ELECTRICAL ENGINEERING
FUNDAMENTALS
second edition

Vincent Del Toro
Professor
Electrical Engineering Department
City University of New York

PRENTICE-HALL, Englewood Cliffs, NJ 07632

Library of Congress Cataloging-in-Publication Data

DEL TORO, VINCENT.
    Electrical engineering fundamentals.

    Bibliography: p.
    Includes index.
    1. Electric engineering. I. Title.
TK145.D368 1986   621.3   85-24378
ISBN 0-13-247131-0

ISBN 0-13-247131-0   01

Editorial/production supervision: Claudia Citarella
Cover design: Photo Plus Art
Manufacturing buyer: Rhett Conklin

PRENTICE-HALL INTERNATIONAL (UK) LIMITED, *London*
PRENTICE-HALL OF AUSTRALIA PTY. LIMITED, *Sydney*
PRENTICE-HALL CANADA INC., *Toronto*
PRENTICE-HALL HISPANOAMERICANA, S.A., *Mexico*
PRENTICE-HALL OF INDIA PRIVATE LIMITED, *New Delhi*
PRENTICE-HALL OF JAPAN, INC., *Tokyo*
PRENTICE-HALL OF SOUTHEAST ASIA PTE. LTD., *Singapore*
EDITORA PRENTICE-HALL DO BRASIL, LTDA., *Rio de Janeiro*
WHITEHALL BOOKS LIMITED, *Wellington, New Zealand*

To *Franca, Peter,* and *Marc*

Contents

preface xiii

_____ chapter one _____

The Fundamental Laws of Electrical Engineering 1

1-1 Units 2   **1-2** Electric Current 4   **1-3** Coulomb's Law 6   **1-4** Ohm's Law 11   **1-5** Faraday's Law of Electromagnetic Induction 13 **1-6** Kirchhoff's Laws 15   **1-7** Ampère's Law 19   Summary Review Questions 20   Problems 21

part one
ELECTRIC CIRCUIT THEORY

_____ chapter two _____

The Circuit Elements 24

2-1 Ideal Independent Current and Voltage Sources 25   **2-2** Reference Directions and Symbols 29   **2-3** Energy and Power 31   **2-4** The

Resistance Parameter 32   **2-5** The Inductance Parameter 38   **2-6** The Capacitance Parameter 44   **2-7** The Operational Amplifier: Its Role as a Circuit Element and as a Dependent (Controlled) Source 49   Summary Review Questions 62   Problems 64

_____chapter three_____
Elementary Network Theory 72

3-1 Series and Parallel Combinations of Resistances 73   **3-2** Series and Parallel Combinations of Capacitances 75   **3-3** Series and Parallel Combinations of Inductances 76   **3-4** Series-Parallel Circuits 80
3-5 The Superposition Theorem 84   **3-6** Network Analysis by Mesh Currents 87   **3-7** Circuit Analysis by Node-Pair Voltages 94
3-8 Thévenin's Theorem 100   **3-9** Norton's Theorem: Conversion of a Voltage Source to a Current Source 104   **3-10** Network Reduction by Δ-Y Transformation 109   Summary Review Questions 112   Problems 113

_____chapter four_____
Circuit Differential Equations: Forms and Solutions 121

4-1 The Differential Operator 122   **4-2** Operational Impedance 123
4-3 The Operator Formulation of Circuit Differential Equations. Equivalency 126   **4-4** General Formulation of Circuit Differential Equations 129   **4-5** The Forced Solution (or Particular Solution) 130
4-6 The Natural Response (or Transient Solution) 137   **4-7** Complete Response of Linear Second-Order Case 142   Summary Review Questions 147   Problems 148

_____chapter five_____
Circuit Dynamics and Forced Responses 155

First-Order Circuits 156   **5-1** Step Response of an _RL_ Circuit 156
5-2 Step Response of an _RC_ Circuit 164   **5-3** Duality 171   **5-4** Pulse Response of the _RC_ Circuit 174   **5-5** The Impulse Response 176
Second-Order Circuits 180   **5-6** Step Response of Second-Order System (_RLC_ Circuit) 180   **5-7** Complete Response of _RL_ Circuit to Sinusoidal Input 195   **5-8** Response of _RLC_ Circuit to Sinusoidal Inputs 199
Summary Review Questions 201   Problems 202

chapter six

The Laplace-Transform Method of Finding Circuit Solutions
211

6-1 Nature of a Mathematical Transform 212 **6-2** The Laplace Transform: Definition and Usefulness 213 **6-3** Laplace Transform of Common Forcing Functions 217 **6-4** Laplace Transform of the Unit-Impulse Function 221 **6-5** Useful Laplace-Transform Theorems 224 **6-6** Example: Laplace-Transform Solution of First-Order Equation 225 **6-7** Inverse Laplace Transformation through Partial-Fraction Expansion 228 **6-8** Step Response of an *RL* Circuit 223 **6-9** Step Response of an *RC* Circuit 237 **6-10** The Impulse Response 240 **6-11** Step Response of Second-Order System (*RLC* Circuit) 241 **6-12** Complete Response of *RL* Circuit to Sinusoidal Input 245 Summary Review Questions 248 Problems 249

chapter seven

Sinusoidal Steady-State Response of Circuits 253

7-1 Sinusoidal Functions—Terminology 254 **7-2** Average and Effective Values of Periodic Functions 256 **7-3** Instantaneous and Average Power. Power Factor 261 **7-4** Phasor Representation of Sinusoids 264 **7-5** Sinusoidal Steady-State Response of Single Elements—*RLC* 273 **7-6** The Series *RL* Circuit 280 **7-7** The Series *RC* Circuit 288 **7-8** The *RLC* Circuit 289 **7-9** Application of Network Theorems to Complex Impedances 291 **7-10** Resonance 296 **7-11** Balanced Three-Phase Circuits 305 **7-12** Fourier Series 314 Summary Review Questions 321 Problems 323

part two
ELECTRONICS

chapter eight

Electron Control Devices: Semiconductor Types 333

8-1 The Boltzmann Relation and Diffusion Current in Semiconductors 335 **8-2** The Semiconductor Diode 346 **8-3** The Transistor (or Semiconductor Triode) 351 **8-4** The Junction Field-Effect Transistor (JFET) 356 **8-5** The Insulated-Gate FET (or MOSFET) 363 **8-6** The Silicon-Controlled Rectifier 367 Summary Review Questions 369 Problems 371

Contents

_____chapter nine_____
Semiconductor Electronic Circuits 375

9-1 Graphical Analysis of Transistor Amplifiers 375   **9-2** Linear Equivalent Circuits 387   **9-3** Biasing Methods for Transistor Amplifiers 395 **9-4** Computing Amplifier Performance 404   **9-5** Frequency Response of *RC* Coupled Transistor Amplifiers 420   **9-6** Integrated Circuits 429 Summary Review Questions 435   Problems 436

_____chapter ten_____
Special Topics and Applications 444

10-1 Electronic-Circuit Applications Involving Diodes 444   **10-2** The Emitter-Follower Amplifier 448   **10-3** The Push-Pull Amplifier 451 **10-4** Modulation 453   **10-5** Amplitude Demodulation or Detection 458 Summary Review Questions 460   Problems 461

part three
DIGITAL SYSTEMS

_____chapter eleven_____
Binary Logic: Theory and Implementation 465

11-1 Binary Logic 465   **11-2** Logic Gates 469   **11-3** Implementation of Digital Logic Gates 472   **11-4** Binary and Other Number Systems 482 **11-5** Conversions between Number Systems 484   **11-6** Applications of Gate Circuits for Data Processing 489   Summary Review Questions 492 Problems 494

_____chapter twelve_____
Simplifying Logical Functions 497

12-1 Boolean Algebra: Postulates and Theorems 498   **12-2** Algebraic Simplification of Logical Functions 504   **12-3** Simplification via Karnaugh Maps 508   **12-4** An Alternative Scheme: Product of Sums 513 Summary Review Questions 517   Problems 518

_____ chapter thirteen_____

Components of Digital Systems 522

Combinational Logic Circuits 522   **13-1** Encoders 524
13-2 Adders 525   **13-3** Subtractors 532   **13-4** Decoders and
Demultiplexers 536   **13-5** Multiplexers (Data Selectors) 540   **13-6** Read-
Only Memory Units (ROMs) 542   Synchronous Sequential Logic
Circuits 544   **13-7** Flip-Flops 545   **13-8** Counters 555
13-9 Registers 567   **13-10** Random Access Memory (RAM) 573
Summary Review Questions 576   Problems 579

_____ chapter fourteen_____

Microprocessor Computer Systems 587

14-1 An Elementary Computer: Basic Architecture 587
14-2 Microprocessors 596   **14-3** Transfer Instructions 599
14-4 Operation Instructions 602   **14-5** Control Instructions 603
14-6 Assembly Language Programming: Examples 606   Summary
Review Questions 610   Problems 612

part four
ELECTROMECHANICAL ENERGY CONVERSION

_____ chapter fifteen_____

Magnetic Theory and Circuits 614

15-1 Ampère's Law—Definition of Magnetic Quantities 615   **15-2** Theory of
Magnetism 623   **15-3** Magnetization Curves of Ferromagnetic
Materials 625   **15-4** The Magnetic Circuit: Concept and Analogies 629
15-5 Units for Magnetic Circuit Calculations 632   **15-6** Magnetic Circuit
Computations 634   **15-7** Hysteresis and Eddy-Current Losses in
Ferromagnetic Materials 639   **15-8** Relays—An Application of Magnetic
Force 643   Summary Review Questions 647   Problems 648

_____ chapter sixteen_____

Transformers 653

16-1 Theory of Operation and Development of Phasor Diagrams 654
16-2 The Equivalent Circuit 664   **16-3** Parameters from No-Load Tests 673
16-4 Efficiency and Voltage Regulation 678   **16-5** Mutual
Inductance 680   Summary Review Questions 683   Problems 684

Contents   ix

_____chapter seventeen_____

Electromechanical Energy Conversion 688

17-1 Basic Analysis of Electromagnetic Torque 689   **17-2** Analysis of Induced Voltages 696   **17-3** Construction Features of Electric Machines 698   **17-4** Practical Forms of Torque and Voltage Formulas 703   Summary Review Questions 709   Problems 711

_____chapter eighteen_____

The Three-Phase Induction Motor 714

18-1 The Revolving Magnetic Field 715   **18-2** The Induction Motor as a Transformer 719   **18-3** The Equivalent Circuit 722   **18-4** Computation of Performance 726   **18-5** Torque-Speed Characteristic. Starting Torque and Maximum Developed Torque 729   **18-6** Parameters from No-Load Tests 731   **18-7** Ratings and Applications of Three-Phase Induction Motors 733   **18-8** Controllers for Three-Phase Induction Motors 733   Summary Review Questions 739   Problems 741

_____chapter nineteen_____

Three-Phase Synchronous Machines 745

19-1 Generation of a Three-Phase Voltage System 746   **19-2** Synchronous Generator Phasor Diagram and Equivalent Circuit 747   **19-3** The Synchronous Motor 750   **19-4** Synchronous Motor Phasor Diagram and Equivalent Circuit 751   **19-5** Computation of Synchronous Motor Performance 752   **19-6** Power-Factor Control 756   **19-7** Synchronous Motor Applications 757   Summary Review Questions 759   Problems 760

_____chapter twenty_____

D-C Machines 763

20-1 D-C Generator Analysis 763   **20-2** D-C Motor Analysis 768   **20-3** Motor Speed-Torque Characteristics. Speed Control 771   **20-4** Speed Control by Electronic Means 775   **20-5** Applications of D-C Motors 781   **20-6** Starters and Controllers for D-C Motors 781   Summary Review Questions 786   Problems 787

_____chapter twenty-one_____

Single-Phase Induction Motors 791

21-1 How the Rotating Field Is Obtained 791   **21-2** The Different Types of Single-Phase Motors 794   **21-3** Characteristics and Typical Applications 797   Summary Review Questions 797   Problems 798

_____chapter twenty-two_____

Stepper Motors 800

22-1 Construction Features 801   **22-2** Method of Operation 801   **22-3** Drive Amplifiers and Translator Logic 807   **22-4** Half-Stepping and the Required Switching Sequence 810   **22-5** The Reluctance-Type Stepper Motor 813   **22-6** Ratings and Other Characteristics 814   Summary Review Questions 815   Problems 816

part five
FEEDBACK CONTROL SYSTEMS

_____chapter twenty-three_____

Principles of Automatic Control 818

23-1 Distinction between Open-Loop and Closed-Loop (Feedback) Control 819   **23-2** Block Diagram of Feedback Control Systems. Terminology 820   **23-3** Position Feedback Control System. Servomechanism 823   **23-4** Typical Feedback Control Applications 826   **23-5** Feedback and Its Effects 832   Summary Review Questions 840   Problems 841

_____chapter twenty-four_____

Dynamic Behavior of Control Systems 844

24-1 Dynamic Response of Second-Order Servomechanism 845   **24-2** Error-Rate Control 858   **24-3** Output-Rate Control 861   **24-4** Integral-Error (or Reset) Control 864   **24-5** Transfer Functions of System Components 866   Summary Review Questions 882   Problems 884

_____ appendices _____

Appendix A: Units and Conversion Factors 891   **Appendix B:** Periodic
Table of the Elements 893   **Appendix C:** Conduction Properties of
Common Metals 894   **Appendix D:** Selected Transistor
Characteristics 895   **Appendix E:** Answers to Selected Problems 899

Index 911

Preface

The purpose of this book is to provide a firm foundation in the study of electrical engineering to both majors and nonmajors alike. It treats each of five principal areas of electrical engineering in sufficient depth to impart real understanding and even perhaps a degree of sophistication. It is not intended to be a survey. For the student expecting to major in electrical engineering this exposure will enable him to better determine areas that interest him most. These can then be pursued with several more advanced courses. For the nonmajor the exposure serves not only to allow him to converse more intelligently with electrical engineers in bringing about solutions to interdisciplinary-oriented problems but also to act independently on relatively routine matters involving electrical engineering.

Emphasis on the fundamental laws of electrical engineering is a theme that is established in the first chapter and referred to often in the development of the subject matter of this book. It is the objective of Chapter 1 to stress the fact that the whole science of electrical engineering is based upon a few experimentally established fundamental laws. Once these principles are grasped and understood, a base is built from which the study of the various areas of electrical engineering follows with considerable ease. Thus electric circuit theory is seen to develop naturally from five basic laws: Coulomb's law, which leads to the concept of capacitance; Ohm's law, which leads to the concept of resistance; Faraday's law, which leads to the concept of inductance; and finally Kirchhoff's voltage and current laws, which provide a systematic formulation of the principles involved in the first three laws. In a similar fashion, the subject matter of electromechanical energy conversion is derived from two basic laws: Faraday's law and Ampère's law. In semiconductor electronics it is the Boltzmann relationship that is important.

The five major parts of this book are: Electric Circuit Theory, Electronics, Digital Systems, Electromechanical Energy Conversion, and Feedback Control Systems.

Part I of the book is devoted to an exposition of the theory of electric circuits. This begins in Chapter 2 with a discussion of circuit parameters as they derive from the fundamental laws. An outstanding feature here is that each parameter is treated from three viewpoints: circuit, energy, and geometrical. In this way a complete rather than a partial understanding of these important aspects of electric circuits emerges. How often does it happen that a student knows how to identify inductance in terms of a description that involves current and voltage (circuit viewpoint) but does not know how to change its value (geometrical viewpoint)? Chapter 2 concludes with a description of the operational amplifier oriented to stess its importance as a circuit element and as a dependent voltage and current source. For simplicity at this stage of the book the treatment of the op-amp is essentially confined to an input-output relationship. Chapter 3 concerns itself with the elementary network theorems. In the interest of placing maximum emphasis on the theorems, networks consisting solely of resistors are used, thereby freeing the treatment of the clutter of complex numbers. The subject matter is developed in a logical and motivated fashion. The response of a network is found first by the more cumbersome methods of simple network reduction and superposition. Improvements in the solution process, through the use of mesh currents and node-pair voltages, are then revealed. When the situation warrants, further simplifications are effected by the use of Thévenin's and Norton's theorems. Also included in this second edition is the matrix formulation of the mesh and nodal methods of circuit analysis. Inclusion of such material is illustrative of the extensions that were made throughout the text to meet the more demanding requirements of EE majors. One can be assured, however, that the omission of such topics by nonmajors will cause no disruption of the continuity of the subject matter. Chapter 4 deals with a description of the classical solution of linear differential equations through use of the differential operator. The treatment is extensive and is generalized for easy application in subsequent chapters dealing with the dynamical behavior of circuits and systems. Considerable attention is focused here to better convey an understanding as to why the dynamical solutions assume the forms they do. Then in Chapter 5 the procedures of Chapter 4 are applied to the circuits that are frequently encountered in electrical engineering. In this revised edition the second-order case is treated in ample detail for the three modes of interest: the overdamped, critically damped, and underdamped responses. In Chapter 6 the Laplace-transform method of finding the solution to circuit problems is introduced and generously illustrated. This chapter surely is appropriate for the major students, although it can be used to advantage by nonmajors as well. All of the circuit problems that are solved by the classical procedure in Chapter 5 are now also solved by use of the Laplace transform in Chapter 6. Consequently, the student has an opportunity to compare the two approaches and thereby to benefit from exposure to both procedures. Of course, elimination of Chapter 6 from a course syllabus will in no way diminish the effectiveness of the classical method in solving circuit problems in the transient state. Part I ends with Chapter 7, which is a comprehensive treatment of the important topic of the sinusoidal steady-state response of circuits.

Electronics is the subject matter of Part II. Here emphasis is directed exclusively to semiconductor devices and integrated circuits. The subject matter is divided into three sections: electronic device description (Chapter 8), semiconductor electronic circuits (Chapter 9), and common applications (Chapter 10). Chapter 8 describes the external characteristics and control capabilities of semiconductor diodes and triodes as well as the silicon-controlled rectifier. Once the theoretical explanations are completed, the devices are then expressed in terms of appropriate parameters and/or sources so that the powerful tools of circuit analysis may be applied to the circuits where they appear.

Chapters 9 and 10 cover the chief topics necessary to an understanding of electronic circuits. The concern here is with the behavior and performance calculations of circuits that use diodes and triodes for control and amplifying purposes. Moreover, wherever appropriate, general expressions are derived for all the current and voltage gains encountered. Included too are the distinctions existing among transistor gain, transistor amplifier gains and the stage gain. The topic of electronic circuits is handled with the systems concept in mind. The amplifier is discussed not only for its own sake, but also as part of an overall system. Accordingly, the two-stage, then the three-stage, and finally the four-stage amplifiers are analyzed with the latter arrangement specifically planned to represent an intercommunication system. In addition a concerted effort was made to present an up-to-date treatment as is apparent from the substantial space allotted to the JFET and MOSFET devices as well as to integrated circuits (IC's).

Digital systems are the concern of Part III. This material is entirely new to this edition and it is included in recognition of the ever-growing importance that digital devices and techniques are exerting in engineering design and practice. The introduction to digital systems is covered in four chapters. Chapter 11 provides the basic definitions associated with binary logic and gates. The up-to-date electronic implementation of digital logic gates is discussed at length with a view toward providing a sense of their operating characteristics as well as to furnish a distinction among the various types. The manner of representing various number systems and the conversion procedures that allow movement from one to another are also described. A chapter on the techniques available for simplifying logical functions is included as well. This is Chapter 12. Although minimization of logical functions is not as important as it was in the early days of the development of digital systems, the material is included primarily because of the background it provides in studying some of the subject matter of Chapter 13. A rather thorough description of the composition and operation of the principal components to be found in digital systems everywhere is described in Chapter 13. Digital devices are divided into two main groups: those that are found in combinational logic circuits and those that make up sequential logic circuits such as flip-flops, counters, registers, and random-access memories. This section on digital systems is concluded in Chapter 14 by treating at some length the operation of a microprocessor computer system, which serves as a forceful application of the notions of the preceding three chapters. Chapter 14 ends with a series of examples that illustrate the technique of assembly language programming. Because the emphasis here is merely to illustrate the technique, no effort was made to use a more recent microprocessor than the Intel 8080.

Part IV is devoted to an exposition of electromechanical energy conversion. The topic begins with a chapter on magnetic theory and circuits thus providing the background needed for the study of electrical machinery. Chapter 16, which follows, then deals with the theory and operation of the transformer. This is a prerequisite to the study of a-c machines. Chapter 17 is devoted to an analysis of the general expression for electromagnetic torque as derived from Ampère's law and an analysis of induced voltage as derived from Faraday's law. The treatment is kept as general as possible to stress that the same basic laws underlie the operation of both a-c and d-c machines. There then follow discussions of the construction features of the various types of electric machines along with the practical forms of the torque and voltage formulas. It is useful to note here that in curricula where relatively little time can be allotted to the study of electromechanical energy conversion, Chapters 15 to 17 provide a germane compendium of the subject matter. In situations where ample time is available, Chapter 17 can be followed appropriately with Chapters 18 to 21 where attention is directed to a description of the operation, performance, characteristics, and applications of the various categories of electrical machines: the three-phase induction motor, the three-phase synchronous motor, the d-c machines, and the single-phase induction motor. To make this treatment of machinery as useful as possible for the nonelectrical majors, particular attention is given to the ratings and applications of electric motors which are conveniently summarized in suitable tables. Controllers for these motors are also discussed quite extensively.

Another noteworthy addition to this second edition is the inclusion of a chapter on the stepper motor. This device is now almost ubiquitous since it is found in many peripheral devices surrounding digital systems in general and digital computers in particular.

Part V deals with feedback control systems. The underlying principles of automatic control are carefully explained and illustrated in Chapter 23 with examples from various fields of engineering. This is followed in Chapter 24 with a general treatment of the dynamic behavior of control systems. Here the student is exposed to systems analysis by dealing directly with the differential equation approach as well as with the transfer function approach. The advantages and the implementation of such procedures as error-rate, output-rate, and integral-error control are also described. The combination of Chapters 23 and 24 should provide a sufficiently strong background in control system theory where the time in the curriculum is limited.

Another feature of this second edition concerns the back matter that immediately follows the text material of each chapter. New is the inclusion of a section entitled Summary Review Questions. These questions are presented to the reader in place of a summary and are designed to elicit a mature understanding and pointed relevance of the subject matter of each section of the chapter. If the student experiences difficulty in answering any of these questions, it is an indication that further study of the chapter is needed.

The section on Summary Review Questions is then followed by a section on Problems, which are now divided into two groups. Group I Problems are designed to provide a straightforward illustration of the subject matter of the chapter. Little or no challenging extension of the subject matter is usually required to obtain solutions to these problems. Accordingly, such problems can best serve as the chief source of assignments for the nonmajor students. On the other hand, the problems that are included in Group II are generally designed to require the student to apply his/her understanding to situations that call for the projection of principles to nonroutine circumstances. Assignments in both groups would certainly be appropriate for the majors who use this text.

Finally, I want to express my gratitude to Mitchell Pfeffer who read the material on Digital Systems in an impressively thorough and meticulous fashion. I am confident that his many suggestions have helped to improve the precision and clarity of the treatment.

Vincent Del Toro

chapter one

The Fundamental Laws of Electrical Engineering

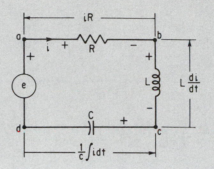

The science of electrical engineering is based on just a few experimentally established fundamental laws. The principles and concepts that underlie the operation and performance of many engineering devices are very often the same in spite of differences in appearance and arrangement. In the interest of stressing the importance of these basic laws, attention is focused on the historical frame of reference as well as the final experimentation which culminated in their strikingly simple formulations. Once the fundamental laws are studied and understood, a considerable amount of perspective will have been gained. In turn, this will facilitate the understanding of those branches of engineering where the appropriate laws provide the foundation.

For example, it will be seen that the field of *electric circuit theory* emanates from the fundamental results achieved by Coulomb (1785), Ohm (1827), Faraday (1831), and Kirchhoff (1857). The dates denote the years in which the laws, which today bear their names, were first published. Likewise, in Part III of this book, it will be seen that the whole subject matter of electromagnetic devices and electromechanical energy conversion can be treated and analyzed by applying just two of the fundamental laws—Ampère's law (1825) and Faraday's law of induction (1831). In electronics a similar situation prevails, although it is not quite as clear-cut. At the turn of the twentieth century the age of electronics was ushered in by the invention of the vacuum tube. In this field fundamental contributions were made by Richardson and Dushman, who were responsible for identifying the properties and characteristics of thermionic emission. This was accompanied by the significant efforts of Langmuir and Child in describing the space-charge behavior of these vacuum devices. As a result of the work of these experimenters it became possible to treat the vacuum device as a circuit element, the external characteristics of which were readily identifiable. In recent years, however, the vacuum tube has slipped to a position of secondary importance in electronics because of the advent of semiconductor devices. Accordingly, considerable attention is devoted here to the principles of operation involved in these latter devices because an entirely new concept is encountered, namely, that of current flow by a diffusion process. The Boltzmann relation is fundamental in describing the external behavior of the semiconductor.

The subject matter of this chapter begins with a discussion of units. The fundamental laws are mathematical expressions of experimentally established conclusions. Hence means must be available for making measurements—and for this a consistent system of units is a prerequisite.

1-1 UNITS

Engineering is an applied science dealing with physical quantities. Accordingly, the establishment of a universally accepted system of units is indispensable if order, communication, and understanding among the engineers and scientists throughout the world are to be achieved. A *unit* of a physical quantity is a standard of measurement regarded as an undivided entity. Once a unit is agreed upon, its application to an unknown quantity merely requires determining how many units make up the whole. Thus, if the unit of length is the meter, then any unknown length can be measured by finding the number of times the unit is taken to make up the whole. To be useful all basic units must be permanent, reproducible, available for use, and adaptable to precise comparison.

The number of different physical quantities with which the electrical engineer must be concerned exceeds thirty in number. Although it is possible to establish a standard unit for each quantity, it is really unnecessary to do so because many of the quantities are functionally related through experiment, derivation, or definition. In the study of mechanics, for example, the units of only three quantities need to be declared arbitrarily as standards not specified in terms of anything else. All other quantities can be expressed in terms of the three arbitrary units by means of the experimental, derived, and defined relationships between the physical quantities. The three quantities of concern are called the *fundamental quantities* and in mechanics are identified as *length, mass,* and *time.* The corresponding units selected for these quantities are referred to as the *fundamental units.*

Two considerations enter into the selection of the fundamental units. First, the basic relationship between the various quantities involved in the study of the given discipline shall involve a minimum number of constants. Second, the measuring units shall be of a practical size. In 1902, G. Giorgi proposed that the meter-kilogram-second (MKS) system of units best meets these needs, and today this comprehensive system is universally accepted.

It is worthwhile at this point to indicate briefly how other quantities in the study of mechanics can be described in terms of the three fundamental units. Consider, for example, velocity. It is referred to as a defined quantity because it is expressed as length per unit time, or, in terms of the fundamental units, as meters/second. Acceleration can be similarly treated. The force quantity can be described in terms of the fundamental units through the aid of an experimental relationship, namely, Newton's law. Thus

$$F = ma \qquad\qquad (1\text{-}1)$$

where m is the mass in kilograms and a is the acceleration in meters/second^2. As a matter of convenience the unit for force in the MKS system is called the *newton*, but note that it is entirely expressible in terms of the fundamental units. Thus, from an inspection of the right side of Eq. (1-1) the dimensional relationship for forces is $[MLT^{-2}]$. The *dimension* of a quantity shows the fundamental units which enter as factors and the exponents. *Work* is still another example of a defined quantity. Since work involves the product of force by length, it can be dimensionally expressed as $[ML^2T^{-2}]$. Note, too, that the simple form of Eq. (1-1), free of constant multiplying factors, resulted from the proper choice of fundamental units. Thus, if the centimeter had been chosen in place of the meter as the fundamental unit of length, and provided other units remain unchanged, Eq. (1-1) would have had to be written as $F = ma \times 10^{-2}$.

Although for the study of mechanics only three fundamental units are required, a fourth fundamental unit must be introduced for the study of electricity and magnetism. For practical reasons the additional fundamental quantity is chosen to be *current* and the corresponding fundamental unit is called the *ampere*. However, from a purely theoretical viewpoint the fourth basic quantity could be charge, having the fundamental unit of the coulomb. One is derivable from the other. The important consideration which led to the selection of the ampere is that the ampere serves as the link between electrical, magnetic, and mechanical quantities and is more readily measured. The definition of the ampere which appears below stresses the matter sufficiently.

With this background, attention may now be directed to the formal definitions of the four fundamental units used in the study of electrical engineering. They are as follows:

The *meter* was defined in 1960 by an international commission as 1,650,763.73 wavelengths of the radiation of the orange-red line of krypton 86. This is a definition in terms of the wavelength of light, which makes it more permanent than the definition previously used.

The *kilogram* is the mass of a platinum-iridium alloy of cylindrical shape kept at the International Bureau of Weights and Measures at Sèvres, France. It is approximately equal to the mass of a cube (measuring 0.1 m on each side) of pure water at 4°C.

The *second* is now defined as 1/31,556,925.9747 of the tropical year 1900.

The *ampere* is that current which when flowing through two infinitely long straight wires of negligible cross-sectional area and placed 1 m apart in a vacuum produces a force per unit length of 2×10^{-7} N/m.

In recent years in electrical engineering, with the advent of very high frequency oscillators, such terms as *pico* and *nano* have been occurring frequently in the literature. The following tabulation is presented as an explanation of these and other terms.

deci (d-, 10^{-1})   deka (da-, 10^{1})
centi (c-, 10^{-2})   hekto (h-, 10^{2})
milli (m-, 10^{-3})   kilo (k-, 10^{3})
micro (μ-, 10^{-6})   mega (M-, 10^{6})
nano (n-, 10^{-9})   giga (G-, 10^{9})
pico (p-, 10^{-12})   tera (T-, 10^{12})
femto (f-, 10^{-15})   exa (E-, 10^{15})
atto (a-, 10^{-18})   peta (P-, 10^{18})

Thus a picosecond is one millionth of one millionth of a second, and a centimeter is one hundredth of a meter. Similarly, a megaton is 1 million tons. A little thought reveals that the MKS system of units offers the additional flexibility in the use of prefixes, which is not so readily accomplished with a system of units such as the English system where inches, feet, and yards are used.

For convenience a table of units is given in Appendix A.

1-2 ELECTRIC CURRENT

Charge has been associated with the idea of electricity from the very beginnings. As early as 600 B.C. it was found that if the fossil resin amber was rubbed with a dry substance, it was possible for the amber to exert a force of attraction on a light material such as a feather or straw. In fact, it is from the Greek word for amber—*elektron*—that the word for electricity is derived. In this situation the amber is described as possessing *frictional electricity* through the accumulation of charge. Up to the end of the eighteenth century this was the only form of electricity known to man. In 1799, however, Allessandro Volta developed the copper-zinc battery, which he showed was capable of producing electricity in a continuously flowing state in wires. Initially this type of electricity was called *current electricity* to distinguish it from the frictional kind. But with further experimentation Volta was able to demonstrate that the two types were identical; he was able to produce the same results. Of course, today our knowledge of the atomic structure of matter bears out the truth of Volta's work.

In terms of the atomic description we know that all matter is composed of atoms. The atom in turn consists of a central body called the *nucleus* and a number of smaller particles, called *electrons,* which move in approximately elliptical orbits around the nucleus. The electron is the smallest indivisible particle of electricity known to man. The American physicist R. A. Millikan in his renowned oil-drop experiment showed the charge of the electron to be 1.602×10^{-19}

coulomb (C),† and it was arbitrarily called a *negative* charge. The nucleus of the atom is made up of two types of particles—the proton and the neutron. The *proton* has a mass 1837 times that of the electron and carries a positive charge equal to the sum of the electron charges of the atom. The *neutron* has the same mass as the proton but no charge. Thus we see that the role of positive and negative charges in the description of matter is an important one.

Electric current is defined as the time rate of change of charge passing through a specified area. The moving charges may be positive or negative; the area may be the cross-sectional area of a wire or some other suitable spatial area where charges are in motion. Expressed mathematically, we can write

$$i = \frac{dq}{dt} \tag{1-2}$$

In this equation i denotes the instantaneous electric current and q represents the net charge, which may be of the positive as well as negative kind. That is,

$$q = N_1 p + N_2 e \tag{1-3}$$

where $N_1 p$ denotes the total positive charge and $N_2 e$ the total negative charge. The positive direction of the current i is arbitrarily taken to be the direction of flow of positive charges.

Since the symbol e is used to denote the charge of an electron we may write

$$e = 1.602 \times 10^{-19} \, C/electron$$

From this result we see that there are 6.24×10^{18} electrons in 1 C, which is a staggering number. Such a large number is used for the coulomb so that smaller numbers can be used for ordinary engineering calculations.

The ability to put charge in motion, i.e., to effect a current flow, is important in many ways in electrical engineering. For example, it is the means by which energy can be transferred from one point to another; as another example, control of the rate at which charge is put into motion makes it possible to transmit intelligence in such communication devices as radio, television, and Telstar.

Circuit.   An electric circuit is a closed path, composed of active and passive elements, to which current flow is confined. Figure 1-1 depicts a typical circuit containing one passive and one active element. An active element is one which supplies energy to the circuit. A passive element is one which receives energy and then either converts it to heat or stores it in an electric or magnetic field. The battery in Fig. 1-1 is the active element.

---

† The coulomb is that unit of charge which when placed 1 m apart from an identical particle of charge is repelled by a force of $10^{-7} c^2$, where c is the velocity of light in meters/second.

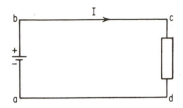

Fig. 1-1 Example of an electric circuit. The current I is shown pointed in the direction of conventional flow. Electrons flow in the opposite direction.

1-3 COULOMB'S LAW

Appearing in Fig. 1-2 are two point positive charges of value Q_1 and Q_2 separated by a distance of r meters. The distance between the charges is assumed to be large compared to the size of the mass on which the charges reside. Charles A. Coulomb was able to show experimentally that a force of repulsion exists between Q_1 and Q_2 which is directly proportional to the magnitude of the charges and inversely proportional to the distance between them. In terms of MKS units, Coulomb showed the magnitude of this force to be given by

$$F = \frac{Q_1 Q_2}{4\pi\varepsilon r^2} \quad \text{N} \qquad (1\text{-}4)$$

where Q_1 and Q_2 are expressed in coulombs and r in meters and ε is a constant related to the surrounding medium. The 4π is a proportionality constant which appears whenever Coulomb's law is expressed in rationalized MKS units. It is this equation which is known as *Coulomb's law*. However, Eq. (1-4) is also frequently written as

$$F = \mathscr{E}Q_2 \qquad (1\text{-}5)$$

where

$$\mathscr{E} \equiv \frac{Q_1}{4\pi\varepsilon r^2} \quad \frac{\text{N}}{\text{C}} \quad \text{or} \quad \text{V/m} \qquad (1\text{-}6)$$

The quantity $\mathscr{E}$ is called the *electric field intensity* and has the units of newtons per coulomb or volts/meter. The former unit is obvious from Eq. (1-5); the latter is discussed presently.

Although Eq. (1-4) bears Coulomb's name, he is not generally credited with having originated the formulation of the law. Long before Coulomb began his experiments it was known that like charges repel and unlike charges attract. It was Benjamin Franklin who proposed inquiries about the behavior of static

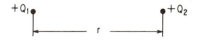

Fig. 1-2 Configuration for measuring electrostatic force of repulsion by Coulomb's law.

charges that finally culminated in Coulomb's successful measurements. While experimenting, Franklin one day discovered that when a conductor was in a charged state but not carrying current (i.e., not part of a complete circuit) all of the charge resided on the surface of the conductor. Franklin subsequently asked Joseph Priestly† to investigate the matter more thoroughly. In the process of doing so Priestly in 1766 was able to deduce the law which bears Coulomb's name. However, it was Coulomb's brilliant invention of a torsion balance which enabled him in 1785 to verify Priestly's deductions with significant accuracy. Coulomb's measurements led directly to Eq. (1-4).

Permittivity. A glance at Eq. (1-4) shows that through measurements it is possible to know all the factors in this equation with the exception of the pro-portionality factor ε, which is a property of the medium in which the experiment is performed. This quantity is called the *permittivity*. When the experiment is performed in a vacuum (free space), the value of the permittivity as obtained from Eq. (1-4) and expressed in MKS units comes out to be

$$\varepsilon_0 = \frac{1}{36\pi \times 10^9} = 8.854 \times 10^{-12} \quad \text{F/m} \tag{1-7}$$

By repeating Coulomb's experiment in oil for the same values of Q_1, Q_2, and r it is found that the resulting force is only about half as much as for air. Accordingly, this means the permittivity of oil is greater than that of air. A convenient way of expressing this difference is to introduce a quantity called *relative permittivity,* which is defined as

$$\varepsilon_r = \frac{\varepsilon}{\varepsilon_0} \tag{1-8}$$

Thus a value of ε_r of 2 reveals that for a given configuration, such as depicted in Fig. 1-2, the resulting force in the given medium is half the amount which is available in free space.

EXAMPLE 1-1   Find the force in free space between two like point charges of 1C each and placed 1m apart.

Solution: From Eqs. (1-4) and (1-7) we have

$$F = \frac{1 \times 1}{4\pi \dfrac{1}{36\pi \times 10^9} \times 1} = 9 \times 10^9 \, \text{N} \approx 1 \text{ megaton} \tag{1-9}$$

Note the gigantic size of the force. This calculation again illustrates that 1C of electricity is an exorbitant quantity which is not available for ordinary engineering computations.

† Priestly is better known for the discovery of oxygen.

Potential Difference. Coulomb's law serves as the starting point for the study of a considerable part of electrical engineering. It is our purpose next to establish additional background that follows naturally from Coulomb's law and that will be useful later on in the book. Refer to Fig. 1-3. The charge Q_1 is

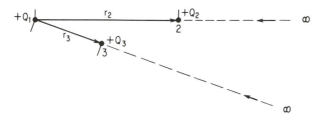

Fig. 1-3 Illustrating the law of potential difference.

assumed located at some fixed point in space. The charge Q_2 is assumed initially to be on a horizontal path from Q_1 but infinitely far away. Hence the force of repulsion existing between Q_1 and Q_2 is zero. Let us consider next what happens as Q_2 is moved from infinity to a point where it is finally located r_2 meters from Q_1. It should be evident that as Q_2 is moved in this direct line from infinity to positions which are closer and closer to Q_1, greater and greater forces are required to accomplish this. Furthermore, since the force exerted on Q_2 acts through a distance which measures from infinity right up to r_2, it follows that work is being done on Q_2. As a matter of fact this work is imparted to Q_2 in the form of *potential energy,* which is recoverable in the form of kinetic energy immediately upon release of the force which keeps Q_2 fixed at a position r_2 meters from Q_1. Keep in mind that there is a force of repulsion between Q_1 and Q_2 which attempts to send Q_2 back to infinity. The fact that work must be done on Q_2 to bring it from infinity to r_2 may also be described in terms of the electric field intensity associated with charge Q_1 and expressed by Eqs. (1-6) and (1-5). Accordingly, owing to Q_1 an electric field of diminishing magnitude exists from r_2 to ∞. In moving Q_2 from ∞ to r_2 work must be done *against* this electric field. It is worthwhile at this point to use a description which stresses not so much the total work imparted but rather the total *work per unit charge,* which is called *voltage.* Viewed dimensionally the work imparted to Q_2 requires applying the length dimension to both sides of Eq. (1-5), which leads to

$$\text{newtons} \times \text{meters} = \frac{\text{newtons}}{\text{coulomb}} \times \text{coulombs} \times \text{meters} \qquad (1\text{-}10)$$

However, to deal in terms of work per unit charge it is necessary to divide both sides of Eq. (1-10) by the unit of charge Q_2. Thus

$$\frac{\text{newton-meters}}{\text{coulomb}} = \frac{\text{newtons}}{\text{coulomb}} \times \text{meters} = \text{voltage} \qquad (1\text{-}11)$$

It follows then that to find the work imparted per coulomb in moving Q_2 from infinity to r_2, it is necessary to perform an integration of electric field intensity

$\mathscr{E}$(newtons/coulomb) with respect to length (meters). We may express this mathematically as

$$\boxed{\begin{aligned} \text{voltage} &= \frac{\text{work}}{\text{charge}} \\ V_2 &= \frac{W_2}{Q_2} \end{aligned}} \tag{1-12}$$

or, more fully,

$$\boxed{V_2 = \frac{W_2}{Q_2} = -\int_{\infty}^{r_2} \mathscr{E}_1 \, dr} \tag{1-13}$$

The minus sign is arbitrarily inserted to indicate that work is done on the charge *against* the action of the electric field produced by Q_1. Since in this case $\mathscr{E}_1$ is given by Eq. (1-6), V_2 may be evaluated as

$$V_2 = -\int_{\infty}^{r_2} \frac{Q_1}{4\pi \varepsilon r^2} dr = \frac{Q_1}{4\pi \varepsilon r_2} \quad \text{V} \tag{1-14}$$

Thus Eq. (1-14) expresses the potential (or voltage) of point 2 in the configuration of Fig. 1-3 *with respect to a point at infinity*. Moreover, note that since voltage is work per unit charge, Eq. (1-14) is independent of Q_2.

In the interest of making clear to the reader the meaning of *potential difference* let us take another charge Q_3 and move it from infinity to r_3. Refer again to Fig. 1-3. Clearly, since r_3 is smaller than r_2, more work per unit charge must be done to place a charge at point 3. Reasoning as before, we may describe the potential of point 3 as

$$V_3 = -\int_{\infty}^{r_3} \frac{Q_1}{4\pi \varepsilon r^2} dr = \frac{Q_1}{4\pi \varepsilon r_3} \quad \text{V} \tag{1-15}$$

Because V_3 is greater than V_2, there exists a *potential difference* between these two points. It should be apparent then that when a unit charge moves from a point of higher potential (point 3) to one of lower potential (point 2) it gives up energy. Conversely, when a unit charge moves from a point of lower potential to one of higher potential it receives energy. This, then, is the meaning to be associated with such terms as potential difference or voltage rise or voltage drop. Return to Fig. 1-1 for an illustration of the application of these terms. Current is assumed flowing in a clockwise direction, which means that either positive charges are moving clockwise or negative charges are moving counterclockwise. As the positive charges move through the voltage rise from *a* to *b* they receive energy so that the potential of *b* is higher than *a*. However, as they move from *c* to *d* they undergo a voltage drop, thereby losing energy to the device appearing between these two points. It is interesting to note that the stream of charge gains energy in one part of the circuit (at the active element) and gives it up at the other part (the passive element). Of course, the total energy remains unchanged in accordance with the law of conservation of energy.

Further Considerations. If the charge Q_1 in Fig. 1-3 is doubled, the electric field intensity by Eq. (1-6) correspondingly doubles. Therefore to move a unit positive charge from infinity to the same points 2 and 3 requires twice the energy. The potential of each of these points becomes twice as large as well, and so too does the voltage or potential difference. One can conclude from these observations that charge, Q, and voltage, V, are related through some proportionality factor which can be reasonably expected to depend upon the medium in which the charges find themselves as well as upon the geometrical configuration prevailing. Calling this proportionality the capacitance C, we can write

$$Q = CV \qquad (1\text{-}16)$$

More is said about this equation and capacitance in Chapter 2.

Gauss's Law. This law is an important consequence of Coulomb's law and provides additional useful knowledge, which is needed in the work that follows in Chapter 2. In this connection consider that a sphere of radius r is put around a point charge Q as depicted in Fig. 1-4. By Eq. (1-6) we know that the value of the electric field intensity at any point on the surface of the sphere is

$$\mathscr{E} = \frac{Q}{4\pi\varepsilon r^2} \qquad (1\text{-}17)$$

Let us now form the product of $\mathscr{E}$ and the surface area of the sphere, which is

$$A = 4\pi r^2 = \text{area of sphere} \qquad (1\text{-}18)$$

Thus

$$\mathscr{E}A = \frac{Q}{4\pi\varepsilon r^2}(4\pi r^2) = \frac{Q}{\varepsilon} \qquad (1\text{-}19)$$

Rewriting Eq. (1-19) yields

$$Q = (\varepsilon\mathscr{E})A = DA \qquad (1\text{-}20)$$

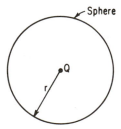

Fig. 1-4 Configuration for deriving Gauss's electric flux law.

where

$$D \equiv \varepsilon \mathscr{E} = \text{electric flux density} \qquad (1\text{-}21)$$

The quantity D is identified as an electric flux density because, as revealed by Eq. (1-20), it is combined with an area term A to yield charge, which is also called *electric flux*.

Equation (1-20) is known as *Gauss's law* and it states that the total electric flux through any closed surface surrounding charge is equal to the amount of charge enclosed. It is discussed here because if we compare Eq. (1-20) with Eq. (1-16), information about the nature of capacitance readily follows. This matter is treated in Sec. 2-6.

1-4 OHM'S LAW

This law is perhaps one of the first things learned about electricity in any elementary course on the subject. It is not at all unlikely that the reader was introduced to Ohm's law in its simplest form in his physics course. This is mentioned by way of further stressing its importance in the study of electrical engineering.

Georg Simon Ohm (1787–1854) was a professor of mathematics and physics who devoted considerable time and effort to experiments dealing with voltaic cells and conductors.† These experiments included the effects of temperature on the resistivity of various metals. In spite of this extensive laboratory work, however, the law that bears his name was the result of a mathematical analysis of the galvanic circuit based on an analogy between the flow of electricity and the flow of heat. This work was described in a pamphlet published in 1827.‡ By following the formulation of Fourier's heat conduction equation and using electric field intensity as analogous to temperature gradient, Ohm was able to show that the current flow in a circuit composed of a battery and conductors can be expressed as

$$I = \frac{A}{\rho}\frac{dv}{dl} \qquad (1\text{-}22)$$

where the derivative term denotes the electric field gradient. In the language used by Ohm v was called the electroscopic force by way of representing the volume density of electricity at a point in the conductor—a terminology consistent with the analogous situation in heat flow through a solid described in terms of

† See William Francis Magie, *A Source Book in Physics* (New York: McGraw-Hill Book Company, 1935).

‡ "Die galvanische Kette mathematisch bearbeitet," i.e., The Galvanic Circuit Investigated Mathematically.

the quantity of heat per unit volume. In the case where a conductor of uniform cross-sectional area is used, Eq. (1-22) may be written as

$$I = \frac{A}{\rho} \frac{V}{L} = \frac{V}{R} \quad \text{A} \qquad (1\text{-}23)$$

where V is the potential difference in volts appearing across the conductor of length L, A is the cross-sectional area, ρ is a property of the material called the *resistivity*, and R is the resistance of the conductor in ohms.

Equation (1-23) is a mathematical description of Ohm's law. It states that the strength of the current in a wire is proportional to the potential difference between its ends. Experimentation aimed at verifying Eq. (1-23) reveals the need for maintaining constant temperature—otherwise the results change accordingly. (More is said about this in Chapter 2.)

Ohm's law may be alternatively expressed as

$$V = IR \qquad (1\text{-}24)$$

In this form it states that for any given potential difference, the amount of current produced is inversely proportional to the resistance, which in turn is dependent upon the composition of the wire. It is interesting to note, therefore, that Ohm's law is not among those basic laws which are independent of materials.

The simplicity of Ohm's law is impressive. It almost commands disbelief; and as a matter of fact this is precisely what happened when Ohm first published his results. He was ridiculed and subsequently resigned from his professorship. Fourteen years later, however, he received the recognition due him, and all ended well.

During Ohm's time experimental verification of his law was achieved exclusively by using direct-current sources, namely the voltaic cells. Since then further experiments have been made to show that Ohm's law is also valid when the potential difference across a linear resistor is time-varying. In this case Ohm's law is written as

$$v = iR \qquad (1\text{-}25)$$

where lowercase letters are used for potential difference and current to emphasize the nonconstant nature of these quantities. This matter is treated at considerable length in Chapter 7.

Devices are often encountered in the study of electrical engineering where the volt-ampere characteristic is not linear, i.e., the current is not related to potential difference by a *constant* proportionality factor. For example, for the incandescent lamp the volt-ampere curve looks as depicted in Fig. 1-5, which is clearly nonlinear. Does Ohm's law apply in this instance? An affirmative answer

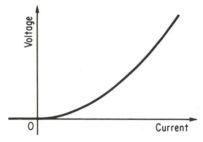

Fig. 1-5 Voltage versus current curve for incandescent lamp.

is possible provided that Ohm's law is expressed in differential form. Thus

$$dv = R\,di \qquad (1\text{-}26)$$

Note that R in this case is the slope of the curve of Fig. 1-5 corresponding to a given point on the curve. As this point is changed, so too is the value of resistance as it applies to that point in the differential sense.

The very simplicity of Ohm's law can lead to difficulty because of the tendency to oversimplify. It is important to know that although Ohm's law applies to many materials—copper, aluminum, iron, brass, bronze, tungsten, etc.—it does not apply to all. For example, it does not apply to nonlinear devices involving step discontinuities such as might be found in zener diodes and voltage regulator tubes.

1-5 FARADAY'S LAW OF ELECTROMAGNETIC INDUCTION

Michael Faraday (1791–1867) distinguished himself as an outstanding experimentalist. He had very little formal academic training. He was born of humble parents and spent seven years as a bookbinder's apprentice. However, Faraday was an avid reader of scientific works and at the age of 20, after listening to a lecture by Sir Humphry Davy,† wrote to Davy requesting that he be taken on as an assistant. Following an interview, Davy agreed to the arrangement and so began a career for Faraday which culminated in fundamental contributions to the science of electricity. His crowning achievement was the establishment of the law of electromagnetic induction which related electricity and magnetism. This was accomplished in 1831 through a series of brilliant experiments.

In the beginning, upon joining Davy at the Royal Institution, Faraday directed his efforts to experiments in chemistry. In 1820, however, the whole scientific world was excited over the discovery by Hans Christian Oersted that a current-

---

† Davy developed the first electric arc lamp. He also isolated sodium and potassium by means of an electric current.

carrying wire could be made to deflect a freely suspended magnet. Faraday repeated Oersted's experiment and then went further to show that not only was it possible for a magnet to move round a current-carrying wire but also for a current-carrying wire to move round a magnet. The reason for the phenomena, however, remained a mystery. This mystery further deepened when in 1825 André Marie Ampère demonstrated that a force also existed between two current-carrying wires. Although Oersted and Ampère both kept searching for the answer in the conductor or the magnet, Faraday did not. Rather, he conceived the forces to be due to tensions in the *medium* in which the conductors and/or magnets were placed. As a matter of fact, it was this attitude which led Faraday to introduce the concept of lines of force. Since he was not a mathematician, he used descriptions which enhanced visualization.

In 1831, Faraday showed that electricity could be produced from magnetism. He demonstrated that induced currents could be made to flow in a circuit whenever (1) current in a neighboring circuit is established or interrupted, (2) a magnet is brought near a closed circuit, and (3) a closed circuit is moved about in the presence of a magnet or other closed current-carrying circuits. Figure 1-6(a) shows one of the experiments made by Faraday. By closing the switch he was able to observe a deflection in the galvanometer† connected to the second circuit. Moreover, upon opening the switch, he again observed a deflection but this time in a reverse direction. Figure 1-6(b) shows schematically still another arrangement used by Faraday to show how electricity is produced from magnetism. If either the magnet was moved toward the circuit or the circuit toward the magnet, the galvanometer registered a deflection.

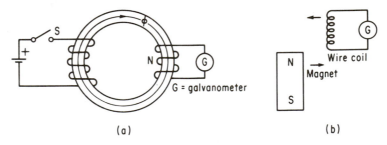

(a)   (b)

Fig. 1-6 Illustrating Faraday's law of induction.

Faraday was a disciplined and meticulous experimenter who kept careful and complete records of his laboratory work. From these descriptions‡ it has been possible to formulate Faraday's law of induction mathematically as follows:

$$e = -\frac{d\lambda}{dt} = -N\frac{d\phi}{dt} \quad \text{V}$$ (1-27)

† A galvanometer is a sensitive instrument used to measure small currents and voltages.

‡ *Experimental Researches in Electricity*, 3 vols. (London, 1839).

The quantity e denotes the electromotive force induced in a closed circuit having a flux linkage of λ weber-turns. In those instances where the magnetic flux ϕ penetrates all the turns of the coil as shown in Fig. 1-6(a), Faraday's law may be expressed by the second form of Eq. (1-27). The negative sign is due to Emil Lenz, who subsequent to Faraday's experiments pointed out that the direction of the induced current is always such as to oppose the action that produced it. This reaction is commonly called *Lenz's law*.

As pointed out in Chapter 17, Faraday's law as embodied in Eq. (1-27) is one of the two basic relationships upon which the entire theory of electromagnetic and electromechanical energy conversion devices is based. In fact, soon after Faraday's work of genius was published in 1831 explanations were at last possible for the phenomena observed by Oersted, Ampère, and other experimenters. The foundation was now established to facilitate the ensuing rapid development of electric motors and generators.

Faraday was also the first experimenter to identify the *emf of self-induction* which manifested itself whenever circuits carrying current in long wires or circuits wound with many turns were disconnected. The American inventor Joseph Henry also independently discovered the current of self-induction but not before Faraday. Both experimenters were able to demonstrate that a changing current produced an emf of self-induction in a coil of wire which varied directly with the time rate of change of current. Expressed mathematically,

$$e = L\frac{di}{dt}$$

(1-28)

where L is a proportionality factor called the *coefficient of self-inductance* which is dependent upon the medium and some physical dimensions (see Chapter 2). This result is equivalent to Faraday's law of induction as expressed in Eq. (1-27). Moreover, Eq. (1-28) is extremely important in the development of electric circuit theory. It serves to identify one of the three basic circuit parameters—*inductance*. The other two parameters, resistance and capacitance, have already been identified with Ohm's and Coulomb's laws, respectively.

1-6 KIRCHHOFF'S LAWS

By the middle of the nineteenth century a considerable amount of experience with electric circuits had been gained by the leading men of science of the time. However, it was Gustav Robert Kirchhoff (1824–1887) who published the first systematic formulation of the principles governing the behavior of electric circuits. He advanced no new experimental facts or concepts but merely restated familiar principles. His work was embodied in two laws—a current and a voltage law—which together are known as *Kirchhoff's laws*. It is upon these laws that electric circuit theory is based.

Kirchhoff's current law states that *the sum of the currents entering or leaving a junction point at any instant is equal to zero.* A junction point is that place in a circuit where two or more circuit elements are joined together. It is often called an *independent node.* In Fig. 1-7 n corresponds to the junction point or node. The current law may be expressed mathematically as

$$\sum_{j=1}^{k} I_j = 0 \qquad (1\text{-}29)$$

where k denotes the number of circuit elements connected to the node in question and Σ is the Greek symbol used to indicate summation. For the circuit of Fig. 1-7 there are five circuit elements joined at node n—four passive elements (the resistors) and one active element (the battery). Applying Eq. (1-29) to node n for the situation where the currents are all assumed to be *entering* the node, we have

$$I_5 - I_1 - I_2 - I_3 - I_4 = 0 \qquad (1\text{-}30)$$

The minus signs are used because these currents are defined as leaving rather than entering the node. Equation (1-30) may be rewritten as

$$I_5 = I_1 + I_2 + I_3 + I_4 \qquad (1\text{-}31)$$

and it states in mathematical terms that the current entering the node is equal to the sum of the currents leaving it.

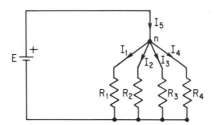

Fig. 1-7 Illustrating Kirchhoff's current law.

In countless experiments performed over the past century Kirchhoff's current law has never been found to be invalid. This is understandable when it is realized that the current law is nothing more than a restatement of the *principle of conservation of charge.* This principle states that the number of electrons passing per second must be the same for all points in the circuit. Accordingly, if the summation of all the currents at a node were not to add up to zero, there would have to occur an accumulation of charge at the node. For the sake of argument suppose the sum of currents at node n were not zero but rather one ampere. Then charge would accumulate at the rate of 1 C for each second. We have already learned, however, that by Coulomb's law a charge accumulation of this kind would produce explosive forces. But experiment shows no evidence of this whatsoever. Any accumulation of charge at the node means an accumulation of mass. Once again experiment fails to reveal that this happens. Therefore, it is proper to conclude that for every charged particle that enters the node there is another charged particle that leaves.

Kirchhoff's voltage law states that *at any time instant the sum of voltages in a closed circuit is zero.* Essentially this law is a restatement of the *law of conservation of energy.* To understand this, refer to Fig. 1-8 which depicts a 12-V battery supplying energy to three series-connected resistors representing a total resistance of 12 Ω. By Ohm's law there exists a current of 1 A, which means that charge flows throughout the circuit at the rate of 1 C per second. This current is denoted by I. The positive direction of current flow is *a-b-c-d-a*. The electrons, of course, are moving in the reverse sense.

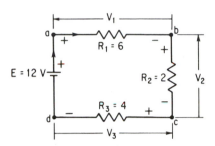

Fig. 1-8 Illustrating Kirchhoff's voltage law.

The constant current which flows in this circuit means that the average rate of movement of charge is constant. However, as electrons traverse the circuit elements R_1, R_2, and R_3, they collide with the atoms in the space lattice configurations of these elements, thereby dissipating energy in the form of heat. Since this energy must be supplied by the battery, the law of conservation of energy permits us to write

$$W = W_1 + W_2 + W_3 \qquad (1\text{-}32)$$

where W denotes the energy supplied by the battery in a specified time interval, W_1 denotes the energy dissipated in R_1 in the same interval of time, and similarly for W_2 and W_3. Furthermore, if we denote by Q the total charge which moves throughout the circuit during this same interval of time, it follows that the work per unit charge may be written as

$$\frac{W}{Q} = \frac{W_1}{Q} + \frac{W_2}{Q} + \frac{W_3}{Q}$$

But because work per unit charge is voltage, this last equation can be expressed as

$$E = V_1 + V_2 + V_3$$

Transposing then yields

$$V_1 + V_2 + V_3 - E = 0 \qquad (1\text{-}33)$$

Note that Eq. (1-33) is a direct mathematical statement of Kirchhoff's voltage law. Due regard must be given to signs. As Eq. (1-33) is written, all voltage drops are treated as positive quantities. Conversely, then, all voltage rises encountered in the loop must be prefixed with a negative sign. A little thought reveals that the reverse convention is equally valid, i.e., voltage rises may be treated as positive quantities while voltage drops are taken as negative.

A generalized formulation of Kirchhoff's voltage law is written as

$$\sum_{j=1}^{k} V_j = 0 \qquad (1\text{-}34)$$

where V_j represents the voltage drop of the jth element in any given closed circuit which is assumed to have k elements. Of course, as the loop is traversed in an assumed direction all voltage rises must carry the minus sign.

In Fig. 1-8 the source voltage was assumed to be a battery of fixed magnitude, thus producing a constant current. However, Kirchhoff's voltage law is general enough so that it applies equally as well when the source voltage is a time-varying quantity producing a time-varying current. Refer to Fig. 1-9. Appearing across d-a is the varying source voltage. Between points a-b is a resistor of value R similar to those used by Ohm in his experiments. If the instantaneous value of the current is denoted as i then the potential difference between the ends of the resistor is iR. Across points b-c appears a coil of the type used by Faraday and across which is developed the emf of self-induction, $L(di/dt)$. Finally, between points c-d a capacitor appears. The voltage drop across the capacitor is readily found from Eqs. (1-16) and (1-2). Thus

$$V = \frac{1}{C}q = \frac{1}{C}\int i\, dt \qquad (1\text{-}35)$$

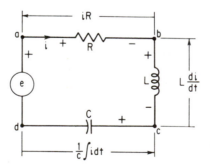

Fig. 1-9 Illustrating Kirchhoff's voltage law for a time-varying source.

Then by Kirchhoff's voltage law we can write

$$e = iR + L\frac{di}{dt} + \frac{1}{C}\int i\, dt \qquad (1\text{-}36)$$

This is an equation concerning which a great deal is said in subsequent chapters.

1-7 AMPÈRE'S LAW

André Marie Ampère (1775–1836) combined extraordinary experimental ability with a genius for mathematics to achieve outstanding contributions in the field of electricity and magnetism. Like all the other leading scientists of the day Ampère was so fascinated by Oersted's discovery of the effect of current in a wire on a magnetic needle that he forthwith directed all his attention to a study of this new phenomenon. A short time afterwards he not only succeeded in extending Oersted's experiments but also came up with a rigorous mathematical theory to describe them. Ampère's most famous experiment in this regard dealt with the force produced between two very long, parallel wires carrying currents. (It will be recalled that this is the configuration which is used today to identify the fundamental unit of the ampere.) Ampère's facility for mathematics enabled him to describe the force in terms of the following equation:

$$F = \frac{\mu I_1}{2\pi r} l I_2 \quad \text{N} \tag{1-37}$$

where I_1 and I_2 are the currents flowing in the two conductors, l is the length of the conductors, and r denotes the distance separating the wires at their center lines. The quantity μ is a property of the medium.

Ampère further showed that the current I_1 could be looked upon as producing a magnetic field given by†

$$B = \frac{\mu I_1}{2\pi r} \tag{1-38}$$

This result was also observed by two other French experimenters of the day— J. B. Biot and F. Savart. Upon substitution of Eq. (1-38) into (1-37) there results that simplified form of Ampère's law which the reader very likely encountered for the first time in an elementary physics course. Thus

$$F = B l I_2 \tag{1-39}$$

In order for Eq. (1-39) to be valid it is necessary for the wires carrying current I_1 and I_2 to lie in the same plane. If, for some reason, the conductor carrying I_2 is inclined by some angle θ with respect to a line perpendicular to the wire carrying I_1, then Ampère's law states that the force is given by

$$F = B l I_2 \sin \theta \tag{1-40}$$

In the interest of covering all cases of this kind as well as those in which more than a single conductor is used to carry I_2, use is made of a generalized formulation of Ampère's law. In equation form it may be expressed as

$$\boxed{F = K_B B I_2 f(\theta) = K_\phi \phi Z_2 I_2 f(\psi)} \tag{1-41}$$

Here B denotes the magnetic field produced by the circuit which carries current I_1, and $f(\theta)$ is some function of the angular displacement existing between the

† See Sec. 15-1 for a further explanation of B.

direction of the B-field and the direction in which I_2 flows. Often $f(\theta)$ is the sinusoidal function, $\sin \theta$. The quantity K_B accounts for such factors as the length of the conductor, the number of poles present in the field, and so forth. An alternate version of Ampère's law finds the force expressed in terms of the magnetic flux field, ϕ, the number of conductors Z_2 which carry I_2, and some function of the angular displacement between the flux and the physical distribution of the conductors Z_2. Often $f(\psi)$ is $\cos \psi$. Refer to Chapter 17 for more details.

On the basis of the information appearing in Eq. (1-41), Ampère's law may be described in words as follows. *To produce an average force (or torque) in an electromagnetic device, there must exist a flux field and an ampere-conductor distribution (which are stationary with respect to each other) and a favorable field pattern (i.e., $f(\psi) \neq 0$).* This statement applies even for those cases in which the flux and current are time-varying quantities. All that is required is that the frequency of variations be the same.

Summary review questions

1. Why is it important to have a universally accepted set of units?
2. Distinguish between the fundamental units and the other units that are encountered in the study and practice of engineering and science.
3. Give a general and inclusive description of electric current and illustrate your statement with appropriate symbols.
4. State Coulomb's law and identify the kind of information that can be deduced from it.
5. What is permittivity? Describe how this quantity can be found for various materials (media).
6. Assume that a circuit consists of a battery connected in series with a resistor. Describe the action that takes place in the circuit as an electron is made to circulate around the circuit.
7. Distinguish among voltage, voltage drop, voltage rise, and potential difference.
8. Does capacitance exist between two thin current-carrying conductors? Explain.
9. What are the units of electric flux density?
10. Describe how the resistance property of a conducting material can be determined using Ohm's law. Can this procedure be applied to nonlinear materials? Explain. Can this procedure be applied to devices that exhibit discontinuities? Explain.
11. Draw a series circuit comprising a battery, a switch, a resistor, and a coil. Describe the effect of the law of electromagnetic induction when the switch is suddenly opened after being allowed to remain closed for a period of time. How does Lenz's law manifest itself in this action?
12. Explain the meaning of inductance and the manner in which it is related to the circuit variable v and the circuit variable i.
13. Does the circuit of Question 11 possess inductance when the circuit is in a quiescent state? Explain.

14. Explain the principle of conservation of charge and its relationship to Kirchhoff's current law.

15. Describe the principle of conservation of energy and its relationship to Kirchhoff's voltage law.

16. Refer to Fig. 1-9 and explain the manifestation of the law of conservation of energy in this circuit. In particular, comment on the specific situation that prevails at each circuit element.

17. State Ampère's law and illustrate its applications to show that the net force is zero in the case where two long, current-carrying conductors are placed in quadrature with one another.

Problems

GROUP I

1-1. In a closed electric circuit the number of electrons that pass a given point is known to be 20 billion billion electrons per second. Find the corresponding current flow in amperes.

1-2. A negative point charge Q_1 of magnitude 2×10^{-12} C is placed in a vacuous medium.
   (a) Find the value of the electric field at a point 1 m from the charge.
   (b) Compute the work done in bringing an electron from a point very far removed from Q_1 to a point 0.5 m away.
   (c) What force is acting on the electron in part (b)?

1-3. A positive point charge of value 10^{-6} C is fixed at a point in a vacuum. An identical point charge is then brought to a distance 10 cm away.
   (a) Compute the potential energy of the second charge.
   (b) How much work per unit charge must be done in moving charges about in the electric field created by the fixed charge source? Is this a constant quantity? Explain.

1-4. Two flat parallel plates, measuring 0.1 m by 0.2 m, are charged by transferring 10^{-6} C from one plate to the other. The plates are separated by 2 cm and oil is placed between them. The relative permittivity of the oil is 2.
   (a) Find the electric flux density between the plates.
   (b) Determine the value of the electric field intensity.
   (c) Compute the potential difference existing between the plates.
   (d) What is the capacitance of the parallel plates?

1-5. For the circuit shown in Fig. P1-5 find the current that flows and also the potential difference across each resistor.

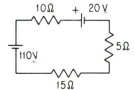

Fig. P1-5

1-6. Four pieces of apparatus are connected as shown in Fig. P1-6. Free electrons circulate at a constant rate in the direction *a-b-c-d*. The quantity of electricity passing a reference point on the circuit in a given time is 50 C. During this time, in A, the electrons receive energy in the amount of 600 J. In B, energy of 200 J is given up by the electrons, and in C, 300 J is supplied to the electrons. Redraw the circuit showing batteries or resistors instead of boxes A, B, C, and D. Mark polarities and compute the emf or voltage drop for each element.

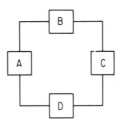

Fig. P1-6

1-7. Determine the instantaneous polarity for the coil configurations shown in Fig. P1-7. The flux is increasing in the direction shown.

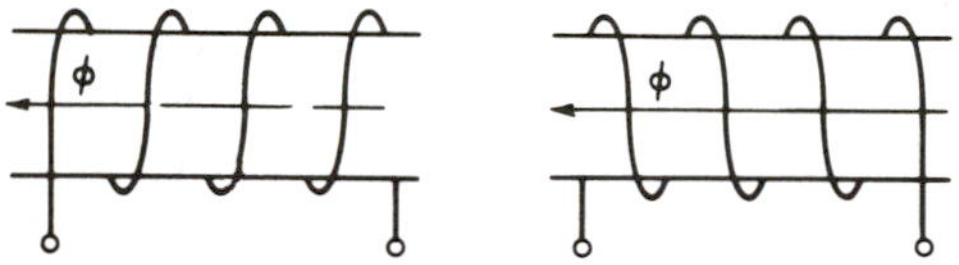

Fig. P1-7

1-8. (a) Find the magnetic flux density in free space at a distance 5 cm from a long straight wire carrying a current of 50 A.

    (b) Calculate the force produced on a *second* wire 1 m long placed 5 cm from the first wire and carrying 100 A.

    (c) Repeat parts (a) and (b) assuming the wires are embedded in iron having a relative permeability of 1500. Use $\mu_0 = 4\pi \times 10^{-7}$.

1-9. A portion of a network has the configuration shown in Fig. P1-9. The voltage drops existing across the three resistors are known to be respectively 20, 40, and 60 V having the polarities indicated. Find the value of R_3.

1-10. In Fig. P1-10 find the voltage drop across R_1 and R_2. The resistance R_3 is not specified.

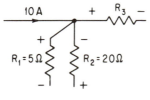

Fig. P1-9

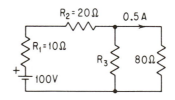

Fig. P1-10

The Fundamental Laws of Electrical Engineering   Chap. 1

GROUP II

1-11. The voltage drop across the 15-Ω resistance in the circuit of Fig. P1-11 is 30 V having the polarity indicated. Find the value of R.

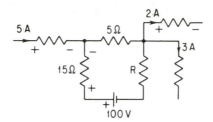

Fig. P1-11

1-12. A stationary magnetic field is produced by the magnets shown in Fig. P1-12. Between the magnets there is placed a cylindrical rotor which supports current-carrying conductors. The direction of current flow in these conductors is represented by dots (out of paper) and crosses (into the paper).
(a) If these current directions are assumed fixed, is there a torque on each conductor?
(b) Does the rotor experience a net torque different from zero, i.e., does the rotor turn?
(c) What recommendation can you make to improve the situation?

Fig. P1-12

1-13. In the configuration shown in Fig. P1-13 the coil has 100 turns and is attached to the rotating member which revolves at 25 rev/s. The magnetic flux is a radial uniform field and has a value of $\phi = 0.002$ weber (Wb). Compute the emf induced in the coil.

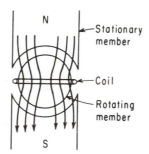

Fig. P1-13

chapter two

The Circuit Elements

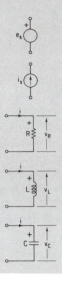

This is the first of four chapters devoted to a treatment of electric circuit theory. It is the purpose of *electric circuit theory* to furnish a means of describing the transfer of energy from a suitable source to appropriate engineering devices which are designed to achieve certain useful results. Through circuit theory, for example, we can readily analyze the generation, transmission, and application of electrical energy for home, commercial, and industrial uses. In like fashion the design, operation, and performance of radio and television circuits, telephone and telegraph circuits, and industrial electronic circuitry are easily determined through the techniques of circuit analysis.

It is important to understand at the outset, however, that *circuit analysis is really a simplified and approximate form of field analysis*. Recall that each of the basic experiments performed by Coulomb, Faraday, and Ohm dealt with the medium of space, involving such quantities as electric field intensity (volts/meter), charge density (coulombs/meter^3), and magnetic flux density (webers/meter^2). The dependence of these quantities on length, volume, and area accordingly identified them as space or field variables. These variables also depend on time variations to which they may be subjected. If all problems in electricity and magnetism were to be solved as straight field problems, they would be quite complex and their solution would be laborious. Fortunately, however, under ordinary circumstances it is permissible to solve these problems by dealing with integrated effects. Thus, for example, instead of working with the space variable for electric field intensity we can work directly with voltage, which is the electric field intensity integrated with respect to distance. Similarly, in place of current density we can deal with current, which we obtain by integrating the current density over an appropriate area such as that of a conducting wire. This approach not only simplifies analysis but also facilitates measurements, because current and voltage are easily measured.

If we are to use circuit analysis rather than field analysis, the loss of energy through radiation must be negligible. Section 1-6 explains how charge receives energy from an active element in a circuit and delivers it to the passive elements. At the passive elements this energy may then be dissipated as heat or stored in magnetic or electric fields. Energy transfer requires motion of charges. Recall, too, that when a charge is accelerated by a field it gains energy, whereas when it is decelerated it gives up energy to the field. Now it can be shown that for engineering devices of dimensions small compared with a wavelength, the energy lost through radiation is indeed negligible unless exorbitant values of charge acceleration are achieved.† In turn this means that throughout the closed path in which the charge particles move, the energy is confined solely to those devices present in the closed path. In such cases integrated effects may be used to describe and analyze problems which are determined in terms of the physical dimensions and properties of the devices as well as the voltage and current existing at a finite number of equipotential surfaces. These surfaces are called the *terminals* of the device. In Fig. 1-8, for example, it is possible to express the energy delivered to the device appearing across terminals *a* and *b* in terms of the integrated voltage between these points and the current flowing into terminal *a*. The subject matter of this book is restricted to those situations where there is negligible radiation loss. Hence circuit analysis (by application of Kirchhoff's laws) is always valid.

This chapter begins with a brief description of ideal current and voltage sources as two important circuit elements which were not introduced in a special way in Chapter 1. Then following a description of the sign conventions and symbols to be used in circuit diagrams, attention is directed to how the energy received by an electric device is expressible in terms of its associated voltage and current. The next three sections focus on the three important circuit elements which resulted from the fundamental investigations of Ohm, Faraday, and Coulomb. These circuit elements are respectively the resistor, the inductor, and the capacitor. The chapter closes with an introduction to the operational amplifier as a circuit element and as a controlled source element as well.

2-1 IDEAL INDEPENDENT CURRENT AND VOLTAGE SOURCES

All electric circuits are driven by sources. These sources may be *independent* of other network variables, as in the case of synchronous generators that are used to supply energy to homes and many industrial and commercial loads; or

---

† Of the order of 10^{16} ft/s^2.

they may be of a *dependent* type, such as is frequently encountered in electronic circuits. Attention here is confined to the independent type. However, independence of a circuit variable such as current or voltage does not imply independence of time. Accordingly, many independent sources are often time varying, as with the synchronous generator whose phase voltage is described by $e(t) = E_m \sin \omega t$, where ω is determined by the speed of rotation of the generator.

The distinguishing feature of an *independent voltage source* is that the value of the voltage is not dependent on either the magnitude or direction of the current flowing through the source. Figure 2-1(a) depicts the model representation of such a two-terminal source. A lowercase letter is used when the character or value of the source is unspecified or time varying. If the voltage source is known to be an ideal battery, an uppercase letter E or V is used to denote its value as illustrated in Fig. 2-1(b). Displayed in Fig. 2-1(c) is the voltage-current characteristic of an ideal constant-voltage source. Observe that the characteristic is horizontal to the current axis, which is entirely consistent with the fact that any value of current magnitude and direction can be associated with the voltage source. Of course, it is assumed that the voltage source is not being operated beyond its energy capability; otherwise, the ideal representation is no longer valid.

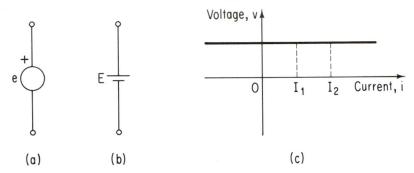

Fig. 2-1 Independent voltage source: (a) symbolic representation; (b) specific example of a battery as a voltage source; (c) the v-i characteristic.

A notable point about an ideal voltage source is its zero internal resistance. This is readily demonstrated by invoking Ohm's law in incremental form as it applies to the v-i characteristic. Thus we can write

$$R = \frac{\Delta V}{\Delta I}$$

where ΔV denotes the change in source voltage associated with a given change in current, $\Delta I = I_2 - I_1$. From Fig. 2-1(c) it is clear that the change in voltage associated with the change in current is zero. Hence

$$R = \frac{\Delta V}{I_2 - I_1} = \frac{0}{I_2 - I_1} = 0$$

This result applies to all ideal voltage sources, even those which are time varying.

The ideal independent *current source* is a two-terminal element which supplies

The Circuit Elements   Chap. 2

its specified current to the circuit in which it is placed independently of the value and direction of the voltage appearing across its terminals. Displayed in Fig. 2-2(a) is the symbol used to represent the current source, and Fig. 2-2(b) shows the v-i characteristic. By applying an analysis similar to that which is used for the voltage source, it is easy to demonstrate that the internal resistance of an ideal current source is infinite. Thus

$$R = \frac{\Delta V}{\Delta I} = \frac{V_2 - V_1}{0} = \infty$$

Accordingly, when one "looks" into the terminals of a current source, one "sees" an open circuit. This is a useful bit of information to keep at hand when doing circuit analysis that involves current sources.

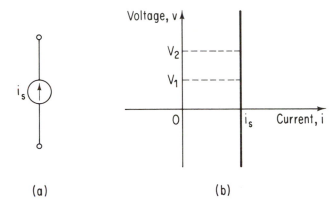

Fig. 2-2 Independent current source: (a) symbolic representation; (b) the v-i characteristic.

EXAMPLE 2-1   An ideal 1-V voltage source is applied to a 10-Ω resistor.

(a) Determine the voltage drop across the resistor as well as the current flow.

(b) An additional 5-Ω resistor is connected in series with the first 10-Ω resistor. Find the resulting current flow and the voltage drop across each resistor.

(c) Repeat parts (a) and (b) for the case where the 1-V voltage source is replaced by a 1-A current source.

Solution: (a) Figure 2-3(a) depicts the circuit arrangement. Clearly, the voltage across the 10-Ω resistor is

$$V_{ab} = 1 \text{ V}$$

The corresponding current flow is

$$I_{ab} = \frac{V_{ab}}{R} = \frac{1 \text{ V}}{10} = 0.1 \text{ A}$$

(b) Figure 2-3(b) shows the new arrangement. It is instructive to note that a voltage source makes its full voltage available to the circuit. Consequently, now the 1-V source is made to appear across the series combination of the 10-Ω and the 5-Ω resistors. This produces a current flow of

$$I = \frac{V}{R_{ab} + R_{bc}} = \frac{1}{10 + 5} = \frac{1}{15} = 0.067 \text{ A}$$

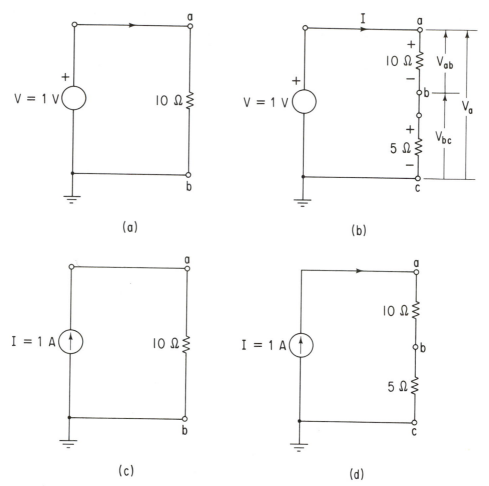

Fig. 2-3 Circuit diagram for Example 2-1: (a) voltage source applied to single resistor; (b) two series-connected resistors; (c) current source applied to single resistor; (d) current source supplying two series-connected resistors.

The corresponding voltage drops are then

$$V_{ab} = IR_{ab} = 0.067(10) = 0.67 \text{ V}$$

and

$$V_{bc} = IR_{bc} = 0.067(5) = 0.33 \text{ V}$$

(c) The circuit arrangement in this case is illustrated in Fig. 2-3(c) and (d). Observe that in each case the total current of the current source is supplied to the connected circuit; that is, the current flow is 1 A whether just the one resistor or both resistors are used. When just the 10-Ω resistor is used, the voltage across the resistor for a 1 A current flow is clearly

$$V_{ab} = IR_{ab} = (1)(10) = 10 \text{ V}$$

When a series connection of the two resistors is used, the voltage across the combined resistors is

$$V_{ac} = V_{ab} + V_{bc} = (1)(10) + (1)(5) = 15 \text{ V}$$

2-2 REFERENCE DIRECTIONS AND SYMBOLS

The selection of a consistent set of reference directions makes it easier to read and draw circuit diagrams. The actual choices are quite arbitrary as long as they are consistent. In the case of current flow it is customary to chose the positive flow direction opposite to the actual flow of the electrons in response to the electric field produced within the body of the conductor. This has the effect of orienting the current flow so that i moves from the terminal of lower potential to the terminal of higher potential for sources and from higher potential to lower potential for passive elements. A glance at Fig. 2-3(b) readily illustrates the point.

The reference directions for two-terminal passive elements such as resistors, inductors, and capacitors are so established that the defined positive current is made to flow into the defined positive terminal of the element. Thus, in Fig. 2-3(b), the voltage drop is assumed to have a polarity that puts terminal a at a higher potential than terminal b. This is the defined positive direction for the voltage V_{ab}. Moreover, when arrows are used to represent such voltages, they are always drawn so that the arrowhead corresponds to the positively defined terminal. It is important to understand that this choice of the positive direction of both the current flow and the voltage drop in the circuit of Fig. 2-3(b) is independent of the particular polarity of the source. Accordingly, if the voltage source in this circuit were to be *reversed*, it follows that for the previously defined positive direction we would get the results $I = -0.067$ A and $V_{ab} = -0.67$ V. The negative sign for current simply recognizes that in this altered circuit the actual current flow is opposite to the assumed positive direction. A similar interpretation applies to V_{ab}.

Electric circuits generally have one point in the circuit that is used as a reference with respect to which the potential of all other points is measured. It is customary to select that terminal of the source which is at the lower potential. The potential of this point is frequently called the *ground potential* and is denoted by the symbol indicated in Fig. 2-3. In house circuits the reference wire is often connected to a pipe of the plumbing system and thereby establishes a true ground connection. In electronic portable devices the reference ground connection is usually the chassis. When the potential of a point in a circuit is to be expressed relative to the ground potential, the second subscript may be omitted. Thus, in the circuit of Fig. 2-3(b), the potential of point a with respect to ground potential may be written simply as V_a.

The reference direction for power must be consistent with the reference directions for voltage and current because power is dependent on the product of these two quantities (see Sec. 2-3). If the actual current and voltage have the same signs as the assumed directions, the power is positive. This means that

TABLE 2-1 SYMBOLS FOR ELEMENTARY CIRCUITS

Circuit element	Symbol for model	Comment
Voltage source		e_s appears in the circuit for all i
Current source		i_s supplied to the circuit for all v
Resistor		$v_R = iR$
Inductor		$v_L = L\dfrac{di}{dt}$ or $i = \dfrac{1}{L}\int v_L\,dt$
Capacitor		$v_c = \dfrac{1}{C}\int i\,dt$ or $i = C\dfrac{dv_c}{dt}$
Joined wires		Connection is made at the dot
Unjoined wires		Wires cross without contact
Open circuit		i is zero for all v
Short circuit		v is zero for all i

the passive element receives energy from the source. When the actual current and voltage are both opposite to the assumed positive directions, the conclusion remains the same because a product of two negative quantities is involved.

Displayed in Table 2-1 is a summary of the symbols that are usually found in elementary electric circuits. Additional items can be included as new devices are discussed. Many electronic devices have symbols for model representation that are particularly pertinent to their functional behavior. This becomes evident as these units are introduced.

2-3 ENERGY AND POWER

Our purpose here is to obtain a relationship between the energy delivered to a device and the integrated effects of the electric field intensity (voltage) and current density (current). Figure 2-4 depicts a battery delivering energy to a circuit element appearing between terminals c and d. It is assumed that the current flowing through the device is i and the potential difference across the terminals is v. From Eq. (1-12) we know that v denotes the amount of energy released by each unit charge as it moves through the circuit element from c to d. Therefore, the total energy delivered to the device by the charge carriers is

$$w = vq \text{ coulomb-volt (or joules)} \tag{2-1}$$

Moreover, the quantity of electricity q associated with a constant current i is

$$q = it \quad \text{C} \tag{2-2}$$

Inserting Eq. (2-2) into (2-1) yields the desired result. Thus

$$\boxed{w = vit} \qquad \text{W-s (J)} \tag{2-3}$$

which shows that the energy absorbed is given by the product of the potential difference, the current, and the time. Although Fig. 2-4 depicts a situation where a constant current flows, Eq. (2-3) is equally applicable for time-varying quantities. Lowercase letters are used to stress this point.

Power is a defined quantity. It is the time rate of doing work. Letting p denote power it can be expressed mathematically as

$$\boxed{p \equiv \frac{w}{t}} \quad \text{J/s} \tag{2-4}$$

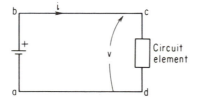

Fig. 2-4 Delivering energy to the circuit element by means of current i.

Upon substituting Eq. (2-3), there results an alternative expression

$$\boxed{p = vi} \quad \text{W J/s} \tag{2-5}$$

Note that power expresses a device's energy-delivering or energy-absorbing capabilities per unit of time. Accordingly, if we compare two devices and we find that one can absorb three times as much energy as the other in any given second, we say that the first has three times the power of the second.

2-4 THE RESISTANCE PARAMETER

Resistance is one of three basic parameters of electric circuit theory. Strictly speaking, every circuit element exhibits some degree of resistance—but often, as in the case of inductors, capacitors, or ordinary wire connections between various circuit elements, it is of minor importance.

In Chapter 1 resistance is introduced as the proportionality factor in Ohm's law relating current to potential difference. Ordinarily, the free electrons associated with the atoms of a conducting material undergo *random* thermal motion and so do not constitute a net current flow. However, the application of a voltage source to such a conductor produces an electric field within the body of the conductor which then imparts a directed velocity component to the random motion of the electrons. As the electrons respond to the electric field by moving in a direction opposite to it, they encounter frequent collisions with the atoms in the lattice structure of the conducting material, which in turn produces an irreversible heat loss. In a general way resistance can be described as that property of a circuit element which offers opposition to the flow of current and in so doing converts electrical energy into heat energy. In the interest of furnishing a complete picture the resistance parameter is treated below from several viewpoints, each of which reveals some useful and interesting facet of the properties and characteristics of this important circuit element.

Circuit Viewpoint. A circuit element is described from the circuit viewpoint when it is expressed in terms of an associated current and potential difference. Since by Ohm's law we have

$$\boxed{R = \frac{v}{i}} \quad \Omega \tag{2-6}$$

it follows that this equation embodies the circuit-viewpoint description of resistance. Equation (2-6) is a linear algebraic expression when the proportionality factor R is independent of the current. Moreover, Eq. (2-6) is valid for negative as well as positive values of i. These remarks are graphically depicted in Fig. 2-5. It is also worth noting that when Ohm's law correctly describes the relation between terminals c and d, the included circuit element is called a *resistor*.

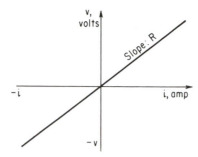

Fig. 2-5 Graphical representation of Ohm's law.

Experiment shows that the resistance of most metallic conductors varies with temperature. Specifically, if the resistance of a metal at temperature T_1 is R_1, then for normal range of temperatures the resistance at temperature T_2 is given by

$$R_2 = R_1[1 + \alpha_1 (T_2 - T_1)] \tag{2-7}$$

where α_1 is the temperature coefficient of resistance at T_1 and temperature is customarily measured in degrees Celsius. The values of α for various metals appear in Appendix C. Investigations reveal that a linear variation of resistance with temperature prevails over the approximate range -50 to $200°C$. Consequently, a more convenient form of Eq. (2-7) is expressed approximately as

$$\frac{R_2}{R_1} = \frac{K + T_2}{K + T_1} \tag{2-8}$$

For copper, the commonly used conductor material, K takes on the value of 234.5. This value is obtained by extrapolating the linear curve to zero resistance as shown in Fig. 2-6. Hence Eq. (2-8) becomes

$$\frac{R_2}{R_1} = \frac{234.5 + T_2}{234.5 + T_1} \tag{2-9}$$

This expression is useful in computing the effect of increased temperature on copper conductors, particularly because copper is almost always used in the temperature range for which the equation is valid. By way of illustration, Eq. (2-9) indicates that if the resistance of copper conductors in a given circuit at an ambient temperature of T_1 equal to $20°C$ is R_1, then the resistance at an

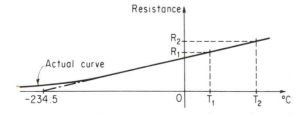

Fig. 2-6 Variation of resistance with temperature for copper.

Sec. 2-4   The Resistance Parameter

operating temperature T_2 equal to 80°C is

$$R_2 = R_1 \left(\frac{234.5 + 80}{234.5 + 20} \right) = 1.235 \, R_1 \tag{2-10}$$

Thus there is a 23.5% increase in resistance at the higher temperature. This increase in resistance occurs in many engineering devices such as motors, generators, and transformers when they operate at rated output power.

Energy Viewpoint. The resistance parameter may also be looked upon in terms of its characteristic property of converting electrical energy into heat energy. If Eqs. (2-5) and (2-6) are combined, it follows that the power absorbed by a resistor is

$$p = vi = (iR)i = i^2R \tag{2-11}$$

This is called *Joule's law* and it states that the number of electrons colliding with the atoms of a circuit element to produce heat is proportional to the square of the current. The corresponding amount of energy converted to heat in the time interval $t_2 - t_1$ is readily found from

$$W = \int_{t_1}^{t_2} Ri^2 \, dt \qquad \text{J} \tag{2-12}$$

The integral form is applicable whenever the current i is a time-varying quantity. When i is a constant quantity I, Eq. (2-12) becomes

$$W = RI^2t \qquad \text{J} \tag{2-13}$$

where

$$t = t_2 - t_1 \qquad \text{s} \tag{2-14}$$

Accordingly, Eq. (2-13) indicates that resistance may also be expressed in terms of the dissipated energy per unit time per unit current squared. Thus

$$\boxed{R = \frac{W}{I^2t}} \qquad \Omega \tag{2-15}$$

All current-carrying conductors and resistors involve a heat loss which occurs at a rate given by I^2R. At first this heat is stored in the body of the material, thereby bringing about a temperature rise. As soon as the body temperature exceeds the ambient temperature, however, heat is transferred to the surrounding medium. Eventually a point is reached where the heat is transferred at the same rate that it is produced. Consequently, heat ceases to be stored in the material, and temperature no longer rises. The ability of a resistor to transfer heat necessarily depends on the area exposed to the surrounding medium. If the power-dissipating capability of a resistor or rheostat is inadequate for a specified rating it can easily burn out. To prevent such occurrences resistors and rheostats are given a specified safe power rating which is determined by test. Thus one may purchase a 2-W, 100-Ω resistor or a 10-W, 100-Ω resistor. Although the resistance values are the same, the physical dimensions of the 10-W resistor are much larger.

EXAMPLE 2-2   A 100-Ω resistor is needed in an electric circuit to carry a current of 0.3 A. The following resistors are available from stock:

$$
\begin{array}{ll}
100\ \Omega & 5\ \text{W} \\
100\ \Omega & 7.5\ \text{W} \\
100\ \Omega & 10\ \text{W}
\end{array}
$$

Which resistor would you specify?

Solution: The power associated with a current of 0.3 A and 100 Ω is

$$P = I^2R = (0.3)^2(100) = 9\ \text{W}$$

Hence select the 10-W resistor to guard against overheating and possible damage.

Geometrical Viewpoint.   The resistance parameter is fundamentally a geometric constant. This fact was discovered by Ohm in his original investigations. In his analogy to Fourier's heat-conduction equation Ohm showed that the resistance of a conductor of uniform dimensions was directly dependent upon the length, inversely proportional to the cross-sectional area, and also dependent upon the physical conduction properties of the material. This information appears in Eq. (1-23) and is repeated here for convenience. Thus

$$\boxed{R = \rho\frac{L}{A}}\quad \Omega \tag{2-16}$$

where ρ = resistivity of the material, Ω

L = length of the conductor, m

A = cross-sectional area, m^2

Equation (2-16) is easily verified experimentally. For example, it is found that for a material of fixed resistivity doubling the length doubles the resistance while doubling the area halves the resistance.

EXAMPLE 2-3   Find the resistance of a round copper conductor having a length of one meter and a uniform cross-sectional area of 1 cm^2. The resistivity of copper is 1.724×10^{-8} Ω-m.

Solution: From Eq. (2-16) there results

$$R = \rho\frac{L}{A} = 1.724 \times 10^{-8}\frac{1\ \text{m}}{10^{-4}\ \text{m}^2} = 1.724 \times 10^{-4}\ \Omega$$

Often the resistivity of copper is expressed in units of micro-ohm-cm in order to avoid the 10^{-8} factor. It is so expressed in Table C-1 of Appendix C. In this table note that the resistivity is also expressed in units of ohm-circular mils/foot. Reference to Eq. (2-16) reveals that this description represents the resistance of a piece of material which has a cross-sectional area of one circular mil and a length of one foot. The *circular mil* is a convenient unit which is used to describe the cross-sectional area of round conductors. By definition, a circular mil is a unit of area and specifically it denotes the area of a circle having a diameter of 0.001 in. or 1 mil. What then is the area of a round wire having a diameter of 0.10 in. expressed in units of circular mils? As is customary in dealing with measurements, we must find the number of times the circle of 1-mil diameter fits into the circle of 100-mil diameter. Since the area of a circle is proportional to the square of the diameter, it follows that the answer is $(100)^2 = 10,000$ circular mils. This clearly leads to the following simple rule: To find the area in circular mils, multiply the diameter expressed in inches by 1000 and square the resulting number.

EXAMPLE 2-4   Find the resistance of a round copper wire having a diameter of 0.1 in. and a length of 10 ft.

Solution:

$$R = \rho \frac{L}{A} = 10.37 \times \frac{10}{10,000} = 0.01037 \ \Omega$$

Commercial wire sizes are standardized in the United States in accordance with the area expressed in circular mils. Each wire size is given an AWG (American Wire Gage) number for identification and each has a maximum allowable current rating which varies with the type of insulation used. Table C-2 in Appendix C lists this information.

Conductance is defined as the reciprocal of resistance. The unit used to measure conductance is the siemens. The symbol for conductance is G. Thus

$$G \equiv \frac{1}{R} \tag{2-17}$$

Hence

$$G = \frac{1}{\rho} \frac{A}{L} = \sigma \frac{A}{L} \tag{2-18}$$

The quantity σ denotes the conductivity and is the reciprocal of resistivity.

Ohm's law may be expressed in terms of the conductance of a circuit element. From Eqs. (2-6) and (2-17) we can write

$$i = \frac{v}{R} = Gv \quad \text{A} \tag{2-19}$$

Thus by multiplying the potential difference by the conductance we obtain the corresponding current. Sometimes this is a preferred procedure in circuit analysis, as is demonstrated later in the book.

Physical Aspects.   The resistors that are found in electric circuits doing many different tasks are of several types as regards to physical construction, power dissipating capability, and tolerance of the resistance value. The *wire-wound* resistor is often found in larger industrial applications where greater power handling capability is required. These resistors are constructed of a nickel-chromium alloy and are wound on a ceramic core. As such they are considerably less sensitive to ambient temperature variation. They can also be constructed to a higher accuracy than is possible with most other types. Deviations of the resistance from its nominally designed value may be as small as 1 to 0.01% and even lower in some instances. The *carbon-type* resistor has found widespread use in electronic circuits, although in recent years it is being replaced increasingly by the diffused

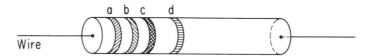

Fig. 2-7 Carbon-type resistor with color-coded bands.

resistor. The physical form of the carbon resistor is illustrated in Fig. 2-7. As expected from the nomenclature, the cylindrical shape is composed of carbon and these units are relatively inexpensive to build. However, they are highly sensitive to temperature variations. In the interest of making the resistance value easy to identify carbon resistors are color coded in accordance with the information that appears in Table 2-2. The color bands are painted near one end of the resistor as indicated in Fig. 2-7. The first three bands (a, b, c) convey information about the nominal value of the resistance consistent with the expression $R = (ab)10^c$. Thus, if band a is black, band b is brown, and band c is green, then the nominal resistance value is $R = (01)10^5 = 100$ kΩ. The fourth band, d, provides information about the manufacturing tolerance of the resistor. Thus, if the band is painted gold, the actual value of a given resistor taken from stock can fall anywhere in the range 95 to 105 kΩ, i.e., 100 kΩ $\pm$ 5% of the nominal value. Carbon resistors are essentially low-wattage resistors. The power dissipating capability of these units ranges from about 0.1 to 2 W and the physical size of the larger units have diameters less than 1 cm.

TABLE 2-2   COLOR CODE FOR CARBON RESISTORS

To determine resistance value in ohms, $R = (ab)10^c$:

Color	Value of a, b, c	Value of c
Black	0	
Brown	1	
Red	2	
Orange	3	
Yellow	4	
Green	5	
Blue	6	
Violet	7	
Gray	8	
White	9	
Gold		-1
Silver		-2

To determine tolerance value:

Color	Tolerance value (%)
Black or no color	± 20
Silver	± 10
Gold	± 5

A third type resistor is the *metal film* resistor. It is constructed by using film deposition techniques to deposit a thick film of resistive material onto an insulating substrate in the manner depicted in Fig. 2-8. The accuracy of these units can be made to approach those of the wire-wound types because they can be trimmed quite accurately with the use of a laser. Finally, there is the *diffused resistor,* which is fabricated by the same procedure used for integrated circuits. These resistors normally carry a tolerance value of ±20% because they cannot be trimmed.

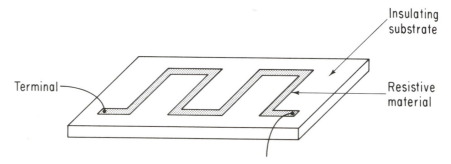

Fig. 2-8 Metal-film resistor.

Standard Resistance Values. In the interest of limiting the number of resistors to be manufactured to meet the needs of the electronics industry, manufacturers agreed to furnish a total of 24 steps for each decade of ohmic resistance. Moreover, each step is to be separated from its adjacent one by approximately ±10%. The number of decades to be covered corresponds to the value of c in Table 2-2. Appearing in Table 2-3 are the standard values as they apply to the first decade with the values ranging from 1 to 9.1 Ω. The standard tolerance value for these resistance values is ±5%. However, all resistance values in the first row are also available as ±10% resistors. All entries on the third row are available with ±20% tolerances as well as ±10%.

TABLE 2-3 STANDARD RESISTANCE VALUES

1.0	1.5	2.2	3.3	3.4	6.8
1.1	1.6	2.4	3.6	5.1	7.5
1.2	1.8	2.7	3.9	5.6	8.2
1.3	2.0	3.0	4.3	6.2	9.1

2-5 THE INDUCTANCE PARAMETER

As described in Sec. 1-5 inductance was first discovered by Faraday in his renowned experiments of 1831. In a general way *inductance* can be characterized as that property of a circuit element by which energy is capable of being stored in a magnetic flux field. A significant and distinguishing feature of inductance, however, is that it makes itself felt in a circuit only when there is a *changing current*. Thus, although a circuit element may have inductance by virtue of its geometrical and magnetic properties, its presence in the circuit is not exhibited unless there is a time rate of change of current. This aspect of inductance is particularly stressed when we consider it from the circuit viewpoint.

Circuit Viewpoint. The current-voltage relationship involving the inductance parameter is expressed by Eq. (1-28) and is repeated here for convenience. Thus

$$v_L = L\frac{di}{dt} \tag{2-20}$$

In general both v_L and i are functions of time. Figure 2-9 depicts the potential difference v_L appearing across the terminals of the inductance parameter when a changing current flows into terminal c. Note that the arrowhead on the v_L quantity is shown at terminal c, indicating that this terminal is instantaneously positive with respect to terminal d. In turn this means that current is increasing in the positive sense, i.e., the slope di/dt in Eq. (2-20) is positive. Any circuit element that exhibits the property of inductance is called an *inductor* and is denoted by the symbolism shown in Fig. 2-9. In the ideal sense the inductor is considered to be resistanceless, although practically it must contain the wire resistance out of which the inductor coil is formed.

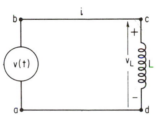

Fig. 2-9 Illustrating the inductance parameter.

It follows from Eq. (2-20) that an appropriate defining equation for inductance is

$$L = \frac{v_L}{di/dt} \quad \frac{\text{volt-seconds}}{\text{ampere}} \text{ or henrys} \tag{2-21}$$

Thus by recording the potential difference at a given time instant across the terminals of an inductor and dividing by the corresponding derivative of the current time function we determine the inductance parameter. Note that the units of inductance are volt-second/ampere. For simplicity this is more commonly called the *henry*.

A *linear inductor* is one for which the inductance parameter is independent of current. As current flows through an inductor it creates a space flux. When this flux permeates air, strict proportionality between current and flux prevails so that the inductance parameter stays constant for all values of current. A plot of the potential difference across the coil as a function of the derivative of the current then appears as shown in Fig. 2-10, which is a plot of Eq. (2-21). Note the similarity to Fig. 2-5, which applies for the resistance parameter. Of course the abscissa is different in each case. When the flux is made to penetrate iron, however, it is possible for large currents to upset the proportional relationship between the current and the flux it produces. In such a case the inductor is called *nonlinear* and a plot of Eq. (2-21) will then no longer be a straight line.

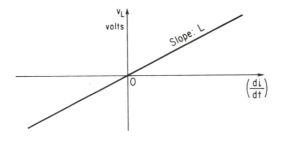

Fig. 2-10 Graphical representation of the inductance parameter L from the circuit viewpoint.

For the resistance parameter Ohm's law can be written to express either voltage in terms of current [see Eq. (1-24)] or current in terms of voltage [see Eq. (1-23)]. The same procedure may be followed for the inductance parameter. Equation (2-20) already expresses the voltage as a function of the current. However, to express the current in terms of the potential difference across the inductor, Eq. (2-20) must be transposed to read as follows:

$$di = \frac{1}{L} v_L \, dt \qquad (2\text{-}22)$$

In integral form this becomes

$$\int_{i(0)}^{i(t)} di = \frac{1}{L} \int_0^t v_L \, dt \qquad (2\text{-}23)$$

or

$$i(t) = \frac{1}{L} \int_0^t v_L \, dt + i(0) \qquad (2\text{-}24)$$

Equation (2-24) thus reveals that the current in an inductor is dependent upon the integral of the voltage across its terminals as well as the initial current in the coil at the start of integration.

An examination of Eqs. (2-20) and (2-24) reveals an important property of inductance: *the current in an inductor cannot change abruptly in zero time*. This is made apparent from Eq. (2-20) by noting that a finite change in current in zero time calls for an infinite voltage to appear across the inductor, which is physically impossible. On the other hand, Eq. (2-24) shows that in zero time the contribution to the inductor current from the integral term is zero so that the

The Circuit Elements   Chap. 2

current immediately before and after application of voltage to the inductor is the same. In this sense, then, we may look upon inductance as exhibiting the property of inertia.

EXAMPLE 2-5   An inductor has a current passing through it which varies in time in the manner depicted in Fig. 2-11(a). Find the corresponding time variation of the voltage drop appearing across the inductor terminals, if it is assumed that the inductance of the coil is 0.1 H.

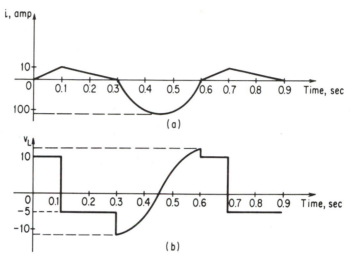

Fig. 2-11   (a) Input current wave shape to an inductor; (b) corresponding voltage variation across inductor terminal.

Solution: The solution appears in Fig. 2-11(b). Note that in the interval from 0 to 0.1 s, $di/dt = 100$ A/s. Hence the voltage across the coil is then a constant given by

$$v_L = L\frac{di}{dt} = (0.1)(100) = 10 \text{ V} \qquad \text{for } 0 < t < 0.1$$

In the time range from 0.1 to 0.3 s the slope of the current curve is -50. Hence the voltage across the coil is -5 v. Finally, in the interval from 0.3 to 0.6 s the current wave slope is sinusoidal so that the corresponding voltage wave is a cosine. In drawing the cosine wave it is assumed that the maximum slope of the sine wave exceeds 100 A/s.

It is interesting to note in Fig. 2-11 that unlike the current in an inductor, the voltage across the inductor is allowed to change discontinuously. Why?

Energy Viewpoint.   Assume that an inductor has zero initial current. Then if a current i is made to flow through the coil across which appears the potential difference v_L, the total energy received in the time interval from 0 to t is

$$W = \int_0^t v_L i \, dt \qquad \text{J} \tag{2-25}$$

Inserting Eq. (2-20) leads to

$$W = \int_0^t \left(L\frac{di}{dt}\right) i \, dt = \int_0^i Li \, di \tag{2-26}$$

or

$$W = \frac{1}{2}Li^2 \quad \text{J} \qquad (2\text{-}27)$$

Continuing with the assumption that the inductor has no winding resistance, Eq. (2-27) states that the inductor absorbs an amount of energy which is proportional to the inductance parameter L as well as the square of the instantaneous value of the current. Thus energy is stored by the inductor in a magnetic field. It is of finite value and retrievable. As the current is increased, so too is the stored magnetic energy. Note, however, that this energy is zero whenever the current is zero. Because the energy associated with the inductance parameter increases and decreases with the current, we can properly conclude that the inductor has the property of being capable of returning energy to the source from which it receives it.

A glance at Eq. (2-27) indicates that an alternative way of identifying the inductance parameter is in terms of the amount of energy stored in its magnetic field corresponding to its instantaneous current. Thus, in mathematical form we can write

$$L = \frac{2W}{i^2} \quad \text{H} \qquad (2\text{-}28)$$

This is an energy description of the inductance parameter.

There is one final point worthy of note. It has already been demonstrated that for a potential difference to exist across the inductor terminals the current must be changing. A constant current results in a zero voltage drop across the ideal inductor. This is not true, however, about the energy absorbed and stored in the magnetic field of the inductor. Equation (2-27) readily verifies this fact. A constant current results in a fixed energy storage. Any attempt to alter this energy state is firmly resisted by the effects of the initial energy storage. This again reflects the inertial aspect of inductance.

Geometrical Viewpoint. The voltage drop across the terminals of an inductor may be expressed from a circuit viewpoint by Eq. (2-20). However, this same voltage drop may be described by Faraday's law in terms of the flux produced by the current and the number of turns N of the inductor coil. Accordingly, we may write

$$v_L = L\frac{di}{dt} = N\frac{d\phi}{dt} \qquad (2\text{-}29)$$

It then follows that

$$L = N\frac{d\phi}{di} \qquad (2\text{-}30)$$

In those cases where the flux ϕ is directly proportional to current i for all values (i.e., for linear inductors), the last expression becomes

$$L = \frac{N\phi}{i} \qquad \frac{\text{Wb-t}}{\text{A}} \tag{2-31}$$

Here the inductance parameter has a hybrid representation because it is in part expressed in terms of the circuit variable i and in part in terms of the field variable ϕ. To avoid this we replace flux by its equivalent,† namely,

$$\phi = \frac{\text{mmf}}{\text{magnetic reluctance}} = \frac{Ni}{\mathcal{R}} \tag{2-32}$$

where mmf denotes the magnetomotive force which produces the flux ϕ in the magnetic circuit having a reluctance $\mathcal{R}$. Appearing in Fig. 2-12 is an inductor consisting of N turns wound about a circular iron core.

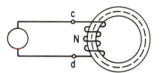

Fig. 2-12 Linear inductor with iron core.

If the core is assumed to have a mean length of l meters and a cross-sectional area of A_m meters^2, then the magnetic reluctance can be shown to be

$$\mathcal{R} = \frac{l}{\mu A_m} \tag{2-33}$$

where μ is a physical property of the magnetic material and is called the permeability. Note the similarity of Eq. (2-33) to Eq. (2-16).

Upon substituting Eqs. (2-32) and (2-33) into (2-31) there results the expression for the inductance parameter of the circuit of Fig. 2-12. Thus

$$L = \frac{N^2 \mu A_m}{l} \tag{2-34}$$

A study of Eq. (2-34) reveals some interesting and useful facts about the inductance parameter which are not readily available when this quantity is defined either from the circuit or the energy viewpoint. Most impressive is the fact that inductance, like resistance, is dependent upon the geometry of physical dimensions and the magnetic property of the medium. This is significant because it tells us what can be done to change the value of L. Thus for the inductor illustrated in Fig. 2-12, the inductance parameter may be increased in any one of four ways: increasing the number of turns, using an iron core of higher permeability, reducing the length of the iron core, and increasing the cross-sectional area of the iron core.

It is interesting to note that neither the circuit viewpoint nor the energy viewpoint could tell us these things, because essentially they deal with the effects

† See Eq. (15-21).

associated with a given inductor geometry. It is emphasized that all three viewpoints are needed to complete the picture of circuit parameters and to give proper perspective.

2-6 THE CAPACITANCE PARAMETER

Attention is next directed to capacitance, the third basic parameter of electric circuit theory. In a general way capacitance can be characterized as that property of a circuit element in which energy is capable of being stored in an electric field. A significant and distinguishing feature of capacitance is that its influence in an electric circuit is manifested only when there exists a *changing potential difference* across the terminals of the circuit element. This aspect of capacitance is easily apparent when treated from a circuit viewpoint.

Circuit Viewpoint. In Sec. 1-3 capacitance is introduced as the proportionality factor relating the charge between two metal surfaces (or conductors) to the corresponding potential difference existing between them. From Eq. (1-16), therefore, we have

$$q = Cv_c \tag{2-35}$$

where q represents the charge and v_c denotes the potential difference. Lowercase letters are used here to stress the instantaneous nature of the quantities. To obtain a definition of capacitance from a circuit viewpoint it is necessary to introduce current into the formulation of Eq. (2-35). This is readily accomplished by substituting Eq. (2-35) into the general expression for current as given by Eq. (1-2). Thus

$$\boxed{i = \frac{dq}{dt} = C\frac{dv_c}{dt}} \quad \text{A} \tag{2-36}$$

This expression shows the manner in which the current flowing through a capacitance parameter is related to the potential difference appearing across it. Any circuit element showing the property of yielding a current which is directly proportional to the rate of change of the voltage across its terminals is called a *capacitor*. A capacitor usually consists of large metal surfaces separated by small distances.

With the establishment of the current-voltage relationship of Eq. (2-36) the definition of capacitance from a circuit viewpoint readily follows. Thus

$$\boxed{C = \frac{i}{dv_c/dt}} \quad \text{F} \tag{2-37}$$

Moreover, from the terms appearing on the right side of Eq. (2-37) it is seen that the unit of capacitance is ampere-second/volt or coulomb/volt. However, for convenience this quantity is defined as the *farad*. Hence the unit of capacitance is the farad. In accordance with Eq. (2-37) the capacitance of a circuit element may be found by taking at any instant the value of the current passing through it and dividing by the corresponding value of the derivative of the voltage appearing across its terminals. Note the similarity in form which this expression for capacitance bears to the corresponding expression for inductance. Refer to Eq. (2-21). The difference lies in the interchange of roles between current and voltage.

Equation (2-36) expresses the capacitor current in terms of the capacitor voltage. It is often necessary to express the capacitor voltage as a function of its current. To do this Eq. (2-36) must be transposed and integrated in the fashion indicated below. Thus

$$dv_c = \frac{1}{C} i \, dt \qquad (2\text{-}38)$$

Integrating both sides, we have

$$\int_{v_c(0)}^{v_c(t)} dv_c = \frac{1}{C} \int_0^t i \, dt \qquad (2\text{-}39)$$

or

$$v_c(t) = \frac{1}{C} \int_0^t i \, dt + v_c(0) \qquad (2\text{-}40)$$

The quantity $v_c(0)$ denotes the initial voltage appearing across the capacitor plates upon start of the integration process. When there is no initial voltage on the capacitor, Eq. (2-40) becomes more simply

$$\boxed{v_c = \frac{1}{C} \int_0^t i \, dt} \quad \text{V} \qquad (2\text{-}41)$$

Figure 2-13 shows a graphical representation of this equation. Compare this with Figs. 2-7 and 2-10 and note the similarity in shapes and the difference in the

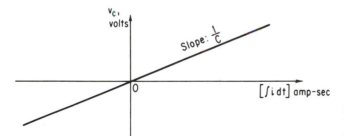

Fig. 2-13 Graphical representation of the capacitance parameter from the circuit viewpoint.

abscissa quantities. Depicted in Fig. 2-14 is the circuit diagram for the capacitor circuit element showing the symbol used to denote the capacitor. As current enters the capacitor, Eq. (2-41) reveals that the potential difference increases, with the upper plate becoming more positive than the lower plate. Electrons are being transferred from the upper plate to the lower plate by the external circuit.

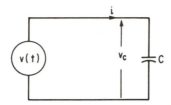

Fig. 2-14   The capacitor circuit.

To indicate this, the arrowhead on v_c is pointed upward. Of course, as i varies and becomes negative for a long enough period of time, it is entirely possible that the net result of integrating the current as called for in Eq. (2-41) will be a negative v_c.

A study of Eqs. (2-36) and (2-41) brings into evidence an important property of capacitance: *The voltage across a capacitor cannot change discontinuously.* Equation (2-36) shows that an abrupt change in capacitor voltage is not admissible because a finite change in v_c in zero time gives a value of infinity for dv_c/dt. Thus the capacitor current is infinite—a physical impossibility. On the other hand, Eq. (2-41) points out that for any finite change in capacitor current, however large, the integral contribution in zero time must necessarily be zero. Hence the capacitor voltage cannot change instantaneously. It is interesting to note here that a step change in current through the capacitor is allowable.

EXAMPLE 2-6   Assume that the current waveshape of Fig. 2-11(a) is made to flow through the capacitor of Fig. 2-14. For a capacitance of 0.01 F, find the capacitor voltage waveshape as a function of time.

Solution: The solution readily follows from an application of Eq. (2-41). In the interval from 0 to 0.1 s the current has a variation given by $i = 100t$. Hence the corresponding voltage variation is

$$v_c = \frac{1}{C} \int_0^t 100t \, dt \qquad 0 < t < 0.1$$
$$= 100 \times 50t^2 = 5000t^2$$

(2-42)

Thus the voltage wave in this interval is parabolic and at $t = 0.1$ s has a value of 50 V. The capacitor voltage can also be obtained by finding the area of triangle Oab and multiplying by $1/C$. By either method the area under the current wave is obtained in the given time interval. This latter procedure was followed to find the key points shown in Fig. 2-15. Note that at time t' the sum of the positive and negative areas of the current wave is zero. Hence v_c is likewise zero. Then a short time later the capacitor voltage becomes negative. Note, too, that in the interval between 0.3 s and t' the capacitor voltage is positive in spite of the fact that the current is negative.

Energy Viewpoint.   Assume that the capacitor of Fig. 2-13 has zero initial voltage across it and that a current i is allowed to flow in the circuit for a time interval t. The energy delivered to the capacitor during this time is given by

$$W = \int_0^t v_c i \, dt$$

(2-43)

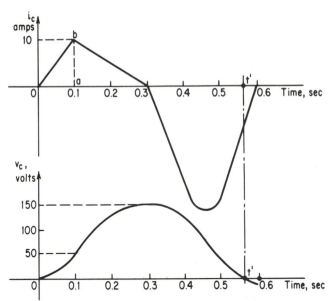

Fig. 2-15 Solution to Example 2-6.

Inserting Eq. (2-36) into the last expression yields

$$W = \int_0^t v_c \left(C \frac{dv_c}{dt} \right) dt = \int_0^v C v_c \, dv_c \qquad (2\text{-}44)$$

or

$$\boxed{W = \frac{1}{2} C v_c^2} \quad \text{J} \qquad (2\text{-}45)$$

Equation (2-45) states that the capacitor absorbs an amount of energy which is proportional to the capacitance parameter and the square of the instantaneous value of the voltage appearing across the capacitor. The absorbed energy in turn is stored by the capacitor in an electric field existing between its two plates. Note that as the capacitor voltage increases the energy increases and as v_c decreases in magnitude the associated energy decreases. Again it is reasonable to conclude therefore that like the inductor, the capacitor has the capacity of interchanging energy with the source. This stands in sharp contrast with the resistor, which can only dissipate energy in the form of heat.

With the derivation of Eq. (2-45) it is now possible to define capacitance from an energy viewpoint. Accordingly, we can write

$$\boxed{C = \frac{2W}{v_c^2}} \quad \text{F} \qquad (2\text{-}46)$$

Therefore capacitance may be identified in terms of the instantaneous value of the energy stored in its electric field and the corresponding value of the potential difference appearing across its terminals.

When the voltage across a capacitor is constant, there can be no current flow. This is required by Eq. (2-36). However, this does not mean that energy cannot be stored. In fact Eq. (2-41) states that for constant capacitor voltage there exists a finite and constant energy stored in its electric field. This situation is analogous to that which prevails in the inductor.

Geometrical Viewpoint. The amount of charge which accumulates on the plates of a capacitor is commonly expressed by Eq. (2-35) as $q = Cv_c$. However, by means of Gauss's flux theorem it is also possible to express this accumulated charge in terms of the electric field quantity $\mathscr{E}$. This description appears in Eq. (1-20) and is repeated here for convenience. Thus

$$q = \varepsilon A \mathscr{E} \tag{2-47}$$

Recall that ε denotes the permittivity (or specific dielectric constant) of the material between the plates of the capacitor, and A represents the area of the plates. For the sake of illustration let us consider the capacitor to have a definite configuration, namely that of two flat, parallel plates separated by a distance of d meters. In such a case the electric field intensity, which has units of volts/meter, is simply described as

$$\mathscr{E} = \frac{v_c}{d} \quad \text{volts/meter} \tag{2-48}$$

Upon insertion of Eq. (2-48) into Eq. (2-47) the latter becomes

$$q = \varepsilon A \frac{v_c}{d} \tag{2-49}$$

Note that now we have an expression for charge which involves the physical dimensions of the capacitor as well as the capacitor voltage.

By equating the two alternative forms, namely Eqs. (2-35) and (2-49), for the charge appearing on the plates of a capacitor for a voltage v_c we have

$$Cv_c = \varepsilon \frac{A}{d} v_c \tag{2-50}$$

or

$$\boxed{C = \varepsilon \frac{A}{d}} \quad \text{F} \tag{2-51}$$

Equation (2-51) thus makes available a definition of capacitance expressed in terms of its geometrical configuration and the physical property of the material lying between its two metal surfaces. Although the foregoing expression is applicable to the specific arrangement of a parallel-plate capacitor, the conclusions relating to the factors upon which capacitance depends are general. Accordingly, it can be said that the capacitance parameter is as a rule directly proportional to the dielectric constant of the material, inversely proportional to the spacing between the metal surfaces, and directly proportional to the metal surface area.

It is interesting to note the striking resemblance which the geometrical descriptions of the resistance, inductance, and capacitance parameters of electric circuit theory bear to one another. Refer to Eqs. (2-16), (2-34), and (2-51). In each instance there is involved the physical property of the material. These are resistivity ρ for resistance, permeability μ for inductance, and permittivity ε for capacitance. Moreover, in each case there appears the ratio of length to area.

2-7 THE OPERATIONAL AMPLIFIER: ITS ROLE AS A CIRCUIT ELEMENT AND AS A DEPENDENT (CONTROLLED) SOURCE

The operational amplifier (op-amp), manufactured with integrated transistors, diodes, resistors and capacitors, is an extremely versatile device that is found doing countless tasks in many electronic circuits. At this point in the book we limit our interests in the op-amp to an external description and to those applications where it serves as a dependent source for important electric circuits as well as a device for performing such important functions as isolation, multiplication, addition, and subtraction. It can also be made to perform the important mathematical function of integration which together with its summing capabilities puts it in a position to model differential equations of physical systems.

As an integrated electronic device the op-amp physically takes the form shown in Fig. 2-16(a). This unit is called an eight-pin dual-in-line package and may contain from one to four op-amps. Figure 2-16(b) depicts a more detailed description for terminal identification of the one-op-amp package. Observe that

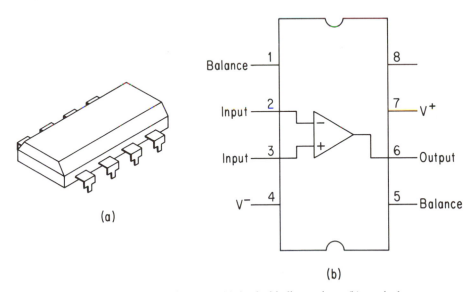

Fig. 2-16 Integrated op-amp: (a) the dual-in-line package; (b) terminal identification. The actual circuitry within the package consists of many integrated resistors, capacitors, diodes, and transistors.

there are two input terminals (pins 2 and 3) and one output terminal (pin 6). Both a positive (pin 7) and negative (pin 4) supply voltage must be furnished. Usually, this voltage is fixed at ± 15 V with respect to ground reference.

The op-amp can accommodate a positive as well as a negative input signal simultaneously or either one separately. The schematic arrangement showing the input signals and the positive and negative power supplies is displayed in Fig. 2-17(a). The signal v_p is applied to pin 3, which produces a noninverting output. That is, the output voltage appearing at pin 6 has the same polarity as the input voltage. When a signal v_n is applied to pin 2, the internal circuitry of the op-amp yields an output signal of inverted (opposite) polarity. In the interest of simplicity the symbol most commonly used to model the op-amp is shown in Fig. 2-17(b). The design of the op-amp is such that the output voltage is related to the two input voltages by

$$v_o = A(v_p - v_n) = Av_d \qquad (2\text{-}52)$$

Keep in mind that v_p denotes the voltage that is applied to the noninverting terminal (pin 3), and v_n is the signal applied to the inverting terminal (pin 2), and v_d is the net differential input voltage. The quantity A is the open loop gain of the op-amp and customarily has values of 100,000. This means that the net differential input voltage is increased at the output terminal (pin 6) by this factor. Clearly, then, if overloads are to be avoided, it is necessary to restrict v_d to those values that will keep v_o at least below the magnitude of the supply voltage. Accordingly, for the case where $A = 10^5$ it follows that v_d must be limited to

$$v_d = \frac{15}{10^5} = 0.00015 \text{ V} = 150 \ \mu\text{V}$$

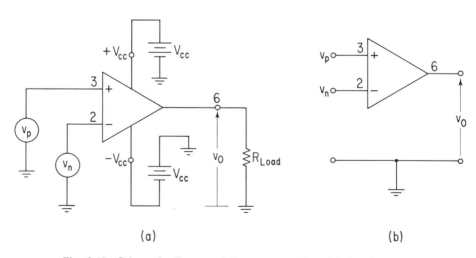

(a)   (b)

Fig. 2-17 Schematic diagram of the op-amp: (a) model showing supply voltages and signal inputs with connection to ground reference; (b) commonly used simplified symbol for modeling the op-amp.

which is obviously a small quantity. Fortunately, there are techniques for increasing this limit to more practical values. These are considered presently.

One of the design characteristics of the op-amp is that saturation of the output voltage sets in rapidly once the differential input signal reaches a threshold value (such as the 150 μV just computed). This feature is illustrated in Fig. 2-18. The saturation value is approximately taken to be equal to the supply voltage V_{cc} for convenience. Actually, it is always less than this quantity. The slope of the curve represents the gain A.

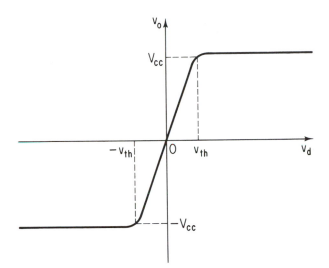

Fig. 2-18 Transfer characteristic of the op-amp. Here v_{th} represents the threshold value of v_d beyond which saturation sets in.

Three notable features characterize the op-amp model of Fig. 2-17 which are very helpful in obtaining solutions to circuit problems that involve op-amps. In the interest of stressing a simplified strategy for analysis, the *idealized* versions of these features are specified. The first is that the resistance presented to a source applied at either input terminal of the op-amp is for all practical purposes infinite. The actual value may be of the order of several megohms but this is tantamount to infinity. This point is important because it means that the presence of the op-amp in the circuit does not change the behavior of the signal sources. In other words, there is no loading effect to be concerned with. The second characteristic feature of the op-amp is that the output resistance is zero. Output resistance is the resistance that appears across the output terminals looking back into the amplifier. The usual value is under 100 Ω, but for the level of load resistor values generally used across the output terminals of the op-amp in applications, this value can be considered to be negligible. Keeping the output resistance small compared to the load resistor is important if most of the voltage generated by the op-amp in response to input signals is to be available at the load resistor, where it does useful work. The output resistance is like the internal resistance of a battery and we know the importance of keeping this quantity as small as possible. The third and final ideal feature of the op-amp is that it is considered to have a direct gain from v_d to v_o of infinity. Although we know this gain to be finite and in the vicinity of 100,000, this figure is high enough so that the practical consequences are the same as if it were treated to be of infinite value.

A careful examination of the ideal model representation depicted in Fig. 2-17(b) reveals an essential difference between the op-amp as a circuit device and those elements that are displayed in Table 2-1. Each of the circuit elements summarized in this table is of the one-port variety. That is, the circuit component has a *single* pair of terminals into or out of which a current flows and across which appears a voltage. In contrast the op-amp is a two-port element. It has a pair of input terminals (say pin 3 and ground or pin 2 and ground) with which a current and voltage are associated and a pair of output terminals (pin 6 and ground) with which a different set of voltage and current is associated. Because the ground terminal is common, this version of the op-amp is alternatively referred to as a three-wire two-port device. We now direct attention to the various ways in which this two-port device may be used to generate convenient controlled sources and to perform the useful operations of multiplication, isolation, addition, and subtraction.

Noninverting Mode. The circuit configuration for use of the op-amp in the noninverting mode is depicted in Fig. 2-19. There is a single external input signal $v_i = v_p$ that is applied to the + pin of the op-amp. A signal also is made to appear at the − input terminal, but this is derived from the output voltage by means of resistors R_1 and R_f. To determine the expression for v_n, we first note that when an output voltage v_o appears at the output terminal of the op-amp we get, by Ohm's law,

$$v_o = I_o (R_f + R_1) \tag{2-53}$$

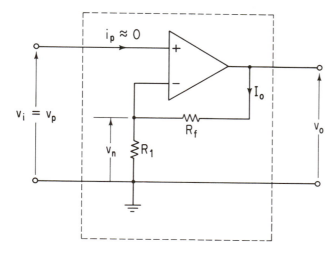

Fig. 2-19 Op-amps as a two-port element used in the noninverting mode.

The connection of the junction point of R_f and R_1 to the negative input pin causes no effect in the circuit involving R_f and R_1 because of the assumption of infinite input resistance for the op-amp. As a matter of fact, it is for this reason too that we can say that the current produced by v_i, namely i_p, is zero. The potential of v_n with respect to ground reference is again, by Ohm's law, simply expressed as

$$v_n = I_o R_1 \tag{2-54}$$

Upon taking the ratio of this last expression with Eq. (2-53), it follows that

$$v_n = \frac{R_1}{R_f + R_1} v_o \tag{2-55}$$

Our interest is to relate the output signal to the input signal and this is conveniently achieved by invoking Eq. (2-52) and using $v_p = v_i$ and Eq. (2-55) for v_n. Thus

$$v_o = v_d A = \left(v_i - \frac{R_1}{R_1 + R_f} v_o \right) A \tag{2-56}$$

Dividing through by A on both sides of Eq. (2-56), we get

$$\left(v_i - \frac{R_1}{R_1 + R_f} v_o \right) = \frac{v_o}{A} \tag{2-57}$$

Next we introduce characteristic information about the op-amp. First, we recall that v_o is a finite quantity and often takes on values less than 15 V. Moreover, the op-amp gain A is so large that we may treat it as infinite. As a result we can confidently set the right side Eq. (2-57) to zero and thereupon obtain an expression for v_o in terms of v_i. This leads to

$$\boxed{v_o = \frac{R_1 + R_f}{R_1} v_i = \left(1 + \frac{R_f}{R_1} \right) v_i} \tag{2-58}$$

Equation (2-58) makes it obvious that the magnitude of the output voltage can now be controlled by the proper selection of resistors R_f and R_1. In fact, this circuit arrangement readily provides us with a *voltage-controlled voltage source* (VCVS). Accordingly, if R_f is set equal to 19 kΩ and R_1 is 1 kΩ, the output voltage of the op-amp will be 20 times the value of the input voltage. Particularly noteworthy too is the fact that with the arrangement of Fig. 2-19 the input voltage can now be made to take on values much larger than the highly restrictive values

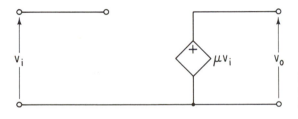

Fig. 2-20 Circuit model for a voltage-controlled voltage source (VCVS). Here μ is a numeric.

associated with v_d when used without resistors R_f and R_1. The circuit model used to represent a voltage-controlled voltage source is shown in Fig. 2-20. The factor μ is a numeric which for the circuit of Fig. 2-19 clearly takes on the value $(1 + R_f/R_1)$. The input circuit feeds into an open circuit in recognition of the fact that the input resistance is assumed to be very high.

The Voltage Follower (Isolation Mode). An interesting result is obtained when the resistor R_f in Fig. 2-19 is set equal to zero as illustrated in Fig. 2-21.

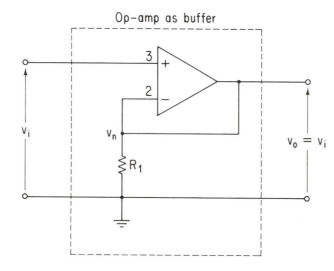

Fig. 2-21 Voltage follower. The output voltage tracks the input voltage in magnitude as well as sign.

A glance at Eq. (2-58) makes it apparent that for such a circuit condition we have

$$v_o = v_i \qquad (2\text{-}59)$$

In other words, the output of the op-amp exactly tracks the input voltage both in sign and magnitude. This is the reason the circuit is called a *voltage follower*.

The result displayed in Eq. (2-59) can also be found from a direct application of Eq. (2-52). Observe in Fig. 2-21 that by the physical connection v_n is equal to v_o. Accordingly, Eq. (2-52) allows us to write

$$v_o = v_d A = (v_i - v_o)A$$

Collecting terms and formulating the ratio of output voltage to input voltage then yields

$$\frac{v_o}{v_i} = \frac{A}{1 + A} = 1 \qquad \text{for } A \gg 1 \qquad (2\text{-}60)$$

The approximation used in this equation to identify a unity gain for the voltage follower is the same one that allowed us to set the right side of Eq. (2-57) to zero, which in turn gave us Eq. (2-58).

The direct connection of the output voltage to the inverting terminal of the

The Circuit Elements   Chap. 2

op-amp represents the case of 100% *negative feedback* of the output to the input. In such cases the output quantity is always related to the input quantity by the intermediate expression in Eq. (2-60). Of course, when the open-loop gain A is allowed to assume very large values compared to unity, the corresponding closed-loop gain becomes approximately unity. However, a unity transmission gain is not the only feature of the voltage follower. A more detailed analysis of the equivalent circuit of the op-amp reveals that the resistance looking into the input terminals (i.e., pin 3 and ground) takes on the value AR_i. It will be recalled that when the op-amp is used in the open loop mode (i.e., without a feedback connection from the output) R_i already has a high value of 1 or 2 MΩ. Hence for a typical value of $A = 10^5$ it follows that the input resistance for the voltage follower is 100,000 MΩ or more. What this means in essence is that the op-amp produces virtually no effect on the input signal voltage source even if the latter should be characterized by a high internal resistance. Detailed analysis of the equivalent circuit of the op-amp also reveals that the resistance R_o looking into the output terminals, which is normally just under 100 ohms, is reduced to R_o/A when 100% feedback is used. Consequently, the output impedance of the op-amp assumes values of the order of 0.001 Ω. This means that on the output side, the op-amp behaves like a voltage source (e.g., a battery) with negligible internal resistance.

Because the voltage follower possesses the three outstanding characteristics of extremely large input resistance, unity transmission gain, and extremely low output resistance, it is the ideal circuit device to serve as a *buffer* (or isolation unit). In this way, it can be made to prevent the disturbance of one part of a circuit on another as might be encountered, for example, in filter design.

Inverting Mode.   Figure 2-22 depicts the circuit configuration of the op-amp in the inverting mode. Observe that the noninverting terminal (pin 3) is tied

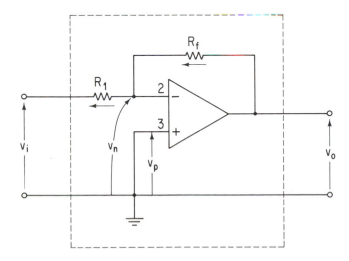

Fig. 2-22 Op-amp as a two-port circuit element configured to operate in the inverting mode.

to the ground reference, which means that $v_p = 0$. Insertion of this fact into the basic voltage equations of Eq. (2-52) yields the result

$$v_o = A(v_p - v_n) = -Av_n \qquad (2\text{-}61)$$

This expression may now be used to derive information about the magnitude of v_n in this particular application of the op-amp. Typical maximum values of v_o in these applications is determined by the value of the supply voltage V_{cc}. [See Fig. 2-17(a).] Thus for a V_{cc} of 15 V and $A = 100,000$, the maximum values to be expected for v_n in normal operation is therefore approximately 0.00015 or 150 μV. This quantity is so small that it is commonly accepted as being at ground potential for all practical purposes. Under these circumstances it is interesting to note that the resistance looking into the port is then simply R_1. Therefore, when the op-amp is used in the inverting mode, the input resistance can be described by the relationship

$$R_i = R_1 \qquad (2\text{-}62)$$

where R_i denotes the resistance encountered upon looking into the input port. Note too that the input resistance for this mode is considerably less than the very high value encountered in the noninverting mode of the op-amp.

How is the output voltage related to the input voltage in this inverting mode of the op-amp? The answer readily follows from an application of Kirchhoff's current law at pin 2 in the configuration of Fig. 2-22. The sum of the currents flowing through resistors R_1 and R_f to pin 2 must add up to zero because in the ideal case the current that flows to pin 2 is zero. (Keep in mind that v_n is now being considered to be at zero potential.) Expressed mathematically, we can write

$$\frac{v_i}{R_1} + \frac{v_o}{R_f} = 0$$

or

$$\boxed{v_o = -\frac{R_f}{R_1} v_i} \qquad (2\text{-}63)$$

Because the magnitude of the output voltage for a given input voltage can be controlled by the selection of resistors R_f and R_1, we have another version of a voltage-controlled voltage source (VCVS). Thus the output of the op-amp in the inverting mode can also be made to serve as a *dependent voltage source*. The corresponding model for this mode is drawn in Fig. 2-23. A comparison of the model displayed in Fig. 2-20 for the noninverting mode indicates that the difference lies chiefly in the input resistance encountered. For the noninverting mode it is so high that the model on the input side shows an open circuit. For the inverting mode the value of the input resistance is finite and is accordingly included in the model.

When using the inverting mode of the op-amp as a dependent voltage source, it is important to compare the internal resistance of the source voltage v_i with

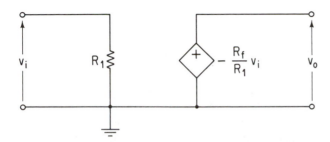

Fig. 2-23 Circuit model for the op-amp used in the inverting mode and revealing its voltage-controlled voltage-source characteristic.

the value used for R_1 in Fig. 2-22. If the source resistance, R_s, is not negligible, then Eq. (2-63) for the output voltage must be modified to

$$V_o = -\frac{R_f}{R_1 + R_s} v_i \tag{2-64}$$

Op-Amp Circuits for Addition and Subtraction. The addition of two quantities can readily be achieved by modifying the circuitry of Fig. 2-22 to that depicted in Fig. 2-24. The purpose of this circuit is to obtain the scaled sum of the two input quantities v_1 and v_2. In turn, these voltages can be arranged to represent through appropriate scale factors a pair of physical variables of whatever composition. The presence of the second resistor R_2 does not affect the conclusion reached concerning v_n in Fig. 2-22, namely, that v_n is essentially at ground potential. Hence by Kirchhoff's current law written at the negative input terminal of the op-amp, we can write

$$\frac{v_1}{R_1} + \frac{v_2}{R_2} + \frac{v_o}{R_F} = 0$$

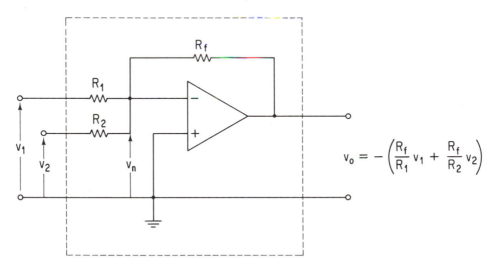

$$v_o = -\left(\frac{R_f}{R_1} v_1 + \frac{R_f}{R_2} v_2\right)$$

Fig. 2-24 Op-amp arranged to provide an output which is the negative of the scaled sum of v_1 and v_2.

or

$$v_o = -\left(\frac{R_f}{R_1}v_1 + \frac{R_f}{R_2}v_2\right) \qquad (2\text{-}65)$$

Thus the output voltage is the negative of the scaled sum of the two input quantities v_1 and v_2. The proper selection of the three resistors permits treating many coefficients that may be associated with the two input variables.

A little thought should make it apparent that it is possible to sum n variables by including n input resistors. In such a case the expression for the output voltage becomes

$$v_o = -\left(\frac{R_f}{R_1}v_1 + \frac{R_f}{R_2}v_2 + \cdots + \frac{R_f}{R_n}v_n\right) \qquad (2\text{-}66)$$

where R_n denotes the summing resistor to which the nth input voltage v_n is applied at the common inverting terminal of the op-amp.

Subtraction is achieved by adding the negative version of an input voltage. For example, suppose that the quantity v_2 should need to be subtracted from the quantity appearing inside the parentheses of Eq. (2-66). To accomplish this the signal v_2 is first applied to an inverter, which is simply the circuit of Fig. 2-22 with $R_f = R_1$. The inverter output, which is now $-v_2$, is then applied as the input at resistor R_2 in the circuit of Eq. (2-66).

It is instructive to observe in Eq. (2-66) that two important mathematical operations are at our disposal in the circuitry associated with this expression. They are multiplication (as represented by R_f/R_n) and addition (or subtraction). Accordingly, it is a routine matter to devise an op-amp circuit model to furnish the solution of n algebraic simultaneous equations. Such a model is customarily called an *analog computer* solution of the problem.

The selection of the input resistors, $R_1, R_2, \ldots, R_n$, in these models is often confined to the range 10 to 100 kΩ. Sometimes the upper limit is extended to a megohm. The lower limit is set in the interest of minimizing the importance of the source internal resistance associated with the input voltages as well as to limit the power dissipation. The upper limit is selected to minimize terminal leakage problems.

A Current-Controlled Current Source (CCCS). In the interest of illustrating the versatility of the op-amp in furnishing various types of dependent sources, attention is next directed to the generation of a dependent current source that is driven by an input current. Refer to Fig. 2-25. The input current I_i is assumed to originate from a suitable current source and it is applied to the inverting input as indicated. For an assumed output voltage v_o Kirchhoff's current law applied to the inverting input terminal yields

$$I_i + \frac{v_o}{R_f} = 0 \qquad (2\text{-}67)$$

or

$$v_o = -I_i R_f \qquad (2\text{-}68)$$

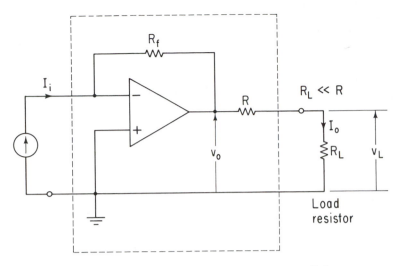

Fig. 2-25 Op-amp arranged to provide a current-controlled current source.

Furthermore, by writing Kirchhoff's voltage law for the output circuit where the current I_o flows, we have

$$v_o = I_o R + I_o R_L = I_o (R + R_L) \qquad (2\text{-}69)$$

The resistor R_L is the model for the load. It represents a device (perhaps the armature of a motor) where a useful task is to be accomplished subject to a controlled current I_o. Upon equating Eqs. (2-68) and (2-69), we get

$$I_o = -\frac{R_f}{R + R_L} I_i \qquad (2\text{-}70)$$

To make this relationship independent of the load resistor R_L, it is customary to design the op-amp circuit so that $R_L \ll R$. With this restriction imposed, Eq. (2-70) becomes simply

$$\boxed{I_o = -\frac{R_f}{R} I_i} \qquad (2\text{-}71)$$

Thus the output current I_o is controllable through the proper selection of R_f and R and dependent on the input current I_i.

The circuit representation (or model) of a dependent current source driven by an input current is shown in Fig. 2-26. Observe that the symbol for a controlled current source is a rhombus enclosing an arrow. The current gain factor is generally represented by β. Clearly, for the case of Fig. 2-25, $\beta = -R_f/R$.

Two additional dependent sources can be generated by op-amp techniques. One is the voltage-controlled current source (VCCS); the other is the current-controlled voltage source (CCVS). In this connection the reader is referred to Probs. 2-40 and 2-41.

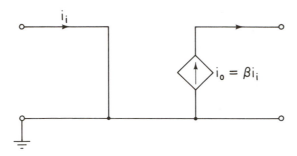

Fig. 2-26 Circuit representation (model) of a current-controlled current source.

i_i

$i_o = \beta i_i$

EXAMPLE 2-7 In the op-amp noninverting circuit displayed in Fig. 2-19 the op-amp is known to have an open loop gain of 10^5 and resistor values of $R_1 = 1 \text{ k}\Omega$ and $R_f = 39$ kΩ.

    (a) Find the actual value of the voltage gain of this op-amp circuit by working with the actual value of A.

    (b) Determine the ideal value of the voltage gain by treating A as infinite.

    (c) Calculate the percent error caused by working with the ideal case.

Solution: (a) By Eq. (2-55) the potential of the negative terminal of the op-amp is

$$v_n = \frac{R_1}{R_1 + R_f} v_o$$

Hence the differential input to the op-amp is given by

$$v_d = v_p - v_n = v_i - \frac{R_1}{R_1 + R_f} v_o$$

Moreover, by invoking the basic voltage expression for the op-amp as described by Eq. (2-52), we can write

$$v_o = Av_d = Av_i - \frac{AR_1}{R_1 + R_f} v_o$$

from which the formula for the actual voltage gain of the circuitry of Fig. 2-19 is found to be

$$\frac{v_o}{v_i} = \frac{A}{1 + \dfrac{AR_1}{R_1 + R_f}} = \frac{10^5}{1 + \dfrac{10^5}{40}} = 39.98 \qquad (2\text{-}72)$$

    (b) For the ideal case the voltage gain is given by Eq. (2-55). Accordingly, we get

$$\frac{v_o}{v_i} = 1 + \frac{R_f}{R_1} = 1 + 39 = 40$$

    (c) The percent error of the ideal value compared to the actual value can be represented by

$$\% \text{ error} = \frac{40 - 39.98}{40} \times 100 = 0.05\%$$

which is clearly a very small error. This helps to explain why the ideal op-amp characteristics are used in the analysis of circuits where the op-amp is a component.

EXAMPLE 2-8 Design a noninverting op-amp circuit that yields a voltage gain of 4. Use the ideal characteristics of the op-amp.

Solution: From Eq. (2-58) we have

$$\frac{v_o}{v_i} = 4 = 1 + \frac{R_f}{R_1}$$

Hence

$$\frac{R_f}{R_1} = 4 - 1 = 3$$

A suitable choice for R_1 is 10 kΩ. Hence $R_f = 30$ kΩ.

EXAMPLE 2-9 Repeat Example 2-8 for the op-amp used in the inverting mode and find the input resistance. Assume negligible internal source resistance. Also, select the design resistors from the preferred resistance group that appears in Table 2-3.

Solution: The voltage gain in this case is given by Eq. (2-63) as it applies to the circuitry of Fig. 2-22. Accordingly,

$$\frac{v_o}{v_i} = -\frac{R_f}{R_1} = -4$$

This voltage gain is easily realized by choosing $R_f = 30$ kΩ and $R_1 = 7.5$ kΩ, both of which are standard values which are readily available.

The input resistance is R_1 and so has a value of 7.5 kΩ.

EXAMPLE 2-10 A physical process is described by the equation $z = 2x - 3y$, where x and y are variables that may be represented by voltages well within the threshold value of the op-amp. Design an op-amp circuit that makes the quantity z available as a voltage. Confine the design to the use of standard resistor values. Assume that x and y originate as positive quantities.

Solution: An inverter is needed in this solution because x and y appear with opposite signs in the process equation. Because the summing node of the op-amp always involves a sign change, it follows that the input x must undergo two sign changes. The circuit arrangement is shown in Fig. 2-27. The negative version of x is obtained from the first

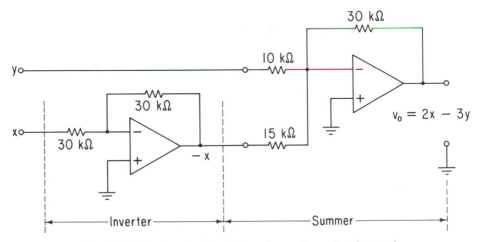

Fig. 2-27 Op-amp circuit solution of $z = 2x - 3y$ where z is generated as the output voltage v_o.

op-amp by choosing the input resistor equal to the feedback resistor. Each resistor is chosen to be 30 kΩ in the interest of keeping power dissipation low. The summer op-amp now has two inputs, y and $-x$. It is the purpose of the resistor of the summer op-amp to furnish the required multiplication called for by the process equation. By choosing a feedback resistor of 30 kΩ and an input resistor of 10 kΩ, the required multiplication factor of 3 is obtained for y. Similarly, by applying $-x$ to an input summing resistor of 15 kΩ, the necessary multiplication factor of 2 is obtained. The output of the summer op-amp v_o now represents the physical variable z as a voltage.

Summary review questions

1. What is the value of the internal resistance of an ideal voltage source? Explain.
2. What is the value of the internal resistance of an ideal current source? Explain.
3. When an ideal current source is placed as part of closed circuit, comment on what factors determine the voltage that appears across the current source terminals.
4. When an ideal voltage source is placed as part of a closed circuit, comment on the factors that determine the magnitude of the current associated with the voltage source.
5. Describe the convention that is used to define the positive direction for a voltage drop that appears across a passive circuit element in a closed circuit.
6. What is a ground reference? What is its purpose in an electric circuit?
7. Can any point in an electric circuit be taken as the reference potential? Explain. What shortcomings can you associate with such a procedure?
8. What is the definition of power? How is it related to energy?
9. Confining your description to energy, explain how a 30-hp motor differs from a 10-hp motor.
10. Distinguish between the terms *resistor* and *resistance*.
11. What is meant when resistance is defined from a circuit's viewpoint? Illustrate.
12. How does temperature affect the value of resistance? Why is it important to know this?
13. The resistance parameter can also be defined from an energy viewpoint. Why is this possible, and how is it accomplished?
14. If heat is constantly being added to a current-carrying resistor, why does the resistor temperature not rise indefinitely?
15. Explain the reason for putting a voltage rating on resistors.
16. Why is it important to know the meaning of resistance from a geometrical viewpoint?
17. Define the terms *resistivity, circular mils,* and *conductivity*.
18. Identify the various resistor types and comment on their corresponding tolerance ratings.
19. Distinguish between an inductor and inductance.
20. Describe the circumstance which permits one to conclude that a circuit is inductive. Be specific.
21. Explain why it is that the current flowing through an inductor is not allowed to change instantaneously.

22. Describe the inductance of a coil from a circuit viewpoint, energy viewpoint and geometrical viewpoint. Identify the important information about inductance that follows from a geometrical description and which cannot readily be obtained from the other two descriptions.

23. Does an inductor make its presence known in a circuit under all circumstances? Explain.

24. Explain why it is poor practice to open a knife switch in an electric circuit that contains a large inductance.

25. By what factor is the inductance of a coil increased when the number of turns is doubled?

26. Describe what is meant by the statement "inductance is a function of the medium in which a coil operates."

27. What is the principal cause of the deviation of inductance from linearity?

28. Distinguish between a capacitor and capacitance.

29. How can one determine that in an electric circuit containing a single element the element is a capacitor? Be specific.

30. Explain why the voltage across the terminals of a capacitor is not allowed to change abruptly.

31. Describe capacitance from a circuit, an energy, and a geometrical viewpoint. Identify the important information about capacitance that follows from a geometrical description and which cannot readily be deduced from the other two descriptions.

32. Does a capacitor make its presence known in a circuit at all times? Explain.

33. By what factor is the capacitance of a pair of parallel conducting plates, separated by a dielectric material, changed when the distance between the plates is doubled and the surface area of the plates is halved?

34. As an active element the op-amp is said to be a two-port device. Explain what this means.

35. Identify the three characteristic features of the ideal op-amp. Describe the important role that each serves.

36. Show how the op-amp may be configured to provide a voltage-controlled voltage source. How is the controlled source made adjustable?

37. Draw a circuit diagram to show how the op-amp may be configured to serve the important role of an isolator unit. What is the order of magnitude of the input resistance and the output resistance for such a circuit?

38. Draw a circuit diagram that indicates how the op-amp may be used to provide a fixed multiplication of an input voltage. Show two methods and draw appropriate distinctions in the operation and performance of these circuits in achieving the stated goal.

39. Draw a circuit diagram that shows how the op-amp may be used to obtain the sum of two input voltages. What precaution must be observed in the use of this circuit?

40. What is the role that negative feedback plays in the circuit arrangement of the voltage follower?

41. When the op-amp is used in the inverting mode, the potential of the summing junction of the input and feedback resistors is said to be at virtual ground potential. Explain.

42. Why is the input pin in the noninverting mode of the op-amp not considered to be at ground potential, whereas the input pin in the inverting mode is so treated?

43. What considerations must be made when selecting the input resistor R_1 in the inverting mode of the op-amp?

44. What is an op-amp inverter?

Problems

GROUP I

2-1. In the circuit of Fig. P2-1 find the value of the current and terminal voltage of the current source. Also, determine the power that is dissipated in the 2-Ω resistor.

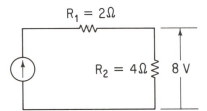

Fig. P2-1

2-2. Refer to Fig. P2-2. Measurement of the voltage across the terminals of the current source yields a value of 22 V. Find the value of R_3 in ohms.

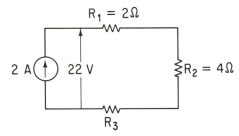

Fig. P2-2

2-3. Refer to Fig. P2-3.
(a) Find the current delivered by the battery.
(b) Determine the resistance of R_2 in ohms.

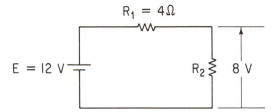

Fig. P2-3

The Circuit Elements   Chap. 2

2-4. In the circuit of Fig. P2-4 the ammeter, which measures current, gives a positive (upscale) deflection when the current enters the ammeter at the + terminal. The voltmeter behaves in a similar manner. The voltage source V is constant.

    (a) What type of circuit element appears in the circuit when both the ammeter and voltmeter read upscale? Explain.

    (b) What is the circuit element when the ammeter is deflected downscale and the voltmeter reads upscale? Explain.

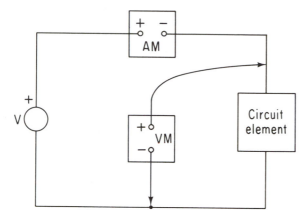

Fig. P2-4

2-5. A constant-voltage source is applied to three resistors in the circuit arrangement illustrated in Fig. P2-5.

    (a) Find the voltage drop across the 3-Ω resistor, V_{ac}.

    (b) Determine the value of the voltage drop V_{ab}.

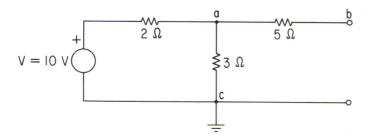

Fig. P2-5

2-6. An electric heater draws 1000 W from a 250-V source. What power does it take from a 208-V source? What is the value of resistance of the heater?

2-7. An electric broiler draws 12 A at 115 V for a period of 3 h.

    (a) If electrical energy costs 5 cents per kilowatt-hour, determine the cost of operating the broiler.

    (b) Find the quantity of electricity in coulombs which passes through the broiler.

    (c) How many electrons are involved?

    (d) What is the rate at which electrical energy is expended?

2-8. Determine the value of the resistance of the electric broiler of Prob. 2-7 in terms of the energy viewpoint. Check this value by finding the resistance also in terms of the circuit viewpoint.

2-9. A resistor is known to dissipate 27.8 kW-hr of energy in 30 min while drawing a current of 10 A.

(a) Obtain the value of resistance from an energy approach.

(b) Determine the resistance by proceeding in terms of a circuit approach.

2-10. A 10-Ω resistor has a voltage rating of 120 V. What is its power rating?

2-11. A coil of standard copper wire having a resistance of 12 Ω at 25°C is imbedded in the core of a large transformer. After the transformer has been in operation several hours, the resistance of the coil is found to be 13.4 Ω. What is the temperature of the transformer core?

2-12. Find the resistance of a bus bar 20 ft long, $\frac{1}{3}$ in. by 3 in. in cross section, and composed of standard annealed copper. Assume temperature is 20°C.

2-13. What must be the length of No. 41 AWG round copper wire so that the resistance will be 0.2 Ω?

2-14. A 400-ft length of round copper wire has a resistance of 0.4 Ω at 20°C. What is the diameter of the wire in inches?

2-15. The coil in the configuration of Fig. 2-12 is equipped with 100 turns. Moreover, the mean length of turn of the magnetic core is known to be 0.2 m and the cross-sectional area is 0.01 m^2. The value of the permeability of the iron is 10^{-3}.

(a) Find the inductance of the coil.

(b) When a d-c voltage is applied to the inductor it is found to draw a current of 0.1 A. How much energy is stored in the magnetic field?

2-16. Find the capacitance of a circuit element in which:

(a) A voltage of 100 V yields an energy storage in an electric field of 0.05 J.

(b) A voltage increases linearly from zero to 100 V in 0.2 s causing a current flow of 5 mA.

(c) Two flat parallel plates are separated by a 0.1-mm layer of mica and have a total area of 0.113 m^2. Assume mica to have a relative permittivity of 10.

2-17. Determine the current which flows through a 1-μF capacitor when

(a) the voltage increases linearly at the rate of 1000 v/s and

(b) the energy storage remains fixed at 0.0005 J.

2-18. Identify those characteristics which the circuit parameters resistance, inductance, and capacitance have in common when described in terms of their geometrical properties.

2-19. A voltage wave having the time variation shown in Fig. P2-19 is applied to a pure inductor having a value of 2 H. Sketch carefully the variation of the current through the inductor over the first 10 s. Indicate all salient values in the current curve.

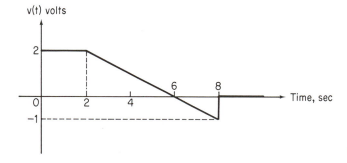

Fig. P2-19

2-20. Design a noninverting mode op-amp circuit that is capable of providing a voltage gain of 10. Assume an ideal op-amp and restrict the selection of resistors to the standard values.

2-21. The op-amp circuitry of Fig. P2-21 is used to provide amplification of an input voltage which is known to be 0.1 V.
  (a) If $A = 10^5$, find the actual value of the output voltage.
  (b) Calculate the value of the differential input to the op-amp.
  (c) In light of the results of parts (a) and (b) comment on the appropriateness of using the ideal characteristics of the op-amp in analysis and design.

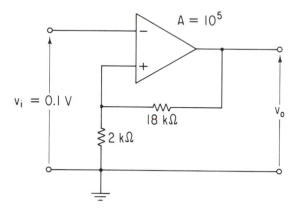

Fig. P2-21

2-22. The op-amp circuitry of Fig. P2-22 is used to obtain a voltage-controlled voltage source (VCVS).
  (a) Draw the corresponding VCVS model with all parameter values clearly indicated. Assume an ideal op-amp.
  (b) If the op-amp supply voltage is known to be ± 10 V, what restriction must be placed on v_i?
  (c) Compare the value found in part (b) with the restriction on the differential input when the op-amp is used without the feedback resistor and $A = 10^5$.

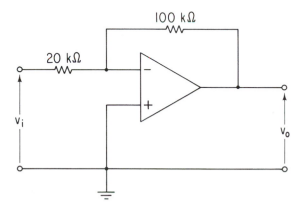

Fig. P2-22

2-23. Refer to the circuitry of Fig. P2-23.
  (a) Find the equation that is being simulated by this circuit.

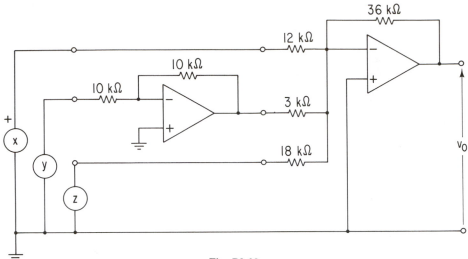

Fig. P2-23

(b) Assume that each input has an internal source resistance of 1 kΩ. Determine the equation that is being simulated under these conditions?

(c) What lesson can be drawn from a comparison of the results found in parts (a) and (b)?

(d) Propose a modification of the circuitry of Fig. P2-23 that serves to eliminate the difficulty encountered in this problem.

2-24. Design an op-amp circuit that simulates the equation $w = -0.5x + 4y - 2z$. Assume an ideal op-amp and negligible source resistances. Use standard resistor values.

GROUP II

2-25. Two current sources and a 2-Ω resistor are arranged as shown in Fig. P2-25.
   (a) Find the power dissipated by the resistor.
   (b) Determine the voltage drop across the parallel combination of the circuit elements.
   (c) What power is associated with the 3-A current source? How is this power distributed?

2-26. In the circuit of Fig. P2-26 the voltage sources have the values $V_1 = -12$ V and $V_2 = 24$ V.
   (a) Find the magnitude and direction of the current flow in the circuit.

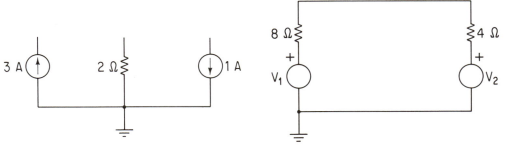

Fig. P2-25

Fig. P2-26

(b) Compute the power associated with each source. Give due regard to signs and state whether the power is being delivered or absorbed by the source.

2-27. Refer to the circuit of Fig. P2-27.

    (a) Assume that both switches are closed. Find the potential of points *a* and *b* with respect to ground.

    (b) Repeat part (a) when S2 is open and S1 remains closed.

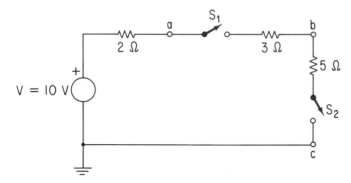

Fig. P2-27

2-28. A current and a voltage source are placed in parallel as indicated in Fig. P2-28. As a practical matter the equivalence of this combination is simply the voltage source. Explain.

2-29. A current source and a voltage source are placed in series as depicted in Fig. P2-29. In essence this arrangement behaves simply as a current source *i*. Explain.

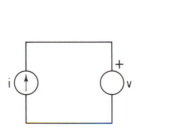

Fig. P2-28

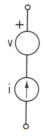

Fig. P2-29

2-30. Find the resistivity expressed in ohm-circular mils per foot of a material that has a resistivity of 4 $\mu\Omega$-cm.

2-31. Find the inductance of a coil in which:

    (a) A current of 0.1 A yields an energy storage of 0.05 J.

    (b) A current increases linearly from zero to 0.1 A in 0.2 s producing a voltage of 5 V.

    (c) A current of 0.1 A increasing at the rate of 0.5 A/s represents a power flow of $\frac{1}{2}$ W.

2-32. When a d-c voltage is applied to the coil of Prob. 2-15, the current is found to vary in accordance with $i = \frac{1}{10}(1 - \varepsilon^{-5t})$ expressed in amperes. Moreover, it is found that after the elapse of 0.2 s, the voltage induced in the coil is 0.16 V. Find the inductance of the coil.

2-33. When a d-c voltage is applied to a capacitor, the voltage across its terminals is found to build up in accordance with $v_c = 150(1 - \varepsilon^{-20t})$. After the elapse of 0.05 s, the current flow is equal to 1.14 mA.

(a) Find the value of the capacitance in microfarads.

(b) How much energy is stored in the electric field at this time?

2-34. A circuit element is placed in a black box. At $t = 0$ a switch is closed and the current flowing through the circuit element and the voltage across its terminals are recorded to have the waveshapes shown in Fig. P2-34. Identify the type of circuit element and its magnitude.

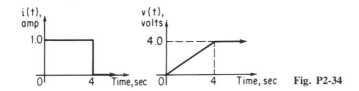

Fig. P2-34

2-35. A voltage waveshape of the form $v(t) = K_1 t^n$ is applied to a deenergized circuit element at time $t = 0$. Determine the current through the element when the element is (a) a resistor, (b) an inductor, and (c) a capacitor.

2-36. The voltage waveshape shown in Fig. P2-36 is applied to a coil having an inductance of $\frac{1}{2}$ H and negligible resistance.

(a) Sketch the corresponding current waveshape as a function of time.

(b) Determine the instantaneous value of the current at the end of $t = 1$ s, $t = 2$ s, and $t = 4$ s.

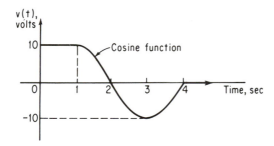

Fig. P2-36

2-37. The voltage waveform shown in Fig. P2-37 is applied separately to a pure capacitor of $\frac{1}{2}$ F and to a pure inductor of $\frac{1}{2}$ H. Carefully sketch the current waveshape for each circuit element over the specified time interval.

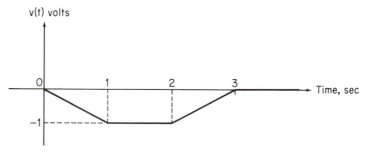

Fig. P2-37

The Circuit Elements   Chap. 2

2-38. In the circuit of Fig. P2-38 if $v_{ab}(0) = 2$ V and $i(t)$ has the variation shown, sketch the waveform $v_{ab}(t)$ on carefully labeled axes. Identify all salient points.

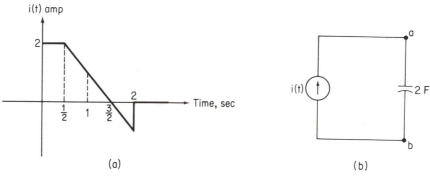

(a)   (b)

Fig. P2-38

2-39. The voltage waveform shown in Fig. P2-39 is applied separately to a pure capacitor of 1 F and to a pure inductor of 1 H. Carefully sketch the current waveshape for each circuit element over the specified time interval. *Hint:* A graphical approach may be found easier than an analytical one.

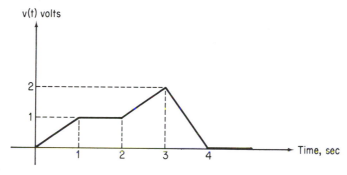

Fig. P2-39

2-40. Propose an op-amp circuit that can serve as a current-controlled voltage source (CCVS). Draw an appropriate model for this dependent source.

2-41. Propose an op-amp circuit that can serve as a voltage-controlled current source (VCCS). Draw an appropriate model for this dependent source.

chapter three

Elementary Network Theory

The network theory treated in this chapter is used extensively in all branches of electrical engineering. It is customary, for example, in analyzing specific engineering devices (such as a transistor, or a polyphase induction motor) to represent the device in terms of an appropriate equivalent circuit. Once this is accomplished, the performance of the device can easily be determined by applying the techniques of circuit theory. It is the purpose of this chapter to describe and develop these techniques.

The first three sections discuss the manner of handling the series and parallel combinations of resistances, capacitances, and inductances. In the remaining sections of the chapter, however, the various analytical techniques of network theory are developed in a framework of circuit elements consisting solely of resistors. This is done to achieve the maximum focusing of attention on the network theory itself by freeing the treatment of the clutter of complex numbers which must be involved† if inductors and capacitors are included. Such an approach in no way causes a loss of generality. In Chapter 7 this same theory is shown to apply with equal validity for circuits involving all three circuit parameters in whatever combinations.

† See Chapter 7.

3-1 SERIES AND PARALLEL COMBINATIONS OF RESISTANCES

Very often in circuit analysis it is necessary to deal with several elements in a closed-loop circuit which exhibit the property of dissipating heat. In the powerline circuits which supply electrical energy to homes and commercial establishments, for example, several resistive elements are frequently combined to carry the same current. Thus in a circuit which supplies electrical power to a lamp, three resistances are found: the internal resistance of the distribution transformer located beneath the street, the lamp resistance, and the resistance of the wires used to conduct the electrical power to the lamp. Similarly, radio and television circuits as well as industrial electronic circuitry employ series combinations of various resistors to achieve specific desirable objectives. To analyze such circuits properly we must know how to treat resistances in series.

By definition, circuit elements that carry the *same* current are said to be in *series*. It is not enough for the circuit elements to carry equal currents, for they can easily do this and yet be physically miles apart in entirely different circuits. The circuit parameters appearing in Fig. 3-1(a) are in series, for it is obvious here that the same current i flows through each circuit element.

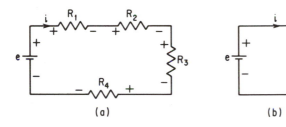

Fig. 3-1 Resistances in series: (a) original configuration; (b) equivalent circuit.

Applying Kirchhoff's voltage law, Eq. (1-34), to the circuit of Fig. 3-1(a) reveals a simple rule for handling resistances in series. Calling all voltage drops positive and voltage rises negative, as the circuit is traversed in the assumed current-flow direction, we can write

$$iR_1 + iR_2 + iR_3 + iR_4 - e = 0 \qquad (3\text{-}1)$$

Rearranging yields

$$e = i(R_1 + R_2 + R_3 + R_4) \qquad (3\text{-}2)$$

The current i is factored out because it is common to each resistance. Consequently, the quantity in parentheses may be replaced by an equivalent resistance which is given by

$$R_s = R_1 + R_2 + R_3 + R_4 \qquad (3\text{-}3)$$

Equation (3-2) may then be written simply as

$$e = iR_s \qquad (3\text{-}4)$$

where R_s denotes the *equivalent series resistance* of the circuit. It follows, too, from this analysis that the original circuit configuration of Fig. 3-1(a) may be replaced by the equivalent circuit shown in Fig. 3-1(b), which is merely a circuit interpretation of Eq. (3-4).

In general, if there are n series-connected resistances in a circuit, *the equivalent series resistance is obtained by taking the sum of the individual resistances.* Expressed mathematically, we have

$$R_s = R_1 + R_2 + \cdots + R_n = \sum_{j=1}^{n} R_j \qquad (3\text{-}5)$$

Circuit elements are also very frequently found in parallel combinations. In the home all electric light bulbs appear in parallel paths with respect to the source voltage. Other circuit elements such as the electric ironer and the electric broiler when used simultaneously are in parallel. By definition circuit elements are said to be in *parallel* when the *same* potential difference appears across their terminals. In accordance with this definition, the resistances R_1, R_2, and R_3 in Fig. 3-2(a) are in parallel.

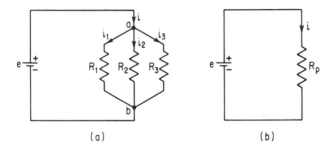

(a)   (b)

Fig. 3-2 Resistances in parallel: (a) original configuration; (b) equivalent circuit.

It is possible, again by means of circuit analysis, to treat this parallel combination of resistances in terms of an equivalent quantity. Kirchhoff's current law states that the current entering terminal a is equal to the sum of currents leaving this terminal. Expressed in equation form we have

$$i = i_1 + i_2 + i_3 \qquad (3\text{-}6)$$

However, from Ohm's law as it relates to each resistance, Eq. (3-6) may be rewritten

$$i = \frac{e}{R_1} + \frac{e}{R_2} + \frac{e}{R_3} \qquad (3\text{-}7)$$

Again by factoring out the common variable, which in this instance is the voltage e, there results

$$i = e \left(\frac{1}{R_1} + \frac{1}{R_2} + \frac{1}{R_3} \right) \qquad (3\text{-}8)$$

Elementary Network Theory   Chap. 3

The expression in parentheses may be replaced by an equivalent quantity defined as

$$\frac{1}{R_p} = \frac{1}{R_1} + \frac{1}{R_2} + \frac{1}{R_3} \qquad (3-9)$$

where R_p denotes the equivalent resistance of the parallel combination of resistances. Upon substituting Eq. (3-9) into Eq. (3-8) we obtain a simplified equation for the circuit. Thus

$$i = \frac{e}{R_p} \qquad (3-10)$$

Figure 3-2(b), which is the circuit representation of Eq. (3-10), may accordingly be considered as the equivalent circuit of the configuration of Fig. 3-2(a).

A general formulation of the foregoing procedure states that *the equivalent resistance of n parallel-connected resistances is the reciprocal of the sum of the reciprocals of the individual resistances.* Expressed in equation form this becomes

$$\boxed{\frac{1}{R_p} = \frac{1}{R_1} + \frac{1}{R_2} + \cdots + \frac{1}{R_n} = \sum_{j=1}^{n} \frac{1}{R_j}} \qquad (3-11)$$

Equation (3-11) deals with the reciprocal of resistance. The unit for this quantity is the siemen, S. On the basis of the definition for conductance, which was introduced in Chapter 2 [see Eq. (2-17)], Eq. (3-11) may also be expressed as

$$G_p = G_1 + G_2 + \cdots + G_n = \sum_{j=1}^{n} G_j \quad S \qquad (3-11a)$$

Examples illustrating the use of Eqs. (3-5) and (3-11) to simplify circuit analysis are given in Sec. 3-4.

3-2 SERIES AND PARALLEL COMBINATIONS OF CAPACITANCES

Series and parallel combinations of capacitances occur quite often in electronic circuitry—much more so than in the other fields of electrical engineering.

Appearing in Fig. 3-3(a) is a circuit involving two series-connected capacitors having capacitances C_1 and C_2, respectively. When a voltage is applied across the combination, the capacitors will have equal displacement of charge. But the potential difference across their terminals will be different provided that C_1 is not equal to C_2. However, by Kirchhoff's voltage law the sum of these two voltages must add up to the source voltage. Thus

$$e = v_1 + v_2 \qquad (3-12)$$

But by Eq. (2-41) for zero initial capacitor voltage

$$v_1 = \frac{1}{C_1} \int_0^t i \, dt \quad \text{and} \quad v_2 = \frac{1}{C_2} \int_0^t i \, dt \qquad (3-13)$$

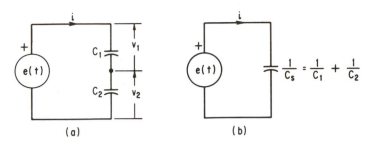

Fig. 3-3 Series-connected capacitances: (a) original circuit; (b) equivalent circuit.

Inserting Eqs. (3-13) into Eq. (3-12) yields

$$e = \frac{1}{C_1} \int_0^t i\, dt + \frac{1}{C_2} \int_0^t i\, dt \tag{3-14}$$

or

$$e = \left(\frac{1}{C_1} + \frac{1}{C_2}\right) \int_0^t i\, dt \tag{3-15}$$

This expression may now be simplified by introducing the substitution

$$\frac{1}{C_s} = \frac{1}{C_1} + \frac{1}{C_2} \tag{3-16}$$

where C_s denotes the equivalent capacitance of two series-connected capacitances. Accordingly, Eq. (3-15) becomes

$$e = \frac{1}{C_s} \int_0^t i\, dt \tag{3-17}$$

It follows then that Fig. 3-3(a) may be replaced by an equivalent circuit which is consistent with Eq. (3-17). This circuit is shown in Fig. 3-3(b).

A general formulation of the equivalent capacitance of n series-connected capacitances is

$$\boxed{\frac{1}{C_s} = \frac{1}{C_1} + \frac{1}{C_2} + \cdots + \frac{1}{C_n} = \sum_{j=1}^{n} \frac{1}{C_j}} \tag{3-18}$$

Alternatively,

$$\frac{1}{C_s} \equiv S_s = S_1 + S_2 + \cdots S_n = \sum_{j=1}^{n} S_j \tag{3-19}$$

where S denotes stiffness expressed in *darafs*. Comparison of Eq. (3-18), which applies for *series-connected capacitances*, with Eq. (3-11), which applies for *parallel-connected resistances*, reveals the expressions to be identical in form. *Therefore, if the expression involves capacitances, series-connected capacitors are treated in the same manner as the resistances of parallel-connected resistors.*

In the light of this conclusion it is reasonable to expect that capacitances in parallel may be treated in the same manner as resistances in series. The validity of this statement is readily established by applying Kirchhoff's current law to the circuitry of Fig. 3-4(a). Therefore,

$$i = i_1 + i_2 \qquad (3\text{-}20)$$

But by Eq. (2-36)

$$i_1 = C_1 \frac{de}{dt} \qquad \text{and} \qquad i_2 = C_2 \frac{de}{dt}$$

Equation (3-20) then becomes

$$i = C_1 \frac{de}{dt} + C_2 \frac{de}{dt} \qquad (3\text{-}21)$$

Collecting terms leads to

$$i = (C_1 + C_2) \frac{de}{dt} \qquad (3\text{-}22)$$

By introducing into Eq. (3-22) the expression

$$C_p = C_1 + C_2 \qquad (3\text{-}23)$$

where C_p is the equivalent capacitance of the parallel combination, we obtain

$$i = C_p \frac{de}{dt} \qquad (3\text{-}24)$$

The circuit representation of this expression appears in Fig. 3-4(b), which is consequently the equivalent circuit of Fig. 3-4(a).

When n capacitances are connected in parallel it follows that the equivalent capacitance may be expressed as

$$\boxed{C_p = C_1 + C_2 + \cdots + C_n = \sum_{j=1}^{n} C_j} \qquad (3\text{-}25)$$

Since Eq. (3-25) is identical in form to Eq. (3-5) we see that the conclusion stated at the outset is indeed valid.

It is interesting to observe in the foregoing analysis that our attention was focused solely on the manner in which the capacitance parameters combined for equivalence. We deliberately avoided discussing the influence of the time variation of current and voltage in the expressions of Eqs. (3-17) and (3-24). Consideration

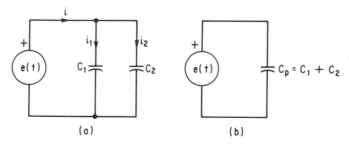

(a)   (b)

Fig. 3-4 Capacitances in parallel: (a) original circuit; (b) equivalent circuit.

will be given to the effects of the integral of a time-varying current and the derivative of a time-varying voltage in Chapter 7, where the subject matter is more appropriate.

3-3 SERIES AND PARALLEL COMBINATIONS OF INDUCTANCES

Depicted in Fig. 3-5(a) is a circuit involving two inductors in series. The source voltage is assumed to be time varying, causing a changing current to flow through the inductors. Then, by Kirchhoff's voltage law,

$$e(t) = L_1\frac{di}{dt} + L_2\frac{di}{dt} \tag{3-26}$$

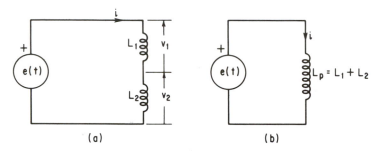

Fig. 3-5 Inductances in series: (a) original circuit; (b) equivalent circuit.

Factoring out the common term leads to

$$e(t) = (L_1 + L_2)\frac{di}{dt} \tag{3-27}$$

Letting

$$L_s = L_1 + L_2 \tag{3-28}$$

and substituting into Eq. (3-27) yields

$$e(t) = L_s\frac{di}{dt} \tag{3-29}$$

where L_s denotes the equivalent inductance for the series connection.

Figure 3-5(b) shows the equivalent circuit of Fig. 3-5(a) based on Eq. (3-29). Note that inductances in series are treated in the same fashion as resistances in series.

In a situation where there are n series-connected inductances in an electric circuit, the resultant equivalent inductance is the sum of the individual inductances. Thus

$$\boxed{L_s = L_1 + L_2 + \cdots + L_n = \sum_{j=1}^{n} L_j} \tag{3-30}$$

A parallel combination of two inductances is shown in Fig. 3-6(a). By Kirchhoff's current law

$$i = i_1 + i_2 \tag{3-31}$$

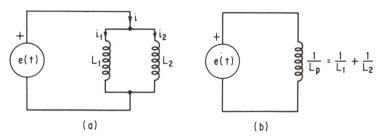

Fig. 3-6 Inductances in parallel: (a) original circuit; (b) equivalent circuit.

Then, in accordance with Eq. (2-24) and assuming zero initial current through each inductance

$$i_1 = \frac{1}{L_1} \int_0^t e(t)\, dt \tag{3-32}$$

and

$$i_2 = \frac{1}{L_2} \int_0^t e(t)\, dt \tag{3-33}$$

Inserting Eqs. (3-32) and (3-33) into Eq. (3-31) then yields

$$i = \left(\frac{1}{L_1} + \frac{1}{L_2} \right) \int_0^t e(t)\, dt \tag{3-34}$$

or

$$i = \frac{1}{L_p} \int_0^t e(t)\, dt \tag{3-35}$$

where

$$\frac{1}{L_p} = \frac{1}{L_1} + \frac{1}{L_2} \tag{3-36}$$

The quantity L_p is the equivalent inductance of the parallel combination of the individual inductances and is depicted in Fig. 3-6(b).

For the general case where there are n parallel-connected inductances, Eq. (3-36) becomes

$$\boxed{\frac{1}{L_p} = \frac{1}{L_1} + \frac{1}{L_2} + \cdots + \frac{1}{L_n} = \sum_{j=1}^{n} \frac{1}{L_j}} \tag{3-37}$$

Equation (3-37) indicates that inductances in parallel are treated in the same way as resistances in parallel.

3-4 SERIES-PARALLEL CIRCUITS

In many practical circuits in electrical engineering there occur situations where a circuit element is in series with a parallel combination of other circuit elements. Although these configurations may involve all three of the circuit parameters, we shall, in the remainder of this chapter, confine our attention exclusively to circuits involving only the resistance parameter. We follow this procedure because it is simpler. Recall that where capacitance and inductance are involved a complete description of the current-voltage relationships is dependent upon a specified variation of the voltage or current applied to each circuit element. Reference to Eqs. (3-24) and (3-29) should make this apparent. The presence of the derivative or integral in these expressions leads to an additional complexity† which can only serve at this stage to obscure the development of the theory as it applies to circuit analysis. There is no loss of generality in the application of the theory and principles herein described. A reading of Sec. 7-9 bears this out. There it is seen that the very same theorems which are treated in this chapter and illustrated solely in terms of resistive circuits are applied with equal validity to circuit configurations involving all three circuit parameters.

The theoretical considerations which are required to solve series-parallel combinations of the resistance parameter have already been studied. They are embodied in Kirchhoff's current and voltage laws, Ohm's law, and Eqs. (3-5) and (3-11) which are a consequence of these laws. The procedure for handling series-parallel circuits is best illustrated by examples.

EXAMPLE 3-1   The circuit of Fig. 3-7(a) involves the combination of three parallel resistors in series with the resistor R_1. For the resistance values shown find the value of current flowing from the voltage source. All resistance values are expressed in ohms.

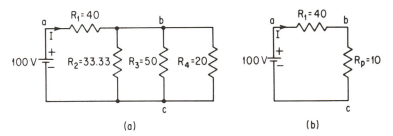

Fig. 3-7   Circuitry for Example 3-1: (a) original configuration; (b) equivalent circuit.

Solution: As the first step in the procedure we find the equivalent resistance of the parallel combination of R_2, R_3, and R_4. Thus, by Eq. (3-11),

$$\frac{1}{R_p} = \frac{1}{R_2} + \frac{1}{R_3} + \frac{1}{R_4} = \frac{1}{33.33} + \frac{1}{50} + \frac{1}{20} = \frac{3 + 2 + 5}{100} = \frac{1}{10} \qquad (3\text{-}38)$$

† By dealing with capacitance and inductance parameters at this point we would have to work with complex numbers rather than real numbers alone. See Chapter 7.

Hence
$$R_p = 10 \ \Omega$$

The original circuit configuration can now be represented by the equivalent circuit of Fig. 3-7(b), which is merely a simple series circuit. Then, by Kirchhoff's voltage law

$$100 = IR_1 + IR_p = I(R_1 + R_p) \tag{3-39}$$

or

$$I = \frac{100}{R_1 + R_p} = \frac{100}{40 + 10} = 2 \text{ A} \tag{3-40}$$

EXAMPLE 3-2   In the circuitry of Fig. 3-7 find the potential difference across terminals *bc*.

Solution: Refer to Fig. 3-7(b). By Ohm's law the voltage drop across *bc* is

$$V_{bc} = IR_p = (2)(10) = 20 \text{ V} \tag{3-41}$$

where I is given by Eq. (3-40).

A glance at Fig. 3-7(b) shows that the current I is common to both resistors. Accordingly it should be possible to obtain the solution without solving for the current specifically. This is readily demonstrated as follows. The voltage across *bc* can be generally written as

$$V_{bc} = IR_p \tag{3-42}$$

Furthermore, by Kirchhoff's voltage law

$$E = I(R_1 + R_p) \tag{3-43}$$

When we divide Eq. (3-42) by Eq. (3-43) the current term cancels, leaving as the expression for the voltage drop across *bc*

$$V_{bc} = \frac{R_p}{R_1 + R_p} E \tag{3-44}$$

Inserting the specified values of the parameters into Eq. (3-44) then gives

$$V_{bc} = \frac{10}{40 + 10}(100) = \frac{1000}{50} = 20 \text{ V} \tag{3-45}$$

Equation (3-44) is used often in circuit analysis. It is worthwhile, therefore, to express it in more general terms as follows: In a circuit composed of n series-connected resistors and a source E the voltage drop V_k appearing across the terminals of resistor R_k is

$$V_k = \frac{R_k}{R_1 + R_2 + \cdots + R_n} E \tag{3-46}$$

Equation (3-46) is frequently referred to as the *voltage-divider rule*.

EXAMPLE 3-3   Determine the equivalent series circuit of the arrangement of circuit elements appearing in Fig. 3-8(a). All resistances are in ohms.

Solution: To obtain the result we follow a process of *network†* reduction which basically involves applying Eqs. (3-5) and (3-11) to series and parallel combinations of resistances starting with the pair of terminals farthest removed from the source voltage. For the case

---

† The word *network* is used synonymously with the word *circuit*.

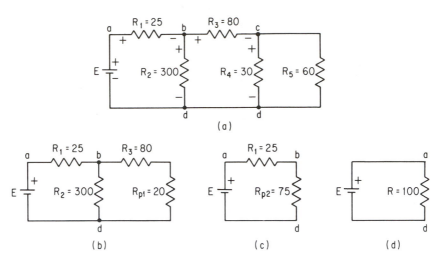

Fig. 3-8 Circuit configuration for Example 3-3: (a) original; (b) intermediate reduction; (c and d) final form.

at hand terminals cd are the farthest removed. Since R_4 and R_5 are in parallel, Eq. (3-11) yields

$$\frac{1}{R_{p1}} = \frac{1}{R_4} + \frac{1}{R_5} \tag{3-47}$$

or

$$R_{p1} = \frac{R_4 R_5}{R_4 + R_5} \tag{3-48}$$

Equation (3-48) is another useful expression to keep in mind because the combination of two parallel resistances occurs often in circuit analysis. Substituting the values for R_4 and R_5 yields

$$R_{p1} = \frac{30(60)}{30 + 60} = 20 \ \Omega \tag{3-49}$$

By means of this computation Fig. 3-8(a) reduces to the form shown in Fig. 3-8(b). Note there that R_3 and R_{p1} are in series, so that by Eq. (3-5) these may be combined into a single resistance of value

$$R_s = R_3 + R_{p1} = 80 + 20 = 100 \ \Omega \tag{3-50}$$

A further reduction (or simplification) can be accomplished by again applying Eq. (3-11) to the two parallel resistances, R_2 and R_s, which appear across terminals bd. Thus

$$R_{p2} = \frac{R_2 R_s}{R_2 + R_s} = \frac{300(100)}{300 + 100} = 75 \ \Omega \tag{3-51}$$

Since R_{p2} is in series with R_1, it follows that Fig. 3-8(b) can now be represented by Fig. 3-8(c) or Fig. 3-8(d), which is the desired result.

EXAMPLE 3-4   In the circuit of Fig. 3-8(a) find the value of E which yields a power dissipation in R_5 of 15 W.

Solution: The solution is readily found by applying the laws of Ohm and Kirchhoff, but first it is necessary to find the current flowing through resistor R_5. Since power is energy per unit time, it follows from Eq. (2-15) that

$$I_5^2 R_5 = 15 \text{ W} \tag{3-52}$$

where I_5 denotes the current flowing through R_5. Hence

$$I_5 = \sqrt{\frac{15}{R_5}} = \sqrt{\frac{15}{60}} = \frac{1}{2} \text{ A} \tag{3-53}$$

Then by Ohm's law the voltage drop across terminals *cd* is

$$V_{cd} = I_5 R_5 = \tfrac{1}{2}(60) = 30 \text{ V} \tag{3-54}$$

Because this same voltage appears across R_4, the current through R_4 is

$$I_4 = \frac{V_{cd}}{R_4} = \frac{30}{30} = 1 \text{ A} \tag{3-55}$$

The current which flows through R_3 is the current entering junction *c*, which is equal to the two currents—I_4 and I_5—which leave junction *c*. Therefore,

$$I_3 = I_4 + I_5 = 1 + \tfrac{1}{2} = \tfrac{3}{2} \text{ A} \tag{3-56}$$

The corresponding potential difference associated with I_3 flowing through R_3 is then

$$V_{bc} = I_3 R_3 = \tfrac{3}{2}(80) = 120 \text{ V} \tag{3-57}$$

To find the current which flows through R_2 we must first determine the voltage across it, V_{bd}. Applying Kirchhoff's voltage law to loop *dbcd* in Fig. 3-8(a), we have

$$V_{bd} = V_{bc} + V_{cd} = 120 + 30 = 150 \text{ V} \tag{3-58}$$

Hence

$$I_2 = \frac{V_{bd}}{R_2} = \frac{150}{300} = \frac{1}{2} \text{ A} \tag{3-59}$$

Since the current flowing through R_1 and into junction *b* splits into two paths yielding currents I_2 and I_3, both of which are now determined, it follows that

$$I_1 = I_2 + I_3 = \tfrac{1}{2} + \tfrac{3}{2} = 2 \text{ A} \tag{3-60}$$

The corresponding voltage drop caused by this current flowing through R_1 is therefore

$$V_{ab} = I_1 R_1 = 2(25) = 50 \text{ V} \tag{3-61}$$

Finally, by applying Kirchhoff's voltage law to loop *abda* we obtain the desired solution.

$$E = V_{ab} + V_{bd} = 50 + 150 = 200 \text{ V} \tag{3-62}$$

In recapitulation note that the solution of this problem was achieved by applying Ohm's law to get the voltage drop across an individual circuit element, Kirchhoff's current law at appropriate junction points such as *b* and *c* in Fig. 3-8(a) to obtain information about the current entering the junction, and also Kirchhoff's voltage law to appropriate loops in order to obtain the proper levels of potential differences.

EXAMPLE 3-5   Appearing in Fig. 3-9(a) is a series-parallel circuit which involves two voltage sources. For the values of the parameters shown in the figure find the current which flows through R_2, the 20-Ω resistor.

Solution: First replace the parallel combination of resistances across *bc* by its equivalent value. Thus

$$R_p = \frac{R_2 R_3}{R_2 + R_3} = \frac{20(60)}{20 + 60} = 15 \; \Omega \tag{3-63}$$

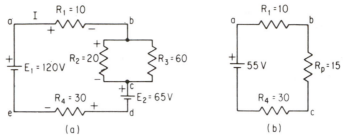

Fig. 3-9 Circuit for Example 3-5, to illustrate superposition: (a) original configuration; (b) equivalent circuit.

Then apply Kirchhoff's voltage law to the loop. Assume that the loop current flows clockwise. Calling voltage drops positive and voltage rises negative, the loop voltage equation becomes

$$IR_1 + IR_p + E_2 + IR_4 - E_1 = 0 \tag{3-64}$$

Combining and rearranging terms leads to

$$I(R_1 + R_p + R_4) = E_1 - E_2 = 120 - 65 = 55 \tag{3-65}$$

The circuit representation of this last expression is shown in Fig. 3-9(b). The value of I consequently is

$$I = \frac{55}{R_1 + R_p + R_4} = \frac{55}{10 + 15 + 30} = 1 \text{ A} \tag{3-66}$$

To find the current through R_2 it is necessary first to compute the voltage across terminals bc. Hence

$$V_{bc} = IR_p = 1(15) = 15 \text{ V} \tag{3-67}$$

Then by Ohm's law the current through R_2 is

$$I_2 = \frac{V_{bc}}{R_2} = \frac{15}{20} = \frac{3}{4} \text{ A} \tag{3-68}$$

which is the desired result.

3-5 THE SUPERPOSITION THEOREM

This theorem has to do with the presence of two or more energy sources acting within a network. Accordingly, in the interest of establishing the proper background, let us return temporarily to the circuit of Fig. 3-9(a) in which appear two distinct voltage sources. By Eq. (3-64) the expression for the net current I existing in this circuit can be written in terms of the effect which each voltage source produces. Thus

$$I = \frac{E_1}{R_1 + R_p + R_4} - \frac{E_2}{R_1 + R_p + R_4} \tag{3-69}$$

An examination of the first term on the right side of this equation reveals that $E_1/(R_1 + R_p + R_4)$ is precisely the current which flows in the circuit in the assumed direction of I when E_2 is removed and replaced with a short circuit between points c and d and E_1 is allowed to act alone. Similarly, the quantity $-E_2/(R_1 + R_p + R_4)$ is the current produced in the same circuit by the voltage source E_2 when E_1 is removed and replaced with a short circuit between points a and e. The minus sign indicates that the current produced by E_2 is in a direction *opposed* to that caused by E_1 acting alone. Furthermore, because the current produced by either voltage is *linearly* related to the voltage in accordance with Ohm's law, the two effects may be superposed to yield the net current I. Performing the operations called for individually in Eq. (3-69) we have as the net current

$$I = \frac{120}{55} - \frac{65}{55}$$

$$= 2.182 - 1.182 = 1.000 \text{ A}$$

(3-70)

which is the same result as Eq. (3-66). Note that $I_1 = 2.182$ A is the solution of the current in the circuit due to E_1 alone, and $I_2 = -1.182$ A is the solution of the current (flowing opposite to the assumed positive direction of current flow for I) in the circuit due to E_2 acting alone.

Accordingly, on the basis of the foregoing discussion the *superposition theorem* may be stated as follows: If I_1 is the response to E_1 and I_2 is the response to E_2, then in any *linear* circuit the response to the combined forcing function $(E_1 + E_2)$ is $(I_1 + I_2)$. This theorem is not limited to circuit theory only. It is applicable in many fields. The sole requirement is that cause and effect bear a linear relationship to one another.

EXAMPLE 3-6   Find the power dissipated in the 10-Ω resistor R_1 in Fig. 3-9(a).

Solution: The power is given by the equation

$$P_1 = I^2 R_1 = I^2(10) = 10 \text{ W}$$

(3-71)

Let us investigate at this point whether this power can be computed by superposition. Proceeding for the moment on the assumption that such a procedure is valid, we write

$$P = I_1^2 R_1 + (-I_2)^2 R_1$$

$$= (2.182)^2 10 + (-1.182)^2 10$$

(3-72)

$$= 47.6 + 13.95$$

$$= 61.55 \text{ W}$$

A comparison of the two results shows a vast inconsistency. In fact the result from superposition gives a power dissipation in R_1 alone which exceeds half the total power delivered by the E_1 source. By Eq. (2-5) this power is $E_1 I = 120(1) = 120$ W. Clearly, then, superposition does not work in this case. A glance at Eq. (3-72) points out the reason. Note that the two variables involved, power and current, are *not linearly related* because current appears to the second power. Therefore, superposition does not apply and so cannot be expected to give correct answers.

In the configuration of Fig. 3-9(a) there is little to be gained by using the principle of superposition to obtain the solution. The use of Kirchhoff's voltage law is indeed simpler. However, this is not true of the circuit arrangement of Fig. 3-10(a). This diagram is obtained from Fig. 3-9(a) by placing E_2 in the R_3 branch.

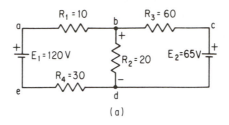

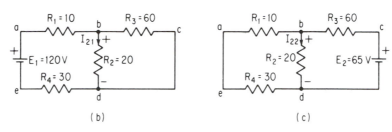

Fig. 3-10 Illustrating superposition (all resistances in ohms): (a) original circuit; (b) response due to E_1 alone; (c) response due to E_2 alone.

EXAMPLE 3-7 By means of the superposition theorem find the current which flows through R_2 in the circuit of Fig. 3-10(a).

Solution: Figure 3-10(b) depicts the circuit with E_1 acting alone. Note that this circuit is identical to Fig. 3-9(b) with the exception that the source voltage is 120 V. The total current supplied by E_1 is therefore

$$I_1 = \frac{E_1}{55} = \frac{120}{55} = 2.182 \text{ A} \tag{3-73}$$

The potential difference across R_2 due to I_1 is then equal to

$$V_{bd1} = I_1 R_p = 2.182(15) = 32.7 \text{ V} \tag{3-74}$$

where the subscript 1 denotes the voltage across bd due to E_1. Hence the current which flows through R_2 as a result of E_1 is

$$I_{21} = \frac{32.7}{R_2} = \frac{32.7}{20} = 1.64 \text{ A} \tag{3-75}$$

where the first subscript denotes that resistor R_2 is involved and the second subscript refers to source E_1.

Appearing in Fig. 3-10(c) is the circuit diagram which allows the effect of E_2 to be found. E_1 is removed. In arm $baed$ the 10-Ω and the 30-Ω resistors may be combined into a 40-Ω resistor. Then between terminals bd there appears the parallel combination of the 40- and the 20-Ω resistors. Hence

$$R_p = \frac{40(20)}{40 + 20} = \frac{40}{3} \Omega \tag{3-76}$$

With this quantity available the potential difference across R_2 caused by E_2 is readily obtained by using the voltage-divider rule. Refer to Eq. (3-46). Thus

$$V_{bd2} = \frac{R_p}{R_p + R_3} E_2 = \frac{\frac{40}{3}}{60 + \frac{40}{3}} (65) = 11.8 \text{ V} \tag{3-77}$$

where the subscript 2 refers to source E_2. Note that the voltage drop across R_2 caused by E_2 is in the *same* direction as that caused by E_1. Therefore, the current produced in R_2 by E_2 is

$$I_{22} = \frac{V_{bd2}}{R_2} = \frac{11.8}{20} = 0.59 \text{ A} \tag{3-78}$$

Accordingly, by superposition the total current flowing through R_2 in the configuration of Fig. 3-10(a) is

$$I_2 = I_{21} + I_{22} = 1.64 + 0.59 = 2.23 \text{ A} \tag{3-79}$$

The plus sign is used here because E_1 and E_2 both cause current to flow from terminal b to d in passing through R_2.

3-6 NETWORK ANALYSIS BY MESH CURRENTS

Network reduction and superposition are only two of several methods which are available to the engineer for determining the response of a circuit to one or more energy sources acting within the circuit. In this section attention is focused on a third method which can be used to find the current flowing through any particular circuit element. However, before proceeding with the treatment, it is useful first to define a network terminology which makes it easier to discuss this method as well as the one described in the next section.

The word *network* is used synonymously with the term *circuit* and refers to any arrangement of passive and/or active circuit elements which form closed paths. Illustrated in Fig. 3-11 is a typical network. Note that this particular circuit is the same as Fig. 3-10(a) except that R_1 and R_4 are combined into a single 40-Ω resistor, which is permitted in accordance with Eq. (3-5). This network is composed of five circuit elements. There are three passive elements, namely resistances R_1, R_2, and R_3, and two active elements, namely the energy sources E_1 and E_2.

A *node* of a network is an equipotential surface at which two or more circuit elements are joined. Thus in Fig. 3-11 terminals a, b, c, and d are nodes.

A *junction* is that point in a network where three or more circuit elements are joined. In the network of Fig. 3-11 there are two junction points—b and d. The junction is sometimes referred to as an *independent node*.

A *branch* is that part of a network which lies between junction points.

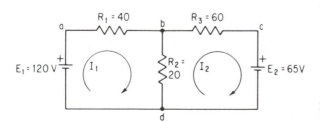

Fig. 3-11 Two-mesh network. This is Fig. 3-10(a) redrawn.

Accordingly, in Fig. 3-11 we see that there are a total of three branches. The branch *dab* consists of the two circuit elements E_1 and R_1. The second branch involves R_3 and E_2, and finally the third branch is merely R_2.

A *loop* is any closed path of the network. Examples of loops in Fig. 3-11 are *abda*, *dbcd*, and *abcda*.

A *mesh* is the most elementary form of a loop. It is a property of a planar† network diagram and must be so identified that it cannot be further divided into other loops. In the circuit of Fig. 3-11 both loops *abda* and *dbcd* qualify as meshes, but *abcda* cannot because it encloses the first two loops.

A *tree* is defined as that part of a network composed of those branches between the junction points which can be drawn without forming a closed path. A tree of the network of Fig. 3-11 is depicted in Fig. 3-12. It consists of the two junction points *b* and *d* and the branch R_2 between these points. Note that if either of the two remaining branches were drawn, a closed path would result. Hence these branches are not permitted and so are not part of the tree. The proper identification of the tree of a network is useful because it facilitates the determination of the minimum number of independent equations needed to solve the circuit completely.

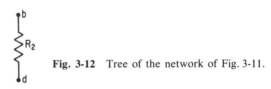

Fig. 3-12 Tree of the network of Fig. 3-11.

We shall consider here the *mesh method* of solving problems in circuitry because it offers some advantages over the methods discussed so far. To illustrate, let us describe the method and then apply it to an example. For convenience the procedure is explained in conjunction with the circuitry of Fig. 3-11. Although in the solution of circuit problems we are often concerned with finding the current which flows in each branch of the network, by the use of loop currents we can often obtain this information with much less effort. As a general rule those loops should be chosen which are simplest in form. In other words, we should work with meshes. Often we can readily identify the meshes to be used, once the tree of the network is drawn. It then simply requires introducing the remaining branches. Each branch so introduced identifies an appropriate mesh. Thus in Fig. 3-12, by adding branch *bad* to the tree a mesh results. Similarly, by introducing branch *bcd* a second mesh results. It is important to note here that as each mesh is formed, at least one branch is included in the newly formed mesh which was not included as part of a loop previously formed. Consequently, by this process we are assured of forming the correct number and composition of the meshes needed to solve the problem. For the network of Fig. 3-11 we have two independent meshes, each of which leads to an independent mesh equation.

† Two-dimensional.

To write the mesh equations we must first introduce the mesh currents. By definition, a *mesh current* is that current which flows around the perimeter of a mesh; for convenience, all mesh currents are drawn in a clockwise direction as shown in Fig. 3-11. Mesh currents may or may not have a direct identification with branch currents. For example, in mesh 1 of Fig. 3-11 the branch current through R_1 is identical to the mesh current. However, the branch current through R_2 is clearly the difference between the two mesh currents, $I_1 - I_2$, because I_1 and I_2 have opposite assumed positive directions. Thus branch currents have a physical identity, and they may be measured. Mesh currents, on the other hand, are fictitious quantities which are introduced because they allow us to solve problems in terms of a minimum number of unknowns. Hence in Fig. 3-11, although there are three unknown branch currents; we can readily find these by determining the two unknown mesh currents I_1 and I_2.

At this point in our study it is appropriate to refer to an expression which relates the number of independent mesh equations to the number of branches and junction points of the network. It reads as follows:

$$\boxed{m = b - (j - 1)} \tag{3-80}$$

where m = number of independent mesh equations

   b = number of branches

   j = number of junction points

The -1 is included because some point in the network is needed as a reference point from which voltage measurements can be made, and a junction point is chosen for this purpose. Applying Eq. (3-80) to the circuitry of Fig. 3-11 yields

$$m = 3 - (2 - 1) = 2$$

Accordingly, this network can be completely analyzed by solving two independent equations. Of course, this is a fact we had already learned in connection with forming the appropriate meshes from the tree. Equation (3-80), however, presents the information in precise mathematical form.

The independent mesh equations referred to in the foregoing discussion are those which are obtained by applying Kirchhoff's voltage law to each independent mesh. By traversing mesh 1 in the clockwise direction and calling voltage drops positive and voltage rises negative, we obtain

$$I_1R_1 + I_1R_2 - R_2I_2 - E_1 = 0 \tag{3-81}$$

Note that the quantity $(-R_2I_2)$ appears because of the effect which mesh current I_2 has in mesh 1 through the mutual resistance R_2. Since I_2 flows in a direction opposite to I_1 through R_2, it produces a rise in voltage for a clockwise traversal in mesh 1.

Rearranging Eq. (3-81) leads to

$$I_1(R_1 + R_2) - I_2R_2 = E_1 \tag{3-81a}$$

With this formulation the coefficient of mesh current I_1 is called the *self-resistance* of mesh 1 and is denoted by R_{11} (read as R one-one). A glance at Fig. 3-11 makes it apparent that R_{11} is simply the sum of all resistances found in mesh 1.

Moreover, the coefficient of mesh current I_2 is called the *mutual resistance* of the network because it is common to meshes 1 and 2. It is represented by the symbol R_{12}. In terms of this standard notation, Eq. (3-81a) can be rewritten as

$$R_{11}I_1 + R_{12}I_2 = E_1 \qquad (3\text{-}82)$$

Equation (3-82) is thus one of the two independent equations needed for a complete determination of the currents existing in the network. This equation involves the two unknown quantities I_1 and I_2. All else is known.

The second required mesh equation is found by applying Kirchhoff's voltage law in a similar manner to mesh 2. Thus

$$I_2R_2 + I_2R_3 + E_2 - R_2I_1 = 0 \qquad (3\text{-}83)$$

Collecting terms yields

$$-R_2I_1 + (R_2 + R_3)I_2 = -E_2 \qquad (3\text{-}84)$$

Again note the presence of a term which involves I_1. This comes about because of the mutual resistance R_2. The coefficient of I_2 is called the self-resistance of mesh 2 and is denoted R_{22}. Hence Eq. (3-84) may be rewritten as

$$+R_{21}I_1 + R_{22}I_2 = -E_2 \qquad (3\text{-}85)$$

This is the second of the two required independent equations. A comparison of Eq. (3-85) with Eq. (3-82) shows each equation to be truly independent because R_{11} and R_{22} involve two different parameters of the network. That is, R_{11} involves R_1 but not R_3, and R_{22} involves R_3 but not R_1.

Mesh Analysis via Matrices. In dealing with networks that involve many meshes it is a distinct advantage to employ the shorthand notation of matrix algebra. The set of algebraic equations that permits the solution of the two-mesh circuit of Fig. 3-11 and is represented by Eqs. (3-82) and (3-85) can be written in compact form as

$$\mathbf{RI} = \mathbf{V} \qquad (3\text{-}86)$$

where $\mathbf{V}$ denotes a column vector representing the net source that acts to produce current in the assumed clockwise direction, $\mathbf{I}$ is a column vector denoting the independent mesh currents, and $\mathbf{R}$ is an $n \times n$ resistance matrix with n representing the number of mesh currents. In the case of the circuit of Fig. 3-11, $\mathbf{V}$ is a 2×1 vector having components $\mathbf{V}_1$ and $\mathbf{V}_2$. Specifically, $\mathbf{V}_1$ is the net voltage acting to produce $\mathbf{I}_1$ and is equal to the right side of Eq. (3-82), which is $\mathbf{E}_1$. Similarly, $\mathbf{V}_2$ is found to be the net voltage rise which acts to sustain the current $\mathbf{I}_2$. A glance at Fig. 3-11 makes it apparent in this case that the quantity $\mathbf{V}_2$ is equal to $-\mathbf{E}_2$. The negative sign is used because the voltage sources encountered in mesh 2 serve to support current flow that is in a direction opposite to the assumed direction of mesh current $\mathbf{I}_2$. The resistance matrix $\mathbf{R}$ is the representation of the self-resistance of the meshes as they appear on the principal diagonal of the matrix as well as all the mutual resistances. In general terms for the two-mesh case $\mathbf{R}$ takes the form

$$\mathbf{R} = \begin{bmatrix} \mathbf{R}_{11} & \mathbf{R}_{12} \\ \mathbf{R}_{21} & \mathbf{R}_{22} \end{bmatrix} \qquad (3\text{-}87)$$

while for the specific situation of Fig. 3-11 this becomes

$$\mathbf{R} = \begin{bmatrix} R_1 + R_2 & -R_2 \\ -R_2 & R_2 + R_3 \end{bmatrix} \tag{3-88}$$

Following the rules of matrix algebra the solution for the current in Eq. (3-86) is found by multiplying both sides of the equation by the inverse resistance matrix. Thus we get for the mesh current vector

$$\mathbf{I} = \mathbf{R}^{-1}\mathbf{V} \tag{3-89}$$

Generally, Eq. (3-89) is an abbreviated way of representing a number of simultaneous algebraic equations equal to the number of independent mesh currents.

To obtain expressions for the mesh currents $\mathbf{I}_1$ and $\mathbf{I}_2$ in the circuit of Fig. 3-11, use is made of *Gauss's elimination* method. Thus, to eliminate $\mathbf{I}_2$ from Eqs. (3-82) and (3-85), we multiply the former by $\mathbf{R}_{22}$ and the latter by $\mathbf{R}_{12}$ and then subtract the latter expression from the former. Hence

$$\mathbf{I}_1\mathbf{R}_{11}\mathbf{R}_{22} + \mathbf{I}_2\mathbf{R}_{12}\mathbf{R}_{22} = \mathbf{V}_1\,\mathbf{R}_{22} \tag{3-90}$$

$$-\mathbf{I}_1\mathbf{R}_{21}\mathbf{R}_{12} - \mathbf{I}_2\mathbf{R}_{22}\mathbf{R}_{12} = -\mathbf{V}_2\,\mathbf{R}_{12} \tag{3-91}$$

or
$$\mathbf{I}_1(\mathbf{R}_{11}\mathbf{R}_{22} - \mathbf{R}_{12}^2) = \mathbf{V}_1\mathbf{R}_{22} - \mathbf{V}_2\mathbf{R}_{12} \tag{3-92}$$

Equation (3-92) is expressed in matrix formulation quite simply as

$$\mathbf{I}_1 = \frac{\begin{vmatrix} \mathbf{V}_1 & \mathbf{R}_{12} \\ \mathbf{V}_2 & \mathbf{R}_{22} \end{vmatrix}}{\begin{vmatrix} \mathbf{R}_{11} & \mathbf{R}_{12} \\ \mathbf{R}_{21} & \mathbf{R}_{22} \end{vmatrix}} = \frac{\begin{vmatrix} \mathbf{V}_1 & \mathbf{R}_{12} \\ \mathbf{V}_2 & \mathbf{R}_{22} \end{vmatrix}}{\mathbf{D}_R} \tag{3-93}$$

where the vertical lines denote evaluation by the rules of determinants. The quantity $\mathbf{D}_R$ is used to denote the *value* of the resistance matrix.

Applying a similar procedure for the solution of the second mesh current $\mathbf{I}_2$ leads to the result

$$\mathbf{I}_2 = \frac{\begin{vmatrix} \mathbf{R}_{11} & \mathbf{V}_1 \\ \mathbf{R}_{21} & \mathbf{V}_2 \end{vmatrix}}{\begin{vmatrix} \mathbf{R}_{11} & \mathbf{R}_{12} \\ \mathbf{R}_{21} & \mathbf{R}_{22} \end{vmatrix}} = \frac{\begin{vmatrix} \mathbf{R}_{11} & \mathbf{V}_1 \\ \mathbf{R}_{21} & \mathbf{V}_2 \end{vmatrix}}{\mathbf{D}_R} \tag{3-94}$$

A study of Eqs. (3-93) and (3-94) leads to an important simplified procedure in writing the solution for the dependent variable (i.e., the mesh currents in the present situation) in the equation set represented by the condensed form of Eq. (3-86). First observe that in each case the denominator is the determinant of the resistance matrix of the network. Then note that to evaluate $\mathbf{I}_1$ it is necessary to use an $n \times n$ determinant in the numerator, where the first column of the resistance matrix is replaced by the elements of the $\mathbf{V}$ matrix, respectively. In a similar manner the solution for $\mathbf{I}_2$ involves a numerator determinant where the elements of the second column of the resistance matrix are replaced by the elements of the $\mathbf{V}$ matrix. This procedure, which can be executed without going through the steps of the Gauss elimination process, is commonly referred to as *Cramer's rule*.

EXAMPLE 3-8   In the network of Fig. 3-11 find the magnitude and direction of each branch current by using mesh analysis.

Solution: Inserting the specified value for R_1, R_2, R_3, E_1, and E_2 into Eqs. (3-82) and (3-84) gives

$$60I_1 - 20I_2 = 120 \tag{3-95}$$

and

$$-20I_1 + 80I_2 = -65 \tag{3-96}$$

Through the use of determinants we obtain directly

$$I_1 = \frac{\begin{vmatrix} 120 & -20 \\ -65 & 80 \end{vmatrix}}{\begin{vmatrix} 60 & -20 \\ -20 & 80 \end{vmatrix}} = \frac{83}{44} = 1.885 \text{ A} \tag{3-97}$$

and

$$I_2 = \frac{\begin{vmatrix} 60 & 120 \\ -20 & -65 \end{vmatrix}}{4400} = \frac{15}{44} = -0.341 \text{ A} \tag{3-98}$$

The minus sign for I_2 indicates that the true direction of flow I_2 is counterclockwise.

The branch currents expressed in terms of the computed mesh currents are then as follows:

$$I_{bd} = \text{current in branch } bd$$

$$= I_1 - I_2 = 1.885 - (-0.341) = 2.23 \text{ A} \tag{3-99}$$

$$I_{ab} = \text{current in branch } dab$$

$$= I_t = 1.885 \text{ A} \tag{3-100}$$

$$I_{cb} = \text{current in branch } dcb$$

$$= -I_2 = 0.341 \text{ A} \tag{3-101}$$

Note that I_{bd} above is the same as the current found through R_2 by using the method of superposition. Refer to Eq. (3-79). It is interesting to note, too, that with the mesh method of solution the currents in the remaining two branches are available immediately as the mesh currents themselves, whereas to find these same branch currents by superposition additional calculations are needed.

In the interest of gaining some additional experience with the mesh method of circuit analysis, let us apply the technique in a general way to the more complicated circuit of Fig. 3-13. An inspection of the network reveals that there are three junction points, b, c, and e. Furthermore, between these junction points there are a total of five branches. Therefore, by Eq. (3-80) it follows that there are three independent mesh equations.

$$m = b - (j - 1) = 5 - (3 - 1) = 3 \tag{3-102}$$

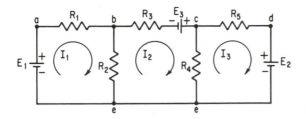

Fig. 3-13 Three-mesh network.

A tree of the network is depicted in Fig. 3-14. Although several other representations of the tree of this network are possible, the one shown in Fig. 3-14 is preferred because it leads directly to a convenient mesh formation. Thus by joining branch *eab* to *be*, mesh 1 is formed, and by introducing branch *bc* into place, mesh 2 is formed, and by inserting branch *cde* into place, mesh 3 is formed.

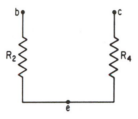

Fig. 3-14 Tree of the network for Fig. 3-13.

Although Kirchhoff's voltage law may be used in each mesh to obtain the three independent equations which permit a complete solution of the circuit, we shall do so instead by introducing self- and mutual resistances. Thus, an examination of mesh 1 shows that the total resistance in this mesh is

$$R_{11} = R_1 + R_2 \qquad (3\text{-}103)$$

where R_{11} denotes the self-resistance of mesh 1. It is the total resistance encountered in traversing the perimeter of mesh 1. Furthermore, since R_2 is common to meshes 1 and 2

$$R_{12} = -R_2 \qquad (3\text{-}104)$$

where R_{12} denotes the mutual resistance between mesh 1 and mesh 2. Also, since there is no circuit element in mesh 1 which is common to mesh 3, it follows that the coupling term between meshes 1 and 3 is zero. Hence

$$R_{13} = 0 \qquad (3\text{-}105)$$

Consequently, the voltage equation as it applies to mesh 1 is

$$(R_1 + R_2)I_1 - R_2 I_2 - 0 = E_1 \qquad (3\text{-}106)$$

The rule for handling voltage sources that appear in a mesh is to use a plus sign for this quantity on the right side of the equation if its acts to produce a current which is in the same direction as the assumed mesh current, and to use a minus sign otherwise.

By the same reasoning which leads to Eq. (3-106) the voltage equation for the second mesh is

$$-R_2 I_2 + (R_2 + R_3 + R_4)I_2 - R_4 I_3 = E_3 \qquad (3\text{-}107)$$

Note that mesh 2 is coupled to mesh current I_1 through R_2 and to mesh current I_3 through R_4.

Similarly, the voltage equation for mesh 3 is

$$0 - R_4I_2 + (R_4 + R_5)I_3 = -E_2 \tag{3-108}$$

Equations (3-106), (3-107), and (3-108) involve three unknown mesh currents. Once these are found, then any one of the five branch currents can readily be determined. Mesh analysis has thereby simplified circuit analysis

3-7 CIRCUIT ANALYSIS BY NODE-PAIR VOLTAGES

It is shown in the preceding section that finding the complete solution (i.e. all the branch currents) of the network of Fig. 3-11 is simpler by the mesh method than by the method of superposition. However, it should not be implied that there is no other, even simpler approach to the problem. For the network under discussion, in fact, there is another method of solution which makes available in a single independent equation the essential information needed to compute all branch currents.

A study of Fig. 3-11 shows why this is possible. For convenience Fig. 3-11 is repeated in Fig. 3-15 with notations which better suit our purposes here. As previously pointed out, the circuit of Fig. 3-15 contains two junction points which in Fig. 3-11 are called b and d but in Fig. 3-15 are identified as 1 and 0. Since the voltages in a circuit must be measured between terminals, it is appropriate and useful to designate one terminal as a reference point. This is a point in the network with respect to which all other levels of potential are measured. In our work we assign that junction marked 0 as the reference point. Often this terminal is called the circuit *ground* and is represented by the symbol shown in Fig. 3-15. A glance at the circuit shows that node a is at a potential E_1 and node c is at a potential E_2 with respect to ground. The only potential which is unknown is that of junction 1. If this potential, V_1, can be found, each branch current can be determined because the voltage drop across each resistor will then be known.

The equation for finding V_1 is obtained by applying Kirchhoff's current law at junction 1. The usual procedure is to assume that all the branch currents leave the junction. Thus

$$I_1 + I_2 + I_3 = 0 \tag{3-109}$$

where these currents are defined as shown in Fig. 3-15. But the voltage across

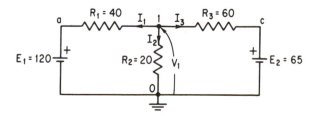

Fig. 3-15 Solution by nodal analysis.

R_1 is the difference in potential between points 1 and a. Hence we can write

$$I_1 = \frac{V_1 - E_1}{R_1} \tag{3-110}$$

In writing the current expression the assumption is made that the node potential is always higher than the other voltages appearing in the equation. If, in fact, it turns out not to be so, a negative value for I_1 will result. This means that I_1 flows into the node and not away from it. The corresponding expressions for the currents in branches 2 and 3 are clearly

$$I_2 = \frac{V_1}{R_2} \tag{3-111}$$

and

$$I_3 = \frac{V_1 - E_2}{R_3} \tag{3-112}$$

Upon substituting these equations into Eq. (3-109), we obtain

$$\frac{V_1 - E_1}{R_1} + \frac{V_1}{R_2} + \frac{V_1 - E_2}{R_3} = 0 \tag{3-113}$$

Equation (3.113) could be written directly from an examination of the circuit diagram. As the reader gains proficiency with these circuits he will learn to do so with ease.

Equation (3-113) is the independent equation which allows V_1 to be determined. All other quantities in the equation are known. For reasons which should be obvious the foregoing procedure is referred to as the *nodal method* of circuit analysis.

EXAMPLE 3-9 By using the values specified in Fig. 3-15 find the magnitude and direction of each of the branch currents.

Solution: From Eq. (3-113) we have

$$V_1 \left(\frac{1}{R_1} + \frac{1}{R_2} + \frac{1}{R_3} \right) = \frac{E_1}{R_1} + \frac{E_2}{R_3} \tag{3-114}$$

$$V_1(\tfrac{1}{40} + \tfrac{1}{20} + \tfrac{1}{60}) = \tfrac{120}{40} + \tfrac{65}{60}$$

$$\therefore \quad V_1 = 44.6 \text{ V}$$

Inserting this value of V_1 into Eqs. (3-110), (3-111), and (3-112) yields the branch currents. Thus

$$I_1 = \frac{V_1 - E_1}{R_1} = \frac{44.6 - 120}{40} = -1.88 \text{ A} \tag{3-115}$$

The minus sign means that I_1 flows into junction 1.

$$I_2 = \frac{V_1}{R_2} = \frac{44.6}{20} = 2.23 \text{ A} \tag{3-116}$$

and

$$I_3 = \frac{V_1 - E_2}{R_3} = \frac{44.6 - 65}{60} = -0.341 \text{ A} \tag{3-117}$$

This current too flows into the junction. Note that these answers compare favorably with those obtained by the mesh method. Note too the smaller amount of effort needed to obtain the solution.

We are now familiar with both the nodal and the mesh methods for finding the branch currents in electrical networks. At this point we may ask: When is one method to be preferred over the other? The answer is surprisingly simple. The general rule is this: Use that method which requires a smaller number of independent equations to obtain the solution. In the nodal method the number of independent node-pair equations needed is one less than the number of junctions in the network. In equation form

$$\boxed{n = j - 1} \tag{3-118}$$

where n denotes the number of independent node equations and j denotes the number of junctions. Of course, the -1 accounts for the fact that one of the junctions is the reference. The number of independent mesh equations to solve a network problem is given by Eq. (3-80). In any given problem, therefore, start by computing m and n. If $m < n$, then the mesh method is generally preferred. Otherwise, it is easier to use the nodal method. In the network of Fig. 3-15 the number of independent mesh equations needed to obtain a complete solution is

$$m = b - (j - 1) = 3 - (2 - 1) = 2 \tag{3-119}$$

whereas the number of independent node equations required is

$$n = j - 1 = 2 - 1 = 1 \tag{3-120}$$

Accordingly, the nodal method is the simpler approach. A comparison of Examples 3-8 and 3-9 bears this out. It should be pointed out, however, that the difference in effort is not overwhelming because we are dealing with a relatively simple problem. But as the complexity increases and the difference between the values of m and n gets large, the difference in effort between the two methods becomes significant. As a general rule the nodal method shows to advantage when the network has many parallel circuits.

It is possible to formulate the defining independent nodal equations of a network in the same manner as was done for the mesh method. To illustrate the procedure let us return to the network of Fig. 3-13, which is repeated in Fig. 3-16 with the modification that the junction points e, b, and c are replaced by the notation 0, 1, and 2, respectively. Here again we have a network for which the solution is easier by the nodal method than by the mesh method. In accordance with Eq. (3-80) there are three independent mesh equations, but by Eq. (3-118) there are only two independent node-pair voltage equations.

Applying Kirchhoff's current law at the independent node 1, and assuming that all branch currents are leaving the junction, we get

$$\frac{V_1 - E_1}{R_1} + \frac{V_1}{R_2} + \frac{V_1 + E_3 - V_2}{R_3} = 0 \tag{3-121}$$

The first two terms in this equation should be self-evident. The third term needs some comment. Keep in mind that the principle involved in the nodal method is to find the potential difference appearing across each circuit element in a branch. In the case of R_3 an inspection of the circuit diagram shows that the left side of R_3 is at a potential of V_1. The potential of the right side of R_3 is found by starting at the reference node and then summing all the voltages encountered in reaching the right side. Thus the potential of the right side of R_3 is $(+V_2 - E_3)$. Consequently, the potential difference across R_3 is $[V_1 - (V_2 -)E_3]$ or $(V_1 + E_3 - V_2)$. Division of this quantity by R_3 then gives the branch current through R_3.

By similar reasoning the current equation at independent node 2 can be written as

$$\frac{(V_2 - E_3) - V_1}{R_3} + \frac{V_2}{R_4} + \frac{V_2 - E_2}{R_5} = 0 \tag{3-122}$$

Thus there are two equations and two unknown quantities—V_1 and V_2.

Collecting terms and rearranging, Eq. (3-121) becomes

$$V_1\left(\frac{1}{R_1} + \frac{1}{R_2} + \frac{1}{R_3}\right) - V_2\frac{1}{R_3} = \frac{E_1}{R_1} - \frac{E_3}{R_3} \tag{3-123}$$

The coefficient of V_1 is a conductance; more specifically it is the *self-conductance* of node 1. It is found by forming the sum of the reciprocal of each of the resistances connected to node 1. It is denoted by the symbol G_{11}. That is,

$$G_{11} = \frac{1}{R_1} + \frac{1}{R_2} + \frac{1}{R_3} = G_1 + G_2 + G_3 \tag{3-124}$$

The coefficient of V_2 is also a conductance. Called the mutual conductance, it is found by taking the reciprocal of the equivalent resistance that lies between nodes 1 and 2. It is denoted by G_{12} and in this case is

$$G_{12} = -\frac{1}{R_3} \tag{3-125}$$

Accordingly, Eq. (3-123) can be written more compactly as

$$G_{11}V_1 + G_{12}V_2 = \frac{E_1}{R_1} - \frac{E_3}{R_3} \tag{3-126}$$

The minus sign always appears on the left side of this formulation because all independent node voltages are considered to be positive with respect to the reference node.

By performing the same operations on Eq. (3-122) we get

$$+G_{21}V_1 + G_{22}V_2 = \frac{E_3}{R_3} + \frac{E_2}{R_5} \tag{3-127}$$

where

$$G_{21} = G_{12} = -\frac{1}{R_3} \tag{3-128}$$

and

$$G_{22} = \frac{1}{R_3} + \frac{1}{R_4} + \frac{1}{R_5} = G_3 + G_4 + G_5 \tag{3-129}$$

Note that the units of the right sides of Eqs. (3-126) and (3-127) are expressed in amperes and involve the source voltages. Hence one may look upon the expressions on the right side of these equations as equivalent current sources applied to nodes 1 and 2, respectively. For further discussion of this matter, refer to Sec. 3-9.

Matrix Formulation of the Nodal Method. Equations (3-126) and (3-127) may be represented in the more general matrix form by writing

$$\begin{bmatrix} G_{11} & G_{12} \\ G_{21} & G_{22} \end{bmatrix} \begin{bmatrix} V_1 \\ V_2 \end{bmatrix} = \begin{bmatrix} I_1 \\ I_2 \end{bmatrix} \tag{3-130}$$

or, even more succinctly, as

$$\mathbf{GV} = \mathbf{I} \tag{3-131}$$

For the nodal method, which is based on the current law, the elements of $\mathbf{V}$ are the unknown quantities to be determined, whereas the elements of $\mathbf{I}$ are the driving forces associated with the sources that account for the various node-pair voltages. For the circuit of Fig. 3-16 I_1 is given by $(E_1/R_1 - E_3/R_3)$ and I_2 is expressed by $(E_3/R_3 + E_2/R_5)$. Observe that at node 1 source E_3 causes a negative sign to appear in these expressions, while at node 2 the sign is positive. A simple rule to use in writing the proper expressions for the net source currents associated with the various nodes in the nodal method is to see whether the positive or negative side of the voltage source is nearer to the node of interest. Accordingly, in Fig. 3-16 at node 1 there is an equivalent current source impinging on this

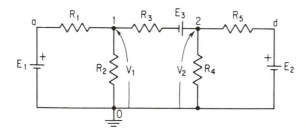

Fig. 3-16   Two node-pair network.

node of value E_1/R_1; it is treated as positive because the positive side of this source voltage is nearer to node 1 than is the negative side. Similarly, source E_3 is associated with an equivalent current source [see Eq. (3-126)] of value $-E_3/R_3$ as it appears at node 1. Here the negative sign is used because in this instance the negative side of the voltage source is nearer to node 1. Thus, by simple inspection of the circuit of Fig. 3-16, it is a routine matter to write the expression for I_1 in Eq. (3-130) without the need to write Kirchhoff's current law at the node of interest. The same technique is applicable to all the other nodes in the circuit.

The solution for the node-pair voltages in Eq. (3-130) by Cramer's rule yields the following expressions:

$$V_1 = \frac{\begin{vmatrix} I_1 & G_{12} \\ I_2 & G_{22} \end{vmatrix}}{\begin{vmatrix} G_{11} & G_{12} \\ G_{21} & G_{22} \end{vmatrix}} = \frac{G_{22}}{D_G} I_1 - \frac{G_{12}}{D_G} I_2 \qquad (3\text{-}132)$$

and

$$V_2 = \frac{\begin{vmatrix} G_{11} & I_1 \\ G_{21} & I_2 \end{vmatrix}}{\begin{vmatrix} G_{11} & G_{12} \\ G_{21} & G_{22} \end{vmatrix}} = -\frac{G_{21}}{D_G} I_1 + \frac{G_{11}}{D_G} I_2 \qquad (3\text{-}133)$$

where D_G is the determinant of the conductance matrix **G** of the two-node pair case. A study of the last two equations reveals that the coefficients of the equivalent source currents I_1 and I_2 are simply the elements of the inverse of the conductance matrix **G**.

EXAMPLE 3-10 The circuit elements in Fig. 3-16 have the following values expressed in ohms and volts: $R_1 = 10$, $R_2 = 2$, $R_3 = 3$, $R_4 = 1$, $R_5 = 5$, and $E_1 = 24$, $E_2 = 12$, and $E_3 = 6$. Find the values of the node-pair voltages V_1 and V_2.

Solution: The general expressions of the solution are displayed in Eqs. (3-132) and (3-133). The actual values of the node-pair voltages are readily determined once the quantities on the right side of these equations are evaluated. Thus, we begin by first finding the elements of the **G** matrix.

$$G_1 = \frac{1}{R_1} + \frac{1}{R_2} + \frac{1}{R_3} = \frac{1}{10} + \frac{1}{2} + \frac{1}{3} = 0.93333$$

$$G_{12} = -\frac{1}{R_3} = -\frac{1}{3} = -0.33333$$

$$G_{22} = \frac{1}{R_3} + \frac{1}{R_4} + \frac{1}{R_5} = 0.33333 + 1.0 + 0.2 = 1.53333$$

$$D_G = \begin{vmatrix} 0.93333 & -\frac{1}{3} \\ -\frac{1}{3} & 1.53333 \end{vmatrix} = 1.32$$

The equivalent current sources are

$$I_1 = \frac{E_1}{R_1} - \frac{E_3}{R_3} = 2.4 - 2 = 0.4 \text{ A}$$

$$I_2 = \frac{E_3}{R_3} + \frac{E_2}{R_5} = 2 + 2.4 = 4.4 \text{ A}$$

Introducing these quantities into the cited equations thus yields

$$V_1 = \frac{G_{22}}{D_G} I_1 - \frac{G_{12}}{D_G} I_2 = \frac{1.53333}{1.32} (0.4) + \frac{0.33333}{1.32} (4.4) = 1.5757 \text{ V}$$

$$V_2 = \frac{G_{21}}{D_G} I_1 + \frac{G_{11}}{D_G} I_2 = \frac{0.33333}{1.32}(0.4) + \frac{0.93333}{1.32}(4.4) = 3.212 \text{ V}$$

3-8 THÉVENIN'S THEOREM

Situations sometimes occur in electrical engineering in which it is desirable to find a particular branch current in a network as the resistance of that branch is varied while all other resistances and sources remain constant. For example, in the circuit configuration of Fig. 3-15 it may be desired to find the current which flows through R_2 for 10 different values of R_2, assuming that E_1, E_2, R_1, and R_3 remain unchanged. We know that by repeatedly applying the superposition method or the mesh method or the nodal method to the network we can find the ten different values of branch current through R_2. We also know that perhaps the least effort is required in this case if we use the nodal method. In circumstances such as these, however, it is possible to resort to yet another technique of circuit analysis which appreciably simplifies the labor required to obtain the solution.

For the moment, so that we can better understand the procedure involved, let us leave Fig. 3-15 and turn attention to Fig. 3-17, which depicts a simple series circuit. The energy source is a battery having an emf E_b and an internal resistance R_i. The battery supplies energy to the load resistor R_L. When R_L is removed from the battery, i.e., the switch S is open, a voltmeter placed across terminals 01 reads an open-circuit voltage V_{oc} which is the battery emf. That is, $V_{oc} = E_b$. Furthermore, the resistance which is "seen" from the output terminals looking to the left into the source is clearly the internal resistance of the battery R_i. Therefore, as far as R_L is concerned, it "sees" appearing at terminals 01 a source voltage of value V_{oc} and an internal impedance R_i. As a matter of fact, even if that part of the circuit to the left of 01 were actually made up of series-parallel combinations of resistances and sources, nevertheless to an observer looking toward the left from the output terminals there appears an open circuit voltage V_{oc} and an equivalent resistance associated with V_{oc} in exactly the same manner as depicted in the simple case of Fig. 3-17. The important thing to realize here is that at the output terminals with S open there can appear only one equivalent voltage V_{oc} and an equivalent resistance. Just how these quantities take on the values they do is irrelevant as far as the output terminals and the load resistor connected to these terminals are concerned. What really matters

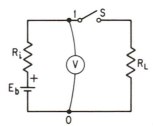

Fig. 3-17 Simple series circuit: battery supplying energy to a load resistor.

to R_L are the terminal quantities themselves V_{oc} and R_i—irrespective of how these are arrived at. On this basis, when the switch S is closed, the resulting current through the load resistor is readily identified as

$$I_L = \frac{V_{oc}}{R_i + R_L} \tag{3-134}$$

Keep in mind that the application of this technique of circuit analysis has relevance in those situations where it is of interest to determine the electrical quantities associated with a particular circuit parameter while all other circuit elements remain invariant. Thus in Eq. (3-134) the circuit parameter of interest is R_L. The actual circuit in which R_L finds itself may be complex or very simple. But whatever the situation, the current, voltage, and power at R_L can be determined by Eq. (3-134) provided that the remainder of the circuit is replaced by an equivalent voltage source, V_{oc}, and an equivalent resistance R_i. The quantity V_{oc} is determined by applying circuit analysis to the original network with R_L removed. Moreover, V_{oc} corresponds to the voltage that appears across the terminals where R_L is to be connected. The quantity R_i is found by measuring the resistance between the terminals where R_L is to be placed. This can usually be achieved by the simple application of network reduction. However, R_i can also be found by replacing R_L with a short circuit and then using circuit analysis on this simplified version of the original circuit to find the short-circuit current. On the assumption that V_{oc} is available, the value of R_i then becomes simply

$$R_i = \frac{V_{oc}}{I_{sc}} \tag{3-135}$$

It is important to keep in mind in using this expression that V_{oc} is found at the terminals of the load resistor but with R_L replaced by an open circuit, whereas I_{sc} is found at these same terminals but with R_L replaced by a short circuit.

Let us now return to the circuit of Fig. 3-15 and rearrange it in the manner shown in Fig. 3-18(a). The drawing of the diagram is modified in order to bring it more into correspondence with Fig. 3-17. Note that R_2 of Fig. 3-18 corresponds to R_L of Fig. 3-17. With sources E_1 and E_2 and resistors R_1 and R_3 remaining

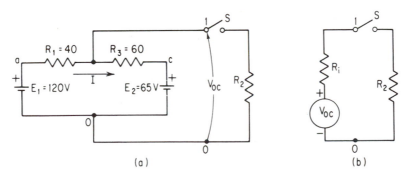

Fig. 3-18 Redrawing Fig. 3-15 to isolate R_2: (a) original network redrawn; (b) equivalent circuit.

unaltered in Fig. 3-18, it should be possible to determine the open-circuit voltage appearing across the output terminals 01 (with switch S open, of course) as well as the equivalent resistance looking into terminals 01 toward the left (or source) side. The open-circuit voltage is readily found as the potential difference between points 0 and 1 as determined by the flow of current in the closed path composed of E_1, R_1, R_3, and E_2. The equivalent internal resistance is merely the parallel combination of R_1 and R_3. Keep in mind that when we are finding the equivalent internal resistance the source voltages are set equal to zero. Figure 3-18(b) depicts the equivalent circuit of Fig. 3-18(a). The procedure is now illustrated by the following example.

EXAMPLE 3-11 In the circuit of Fig. 3-15 find the branch current which flows through R_2 when R_2 has the following values: 11, 19, 46, and 176 Ω.

Solution: Refer to Fig. 3-18(a). Let us first find the open-circuit voltage $V_{oc} = V_{01}$. Thus we need to determine the potential of point 1 relative to 0, which can be considered the reference or ground point. Writing Kirchhoff's voltage law for circuit $ac0a$ yields

$$IR_1 + IR_3 + 65 - 120 = 0 \tag{3-136}$$

where I is the current flowing in loop $ac0$ with switch S open. For the specified values of R_1 and R_3 this current becomes

$$I = \frac{55}{R_1 + R_3} = \frac{55}{40 + 60} = 0.55 \text{ A} \tag{3-137}$$

Therefore,

$$V_{0c} = V_{01} = E_1 - IR_1 = 120 - (0.55)(40) = 98 \text{ V} \tag{3-138}$$

The equivalent resistance appearing between the output terminals 01 is then

$$R_i = \frac{R_1 R_3}{R_1 + R_3} = \frac{40(60)}{40 + 60} = 24 \ \Omega \tag{3-139}$$

To find R_i by application of Eq. (3-135), the circuit of Fig. 3-18 is altered by replacing R_L with a short circuit and then finding the current that flows through the short circuit. Placing a short circuit between terminals 1 and 0 in Fig. 3-18 means that each voltage source produces a current in the short circuit in the same direction. The total short-circuit current is given by

$$I_{sc} = \frac{E_1}{R_1} + \frac{E_2}{R_3} = \frac{120}{40} + \frac{65}{60} = 3 + 1.0833 = 4.0833 \text{ A} \tag{3-140}$$

Hence the resistance looking back into terminals 0 and 1 from R_L is

$$R_i = \frac{V_{oc}}{I_{sc}} = \frac{98}{4.0833} = 24 \ \Omega \tag{3-141}$$

Having established V_{oc} and R_i, it then follows from Fig. 3-18(b) that the branch current through R_2 is

$$I_2 = \frac{V_{oc}}{R_i + R_2} = \frac{98}{24 + R_2} \tag{3-142}$$

Accordingly,

when $R_2 = 11 \ \Omega$, $I_2 = \frac{98}{35} = 2.8$ A

when $R_2 = 19 \ \Omega$, $I_2 = \frac{98}{43} = 2.28$ A

when $R_2 = 46 \ \Omega$, $I_2 = \frac{98}{70} = 1.4$ A

when $R_2 = 176 \ \Omega$, $I_2 = \frac{98}{200} = 0.49$ A

Note that once the equivalent source—V_{oc} and R_i—is determined, Eq. (3-142) allows the current through R_2 to be found with very little effort for each value of R_2. The same cannot be said for the superposition, mesh, and nodal methods.

A French engineer, M. L. Thévenin, was the first person to discover the equivalence in networks such as those depicted in Fig. 3-18. This technique of circuit analysis was first published by him in 1883. Known today as Thévenin's theorem, it may be stated as follows (for the resistive network case):

> Any two terminals of a network composed of linear passive and active circuit elements may be replaced by an equivalent voltage source and an equivalent series resistance. The voltage source is equal to the potential difference between the two terminal points caused by the active network with no external elements connected to these terminals. The series resistance is the equivalent resistance looking into the two terminal points with all power sources within the terminal pair inactive.

EXAMPLE 3-12   A network has the configuration shown in Fig. 3-19(a). All resistance values are expressed in ohms.

   (a) Find the current through R_L when it takes on values of 10, 50, and 200 Ω.

   (b) Determine the value of R_L corresponding to which there is a maximum power transfer to the load resistor. Compute this maximum power.

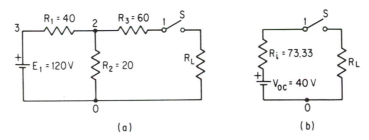

Fig. 3-19 Circuit configuration for Example 3-11: (a) original network with resistances expressed in ohms; (b) Thévenin equivalent circuit.

Solution: Although the solution to part (a) can be found by applying the technique of network reduction, it is simpler to find the Thévenin equivalent circuit and then use Eq. (3-134) directly.

   (a) The open-circuit voltage appearing at terminal 1 is the same as that at terminal 2. Applying the voltage-divider rule to the closed path 032 we get

$$V_{oc} = V_{02} = V_{01} = \frac{R_2}{R_1 + R_2} E_1 = \frac{20}{40 + 20}(120) = 40 \text{ V} \tag{3-143}$$

Also the equivalent resistance looking into the network from terminal pair 01 toward the left is R_3 in series with the parallel combination of R_1 and R_2. Thus

$$R_1 = R_3 + \frac{R_1 R_2}{R_1 + R_2} = 60 + \frac{40(20)}{40 + 20} = 60 + \frac{40}{3} = 73.33 \ \Omega \tag{3-144}$$

Hence the Thévenin equivalent circuit of Fig. 3-19(a) appears as shown in Fig. 3-19(b). Accordingly, the corresponding load currents are

$$\text{for } R_L = 10, \quad I_L = \frac{40}{73.33 + 10} = 0.480 \text{ A}$$

$$\text{for } R_L = 50, \quad I_L = \frac{40}{73.33 + 50} = 0.324 \text{ A}$$

$$\text{for } R_L = 200, \quad I_L = \frac{40}{73.33 + 200} = 0.146 \text{ A}$$

(b) The general expression for the power in the load resistor is

$$P_L = I_L^2 R_L = \left(\frac{V_{oc}}{R_i + R_L} \right)^2 R_L = \frac{V_{oc}^2 R_L}{(R_i + R_L)^2} \tag{3-145}$$

In order to find the value of R_L for which P_L is a maximum, it is necessary to differentiate Eq. (3-145) with respect to R_L and set the result equal to zero. Thus

$$\frac{dP_L}{dR_L} = V_{oc}^2 \left[\frac{-2R_L}{(R_i + R_L)^3} + \frac{1}{(R_i + R_L)^2} \right] = 0$$

$$= \frac{-2R_L + R_i + R_L}{(R_i + R_L)^3} = 0 \tag{3-146}$$

From which it follows that

$$\boxed{R_i = R_L} \tag{3-147}$$

Hence for the problem at hand

$$R_L = R_i = 73.33 \ \Omega$$

The maximum power to R_L is then

$$P_{L\,max} = \frac{V_{oc}^2}{4R_L^2} R_L = \frac{V_{oc}^2}{4R_L} = \frac{(40)^2}{4(73.33)} = 5.45 \text{ W} \tag{3-148}$$

Equation (3-147) is a useful result. It states that the maximum transfer of power to a load resistor occurs when the load resistor has a value equal to the series equivalent or internal resistance of the source. Note that the efficiency at maximum power transfer is 50%. This result brings into evidence another advantage of the Thévenin equivalent representation of a network. Once R_i is computed, it shows at a glance the condition for maximum power transfer. The Thévenin equivalent circuit conveys other useful information as well. Thus in Fig. 3-19(b) it is immediately apparent that the maximum voltage that can ever appear at the load resistor for the circuit configuration of Fig. 3-19(a) is 40 V. This conclusion is not so obvious in the original circuit.

3-9 NORTON'S THEOREM: CONVERSION OF A VOLTAGE SOURCE TO A CURRENT SOURCE

Let us look further into the description of the circuit of Fig. 3-17. Note that when R_L is made equal to zero, a current flows through the output terminals (from 1 to 0) which is the maximum possible current that can be supplied by the given voltage source. In this sense, then, one can treat the voltage source in terms of a source which is capable of supplying current up to a maximum limited by the series equivalent resistance R_i. In other words, the voltage source may be considered in terms of a current source of magnitude $I = E_b/R_i$. However, in order for a description from this viewpoint to be completely consistent with the conditions as they prevail in Fig. 3-17 between the terminal pair 01 at open circuit, the current source must appear across a resistor placed between points 0 and 1 and having such a value that the equivalent open-circuit voltage of the original circuit appears across it. Since we already know the value of the current source as $I = E_b/R_i$, we must now determine that value of shunt† resistance which when placed across the current source yields E_b. Expressed mathematically, we have

$$IR_x = E_b \qquad (3\text{-}149)$$

But

$$I = \frac{E_b}{R_i} \qquad (3\text{-}150)$$

Therefore,

$$R_x = R_i \qquad (3\text{-}151)$$

Accordingly, as far as the network terminal pair 01 is concerned at open-circuit, the part to the left of these terminals may be represented either by a voltage source and a series internal resistance or by a current source and a shunt resistance of value R_i. The current-source equivalent of the voltage source is illustrated in Fig. 3-20. The symbol used to denote it is a circle-enclosed arrow indicating the direction of current flow assumed as positive. In dealing with current sources it is important to understand that the total current of the current source is supplied to the circuit elements placed across its terminals. If there is just one path available, as in Fig. 3-20 when the switch S is open, then that one circuit element draws the full current. If more than one circuit element is placed across the current source, then together they draw the full current of the source.

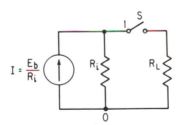

Fig. 3-20 Current source equivalent of the circuit of Fig. 3-19.

A current source always delivers constant current irrespective of the particular configuration of the circuit to which it is connected. This is not unlike the behavior of a voltage source. Thus in the circuit of Fig. 3-17 the battery delivers its full voltage to the circuit at all times. This continues to be true even if additional series-parallel circuit elements are placed in the circuit.

Why should it be at all desirable to work with equivalent current sources?

---

† This is called a shunt resistance because it is placed across the output terminals in parallel with the current source.

What are the advantages? The biggest advantage is that the analysis of some very important engineering devices is much easier to perform. This is especially true of those engineering devices, such as the transistor and the vacuum-tube pentode, which are characteristically current sensitive in their behavior and operation. Furthermore, in those instances of circuit analysis which can be most easily handled by the nodal method, the idea of a current-source equivalent of voltage sources is a natural consequence of the mathematical formulation associated with the method. This is readily borne out by referring to the right side of Eqs. (3-126) and (3-127). Here the quantities E_1/R_1, E_3/R_3, and E_2/R_5 are nothing more than the current-source equivalences of the corresponding voltage sources.

To help complete our understanding of the current-source equivalence let us investigate the behavior of the circuits shown in Figs. 3-17 and 3-20 when the switch S is closed. In Fig. 3-17, by Kirchhoff's voltage law, the expression for the output voltage appearing across terminal-pair 01 is

$$E_b - I_L R_i = I_L R_L = V_{01} \qquad (3\text{-}152)$$

Dividing each term through by R_i yields

$$\frac{E_b}{R_i} - I_L = \frac{V_{01}}{R_i} \qquad (3\text{-}153)$$

Inserting Eq. (3-150) and rearranging leads to

$$I_L = I - \frac{V_{01}}{R_i} \qquad (3\text{-}154)$$

This expression states that the load current is equal to the equivalent source current less the amount drawn by the shunt resistance R_i. It is interesting to note that Eq. (3-154) can be written immediately by referring to the current-source equivalent circuit of Fig. 3-20 and writing Kirchhoff's current law at node 1.

Another interesting result pertaining to the current-source equivalent circuit can be obtained by rewriting Eq. (3-154) and using the fact that $I_L = V_{01}/R_L$. Thus

$$I = I_L + \frac{V_{01}}{R_i} = \frac{V_{01}}{R_L} + \frac{V_{01}}{R_i} = V_{01}\left(\frac{1}{R_L} + \frac{1}{R_i}\right) \qquad (3\text{-}155)$$

But the quantity in parentheses is nothing more than the parallel combination of R_i and R_L in Fig. 3-20. Hence by setting

$$\frac{1}{R_p} = \frac{1}{R_L} + \frac{1}{R_i} \qquad (3\text{-}156)$$

it follows that the expression for the voltage across the output terminals 01 is simply

$$\boxed{V_{01} = I R_p} \qquad (3\text{-}157)$$

where I is the equivalent current source. Equation (3-157) provides a very simple and direct way of finding a terminal voltage, especially where many parallel elements are involved.

The current equivalent of a voltage source was first developed by E. L. Norton, an engineer employed by the Bell Telephone Laboratories. Today this equivalence is referred to as *Norton's theorem* and is described as follows:

> Any two terminals of a network composed of linear passive and active circuit elements may be replaced by an equivalent current source and a parallel resistance. The current of the source is the current measured in the short circuit placed across the terminal pair. The parallel resistance is the equivalent resistance looking into the terminal pair with all independent power sources inactive.

Note that this theorem does not require that the Thévenin equivalent circuit be found first. The short-circuit current and the parallel resistance may be found directly as illustrated in the examples which follow.

EXAMPLE 3-13 By means of Norton's current-source equivalent circuit of the network shown in Fig. 3-21(a) find the current which flows through R_2 when it has a value of 11 Ω. All resistance values in Fig. 3-21 are expressed in ohms.

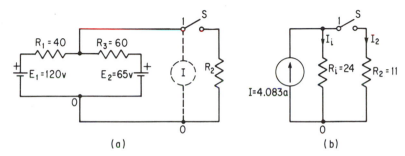

Fig. 3-21 Network configuration for Example 3-12: (a) original circuit; (b) Norton's current source equivalent.

Solution: To find the current-source equivalent circuit associated with the network terminal pair 01, it is necessary to find the current which flows through a short circuit placed across the terminals, and the resistance looking into these same terminals when the short circuit is removed. In Fig. 3-21(a) the short circuit connection is represented by a dashed line leading to a circle enclosing the letter I. This symbol can be thought of as depicting an instrument for measuring current, such as an ammeter. An examination of the circuit lying to the left of terminal pair 01 indicates that the current I is composed of two components. There is a current contribution from source E_1 which takes a path through R_1 through the short circuit and then back to E_1. This component of I is thus $E_1/R_1 = 120/40 = 3$ A. The contribution from source E_2 takes a path through R_3, then through the short circuit and back to E_2. The magnitude of this current is $E_2/R_3 = 65/60 = 1.083$ A. Thus

$$I = \frac{E_1}{R_1} + \frac{E_2}{R_3} = 3 + 1.083 = 4.083 \text{ A} \tag{3-158}$$

Since the parallel resistance is the equivalent resistance looking into the network at the specified terminal pair 01 with all source voltages set to zero, it is given by

$$R_i = \frac{R_1 R_3}{R_1 + R_3} = \frac{40(60)}{40 + 60} = 24 \ \Omega \tag{3-159}$$

Hence the Norton equivalent circuit takes on the form shown in Fig. 3-21(b).

When the switch S is closed, the source current I in Fig. 3-21(b) splits into two paths as I_i and I_2. Because R_i and R_2 are in parallel, the potential difference across each resistor is the same. Consequently, we can write

$$I_2 R_2 = I_i R_i \tag{3-160}$$

But

$$I_i = I - I_2 \tag{3-161}$$

Therefore,

$$I_2 R_2 = IR_i - I_2 R_i$$

or

$$\boxed{I_2 = \frac{R_i}{R_i + R_2} I} \tag{3-162}$$

Equation (3-162) is useful when circuit analysis is being performed with current sources. It is analogous to the voltage-divider rule for voltage sources. It states that for two parallel branches the current in one branch is to the total current as the resistance of the second branch is to the sum of the resistances of the two branches. For the example at hand the current in branch I_2 is found to be

$$I_2 = \frac{R_i}{R_i + R_2} I = \frac{24}{24 + 11}(4.083) = 2.8 \text{ A} \tag{3-163}$$

which checks with the value found by using Thévenin's theorem in Example 3-10.

EXAMPLE 3-14   The small-signal equivalent circuit of an electronic amplifier has the configuration depicted in Fig. 3-22(a). The indicated circuit parameters have the following values:

$$\mu = 99 \qquad R_1 = 147{,}000 \ \Omega$$
$$r_p = 66{,}000 \ \Omega \qquad R_2 = 100{,}000 \ \Omega$$

Find the value of the output voltage for $E_g = 0.4$ V.

Solution: The current which flows in the short circuit placed across the output terminals 01 is

$$I = \frac{\mu E_g}{r_p} = 1.5 \times 10^{-3} E_g \quad \text{A} \tag{3-164}$$

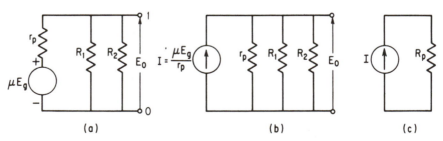

Fig. 3-22 Network for Example 3-13: (a) original circuit; (b) current source equivalent; (c) simplified equivalent.

Note that the short circuit across 01 removes R_1 and R_2 from any consideration in determining the current source I. Thus the current-source equivalent circuit may be represented as shown in Fig. 3-22(b). By finding the parallel combination of the three parallel resistances we can depict the equivalent circuit as shown in Fig. 3-22(c), where

$$\frac{1}{R_p} = \frac{1}{r_p} + \frac{1}{R_1} + \frac{1}{R_2} = \frac{1}{31,200} \tag{3-165}$$

The output voltage is equal to the source current multiplied by R_p. Thus

$$E_0 = IR_p = \frac{\mu}{r_p} E_g R_p = 1.5 \times 10^{-3}(0.4)(31.2)10^{+3} = 18.72 \text{ V} \tag{3-166}$$

3-10 NETWORK REDUCTION BY Δ-Y TRANSFORMATION

The techniques of series-parallel network reduction are not always applicable. A case in point is illustrated in Fig. 3-23. Examination of the diagram shows that although two parallel paths may be traced on leaving terminal a, yet the paths from b and c are neither of the series nor parallel kind. Of course one can argue here that the techniques of mesh and nodal analysis certainly lend themselves to a solution of the branch currents that flow in such a circuit configuration. A little thought, however, indicates that in either case it is necessary to solve three independent equations in order to arrive at a solution. In situations of the type under consideration the Δ-to-Y transformation simplifies the solution procedure considerably. As a matter of fact it often makes it unnecessary to solve simultaneous mesh or nodal equations by modifying the circuit in such a way as to permit the standard series-parallel network reduction technique to be applied directly.

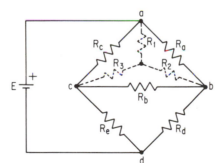

Fig. 3-23 The original circuit configuration appears drawn with solid lines. The Y-equivalent network of the abc delta network is drawn with dashed lines.

The source of difficulty in applying the procedures of series-parallel reduction to the original configuration of Fig. 3-23 essentially lies in the three-terminal character of a circuit such as abc, which is often called a delta (Δ) network. It is significant to note that if this network could be replaced by an equivalent network which effectively removes the cross arm cb, then combinations of conventional series and parallel resistances could be readily identified and the familiar techniques of Sec. 3-4 applied. A network which possesses this capability is the wye (Y) network. It is depicted in Fig. 3-23 with a dashed-lined drawing. With such an equivalent circuit note that R_2 and R_d are in series and so, too, are R_3 and R_e. Moreover, these series combinations in turn are in parallel with each other. Accordingly, we find that the equivalent resistance between points ad can be written quite simply and directly as

$$R_{ad} = R_1 + \frac{(R_3 + R_e)(R_2 + R_d)}{R_3 + R_e + R_2 + R_d} \tag{3-167}$$

The current delivered by the source in the circuit of Fig. 3-23 is then easily determined.

We turn attention next to the procedure which describes how the equivalent Y network can be computed from the known arms of the given Δ network, which in Fig. 3-23 are R_a, R_b, and R_c. This part of the circuit is repeated for clarity's sake in Fig. 3-24. A study of Fig. 3-24 should make it apparent that for the Y and Δ networks to be equivalent it is necessary that the resistance appearing between each of the three pairs of terminals—ab, bc, and ca—be the same. Thus, looking into the terminals ab, the equivalent resistance for the Y arrangement is merely R_1 in series with R_2. On the other hand, for the Δ arrangement the equivalent resistance between ab is R_a in parallel with the series combination of R_b and R_c. Expressing this mathematically, we can write

$$[R_{ab}]_Y = [R_{ab}]_\Delta \tag{3-168}$$

or

$$R_1 + R_2 = \frac{R_a(R_b + R_c)}{R_a + R_b + R_c} \tag{3-169}$$

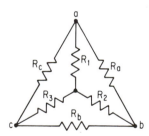

Fig. 3-24 Three-terminal Δ and Y networks.

Similarly, for terminals bc we have

$$[R_{bc}]_Y = [R_{bc}]_\Delta \tag{3-170}$$

or

$$R_2 + R_3 = \frac{R_b(R_c + R_a)}{R_a + R_b + R_c} \tag{3-171}$$

and for terminals ca

$$[R_{ca}]_Y = [R_{ca}]_\Delta \tag{3-172}$$

Elementary Network Theory   Chap. 3

or

$$R_3 + R_1 = \frac{R_c(R_a + R_b)}{R_a + R_b + R_c} \tag{3-173}$$

Equations (3-169), (3-171), and (3-173) represent three equations involving the three unknown resistances of the equivalent Y network. Solving these equations simultaneously leads to the following useful results:

$$R_1 = \frac{R_a R_c}{R_a + R_b + R_c} \tag{3-174}$$

$$R_2 = \frac{R_a R_b}{R_a + R_b + R_c} \tag{3-175}$$

$$R_3 = \frac{R_b R_c}{R_a + R_b + R_c} \tag{3-176}$$

A comparison of these equations with the circuitry of Fig. 3-24 reveals that any given arm of the Y network is found by taking the product of the two *adjacent* arms of the Δ network and dividing by the sum of the Δ-network arms. This is an easy rule for remembering the foregoing equations.

EXAMPLE 3-15   In the original circuit configuration of Fig. 3-23 assume that the resistances have the following values expressed in ohms:

$$R_a = 20 \quad R_d = 24$$
$$R_b = 30 \quad R_e = 5$$
$$R_c = 50$$

Find the current delivered by the source to the network. Assume that $E = 220$ V.

Solution: Replace the *abc* delta network by an equivalent wye which by Eqs. (3-174) to (3-176) has arm values as follows:

$$R_1 = \frac{R_a R_c}{R_a + R_b + R_c} = \frac{20(50)}{20 + 30 + 50} = 10 \ \Omega$$

$$R_2 = \frac{R_a R_b}{R_a + R_b + R_c} = \frac{20(30)}{20 + 30 + 50} = 6 \ \Omega$$

$$R_3 = \frac{R_b R_c}{R_a + R_b + R_c} = \frac{30(50)}{20 + 30 + 50} = 15 \ \Omega$$

Then by Eq. (3-167) the equivalent resistance in ohms between terminals *ad* is

$$R_{ad} = R_1 + \frac{(R_2 + R_d)(R_3 + R_e)}{R_2 + R_3 + R_d + R_e} = 10 + \frac{(6 + 24)(15 + 5)}{50} = 10 + 12 = 22$$

Hence

$$I = \frac{E}{R_{ad}} = \frac{220}{22} = 10 \ \text{A}$$

1. What is the test for determining whether or not two circuit elements are in series? In parallel?

2. Which has the greater equivalent resistance; two equal resistors in series or in parallel? Explain.

3. Which has the greater equivalent capacitance, two equal capacitors in series or in parallel? Explain.

4. Which has the greater equivalent inductance, two equal inductors in series or in parallel? Explain.

5. Identify that property of a capacitor which when connected in series with another capacitor is responsible for treating it in the same manner as two parallel resistors when the equivalent series capacitance is required.

6. Describe and illustrate what is meant by the term *network reduction*.

7. Describe and illustrate the voltage-divider principle.

8. Describe and illustrate the current-divider principle.

9. What is superposition, and why is it important? Is superposition peculiar to electrical networks alone? Explain.

10. Distinguish among a mesh current, a branch current, and a loop current.

11. Why are mesh currents important in circuit analysis?

12. Distinguish between a node and an independent node.

13. Distinguish between a mesh and a loop of a circuit.

14. Can a branch contain more than one circuit element? Explain and illustrate.

15. How is the tree of a circuit defined? Describe its importance in circuit analysis.

16. Can complete knowledge of the mesh currents in a network permit the evaluation of the current that flows in *any* branch of the network? Explain.

17. How is the number of independent equations needed to competely solve a circuit problem determined? Illustrate.

18. Name the basic circuit law upon which the mesh method of circuit analysis is based.

19. Name the basic circuit law upon which the nodal method of circuit analysis is based.

20. Is the number of independent equations needed to solve a given circuit's problem fixed? Explain.

21. In a given network it is possible to identify four meshes and four independent nodes. Which method of circuit analysis is to be preferred? Why?

22. A network is found to have a total of seven branches and four junction points (i.e., independent nodes). Which method is to be preferred in analyzing the circuit? Explain.

23. State the circumstances for which a Thévenin equivalent circuit is useful.

24. Describe the basic idea that underlies the Thévenin equivalent circuit.

25. How is the Norton equivalent circuit related to the Thévenin equivalent circuit?

26. If the Thévenin equivalent of a network is represented by V_{oc} and R_i, show why R_i is correctly given by V_{oc}/I_{sc} when I_{sc} is the current that flows when the Thévenin equivalent is shorted.

27. When is the Δ-Y transformation useful in network reduction? Illustrate.

Problems _____

GROUP I

3-1. Three resistances having values of 10, 20, and 30 Ω are connected in series to form an electric circuit. Find the equivalent series resistance.

3-2. The three resistances of Prob. 3-1 are placed in parallel across a voltage source. Find the equivalent parallel resistance.

3-3. Four resistances of values 50,000 Ω, 250 kΩ, 1 MΩ, and 500 kΩ are placed in parallel. Compute the equivalent parallel resistance.

3-4. The capacitance values of three capacitors are 10, 20, and 40 μF. When these are placed in parallel across a 200-V source, find:
(a) The equivalent capacitance.
(b) The total charge residing on the capacitors.
(c) The charge on each capacitor.

3-5. The three capacitors of Prob. 3-4 are placed in series across a 350-V source.
(a) Compute the equivalent capacitance.
(b) Find the charge on each capacitor.
(c) Determine the voltage drop across each capacitor.

3-6. In a configuration of three series-connected capacitors the voltage drops across their terminals are 20, 30, and 50 V and the charge on each capacitor is 300 μC.
(a) Find the capacitance of each capacitor.
(b) What is the series equivalent capacitance?
(c) Compute the total stored energy.

3-7. Three series-connected inductor coils have voltages of 20, 30, and 50 V appearing across their terminals when the circuit current is changing at a rate of 100 A/s. Determine the equivalent series inductance.

3-8. Three inductors are arranged as shown in Fig. P3-8.
(a) Find the equivalent inductance as seen by the source.
(b) Determine the value of the emf of self-inductance in coil L_2 when the current in coil L_1 is changing at a rate of 1500 A/s.

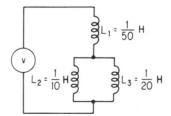

$L_1 = \frac{1}{50}$ H

$L_2 = \frac{1}{10}$ H

$L_3 = \frac{1}{20}$ H

Fig. P3-8

3-9. Find the equivalent resistance appearing between terminals ab for the circuit configuration shown in Fig. P3-9. All resistance values are expressed in ohms.

3-10. In the circuit configuration of Fig. P3-10 determine the voltage drop across each circuit element as well as the power dissipated. All resistances are in ohms.

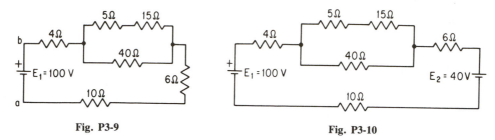

Fig. P3-9   Fig. P3-10

3-11. For the circuit of Fig. P3-11, find the current *I*. All resistance values are in ohms.

3-12. In the configuration of Fig. P3-12, find the value of *E* which permits a power dissipation of 180 W in the 20-Ω resistor. All values are expressed in ohms.

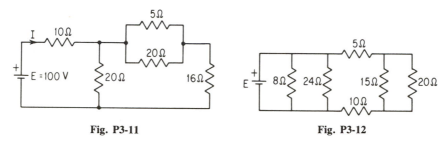

Fig. P3-11   Fig. P3-12

3-13. Determine the equivalent resistance appearing at the battery terminals in the circuit of Fig. P3-12.

3-14. Find the voltage drop appearing across the 18-Ω resistor in the network of Fig. P3-14. All values are in ohms.

3-15. Determine the equivalent resistance between points *ab* in the circuit of Fig. P3-15. All resistances are in ohms.

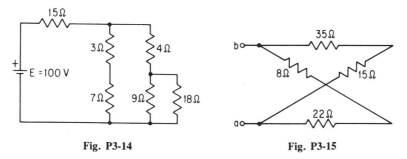

Fig. P3-14   Fig. P3-15

3-16. In the circuitry of Fig. P3-15, determine:
   (a) The voltage needed across *ab* so that the voltage drop across the 15-Ω resistor is 45 V.
   (b) The corresponding voltage across the 8-Ω resistor for this condition.

3-17. Find the equivalent resistance between terminals *ab* for the circuit of Fig. P3-17. All resistance values are in ohms.

3-18. In Fig. P3-17 compute the voltage required between terminals *ab* so that a voltage drop of 45 V occurs across the 15-Ω resistor.

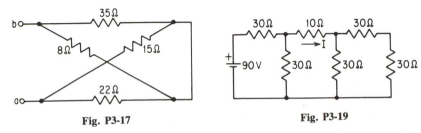

Fig. P3-17   **Fig. P3-19**

3-19. Find the current I which flows through the 10-Ω resistor in the circuit of Fig. P3-19. All resistances are in ohms.

3-20. In the circuit of Fig. P3-20 find the value of R. All resistances are in ohms.

3-21. By means of superposition find the current which flows in $R_2 = 20\ \Omega$ for the circuit of Fig. P3-21.

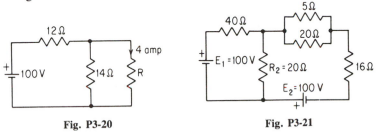

Fig. P3-20   **Fig. P3-21**

3-22. Find the current and the power dissipation in each resistor of the circuit shown in Fig. P3-22.

3-23. Determine the current which flows through each resistor in the circuit of Fig. P3-23. All resistance values are in ohms. Use superposition.

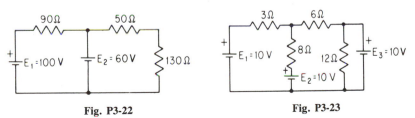

Fig. P3-22   **Fig. P3-23**

3-24. In the circuit configuration shown in Fig. P3-24 determine the number of (a) circuit elements, (b) nodes, (c) junction points, (d) branches, and (e) meshes. Draw that tree of the network that leads directly to the mesh equations.

3-25. Repeat Prob. 3-24 for the circuit shown in Fig. P3-25.

3-26. Repeat Prob. 3-24 for the network depicted in Fig. P3-26.

3-27. Find the current which flows through the 15-Ω resistor in the circuit of Fig. P3-27 by using the mesh method. Check your solution by using the superposition theorem.

3-28. Solve Prob. 3-27 by using the nodal method of analysis.

3-29. In the network of Fig. P3-29 find the current through the 90-Ω resistor.

3-30. Determine the current which flows through the 30-Ω resistor in the circuit of Fig. P3-30.

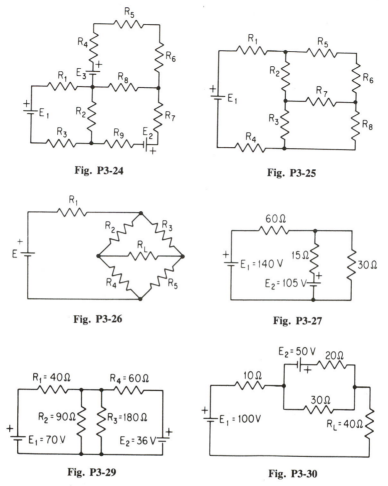

Fig. P3-24

Fig. P3-25

Fig. P3-26

Fig. P3-27

Fig. P3-29

Fig. P3-30

3-31. In the circuit of Fig. P3-29 find the current through R_2 when it is made to take on the following values: R_2 = 62.5, 106, 209, and 784 Ω. Use the method requiring the least effort to yield the answers.

3-32. Assuming that R_L in Fig. P3-30 is the load resistor, find the Thévenin equivalent circuit.

3-33. Solve Prob. 3-22 by the mesh method.

3-34. Solve Prob. 3-23 by the nodal method.

3-35. A circuit has an arrangement of circuit elements as depicted in Fig. P3-35.

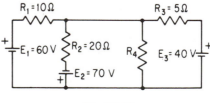

Fig. P3-35

(a) Find the Thévenin equivalent circuit considering R_4 as the variable load resistance.

(b) Find the current through R_4 when it has values of $\frac{4}{7}$ and $\frac{40}{7}$ Ω.

3-36. In the circuit of Fig. P3-36 it is desirable to obtain information about each of the branch circuits.

(a) By using the mesh method write (but do not solve) the necessary independent equations which will allow all branch currents to be determined.

(b) By using the nodal method, write (but do not solve) the independent equations from which all the branch currents can be found.

3-37. In the circuit of Fig. P3-37 find the current delivered by the battery.

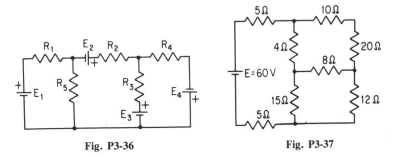

Fig. P3-36   Fig. P3-37

3-38. Use Thévenin's theorem to replace the three-loop circuit of Fig. P3-38 by a single-loop equivalent circuit in which the identity of R_L is preserved.

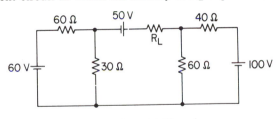

Fig. P3-38

3-39. In the circuit shown in Fig. P3-39 find the current in the 90-Ω resistor using the mesh method.

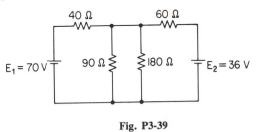

Fig. P3-39

GROUP II

3-40. Three capacitors are connected across a 100-V source in the manner shown in Fig. P3-40.

(a) Find the charge on each capacitor.

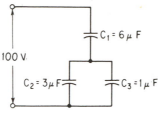

Fig. P3-40

(b) Compute the total stored energy.

(c) Compute the energy stored in each capacitor.

3-41. The capacitors of Prob. 3-6 are disconnected in the charged condition and then placed in parallel. However, one of the capacitors is connected with its polarity reversed from that of the other two.

    (a) Find the resultant voltage of the parallel combination.

    (b) Determine the charge on each capacitor.

    (c) Compute the stored energy.

3-42. The equivalent inductance of two inductors placed in parallel is 4 H. When series-connected, the equivalent inductance is 20 H. Find the values of the individual inductances.

3-43. Set up the equations which will allow you to determine all the branch currents in the circuits of Figs. P3-24 and P3-26. Give reasons to justify the method you have chosen to solve the problem.

3-44. Solve for the power delivered to the 10-Ω resistor in the circuit shown in Fig. P3-44. All resistances are in ohms.

3-45. Use Thévenin's theorem to find the power delivered to the 3-Ω resistance in the circuit of Fig. P3-45. All resistances are in ohms.

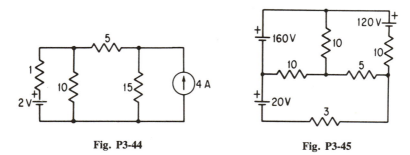

Fig. P3-44　　　　　　　　**Fig. P3-45**

3-46. Use Thévenin's theorem to replace the three-loop equivalent circuit of Fig. P3-46 by a single-loop equivalent circuit in which the identity of R_L is preserved. All resistances are expressed in ohms.

3-47. Derive the appropriate expressions which will permit converting a given wye arrangement of resistances to an equivalent delta arrangement.

3-48. Refer to Fig. P3-48.

    (a) Draw an appropriate tree of the network.

    (b) How many equations must be solved to specify completely the current in each branch of the circuit assuming all source and resistance values are known?

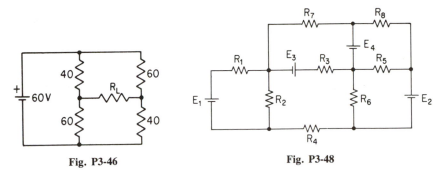

Fig. P3-46

Fig. P3-48

3-49. The circuit appearing in Fig. P3-49 is a typical potentiometer arrangement which is used to measure voltage.
   (a) Find the voltage at point A with respect to the ground terminal.
   (b) Find the Thévenin equivalent circuit for terminals 0A.
   (c) Calculate the galvanometer current.

3-50. Refer to the circuit shown in Fig. P3-50. Consider that $0 \leqslant A \leqslant 1$.
   (a) If I is a constant current source, find the output voltage E_0 in terms of I and the circuit parameters shown.
   (b) Assume that the current source is ideal (i.e., has infinite resistance). Compute the resistance looking into terminals AB.
   (c) If $R_L = 0$, compute the output current I_L.

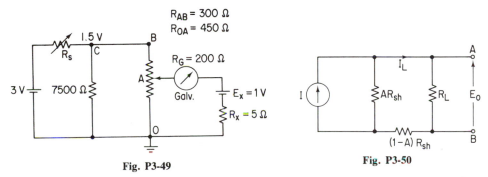

Fig. P3-49

Fig. P3-50

3-51. In the circuit of Fig. P3-51 find the Thévenin equivalent circuit existing between points a and b.

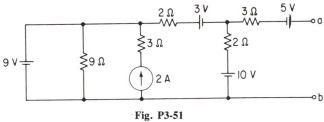

Fig. P3-51

3-52. By direct application of Norton's theorem find the Norton equivalent circuit as seen by R_L in the circuit of Fig. P3-52.

3-53. For the circuit shown in Fig. P3-53 find the Thévenin equivalent circuit as seen by the 20-Ω resistor appearing between points a and b.

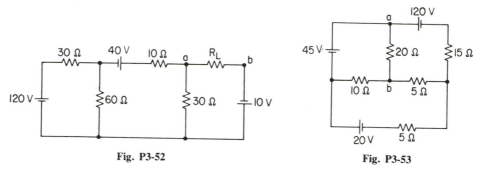

Fig. P3-52

Fig. P3-53

3-54. In the circuit of Fig. P3-54, answer the following questions as they relate to the mesh method of analysis.
(a) How many meshes does this circuit have?
(b) What are the mesh currents for the inner and outer meshes?
(c) Find all mesh currents.
(d) Determine voltages V_1 and V_2.

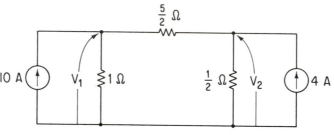

Fig. P3-54

3-55. Find the node-pair voltages V_1 and V_2 in Fig. P3-54 using the nodal method.

3-56. Determine the voltages V_1 and V_2 in Fig. P3-54 using network reduction that involves replacing the current sources with voltage sources.

chapter four

Circuit Differential Equations: Forms and Solutions

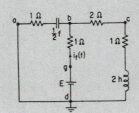

$$i_f(t) = \frac{2p^2 + 4p + 2}{4p^2 + 11p + 8} E$$

$$4\frac{d^2 i_f}{dt^2} + 11\frac{di_f}{dt} + 8i_f = 0 + 0 + 2E$$

In dealing with the subject matter of electrical engineering, it is frequently necessary to determine the solutions of nonhomogeneous linear differential equations with constant coefficients. In this way information is obtained about the dynamic and forced responses of a circuit, device, or system to a driving force. The method of solution described here is the classical one. Use is made of the differential operator. Its introduction permits handling integrodifferential equations in an algebraic form, which in turn leads to the notion of operational impedance and thence to operational network functions.

The classical solution of linear differential equations can be identified as a three-part procedure. First, the forced solution is found. Mathematicians like to call this part the particular solution. Next, the transient solution is determined. This is nothing more than the solution of the *homogeneous* linear differential equation. Mathematicians call this part the complementary function for the simple reason that when it is combined with the particular solution a total solution results which completely satisfies the nonhomogeneous linear differential equation for *all* time after application of the forcing function. In the third and final step of the procedure the constants of integration are evaluated from the initial conditions. Of course if the linear differential equation is of nth order, then n constants of integration will need to be determined.

In studying this chapter it is helpful to keep in mind that the primary objective is to describe the classical method of solving linear differential equations.

Applications of these techniques to formal engineering situations is reserved for Chapter 5.

4-1 THE DIFFERENTIAL OPERATOR

Situations often occur in engineering where one or more energy-storing elements are subjected to a time-varying forcing function. An example of this is shown in Fig. 1-9. A glance at Eq. (1-36), which is the describing differential equation for the circuit, reveals the presence of both an integral as well as a derivative of time. For this reason, it is called an integrodifferential equation. As it stands here, this equation is in an inconvenient form to manipulate in seeking out a solution. To help facilitate the solution process therefore, the notation of operational calculus is employed. We define

$$p \equiv \frac{d}{dt} \tag{4-1}$$

where p denotes differentiation with respect to time and is called the *differential operator*. Accordingly, the derivative of current with respect to time appearing in Eq. (1-36) can now be represented in the operator form as pi. That is,

$$\frac{di}{dt} = pi \tag{4-2}$$

Moreover, in situations where a second derivative is involved, we can write

$$\frac{d^2i}{dt^2} = \frac{d}{dt}\left(\frac{di}{dt}\right) = p(pi) = p^2i \tag{4-3}$$

In general the operator form of an nth derivative can be expressed as

$$\boxed{\frac{d^n}{dt^n} \equiv p^n} \tag{4-4}$$

The integral term appearing in Eq. (1-36) can be handled in a similar manner. However, it is useful to keep in mind that integration is the inverse operation of differentiation and, in addition, involves a constant of integration. To illustrate this point recall that the current flow through a capacitor is given by

$$i = C\frac{dv_c}{dt} \tag{4-5}$$

where v_c denotes the voltage appearing across the terminals of the capacitor. The capacitor voltage can be found explicitly by integrating Eq. (4-5). Accordingly,

$$v_c = \frac{1}{C}\int_0^t i\, dt + V_0 \tag{4-6}$$

where V_0 is a constant of integration and specifically represents the voltage across the capacitor that is associated with an initial charge appearing across the capacitor plates before applying the forcing function. The term involving integration with time can be represented in operator form by

$$\int_0^t i \, dt = \frac{1}{p} i \tag{4-7}$$

from which it follows that

$$\frac{1}{p}(\quad) \equiv \int_0^t (\quad)dt \tag{4-8}$$

Note that the inverse p operator is defined by the definite integral only. Accordingly, whenever the p-operator notation is used to denote integration such matters as constants of integration arising from initial conditions must be treated separately.

Finally, it is worthwhile to note that differentiation followed by integration of a function i or integration followed by differentiation of the function i yields i for zero initial conditions. In operator form this is expressed as

$$\frac{1}{p}pi = p\frac{1}{p}i = i \tag{4-9}$$

The algebraic manipulation of the p operator in this equation is indeed significant.

4-2 OPERATIONAL IMPEDANCE

When a time-varying voltage e is applied to a simple resistor R, the current is given by Ohm's law as

$$i = \frac{e}{R} \tag{4-10}$$

If this same voltage is applied to a simple inductor, the relation between e and i is given by Faraday's law as

$$e = L\frac{di}{dt} \tag{4-11}$$

Introducing the differential operator in Eq. (4-11) allows writing

$$e = pLi$$

or

$$i = \frac{e}{pL} \tag{4-12}$$

Thus the use of the differential operator permits the differential equation to be written as an algebraic expression, where pL plays a role for inductance that is the same as R for resistance. The inductor is said to possess an operational impedance of pL.

When e is applied to a simple capacitor the pertinent relationship is

$$C\frac{de}{dt} = i \tag{4-13}$$

or, in operator form,

$$Cpe = i$$

from which it follows that

$$i = \frac{e}{1/pC} \qquad (4\text{-}14)$$

Then, by analogy to Eqs. (4-10) and (4-12), the capacitor can be said to present to the forcing function e an operational impedance of $1/pC$. It is interesting to note that when e is made to assume a fixed value, corresponding to which the differential operator p is zero, the operational impedance of the capacitor is infinite. This means that the ensuing current is zero, and the result is indeed familiar.

The term *operational impedance* can be applied to single elements or to several elements placed in series. In this connection consider the series RLC circuit of Fig. 4-1. Assume that neither L nor C has any initial energy, i.e., the circuit is initially passive. By Kirchhoff's voltage law the describing differential equation for the circuit is

$$e = Ri + L\frac{di}{dt} + \frac{1}{C}\int i \, dt \qquad (4\text{-}15)$$

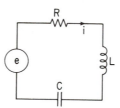

Fig. 4-1 Series RLC circuit with zero initial energy.

In terms of the p operator the expression becomes

$$e = Ri + pLi + \frac{1}{pC}i \qquad (4\text{-}16)$$

or

$$e = \left(R + pL + \frac{1}{pC}\right)i = Z(p)i \qquad (4\text{-}17)$$

where

$$Z(p) \equiv R + pL + \frac{1}{pC} \qquad (4\text{-}18)$$

and denotes the operational impedance of the series RLC circuit. In the case of this simple circuit Eq. (4-18) also qualifies as the *driving-point operational impedance* of the circuit. In general, the driving-point operational impedance is a network function that expresses the ratio of voltage to current taken at the *same* terminal pairs of a network.

EXAMPLE 4-1 Write the expression for the operational impedance for each branch of the circuit shown in Fig. 4-2 and determine the expression for the driving-point impedance appearing at terminal pair *a-d*.

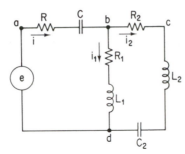

Fig. 4-2 Circuit for Example 4-1.

Solution: There are three branches: *ab, bd,* and *bcd.* The corresponding operational impedances are

$$Z_{ab} = R + \frac{1}{pC} \tag{4-19}$$

$$Z_{bd} = R_1 + pL_1 \tag{4-20}$$

$$Z_{bcd} = R_2 + pL_2 + \frac{1}{pC_2} \tag{4-21}$$

The driving-point impedance appearing between terminals *a-d* represents the ratio of the voltage *e* applied at these terminals and the current *i* that flows into and out of these same terminals. Applying the reduction technique described in the preceding chapter, we obtain the desired result:

$$Z_{ad} = \frac{e}{i} = Z_{ab} + \frac{Z_{bd}Z_{bcd}}{Z_{bd} + Z_{bcd}} \tag{4-22}$$

$$= R + \frac{1}{pC} + \frac{(R_1 + pL_1)(R_2 + pL_2 + 1/pC_2)}{R_1 + R_2 + p(L_1 + L_2) + 1/pC_2}$$

Occasions often arise in electrical engineering when it is desirable to relate an electrical quantity in one part of a circuit or system to an appropriate electrical quantity in another part of the circuit or system. For example, in the circuit of Fig. 4-1, it may be desirable to express the ratio of voltage appearing across the inductor to the source voltage. This is readily and conveniently accomplished by the use of the operator form of impedances together with the voltage-divider relationship. Thus, if v_L is used to denote the voltage developed across the inductor, the desired result can then be written as

$$\frac{v_L}{e} = \frac{pL}{R + pL + 1/pC} \tag{4-23}$$

Such an expression is commonly called a *transfer function.* The transfer function relates two quantities that are identified at *different* terminal pairs. Moreover, the quantities involved are not restricted in the way they are for the driving-point impedance.† Transfer functions may be used to relate ratios of voltages, ratios of currents, ratios of current to voltage, or ratios of voltage to current, provided that the quantities are not identified at the same terminal pairs.

4-3 THE OPERATOR FORMULATION OF CIRCUIT DIFFERENTIAL EQUATIONS. EQUIVALENCY

It is important at the outset to feel satisfied that the p-operator form of a circuit differential equation is in all respects equivalent to the original equation that it represents. An example is now considered that shows not only this equivalency but the ease of formulation as well. First the problem is solved without the use of the p operator.

In the circuit of Fig. 4-3 let it be desired to find the describing differential equation that relates the voltage across the inductor, v_L to the source voltage e. By Kirchhoff's voltage law for zero initial energy, we can write

$$e = Ri + L\frac{di}{dt} + \frac{1}{C}\int_0^t i\, dt$$

or

$$e = Ri + v_L + \frac{1}{C}\int_0^t i\, dt \tag{4-24}$$

where

$$v_L = L\frac{di}{dt} \tag{4-25}$$

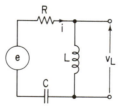

Fig. 4-3 Circuit to illustrate equivalency between p-operator algebraic equation and the circuit differential equation.

Equations (4-24) and (4-25) are a pair of simultaneous equations relating v_L and i. By employing Eq. (4-25) we can eliminate i from Eq. (4-24). Thus

$$i = \frac{1}{L}\int_0^t v_L\, dt \tag{4-26}$$

and

$$e = \frac{R}{L}\int_0^t v_L\, dt + v_L + \frac{1}{LC}\int_0^t\int_0^t v_L\, dt \tag{4-27}$$

† If the quantities in the ratio e/i (which is used to define a driving-point impedance) are inverted, the ensuing result is the driving-point admittance.

Differentiating Eq. (4-27) once with respect to time yields

$$\frac{de}{dt} = \frac{R}{L}v_L + \frac{dv_L}{dt} + \frac{1}{LC}\int_0^{t} v_L\, dt \tag{4-28}$$

The last remaining integral may be eliminated by differentiating with respect to time once again. Accordingly, there results

$$\frac{d^2e}{dt^2} = \frac{R}{L}\frac{dv_L}{dt} + \frac{d^2v_L}{dt^2} + \frac{1}{LC}v_L \tag{4-29}$$

which is the desired relationship.

The p-operator formulation of the solution to this problem leads more directly to the result because of the algebraic feature involved. By the voltage-divider procedure, we can immediately relate v_L to e. Thus

$$v_L = \frac{pL}{R + pL + 1/pC}e = \frac{p^2LC}{RCp + LCp^2 + 1}e \tag{4-30}$$

or

$$RCpv_L + LCp^2v_L + v_L = LCp^2e \tag{4-31}$$

or

$$p^2e = \frac{R}{L}pv_L + p^2v_L + \frac{1}{LC}v_L \tag{4-32}$$

Now that the algebraic manipulations have led to the appropriate relation between e and v_L, the p operator may be replaced by the corresponding time derivatives, thus leading to the required differential equation

$$\frac{d^2e}{dt^2} = \frac{R}{L}\frac{dv_L}{dt} + \frac{d^2v_L}{dt^2} + \frac{1}{LC}v_L \tag{4-33}$$

Comparison of Eq. (4-33) with Eq. (4-29) shows them to be identical. It should not be difficult to see that for more complicated circuits the advantage of the algebraic manipulations associated with the p notation can be appreciable indeed. This point is illustrated by the example that follows.

EXAMPLE 4-2   In the circuit of Fig. 4-4 find the differential equation that relates the voltage v_0 to the source voltages e_1 and e_2.

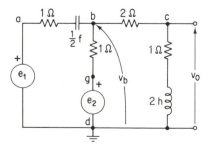

Fig. 4-4   Circuit for Example 4-2. Circuit parameter values are chosen for convenience.

Solution: (It is suggested to the reader that an attempt first be made to obtain the desired result by working entirely with integrodifferential equations. The ensuing inconvenience of such an approach will become apparent rather soon.)

The algebraic nature of the p operator allows us to exploit our knowledge of circuit theory immediately by employing that method of analysis that leads to the desired result with the least effort. Applying the nodal method of analysis to this circuit reveals that there is only one independent unknown, namely, the voltage at junction b, v_b. Kirchhoff's current law at b leads to

$$\frac{v_b - e_1}{Z_{ab}} + \frac{v_b - e_2}{Z_{gb}} + \frac{v_b}{Z_{bcd}} = 0 \tag{4-34}$$

where

$$Z_{ab} = 1 + \frac{2}{p} = \frac{2 + p}{p} \tag{4-35}$$

$$Z_{gb} = R_{gb} = 1 \tag{4-36}$$

$$Z_{bcd} = 3 + 2p \tag{4-37}$$

Equation (4-34) can be rewritten as

$$v_b \left(\frac{p}{2 + p} + 1 + \frac{1}{3 + 2p} \right) = \frac{p}{2 + p} e_1 + e_2 \tag{4-38}$$

or

$$v_b = \frac{p(3 + 2p)e_1 + (2 + p)(3 + 2p)e_2}{4p^2 + 11p + 8} \tag{4-39}$$

In passing it should be noted that this last equation relates the nodal voltage at terminal b to the source voltages e_1 and e_2. The corresponding differential equation is found by replacing the p operators by their equivalent time derivatives.

To obtain the result called for in the statement of this problem, however, it is necessary merely to write v_0 in terms of v_b by applying the voltage-divider relationship. Thus

$$v_0 = \frac{Z_{cd}}{Z_{bcd}} v_b = \frac{1 + 2p}{3 + 2p} v_b \tag{4-40}$$

Inserting Eq. (4-39) into (4-40) yields

$$(4p^2 + 11p + 8)v_0 = (2p^2 + p)e_1 + (2p^2 + 5p + 2)e_2 \tag{4-41}$$

The corresponding differential equation is then

$$4\frac{d^2v_0}{dt^2} + 11\frac{dv_0}{dt} + 8v_0 = 2\frac{d^2e_1}{dt^2} + \frac{de_1}{dt} + 2\frac{d^2e_2}{dt^2} + 5\frac{de_2}{dt} + 2e_2 \tag{4-42}$$

An examination of this equation shows that the left side consists of a linear combination of the response function v_0 and its derivatives. The right side constitutes *the total forcing function* that causes the response v_0. Note that the right side is made up of a linear combination of the applied sources e_1 and e_2 and their derivatives. Equation (4-42) is the nonhomogeneous linear differential equation with constant coefficients that describes the response function v_0 for this circuit. It is important here to appreciate the ease with which this result was obtained employing the p operator and conventional circuit theory.

EXAMPLE 4-3   Refer to Fig. 4-4. Determine the expressions for the operational driving-point impedances appearing at terminals *ad* and *dg*.

Solution: At terminal pair *ad* we have

$$Z_{ad}(p) = Z_{ab} + \frac{Z_{bg}Z_{bcd}}{Z_{bg} + Z_{bcd}} = 1 + \frac{2}{p} + \frac{1(3 + 2p)}{4 + 2p}$$

$$= \frac{4p^2 + 11p + 8}{2p^2 + 4p} \qquad (4\text{-}43)$$

Similarly, at terminal pair *gd* we can write

$$Z_{gd}(p) = Z_{bg} + \frac{Z_{ab} + Z_{bcd}}{Z_{ab}Z_{bcd}} = 1 + \frac{\dfrac{p + 2}{2}(3 + 2p)}{\dfrac{p + 2}{p} + 3 + 2p}$$

$$= \frac{4p^2 + 11p + 8}{2p^2 + 4p + 2} \qquad (4\text{-}44)$$

Inspection of Eqs. (4-43) and (4-44) shows that the numerator expressions for the driving-point impedances at different terminal pairs of the circuit are identical. In fact, a glance at Eq. (4-41), relating the response voltage v_0 to e_1 and e_2, indicates the presence of the same quadratic equation in p. This occurrence is by no means an accident or a coincidence. The quadratic expression is intimately associated with the values of the circuit parameters appearing in the prescribed circuit configurations and determines the character of the dynamic response of this circuit to applied forces. This matter is treated at some length in the following sections and in the next chapter.

4-4 GENERAL FORMULATION OF CIRCUIT DIFFERENTIAL EQUATIONS

Attention is directed in this and the next two sections to the components that are needed to identify a complete solution to a nonhomogeneous linear differential equation that relates a response function to applied source functions. A typical example of such an equation encountered in electrical engineering is illustrated by Eq. (4-42). For ease of handling let us simplify this equation by eliminating e_1 and replacing it with a short circuit. The voltage v_0 is then related to e_2 in Fig. 4-4 by

$$4\frac{d^2v_0}{dt^2} + 11\frac{dv_0}{dt} + 8v_0 = 2\frac{d^2e_2}{dt^2} + 5\frac{de_2}{dt} + 2e_2 \qquad (4\text{-}45)$$

An equation of this type is often referred to as an *equilibrium equation.* It relates linear combinations of the response function and its derivatives to linear combinations of the source function and its derivatives. Moreover, as demonstrated in the preceding section, this equilibrium equation is readily obtained from the circuit configuration through the use of circuit theory and operational impedances. The operator version of Eq. (4-45) is clearly

$$v_0 = \frac{2p^2 + 5p + 2}{4p^2 + 11p + 8}e_2 \qquad (4\text{-}46)$$

More generally, we can write

$$v_0 = G(p)e_2 \qquad (4\text{-}47)$$

where $G(p)$ is called the operational network function.

In any circuit configuration the foregoing procedure can be generalized to the form

response function = (network function) × (applied source function)   (4-48)

or, more succinctly,

$$y(t) = G(p)f(t) = \frac{N(p)}{D(p)}f(t) \qquad (4\text{-}49)$$

where $y(t)$ denotes any appropriate response function, such as current or voltage in a particular part of a circuit, $G(p)$ denotes the total network function $N(p)$ is the numerator polynomial in p of the network function, $D(p)$ is the denominator polynomial in p of the network function, and $f(t)$ denotes the applied source function. A preferred way of writing this last equation is

$$\boxed{D(p)y(t) = N(p)f(t)} \qquad (4\text{-}50)$$

Equation (4-50) is the nonhomogeneous linear differential equation that relates the response $y(t)$ in operator form. This version of the equilibrium equation brings into evidence the fact that the total forcing function in the network that is responsible for establishing the response $y(t)$ is generally a linear combination of the source function and its derivatives, in other words, the right side of Eq. (4-50). Of course in the special case where $N(p)$ is unity, the forcing function and the source function are the same.

4-5 THE FORCED SOLUTION (OR PARTICULAR SOLUTION)

If $f(t)$ in Eq. (4-50) is an applied source function, then $y_f(t)$ is said to be a solution to the homogeneous linear differential equation provided that it satisfies the equation. Experience shows that if $f(t)$ is a sinusoidal function, then the form $y_f(t)$ must also be sinusoidal. In general $y_f(t)$ will differ from the source function in amplitude and phase and it will continue to exist in the circuit so long as the source function remains applied. This is the reason why $y_f(t)$ is referred to as the forced solution. It is also frequently referred to as the *steady-state solution* because it remains in the circuit long after the transient terms disappear. Another commonly used description for this part of the solution is the term *particular solution*. This is in obvious reference to the fact that the form of $y_f(t)$ is particularized to the nature of the source function. When the source function is an exponential, the particular solution will likewise be exponential. Similarly, when a constant source is used, the resulting forced solution is constant albeit of different amplitude. A little thought about this matter should make it plain that such results are to be entirely expected in view of the *linear* nature of the differential equation. Let us now illustrate these statements with specific analyses and examples.

Response to Constant Sources. Refer to Fig. 4-5, which is the circuit of Fig. 4-4 modified so that e_1 is eliminated and e_2 is the fixed battery voltage E. Let it be desired to find the forced battery current $i_f(t)$ corresponding to a source voltage E.

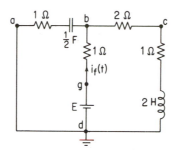

Fig. 4-5 Circuit of Fig. 4-4 with $e_1 = 0$ and $e_2 = E$. All resistance parameter values are expressed in ohms.

Before proceeding with the solution in this case, a word is in order concerning the general procedure to be followed. (1) Obtain the equilibrium equation by applying circuit theory to the configuration of circuit elements described in terms of their operational impedances. (2) Put the resulting expression in the form of Eq. (4-50). Then examine the right side of this equation to determine the exact form needed by the forced solution to qualify as a solution to the equilibrium equation. (3) Finally, solve for the unknown quantities.

To find the expression for the current $i_f(t)$ in the circuit of Fig. 4-5, we need to know the driving-point impedance, Z_{gd}. Keep in mind that this is the impedance that is seen looking into the circuit at terminals g and d. The response function $i_f(t)$ is then related to the source function E by

$$i_f(t) = \frac{1}{Z_{gd}} E \tag{4-51}$$

This driving-point impedance was calculated in Example 4-3 and is given by Eq. (4-44). Thus

$$i_f(t) = \frac{2p^2 + 4p + 2}{4p^2 + 11p + 8} E \tag{4-52}$$

or

$$(4p^2 + 11p + 8)i_f(t) = (2p^2 + 4p + 2)E \tag{4-53}$$

The corresponding differential equation is then

$$4\frac{d^2 i_f}{dt^2} + 11\frac{di_f}{dt} + 8i_f = 0 + 0 + 2E \tag{4-54}$$

Examination of this equilibrium equation reveals that because there are no derivative terms on the right side of the equation (which constitutes the total forcing function), the derivative terms of the response function must be identically equal to zero.

Hence we conclude that

$$i_f(t) = I_0 \tag{4-55}$$

Inserting this expression into Eq. (4-54) leads to the value that I_0 assumes in order to satisfy the equilibrium equation. Accordingly,

$$I_0 = \frac{E}{4} \tag{4-56}$$

This is the forced or steady-state solution.

Examination of the circuit of Fig. 4-5 shows that this result is entirely expected on the basis of physical considerations. The current $i_f(t)$ finds a closed path only in loop *gbcdg*. The left-hand loop is obviously an open circuit in the steady state because of the presence of the capacitor. The total resistance in the right-hand loop is 4 Ω. Moreover, in the steady state the voltage across the 2-H inductor is zero because di/dt is zero. Accordingly, the current is simply a d-c current of magnitude $E/4$.

A little reflection concerning the procedure involved in identifying the steady-state solution leads to the conclusion that the result can be found directly from the operator form of the differential equation by merely setting $p = d/dt = 0$ for constant sources. For the case at hand we have

$$i_f(t) = [G(p)]_{p=0} E = \left[\frac{2p^2 + 4p + 2}{4p^2 + 11p + 8} \right]_{p=0} E = \frac{E}{4} \tag{4-57}$$

In general, the forced response due to a constant source may be found from

$$\boxed{y_f(t) = [G(p)]_{p=0} f(t)} \tag{4-58}$$

Response to Exponential Sources. The exponential function occupies a unique place in the realm of mathematics. It is the only function which preserves its form after being subjected to operations of differentiation and integration. This unusual property accounts for its presence in many mathematical situations and most particularly in the solutions to homogeneous linear differential equations.

Consider that $f(t)$ in Eq. (4-50) is a function that varies exponentially with time, that is,

$$f(t) = A\varepsilon^{st} \tag{4-59}$$

Differentiating with respect to time yields

$$\frac{df(t)}{dt} = sA\varepsilon^{st} = sf(t) \tag{4-60}$$

In terms of the operator notation we can write

$$pf(t) = sf(t) \tag{4-61}$$

It is important to note in Eq. (4-60) that *differentiation with respect to time is equivalent to the algebraic operation of multiplying the original function by s*, which is the coefficient of t in the exponent of ε. In the operator notation of this process, as represented by Eq. (4-61), the distinction between p and s should be kept clearly in mind. Here p denotes differentiation with respect to time, whereas s is merely an *algebraic multiplier* which appears because of the unique property of the exponential function.

The integration process leads to a similar result. Thus, if we assume no initial energy, we can write

$$\int_0^t f(t)dt = \frac{1}{s}A\varepsilon^{st} = \frac{1}{s}f(t) \tag{4-62}$$

or, in operator form,

$$\frac{1}{p}f(t) = \frac{1}{s}f(t) \tag{4-63}$$

Accordingly, integration of the exponential function amounts to *division by s*. Note the consistent inverse relationship this bears to differentiation. *In all succeeding work throughout this textbook we shall adhere to the policy of using the symbol s wherever the derivatives and integrals of the exponential function are involved.*

We focus attention next onto an example to illustrate the procedure involved in finding the forced solution associated with an exponential source function. The circuit diagram appears in Fig. 4-6. It is the same configuration as appears in Fig. 4-4 with the change that e_2 is eliminated. From our knowledge of circuit theory it follows that the expression for the forced current produced by the exponential source is

$$i_f(t) = \frac{1}{Z_{ad}} e_1(t) \tag{4-64}$$

where Z_{ad} denotes the driving-point impedance at terminal pair ad. This quantity was previously calculated, and is described by Eq. (4-43). It therefore follows that

$$i_f(t) = \frac{2p^2 + 4p}{4p^2 + 11p + 8}10\varepsilon^{-0.2t} \tag{4-65}$$

Rearranging into a form consistent with Eq. (4-50) there results

$$(4p^2 + 11p + 8)i_f(t) = (2p^2 + 4p)10\varepsilon^{-0.2t} \tag{4-66}$$

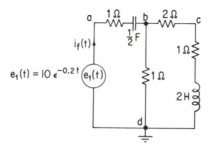

Fig. 4-6 Circuit to illustrate calculation of forced solution to an exponential source function. All resistance values are in ohms.

The total forcing function in this circuit that establishes the forced solution $i_f(t)$ is clearly the entire right side of Eq. (4-66), which in this case takes on the value

$$(2p^2 + 4p)10\varepsilon^{-0.2t} = 20p^2\varepsilon^{-0.2t} + 40p\varepsilon^{0.2t}$$

$$= 20\frac{d^2}{dt^2}\varepsilon^{-0.2t} + 40\frac{d}{dt}\varepsilon^{-0.2t} \qquad (4\text{-}67)$$

$$= 0.8\varepsilon^{-0.2t} - 8\varepsilon^{-0.2t}$$

$$= -7.2\varepsilon^{-0.2t}$$

Hence Eq. (4-66) becomes

$$(4p^2 + 11p + 8)i_f(t) = -7.2\varepsilon^{-0.2t} \qquad (4\text{-}68)$$

Because the right side of Eq. (4-68) contains just an exponential function, it follows that to satisfy this equation it is merely necessary for the forced solution to be of the form

$$i_f(t) = A\varepsilon^{-0.2t} \qquad (4\text{-}69)$$

The amplitude A is determined in a way to assure that the left side of Eq. (4-68) equals the right side. Therefore, insert Eq. (4-69) into Eq. (4-68) and carry out the differentiation process called for by p^2 and p, respectively, and obtain

$$0.16A\varepsilon^{-0.2t} - 2.2A\varepsilon^{-0.2t} + 8A\varepsilon^{-0.2t} = -7.2\varepsilon^{-0.2t} \qquad (4\text{-}70)$$

from which it follows that

$$A = -\frac{7.2}{5.96} = -1.21 \qquad (4\text{-}71)$$

Consequently, the final expression for the forced solution becomes

$$i_f(t) = -1.21\varepsilon^{-0.2t} \qquad (4\text{-}72)$$

There is a quicker way to find the forced solution which exploits the fact that differentiation of the exponential function leads to algebraic multiplications by s. As a result the original expression for the forced solution as represented by Eq. (4-65) may be written as

$$i_f(t) = \frac{2s^2 + 4s}{4s^2 + 11s + 8}10\varepsilon^{-0.2t} \qquad (4\text{-}73)$$

In replacing p by s here there is a significant point that we should not allow to escape us. Whereas the presence of p made it necessary to carry out differentiation in the usual manner, now with the permissible substitution of s all that we need do is treat s as an algebraic quantity whose value is $s = -0.2$. Accordingly, the forced solution becomes immediately

$$i_f(t) = \left. \frac{2s^2 + 4s}{4s^2 + 11s + 8} \right|_{s=-0.2} 10\varepsilon^{-0.2t} \qquad (4\text{-}74)$$

$$= \frac{-0.08 - 0.8}{-0.16 - 1.1 + 8}10\varepsilon^{-0.2t} = -1.21\varepsilon^{-0.2t}$$

On the basis of the foregoing experience we can generalize that whenever exponential source functions $f(t) = F\varepsilon^{s_g t}$ are used the corresponding response function $y_f(t)$ can be immediately determined from

$$y_f(t) = G(p)f(t) = |G(s)|_{s=s_g} F\varepsilon^{s_g t} \tag{4-75}$$

where the subscript g denotes generator source and F is the amplitude of the exponential function. A comparison of this last equation with Eq. (4-58) reveals that the generalized procedure for finding the forced solution to a constant source is identical. This conclusion is entirely expected, however, because the constant source function is actually that special case of the exponential function where $s_g = 0$.

Response to Sinusoidal Sources. Refer to Fig. 4-6 and assume now that $e_1(t) = 10 \sin 3t$. Let it be desired to find the forced current response.

The appropriate equilibrium equation is found from the relationship

$$i_f(t) = \frac{1}{Z_{ad}} e_1(t) = \left(\frac{2p^2 + 4p}{4p^2 + 11p + 8} \right) 10 \sin 3t \tag{4-76}$$

Rearranging to conform with Eq. (4-50), we have

$$(4p^2 + 11p + 8)i_f(t) = (2p^2 + 4p)10 \sin 3t \tag{4-77}$$

Once again the complete right-side expression represents the total forcing function that is effective in establishing the forced response. This has the value

$$(2p^2 + 4p)10 \sin 3t = -180 \sin 3t + 120 \cos 3t \tag{4-78}$$

This expression now clearly indicates that the assumed form of the forced solution must contain both a sine and a cosine component; otherwise, it will be impossible for Eq. (4-77) to be balanced. Accordingly, we must take

$$i_f(t) = A \sin 3t + B \cos 3t \tag{4-79}$$

Moreover,

$$pi_f(t) = 3A \cos 3t - 3B \sin 3t \tag{4-80}$$

and

$$p^2 i_f(t) = -9A \sin 3t - 9B \cos 3t \tag{4-81}$$

Inserting the last three equations into Eq. (4-77) and collecting terms leads to

$$(-28A - 33B) \sin 3t + (33A - 28B) \cos 3t = -180 \sin 3t + 120 \cos 3t \tag{4-82}$$

Setting the coefficients of like terms equal to one another leads to two simultaneous equations which in turn are readily solved to furnish the values of A and B. Thus

$$28A + 33B = 180 \tag{4-83}$$

$$33A - 28B = 120 \tag{4-84}$$

from which it is found that

$$A = 4.8 \quad \text{and} \quad B = 1.38 \tag{4-85}$$

The complete forced solution is then

$$i_f(t) = 4.8 \sin 3t + 1.38 \cos 3t \tag{4-86}$$

A considerable saving of effort is possible in obtaining the foregoing forced solution if it is recalled that the sinusoidal function is expressible in terms of an exponential function by use of Euler's formula. Accordingly,

$$10 \sin 3t = Im (10\varepsilon^{j3t}) \tag{4-87}$$

$$= Im (10 \cos 3t + j10 \sin 3t)$$

where Im denotes "imaginary part of . . ." and $\varepsilon^{j3t} = \cos 3t + j \sin 3t$. Moreover, in Chapter 6 it is shown that analyses involving sinusoidal functions may actually be carried out using the entire exponential function ε^{j3t} provided we remember at the end to take the imaginary part of the resulting exponential solution. The procedure is readily illustrated by the example at hand.

By means of Eq. (4-75) we can write forthwith

$$i_f(t) = \left| \frac{2p^2 + 4p}{4p^2 + 11p + 8} \right|_{s=s_g} 10\varepsilon^{j3t} = \left| \frac{2s^2 + 4s}{4s^2 + 11s + 8} \right|_{s=j3} 10\varepsilon^{j3t} \tag{4-88}$$

Evaluating the network function for $s = j3$ then yields

$$i_f(t) = \left| \frac{-18 + j12}{-28 + j33} \right| 10\varepsilon^{j3t} = \frac{21.6 \underline{/146.3°}}{43.3 \underline{/130.3°}} 10\varepsilon^{j3t}$$

$$= 5\underline{/16°}\, \varepsilon^{j3t} = 5\varepsilon^{j(3t + 16°)} \tag{4-89}$$

The actual solution corresponding to the applied sinusoidal function is obtained as the imaginary part of Eq. (4-89). Thus

$$i_f(t) = Im(5\varepsilon^{j(3t + 16°)}) = 5 \sin (3t + 16°) \tag{4-90}$$

which is the desired result. Note that the forced response has the same form as the source function (i.e., a sinusoid) but it differs in amplitude and phase.

By applying the expansion formula for the sine of the sum of two angles an expression results for Eq. (4-90) which is identical to that of Eq. (4-86). Thus

$$i_f(t) = 5 \sin (3t + 16°) = 5 \sin 3t \cos 16° + 5 \cos 3t \sin 16° \tag{4-91}$$

$$= 4.8 \sin 3t + 1.38 \cos 3t$$

Response to Polynomial Sources.   Although the general form of the polynomial function

$$f(t) = a_0 + a_1 t + a_2 t^2 + a_3 t^3 + \cdots \tag{4-92}$$

contains many higher-order terms in t, we shall confine our attention here to a source function that consists solely of the second term, i.e., $f(t) = a_1 t$. We already know how to find the forced solution when the source is a constant. In situations where there exists both the first and second terms, it is merely necessary to superpose the individual results in order to obtain the total forced response. The linearity feature of the circuit permits this procedure.

To treat a specific case, refer to the circuit of Fig. 4-7, which is our familiar configuration; but note that now the source function is

$$e_2(t) = 2t \tag{4-93}$$

This function varies linearly with time and has a slope of 2. Because of its sloping graphical representation, Eq. (4-93) is frequently referred to as a *ramp function*.

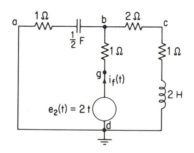

Fig. 4-7 Circuit to illustrate calculations of forced response to a ramp function $e_2(t) = 2t$. All resistance values are in ohms.

Again circuit theory leads us to write in this case

$$i_f(t) = \frac{1}{Z_{gd}} e_2(t) = \frac{2p^2 + 4p + 2}{4p^2 + 11p + 8}(2t) \qquad (4\text{-}94)$$

Rearranging to place the response function $i_f(t)$ alone on one side of the equation yields

$$(4p^2 + 11p + 8)i_f(t) = (2p^2 + 4p + 2)(2t) \qquad (4\text{-}95)$$

The effective forcing function thus takes the form shown in Eq. (4-96):

$$(2p^2 + 4p + 2)2t = 8 + 4t \qquad (4\text{-}96)$$

Accordingly, the forced response can be written as the sum of a constant term plus a ramp. Thus

$$i_f(t) = A + Bt \qquad (4\text{-}97)$$

Substituting the last two equations into Eq. (4-95) leads to

$$11B + 8A + 8Bt = 8 + 4t \qquad (4\text{-}98)$$

Equating like coefficients and solving for the unknown quantities, we get

$$A = \tfrac{5}{16} \qquad \text{and} \qquad B = \tfrac{1}{2} \qquad (4\text{-}99)$$

so that the complete expression for the forced solution becomes

$$i_f(t) = \tfrac{5}{16} + \tfrac{1}{2}t \qquad (4\text{-}100)$$

When the source polynomial function contains higher-order terms, a similar procedure is followed. However, a little thought should make it apparent that, as more terms appear in the source polynomial, the greater will be the number of simultaneous equations that must be solved to obtain the explicit forced solution.

4-6 THE NATURAL RESPONSE (OR TRANSIENT SOLUTION)

The foregoing pages were devoted to a detailed description of the manner in which the forced solution to a circuit linear differential equation is found in response to various types of applied source functions. The forced solution is the solution that is found to exist when the circuit has "settled down" in its response to the disturbing effect of an applied source function. For this reason it is frequently referred to as the *steady-state solution*. It can always be shown that this solution does satisfy the defining differential equation; but, in general, it is not a valid solution over the entire time domain. This problem can best be described by a simple example. Consider the circuit depicted in Fig. 4-8. Let it be desired to find the complete solution for the voltage v appearing across the resistor for all time t after the switch is closed. To determine the equation that relates v to the source voltage E use is made of voltage division. Thus

$$\frac{v}{E} = \frac{R}{R + pL} = \frac{1}{1 + p(L/R)} \tag{4-101}$$

or

$$\frac{L}{R}\frac{dv}{dt} + v = E \tag{4-102}$$

Equation (4-102) is the nonhomogeneous linear differential equation that relates the response function v to the source function E.

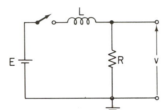

Fig. 4-8 Simple *RL* circuit to illustrate the need for a transient component to the total solution.

The forced solution to this differential equation is readily obtained from Eq. (4-101) upon inserting $p = 0$. Thus

$$v_f = E \tag{4-103}$$

Substituting Eq. (4-103) back into Eq. (4-102) shows that it does satisfy the equation and so rightly qualifies as a solution. But care must be used here, for although Eq. (4-103) qualifies as a solution it does so only after the circuit has settled, i.e., has reached its steady state in response to the source function, which after all was the premise on which the forced solution was determined. However, it is important to note that the forced solution does not qualify as a solution of the governing differential equation in the period immediately following the closing of the switch in Fig. 4-8. This fact is readily demonstrated by noting that at time $t = 0^+$ the current is zero because of the presence of the inductor.† Hence v, which is iR, is also zero. Equation (4-103) therefore cannot be taken to be a complete description of the voltage across the resistor because it fails to describe the behavior of v from time 0^+ to the time when v assumes the steady value of E. In essence, then, there arises, in this period immediately following the application of the source function, a need to add a *complementary function* to the forced solution so that the description would be complete for *all* time after switching. On the basis of what has been said, we should expect too that this complementary function will disappear as steady state is reached, thereby

† See p. 40.

leaving the forced solution alone. It is the purpose of the complementary function to provide a smooth transition from the initial state of the response in the presence of energy-storing elements (the inductor in this case) to the final state. The complementary function serves a twofold purpose: (1) it provides reconciliation between the final state of the solution and that prevailing at $t = 0^+$, and (2) it permits this transition to occur in a smooth fashion and then disappears as the steady state is reached. Because of these features, the complementary function is often called the *transient* solution.

A discussion from an energy viewpoint can help to emphasize further the role of the complementary function. Before closing the switch in the circuit of Fig. 4-8 the current through the inductor is obviously zero. Immediately after switching, at time $t = 0^+$, the current is still zero, as demonstrated in Chapter 2. Consequently the initial energy storage in the inductor is zero. However, in the steady state, the current reaches a level of E/R amperes. The corresponding energy stored in the magnetic field of the inductor then becomes

$$ W = \frac{1}{2} L \left(\frac{E}{R} \right)^2 $$

Accordingly, after the switch is closed, the source function must increase the energy level of the inductor from zero to the aforementioned value. Such a finite change of energy cannot occur in infinitesimal time in the presence of a finite force, i.e., a step change in energy level is not admissible. Rather, the change must take place in a smooth and gradual fashion. Indeed, it is the very task of the complementary function to allow this transition to take place smoothly and gradually.

How do we go about identifying a complementary function to fulfill these requirements? A clue concerning the procedure to follow can be had by resorting to some physical reasoning. When the source voltage E is first applied in the circuit of Fig. 4-8, the source attempts to impose upon the circuit the steady-state solution. However, the circuit, with its energy-storing inductor, immediately reacts to prevent this because it will not tolerate a finite change in energy in the presence of a mere finite force. It is important to realize here that this same reaction of the circuit occurs irrespective of the type of source function applied.† The circuit reaction is intimately related to the presence of the energy-storing elements (and incidentally with the associated resistors). It is therefore *a characteristic of the circuit configuration* in which the energy-storing elements are placed. For this reason this reaction is also known as the *natural response* of the circuit. With this background, then, it should not be difficult to see that the complementary function (or natural response) is obtained as the solution to the homogeneous version of the governing differential equation. For the example at hand the complementary function is the solution to Eq. (4-102) with the source function set equal to zero. Thus the homogeneous differential equation is

$$ \frac{L}{R} \frac{dv}{dt} + v = 0 \tag{4-104} $$

† An exception is the impulse function which has an amplitude approaching infinity.

How do we find a solution to this equation? Examination of the equation should make it clear that the solution must satisfy the condition that the function, v, and its derivative, dv/dt, must be of the same form; otherwise, it will be impossible to make the left side of the equation zero. We already know that there is only one function in all of mathematics that so qualifies—the exponential function. Accordingly, we assume the solution to Eq. (4-104) to be of the form

$$v_t = K\varepsilon^{st} \qquad (4\text{-}105)$$

where v_t denotes transient solution (or complementary function or natural response); K is an amplitude quantity which is to be determined from the initial conditions prevailing at $t = 0^+$ so that a reconciliation between the initial and final energy states can be effected; and s denotes a quantity bearing units of inverse seconds (i.e., frequency).

Substituting the last equation into Eq. (4-104) gives

$$\left(\frac{L}{R}s + 1\right)K\varepsilon^{st} = 0 \qquad (4\text{-}106)$$

Since $K = 0$ leads to a trivial solution, we disregard it. Instead we focus attention on that value of s that makes the parenthetical expression zero. In other words, we solve for s in the equation

$$\frac{L}{R}s + 1 = 0 \qquad (4\text{-}107)$$

which leads to

$$s = -\frac{R}{L} \qquad (4\text{-}108)$$

so that the complementary function becomes

$$v_t = K\varepsilon^{-(R/L)t} \qquad (4\text{-}109)$$

We are now in a position to write a complete solution to the original nonhomogeneous differential equation. Thus

$$v = v_f + v_t = E + K\varepsilon^{-(R/L)t} \qquad (4\text{-}110)$$

The quantity K is found from the initial condition which requires that v be zero at $t = 0^+$. Accordingly, Eq. (4-110) becomes at $t = 0^+$,

$$0 = E + K$$

or

$$K = -E \qquad (4\text{-}111)$$

Hence the complete expression for the voltage across the resistor for *all* time after applying the source function is

$$v = E(1 - \varepsilon^{-(R/L)t}) \qquad (4\text{-}112)$$

A graph of this equation appears in Fig. 4-9. Note how the presence of the transient term permits the voltage at zero plus time to be zero while allowing the steady-state value to be E. The transient component of the solution allows a smooth transition to take place from the initial to the final state.

The rapidity with which this transition takes place is entirely dependent

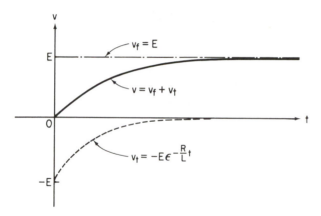

Fig. 4-9 Complete solution of voltage across the resistor in the circuit of Fig. 4-8.

upon the value of s. Equation (4-108) shows that when R is large (or L is small) the transition occurs quickly because the transient term dies out in little time. It is reasonable then to say that the value of s *characterizes* the mode of the response independently of the forcing function applied. In view of the fact then that s is obtained from Eq. (4-107), which in turn derives from Eq. (4-106), the former equation is called the *characteristic equation* of the circuit. A comparison of Eq. (4-106) with Eq. (4-104) reveals that the characteristic equation is readily found from the homogeneous differential equation by replacing time derivatives by s, factoring out the response function, and setting the ensuing parenthetical expression involving the s terms equal to zero.

Generalized Procedure for Determining the Transient Solution. The operator form of a circuit nonhomogeneous differential equation relating a desired response function $y(t)$ to a source function $f(t)$ is given by Eq. (4-50), which is repeated here for convenience:

$$D(p)y(t) = N(p)f(t)$$

Keep in mind that $N(p)$ and $D(p)$ are, respectively, the numerator and denominator polynomial functions in p of the appropriate netwcrk function $G(p)$ that relates $y(t)$ to $f(t)$. The homogeneous version of this differential equation results upon setting $f(t)$ equal to zero. Thus

$$D(p)y(t) = 0 \tag{4-113}$$

In view of the fact that only the exponential function qualifies as a solution to this differential equation, the differential operator p may be replaced by the algebraic multiplier s so that Eq. (4-113) can be rewritten as

$$\boxed{D(s)y(t) = 0} \tag{4-114}$$

Now $D(s)$ is the denominator polynomial of the network function and when it is set equal to zero, it identifies forthwith the characteristic equation of the network. Thus

$$\boxed{D(s) = a_n s^n + a_{n-1} s^{n-1} + \cdots + a_1 s + a_0 = 0} \tag{4-115}$$

The dynamic response of a network to externally applied source functions is completely determined by the roots of this equation. Each specific value of s that satisfies Eq. (4-115) establishes a natural mode of response. The value of n in this equation is equal to the number of independent energy-storing elements in the network. If the n solutions for s in Eq. (4-115) are found to be real and distinct, then the total expression for the natural response of the network is readily shown to be

$$y_t(t) = \sum_{k=1}^{n} K_k \varepsilon^{s_k t} \tag{4-116}$$

Note that there are n values of K that need to be determined from the n initial conditions associated with the n energy-storing elements in the network. When n is 2 or greater, it is also possible to find solutions for s that involve either complex conjugate values or real and equal roots.

4-7 COMPLETE RESPONSE OF LINEAR SECOND-ORDER CASE

The material in the preceding section describes in considerable detail how both the forced and the transient solutions of a homogeneous linear differential equation with constant coefficients are found employing classical techniques. In the interest of providing more experience with those procedures as well as to illustrate the calculation of the coefficients appearing in the transient solution for a situation involving more than a single energy-storing element, attention is now directed to the second-order case where two independent energy-storing elements are present. Specifically, two situations are investigated. In the first case the circuit parameters are assigned such values that the roots of the characteristic equation are real and distinct. In the second case the parameter values lead to a characteristic equation whose roots are complex conjugate.

Real and Distinct Roots. In the circuit of Fig. 4-10, which is assumed initially deenergized, let it be desired to find the complete solution for the voltage across the capacitor, v, for all time after closing the switch. A little physical reasoning before proceeding with the mathematical analysis is always a useful aid. Before closing the switch, the voltage across the capacitor is zero. Moreover, after the elapse of a long period of time (i.e., when steady state is reached) the

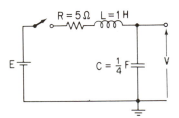

Fig. 4-10 Series *RLC* circuit with parameters adjusted for an overdamped response.

circuit current is known to be zero and so the full source voltage must appear across the capacitor in conformity with Kirchhoff's voltage law. Accordingly, because the initial and final value of the response function v are not the same, we have reason to expect that the circuit will react to provide a smooth transition from the original value to the final value of v. In other words, there exists the need for a transient component to the total solution.

To find the governing differential equation that relates v to the source E, we use the voltage-divider relationship. Thus

$$v = \frac{1/pC}{R + pL + 1/pC} E = \frac{1}{p^2 LC + pRC + 1} E$$

$$= \frac{1/LC}{p^2 + \frac{R}{L}p + \frac{1}{LC}} E = \frac{4}{p^2 + 5p + 4} E \qquad (4\text{-}117)$$

The associated nonhomogeneous differential equation is then

$$\frac{d^2v}{dt^2} + 5\frac{dv}{dt} + 4v = 4E \qquad (4\text{-}118)$$

The forced solution is given by

$$v_f = \left| \frac{4}{p^2 + 5p + 4} \right|_{p=0} E = E \qquad (4\text{-}119)$$

which corroborates our physical reasoning.

The highest order of p in the denominator polynomial of Eq. (4-117) is 2. This is consistent with the fact that the circuit contains two independent energy-storing elements (L and C). Accordingly, two values of s must be expected, each of which establishes a characteristic mode of response. These values are readily obtained from a solution of the characteristic equation, which in this case is

$$D(p) = D(s) = s^2 + 5s + 4 = (s + 4)(s + 1) = 0 \qquad (4\text{-}120)$$

Thus

$$s_1 = -4 \quad \text{and} \quad s_2 = -1 \qquad (4\text{-}121)$$

and so the complementary function becomes

$$v_t = K_1\varepsilon^{-4t} + K_2\varepsilon^{-t} \qquad (4\text{-}122)$$

The complete solution then assumes the form

$$v = v_f + v_t = E + K_1\varepsilon^{-4t} + K_2\varepsilon^{-t} \qquad (4\text{-}123)$$

where K_1 and K_2 must be calculated from known or derived initial conditions involving the response function v. The evaluation of these coefficients constitutes the third and final phase of the three-step solution procedure of the classical method of solving nonhomogeneous linear differential equations.

Since there are two unknown coefficients, two bits of information involving v at time 0^+ are needed to evaluate K_1 and K_2. The first is that v has a zero value at time $t = 0^+$. Inserting this information into Eq. (4-123) yields

$$v(0^+) = 0 = E + K_1 + K_2 \qquad (4\text{-}124)$$

The second bit of information must be derived. In this connection let us differentiate the total solution of Eq. (4-123) with respect to time. This leads to

$$\frac{dv}{dt} = -4K_1\varepsilon^{-4t} - K_2\varepsilon^{-t} \qquad (4\text{-}125)$$

We must now apply our knowledge of circuit theory to involve the capacitor voltage v in such a relationship that its first derivative can be identified in a manner that readily furnishes the information needed. A good starting point in this instance is to write the equation relating the voltage and current of the capacitor. Thus

$$v = \frac{1}{C}\int_0^t i\,dt \qquad (4\text{-}126)$$

Differentiating once, we get

$$\frac{dv}{dt} = \frac{i}{C} \qquad (4\text{-}127)$$

Since we know that i at $t = 0^+$ is zero, because of the presence of the inductor, it follows that

$$\left(\frac{dv}{dt}\right)_{t=0^+} = \frac{i(0^+)}{C} = 0 \qquad (4\text{-}128)$$

Inserting this result into Eq. (4-125) yields the second of the two simultaneous equations involving K_1 and K_2. Hence

$$\left(\frac{dv}{dt}\right)_{t=0^+} = 0 = -4K_1 - K_2 \qquad (4\text{-}129)$$

Equations (4-124) and (4-129) then lead to

$$K_1 = \frac{E}{3} \quad \text{and} \quad K_2 = -\frac{4}{3}E \qquad (4\text{-}130)$$

Substituting these values into Eq. (4-123) and collecting terms, the final expression for the complete solution of the series RLC circuit to a constant source is obtained. Thus

$$v = E(1 + \tfrac{1}{3}\varepsilon^{-4t} - \tfrac{4}{3}\varepsilon^{-t}) \qquad (4\text{-}131)$$

It is interesting to note here that the dynamic response (as described by the transient component of the solution) consists of two exponentially damped natural modes. A graphical representation appears in Fig. 4-11. The exponential term associated with the larger value of s (namely, -4) also has the smaller coefficient. Consequently, the total effect of this term is small and short-lived.

Complex Roots. In this next situation the circuit depicted in Fig. 4-12 is used. Although the configuration is the same as that of Fig. 4-10, a comparison reveals that the value of the resistance is reduced from 5 to 2 Ω. Again, let it be desired to find a complete description of the behavior of the voltage of the initially deenergized capacitor for all time following the application of the constant

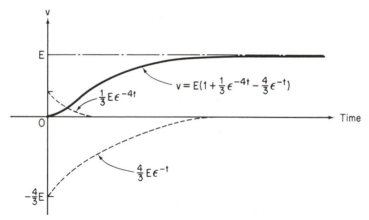

Fig. 4-11 Graphical representation of Eq. (4-131).

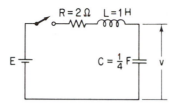

Fig. 4-12 Series *RLC* circuit with parameters selected for an underdamped response.

source function E. The operator form of the nonhomogeneous linear differential equation relating v to E is found by voltage division to be

$$v = \frac{4}{p^2 + 2p + 4}E \tag{4-132}$$

from which it follows that the forced component of the complete solution is

$$v_f = \left. \left| \frac{4}{p^2 + 2p + 4} \right| \right|_{p=0} E = E \tag{4-133}$$

To find the transient component of the total solution we must first find the values of s that satisfy the characteristic equation. Thus

$$D(p) = D(s) = s^2 + 2s + 4 = 0 \tag{4-134}$$

which has the solutions

$$s_{1,2} = -1 \pm j\sqrt{3} \tag{4-135}$$

It is important to note here that the roots are no longer real and distinct but are rather complex conjugates. This should lead us to expect some essential difference in the behavior of the circuit as it reacts to the application of the source function. This difference is readily borne out by writing the expression for the transient response in terms of the natural modes associated with the complex conjugate roots of the characteristic equation. Hence

$$v_t = K_1 \varepsilon^{s_1 t} + K_2 \varepsilon^{s_2 t} = \varepsilon^{-t}(K_1 \varepsilon^{j\sqrt{3}t} + K_2 \varepsilon^{-j\sqrt{3}t}) \tag{4-136}$$

Applying Euler's formula ($\varepsilon^{j\theta} = \cos\theta + j\sin\theta$) to the last expression and collecting terms yields

$$v_t = \varepsilon^{-t}[(K_1 + K_2)\cos\sqrt{3}t + j(K_1 - K_2)\sin\sqrt{3}t] \qquad (4\text{-}137)$$

Equation (4-137) can be written in a simpler manner still. Keep in mind that the left side of this equation is a real quantity. Moreover, v_t is associated with a differential equation that has real coefficients. Therefore, the right side of Eq. (4-137) must also necessarily have real coefficients so that the presence of a term such as $j(K_1 - K_2)$ merely means that $K_1 - K_2$ must assume such a value as to lead to a real coefficient. For ease of handling, then, Eq. (4-127) may be rewritten more simply as

$$v_t = \varepsilon^{-t}(k_1\cos\sqrt{3}t + k_2\sin\sqrt{3}t) \qquad (4\text{-}138)$$

where k_1 and k_2 denote real numbers that can be evaluated directly from the initial conditions.

The complete expression for the total solution thus becomes

$$v = v_f + v_t = E + \varepsilon^{-t}(k_1\cos\sqrt{3}t + k_2\sin\sqrt{3}t) \qquad (4\text{-}139)$$

Upon applying the initial conditions to Eq. (4-139) as well as to appropriate manipulated forms of it, evaluating k_1 and k_2 becomes straightforward. Thus at time $t = 0^+$, Eq. (4-139) becomes

$$v(0^+) = 0 = E + k_1$$

or

$$k_1 = -E \qquad (4\text{-}140)$$

To find k_2 it is necessary to impose a second initial condition. Since the circuit of Fig. 4-12 differs from that of Fig. 4-10 only in the value of the resistance, the initial value of the change in voltage across the capacitor continues to remain valid. In other words Eq. (4-128) applies here, too. However, before we can use this information, the expression for the derivative of the total solution is needed. Differentiating Eq. (4-139) leads to

$$\frac{dv}{dt} = \varepsilon^{-t}[(\sqrt{3}k_2 - k_1)\cos\sqrt{3}t - (\sqrt{3}k_1 + k_2)\sin\sqrt{3}t] \qquad (4\text{-}141)$$

At $t = 0^+$,

$$\frac{dv}{dt} = 0 = \sqrt{3}k_2 - k_1 \qquad (4\text{-}142)$$

from which it follows that

$$k_2 = \frac{k_1}{\sqrt{3}} = -\frac{E}{\sqrt{3}} \qquad (4\text{-}143)$$

Upon inserting Eqs. (4-140) and (4-143) into Eq. (4-139), the final expression for the complete solution results. Hence

$$v = E\left[1 - \varepsilon^{-t}\left(\cos\sqrt{3}t + \frac{1}{\sqrt{3}}\sin\sqrt{3}t\right)\right] \qquad (4\text{-}144)$$

An alternate expression follows from combining the cosine and sine terms into a single cosine term. The result is

$$v = E\left[1 - \frac{2}{\sqrt{3}}\varepsilon^{-t}\cos(\sqrt{3}t - 30°)\right]$$ (4-145)

Fig. 4-13 shows a plot of this equation.

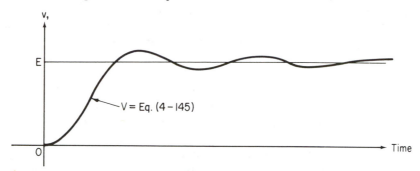

Fig. 4-13 Graphical representation of the underdamped response of a series *RLC* circuit to a constant applied source.

A comparison of the response of the series *RLC* circuit to an applied constant source for the case of complex roots with that found for real and distinct roots reveals that the essential difference lies in the *oscillatory* nature of the response associated with the complex roots. What happens physically in such a case is that during the transient state an interchange of energy occurs between the two independent energy-storing elements. Such an interchange can be shown to take place whenever the damping in the circuit (here represented by the value of resistance) is below a critical value determined by the circuit parameters of the energy-storing elements. A more detailed analysis is presented in the next chapter.

Summary review questions

1. Describe the three parts that are associated with the classical solution of linear integrodifferential equations.
2. Describe the role of the differential operator in effecting the solution of a linear integrodifferential equation.
3. Define operational impedance. What is the operational impedance of a capacitor? Of an inductor?
4. Distinguish between operational impedance and driving-point impedance.
5. Draw a network of circuit elements and illustrate the meaning of a transfer function that relates (a) voltage to voltage, (b) current to voltage, (c) voltage to current, and (d) current to current.
6. The presence of a source function in a network causes a response to occur in the various parts of the network. Distinguish between the source function and the forcing function in producing the network response.

7. Describe the procedure that is involved in determining a specific operational network function.

8. What is meant by the forced solution of a linear network? How is it characterized generally?

9. What is the transient solution of a linear network? What are the factors that enter into the characterization of this solution?

10. In a linear network identify the general form of the network response function (say, for current or voltage) to a constant source function.

11. In a linear network identify the general form of the network response function to an exponential source. Repeat for a sinusoidal source.

12. What is the role of the transient part of the total solution to the integrodifferential equation of a network driven by a source function?

13. Is a transient solution component always required when a source function is applied to a network that contains energy storing elements? Explain.

14. It is said that the transient solution of a network with energy-storage elements has its origin in these circuit elements. Explain the meaning of such a statement.

15. Explain why the transient solution of a linear network is always characterized by the exponential function.

16. What is the characteristic equation of a circuit? Is it necessary for the circuit to contain energy-storing elements to obtain the characteristic equation?

17. Describe the general procedure for finding the characteristic equation of a network that is described by a linear integrodifferential equation.

18. A linear network contains two energy-storing elements. Sketch the general behavior of the transient solution when the circuit parameters are set to yield an overdamped response.

19. What is the general form of the transient solution of a linear second-order network which is underdamped?

20. Discuss the relationship between the *differential operator p* and the *algebraic multiplier s*.

Problems

GROUP I

4-1. Find the driving-point operational impedance appearing between terminal pair *ab* in the circuit depicted in Fig. P4-1.

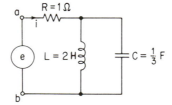

Fig. P4-1

Circuit Differential Equations: Forms and Solutions   Chap. 4

4-2. An applied source function e causes a response i to flow in a network. The governing differential equation that relates i to e is

$$2\frac{d^2i}{dt^2} + 6\frac{di}{dt} + 3i = 2\frac{d^2e}{dt^2} + 3e$$

    (a) Write this equation in operator form.

    (b) Obtain the expression for i entirely in terms of e and the differential operator.

4-3. Refer to the circuit of Fig. P4-1.

    (a) Working entirely in terms of Kirchhoff's circuit laws, determine the differential equation that relates the current in the inductor, i_L, to the source voltage e. Assume that e is a time-varying function.

    (b) Repeat part (a) but this time work in terms of the p operator and appropriate operational impedances.

4-4. Refer to the circuit of Fig. P4-1.

    (a) Apply Kirchhoff's laws and obtain the differential equation that relates the current in the capacitor, i_C, to the source voltage e. Assume that e is a time-varying function.

    (b) Obtain the result of part (a) by working with the p operator and the appropriate operational impedances of the circuit.

4-5. Refer to the circuit depicted in Fig. P4-1.

    (a) Apply Kirchhoff's laws and obtain the differential equation relating the response current i to the source voltage e, where e is time varying.

    (b) Repeat part (a) but now work with the p operator and appropriate operational impedances.

4-6. The circuit of Fig. P4-6 is initially deenergized and the source voltages e_1 and e_2 are time varying.

    (a) Obtain the differential equation that relates the voltage v_b to e_1 and e_2.

    (b) Identify the expression for the total forcing function that produces the response v_b.

4-7. The time-varying voltages e_1 and e_2 are applied as shown in Fig. P4-7.

    (a) Write the expressions for the operational impedances Z_{bd} and Z_{bc}.

    (b) Find the driving-point operational impedances Z_{ad} and Z_{cd}.

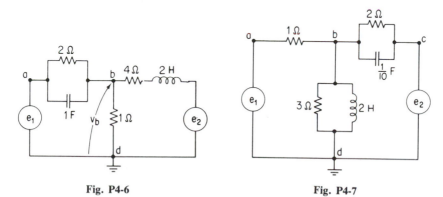

Fig. P4-6   **Fig. P4-7**

4-8. Refer to Fig. P4-7 and consider that e_2 is removed and replaced with a short circuit. Determine the differential equation that relates the current that flows through the 1-ohm resistor to the source function e_1.

4-9. Consider that e_1 in Fig. P4-7 is removed and replaced with a short circuit. Derive the differential equation that relates the current, i_2, to the source e_2.

4-10. Refer to the circuit depicted in Fig. P4-10.
  (a) Find the operational network function that relates v to e.
  (b) In part (a) write the expression for the differential equation relating v to e.
  (c) Determine the operational network function that relates the capacitor current to the inductor current.
  (d) Find the differential equation that relates the capacitor current to the inductor current.

4-11. Refer to Fig. P4-11.
  (a) Find the differential equation relating v to e.
  (b) Find the differential equation that describes the variation of the resistor current to the capacitor current.

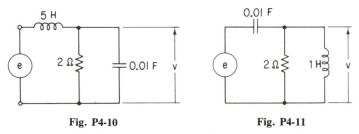

Fig. P4-10   Fig. P4-11

4-12. In the circuit depicted in Fig. P4-12 determine the differential equation that relates the voltage across the capacitor v to the source voltage e.

4-13. In the circuit of Fig. P4-13 obtain the differential equation that relates v to the source voltage e.

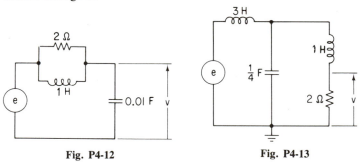

Fig. P4-12   Fig. P4-13

4-14. Obtain the differential equation that describes the variation of the current i in the circuit of Fig. P4-14 for all time after applying the source voltage e.

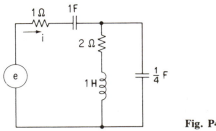

Fig. P4-14

4-15. Refer to the circuit appearing in Fig. P4-15. Determine the describing differential equation that relates the current i_2 to the source current i for all time following application of the source to the circuit.

4-16. A source current I is applied in the circuit of Fig. P4-16.

(a) Show that i_1 is related to the source by the differential equation

$$2\frac{d^2i_1}{dt^2} + 13\frac{di_1}{dt} + 25i_1 = 2\frac{d^2I}{dt^2} + 12\frac{dI}{dt} + 20I$$

(b) Obtain the differential equation that relates i_2 to I.

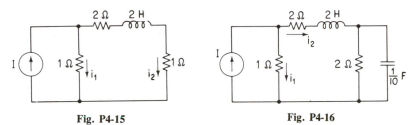

Fig. P4-15   Fig. P4-16

4-17. A constant battery source E is applied to the circuit configuration shown in Fig. P4-17.

(a) Obtain the differential equation that relates the response current i to the source E.

(b) Employ the procedure represented by Eq. (4-58) and find the forced response i_f caused by E.

(c) Justify the validity of the solution found in part (b) from physical considerations relating to the steady-state behavior of circuit elements.

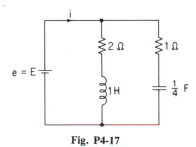

Fig. P4-17

4-18. Refer to the circuit of Fig. P4-10.

(a) Find the forced solution for the voltage across the capacitor when $e = E =$ constant source. Use the technique represented by Eq. (4-58).

(b) Determine the forced solution for the current that flows through the 2-Ω resistor when $e = E =$ constant.

4-19. The source voltage in Fig. P4-11 is a constant quantity of 12 V.

(a) Find the differential equation that relates v to the source voltage.

(b) What is the value of the forced solution?

(c) Is the result in part (b) consistent with physical arguments?

4-20. The source voltage in the circuit of Fig. P4-13 has the constant value E.

(a) Determine the differential equation that relates v to E.

(b) Find the value of the forced solution for v.

(c) Justify the result of part (b) on the basis of physical arguments.

4-21. The source current in Fig. P4-16 has a constant value of 5 A.
   (a) Obtain the differential equation that relates i_1 to I.
   (b) What is the value of the forced solution for i_1?

4-22. Refer to the circuit depicted in Fig. P4-7. Consider e_1 to be short circuited and e_2 to be equal to a constant E.
   (a) Show that the differential equation relating the source current to the source voltage is

$$6\frac{d^2i}{dt^2} + 110\frac{di}{dt} + 30i = 15E$$

   (b) Find the particular solution of the foregoing differential equation.
   (c) Check the validity of the solution of part (b) from a physical consideration of the properties of circuit elements.

4-23. Refer to Fig. P4-10. Assume that $e = \varepsilon^{-t}$.
   (a) Find the differential equation that relates v to e.
   (b) Determine the expression for the forced component of the solution to the equation of part (a).
   (c) Repeat part (b) for $e = \varepsilon^{-10t}$.
   (d) Repeat part (b) for $e = \varepsilon^{+10t}$.

4-24. A source function e produces a response i in accordance with the differential equation

$$2\frac{d^2i}{dt^2} + 6\frac{di}{dt} + 2i = 2\frac{d^2e}{dt^2} + 2e$$

   (a) Write the expression for the homogeneous differential equation.
   (b) Obtain the characteristic equation.
   (c) Determine the expression for the complementary solution and identify each natural mode.

4-25. A constant source voltage $e = 12$ V is applied to a series combination of R and C. The capacitor is initially deenergized.
   (a) Write the expression for the operational input impedance.
   (b) Obtain the differential equation from which the total solution for the response current i can be determined.
   (c) Find the forced current response.
   (d) Find the complete solution for the current response.

4-26. The capacitor in the circuit of Fig. P4-26 has an initial charge of 10 V with the polarity indicated.
   (a) Obtain the complete solution for the current response i as a function of time after the switch is closed.
   (b) Plot the result of part (a).
   (c) What is the expression that describes the variation of the capacitor voltage with time after the switch is closed?

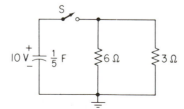

Fig. P4-26

4-27. For the circuit depicted in Fig. P4-11 show that the characteristic equation for describing the dynamic behavior of v to the source function e is identical to the one that describes the dynamic behavior of the capacitor current to the source function e.

4-28. Refer to the circuit of Fig. P4-6 and determine whether or not the complementary function is characterized by an oscillatory response. Identify the natural modes specifically.

4-29. A response i is related to a source function e by the following network function:

$$i = \frac{16p^2 + 80p + 30}{12p^2 + 220p + 60} e$$

Does the transient solution exhibit an oscillatory response? Explain.

GROUP II

4-30. Refer to the circuit of Fig. P4-6. Find the expression for the forced solution for v_b when $e_1 = 100\varepsilon^{-0.5t}$ and $e_2 = 50\varepsilon^{-2t}$.

4-31. Refer to the circuit of Fig. P4-11 and assume that $e = 5 \sin 10t$. Find the expression for the steady-state solution of v.

4-32. A response function v is related to a source function e by the differential equation

$$3\frac{d^3v}{dt^3} + 6\frac{d^2v}{dt^2} + 16\frac{dv}{dt} + 8v = 8e$$

Find the particular solution to this equation when $e = 10 \sin t$.

4-33. Obtain the expression for the forced solution for the current i_2 in the circuit of Fig. P4-16 when $I = 4 \sin 2t$.

4-34. In the circuit of Fig. P4-7 assume that e_2 is removed and replaced by a short circuit and $e_1 = 5 \sin 5t$. Obtain the expression for the forced current response that flows through the 1-Ω resistor.

4-35. The source current in the circuit of Fig. P4-15 is known to be a ramp function, $I = 10t$.
  (a) Find the expression for the forced solution for the current i_1.
  (b) Find the expression for the forced solution for the current i_2.

4-36. Repeat Prob. 4-35 for the case where $I = 5 + 10t$.

4-37. A response function v is related to a source function e by the differential equation

$$3\frac{d^3v}{dt^3} + 6\frac{d^2v}{dt^2} + 16\frac{dv}{dt} + 8v = 8e$$

where $e = 4t$. Obtain the expression for the forced solution.

4-38. Both energy-storing elements are initially deenergized in the circuit of Fig. P4-38. The switch is closed at $t = 0$. Obtain the following:
  (a) The initial current through the capacitor.
  (b) The initial current through the inductor.
  (c) The value of di_1/dt at $t = 0^+$.
  (d) The value of di_2/dt at $t = 0^+$.
  (e) The linear differential equation that relates the current i to the source E.
  (f) The forced current response.
  (g) The characteristic equation of the circuit.

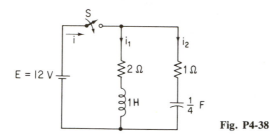

Fig. P4-38

(h) The complete current response.

(i) A carefully drawn plot of part (h).

4-39. Repeat Prob. 4-38 for the case where the battery is replaced by the source $e = 12\varepsilon^{-0.1t}$.

4-40. A ramp function $e = 2t$ is applied to the circuit of Fig. P4-40 at time $t = 0$. The capacitor and inductor have zero initial energy.

(a) Find the forced current response component of i.

(b) Compute the natural modes of the current response.

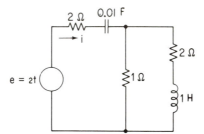

Fig. P4-40

chapter five

Circuit Dynamics and Forced Responses

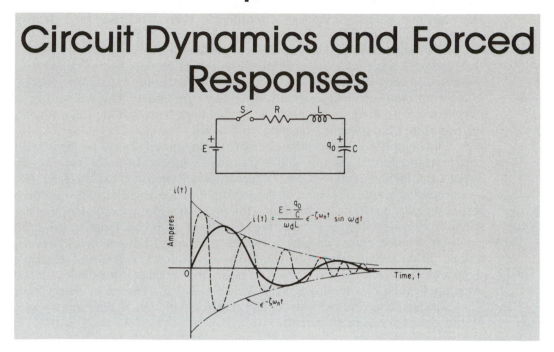

$$i(t) = \frac{E - \dfrac{q_0}{C}}{\omega_d L} \epsilon^{-\zeta \omega_n t} \sin \omega_d t$$

The study of the dynamic behavior of linear circuits and systems containing one or more energy-storing elements is of considerable importance to the engineer for two reasons. First, he often wants to know how long it takes for the circuit to respond to applied source functions. Recall that the energy in an energy-storing element such as an inductor or capacitor cannot change instantaneously. Accordingly, if an applied forcing function demands an increase in the amount of energy storage in a particular part of a circuit, this increase must occur gradually. The change cannot take place in a discontinuous manner, because then infinite forces would be needed. The period of adjustment during which the stored energy changes from some initial level to a new, commanded, final level is called the *settling time* of the circuit (or system). In many engineering applications it is important to keep the response time within tolerable limits.

Second, in situations where there are two or more energy-storing elements the engineer must be able to predict the occurrence of severe oscillations as the circuit (or system) changes from one energy state to another. In electric circuits such oscillations can readily cause ruinous voltages or currents. In electromechanical systems, such as a servomechanism, these oscillations can cause excessively high torques which in turn may damage mechanical parts. In fact, in some extreme cases, the interchange of energy during the transient state is such that the oscillations started by the application of the forcing function do not cease. Accordingly, the new steady-state level is never reached. The system or circuit is then described as being in a state of sustained oscillations. Whether or not this is a desirable state of affairs depends upon the objective which the engineer has in mind. If his intention is to build an oscillator, then his goal is achieved. However, if his intention is to build a follow-up control system which reaches its new steady-state position after the elapse of a reasonable amount of time, then clearly a sustained oscillation must be avoided at all costs.

It is the purpose of this chapter to furnish the background which will allow the reader to determine the complete response of circuits and systems, when subjected to conventional forcing functions. One commonly used source function is the step input of voltage or current or force or torque. The sinusoid is another frequently used source function. In addition, there are the ramp function $[f(t) = t]$ and the parabolic function $[f(t) = t^2]$.

The complete response of a circuit to an input can always be identified in two parts—a forced solution and a transient solution. In linear systems and circuits the forced solution is readily recognizable because it has the same form as the forcing function. Hence, if the forcing function is a sinusoid, the forced response must itself be a sinusoid. However, the character of the transient solution remains the same irrespective of the type of forcing function used. The transient solution makes possible a smooth transition from one energy level to another in the circuit or system. The transient solution is a means of describing the manner in which the circuit reacts to satisfy the boundary (or initial) conditions upon application of a forcing function which calls for a change in energy state. Thus the characteristic behavior of the transient solution is determined solely by the circuit (or system) parameters.

First-order circuits

5-1 STEP RESPONSE OF AN *RL* CIRCUIT

Any circuit composed of resistance and inductance represents a situation in which there occurs a dissipation of energy as well as a storage of energy in a magnetic field. Since we now know how to deal with both voltage and current sources, the step response of the *RL* circuit is found for each source type. Attention is directed first to the step-voltage response.

Appearing in Fig. 5-1 is the circuit arrangement of an initially deenergized inductor in series with a resistor. It is desired to find the complete solution for

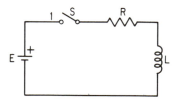

Fig. 5-1 Series *RL* circuit.

the current when the switch S is closed. The governing differential equation results upon applying Kirchhoff's voltage law to the circuit of Fig. 5-1. Thus

$$E = Ri + L\frac{di}{dt} \tag{5-1}$$

The current i is the solution being sought and E is the applied forcing function which causes the response i to exist.

Replacing the time derivative by the operator p, Eq. (5-1) can be written as

$$E = Ri + Lpi = i(R + pL) \tag{5-2}$$

so that the expression for the response becomes

$$i = \frac{1}{R + pL}E = \left(\frac{1}{1 + p\dfrac{L}{R}}\right)\frac{E}{R} \tag{5-3}$$

In accordance with Eq. (4-58), the forced solution is readily determined to be

$$i_f = \left(\frac{1}{1 + p\dfrac{L}{R}}\right)_{p=0}\frac{E}{R} = \frac{E}{R} \tag{5-4}$$

which is entirely in keeping with expectations resulting from a physical examination of this circuit.

The transient solution of Eq. (5-1) is found by employing the procedure described in Sec. 4-6. The transient solution, which is the solution to the homogeneous version of the governing differential equation, must be an exponential of the form $i_t = K\varepsilon^{st}$. The quantity s is determined from the characteristic equation, which results upon inserting the assumed form of the transient solution into the differential equation. Thus

$$L\frac{di_t}{dt} + Ri_t = 0 \tag{5-5}$$

or

$$\begin{aligned} LsK\varepsilon^{st} + RK\varepsilon^{st} &= 0 \\ (sL + R)K\varepsilon^{st} &= 0 \end{aligned} \tag{5-6}$$

A nontrivial solution results when

$$sL + R = s\frac{L}{R} + 1 = 0 \tag{5-7}$$

This is the characteristic equation for the circuit of Fig. 5-1. Note that the characteristic equation could have been written directly from an inspection of Eq. (5-3) by merely setting the denominator expression involving p equal to zero. Thus

$$D(p) = D(s) = 1 + s\frac{L}{R} = 0 \tag{5-8}$$

Recall that the differential operator p may be replaced by the algebraic multiplier

s whenever we are dealing with exponential functions. The pertinent value of s clearly is

$$s = -\frac{R}{L} \tag{5-9}$$

so that the transient solution becomes

$$i_t = K\varepsilon^{-(R/L)t} \tag{5-10}$$

Combining this result with the forced solution of Eq. (5-4) yields the complete solution. Hence

$$i = i_f + i_t = \frac{E}{R} + K\varepsilon^{-(R/L)t} \tag{5-11}$$

There now remains the task of finding K from the initial condition.

Evaluation of initial conditions. Information concerning the initial value of the response, and its derivatives when required, is always derivable from the defining differential equation. Sometimes there is need for appropriate algebraic manipulation of the differential equation before the specific information can be deduced, but the important thing is that this equation can always be made to yield the required information. In the case at hand, for example, knowledge of the value of the response immediately after switching, $i(0^+)$, is obtained by integrating both sides of Eq. (5-1) with respect to time from 0^- (the time just before switching) to 0^+ (the time just after switching). Thus

$$\int_{0^-}^{0^+} E\,dt = \int_{0^-}^{0^+} Ri\,dt + L\int_{0^-}^{0^+} \frac{di}{dt}\,dt \tag{5-12}$$

Now note that the quantity

$$\int_{0^-}^{0^+} E\,dt = 0$$

because for a constant integrand the net contribution of the integral from 0^- to 0^+ is infinitesimally small. A similar reasoning applies for the first term on the right side for a finite value of i. Accordingly, Eq. (5-12) may be written as

$$0 = 0 + L\int_{0^-}^{0^+} di \tag{5-13}$$

or

$$0 = L[i(0^+) - i(0^-)] \tag{5-14}$$

It follows then that

$$i(0^+) = i(0^-) \tag{5-15}$$

Therefore Eq. (5-15) states that the current flowing through the *RL* circuit immediately after the switch is closed is equal to the current existing before switching. This is another way of stating that the current through an inductance cannot change instantaneously—a result which we encountered in Chapter 2. Of course we chose to arrive at Eq. (5-15) as described above in order to illustrate the manner in which information about initial conditions can be determined from the appropriate differential equation. A little thought should make it apparent that Eq. (5-1) was integrated with respect to time between 0^- and 0^+ just so that the quantities of interest, $i(0^+)$ and $i(0^-)$, could be brought into evidence.

Because the switch in Fig. 5-1 is initially open, it follows that $i(0^-) = 0$. Hence, by Eq. (5-15), $i(0^+)$ is also zero so that Eq. (5-11) may now be explicitly written as

$$i(0^+) = 0 = \frac{E}{R} + K \tag{5-16}$$

from which it follows that

$$K = -\frac{E}{R} \tag{5-17}$$

Upon inserting this value into Eq. (5-11), there results

$$\boxed{i(t) = \frac{E}{R} - \frac{E}{R}\varepsilon^{-(R/L)t}} \qquad \text{for } t \geqslant 0 \tag{5-18}$$

Equation (5-18) is the complete solution of Eq. (5-1), which is the governing differential equation for the circuit of interest. Keep in mind that it consists of two parts. One part is E/R, which is the *forced* solution. It depends directly upon the forcing function E and has the same form as E—that is, a constant. The forced solution is often called the *steady-state* solution because, after all transient terms have virtually disappeared, it is the only current which remains for the circuit of Fig. 5-1 with the switch closed. Mathematicians prefer to call this component of the solution the *particular* solution. The second part of the total solution is one which decays with time. Although it requires infinite time for this term to reach zero, in practice it takes a relatively short time for it to become negligibly small. As a matter of fact, when the power of ε has a value of 5, this term is reduced to less than 1 per cent of its maximum value. In most situations this is equivalent to saying that the exponentially decaying term is for all intents and purposes equal to zero. An examination of Eq. (5-18) reveals that the presence of the transient term in the solution is demanded by the need to satisfy the boundary conditions immediately upon application of the forcing function. That is, the forcing function E wants to establish the forced solution E/R, but the presence of the inductance prevents this from occurring instantly. As a result the circuit reacts in such a way as to assure a smooth transition from the initial to the final energy state. This is achieved through the generation of a component of the solution, which not only satisfies the boundary conditions at $t = 0^+$, but also permits the forced solution to exist after a suitable period of adjustment. Examination of Eq. (5-10) shows that at $t = 0^+$, i_t has that value $(-E/R)$ which it must have so that the total current at this time instant be zero as called for by the boundary condition. Equation (5-18) makes this obvious. However, note that as time elapses the transient term decays to zero, leaving just the forced solution. Of course, as i_t decays, time is allowed for transferring energy to the inductance. When steady state is virtually reached, the energy stored in the inductor is

$$W = \frac{1}{2}L\left(\frac{E}{R}\right)^2 \qquad \text{J} \tag{5-19}$$

It is because this energy cannot be transferred instantaneously that the need for the transient term arises.

Inspection of Eq. (5-10) makes it clear that the decay of the transient term is solely dependent upon the circuit parameters—R and L—and entirely independent of the source function. This is certainly not unexpected when it is recalled that the transient term is really a description of the manner in which the circuit reacts to external disturbances, whatever their origin or nature. The only influence that the forcing function has on the transient term is in determining the magnitude of the coefficient of the exponential, but it in no way influences the rate of decay of the transient term.

Time Constant. A plot of Eqs. (5-18) and (5-10) appears in Fig. 5-2. Note that at $t = 0^+$ the steady-state and the transient solutions have equal and opposite values, thus yielding a zero value for the total current as called for by the boundary condition. Note also that as the transient term decays to zero, the total current reaches its forced value. In the interest of establishing a convenient measure of the duration of the transient, let us look more closely at the exponential function of Eq. (5-10). A glance at the power of the exponential, which must be a numeric, reveals that the quantity L/R must in turn bear units of time. Because of this fact and because the quantity is determined solely in terms of the circuit parameters, it is called the *time constant* of the circuit and is denoted by T.

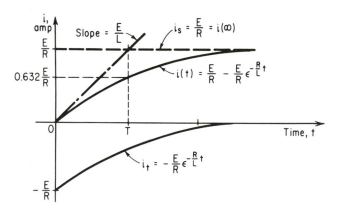

Fig. 5-2 Buildup of current in a series RL circuit.

Thus

$$\boxed{T \equiv \frac{L}{R}} \qquad \text{s} \tag{5-20}$$

A verification that L/R has units of seconds follows directly from the units for

Circuit Dynamics and Forced Responses   Chap. 5

L and R. Thus

$$T = \frac{L\left(\dfrac{\text{volts} \times \text{seconds}}{\text{ampere}}\right)}{R\left(\dfrac{\text{volts}}{\text{ampere}}\right)} = \frac{L}{R}(\text{seconds}) \tag{5-21}$$

To understand how the time constant serves as a convenient measure of the duration of the transient, one need merely observe that at $t = T$ the value of the transient term from Eq. (5-10) is

$$i_t(T) = -\frac{E}{R}\varepsilon^{-T/T} = -\frac{E}{R}\varepsilon^{-1} = -0.368\frac{E}{R} \tag{5-22}$$

Therefore in a period equal to one time constant the transient term has decayed to 36.8% of its initial value. Upon the elapse of a period equal to two time constants, the transient term has a value of

$$i_t(2T) = -\frac{E}{R}\varepsilon^{-2} = -0.135\frac{E}{R} \tag{5-23}$$

Similarly,

$$i_t(3T) = -\frac{E}{R}\varepsilon^{-3} = -0.0498\frac{E}{R} \tag{5-24}$$

and

$$i_t(5T) = -\frac{E}{R}\varepsilon^{-5} = -0.0067\frac{E}{R} \tag{5-25}$$

Thus after the elapse of three time constants the transient term is less than 5% of its initial value and after five time constants it is less than 1% of its initial value. Another way of describing the latter is to say that the response has reached more than 99% of its final value, which in most instances is considered to be steady state.

Another useful interpretation of the time constant follows from an investigation of the slope of the response curve at $t = 0^+$. Differentiating Eq. (5-18) with respect to time yields the general expression for the slope as a function of time. Thus

$$\frac{di}{dt} = \frac{E}{R}\left(\frac{R}{L}\right)\varepsilon^{-(R/L)t} = \frac{E}{R}\left(\frac{1}{T}\right)\varepsilon^{-(R/L)t} \tag{5-26}$$

The initial value of the slope is found upon substituting $t = 0^+$ into Eq. (5-26). Hence

$$\left.\frac{di}{dt}\right\}_{t=0^+} = \frac{E}{R}\left(\frac{1}{T}\right) \tag{5-27}$$

An examination of this expression discloses that, if the current were to continue to change at the rate shown by Eq. (5-27), it would require only one time constant to reach steady state. This statement is illustrated graphically in Fig. 5-2.

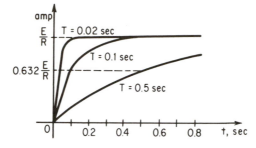

Depicted in Fig. 5-3 are several response curves which are drawn for various values of the time constant. Note the rapid rise to the forced solution for the case where $T = 0.02$ s. By using five time constants as a suitable measure of the duration of the transient term, it follows that the steady state is reached in about 0.1 s. When $T = 0.5$ s, it requires 2.5 s to reach steady state.

EXAMPLE 5-1 In the circuit depicted in Fig. 5-4, the battery has been applied for a long time. For the values of the parameters indicated find the complete expression for the current after closing the switch which removes R_1 from the circuit.

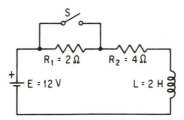

Fig. 5-4 Circuit arrangement for Example 5-1.

Solution: The governing differential equation that applies upon closing the switch is

$$E = iR_2 + L\frac{di}{dt} \tag{5-28}$$

In operator form this becomes

$$E = i(R_2 + pL) \tag{5-29}$$

where the expression in parentheses represents the operational impedance of the circuit. The current response is then

$$i = \left(\frac{1}{1 + p\dfrac{L}{R_2}}\right)\frac{E}{R_2} \tag{5-30}$$

The forced solution is obtained upon setting $p = 0$. Accordingly,

$$i_f = \left(\frac{1}{1 + p\dfrac{L}{R_2}}\right)_{p=0}\frac{E}{R_2} = \frac{E}{R_2} = \frac{12}{4} = 3 \text{ A} \tag{5-31}$$

The transient solution has the form $i_t = K\varepsilon^{st}$ where s is found from the characteristic equation, which in this case is

$$1 + p\frac{L}{R_2} = 1 + s\frac{L}{R_2} = 0 \tag{5-32}$$

Hence

$$s = -\frac{R_2}{L} = -\frac{4}{2} = -2 \tag{5-33}$$

so

$$i_t = K\varepsilon^{-(R_2/L)t} = K\varepsilon^{-2t} \tag{5-34}$$

The complete solution then becomes

$$i = i_f + i_t = 3 + K\varepsilon^{-2t} \tag{5-35}$$

To evaluate K, it is necessary to establish the initial condition prevailing just prior to switching. A glance at Fig. 5-4 shows that the current just before closing the switch is simply

$$i(0^-) = \frac{E}{R_1 + R_2} = \frac{12}{2 + 4} = 2 \text{ A} \tag{5-36}$$

Recalling that the current in an inductor cannot change instantaneously, it follows that

$$i(0^+) = i(0^-) = 2 \text{ A} \tag{5-37}$$

Inserting this result into Eq. (5-35) yields

$$i(0^+) = 2 = 3 + K \tag{5-38}$$

from which

$$K = -1 \tag{5-39}$$

The complete solution for the current thus becomes

$$i(t) = 3 - \varepsilon^{-2t} \tag{5-40}$$

As a check note that at $t = 0^+$ the value of $i(t)$ is 2 A and upon full decay of the transient the current assumes a value of 3 A.

The time constant for this circuit is

$$T = \frac{L}{R} = \frac{2 \text{ H}}{4 \, \Omega} = \frac{1}{2} \text{ s} \tag{5-41}$$

Thus after 2.5 s the transient term has a value less than 1% of its initial value of 1 A.

Step-Current Response of an RL Circuit. Attention is next turned to finding the complete response of an *RL* circuit to which is applied a step-current forcing function. Because we are now involved with current sources, the resistance and inductance are assumed placed in parallel with one another as illustrated in Fig. 5-5. Recall that, it is characteristic of a current source to deliver to the circuit elements connected to it, the total value of its current. Accordingly, to simulate the application of a step current to the *RL* combination, it is merely necessary to open switch S in Fig. 5-5. Clearly, when S is closed the short circuit draws all of the current from the source, leaving no current to flow through *R* and *L*.

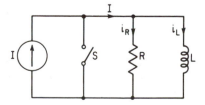

Fig. 5-5 Circuit for finding the step-current response of a parallel *RL* circuit.

When switch S is opened, the source current flows through the parallel combination of resistance and inductance. By Kirchhoff's current law we then have

$$I = i_R + i_L \tag{5-42}$$

where i_R denotes the current which flows through the resistance and i_L denotes the current flowing through the inductance. Also, because of the parallel arrangement, the potential differences appearing across R and L are the same. Therefore, we may further write

$$Ri_R = L\frac{di_L}{dt} \tag{5-43}$$

Inserting Eq. (5-42) into Eq. (5-43) leads to

$$RI = Ri_L + L\frac{di_L}{dt} \tag{5-44}$$

This expression is the governing differential equation for the circuit. Since R and I are both known, the left side of Eq. (5-44) is thereby known and effectively represents the forcing function. In fact the sole unknown in this equation is the current i_L. A comparison of Eq. (5-44) with Eq. (5-1) reveals them to be identical in form so that the solution may be taken to be Eq. (5-18) with the one modification that E is replaced by IR. Hence the complete solution can be written forthwith as

$$i_L = \frac{IR}{R}(1 - \varepsilon^{-(R/L)t}) = I(1 - \varepsilon^{-(R/L)t}) \tag{5-45}$$

The expression for the current through the resistor follows from Eq. (5-42):

$$i_R = I - i_L = I\varepsilon^{-(R/L)t} \tag{5-46}$$

A study of Eq. (5-45) shows that at $t = 0^+$ there is no current flowing through the energy-storing element L, which is consistent with the boundary condition. Keep in mind that with S closed $i(0^-) = 0$. At the same time, however, the expression for i_R shows that all of the source current flows through R. This can take place instantly because there is no energy storage in a resistive element. As time progresses the current I transfers from a path through R to a path solely through L. This is again borne out by Eqs. (5-45) and (5-46). After the elapse of five time constants, for all intents and purposes $i_L = I$ and $i_R = 0$. In the steady state the energy stored in the inductance is $\frac{1}{2}LI^2$.

5-2 STEP RESPONSE OF AN *RC* CIRCUIT

Appearing in Fig. 5-6 is the circuit arrangement of a resistor in series with a capacitor. The capacitor is assumed to have an initial charge of value q_0 owing to the flow of a previously applied current. Let us now find the complete expression for the current after the switch S is closed. The governing differential equation is obtained upon applying Kirchhoff's voltage law to the closed circuit. Thus,

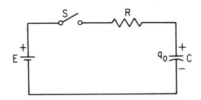

Fig. 5-6 Series RC circuit. The capacitor has an initial charge q_0.

$$E = Ri + \frac{1}{C} \int i \, dt \tag{5-47}$$

The indefinite integral in the last equation may be written as consisting of two parts:

$$\int i \, dt = \int_{-\infty}^{0} i' \, dt + \int_{0}^{t} i \, dt = q_0 + \int_{0}^{t} i \, dt \tag{5-48}$$

where the first term in the middle expression represents the accumulation of charge on the capacitor brought about by the flow of current (i') due to a previously applied action, and the second term denotes the further accumulation of charge due to the present action of closing the switch. This first term may be conveniently replaced by the initial charge q_0. Equation (5-47) can then be rewritten as

$$E = Ri + \frac{q_0}{C} + \frac{1}{C} \int_{0}^{t} i \, dt \tag{5-49}$$

The quantity q_0/C denotes the initial voltage appearing across the capacitor before closing the switch. By placing this quantity on the left side, it combines with the applied source voltage to yield a net driving force for the circuit which is $E - q_0/C$. Thus

$$E - \frac{q_0}{C} = Ri + \frac{1}{C} \int_{0}^{t} i \, dt \tag{5-50}$$

Expressed in terms of the differential operator, Eq. (5-50) becomes

$$E - \frac{q_0}{C} = Ri + \frac{1}{pC} i = i \left(R + \frac{1}{pC} \right) \tag{5-51}$$

where the parenthetical expression denotes the operational impedance of a series RC circuit. The expression for the current response is then

$$i = \frac{E - \dfrac{q_0}{C}}{R + \dfrac{1}{pC}} = \frac{E - \dfrac{q_0}{C}}{R(1 + 1/pRC)} \tag{5-52}$$

Again, by employing the procedures described in detail in Secs. 4-5 and 4-6, the forced and natural responses are readily found from the last equation. Accordingly, for the forced solution

$$i_f = \left(\frac{E}{R} - \frac{q_0}{RC} \right) \left(\frac{1}{1 + 1/pRC} \right)_{p=0} = 0 \tag{5-53}$$

Moreover, by setting the denominator expression of Eq. (5-52) equal to zero, the circuit characteristic equation results. Thus

$$1 + \frac{1}{pRC} = 1 + \frac{1}{sRC} = 0 \tag{5-54}$$

Hence

$$s = -\frac{1}{RC} \tag{5-55}$$

The expression for the transient solution thus becomes

$$i_t = K\varepsilon^{-t/RC} \tag{5-56}$$

Combining Eqs. (5-53) and (5-56) yields the complete solution for the current response:

$$i = i_f + i_t = i_t = K\varepsilon^{-t/RC} \tag{5-57}$$

To evaluate K, it is necessary to know the initial value of the current at $t = 0^+$. A physical argument will now be used for this determination. Refer to Fig. 5-6. When the switch is closed, the source voltage E is applied to the series combination of R and C. The voltage across C, however, cannot change instantaneously because this requires a finite change in energy. Hence it remains at the value q_0/C. The excess voltage† between E and q_0/C must therefore appear across R and this in turn produces a current limited by the value of R. Consequently,

$$i(0^+) = \frac{E - \dfrac{q_0}{C}}{R} = \frac{E}{R} - \frac{q_0}{RC} \tag{5-58}$$

Inserting the result of Eq. (5-58) into Eq. (5-57) yields

$$K = \frac{E}{R} - \frac{q_0}{RC} \tag{5-59}$$

The complete expression for the current response is then

$$\boxed{i(t) = \left(\frac{E}{R} - \frac{q_0}{RC}\right)\varepsilon^{-t/RC}} \tag{5-60}$$

A plot of this equation as a function of time is depicted in Fig. 5-7. Note that at $t = 0^+$ there occurs a step change in current in the circuit. Its magnitude is given by Eq. (5-58). As time elapses this current gradually decays to zero.

An inspection of the exponential term of Eq. (5-60) reveals that the time constant of an RC circuit is given by

$$\boxed{T = RC} \quad \text{s} \tag{5-61}$$

† It is instructive to note here that if the source voltage E were equal to q_0/C the initial and final values of current would be the same (i.e., zero), and so there would be no need for a transient current.

Circuit Dynamics and Forced Responses   Chap. 5

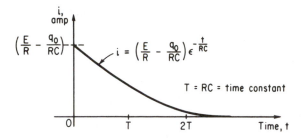

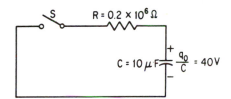

Fig. 5-7 Exponential decay of the current in an RC circuit.

The complete current solution in the case of the simple RC circuit consists merely of the transient term because the capacitor presents an open circuit to the battery source at steady state. Hence the forced solution is zero in this instance.

EXAMPLE 5-2   In the circuit of Fig. 5-8 the capacitor has an initial charge of such value that the voltage appearing across its plates is 40 V. For the values of the parameters indicated find the expression for the current which flows when the switch is closed. Also, find the total energy dissipated in the resistor during the transient state and compare it with the energy initially stored in the capacitor.

Fig. 5-8   Circuit configuration for Example 5-2.

Solution: Taking voltage drops in a clockwise direction as positive, Kirchhoff's voltage law applied to the circuit with S closed yields

$$0 = Ri + \frac{1}{C} \int i \, dt \tag{5-62}$$

A comparison of Eq. (5-62) with Eq. (5-50) discloses that it is identical except for the fact that E is zero. Accordingly, the complete current response for this case follows from Eq. (5-60) upon setting E to zero. Thus

$$i = -\frac{q_0}{RC}\varepsilon^{-t/RC} = -\frac{40}{R}\varepsilon^{-t/RC} \tag{5-63}$$

The time constant is

$$RC = 0.2 \times 10^6 \times 10 \times 10^{-6} = 2 \text{ s} \tag{5-64}$$

The current solution may therefore be written as

$$i(t) = -\frac{40}{0.2 \times 10^6}\varepsilon^{-t/2} = -2 \times 10^{-4}\varepsilon^{-t/2} \text{ A} \tag{5-65}$$

The negative sign indicates that the actual current flow is counterclockwise rather than clockwise as initially assumed.

The amount of energy dissipated in the resistor is readily found from

$$W_{\text{diss}} = \int_0^\infty i^2(t)R \, dt = \int_0^\infty (4 \times 10^{-8})2 \times 10^5 \varepsilon^{-t} \, dt$$

$$= -8 \times 10^{-3}[\varepsilon^{-t}]_0^\infty = 8 \times 10^{-3} \text{ J} \tag{5-66}$$

The amount of energy initially stored in the electric field of the capacitor is given by

$$W_e = \tfrac{1}{2}CV_0^2 = \tfrac{1}{2}10^{-5}(40)^2 = 8 \times 10^{-3} \text{ J} \tag{5-67}$$

A comparison of Eqs. (5-66) and (5-67) shows that all the capacitor energy is dissipated as heat after the switch is closed.

It does not always follow that all of the energy stored on a capacitor will be dissipated in a resistor appearing in series with it. This is illustrated by the following example.

EXAMPLE 5-3 In the circuit depicted in Fig. 5-9 assume that C_1 has an initial charge of such a magnitude that the potential difference appearing across its terminals is equal to 100 V with a polarity which makes the upper plate positive relative to the lower plate. There is no initial charge on C_2. Determine the voltage which appears across each capacitor at steady state.

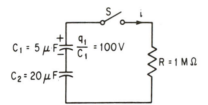

Fig. 5-9 Circuit configuration for Example 5-3.

Solution: Assuming a clockwise flow of current and writing Kirchhoff's voltage law for the circuit we have

$$0 = Ri + \frac{1}{C_2}\int_0^t i \, dt + \frac{1}{C_1}\int i \, dt \tag{5-68}$$

The last term is written as an indefinite integral to remind us that C_1 has an initial condition associated with it. Specifically this quantity can be expressed as

$$\frac{1}{C_1}\int i \, dt = -100 + \frac{1}{C_1}\int_0^t i \, dt \tag{5-69}$$

The negative sign appears because the assumed positive flow of current through C_1 is in a direction which brings about a charge displacement on the plates of C_1 having a polarity opposed to that of the initial charge. Inserting Eq. (5-69) into Eq. (5-68) and transposing the initial condition voltage to the left side of the equation yields

$$100 = Ri + \frac{1}{C_2}\int_0^t i \, dt + \frac{1}{C_1}\int_0^t i \, dt \tag{5-70}$$

Introducing the differential operator enables us to write

$$100 = Ri + \frac{1}{p}\left(\frac{1}{C_1} + \frac{1}{C_2}\right)i = i\left(R + \frac{1}{pC}\right) \tag{5-71}$$

where

$$\frac{1}{C} = \frac{1}{C_1} + \frac{1}{C_2} = \frac{1}{5} + \frac{1}{20} = \frac{1}{4} \tag{5-72}$$

or the equivalent series capacitance is

$$C = 4 \, \mu F \tag{5-73}$$

The nonhomogeneous linear differential equation in this example may then be rewritten in operator form as

$$(1 + pRC)i = p\frac{100}{R} \tag{5-74}$$

The forced response immediately follows from Eq. (5-74) upon setting p equal to zero. This yields

$$i_f = 0 \tag{5-75}$$

which is consistent with physical reasoning.

The homogeneous linear differential equation in operator form can be written by setting the right side of Eq. (5-74) to zero. Thus

$$(1 + pRC)i = 0 \tag{5-76}$$

The corresponding characteristic equation is then

$$1 + pRC = 1 + sRC = 0 \tag{5-77}$$

which leads to

$$s = -\frac{1}{RC} = -\frac{1}{4} \tag{5-78}$$

Accordingly, the associated transient response becomes

$$i_t = K\varepsilon^{-t/RC} = K\varepsilon^{-t/4} \tag{5-79}$$

The complete solution in this case is the same as the transient response because the forced response is zero. Hence

$$i = i_t = K\varepsilon^{-t/4} \tag{5-80}$$

When the switch is closed in the circuit of Fig. 5-9, the entire initial-condition voltage of 100 V appears across the 1-MΩ resistor because the voltage across the capacitors cannot change instantaneously and Kirchhoff's voltage law must be satisfied. Hence the current at time $t = 0^+$ is expressed by

$$i(0^+) = \frac{100}{R} = 10^{-4} \, A \tag{5-81}$$

Upon inserting this result into Eq. (5-80) it follows that the value of K is the same as $i(0^+)$. Accordingly, the complete expression for the clockwise-flowing current is

$$i(t) = 10^{-4} \, \varepsilon^{-t/4} \tag{5-82}$$

The net voltage appearing across C_1 is given by the expression

$$v_1(t) = -100 + \frac{1}{C_1} \int_0^\infty i \, dt = -100 + \frac{10^6}{5} 10^{-4} \int_0^\infty \varepsilon^{-t/4} \, dt \tag{5-83}$$

$$= -100 - \frac{4}{5}100 \left[\varepsilon^{-t/4} \right]_0^\infty = -100 + 80 = -20$$

The minus sign is taken with respect to the clockwise direction of current. Therefore, it means that a rise in voltage occurs as C_1 is traversed in the direction of i. In other words, the upper plate is 20 V positive with respect to the lower plate.

The potential difference appearing across C_2 is expressed in a similar fashion by

$$v_2(t) = \frac{1}{C_2} \int_0^t i \, dt = \frac{10^6}{20} \int_0^t 10^{-4} \varepsilon^{-t/4} \, dt$$

$$= 20(1 - \varepsilon^{-t/4})$$

(5-84)

Clearly at steady state (i.e., $t \rightarrow \infty$) the value of this voltage is

$$v_2(\infty) = 20 \text{ V}$$

(5-85)

The plus sign here means that the lower plate of C_2 is positive with respect to the upper plate.

The potential difference across each capacitor is 20 V with opposite signs. Consequently under these conditions the net potential difference across the resistor is zero, so that no further dissipation of the stored electric-field energy can take place. The energy is "trapped" by the action of the equal and opposite capacitor voltages.

Another interesting aspect of this example is revealed from a glance at the final expressions for v_1 and v_2. These equations show that the transient terms decay in accordance with an equivalent time constant which is determined by the circuit resistance and the equivalent capacitance as represented by Eq. (5-72).

Step-Current Response of an RC Circuit. Again in dealing with a current source the resistance and capacitance parameters are considered to be in parallel as illustrated in Fig. 5-10. As previously explained, a step-current forcing function is applied to the parallel combination by opening the switch. Then by Kirchhoff's current law we can write

$$I = i_R + i_C$$

(5-86)

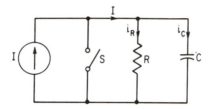

Fig. 5-10 Parallel RC circuit with step-current forcing function.

Moreover, the same potential difference appears across R and C. Hence

$$Ri_R = \frac{\int_0^t i_C \, dt}{C}$$

(5-87)

There can exist no initial charge on the capacitor because the switch S is initially assumed closed. Inserting Eq. (5-86) into Eq. (5-87) yields

$$RI = Ri_C + \frac{1}{C} \int_0^t i_C \, dt$$

(5-88)

A study of the last equation can be made to reveal information about the steady-state value of i_C without resorting to the operator form of the governing differential equation and setting p equal to zero. First note that the left side of the equation is a finite quantity, IR. Also note that, as time increases after opening the switch, the contribution from the integral term becomes larger and

Circuit Dynamics and Forced Responses   Chap. 5

larger. As time gets very large and approaches infinity, the only way the right side of the equation can remain finite so as to balance the left side is for the current to approach zero.

Our experience with differential equations having the form of Eq. (5-88) tells us that the capacitor carries a transient current given by

$$i_C = i_t = K\varepsilon^{-t/RC} \tag{5-89}$$

To evaluate K, we need the initial value of i_C. This too can be deduced from Eq. (5-88) upon inserting $t = 0^+$. In this instance the contribution from the integral term is infinitesimal so that it follows that

$$i_C(0^+) = I \tag{5-90}$$

Therefore the complete expression for the capacitor current is

$$i_C = I\varepsilon^{-t/RC} \tag{5-91}$$

The time solution for the current through the resistor is found from Eqs. (5-86) and (5-89). Thus

$$i_R = I - i_C = I(1 - \varepsilon^{-t/RC}) \tag{5-92}$$

In the configuration of Fig. 5-10 the total source current initially flows entirely through the capacitor and then, as time elapses, it gradually transfers to the resistor. At steady state all the current I flows through R and none through C.

5-3 DUALITY

Two circuits are said to be duals when the *mesh* equations that describe the behavior of one circuit are found to be identical in form to the *nodal* equations that describe the other. Duality therefore is a property of the circuit equations. There are times when the construction of the dual of a circuit is helpful in reducing the effort needed in analyzing simple, conventional circuits.

The origin of the principle of duality can be traced back to the laws of Ohm, Faraday, and Coulomb. In the case of a voltage e applied to a series resistor, R, we have, by Ohm's law,

$$e = Ri \tag{5-93}$$

However, an alternative form of this equation states that the current i is expressed as the product of the conductance G and the voltage e appearing across G. That is,

$$i = Ge \tag{5-94}$$

A comparison of Eqs. (5-93) and (5-94) makes it plain that the roles of voltage and current are interchanged. Moreover, when the roles of current and voltage are interchanged, then the resistance R is replaced by the conductance G.

Similarly, by Faraday's law we have the voltage and current of an inductor related by

$$e = L\frac{di}{dt} \tag{5-95}$$

For the capacitor the relationship is

$$i = C\frac{de}{dt} \tag{5-96}$$

Equation (5-95) is a mesh-oriented equation since it is an expression for voltage. On the other hand, Eq. (5-96) is a node-oriented equation because it is an expression for current. Again note how one may proceed from the mesh orientation to the node orientation by merely interchanging the roles for voltage and current in the describing equations, and also replacing the inductance parameter by the capacitance parameter. The reverse procedure is also valid. One useful feature of this duality property is that one can readily study, for example, the manner in which the voltage across a capacitor varies in response to a specific current source by simply investigating the variation of the current in an inductor subject to a voltage source which is the same type as the current source.

Further examples of the usefulness of duality can be cited. For example, it has already been demonstrated on several occasions *that the current through an inductor cannot change instantaneously*. A moment's thought should make it apparent that the dual statement is that *voltage across the capacitor cannot change instantaneously*. Here the dual words are *voltage* and *current, through* and *across, inductance* and *capacitance*. Similarly, we know that the voltage across an inductor can change instantly. The dual statement is that the current through a capacitor can change instantly—a result we are quite familiar with by now.

To consider the matter of duality further, consider the simple series *RL* circuit depicted in Fig. 5-11. The mesh equation for this circuit is clearly

$$e = Ri + L\frac{di}{dt} \tag{5-97}$$

Now, on the basis of the interchangeability of the roles of voltage and current, resistance and conductance, inductance and capacitance, we can write directly from Eq. (5-97) the current (or nodal) equation that identifies the dual circuit of Fig. 5-11. Thus

$$i = Ge + C\frac{de}{dt} \tag{5-98}$$

A circuit interpretation of this equation leads to the configuration appearing in Fig. 5-12. When the magnitude of *G* is selected equal to *R* and the magnitude of *C* is chosen equal to *L*, an *exact dual* circuit results. The solution for *e* in

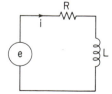

Fig. 5-11   Series *RL* circuit.

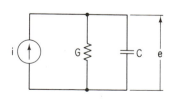

Fig. 5-12   Dual of the circuit shown in Fig. 5-11.

Eq. (5-98) becomes exactly the solution for i in Eq. (5-97). If i in Eq. (5-97) has been previously found, then by this principle of duality the solution for the variation of e to a current source in the circuit of Fig. 5-12 is also known. To illustrate this point, let it be desired that the complete expression for the voltage e be found corresponding to the application of a step current $i = I$ in the circuit of Fig. 5-12. The solution for the current response in Fig. 5-11 to a step voltage $e = E$ was determined in Sec. 5-1 and is given by Eq. (5-18). By the principle of duality then, the required solution may be written forthwith as simply

$$e(t) = \frac{I}{G}(1 - \varepsilon^{-(G/C)t}) \tag{5-99}$$

The identification of the dual of a circuit need not proceed in the manner described in the foregoing. Rather, a direct graphical procedure can be employed. Thus, to draw the dual of the circuit of Fig. 5-11 directly, begin by drawing a dashed line completely encircling the mesh circuit as illustrated in Fig. 5-13(a). This closed loop represents the reference node of the nodal method. Next, place a dot at the center of each mesh of the planar network. Each dot denotes a node of the dual circuit. Obviously, for the circuit of Fig. 5-11, there is just one node required. Then identify appropriate branches between the node points by using a dual element for each element that appears in the original mesh circuit. Accordingly, the voltage source e is replaced by the current source i, the resistance R is replaced by the conductance G, and the inductance L is replaced by the capacitance C. The resulting dual circuit is more clearly depicted in Fig. 5-13(b). The configuration is identical to that of Fig. 5-12.

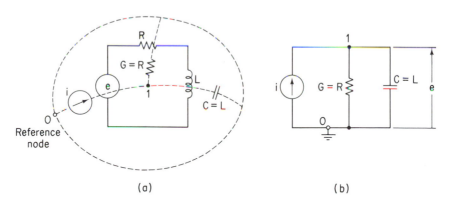

(a)   (b)

Fig. 5-13 (a) Illustration of graphical procedure for identifying the dual circuit; (b) the resulting dual circuit drawn in conventional form.

EXAMPLE 5-4 Find the exact dual of the two-mesh circuit depicted in Fig. 5-14. Use the graphical procedure.

Solution: Draw a dashed line enclosing both meshes and call this the reference node 0. Place two dots in each mesh and call these the nodes 1 and 2. Replace each element of the original circuit by its dual counterpart. The final result is shown in Fig. 5-15.

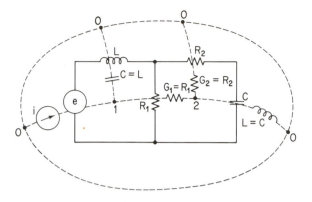

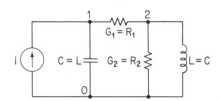

Fig. 5-14 Original two-mesh circuit of Example 5-4 together with graphical solution procedure for finding the dual circuit shown in dashed lines.

Fig. 5-15 Dual circuit of Fig. 5-14.

5-4 PULSE RESPONSE OF THE *RC* CIRCUIT

An interesting application of the theory of transients in *RC* circuits occurs when a rectangular pulse is applied to a series combination of *R* and *C* as depicted in Fig. 5-16. The capacitor is initially deenergized. Let it be desired to find the expressions describing the behavior of the current as well as the capacitor voltage for all time following the application of the pulse.

It is helpful at the outset to understand that any rectangular pulse can be considered as consisting of two step functions—one delayed with respect to the other. A glance at Fig. 5-17 makes this plain. The step function e_1 is assumed originating at time zero with magnitude E and existing to time infinity. The step function e_2 is assumed originating at time $t = T_1$ with amplitude $-E$ and then persisting for all time thereafter. The superposition of these two functions yields a resultant function which is a rectangular pulse having a magnitude E for the time interval $0 < t < T_1$ and zero for all other times.

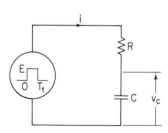

Fig. 5-16 Rectangular pulse applied to *RC* circuit.

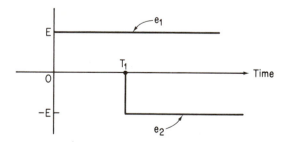

Fig. 5-17 Generation of the rectangular pulse by the superposition of two step functions—one delayed with respect to the other.

The solution for the current in the time interval from zero to T_1 is clearly the step response of the series RC circuit. This result is described by Eq. (5-60) upon setting q_0 equal to zero. Thus

$$i(t) = \frac{E}{R}\varepsilon^{-t/RC} \qquad \text{for } 0 < t < T_1 \qquad (5\text{-}100)$$

The voltage across the capacitor during this same time interval is

$$v_C = \frac{1}{C}\int_0^t i\, dt = \frac{1}{C}\int_0^t \frac{E}{R}\varepsilon^{-t/RC} dt = E(1 - \varepsilon^{-t/RC}) \qquad \text{for } 0 \le t \le T_1 \quad (5\text{-}101)$$

Moreover, the voltage across the resistor is simply described by

$$v_R = iR = E\varepsilon^{-t/RC} \qquad (5\text{-}102)$$

for the same period.

As time elapses and t becomes equal to T_1, voltage abruptly changes from $+E$ to 0. The voltage across the capacitor just prior to T_1, i.e., at $t = T_{1-}$, is given by Eq. (5-101) as

$$v_C(T_1-) = E(1 - \varepsilon^{-T_1/RC}) \qquad (5\text{-}103)$$

As the pulse voltage drops to zero, this capacitor voltage cannot change. Hence

$$v_C(T_1+) = v_C(T_1-) = E(1 - \varepsilon^{-T_1/RC}) \qquad (5\text{-}104)$$

However, since the resistor is not an energy-storing element, the voltage across it is allowed to change abruptly. In fact at time T_1+ the value of this voltage must be the negative of the capacitor voltage in order to satisfy Kirchhoff's voltage law. Thus

$$v_R(T_1+) = -v_C(T_1+) = -E(1 - \varepsilon^{-T_1/RC}) \qquad (5\text{-}105)$$

The resultant change in voltage that occurs across R in the time interval from T_{1-} to T_{1+} is readily computed by subtracting Eq. (5-102) evaluated at T_{1-} from Eq. (5-105). Thus

$$\Delta v_R|_{t=T_1} = v_R(T_1+) - v_R(T_1-) \qquad (5\text{-}106)$$
$$= -E(1 - \varepsilon^{-T_1/RC}) - E\varepsilon_1^{-T_1/RC} = -E$$

This result is entirely consistent with the fact that the change that occurs in the circuit when the pulse drops from $+E$ to 0 at T_1 must all take place across the resistor because the voltage across the capacitor cannot assume any part of this change.

The value of the current prevailing at $t = T_1+$ is found from Kirchhoff's voltage equation written for the circuit at this time instant. Thus

$$0 = i(T_1+)R + v_C(T_1+) \qquad (5\text{-}107)$$

The zero appears on the left side because the pulse value has dropped to zero at T_1+. Moreover, the capacitor has a value of voltage given by Eq. (5-104). Accordingly, it follows that

$$i(T_1+) = -\frac{v_C(T_1+)}{R} = -\frac{E}{R}(1 - \varepsilon^{-T_1/RC}) \qquad (5\text{-}108)$$

The negative sign here denotes that the current now flows in a direction opposite to that assumed in Fig. 5-16. In other words, the capacitor is discharging through the resistor with current flowing counterclockwise in Fig. 5-16.

As time elapses beyond T_1+, the magnitude of current represented by Eq. (5-108) merely decays in the usual manner as determined by the time constant of the circuit. If we introduce a new time variable t' that has its zero value at T_1, the current response can be expressed as

$$i(t') = -\frac{E}{R}(1 - \varepsilon^{-T_1/RC})\varepsilon^{-t'/RC} \qquad (5\text{-}109)$$

where

$$t' = t - T_1 \qquad (5\text{-}110)$$

and

$$t > T_1$$

The foregoing analysis thus makes it apparent that the current response over the total time domain must be described in two parts. For the time interval $0 < t < T_1$, Eq. (5-100) is valid. For $t > T_1$ it is Eq. (5-109) that provides the appropriate description.

In a similar manner the capacitor voltage for the time interval $0 \leq t \leq T_1$ is given by Eq. (5-101), whereas for time beyond this interval the expression is

$$v_C(t') = E(1 - \varepsilon^{-T_1/RC})\varepsilon^{-t'/RC} \qquad (5\text{-}111)$$

where again

$$t' = t - T_1$$

and

$$t \geq T_1$$

A plot of these results appears in Fig. 5-18. These curves are drawn on the assumption that T_1 is comparable to the circuit time constant RC.

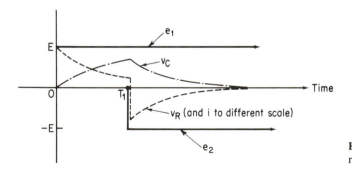

Fig. 5-18 Response of RC circuit to rectangular pulse.

5-5 THE IMPULSE RESPONSE

As a driving force, the impulse is a term often used to describe a situation where an extraordinarily large forcing function is applied to a circuit or system over a very small period of time. The impulse can readily be derived as the limiting case of a pulse. In this connection consider that the pulse depicted in Fig. 5-19(a) has its amplitude quadrupled and its period reduced by one-quarter as shown in Fig. 5-19(b). Note that the area is kept constant. Figure 5-19(c) shows the same pulse area but the amplitude is now further increased by a factor of 5. In the limiting case the height of the pulse is made to approach infinity while the pulse duration is made to approach zero for constant area. Such a function is called the *impulse function* and is represented by the symbolism of Fig. 5-19(d). In each case illustrated in Fig. 5-19 the area of the pulse is given by

$$A = ET_1 \qquad \text{V-s} \tag{5-112}$$

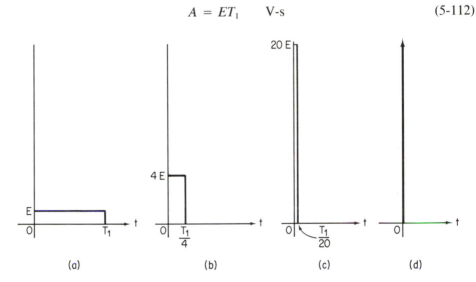

Fig. 5-19 Illustrating the impulse as the limiting case of the pulse: (a) normal pulse; (b) and (c) increased height and diminished pulse period for constant area; (d) the impulse function where height approaches infinity and duration approaches zero for constant area.

When this area is associated with the impulse function itself, it is often called the *strength* of the impulse. The *unit impulse function* results whenever A assumes a value of unity. On the basis of the foregoing description the following mathematical definition of the unit impulse function can be written:

$$\delta(t) = 0 \qquad \text{for } t \neq 0$$

and

$$\int_{-\infty}^{\infty} \delta(t)dt = \int_{0^-}^{0^+} \delta(t)dt = 1 \tag{5-113}$$

There are a number of areas in electrical engineering where waveshapes approaching the impulse function can be found serving practical purposes. This is particularly true in electronic circuits employing sweeps such as are found in television and radar applications. Similar high-force, low-duration forces can be cited in other areas in engineering—for example, the hammer blow. However, knowledge of the impulse function and its properties are also useful from a purely theoretical point of view. One such application is the use of the impulse function to characterize the dynamical response of circuits and systems generally.

To focus attention on a specific problem, let it be desired to find the current response of a series RC circuit to a driving force which is the impulse function of strength A. Refer to Fig. 5-20. The symbol used to denote the impulse function is $\delta(t)$. Since the impulse function is derived as a limiting case of the pulse function, we can reasonably expect that the results obtained for the pulse response can be employed provided that appropriate modifications are made. Thus the voltage appearing across the capacitor plates shortly after application of the impulse function should be derivable from Eq. (5-104). We need merely observe that T_1, in the case of the impulse, is negligibly small compared to the circuit time constant $T = RC$. That is,

$$T_1 \ll T = RC \tag{5-114}$$

Repeating Eq. (5-104) for convenience, we have

$$v_C(T_1) = E(1 - \varepsilon^{-T_1/T}) \tag{5-115}$$

At this point it is helpful to observe further that when Eq. (5-114) prevails, the exponential term in Eq. (5-115) may be replaced by

$$\varepsilon^{-T_1/T} = 1 - \frac{T_1}{T} \tag{5-116}$$

Consequently, Eq. (5-115) becomes simply

$$v_C(T_1) = \frac{ET_1}{T} = \frac{\text{pulse area}}{T} = \frac{A}{T} \tag{5-117}$$

where A clearly carries the units of volt-seconds and T is in seconds.

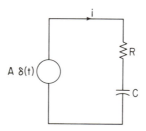

Fig. 5-20 Series RC circuit with an implied impulse function.

A close examination of Eq. (5-117) provides some interesting conclusions. For the case where the pulse is modified to take on the properties of the impulse, T_1 obviously assumes an infinitesimal value which we denote by 0^+. Certainly, in this instance, Eq. (5-116) is even more valid. Accordingly, Eq. (5-117) may be simply rewritten as

$$v_C(0^+) = \frac{\text{pulse area}}{T} = \frac{A}{T} \tag{5-118}$$

By this result we must conclude that, an infinitesimal time after the application of the impulse, the voltage across the capacitor assumes the value A/T. At first this may appear to be a surprising result, for we learned previously that the voltage across a capacitor cannot change instantaneously. However, this statement had always been made subject to the condition "in the presence of a finite force," which is not true here. The impulse function has an amplitude approaching infinity so that it is capable of bringing about a sudden change.

It is now fair to speculate that the change in the capacitor voltage in 0^+ time must be the result of a sudden burst of current. In other words there appears to exist an impulse of current that places the full charge on the capacitor plates consistent with the voltage A/T. To verify that this is indeed what happens, we begin by writing Kirchhoff's voltage law for the circut of Fig. 5-20. Thus

$$A \, \delta(t) = Ri + \frac{1}{C} \int i \, dt \tag{5-119}$$

Integrating both sides of the equation in the infinitesimal time interval from 0^- to 0^+ yields

$$A \int_0^{0^+} \delta(t) dt = R \int_{0^-}^{0^+} i \, dt + \frac{1}{C} \int_{0^-}^{0^+} \int_{0^-}^{0^+} i \, dt \tag{5-120}$$

Examination of Eq. (5-120) discloses that the term involving the double integral cannot possibly make any contribution to the equation balance. This follows from the fact that if the result of the first integration yields a finite value,[†] a second integration of this finite quantity over the infinitesimal time interval still gives a negligible quantity. Hence Eq. (5-120) reduces to

$$A \int_{0^-}^{0^+} \delta(t) dt = R \int_{0^-}^{0^+} i \, dt \tag{5-121}$$

In view of the fact that A and R are both finite numbers, it must next be concluded that the only way for the impulse on the left side to be balanced on the right side is for a current impulse to exist on the right side. Accordingly, the waveshape for i in Eq. (5-121) must be one that has an amplitude approaching infinity and a duration approaching zero with the area fixed. Since the area of such a curve carries the unit of ampere-seconds, which is the dimension of charge, we can properly replace the integrand on the right side of Eq. (5-121) by the impulse function $Q\delta(t)$, where Q denotes the area of the current pulse. Equation (5-121) thus becomes

$$A \int_{0^-}^{0^+} \delta(t) \, dt = R \int_{0^-}^{0^+} Q \, \delta(t) \, dt = RQ \int_{0^-}^{0^+} \delta(t) dt \tag{5-122}$$

By the definition of the impulse function as described in Eq. (5-113), the last equation then reads simply as

$$A = RQ$$

or

$$Q = \frac{A}{R} \tag{5-123}$$

[†] This result is possible only if the integrand is an impulse function.

Equation (5-123) represents the charge placed on the capacitor by the current impulse. The corresponding voltage is then

$$v_C(0^+) = \frac{Q}{C} = \frac{A}{RC} = \frac{A}{T} \tag{5-124}$$

which agrees entirely with Eq. (5-118).

The solution for the current response after time 0^+ is straightforward. It is described simply by the action of "the initial condition" voltage on the capacitor discharging through the RC circuit. Accordingly,

$$i(t) = -\frac{A}{RT}\varepsilon^{-t/T} \qquad \text{for } 0^+ < t < \infty \tag{5-125}$$

The minus sign appears because the actual flow of current is opposite to the assumed positive direction shown in Fig. 5-20.

Second-order circuits

5-6 STEP RESPONSE OF SECOND-ORDER SYSTEM (*RLC* CIRCUIT)

The behavior of a circuit or system which contains two independent energy-storing elements is completely described by a second-order differential equation. Because the second-order system occurs frequently in engineering situations, considerable attention is given here to its analysis, especially with a view to putting the results in a universal form. In this way the results can be applied to second-order systems generally, independently of their particular composition. Thus the independent energy-storing elements may consist of inductance and capacitance or of mass and spring constant and so forth. The study of the second-order system is important also because the behavior of many higher-order systems can frequently be described in terms of an equivalent second-order system. Consequently the conclusions developed here can frequently be applied to such systems with satisfactory results.

Let it be desired to find the complete current response in the circuit of Fig. 5-21 after the switch is closed. Assume that the capacitor has an initial charge of q_0. The governing differential equation for the circuit is found upon applying Kirchhoff's voltage law. Thus

$$E = Ri + L\frac{di}{dt} + \frac{1}{C}\int i \, dt \tag{5-126}$$

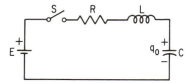

Fig. 5-21 Series *RLC* circuit with an initial charge on *C*.

Equation (5-126) is a second-order nonhomogeneous linear differential equation. Upon introducing the initial condition voltage we can then write

$$E = Ri + L\frac{di}{dt} + \frac{q_0}{C} + \frac{1}{C}\int_0^t i\, dt \tag{5-127}$$

Rearranging and employing the differential operator, the expression becomes

$$E - \frac{q_0}{C} = \left(R + pL + \frac{1}{pC}\right)i \tag{5-128}$$

or

$$i = \left(\frac{1}{R + pL + 1/pC}\right)\left(E - \frac{q_0}{C}\right) \tag{5-129}$$

The forced solution is found upon setting p equal to zero in the last equation which clearly leads to

$$i_f = \text{forced solution} = 0 \tag{5-130}$$

Since the capacitor represents an open circuit to a constant forcing function, this result is entirely expected.

The transient component of the solution to the governing differential equation is found in the fashion described in Sec. 4-7. The first step consists of finding the characteristic equation which here is readily determined by setting the denominator of Eq. (5-129) equal to zero. After rearranging terms, we get

$$\boxed{D(p) = D(s) = s^2 + \frac{R}{L}s + \frac{1}{LC} = 0} \tag{5-131}$$

It is worth noting that this equation also results when the left side of Eq. (5-128) is set equal to zero, thus yielding the homogeneous differential equation in operator form. Moreover, note that each of the three circuit parameters appears and that the highest order of s is 2, which is consistent with the presence of two independent energy-storing elements. If s_1 and s_2 denote the roots of this equation, the formal expression for the response can be written as

$$i(t) = K_1\varepsilon^{s_1 t} + K_2\varepsilon^{s_2 t} \tag{5-132}$$

Although Eq. (5-132) presents a formal solution of the problem at hand, we are interested in writing this solution in a more useful and significant form. In this connection, then, let us return to Eq. (5-131) and write the specific expressions for the roots. Thus

$$s_{1,2} = -\frac{R}{2L} \pm \sqrt{\left(\frac{R}{2L}\right)^2 - \frac{1}{LC}} \tag{5-133}$$

Depending upon the expression under the radical the transient response can be any one of the following: (1) overdamped if $(R/2L)^2 > 1/LC$; (2) critically damped if $(R/2L)^2 = 1/LC$; (3) underdamped if $(R/2L)^2 < 1/LC$.

Complex Roots. Because the underdamped case is the most interesting as well as the most frequently encountered case, attention is confined to it throughout the remainder of this section. The term damping is an appropriate one to use in our description because, as previously demonstrated, it is characteristic of the resistance parameter to dissipate energy. Consequently, its presence serves to prevent an uninterrupted interchange of energy between the two energy-storing elements. Such an uninterrupted interchange would constitute a sustained (or undamped) oscillation. For the underdamped case, then, the expression for the roots of the characteristic equation becomes

$$s_{1,2} = -\frac{R}{2L} \pm j \sqrt{\frac{1}{LC} - \left(\frac{R}{2L}\right)^2} \tag{5-134}$$

This result is obtained by factoring out the minus sign under the radical and recalling that $j = \sqrt{-1}$.

The critical value of damping for fixed values of L and C corresponds to that value of R which makes the radical term go to zero. Hence

$$\left(\frac{R_c}{2L}\right)^2 = \frac{1}{LC} \tag{5-135}$$

or

$$R_c = 2\sqrt{\frac{L}{C}} \tag{5-136}$$

where R_c denotes the value of resistance which yields a critically damped transient response. In order to express the roots of the characteristic equation in a manner which makes the results applicable to all linear second-order systems as well as to provide a quick and convenient way of identifying the dynamic response, two figures of merit are introduced and defined. The first is the *damping ratio*, which is denoted by the Greek letter zeta (ζ) and defined as follows:

$$\boxed{\zeta \equiv \frac{\text{actual damping}}{\text{critical damping}} \equiv \frac{R}{2\sqrt{L/C}}} \tag{5-137}$$

The second figure of merit is the natural radian frequency of the circuit and is defined by

$$\boxed{\omega_n \equiv \frac{1}{\sqrt{LC}}} \tag{5-138}$$

Accordingly, the real part of the roots given by Eq. (5-134) may be expressed in terms of ζ and ω_n as

$$\frac{R}{2L} = \frac{\zeta R_c}{2L} = \frac{\zeta 2\sqrt{L/C}}{2L} = \zeta\frac{1}{\sqrt{LC}} = \zeta\omega_n \tag{5-139}$$

Also, by means of the results in Eqs. (5-138) and (5-139), the radical term may be expressed as

$$\sqrt{\frac{1}{LC} - \left(\frac{R}{2L}\right)^2} = \sqrt{\omega_n^2 - \zeta^2\omega_n^2} = \omega_n\sqrt{1 - \zeta^2} \tag{5-140}$$

Therefore, Eq. (5-134) may be written

$$s_{1,2} = -\zeta\omega_n \pm j\omega_n\sqrt{1 - \zeta^2} \qquad (5\text{-}141)$$

Introducing the substitution

$$\omega_d \equiv \omega_n\sqrt{1 - \zeta^2} = \text{damped frequency of oscillation} \qquad (5\text{-}142)$$

simplifies Eq. (5-141) to

$$\boxed{s_{1,2} = -\zeta\omega_n \pm j\omega_d} \qquad (5\text{-}143)$$

A study of Eq. (5-143) reveals that both the real and the j parts of the complex roots have the units of inverse seconds, which is frequency. This in turn gives rise to the term *complex frequency*. It is shown presently that both parts of the complex frequency play important roles in establishing the character of the transient response.

By using the results appearing in Eq. (5-143) for s_1 and s_2, it is possible to write the characteristic equation entirely in terms of ζ and ω_n. Thus

$$\boxed{s^2 + \frac{R}{L}s + \frac{1}{LC} = s^2 + 2\zeta\omega_n s + \omega_n^2 = (s + \zeta\omega_n)^2 + \omega_d^2} \qquad (5\text{-}144)$$

Inserting Eq. (5-143) into Eq. (5-132) the expression for the current response becomes

$$i = \varepsilon^{-\zeta\omega_n t}(K_1\varepsilon^{j\omega_d t} + K_2\varepsilon^{-j\omega_d t}) \qquad (5\text{-}145)$$

$$= \varepsilon^{-\zeta\omega_n t}[(K_1 + K_2)\cos\omega_d t + j(K_1 - K_2)\sin\omega_d t]$$

Now since the left side is a real quantity, each term on the right side must be real. Hence Eq. (5-145) may be rewritten more conveniently as

$$i = \varepsilon^{-\zeta\omega_n t}(k_1\cos\omega_d t + k_2\sin\omega_d t) \qquad (5\text{-}146)$$

where k_1 and k_2 are coefficients that are to be evaluated from initial conditions.

In this connection the presence of the inductor in the circuit of Fig. 5-21 means that the current cannot change instantaneously. Inserting $t = 0^+$ in Eq. (5-146) then gives

$$i(0^+) = k_1 \qquad (5\text{-}147)$$

To find k_2, a second initial condition must be known. This can readily be deduced from the original differential equation, which is given by Eq. (5-127). Evaluating this equation at $t = 0^+$ we have

$$E - \frac{q_0}{C} = Ri(0^+) + L\left(\frac{di}{dt}\right)_{t=0^+} + \frac{1}{C}\int_0^{0^+} i\,dt \qquad (5\text{-}148)$$

Clearly the integral term offers a negligible contribution and by Eq. (5-147) the first term on the right side is also zero. Hence it follows that the initial value of the rate of change of current is

$$\left(\frac{di}{dt}\right)_{t=0^+} = \frac{E - q_0/C}{L} \qquad (5\text{-}149)$$

This result must now be inserted into the differentiated expression for the current as derived from Eq. (5-146). The result is

$$k_2 = \frac{E - q_0/C}{\omega_d L} \tag{5-150}$$

Accordingly, the final complete solution for the current produced by a step voltage source applied to the series combination of R, L, and C is

$$i(t) = \frac{E - q_0/C}{\omega_d L} \varepsilon^{-\zeta \omega_n t} \sin \omega_d t \tag{5-151}$$

The solution is this case involves only a transient term because, as already pointed out, the forced solution is zero.

A plot of the response function, Eq. (5-151), appears in Fig. 5-22. The response is a damped sinusoid which is confined to an envelope determined by the exponential function $\varepsilon^{-\zeta \omega_n t}$. It is interesting to note that dealing in terms of the envelope alone it is possible to identify a time constant for the second-order system. Thus

$$\boxed{T = \frac{1}{\zeta \omega_n} = \frac{2L}{R}} \tag{5-152}$$

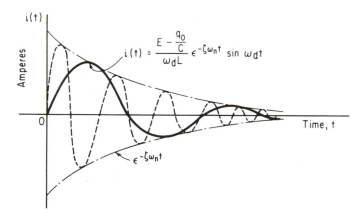

Fig. 5-22 Graphical representation of the current response in an *RLC* circuit for underdamped conditions. The dashed-line curve is for a lower value of *C*.

Accordingly, the meaning which can be attached to the real part of the complex frequency of Eq. (5-143) is that it gives a measure of the response time of the transient. It effectively describes the period of duration of the transient. Initially ω_n was introduced as a figure of merit. By its nature any figure of merit, if it is to be useful, must convey information about a specific performance feature. In the case of a circuit's natural frequency, it yields information about *settling time*,† i.e., the time needed to reach steady state. *The larger the ω_n of a circuit*

† This is also called the *response time*.

the smaller will be the settling time. Thus if two *RLC* circuits are compared and each has the same damping ratio but one has twice the natural frequency of the other, the transients in the circuit with the larger natural frequency will decay twice as fast. This conclusion follows from Eq. (5-152).

The meaning of the *j* part of the complex frequency of Eq. (5-143) is that it conveys information about the *actual* frequency at which the oscillations decay. This conclusion easily follows from Eq. (5-151).

For fixed values of *R* and *L* what is the effect on the transient response of decreasing the capacitance parameter *C*? Equation (5-152) tells us that the envelope is in no way affected by this change, since it is independent of *C*. In other words the time constant and therefore the settling time remain essentially unaltered. Equations (5-137) and (5-138), however, state, respectively, that the damping ratio decreases and the natural frequency increases. The resulting response is then characterized by a higher frequency of oscillation as well as a higher peak oscillation, as illustrated by the dashed-line curve of Fig. 5-22. In Example 5-5 it is shown that the damping ratio is a figure of merit which yields information about maximum overshoot. The smaller the value of ζ the greater will be the peak values attained during the transient period.

Step-Current Response of a Parallel GLC Circuit. The circuit configuration is depicted in Fig. 5-23. All initial conditions are assumed equal to zero. A step-current forcing function is applied to the parallel combination of the circuit elements by opening the switch. The lower of the two nodes of the circuit is taken as the reference node and is shown grounded to stress the point.

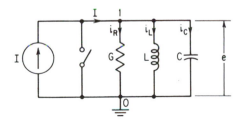

Fig. 5-23 Parallel *GLC* circuit subjected to a step-current forcing function.

A little reflection here reveals that this parallel arrangement of circuit elements is the dual of the series *RLC* circuit. In fact an exact dual occurs if it is assumed that $G = R$, $L = C$, and $C = L$. On this assumption then and by comparison with Eq. (5-126) the governing equation for the circuit of Fig. 5-23 is simply

$$I = Ge + C\frac{de}{dt} + \frac{1}{L}\int e\, dt \qquad (5\text{-}153)$$

Note that the roles of voltage and current have been interchanged. The corresponding solution for the output voltage *e* can be written forthwith from Eq. (5-151). Accordingly,

$$e = \frac{I}{\omega_d C}\varepsilon^{-\zeta\omega_n t}\, sin\, \omega_d t \qquad (5\text{-}154)$$

where

$$\zeta = \frac{G}{2\sqrt{C/L}} \tag{5-155}$$

and

$$\omega_n = \frac{1}{\sqrt{LC}} \tag{5-156}$$

Once the node voltage is determined, finding the expressions for the currents flowing through the individual elements is a routine matter. For example, the current through the resistor is simply given by

$$i_R = \frac{e}{R} = \frac{I}{\omega_d RC} \varepsilon^{-\zeta\omega_n t} \sin \omega_d t \tag{5-157}$$

To find the time function for the current through the inductor, more effort is required because it involves the integration of the product of two time functions. That is,

$$i_L = \frac{I}{L} \int_0^t e \, dt = \frac{I}{\omega_d LC} \int_0^t \varepsilon^{-\zeta\omega_n t} \sin \omega_d t \, dt \tag{5-158}$$

After performing the integration and substituting the limits, the final expression becomes

$$i_L = \frac{I}{\omega_d LC} \left[\frac{\omega_d}{\omega_n^2} - \frac{\varepsilon^{-\zeta\omega_n t}}{\omega_n} \sin(\omega_d t + \gamma) \right] \tag{5-159}$$

where

$$\gamma = \tan^{-1} \frac{\omega_d}{\zeta\omega_n} = \tan^{-1} \frac{\sqrt{1 - \zeta^2}}{\zeta} \tag{5-160}$$

A check at $t = \infty$ shows that the current through the inductor is equal to the source current. This is in accord with the physical situation, which indicates that at steady state the inductance behaves as a short circuit and thereby carries the total source current. The time solution for the capacitor current is found from

$$i_c = C \frac{de}{dt} \tag{5-161}$$

where e is given by Eq. (5-154). It is left as an exercise for the reader to determine the final expression for i_c.

EXAMPLE 5-5   A step voltage of magnitude E is applied to an initially deenergized series *RLC* circuit for which the parameter values are

$$R = 40 \, \Omega \qquad L = 0.2 \, \text{H} \qquad C = 100 \, \mu\text{F}$$

Find the complete solution for the charge on the capacitor and show a plot of the response.

Solution: The solution for the current is already determined and is given by Eq. (5-151). The only modification needed is to set q_0 equal to zero because of the statement of the problem. Thus

$$i(t) = \frac{E}{\omega_d L} \varepsilon^{-\zeta\omega_n t} \sin \omega_d t \tag{5-162}$$

The time solution for the charge thus follows from

$$q(t) = \int_0^t i(t)dt = \frac{E}{\omega_d L} \int_0^t \varepsilon^{-\zeta\omega_n t} \sin \omega_d t \, dt \qquad (5\text{-}163)$$

After integrating, this becomes

$$q(t) = \frac{E}{\omega_d L} \left[\frac{\varepsilon^{-\zeta\omega_n t}(-\zeta\omega_n \sin \omega_d t - \omega_d \cos \omega_d t)}{\omega_n^2} \right]_0^t \qquad (5\text{-}164)$$

Substituting the upper and lower limits then yields

$$q(t) = \frac{E}{\omega_d L} \left[\frac{\omega_d}{\omega_n^2} - \frac{\varepsilon^{-\zeta\omega_n t}}{\omega_n} \sin (\omega_d t + \gamma) \right] \qquad (5\text{-}165)$$

where

$$\gamma = \tan^{-1} \frac{\omega_d}{\zeta\omega_n} \qquad (5\text{-}166)$$

To put $q(t)$ in a more condensed form we must have the values of ζ, ω_d, and ω_n. These are readily obtained from the characteristic equation, which can be written as

$$s^2 + \frac{R}{L}s + \frac{1}{LC} = s^2 + \frac{40}{0.2}s + \frac{1}{0.2 \times 10^{-4}} = s^2 + 200s + 50{,}000 = 0 \qquad (5\text{-}167)$$

or, more conveniently,

$$s^2 + 2\zeta\omega_n s + \omega_n^2 = s^2 + 200s + 50{,}000 = 0 \qquad (5\text{-}168)$$

The left side is the general form of the characteristic equation expressed in terms of ζ and ω_n. A comparison of the constant terms in Eq. (5-168) readily shows that

$$\omega_n = \sqrt{50{,}000} = 100\sqrt{5} = 224 \text{ rad/s} \qquad (5\text{-}169)$$

Equating the coefficients of the s terms gives

$$\zeta\omega_n = 100 \qquad (5\text{-}170)$$

or

$$\zeta = \frac{100}{\omega_n} = \frac{100}{100\sqrt{5}} = \frac{1}{\sqrt{5}} = 0.447 \qquad (5\text{-}171)$$

The damped frequency of oscillation is then

$$\omega_d = \omega_n \sqrt{1 - \zeta^2} = 224\sqrt{1 - \tfrac{1}{5}} = 200 \text{ rad/s} \qquad (5\text{-}172)$$

Accordingly, Eq. (5-165) can now be rewritten

$$q(t) = E\left[C - \frac{C}{\sqrt{1 - \zeta^2}} \varepsilon^{-100t} \sin (200t + \tan^{-1}2) \right] \qquad (5\text{-}173)$$

or

$$q(t) = EC[1 - 1.12\varepsilon^{-100t} \sin (200t + 63.4°)] \qquad (5\text{-}174)$$

The expression for the voltage across the capacitor readily follows from Eq. (5-174) upon dividing both sides by the capacitance parameter C. Thus we get

$$v_C = E[1 - 1.12 \, \varepsilon^{-100t} \sin (200t + 63.4°)] \qquad (5\text{-}175)$$

which applies for zero initial charge.

A plot of Eq. (5-175) appears in Fig. 5-24. Note that E is the steady-state value of the voltage that appears on the capacitor which is consistent with the potential difference E appearing across the capacitor terminals after the transient

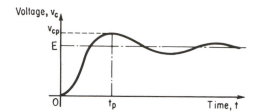

Fig. 5-24 Oscillatory buildup in a capacitor in a series *RLC* circuit with a step-voltage input.

has decayed. Note too that the oscillations decay at a frequency of 200 rad/s, which is the value of ω_d. Inspection of Fig. 5-24 also shows that the first overshoot beyond the steady-state value represents the maximum overshoot associated with the transient. A quantitative measure of this quantity can be found by setting the derivative of $v_c(t)$ equal to zero. This yields the time at which the peak overshoot occurs. Specifically, it is given by the expression

$$t_p = \frac{\pi}{\omega_d} \qquad (5\text{-}176)$$

where t_p denotes the time for the peak overshoot. The corresponding value of the peak overshoot expressed as a percent of the steady-state value can be shown to be given by

$$\text{maximum percent overshoot} = 100\,\frac{v_{cp} - E}{E} = 100\varepsilon^{-\zeta\pi/\sqrt{1-\zeta^2}} \qquad (5\text{-}177)$$

The right side of Eq. (5-177) reveals that the maximum percent overshoot is solely dependent upon the value of the damping ratio ζ. *It is for this reason that ζ is looked upon as a figure of merit which conveys information about maximum overshoot of the transient response.* Since for our example $\zeta = 0.447$, its insertion into Eq. (5-177) reveals that the maximum overshoot is 20% of the steady-state value. Moreover, if it is assumed that steady state is reached after five time constants, that is, $5(\frac{1}{100})$, then clearly the settling time is 0.05 s. It is interesting to note that these two bits of information—settling time and maximum percent overshoot—are really all that is needed to describe the dynamic response of the circuit. Knowledge of these quantities, which are directly obtainable from the characteristic equation and Eqs. (5-152) and (5-177), often makes it unnecessary to determine the formal solution of the response such as that represented by Eq. (5-175).

Because of the considerable importance of Eq. (5-177) in estimating one of the two important features that characterizes the dynamic behavior of linear second-order circuits, it is presented in graphical form in Fig. 5-25. This is the maximum percent overshoot that occurs for various values of the damping ratio in situations that require a fixed steady-state solution. A constant steady-state solution is often associated with constant values of the step function. In fact, because the dynamic response is solely determined by the circuit parameters and is independent of the type of source function that is used in a particular

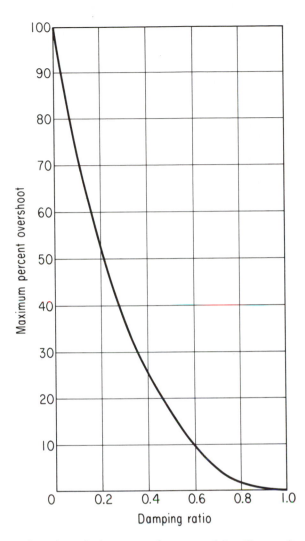

Fig. 5-25 Maximum percent overshoot versus damping ratio for linear second-order systems.

situation, it is convenient to arbitrarily apply the step function whenever it is desired to get a measure of the oscillatory nature of an underdamped response of a linear second-order circuit. The situation is illustrated by the next example.

EXAMPLE 5-6 Refer to the *RLC* circuit shown in Fig. 5-26, where the value of the circuit parameters are indicated and the capacitor is assumed to have zero charge initially.

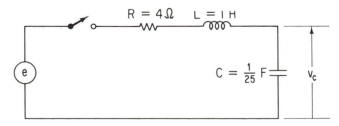

Fig. 5-26 Circuit for Example 5-6. Initial charge on *C* is zero.

Describe the dynamic behavior of this circuit when it is disturbed by an applied source function.

Solution: For convenience we assume that the source function e is a constant and that our interest is to describe the buildup of voltage across the capacitor from its initially deenergized state. The dynamic behavior of a linear circuit is virtually described once the settling time and the maximum percent overshoot are known. Of course, the maximum percent overshoot is determined upon finding the value of the damping ratio and the settling time is found from knowledge of $\zeta\omega_n$. Moreover, as a glance at Eq. (5-144) readily discloses, this information is quickly available from the characteristic equation. Since the damping ratio is customarily restricted for use in underdamped situations, let's ascertain this fact by resorting to Eq. (5-136) to find the critical resistance. Thus

$$R_c = 2\sqrt{\frac{L}{C}} = 1\sqrt{\frac{1}{1/25}} = 10\ \Omega \tag{5-178}$$

Since the circuit $R = 4$ and is less than R_c, it follows that this circuit exhibits an oscillatory response when disturbed. In fact the damping ratio is found to be

$$\zeta = \frac{R}{R_c} = \frac{4}{10} = 0.4 \tag{5-179}$$

On entering Fig. 5-25 we find the maximum overshoot to be 25%.

Alternatively, the damping ratio as well as the natural frequency can be found directly from a comparison of the characteristic equation for the case at hand with that of the universal form. Accordingly, here we have

$$s^2 + 2\zeta\omega_n s + \omega_n^2 = s^2 + \frac{R}{L}s + \frac{1}{LC} = s^2 + 4s + 25 \tag{5-180}$$

Comparing coefficients yields

$$\omega_n = 5 \quad\text{and}\quad 2\zeta\omega_n = 4 \tag{5-181}$$

Hence

$$\zeta = 0.4$$

The time it takes for the transients in this circuit to reduce to less than 1% of the initial values is given by five times the time constant displayed in Eq. (5-152). Thus

$$t_s = 5T = \frac{5}{\zeta\omega_n} = \frac{5}{2} = 2.5\ \text{s} \tag{5-182}$$

A sketch of the variation of the voltage across the capacitor can be fairly accurately displayed for a constant source voltage on the basis of the foregoing information. This is depicted in Fig. 5-27. Observe that use is also made of Eq. (5-176), which serves to place the peak value of the capacitor voltage at time

$$t_p = \frac{\pi}{\omega_d} = \frac{\pi}{\omega_n\sqrt{1-\zeta^2}} = 0.686\ \text{s} \tag{5-183}$$

Critically Damped Response (Real and Equal Roots). Critical damping in the *RLC* circuit occurs when the resistor has the value given by Eq. (5-136), which is $R = 2\sqrt{L/C}$. In such a situation the roots of the characteristic equation are real and equal, as a glance at Eq. (5-133) easily reveals. With the circuit resistance equal to the critical value the damping ratio ζ becomes unity and ω_d assumes a zero value. In other words the underdamped response is at the limit of its oscillatory range and at this limit the response no longer exhibits an overshoot.

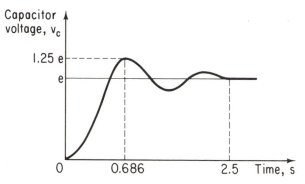

The insertion of $\zeta = 1$ and $\omega_d = 0$ into the general solution of the underdamped case as represented by Eq. (5-175) after appropriate analytical manipulations leads to the expression.

$$v_c = E[1 - (1 + 224t)\varepsilon^{-224t}] \tag{5-184}$$

More generally, Eq. (5-184) can be written as

$$v_c = E[1 - (1 + \omega_n t)\varepsilon^{-\omega_n t}] \tag{5-185}$$

where ω_n is the natural frequency.

It is instructive to observe that the double-root condition of the characteristic equation has caused a time factor to appear in the solution. The form of Eq. (5-185) can be shown to be consistent with a formulation of the transient response, which is expressed as

$$v_t(t) = K_1\varepsilon^{st} + K_2 t\varepsilon^{st} \tag{5-186}$$

where s denotes the value of the double root that results from the characteristic equation. The constants K_1 and K_2, as is customary, are evaluated from the initial conditions. These notions are illustrated in the example that follows.

EXAMPLE 5-7   Obtain the critically damped response for the circuit of Fig. 5-26 assuming that $R = 10\ \Omega = R_c$. Assume too that $e = 12$ V and no initial charge on the capacitor.

Solution: Attention is first directed to a formal solution to show correspondence with Eq. (5-185), after which the abbreviated procedure that was employed in Example 5-6 is applied for comparison purposes.

On closing the switch, Kirchhoff's voltage law yields

$$E = Ri + L\frac{di}{dt} + \frac{1}{C}\int_0^t i\,dt \tag{5-187}$$

Since interest is focused here on the capacitor voltage, it is useful to introduce it (v_c) as the dependent variable in place of i. This is readily accomplished by introducing

$$i = C\frac{dv_c}{dt}$$

into Eq. (5-187). The result is

$$\frac{d^2v_c}{dt^2} + \frac{R}{L}\frac{dv_c}{dt} + \frac{1}{LC}v_c = \frac{E}{LC} \tag{5-188}$$

The operator form of this expression then becomes

$$p^2 v_c + \frac{R}{L} p v_c + \frac{1}{LC} v_c = \frac{E}{LC}$$

or

$$\left(p^2 + \frac{R}{L} p + \frac{1}{LC}\right) v_c = \frac{E}{LC}$$

Hence

$$v_c = \left(\frac{1}{p^2 + \frac{R}{L} p + \frac{1}{LC}}\right) \frac{E}{LC} \tag{5-189}$$

The forced solution readily follows on inserting $p = 0$, thus yielding

$$v_f = E \tag{5-190}$$

The transient solution is determined by the roots of the characteristic equation, which in this case is (replacing p with s)

$$s^2 + \frac{R}{L} s + \frac{1}{LC} = s^2 + 10s + 25 = (s + 5)(s + 5) = 0 \tag{5-191}$$

Clearly, then, there exists a double root at -5. Hence the general form of the transient solution is given by Eq. (5-186), which when combined with the forced solution yields the complete solution, namely,

$$v_c = v_f + v_t = 12 + K_1 \varepsilon^{st} + K_2 t \varepsilon^{st} \tag{5-192}$$

The constants K_1 and K_2 are now evaluated from the initial conditions. At $t = 0^+$ Eq. (5-192) gives

$$v_c(0^+) = 0 = 12 + K_1 \tag{5-193}$$

or

$$K_1 = -12 \tag{5-194}$$

because $v_c(0^+) = 0$ by the problem statement. Since this is a second-order circuit we need initial information about a second variable. We know that the presence of the inductor in the circuit of Fig. 5-26 prevents the current from changing immediately on closing the line switch. Hence $i(0^+) = 0$. In turn, this means that

$$\left(\frac{dv_c}{dt}\right)_{t=0^+} = \frac{i(0^+)}{c} = 0 \tag{5-195}$$

Thus a second initial condition is available and is used after differentiating Eq. (5-192), which leads to

$$\frac{dv_c}{dt} = sK_1 \varepsilon^{st} + K_2(st \varepsilon^{st} + \varepsilon^{st}) \tag{5-196}$$

Evaluating at $t = 0^+$ and inserting Eq. (5-195) and $s = -\omega_n = -5$ yields

$$\left(\frac{dv_c}{dt}\right)_{t=0^+} = 0 = sK_1 + K_2 \tag{5-197}$$

Therefore,

$$K_2 = \omega_n K_1 = 5K_1 = -12\omega_n \tag{5-198}$$

The total solution for the capacitor voltage results upon inserting Eqs. (5-194) and (5-198) into Eq. (5-192). Thus

$$v_c = v_f + v_t = 12 - 12\varepsilon^{-\omega_n t} - 12\omega_n t \varepsilon^{-\omega_n t}$$ (5-199)

$$= 12[1 - (1 + \omega_n t)\varepsilon^{-\omega_n t}]$$

When $\omega_n = 5$ is inserted, the final result becomes

$$v_c = 12[1 - (1 + 5t)\varepsilon^{-5t}]$$ (5-200)

A plot of this equation reveals no overshoots and also permits the settling time to be found.

The abbreviated procedure uncovers these conclusions much more directly. Comparing the actual characteristic equation with the universal form for second-order circuits, we get

$$s^2 + 10s + 25 = s^2 + 2\zeta\omega_n s + \omega_n^2$$ (5-201)

Hence,

$$\omega_n = \sqrt{25} = 5$$

$$2\zeta\omega_n = 10$$

$$\zeta\omega_n = 5$$

$$\zeta = \frac{5}{\omega_n} = 1 \qquad \therefore \text{ critical damping}$$

Of course, this last result means that there are no overshoots in the transient response. Moreover,

$$\omega_d = \omega_n\sqrt{1 - \zeta^2} = 5\sqrt{1 - 1} = 0$$ (5-202)

which in turn means no oscillations because the damped frequency of oscillation is zero. Finally, the settling time can be estimated by using five time constants. Thus

$$t_s = \frac{5}{\zeta\omega_n} = \frac{5}{5} = 1 \text{ s}$$ (5-203)

Observe that the critically damped response reaches the same steady-state value used in Example 5-6, but it does so in much less time: 1 s compared to 2.5 s. Clearly, the oscillations consume time.

Overdamped Response (Real and Distinct Roots). An overdamped response results in a second-order circuit when the roots of the characteristic equation are both real and distinct. If we identify these roots as s_1 and s_2 and the dependent variable of interest as voltage v_c, the *transient* solution is readily represented as

$$v_t = K_1\varepsilon^{s_1 t} + K_2\varepsilon^{s_2 t}$$ (5-204)

Moreover, by calling the forced solution v_f and evaluating it in the usual manner, the expression for the complete solution can then be written generally as

$$v_c = v_f + v_t = v_f + K_1\varepsilon^{s_1 t} + K_2\varepsilon^{s_2 t}$$ (5-205)

The constants K_1 and K_2 are then evaluated in the customary way from initial condition considerations.

EXAMPLE 5-8 Repeat Example 5-7 for the case where the resistance R is increased to 17.2 Ω. All other conditions remain the same.

Solution: Because the governing differential equation continues to be validly represented by Eq. (5-188), it follows that the same forced solution prevails at steady state, i.e., $v_f = 12$ V. The transient solution differs, however, since new roots of the characteristic equation result from the increased circuit resistance. In this case the characteristic equation becomes

$$s^2 + \frac{R}{L}s + \frac{1}{LC} = s^2 + 17.2s + 25 = 0 \qquad (5\text{-}206)$$

which leads to the roots

$$s_1 = -1.6 \quad \text{and} \quad s_2 = -15.6 \qquad (5\text{-}207)$$

Hence the complete solution for the behavior of the capacitor voltage to a step voltage command of 12 V is

$$v_c = 12 + K_1 \varepsilon^{-1.6t} + K_2 \varepsilon^{-15.6t} \qquad (5\text{-}208)$$

The initial conditions employed in the solution to Example 5-7 are also applicable here. Evaluation of v_c and its first derivative at time $t = 0^+$ leads to

$$K_1 = -13.37 \quad \text{and} \quad K_2 = 1.37 \qquad (5\text{-}209)$$

The corresponding expression for the complete solution of the capacitor voltage then becomes

$$v_c = 12 - 13.37\varepsilon^{-1.6t} + 1.37\varepsilon^{-15.6t} \qquad (5\text{-}210)$$

A plot of this function is shown in Fig. 5-28.

Several points are worth noting in this example because these results are not unusual for overdamped circuits. First note the wide difference in the values of the roots of the characteristic equation. The natural mode associated with the larger root (-15.6) damps out very rapidly, as indicated in Fig. 5-28. Moreover, the large root value also carries with it an amplitude for the transient (1.37 in this case) which is correspondingly small compared to the magnitude of the natural mode that is associated with the smaller root value. The net result of these considerations is that the overdamped second-order system can be treated as essentially a first-order system with a dominant characteristic root of -1.6. Consequently, this overdamped second-order system can be considered to exhibit

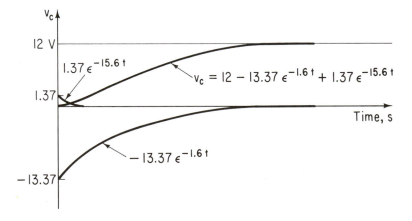

Fig. 5-28 Sketch of capacitor voltage in the circuit of Fig. 5-26 to a step voltage for the over-damped case. Individual transient terms are also indicated.

Circuit Dynamics and Forced Responses   Chap. 5

a time constant of $1/1.6 = 0.625$ s, so that the approximate value of time for the capacitor voltage to reach beyond 99% of its final value is

$$t_s = 5\left(\frac{1}{1.6}\right) = 5(0.625) = 3.125 \text{ s}$$

Note how much longer this time is than the settling time achieved for critical damping. This settling time even exceeds that which was found for the underdamped case of Example 5-6.

5-7 COMPLETE RESPONSE OF *RL* CIRCUIT TO SINUSOIDAL INPUT

So far in the chapter attention has been confined to just one type of input, the step forcing function. In this section we now determine the total response of a series *RL* circuit when a second type of input, the sinusoidal function, is applied to the circuit terminals. The circuit diagram is depicted in Fig. 5-29.

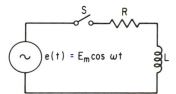

Fig. 5-29 Series *RL* circuit with sinusoidal forcing function.

The voltage source is assumed to be varying in a cosinusoidal fashion. When the switch is closed, the governing differential equation for the circuit becomes

$$E_m \cos \omega t = Ri + L\frac{di}{dt} \tag{5-211}$$

In operator form the response is

$$i = \left[\frac{1}{R + pL}\right]E_m \cos \omega t \tag{5-212}$$

The forced solution is found directly from this equation. Before doing so, recall that a considerable saving of effort results when the cosinusoidal function is replaced by the corresponding exponential function which contains it. Refer to Sec. 4-5. Accordingly, we have

$$i_f^* = \left[\frac{1}{R + pL}\right]E_m \varepsilon^{j\omega t} = \left[\frac{1}{R + sL}\right]_{s = j\omega} E_m \varepsilon^{j\omega t} \tag{5-213}$$

where

$$i_f = Re[i_f^*] \tag{5-214}$$

The real part of i_f^* is the solution we are seeking. This follows from the fact that the cosine is the real part of $\varepsilon^{j\omega t}$. Moreover, note that p and s are interchangeable in Eq. (5-213) as a consequence of the presence of the exponential functions. Performing the substitution called for in Eq. (5-213) yields

$$i_f^* = \frac{E_m}{R + j\omega L} \varepsilon^{j\omega t} = \frac{E_m}{Z\underline{/\theta}} \varepsilon^{j\omega t} = \frac{E_m}{Z} \varepsilon^{-j\theta} \varepsilon^{j\omega t}$$

$$= \frac{E_m}{Z} \varepsilon^{j(\omega t - \theta)} \tag{5-215}$$

where

$$Z = \sqrt{R^2 + \omega^2 L^2} \tag{5-216}$$

and

$$\theta = \tan^{-1} \frac{\omega L}{R} \tag{5-217}$$

Upon applying Eq. (5-214) to Eq. (5-215), the forced or steady-state solution can be identified simply as

$$i_f = Re\left[\frac{E_m}{Z} \varepsilon^{j(\omega t - \theta)}\right] = \frac{E_m}{Z} \cos(\omega t - \theta) \tag{5-218}$$

It is instructive here to note that the forced solution has the same form as the forcing function but that it differs in two respects—amplitude and phase. The amplitude of the current is modified from that of the source voltage by the factor $1/\sqrt{R^2 + \omega^2 L^2} = 1/Z$. The phase or argument† is altered by the angle $-\theta$.

To find the transient component of the solution, we first obtain the characteristic equation. This follows upon setting the denominator of Eq. (5-212) equal to zero. Thus

$$R + pL = R + sL = 0 \tag{5-219}$$

from which

$$s = -\frac{R}{L} \tag{5-220}$$

This is a familiar result and leads to an expression for the transient solution that is the same as was found for the case of the step source function. See Eqs. (5-9) and (5-10). Thus

$$i_t = K\varepsilon^{-(R/L)t} \tag{5-221}$$

The complete solution is obtained by adding Eq. (5-218) to Eq. (5-221), which yields

$$i = i_f + i_t = \frac{E_m}{Z} \cos(\omega t - \theta) + K\varepsilon^{-(R/L)t} \tag{5-222}$$

The presence of the inductor means that $i(0^+) = 0$. Inserting this result into Eq. (5-222) for $t = 0^+$ identifies the value of K as

$$K = \frac{E_m}{Z} \cos(-\theta) = \frac{E_m}{Z} \cos\theta \tag{5-223}$$

† See p. 255 for a definition of this term.

Circuit Dynamics and Forced Responses   Chap. 5

The final expression for the response thus becomes

$$i(t) = \frac{E_m}{Z}[\cos(\omega t - \theta) + \varepsilon^{-(R/L)t}\cos\theta] \qquad (5\text{-}224)$$

A graphical representation of Eq. (5-224) appears in Fig. 5-30. The transient component of the solution behaves as expected. Initially it provides a magnitude of current which is equal and opposite to the instantaneous value of the steady-state current at the instant of switching. In Fig. 5-30 it is assumed that the switch is closed at the instant when the voltage has its positive maximum value. Note that the corresponding steady-state current is not zero at this instant. In Fig. 5-30 it is shown to have a value $0a$. Since the boundary condition demands that the current at this instant be zero, it is necessary for the transient term to have a value equal and opposite. This is depicted as $-0a$ in Fig. 5-30. As time elapses after switching, the transient term decays to zero at a rate determined by the circuit time constant, and the total current in Fig. 5-30 then becomes identical to the steady-state current.

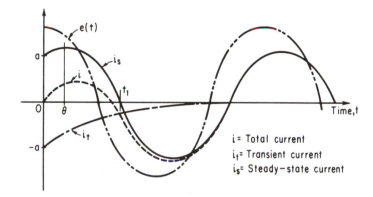

Fig. 5-30 Current response of an *RL* circuit to a sinusoidal forcing function.

The presence of a transient component in the total solution for the current is called for whenever the initial value of current in the *RL* circuit is at variance with the value of the steady-state current at the instant of switching. As an illustration, if in the problem just discussed the initial current flowing through *L* were equal to $0a$ there would be no need for a transient term because the ensuing current would proceed directly into the steady state. As a result of the correspondence of the currents before and after switching, there is no force attempting to bring about an abrupt change in the amount of energy stored in the inductor. This behavior can be described in another way. Let t_1 in Fig. 5-30 denote the time when the steady-state current is zero. If the switch in Fig. 5-29 is closed at this instant, no violation of the boundary condition occurs; consequently there is no need for a transient current. In such a case the response proceeds directly into the steady state. Incidentally, note that although the steady-state current is zero the corresponding value of the applied voltage is not. This is characteristic of the *RL* circuit for sinusoidal forcing functions.

EXAMPLE 5-9 A sinusoidal forcing function $e(t) = 141 \sin 377t$ is applied to an initially de-energized series RL circuit in which $R = 100 \ \Omega$ and $L = \frac{1}{2}$ H.

(a) If the switch which applied the voltage to the RL circuit is closed at the instant when $e(t)$ is passing through zero with a positive slope, determine the initial value of the transient current.

(b) Write the complete expression for the transient solution.

(c) Write the expression for the complete solution of the current response.

(d) At what instantaneous value of the applied voltage will the closing of the switch result in no transient component of current?

Solution: (a) The general form of the steady-state solution is

$$i_f = \frac{E_m}{\sqrt{R^2 + \omega^2 L^2}} \sin(\omega t - \theta) \tag{5-225}$$

This follows directly from Eq. (5-218) except that the sine function is used in order to make it consistent with the assumed sinusoidal forcing function. Since $\omega = 377$ it follows that

$$\omega L = 377(\tfrac{1}{2}) = 188.5 \ \Omega \tag{5-226}$$

and

$$\sqrt{R^2 + \omega^2 L^2} = \sqrt{100^2 + 188.5^2} = 213.5 \ \Omega \tag{5-227}$$

Also,

$$\theta = \tan^{-1} \frac{\omega L}{R} = \tan^{-1} 1.885 = 62° \tag{5-228}$$

Therefore, the value of the steady-state current at $t = 0^+$, which is the time immediately following application of the forcing function, is

$$i_f(0^+) = \frac{E_m}{\sqrt{R^2 + \omega^2 L^2}} \sin(-\theta) = \frac{141}{213.5} \sin(-62°) = -0.584 \ \text{A} \tag{5-229}$$

However, in accordance with the boundary condition the initial value of the current through the inductance must be zero. That is,

$$i(0^+) = 0 = i_f(0^+) + i_t(0^+) \tag{5-230}$$

or

$$0 = -0.584 + i_t(0^+)$$

Hence the initial value of the transient current is

$$i_t(0) = 0.584 \ \text{A} \tag{5-231}$$

(b) The rate of decay of the transient term is determined by the time constant of the circuit, which here is

$$T = \frac{L}{R} = \frac{\frac{1}{2}}{100} = \frac{1}{200} \ \text{s} \tag{5-232}$$

Therefore, the complete expression for the transient term becomes

$$i_t = 0.584\varepsilon^{-200t} \tag{5-233}$$

(c) By adding Eqs. (5-225) and (5-233) the equation for the total solution results.

$$i(t) = 0.61 \sin(377t - 62°) + 0.584\varepsilon^{-200t} \tag{5-234}$$

(d) In order that there be no transient current in the solution, it is necessary to close the switch at that time instant for which i_f is zero. A glance at Eq. (5-225) shows that this occurs whenever

$$(\omega t - \theta) = 0, \pi, 2\pi, \ldots \tag{5-235}$$

Choosing the first value, it follows that

$$\omega t = \theta = 62° \tag{5-236}$$

The corresponding instantaneous value of the forcing function is then

$$e = 141 \sin 62° = 124.2 \text{ V} \tag{5-237}$$

Therefore, if the switch is closed when the applied voltage has an instantaneous value of 124.2 volts, there is no need for a transient term and so none exists.

5-8 RESPONSE OF RLC CIRCUIT TO SINUSOIDAL INPUTS

The treatment of this case presents nothing new in the way of concepts about transients; therefore, the development of the solution is kept brief. The circuit configuration is shown in Fig. 5-31. With the switch closed, the differential equation for the circuit is

$$E_m \cos \omega t = Ri + L\frac{di}{dt} + \frac{1}{C}\int_0^t i\,dt \tag{5-238}$$

All initial conditions are assumed to be zero.

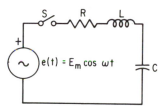

Fig. 5-31 Series RLC circuit with a sinusoidal forcing function. All initial conditions zero.

Proceeding as outlined in Sec. 5-7, we can write the expression for the forced solution directly as

$$i_f^* = \left[\frac{1}{R + pL + \dfrac{1}{pC}}\right]_{p=j\omega} E_m \varepsilon^{j\omega t} = \left[\frac{1}{R + sL + \dfrac{1}{sC}}\right]_{s=j\omega} E_m \varepsilon^{j\omega t} \tag{5-239}$$

Performing the indicated operation leads to

$$i_f^* = \frac{E_m}{Z}\varepsilon^{j(\omega t - \theta)} \tag{5-240}$$

where now

$$Z = \sqrt{R^2 + \left(\omega L - \frac{1}{\omega C}\right)^2} \tag{5-241}$$

and

$$\theta = \tan^{-1} \frac{\omega L - \dfrac{1}{\omega C}}{R} \qquad (5\text{-}242)$$

Then by Eq. (5-214) we find the actual steady-state solution to be

$$i_f = Re[i_f^*] = \frac{E_m}{Z} \cos{(\omega t - \theta)} \qquad (5\text{-}243)$$

Again note that this forced solution has the same form as the source function but differs in its amplitude and argument.

By setting the denominator of Eq. (5-239) equal to zero, the characteristic equation results. Thus

$$R + pL + \frac{1}{pC} = R + sL + \frac{1}{sC} = 0 \qquad (5\text{-}244)$$

Rearranged, this becomes

$$s^2 + \frac{R}{L}s + \frac{1}{LC} = 0 \qquad (5\text{-}245)$$

which we know has two roots, s_1 and s_2. If we assume that the circuit is underdamped, the characteristic equation may be rewritten as

$$s^2 + 2\zeta\omega_n s + \omega_n^2 = 0 \qquad (5\text{-}246)$$

where ζ and ω_n have the definitions given by Eqs. (5-137) and (5-138). Accordingly, the transient component takes the form

$$i_t = \varepsilon^{-\zeta\omega_n t}(K_1 \cos{\omega_d t} + K_2 \sin{\omega_d t}) \qquad (5\text{-}247)$$

as described on p. 183.

The expression for the total solution then becomes

$$i(t) = i_f + i_t = \frac{E_m}{Z} \cos{(\omega t - \theta)} + \varepsilon^{-\zeta\omega_n t}(K_1 \cos{\omega_d t} + K_2 \sin{\omega_d t}) \qquad (5\text{-}248)$$

a plot of which is depicted in Fig. 5-32.

Comparing Eq. (5-248), the current response of the *RLC* case, with Eq. (5-224), the current response for the *RL* case, shows that the responses differ essentially in the character of the transient terms. Because the *RL* configuration

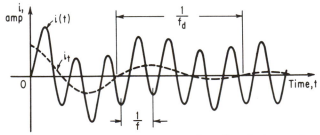

i_t = Transient current
$i(t)$ = Total current

Fig. 5-32 Plot of current response to the configuration of Fig. 5-31. Here f_d is the damped frequency of oscillation in hertz of the transient term and f is the frequency of the applied forcing function.

     Circuit Dynamics and Forced Responses   Chap. 5

involves one energy-storing element and is described by a first-order differential equation there occurs just a simple exponential decay of the transient. The steady-state solution merely rides on this exponential decay as illustrated in Fig. 5-30. For the underdamped *RLC* configuration, however, the transient solution itself involves an exponentially damped oscillation whose frequency is ω_d. Here the steady-state term can be looked upon as riding on the damped oscillation as depicted in Fig. 5-32. It is basically in this respect that the two responses differ.

Summary review questions _____

1. In a linear network what is the meaning of settling time?
2. What is meant by the time constant of a first-order linear circuit?
3. How are settling time and time constant related in a first-order linear circuit?
4. Can you speak of a time constant of a higher-order circuit? Explain.
5. Give a general description of the relationship between the time constants and the settling time of a second-order linear circuit.
6. What is the general technique to be used to gain information about the initial values of the dependent variable in a linear circuit?
7. Describe the role that is served by the transient component in the complete solution to a first-order linear circuit driven by a constant source function.
8. Is the nature of the transient solution dependent on the type of source function (i.e., step, ramp, sinusoid, etc.)? Explain.
9. A sinusoidal source function is applied to an initially deenergized series *RL* circuit. The switch that applies the sinusoidal source to the circuit is closed at that instant in its cycle when the source voltage passes through zero in moving from the negative to the positive range of voltages. Is a transient solution generated? Explain.
10. To what percentage of its initial value does the transient component of a response function in a first-order circuit reach when the settling time is measured in terms of three time constants; in terms of four time constants; in terms of five time constants?
11. A constant source voltage is applied to a series *RC* circuit with zero initial charge. Can the current change abruptly at $t = 0^+$ following application of the source voltage? Explain.
12. Explain the effect, if any, of an initial charge on the capacitor in response to Question 11.
13. Define the dual of a circuit and illustrate your explanation.
14. State the duals of the following terms as they apply to electric circuits: resistance, inductance, voltage, through.
15. Describe and illustrate a graphical procedure for finding the dual of a network.
16. Describe the basic notion involved in finding the pulse response of the series *RC* circuit. Assume that the dependent variable of interest is the voltage across the capacitor.
17. Repeat Question 16 for the voltage across the resistor.
18. Define the impulse function mathematically and illustrate graphically.

19. Explain why the application of an impulse function to a series *RC* circuit gives the appearance of voltage changing abruptly in a capacitor.

20. Why does the impulse response of a circuit appear to be "free" of a forced solution?

21. Identify the various cases of transient responses that can arise in a network that has two independent energy-storing elements.

22. How is the damping ratio of an underdamped second-order circuit defined? What is the specific information about the character of the transient response that is conveyed by this ratio? Illustrate.

23. How is the equivalent time constant of an underdamped second-order circuit defined? What is the specific information about the character of the ensuing transient response that is conveyed by this constant? Is this information likely to be slightly pessimistic or optimistic? Explain.

24. In a series *RLC* circuit define the critical resistance and describe its importance.

25. How is the natural frequency of an *RLC* network determined? Offer a physical explanation of why the expression for the natural frequency involves the particular circuit parameters cited.

26. Repeat Questions 24 and 25 for the parallel *GLC* circuit which is chosen as the exact dual of the series *RLC* circuit.

27. Identify the general form of the forced solution of a series *RL* circuit when it is subject to a sinusoidal source function. Indicate where the response differs and is similar to the source function.

28. Does the character of the transient response in a linear circuit subject to a step function differ from that obtained for a sinusoidal source function? Explain.

29. When a sinusoidal source is applied to a deenergized series *RL* circuit, is it possible to close the switch that connects the source to the circuit without incurring a transient current? Explain.

Problems

GROUP I

5-1. In the circuit of Fig. P5-1, determine the expression for the source current for all time after closing the switch. Assume zero current through the coil when S is closed.

5-2. (a) In the circuit of Fig. P5-2 find the complete expression of the current which flows through the coil when switch S is closed.

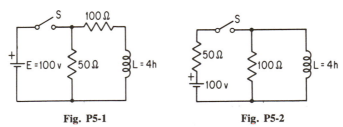

Fig. P5-1   **Fig. P5-2**

(b) What is the final value of the coil current?

(c) How long does it take for the coil current to reach 95% of its final value?

5-3. A mass M is freely suspended and at rest in a viscous material having a viscous coefficient D equal to 20 N-s/m. Neglect gravity effects. A constant force of 10 N is applied to this mass. Determine:

   (a) The expression for the mass velocity for all time after application of the force. Assume $M = 5$ N-s^2/m.

   (b) The final velocity in meters/second.

   (c) The time it takes for the mass to reach within 1% of its final velocity.

5-4. The plot of the current response to a step forcing function of 100 V has the variation depicted in Fig. P5-4. Find the values of resistance and inductance which apply for the circuit.

5-5. Assuming the coil initially deenergized and the current source suddenly applied to the circuit of Fig. P5-5, find the total expression for the current through the energy-storing element.

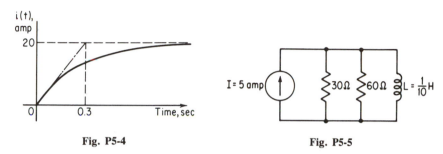

Fig. P5-4	**Fig. P5-5**

5-6. In the circuit of Fig. P5-6 switch S (denoted by the arrow) has been placed at a for a long time. It is then quickly moved to position b along the contact points shown as heavy lines.

   (a) Find the expression for the current through the 30-Ω resistor.

   (b) Compute the energy dissipated in this resistor.

   (c) Compare the result found in (b) with the energy initially stored in the coil.

5-7. Determine the time constant at which the transients in the circuit of Fig. P5-7 decay.

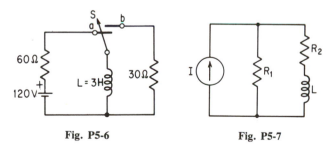

Fig. P5-6	**Fig. P5-7**

5-8. The configuration of an RC circuit is as shown in Fig. P5-8. The initial-condition voltage on the capacitor is zero. Switch S is then put to terminal a.

   (a) Find the expression for the current through the capacitor.

   (b) What is the time constant of the charging circuit?

   Ten seconds after switch S has been at terminal a it is placed at terminal b.

(c) Find the current which flows through the 2-MΩ resistor.

(d) Compute the energy dissipated in this resistor after 2 s.

5-9. In the circuit of Fig. P5-9 switch S has been in position 1 for a long time.

(a) Find the complete solution for the current in the circuit when S is put to position 2.

(b) If it requires five time constants for the transient to disappear, find the time in seconds.

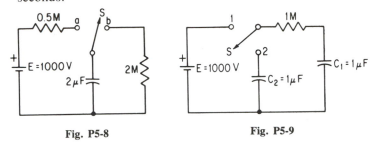

Fig. P5-8   Fig. P5-9

5-10. The circuit of Fig. P5-10 has been in steady state for a long time with the switch S open. Determine the complete time expression for the current when switch S is closed.

5-11. Assuming the circuit of Fig. P5-11 is in a steady-state condition with switch S open, compute the complete expression for the battery current when switch S is closed. What is the time constant of this circuit?

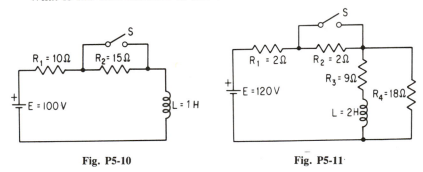

Fig. P5-10   Fig. P5-11

5-12. The circuit of Fig. P5-12 is initially deenergized. The switch is then closed.

(a) Find the expression for the current through R_2 as a function of time after the switch is closed.

(b) Repeat (a) for the current through C.

(c) What is the time constant of this circuit?

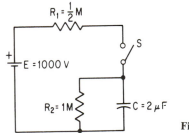

Fig. P5-12

5-13. In the circuit of Fig. P5-13 the switch S is suddenly closed.

(a) Give the initial value of the current in each branch of the circuit.

(b) By replacing the portion of the circuit to the left of *cd* by its Thévenin equivalent, find the equation for the current in the inductance as a function of time.

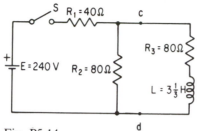

Fig. P5-13

5-14. Refer to Fig. P5-14.

(a) Find the complete expression for the charging capacitor current when switch S is put to position *a*.

(b) After a long time switch S is placed at position *b*. Determine the expression for the current which flows through the 0.1-MΩ resistor.

(c) Find the time constant of the circuit of part (b).

5-15. For the circuit shown in Fig. P5-15 determine:

(a) The characteristic equation.

(b) The time constants at which transients decay in this network.

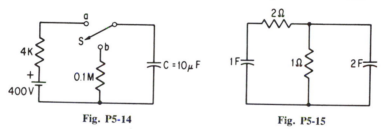

Fig. P5-14   **Fig. P5-15**

5-16. A single rectangular pulse of magnitude E and duration T_1 is applied to a series *RL* circuit. Find the expression for the current for (a) $0 \leqslant t \leqslant T_1$ and (b) $t \geqslant T_1$.

5-17. A single rectangular pulse of current having magnitude *I* and duration T_1 is applied to a parallel *RC* circuit. Obtain the result for the voltage across the capacitor by employing the principle of duality.

5-18. A pulse having an amplitude of 10 volts and a period of duration of $T_1 = \pi/1000$ s is applied to an initially deenergized series *LC* circuit. Find the value of the coil current immediately upon the expiration of the pulse period. Assume that $L = 40$ mH and $C = 100 \, \mu$F.

5-19. The circuit of Fig. P5-19 has been in the condition shown for a long time. The switch is then suddenly closed.

(a) What is the value of v before the switch is closed?

(b) What is the value of v immediately after the switch is closed?

(c) Find the complete expression for v after the switch is closed.

(d) What is the value of the time constant of the transient term?

5-20. The switch S is initially in position *b* with zero initial conditions on *C* and *L*. The switch is then put to terminal *a* moving along the contacts (heavy lines) in Fig. P5-20.

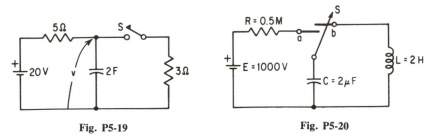

Fig. P5-19   Fig. P5-20

(a) What is the initial value of the voltage appearing across the capacitor when S is switched to *a?* Explain.

(b) Find the time expression for the current through the capacitor.

(c) How long does it take in seconds for the transient current to reach within 1% of its initial value? An approximate answer is acceptable.

After a long time the switch S is quickly moved back to position *b.*

(d) Obtain the *p*-domain solution for the current which flows in the *LC* circuit.

5-21. The response of a circuit to a step forcing function is known to be given by

$$i(t) = 10\varepsilon^{-8t} \sin (6t + \theta)$$

Find the value of the damping ratio and the natural frequency of the current.

5-22. Refer to the circuit of Fig. P5-22. The switch is closed.

(a) Describe what happens at the capacitor.

(b) Find the battery current as a function of time.

5-23. After the circuit shown in Fig. P5-23 has been energized for a long time, switch S is suddenly opened.

(a) Find the expression for the field winding current as a function of time.

(b) What is the voltage across the F_1–F_2 terminals at the instant the switch is opened?

(c) What would be the voltage computed in part (b) if R_1 were increased by 10 times?

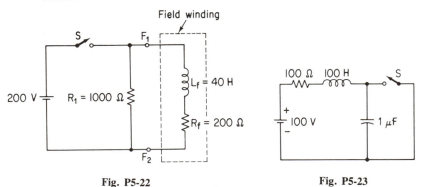

Fig. P5-22   Fig. P5-23

5-24. Refer to Fig. P5-24.

(a) With S_1 closed and S_2 open find the time constant of the circuit.

(b) With S_2 open and S_1 suddenly closed, find the time it takes for the current to reach 0.05 A.

(c) Assume that S_1 has been closed a long time and that S_2 is then suddenly closed, find the relay current as a function of time.

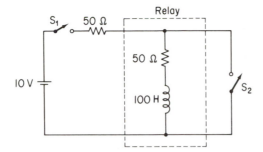

Fig. P5-24

(d) In part (c) how long will it take for the current in the relay to fall to 0.05 A? Compare this with the result obtained in part (b).

5-25. The inductor and capacitor in the circuit of Fig. P5-25 are initially deenergized.
  (a) Find the time it takes for the battery current to reach within 99% of its final value after the switch is closed.
  (b) What is the maximum value of the coil current during the transient state?

5-26. In the circuit shown in Fig. P5-26 the switch has been at a for a long time.
  (a) Derive the expression for the current in the inductor when the switch is placed from a to b.
  (b) How long does it take the current to reach steady state?

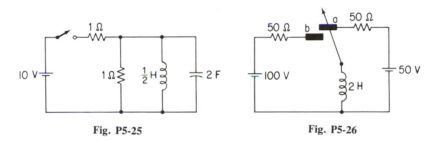

Fig. P5-25   Fig. P5-26

5-27. The behavior of a second-order system is described by

$$\frac{d^2c}{dt^2} + 6.4\frac{dc}{dt} + 64c = 160r$$

  (a) What is the damping ratio of the system?
  (b) Compute the percent maximum overshoot for $r = 0.4$.
  (c) At what frequency does the system response oscillate during the transient state?
  (d) How many seconds does it take for the response to settle within 5% of its final value after the step forcing function r is applied? An approximate answer is acceptable.

5-28. A sinusoidal voltage $e(t) = 150 \cos 20t$ is applied at $t = 0$ to an initially deenergized series RL circuit in which $R = 45\ \Omega$ and $L = 3$ H.
  (a) What is the expression for the forced current response?
  (b) Calculate the maximum value of the transient term.
  (c) Write the time expression for the transient current.
  (d) Why does a transient-current term exist?

5-29. Repeat Prob. 5-28 for the case where the sinusoidal forcing function is described by $e(t) = 150 \cos(20t - 30°)$.

Chap. 5   Problems   **207**

5-30. Repeat Prob. 5-28 for the case where $e(t) = 150 \cos (20t + 30°)$.

5-31. In the circuit of Fig. P5-31 switch S has been at position a for a long time. The switch is then put to position b at an instant when the a-c voltage is at its positive peak value. Determine the complete time expression for the current response when S is put to position b.

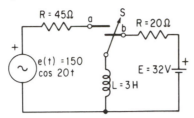

Fig. P5-31

GROUP II

5-32. A circuit containing two independent energy-storing elements is defined by the following differential equation:

$$\frac{d^2x}{dt^2} + 2\frac{dx}{dt} + 2x = f(t)$$

The circuit is initially deenergized so that $x(0^-) = \dot{x}(0^-) = 0$. Find $x(0^+)$, $\dot{x}(0^+)$, and $\ddot{x}(0^+)$ when $f(t)$ is a unit step function.

5-33. For the circuit of Fig. P5-33 determine the two time constants at which the transient terms decay when the network is subjected to a forcing function.

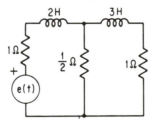

Fig. P5-33

5-34. The forcing function in the circuit of Fig. P5-34(a) is the pulse depicted in Fig. P5-34(b).
(a) Find the expression for the current.
(b) Find the expression for the voltage which appears across the capacitor terminals.

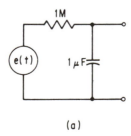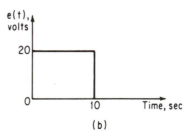

(a)   (b)

Fig. P5-34

5-35. For the circuit of Fig. P5-35 determine whether oscillations occur when a forcing function is applied. If oscillations do occur, find the value of the damping ratio and the damped frequency of response.

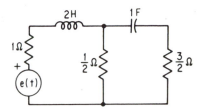

Fig. P5-35

5-36. Refer to the circuit depicted in Fig. P5-36.
  (a) Write the mesh equations for this circuit.
  (b) From the results of part (a) write the corresponding equations that apply to the dual circuit.
  (c) Employ the graphical procedure and find the exact dual circuit.
  (d) Show that the results of (b) and (c) are consistent.

5-37. Repeat Prob. 5-36 for the circuit illustrated in Fig. P5-37.

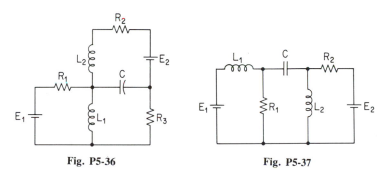

Fig. P5-36   Fig. P5-37

5-38. An impulse function of strength A volt-seconds is applied to the series arrangement of a resistor R and inductor L, which is initially deenergized.
  (a) Find the inductor current at time $t = 0^+$.
  (b) Find $i(t)$ for $0^+ \leq t \leq \infty$.

5-39. An impulse function of strength Q amp-seconds is applied to the parallel arrangement of a conductance G and capacitor C, which is initially deenergized.
  (a) Find the capacitor voltage at time $t = 0^+$.
  (b) Find the expression for the capacitor voltage for $0^+ \leq t \leq \infty$.

5-40. An engineering device contains two independent energy-storing elements. When subjected to a forcing function the governing differential equation becomes

$$2\frac{di}{dt} + 36i + 450 \int i \, dt = 900$$

Just prior to the application of the forcing function the initial value of the response i is known to be 10 and the initial value of the integrating device is known to be 70, i.e., $\int_{-\infty}^{0} i \, dt = 70$.
  (a) Does the response have a steady-state value different from zero in the time domain?
  (b) Find the expression for the complete solution in the time domain.

5-41. The dynamic behavior of an electric circuit is described by

$$\frac{d^2 i}{dt^2} + 4\frac{di}{dt} + 4i = \varepsilon^{-t}$$

All initial conditions are zero. Find the complete solution for the current in response to the source function.

5-42. Refer to the circuit of Fig. 5-26. Assume an initial charge on the capacitor of q_0 corresponding to which is an initial capacitor voltage V_0.
  (a) Determine the forced solution.
  (b) If $V_0 = 4$ V, sketch the complete response of the capacitor voltage as a function of time. Indicate all salient values. Assume e = 10 V.

5-43. For the modified condition described in Prob. 5-42 for the circuit of Fig. 5-26 determine the following:
  (a) The value of di/dt at $t = 0^+$.
  (b) The value of $d^2 v_c / dt^2$ at $t = 0^+$.

5-44. The two energy-storing elements of Fig. P5-44 are initially deenergized.
  (a) Find the characteristic equation.
  (b) Is the transient response characterized by overshoots? If so, compute the maximum percent overshoot.
  (c) Determine the time it takes for the transients to settle to less than 1 percent of the initial values.

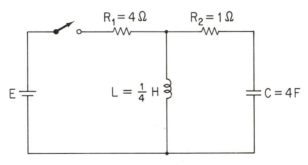

5-45. Repeat Prob. 5-44 for the case where $R_1 = 10\Omega$, $R_2 = 1\Omega$, $L = 1$H, and $C = 0.1$F.

5-46. Repeat Prob. 5-44 for the following values of the parameters: $R_1 = 1\Omega$, $R_2 = 3\Omega$, $L = 1$H, and $C = 1$F.

5-47. A sinusoidal voltage $e(t) = 100 \sin 20t$ is applied to a series RL circuit. The magnitude of the transient term immediately upon switching at $t = 0$ is found to be 0.4 A. Moreover, the transient term is observed to diminish with a time constant of 0.5 s.
  (a) Compute the values of R and L.
  (b) Is the initial value of the transient current positive or negative? Explain.
  (c) At what point in the voltage cycle must the switch be closed for the voltage across the inductance to be a maximum? What is the value of this voltage?

5-48. A sinusoidal voltage of $e(t) = E_m \cos \omega t$ is applied to an initially deenergized series combination of capacitance C and resistor R. Determine the complete expression for the current response.

chapter six

The Laplace-Transform Method of Finding Circuit Solutions

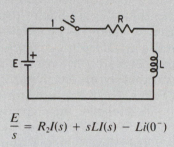

$$\frac{E}{s} = R_2I(s) + sLI(s) - Li(0^-)$$

In dealing with the subject matter of electrical engineering, it is frequently necessary to determine the solutions of linear differential equations. In this way information is obtained about the dynamic and forced responses of a circuit, device, or system to a driving function. Moreover, the development of methods of analysis, design, and synthesis as well as the interpretation of associated results are very often influenced by the particular mathematical formulations used to generate these solutions. It is in the interest of furnishing the reader with the most consistent, systematic, and readily interpretable mathematical tool that attention is given in this chapter to the Laplace-transform method of solving linear differential equations. Certainly no more time is needed to learn the mechanics of the Laplace-transform method than the classical method. Moreover, the former has some very significant and important advantages to offer.

In the first place the Laplace-transform method of finding the solution to a defining differential equation involves a one-step systematic formulation which yields complete information about the sustained and transient solutions. This stands in sharp contrast to the three-part solution procedure of the classical method. In the latter, it is necessary first to find the particular solution associated with the forcing functions, then to find the complementary function which is the solution to the homogeneous differential equation, and finally to evaluate the constants of integration from the initial conditions. Although two aspects of the procedure remain the same, namely finding the roots of the characteristic equation and evaluating the initial conditions, it often happens that where the governing differential equation is of an order higher than the second the amount of labor required to get the solution by the Laplace-transform method is less.

However, by far the more important factor is the systematic formulation provided by the Laplace-transform method. It permits the treatment of source functions and disturbances in the same fashion as initial conditions. This means that the same basic procedure is used to evaluate the forced solution as is used to evaluate the transient solution. This feature is particularly useful because it allows a unifying approach in the analysis and design of circuits, devices, and systems which cannot be otherwise achieved. Thus, for example, in the study of electric circuits it is possible to treat the more general case of the transient response to deterministic inputs before the steady-state sinusoidal response. In addition, greater stress is placed on transform impedance as an operator which converts forcing function or initial condition to response.

Finally, the mathematical formulation of the Laplace transform permits a uniformity of analysis and of interpretation of results in situations which we find constantly recurring in all areas of electrical engineering and many other branches of engineering and science. Thus with this approach we will use the same techniques in arriving at answers to problems irrespective of whether they deal with circuits, amplifiers, feedback devices, control systems, or analog computers.

In the treatment which follows, the emphasis is put on the mechanics of the Laplace-transform method. No attempt is made to present a rigorous exposition. This is left to books devoted primarily to that objective.† Accordingly, a matter such as obtaining the final solution to a problem after applying the Laplace transform is accomplished through the use of tables rather than by means of integration in the complex plane, which is considered beyond the requirements for students of this book. When we accept the use of tables as a valid means of performing the inverse Laplace transformation, there is every reason to use the Laplace transform as a means of solving linear differential equations.

6-1 NATURE OF A MATHEMATICAL TRANSFORM

It is the nature of a transform to simplify the analytical procedure leading to the solution of a problem irrespective of whether the problem is concerned with numbers, functions, or the calculus of differential equations. Specifically the Laplace transform is a mathematical tool for transforming functions. In linear analysis it also serves to convert operation in time, such as differentiation and integration, into simpler algebraic functions of an intermediate variable such as multiplication and division.

The idea of using transforms to simplify the procedure leading to a solution should not be new to the reader. It is assumed that he has used logarithms. Essentially the logarithm is a number transform which permits multiplication to be performed by means of the simpler operation of addition. An example readily

---

† See M. F. Gardner and J. L. Barnes, *Transients in Linear Systems* (New York: John Wiley & Sons, Inc., 1942).

illustrates the method. Let it be desired to find the product of

$$P = (35.6)\,(7.23)\,(181.47)$$

by using a number transform—the logarithm. The first step in the procedure is to find the transform of each number. The use of a logarithm table to the base 10 yields the results shown below.

original number		number transform
35.60	$\longrightarrow$	1.55145
7.23	$\longrightarrow$	0.85914
181.47	$\longrightarrow$	2.25880
46,708.2	$\longleftarrow$	4.66939

The solution in the transformed domain involves next an addition yielding 4.66939. Finally, the solution in the original system of numbers is obtained by consulting the logarithm table. This last step is known as *performing the inverse transformation through the use of tables*. For the example the inverse operation yields the answer of 46,708.2. Thus multiplication, the desired operation, has actually been performed by addition, a simpler operation, in a transformed domain.

The Laplace transform achieves a similar result but on a different plane. Instead of simplifying the multiplication of numbers, it is concerned with functions and the simplification of operations of the calculus, such as differentiation and integration, into the easier operations of algebra such as multiplication and division. Note especially, however, that the method of procedure is identical. Thus to solve an integrodifferential equation the same basic steps are involved. (1) The Laplace transform is used to convert the integrodifferential equation into an algebraic equation. The algebraic equation is expressed in terms of a new (or transformed) variable. (2) The simpler operation of solving the algebraic equation is performed in the transformed domain. (3) The solution in the original domain is obtained by consulting appropriate tables.

6-2 THE LAPLACE TRANSFORM: DEFINITION AND USEFULNESS

A typical form of the differential equations encountered in the study of electrical engineering is illustrated by Eq. (6-1):

$$a_2\frac{d^2i(t)}{dt^2} + a_1\frac{di(t)}{dt} + a_0i(t) - a_{-1}\int i(t)\,dt = f(t) \tag{6-1}$$

where $f(t)$ refers to the applied source function, the coefficients a denote constants, and $i(t)$ is the solution or desired response. A test of the usefulness of the Laplace transform is that it must succeed in converting this integrodifferential equation to a form which is easier to handle in finding the solution. Since Eq. (6-1) involves derivatives as well as an integral, and because of the importance of preserving the identity of the response function in spite of these operations of differentiation and integration, it can reasonably be concluded that the Laplace transform must in some way involve the exponential function. It will be recalled that the exponential function has the property that differentiation and integration always results in a new function which preserves the exponential character.

The direct Laplace transform of a function $f(t)$† is defined as follows:

$$F(s) \equiv \mathscr{L}f(t) \equiv \int_0^\infty f(t)\varepsilon^{-st}\, dt \qquad (6\text{-}2)$$

where s is the intermediate or transformed variable. Note that s is part of the exponential function. Equation (6-2) states that to obtain the Laplace transform of a function $f(t)$, it must be multiplied by ε^{-st} and then integrated from $t = 0$ to $t = \infty$. Furthermore, in order that $F(s)$ be meaningful, it is necessary that the integral converge and that $f(t)$ be defined for $t > 0$ and equal to zero for $t \leqslant 0$. In general the variable s is complex. Accordingly, for most of the functions encountered in electrical engineering, convergence is ensured by imposing the condition that the real part of s be positive, i.e., Re $[s] > 0$. Moreover, because t is used in this book exclusively to represent *time*, it follows that s must have the dimensions of inverse time, which is frequency. It is for these reasons, then, that the transformed variable is described as a *complex frequency*. An examination of Eq. (6-2) then reveals that after integration many time functions can be expressed in terms of the complex frequency s, thus emphasizing that the Laplace transform is a mathematical tool which permits solving time-domain problems in the frequency domain.

To solve the integrodifferential equation by the Laplace transform, it is necessary to apply the Laplace integral to each term of the equation. This converts the original time-domain equation to one expressed entirely in terms of the intermediate variable s. In Eq. (6-1) the unknown response $i(t)$ is identified in the transform domain as $I(s)$, i.e., $\mathscr{L}i(t) = I(s)$. Also, a glance at Eq. (6-2) shows that $\mathscr{L}a_0 i(t) = a_0 I(s)$. Thus one of the five terms of Eq. (6-1) is transformed. We next turn attention to the first derivative term.

How is the Laplace transform of a derivative term such as $a_1\, di(t)/dt$ found? The answer to this question is crucial in establishing whether or not the Laplace transform is useful in simplifying the solution of integrodifferential equations. To demonstrate that the Laplace transform does, in fact, achieve this objective, let us start with the expression for the Laplace transform of $i(t)$. Thus

$$I(s) = \int_0^\infty i(t)\varepsilon^{-st}\, dt \qquad (6\text{-}3)$$

Although we do not as yet know the specific form of $i(t)$, we are assuming that it is a continuous function for $t > 0$ and that its Laplace transform does exist and is given by $I(s)$. The right side of Eq. (6-3) is readily recognized as the integral of the product of two variables. From the calculus we know that

$$\int u\, dv = uv - \int v\, du \qquad (6\text{-}4)$$

---

† The notation procedure in dealing with Laplace transforms is to use lowercase letters to denote the time functions and capital letters for the corresponding Laplace-transformed function.

Hence, letting

$$u = i(t) \qquad dv = \varepsilon^{-st}\, dt$$

$$du = i'(t)\, dt \qquad v = -\frac{1}{s}\varepsilon^{-st}$$

(6-5)

and applying integration by parts yields

$$I(s) = -\frac{i(t)}{s}\varepsilon^{-st}\Big]_0^\infty + \frac{1}{s}\int_0^\infty i'(t)\varepsilon^{-st}\, dt = \frac{i(0)}{s} + \frac{1}{s}\int_0^\infty i'(t)\varepsilon^{-st}\, dt$$

Rearranging leads to

$$\int_0^\infty i'(t)\varepsilon^{-st}\, dt = sI(s) - i(0)$$

(6-6)

An examination of the left side of Eq. (6-6) reveals that this is precisely the expression for the Laplace transform of the derivative of $i(t)$. Thus

$$\mathscr{L}\frac{di(t)}{dt} = \mathscr{L}i'(t) = \int_0^\infty i'(t)\varepsilon^{-st}\, dt$$

(6-7)

Therefore Eq. (6-6) may be rewritten as

$$\boxed{\mathscr{L}i'(t) = sI(s) - i(0)}$$

(6-8)

Equation (6-8) is significant because it shows that the Laplace transform of a derivative in the original time domain carries over as the simpler operation of multiplication by the intermediate variable s in the transform domain. This result bears out the usefulness of the Laplace transform, for in transforming differentiation it preserves the transform of the original time function except for an algebraic operation. In addition, the term $i(0)$ describes in a direct and formal fashion the manner in which the initial condition is to be treated.

Extension of the Laplace transform to derivatives of higher order is readily accomplished by repeated application of the general procedure implied in Eq. (6-8). Thus the Laplace transform of the second derivative is

$$\mathscr{L}i''(t) = s[sI(s) - i(0)] - i'(0)$$
$$= s^2 I(s) - si(0) - i'(0)$$

(6-9)

For the nth derivative the Laplace transform is given by

$$\boxed{\mathscr{L}i^n(t) = s^n I(s) - s^{n-1}i(0) - s^{n-2}i'(0) - \cdots - i^{n-1}(0)}$$

(6-10)

Returning to Eq. (6-1) we see that we are now in a position to Laplace transform each term in the equation with the exception of the term involving the integral. By following a procedure similar to that used for differentiation it can be shown that the Laplace transform of an integral of time may be expressed as

$$\boxed{\mathscr{L}i^{-1}(t) = \int_0^\infty \left[\int i(t)\, dt\right]\varepsilon^{-st}\, dt = \frac{I(s)}{s} + \frac{i^{-1}(0)}{s}}$$

(6-11)

where $i^{-1}(0)$ denotes the initial value of the integral term just before applying the source function. Note that integration in the time domain is transformed to the simpler operation of division by s in the frequency domain. Note too the presence of the initial-condition term associated with the energy-storing element which gives rise to the integral term.

For an n-order integral the result may be expressed as

$$\mathscr{L}i^{-n}(t) = \frac{I(s)}{s^n} + \frac{i^{-1}(0)}{s^n} + \frac{i^{-2}(0)}{s^{n-1}} + \cdots + \frac{i^{-n}(0)}{s} \qquad (6\text{-}12)$$

Equations (6-10) and (6-12) illustrate another advantage of using the Laplace-transform method of solving differential equations. In situations where there are many energy-storing elements present (i.e., the order of the differential equation is high), this method indicates precisely how each initial-condition term enters into the expression for the derivatives and integrals.

With the results developed so far it is now possible to transform the integrodifferential expression of Eq. (6-1) into a completely algebraic equation in the s domain. Thus, by means of Eqs. (6-3), (6-8), (6-9), and (6-11) the integrodifferential equation becomes

$$a_2[s^2 I(s) - si(0) - i'(0)] + a_1[s I(s) - i(0)] + a_0 I(s) +$$
$$a_{-1}\left[\frac{I(s)}{s} + \frac{i^{-1}(0)}{s}\right] = F(s) \qquad (6\text{-}13)$$

where

$$F(s) = \mathscr{L}f(t) = \int_0^\infty f(t)\varepsilon^{st}\, dt \qquad (6\text{-}14)$$

The particular form of $F(s)$ depends upon the nature of $f(t)$. This matter is discussed in the next section.

It is significant to note that Eq. (6-13) is an algebraic expression which involves only one unknown quantity, namely, $I(s)$. Thus the solution to the differential equation as it is found in the transform domain (also called the intermediate result) is

$$I(s) = \frac{F(s) + a_2 si(0) + a_2 i'(0) + a_1 i(0) - a_{-1} i^{-1}(0)/s}{a_2 s^2 + a_1 s + a_0 + a_{-1}/s} \qquad (6\text{-}15)$$

Once $I(s)$ is found, the corresponding expression in the time domain is obtained from suitable tables as is done with logarithms. It is interesting to note in Eq. (6-15) that the initial-condition terms play a role similar to that of the driving function $F(s)$.

6-3 LAPLACE TRANSFORM OF COMMON FORCING FUNCTIONS

A formal solution to a linear integrodifferential equation is readily obtained in an intermediate form by Laplace transforming each term. This procedure leads to a result which is typically represented by Eq. (6-15). Here all the circuit parameters—a_1, a_2, and a_3—as well as all initial conditions are known. The only unknown term is $F(s)$. However, as Eq. (6-14) reveals, the specific form of $F(s)$ depends upon the nature of $f(t)$. Consequently, we next direct attention to the evaluation of $F(s)$ for those time functions which are commonly found in the study of electrical engineering. Once the Laplace transform $F(s)$ is found for a specific time function $f(t)$, it is tabulated for future reference just as is done for logarithms or integrals. Such a tabulation is particularly useful because, as can be shown, the Laplace transform of a time function, if it exists, is unique.

The Unit-Step Function u(t). This function is one of the most commonly used driving functions in electrical engineering. It frequently represents the closing of a switch which applies a constant voltage to a circuit or system. Mathematically, it is defined as

$$u(t) = \begin{cases} 0 & t \le 0 \\ 1 & t > 0 \end{cases} \tag{6-16}$$

and is plotted in Fig. 6-1. To determine the Laplace transform, we perform the integration called for by Eq. (6-2). Accordingly

$$F(s) = \mathcal{L}u(t) = \int_0^\infty \varepsilon^{-st}\, dt = \frac{1}{s} \tag{6-17}$$

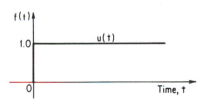

Fig. 6-1 Unit-step function occurring at $t = 0$.

Note that the discontinuous step function in the time domain is converted to a well-behaved analytic function in the s domain. The time function $u(t)$ and its corresponding Laplace transform $1/s$ constitute a unique transform pair, and they are placed in Table 6-1 for ready reference. For situations where the step function has a magnitude F_0 different from one, it follows from Eq. (6-17) that the corresponding Laplace transform is F_0/s. That is, $\mathcal{L}F_0 u(t) = F_0/s$.

The Exponential Function, f(t) = ε⁻ᵃᵗ. Direct application of Eq. (6-2) yields

$$F(s) = \mathcal{L}\varepsilon^{-\alpha t} = \int_0^\infty \varepsilon^{-\alpha t}\varepsilon^{-st}\, dt = \frac{1}{s + \alpha} \tag{6-18}$$

This result is identified as the second Laplace-transform pair of Table 6-1. Keep in mind that t is restricted to positive values only.

TABLE 6-1 LAPLACE-TRANSFORM PAIRS

$f(t)$	$f(t)$ vs. t, where $f(t) = 0$ for $t \leq 0$	$F(s)$	Pole location in s plane
1. $u(t)$		$\dfrac{1}{s}$	
2. ε^{-at}		$\dfrac{1}{s + \alpha}$	
3. $\sin \omega t$		$\dfrac{\omega}{s^2 + \omega^2}$	
4. $\cos \omega t$		$\dfrac{s}{s^2 + \omega^2}$	
5. t		$\dfrac{1}{s^2}$	Double pole at $S = 0$
6. t^2		$\dfrac{2}{s^3}$	Triple pole at $S = 0$
7. t^n		$\dfrac{n!}{s^{n+1}}$	$(n+1)$ multiple pole at $S = 0$
8. $\varepsilon^{-\alpha t} t^n$		$\dfrac{n!}{(s + \alpha)^{n+1}}$	$(n+1)$ multiple pole at $S = -\alpha$

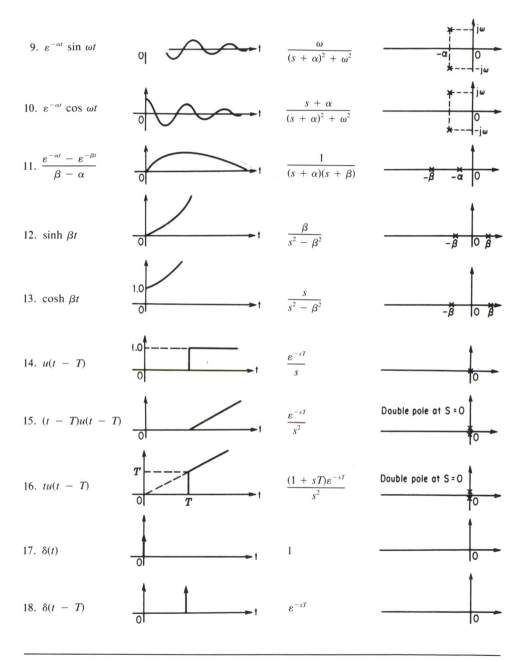

$f(t)$	$f(t)$ vs. t, where $f(t) = 0$ for $t \leqslant 0$	$F(s)$	Pole location in s plane
9. $\varepsilon^{-\alpha t} \sin \omega t$		$\dfrac{\omega}{(s + \alpha)^2 + \omega^2}$	
10. $\varepsilon^{-\alpha t} \cos \omega t$		$\dfrac{s + \alpha}{(s + \alpha)^2 + \omega^2}$	
11. $\dfrac{\varepsilon^{-\alpha t} - \varepsilon^{-\beta t}}{\beta - \alpha}$		$\dfrac{1}{(s + \alpha)(s + \beta)}$	
12. $\sinh \beta t$		$\dfrac{\beta}{s^2 - \beta^2}$	
13. $\cosh \beta t$		$\dfrac{s}{s^2 - \beta^2}$	
14. $u(t - T)$		$\dfrac{\varepsilon^{-sT}}{s}$	
15. $(t - T)u(t - T)$		$\dfrac{\varepsilon^{-sT}}{s^2}$	Double pole at S = 0
16. $tu(t - T)$		$\dfrac{(1 + sT)\varepsilon^{-sT}}{s^2}$	Double pole at S = 0
17. $\delta(t)$		1	
18. $\delta(t - T)$		ε^{-sT}	

The value of s in Eq. (6-18) for which the function becomes infinite is called a pole of $F(s)$. Hence it follows that $s = -\alpha$ is a pole of the Laplace transform of the exponential function. Poles are of particular importance to us because they characterize the nature of the time function associated with the solution in the s domain. This matter is discussed at length in Sec. 6-6.

The Sinusoidal Function, $f(t) = \sin \omega t$. For convenience in evaluating the Laplace integral the exponential form of $\sin \omega t$ is used. Thus

$$f(t) = \sin \omega t = \frac{\varepsilon^{j\omega t} - \varepsilon^{-j\omega t}}{2j} \tag{6-19}$$

Upon inserting into Eq. (6-2), we get

$$\mathcal{L} \sin \omega t = \frac{1}{2j} \int_0^\infty [\varepsilon^{-(s-j\omega)t} - \varepsilon^{-(s+j\omega)t}] \, dt \tag{6-20}$$

$$= \frac{1}{2j} \left[\frac{1}{s - j\omega} - \frac{1}{s + j\omega} \right] = \frac{\omega}{s^2 + \omega^2}$$

This appears as Laplace-transform pair 3 in Table 6-1. Here note that there are two poles of the Laplace-transform function located at $\pm j\omega$ on the imaginary axis.

The Cosinusoidal Function, $f(t) = \cos \omega t$. By recognizing that

$$\cos \omega t = \frac{1}{\omega} \frac{d}{dt} \sin \omega t$$

and applying the differentiation theorem we can write

$$\mathcal{L} \cos \omega t = \frac{1}{\omega} [sF(s) - f(0)]$$

where $F(s) = \mathcal{L} \sin \omega t = \omega/(s^2 + \omega^2)$. Therefore, it follows that

$$\mathcal{L} \cos \omega t = \frac{1}{\omega} \left[\frac{s\omega}{s^2 + \omega^2} - 0 \right] = \frac{s}{s^2 + \omega^2} \tag{6-21}$$

This result appears as transform pair 4 in Table 6-1.

The Ramp Function, $f(t) = t$. Applying the Laplace integral to the time function yields

$$F(s) = \int_0^\infty t\varepsilon^{-st} \, dt \tag{6-22}$$

Integration by parts then leads to

$$F(s) = \frac{-t\varepsilon^{-st}}{s} \Bigg]_0^\infty + \int_0^\infty \frac{1}{s}\varepsilon^{-st} \, dt = \frac{1}{s^2} \tag{6-23}$$

Equation (6-23) reveals that the Laplace transform of the ramp function involves a double pole at the origin of the s domain.

The Parabolic Function, $f(t) = t^2$. Although a direct application of the Laplace integral can be used, it is more convenient here to use the integration theorem. However, it is first necessary to write t^2 in terms of an integral. Thus

$$\mathscr{L}t^2 = \mathscr{L}\int 2t\,dt \tag{6-24}$$

Then by the integration theorem

$$\mathscr{L}\int 2t\,dt = \frac{F_1(s)}{s} + \frac{f_1^{-1}(0)}{s} = \frac{F_1(s)}{s} + 0 \tag{6-25}$$

where $F_1(s) = \mathscr{L}2t = 2/s^2$. Inserting this result into Eq. (6-25) gives

$$\mathscr{L}t^2 = \frac{2}{s^3} \tag{6-26}$$

It is interesting to note that the square function in time involves a cubic inverse function of s. As a matter of fact, an extension of the foregoing analysis leads to the following general result:

$$\mathscr{L}t^n = \frac{n!}{s^{n+1}} \tag{6-27}$$

This result appears as item 7 in Table 6-1.

6-4 LAPLACE TRANSFORM OF THE UNIT-IMPULSE FUNCTION

The unit impulse can be of considerable importance in analyzing the transient behavior of systems. For example, the transient response of a system to an arbitrary forcing function can be determined once the system's response to the unit impulse is established. This is an especially useful tool in light of the fact that the unit-impulse response of linear systems is related in a simple way to its transfer function.

One mathematical definition[†] of the impulse function is

$$\delta(t) \equiv \lim_{\Delta t \to 0} \frac{u(t) - u(t - \Delta t)}{\Delta t} \tag{6-28}$$

which will be recognized as the derivative of the unit-step function. In other words,

$$\delta(t) \equiv \frac{d}{dt}u(t) \tag{6-29}$$

Strictly speaking, therefore, Eq. (6-29) has the value zero for $t > 0$ and infinity at $t = 0$. Now, because the infinity value at $t = 0$ is unrealistic, we introduce a new function.

$$g(t) = 1 - \varepsilon^{-at} \tag{6-30}$$

which can be made to represent the unit-step function very closely by choosing α very large (see Fig. 6-2). The use of $g(t)$ offers the advantage of dealing with

[†] For other definitions, see S. Goldman, *Transformation Calculus and Electrical Transients* (Englewood Cliffs, N.J.: Prentice-Hall, Inc., 1953), p. 101.

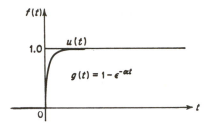

Fig. 6-2 Approximation of the unit-step function.

a function which has a continuous first derivative. Specifically, its value is

$$g'(t) = \frac{d}{dt}g(t) = \alpha \varepsilon^{-\alpha t} \tag{6-31}$$

An important property of this last equation is that

$$\int_0^\infty g'(t)\, dt = \int_0^\infty \alpha \varepsilon^{-\alpha t}\, dt = 1 \tag{6-32}$$

regardless of the value of α. Equations (6-30) to (6-32) can now be used to lead us to a more realistic interpretation of the impulse function. It will be observed that, by making $\alpha \to \infty$, $g(t)$ approaches $u(t)$. Furthermore, by Eq. (6-31) the ordinate intercept approaches infinity, and by Eq. (6-32) the area enclosed by $g'(t)$ remains equal to 1. It therefore follows that, as the ordinate intercept approaches infinity, the abscissa intercept approaches zero with the enclosed area always equal to unity. The impulse function therefore refers to any quantity which has so large a value occurring in so negligibly small an interval of time that it causes a finite change in the state of the system.

The direct Laplace transform of the unit impulse is readily established from Eq. (6-31) as α is made to approach infinity. Thus

$$\mathcal{L}\,\delta(t) = \lim_{\alpha \to \infty} \mathcal{L}\,g'(t) = \lim_{\alpha \to \infty} \mathcal{L}\,\alpha \varepsilon^{-\alpha t} = \lim_{\alpha \to \infty} \frac{\alpha}{s + \alpha} = 1$$

The Laplace transform of a unit impulse may also be found by letting the height of a unit rectangular pulse approach infinity and the width approach zero in such a way as to maintain the enclosed area equal to unity as required by Eq. (6-32). This is illustrated in Fig. 6-3. The expression in the time domain for such a pulse is

$$f(t) = \frac{1}{h}[u(t) - u(t - h)] \tag{6-33}$$

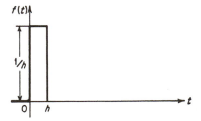

Fig. 6-3 Prelimit form of unit impulse in terms of a rectangular pulse.

The corresponding Laplace transform is

$$\mathscr{L} \text{ pulse} = \frac{1}{h}\left(\frac{1}{s} - \frac{\varepsilon^{-sh}}{s}\right) = \frac{1}{sh}(1 - \varepsilon^{-sh}) \tag{6-34}$$

Inserting the power-series equivalent of ε^{-sh} yields

$$\mathscr{L} \text{ pulse} = \frac{1}{sh}\left[1 - \left(1 - sh + \frac{s^2h^2}{2!} - \frac{s^3h^3}{3!} + \cdots + \right)\right]$$

$$= \left(1 - \frac{sh}{2!} + \frac{s^2h^2}{3!} - \cdots\right)$$

Accordingly,

$$\mathscr{L} \text{ unit impulse} = \mathscr{L} \delta(t) = \mathscr{L}_{h\to 0} \text{ pulse} = 1 \tag{6-35}$$

Consider next the evaluation of the Laplace transform of the unit impulse by yet another method—the use of Eq. (6-8). Since by Eq. (6-29) the impulse function is the derivative of the step function, application of Eq. (6-8) permits us to write

$$\mathscr{L} \delta(t) = sF(s) - u(0) \qquad \text{where } F(s) = \frac{1}{s} \tag{6-36}$$

Accordingly, we have

$$\mathscr{L} \delta(t) = s\frac{1}{s} - u(0) = 1 - u(0) \tag{6-37}$$

A little thought reveals a point of difficulty because of the ambiguity associated with the term $u(0)$. This difficulty indicates that a modification of the definition of the Laplace transform integral is clearly in order. A reasonable modification appears to be to change the lower limit in the integral definition of Eq. (6-2) to either 0^+ or 0^-. It is instructive to note that if 0^+ is used, then $u(0)$ in Eq. (6-37) becomes $u(0^+) = 1$. This leads to a result for the Laplace transform of the unit impulse contrary to that already derived by other means. On the other hand, if the lower limit of the integral is chosen to be 0^-, a perfectly consistent result ensues.

Consequently, it will be our practice from here on in this book to use an integral definition of the Laplace transform that employs 0^- as the lower limit. Thus, repeating the definition of Eq. (6-2), we now write

$$\boxed{F(s) \equiv \int_0^\infty f(t)\varepsilon^{-st}\, dt} \tag{6-38}$$

Correspondingly, the Laplace transforms of all integral and derivative terms will now involve 0^- rather than 0. Thus

$$\mathscr{L} f'(t) = sF(s) - f(0^-) \tag{6-39}$$

and so on for higher derivative terms.

6-5 USEFUL LAPLACE-TRANSFORM THEOREMS

There are four Laplace-transform theorems which when combined with the results of the preceding section provide a sufficient number of Laplace-transform pairs to take care of the vast majority of situations encountered in the study of electrical engineering. These are now described without proof.

Time-Displacement Theorem (Real Translation). This theorem states that if a function $f(t)$ is Laplace-transformable and if the $\mathscr{L}f(t) = F(s)$, then

$$\mathscr{L}f(t - T) = \varepsilon^{-sT}F(s) \tag{6-40}$$

The function $f(t - T)$ is the function $f(t)$ displaced by T. Hence by this theorem displacement in the time domain becomes multiplication by ε^{-sT} in the s domain. A useful application of this theorem is illustrated by the following example.

EXAMPLE 6-1 Find the Laplace transform of the pulse depicted in Fig. 6-4.

Solution: Note that the pulse may be considered as composed of two delayed unit steps. Accordingly, we may write as the time expression for the pulse

$$\text{pulse} = u(t - T_1) - u(t - T_2)$$

then

$$\mathscr{L}(\text{pulse}) = \mathscr{L}u(t - T_1) - \mathscr{L}u(t - T_2) = \frac{\varepsilon^{-sT_1}}{s} - \frac{\varepsilon^{-sT_2}}{s} \tag{6-41}$$

Again note that the Laplace transform converts a discontinuous time function into an expression which is analytical[†] in the s domain.

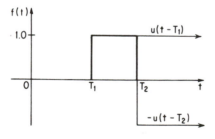

Fig. 6-4 Pulse as the superposition of two delayed steps.

Complex Translation. Of particular concern to us in our study throughout the book is the meaning attached to transform functions which involve a displacement in the complex variable s. We already know how to interpret any of the $F(s)$ functions treated in the preceding section. However, suppose now that for any one of these expressions the variable s is replaced by $(s + \alpha)$. How does this influence the corresponding time function? The answer is furnished by the complex translation theorem, which states that if $f(t)$ is Laplace-transformable and has the transform $F(s)$, then multiplication of $f(t)$ by the exponential time function $\varepsilon^{-\alpha t}$ becomes a translation in the s domain, and vice versa. Expressed

[†] That is, all derivatives of Eq. (6-41) exist in the s-plane except at $s = 0$.

mathematically, we can write

$$\mathscr{L}\varepsilon^{-\alpha t}f(t) = F(s + \alpha) \qquad (6\text{-}42)$$

where $\mathscr{L}f(t)$ is $F(s)$.

By applying this theorem to known Laplace-transform pairs we may readily derive additional Laplace-transform pairs, as illustrated by the following two examples.

EXAMPLE 6-2   Find the Laplace transform of the exponentially damped cosine function.

Solution: In accordance with the complex translation theorem this merely requires replacing s by $s + \alpha$ in the expression for the Laplace transform of the pure cosine function. Thus

$$\mathscr{L}\varepsilon^{-\alpha t}\cos \omega t = \frac{s + \alpha}{(s + \alpha)^2 + \omega^2} \qquad (6\text{-}43)$$

This result appears as item 10 in Table 6-1.

EXAMPLE 6-3   Find the Laplace transform of $\varepsilon^{-\alpha t}t^n$.

Solution: Applying the complex translation theorem to Eq. (6-27) yields

$$\mathscr{L}\varepsilon^{-\alpha t}t^n = \frac{n!}{(s + \alpha)^{n+1}} \qquad (6\text{-}44)$$

This is a very useful result as is demonstrated later in the book.

Final-Value Theorem.   This theorem states that if $f(t)$ and its first derivative $f'(t)$ are Laplace-transformable and the $\mathscr{L}f(t) = F(s)$ and if the poles of $sF(s)$ lie *inside* the left half of the s-plane, then

$$\lim_{s \to 0} sF(s) = \lim_{t \to \infty} f(t) \qquad (6\text{-}45)$$

Equation (6-45) is useful in those situations where the transform solution of a problem is available and all that is needed is information about the final or steady-state solution. By performing the operation called for on the left side of Eq. (6-45), we obtain information about the final value without ever having to evaluate the complete solution in the time domain.

Initial-Value Theorem.   This theorem states that if $f(t)$ and $f'(t)$ are Laplace-transformable, $\mathscr{L}f(t) = F(s)$, and the $\lim sF(s)$ exists, then

$$\lim_{s \to \infty} sF(s) = \lim_{t \to 0} f(t) \qquad (6\text{-}46)$$

Equation (6-46) permits us to evaluate the initial value of the time-domain solution $f(t)$ without ever having to determine $f(t)$ formally. One can work directly with the transform version of the solution to obtain initial values by performing the operation called for on the left side of Eq. (6-46).

6-6 EXAMPLE: LAPLACE-TRANSFORM SOLUTION OF FIRST-ORDER EQUATION

In the interest of bringing the material treated so far into sharper focus and to provide background for the next section, we turn attention next to the Laplace method of solving a linear, first-order differential equation. Refer to Fig. 6-5, which depicts a mechanical system comprised of a spring constant K and a viscous damper D. Mass is negligible. Let it be desired to describe as a function of time the velocity of the lower end of the spring, point p, when it is subjected to a constant downward force of magnitude F. Also, because initial conditions can be handled with ease in a direct and formal way by the Laplace-transform method, let us not hesitate to include this in the analysis by assuming that the spring is already stretched by an amount y_0 in the downward direction owing to a previous excitation. The equation which defines the velocity of point p is readily obtained by equating the applied force F to the sum of the opposing forces. Thus

$$Fu(t) = Dv(t) + K \int v(t)\, dt \qquad (6\text{-}47)$$

Note that the equation has been written to bring into evidence the velocity $v(t)$ which is the variable of interest. The first term on the right side refers to the opposing torque associated with the viscous friction disk. The viscous friction coefficient D has units of force/velocity. The second term involves the integral of velocity which is displacement, and this displacement multiplied by the spring constant K, having units of force/displacement, yields the opposing force represented by the further stretching of the spring. The applied-force term F has

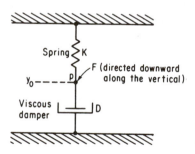

Fig. 6-5 Spring-damper arrangement. Force applied to lower end of spring at point p.

the symbol $u(t)$ attached in order to emphasize that the force is applied at $t = 0$.

As the first step in the systematic Laplace-transform solution of this equation, it is necessary to identify the desired solution $v(t)$ in terms of its s-domain counterpart. Accordingly, we have

$$\mathscr{L}v(t) = V(s) \qquad (6\text{-}48)$$

It is customary to use lower-case letters to denote the dependent variable in the time domain, and capital letters to denote the dependent variable in the s domain. Next we Laplace transform each term of Eq. (6-47). Thus

$$F\mathscr{L}u(t) = D\mathscr{L}v(t) + K\mathscr{L} \int v(t)\, dt \qquad (6\text{-}49)$$

Then from item 1 in Table 6-1

$$F\mathscr{L}u(t) = \frac{F}{s} \qquad (6\text{-}50)$$

Furthermore, from Eq. (6-11) the Laplace transform of the integral term is

$$K\mathscr{L}\int v(t)\,dt = K\left[\frac{V(s)}{s} + \frac{\int_{-\infty}^{0} v_0\,dt}{s}\right] = K\left[\frac{V(s)}{s} + \frac{y_0}{s}\right] \tag{6-51}$$

where $\int_{-\infty}^{0} v_0\,dt$ denotes the initial displacement y_0 stored in the spring and brought about by an assumed previously applied velocity v_0.

Inserting Eqs. (6-48), (6-50), and (6-51) into Eq. (6-49) yields the Laplace-transformed differential equation. Thus

$$\frac{F}{s} = DV(s) + K\left[\frac{V(s)}{s} + \frac{y_0}{s}\right] \tag{6-52}$$

Since this equation involves only the single unknown quantity $V(s)$, the solution is immediately available in the s domain as

$$V(s) = \frac{F}{D}\frac{1}{s + K/D} - \frac{Ky_0}{D}\frac{1}{s + K/D} \tag{6-53}$$

It is important here to understand that Eq. (6-53) is the complete solution to the differential equation as it appears in the transform domain. To find the corresponding time solution, it is necessary to identify the meaning of terms such as $1/(s + K/D)$. If we continue to draw upon our experience of doing multiplication by logarithms, we can be led to expect that the inversion process of going from the s domain to the time domain can effectively be accomplished by the use of tables. Of course we understand that we must first prove that such a procedure is valid. We can accomplish this proof by means of the theory of functions of a complex variable; therefore Table 6-1 can be used in a bilateral fashion. On this basis, then, the term $1/(s + K/D)$ has a unique counterpart in the time domain, which by item 2 in Table 6-1 is identified as the exponential function $\varepsilon^{-(K/D)t}$. It therefore follows that the time solution which corresponds to the s-domain solution of Eq. (6-53) is

$$v(t) = \frac{F}{D}\varepsilon^{-(K/D)t} - \frac{Ky_0}{D}\varepsilon^{-(K/D)t} \tag{6-54}$$

This is the complete solution to the problem described by Eq. (6-47) with the specified initial condition. Note that the initial condition is responsible for the second term on the right side of Eq. (6-54).

The foregoing example may also be used to illustrate the final- and initial-value theorems. A glance at Eq. (6-54) indicates that the final value (i.e., $t \to \infty$) of the velocity of point p is zero. This result can also be obtained by applying the final-value theorem to the transformed solution. Thus

$$v_{\text{final}} = \lim_{s \to 0} s[V(s)] \tag{6-55}$$

$$= \lim_{s \to 0} s\left[\frac{F}{D}\frac{1}{s + K/D} - \frac{Ky_0}{D}\frac{1}{s + K/D}\right] = 0$$

Note that there is no need to deal with the solution as a time function in order to obtain information about the final value of the solution. It can be done solely in terms of the s-plane expression, provided the specified conditions on the use of the theorem are satisfied.

Inspection of Eq. (6-54) also reveals that the initial value of the velocity immediately after application of the force is given by

$$v_{initial} = \frac{F}{D} - \frac{Ky_0}{D} \tag{6-56}$$

That this same result is obtained from the initial-value theorem is readily demonstrated as follows:

$$v_{initial} = \lim_{s \to \infty} sV(s) = \lim_{s \to \infty} s \left[\frac{F/D}{s + K/D} - \frac{Ky_0/D}{s + K/D} \right]$$

$$= \lim_{s \to \infty} \left[\frac{F/D}{1 + \dfrac{1}{s}\dfrac{K}{D}} - \frac{Ky_0/D}{1 + \dfrac{1}{s}\dfrac{K}{D}} \right] = \frac{F}{D} - \frac{Ky_0}{D} \tag{6-57}$$

Although a comparison of the two methods of finding the initial value for the example at hand shows the initial-value theorem at a slight disadvantage in terms of the effort involved, this is not true when the s-domain solution is more complicated.

The process of finding the time solution by going from the s domain to the time domain through the use of appropriate tables can be described as *performing the inverse Laplace transformation*. In the next section we consider the matter of manipulating complicated expressions of transformed solutions into forms which are available in our listing of Laplace-transform pairs as presented in Table 6-1.

6-7 INVERSE LAPLACE TRANSFORMATION THROUGH PARTIAL-FRACTION EXPANSION

In Sec. 6-2 it is shown that Eq. (6-15) is the transformed solution of the integrodifferential equation (6-1). In the case where $F(s)$ in Eq. (6-15) corresponds to a unit-impulse function the expression for the transformed solution may be written as†

$$I(s) = \frac{a_2 s^2 + a_1 s + a_0}{s(s^3 + b_2 s^2 + b_1 s + b_0)} \tag{6-58}$$

Now if this form of the s-plane solution were available in a table of Laplace-transform pairs, the time solution could be written forthwith. A little thought reveals, however, that such a compilation is impractical because to cover all cases the table would have to be endless. Therefore it is customary to make available only a limited number of the basic forms and then to treat all other

---

† Note that the a-coefficients in Eq. (6-58) are not the same as those appearing in Eq. (6-15).

cases by reducing them through partial-fraction expansion into the basic forms. To accomplish this in the case of Eq. (6-58), it is necessary first to find the roots of the cubic equation by any of the standard methods of algebra.‡ For the sake of illustration assume these roots are found to be s_1, s_2, and s_3 and that each is real. Then Eq. (6-58) may be rewritten in terms of a partial-fraction expansion as

$$I(s) = \frac{a_2 s^2 + a_1 s + a_0}{s(s - s_1)(s - s_2)(s - s_3)} = \frac{K_0}{s} + \frac{K_1}{s - s_1} + \frac{K_2}{s - s_2} + \frac{K_3}{s - s_3} \qquad (6\text{-}59)$$

Now since each term on the right side of this expression is identifiable in Table 6-1, a complete description of the time solution merely requires an evaluation of the K coefficients.

Before proceeding further with the partial-fraction expansion as a means of facilitating the evaluation of the inverse Laplace transformation, let us formulate the transformed solution of Eq. (6-59) in completely general terms. Accordingly, we may write

$$I(s) = \frac{a_m s^m + a_{m-1} s^{m-1} + \cdots + a_1 s + a_0}{s^n + b_{n-1} s^{n-1} + \cdots + b_1 s + b_0} = \frac{N(s)}{D(s)} \qquad (6\text{-}60)$$

Since in the vast majority of physical situations the order of the denominator polynomial is very often greater than that of the numerator, i.e., $n > m$, the partial-fraction expansion is directly applicable. In those rare cases where $n = m$ it is necessary to do longhand division in order that $I(s)$ be in the form of a proper fraction. Because of its infrequent occurrence this case will receive no further attention. Hence in the material that follows $I(s)$ is assumed to be expressible as a proper fraction so that partial-fraction expansion is immediately applicable. However, once this is done, the particular manner of evaluating the coefficients of the expansion is dependent upon the character of the roots of $D(s)$ in Eq. (6-60). Attention is now directed to the three cases of interest which arise.

I(s) Contains First-Order Poles Only. Assume that the roots of the denominator polynomial

$$D(s) = s^n + b_{n-1} s^{n-1} + \cdots + b_1 s + b_0 = 0 \qquad (6\text{-}61)$$

are all real and distinct. Then Eq. (6-60) may be rewritten in terms of a partial-fraction expansion as

$$\begin{aligned} I(s) &= \frac{N(s)}{(s - s_1)(s - s_2) \cdots (s - s_n)} \\ &= \frac{K_1}{s - s_1} + \frac{K_2}{s - s_2} + \cdots + \frac{K_k}{s - s_k} + \cdots + \frac{K_n}{s - s_n} \end{aligned} \qquad (6\text{-}62)$$

To evaluate any one of the coefficients it is necessary to isolate it in Eq. (6-62). This is readily accomplished by multiplying both sides of the equation by the

‡ This part of the solution procedure is also necessary when solving differential equations by the classical method.

root factor of the coefficient of interest. Thus, to find K_k requires multiplying through by $(s - s_k)$ and then setting s equal to s_k. This leads to the following general formulation for determining the coefficients.

$$K_k = \left[(s - s_k)\frac{N(s)}{D(s)} \right]_{s=s_k} \qquad (6\text{-}63)$$

Once the K_k coefficient is found, the contribution of the Kth term in the expansion is obtained by recognizing that the inverse Laplace transform of $1/(s - s_k)$ is $\varepsilon^{s_k t}$. It follows then that the complete time solution corresponding to Eq. (6-62) is

$$i(t) = \mathcal{L}^{-1}I(s) = \sum_{k=1}^{n} K_k \varepsilon^{s_k t} \qquad \text{for } t>0 \qquad (6\text{-}64)$$

where K_k is given by Eq. (6-63) and $\mathcal{L}^{-1}$ is the symbol which denotes the inverse Laplace transformation.

EXAMPLE 6-4 Find the inverse Laplace transform of

$$I(s) = \frac{s + 3}{s(s + 1)(s + 2)}$$

Solution: By a partial-fraction expansion we can write

$$I(s) = \frac{s + 3}{s(s + 1)(s + 2)} = \frac{K_1}{s} + \frac{K_2}{s + 1} + \frac{K_3}{s + 2}$$

Applying the operation called for in Eq. (6-63) yields

$$K_1 = [sI(s)]_{s=0} = \left[\frac{s + 3}{(s + 1)(s + 2)} \right]_{s=0} = \frac{3}{2}$$

$$K_2 = [(s + 1)I(s)]_{s=-1} = \left[\frac{s + 3}{s(s + 2)} \right]_{s=-1} = -2$$

$$K_3 = [(s + 2)I(s)]_{s=-2} = \left[\frac{s + 3}{s(s + 1)} \right]_{s=-2} = \frac{1}{2}$$

Hence

$$i(t) = \mathcal{L}^{-1}I(s) = \mathcal{L}^{-1}\frac{3}{2}\frac{1}{s} - 2\mathcal{L}^{-1}\frac{1}{s + 1} + \frac{1}{2}\mathcal{L}^{-1}\frac{1}{s + 2} = \frac{3}{2} - 2\varepsilon^{-t} + \frac{1}{2}\varepsilon^{-2t}$$

I(s) Contains a Pair of Complex Poles Plus One First-Order Pole. In this case $I(s)$ may be written as

$$I(s) = \frac{N(s)}{(s - s_1)(s - s_1^*)(s - s_2)} = \frac{K_1}{s - s_1} + \frac{K_1^*}{s - s_1^*} + \frac{K_2}{s - s_2} \qquad (6\text{-}65)$$

where s_1 and s_1^* are the complex conjugate roots and s_2 is the real root. Assume

$$s_1 = -\alpha + j\omega \qquad \text{and} \qquad s_1^* = -\alpha - j\omega$$

Then the coefficients K_1 and K_1^* are also, in general, complex conjugate. Hence we may write

$$K_1 = A + jB \quad \text{and} \quad K_1^* = A - jB$$

Inserting these quantities into Eq. (6-65) permits us to write it as

$$I(s) = \frac{A + jB}{(s + \alpha) - j\omega} + \frac{A - jB}{(s + \alpha) + j\omega} + \frac{K_2}{s - s_2} \tag{6-66}$$

Combining the first two terms on the right side and simplifying leads to

$$I(s) = \frac{2A(s + \alpha) - 2B\omega}{(s + \alpha)^2 + \omega^2} + \frac{K_2}{s - s_2} \tag{6-67}$$

The values of A and B are found by applying Eq. (6-63). Thus

$$K_1 \equiv A + jB = \left[\frac{N(s)}{D(s)}(s - s_1)\right]_{s = -\alpha + j\omega} \tag{6-68}$$

Since evaluation of the right side will in general be a complex number, it follows that

$$A \equiv Re(K_1) \quad \text{and} \quad B \equiv \mathcal{I}m(K_1) \tag{6-69}$$

where $Re(K_1)$ denotes the real part of K_1, and $\mathcal{I}m(K_1)$ denotes the imaginary part of K_1. Keep in mind, too, that K_2 is evaluated in the usual fashion.

Equation (6-67) is in a form which permits the corresponding time functions to be identified from the standard types appearing in Table 6-1. Thus

$$i(t) = 2A\mathcal{L}^{-1}\frac{s + \alpha}{(s + \alpha)^2 + \omega^2} - 2B\mathcal{L}^{-1}\frac{\omega}{(s + \alpha)^2 + \omega^2} + K_2\mathcal{L}^{-1}\frac{1}{s - s_2} \tag{6-70}$$

Then by transform pairs 2, 9, and 10 the time solution becomes

$$i(t) = 2A\varepsilon^{-\alpha t}\cos\omega t - 2B\varepsilon^{-\alpha t}\sin\omega t + K_2\varepsilon^{s_2 t} \quad t > 0 \tag{6-71}$$

I(s) Contains Multiple Poles Plus One First-Order Pole. In the interest of illustrating the procedure more fully, a root of $D(s)$ of multiplicity three is assumed. If we call this root s_0, the expression for $I(s)$ becomes

$$I(s) = \frac{N(s)}{(s - s_0)^3(s - s_1)} = \frac{K_{01}}{(s - s_0)^3} + \frac{K_{02}}{(s - s_0)^2} + \frac{K_{03}}{s - s_0} + \frac{K_1}{s - s_1} \tag{6-72}$$

Note that when a multiple root is involved, the rules of the partial-fraction expansion require that there be as many coefficients associated with the multiple root as the order of the multiplicity. Moreover, these coefficients must be associated with fractions involving the multiple root which differ in power by one in the manner exhibited on the right side of Eq. (6-72). The general technique for evaluating the coefficients remains the same; it requires isolation through appropriate algebraic manipulation. In the case of K_{01} this merely necessitates multiplying both sides of Eq. (6-72) by $(s - s_0)^3$.

$$(s - s_0)^3 \frac{N(s)}{D(s)} = K_{01} + (s - s_0)K_{02} + (s - s_0)^2 K_{03} + (s - s_0)^3 \frac{K_1}{s - s_1} \quad (6\text{-}73)$$

Upon insertion of $s = s_0$ all terms on the right side drop out with the exception of K_{01}. Hence

$$K_{01} = \left[(s - s_0)^3 \frac{N(s)}{D(s)} \right]_{s=s_0} \quad (6\text{-}74)$$

To evaluate K_{02} requires that it, too, be made to stand alone. Inspection of Eq. (6-73) indicates that the most expedient way of accomplishing this is to differentiate both sides of the equation. Thus

$$\frac{d}{ds}\left[(s - s_0)^3 \frac{N(s)}{D(s)} \right]$$
$$= 0 + K_{02} + 2(s - s_0)K_{03} + 3(s - s_0)^2 \frac{K_1}{s - s_1} - (s - s_0)^3 \frac{K_1}{(s - s_1)^2} \quad (6\text{-}75)$$

Again upon inserting $s = s_0$ we have

$$K_{02} = \left\{ \frac{d}{ds}\left[(s - s_0)^3 \frac{N(s)}{D(s)} \right] \right\}_{s=s_0} \quad (6\text{-}76)$$

A similar procedure makes available the expression for K_{03}. Differentiating Eq. (6-75) and setting $s = s_0$ yields

$$K_{03} = \left\{ \frac{1}{2}\frac{d^2}{ds^2}\left[(s - s_0)^3 \frac{N(s)}{D(s)} \right] \right\}_{s=s_0} \quad (6\text{-}77)$$

On the basis of Eqs. (6-76) and (6-77) a general expression for the coefficients associated with a multiple pole of $I(s)$ is readily seen to be

$$\boxed{K_{0m} = \left\{ \frac{1}{(m-1)!}\frac{d^{m-1}}{ds^{m-1}}\left[(s - s_0)^p \frac{N(s)}{D(s)} \right] \right\}_{s=s_0}} \quad (6\text{-}78)$$

where p denotes the order of multiplicity and $m = 1, 2, \ldots, p$.

The last remaining coefficient K_1 in Eq. (6-72) is determined by Eq. (6-63). Accordingly the time solution may now be expressed as

$$i(t) = K_{01}\mathscr{L}^{-1}\frac{1}{(s - s_0)^3} + K_{02}\mathscr{L}^{-1}\frac{1}{(s - s_0)^2}$$
$$+ K_{03}\mathscr{L}^{-1}\frac{1}{s - s_0} + K_1\mathscr{L}^{-1}\frac{1}{s - s_1} \quad (6\text{-}79)$$

By use of transform pairs 2 and 8 in Table 6-1 the final time solution becomes

$$\boxed{i(t) = \frac{K_{01}}{2}t^2\varepsilon^{s_0 t} + K_{02}t\varepsilon^{s_0 t} + K_{03}\varepsilon^{s_0 t} + K_1\varepsilon^{s_1 t}} \qquad t > 0 \quad (6\text{-}80)$$

Therefore by means of a partial-fraction expansion, Eq. (6-78), and a table of Laplace-transform pairs the time solution for a transformed solution involving multiple poles is found in a direct and systematic fashion.

6-8 STEP RESPONSE OF AN *RL* CIRCUIT

Appearing in Fig. 6-6 is the circuit arrangement of an initially deenergized inductor in series with a resistor. It is desired to find the complete solution for the current when the switch is closed. The governing differential equation results on applying Kirchhoff's voltage law to the circuit of Fig. 6-6. Hence

$$E = Ri + L\frac{di}{dt} \tag{6-81}$$

The current i is the solution being sought and E is the applied source function which causes the response i to exist. To emphasize that E exists only when the switch in Fig. 6-6 is closed, it is helpful to attach the unit step function $u(t)$ to the voltage source E in Eq. (6-81). Hence a more complete formulation of the governing differential equation of the circuit is

$$Eu(t) = Ri + L\frac{di}{dt} \tag{6-82}$$

Equation (6-82) is a linear differential equation of first order. It contains a single energy-storing element as evidenced by the first derivative term.

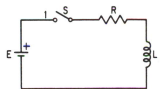

Fig. 6-6 Series *RL* circuit.

The first step in solving Eq. (6-82) by the Laplace-transform method is to Laplace transform both sides of the equation. Thus

$$E\mathscr{L}u(t) = R\mathscr{L}i(t) + L\mathscr{L}\frac{di(t)}{dt} \tag{6-83}$$

The desired solution in the time domain is identified as $i(t)$. The corresponding unknown function in the s domain is called $I(s)$. That is,

$$I(s) = \mathscr{L}i(t) \tag{6-84}$$

Inserting Eqs. (6-84), (6-39), and (6-17) into Eq. (6-83) then yields the transformed algebraic counterpart of the governing differential equation (6-82)

$$\frac{E}{s} = RI(s) + L[sI(s) - i(0^-)] \tag{6-85}$$

Here $i(0^-)$ is used to emphasize that information about the initial value of current immediately before application of the forcing function E is needed before a formal solution for the transformed response $I(s)$ can be written.

Because the switch in Fig. 6-6 is initially open, it follows that $i(0^-) = 0$. Hence Eq. (6-85) may now be explicitly written as

$$\frac{E}{s} = I(s)[R + sL] \tag{6-86}$$

Therefore, the transformed solution for the response becomes

$$I(s) = \frac{E/L}{s(s + R/L)} \tag{6-87}$$

To obtain the time solution which corresponds to Eq. (6-87) we must perform the inverse Laplace transformation, which requires putting $I(s)$ into a form whose terms are readily identifiable in Table 6-1. This result is achieved by using a partial-fraction expansion. Accordingly,

$$I(s) = \frac{E/L}{s(s + R/L)} = \frac{K_0}{s} + \frac{K_1}{s + R/L} \tag{6-88}$$

Then by the application of the operation called for in Eq. (6-63) there results

$$K_0 = \left[\frac{E/L}{s + R/L}\right]_{s=0} = \frac{E}{R} \tag{6-89}$$

and

$$K_1 = \left[\frac{E/L}{s}\right]_{s=-R/L} = -\frac{E}{R} \tag{6-90}$$

The transformed solution may now be written more conveniently as

$$I(s) = \frac{E}{R}\frac{1}{s} - \frac{E}{R}\left[\frac{1}{s + R/L}\right] \tag{6-91}$$

From Table 6-1 we know the equivalent of $1/s$ to be a unit step in the time domain. Also the equivalent of $1/(s + R/L)$ is $\varepsilon^{-(R/L)t}$ in the time domain. Hence the time solution which corresponds to the transformed solution of Eq. (6-91) is

$$i(t) = \mathscr{L}^{-1}I(s) = \frac{E}{R}\mathscr{L}^{-1}\frac{1}{s} - \frac{E}{R}\mathscr{L}^{-1}\frac{1}{s + R/L} \tag{6-92}$$

or

$$\boxed{i(t) = \frac{E}{R} - \frac{E}{R}\varepsilon^{-(R/L)t}} \qquad \text{for } t > 0 \tag{6-93}$$

In Eq. (6-92) the symbol $\mathscr{L}^{-1}$ is used to denote the inverse Laplace transformation. By means of Table 6-1 this operation is carried out by proceeding from the s-function column to the corresponding time-function column. Recall that the direct Laplace transformation involves going from the time function to the s-function column.

The Concept of Operational Impedance.   The governing differential equation for the *RL* circuit is given by Eq. (6-81). It is pointed out in Sec. 6-2 that initial-condition terms are treated in a manner which is similar to that of the source function. Hence for convenience we now assume that all initial conditions are zero. Laplace transforming Eq. (6-81) then leads to

$$E(s) = I(s)[R + sL] \qquad (6\text{-}94)$$

Here $E(s)$, the general representation of the Laplace transform of the forcing function, is used without identifying a specific form such as a step voltage. $E(s)$ can be the Laplace transform of any deterministic forcing function. Rearranging Eq. (6-94) yields

$$\frac{E(s)}{I(s)} = R + sL \equiv Z(s) \qquad (6\text{-}95)$$

Since the left side of Eq. (6-95) involves the ratio of a voltage to a current, the equivalent expression on the right side is called the *operational impedance* and is denoted by $Z(s)$. The term impedance is preferred in this description because both resistance and inductance are involved. Furthermore, the adjective operational is used because it is through the operation of differentiation that the factor s appears in Eq. (6-95) as a multiplier of the inductance parameter. The concept of operational impedance is useful because by means of it one can write the response $I(s)$ caused by a forcing function $E(s)$ by merely applying Ohm's law in a more general form. Thus the response as expressed in the s domain may be written directly as

$$I(s) = \frac{E(s)}{Z(s)} = \frac{E(s)}{R + sL} \qquad (6\text{-}96)$$

An interesting and very significant aspect of this formulation is that both the forced and the transient solutions are obtainable from Eq. (6-96) for zero initial conditions. For the sake of illustration consider the case where the forcing function is a step voltage of magnitude E. Accordingly, $E(s) = E/s$ so that Eq. (6-96) then becomes

$$I(s) = \frac{E}{s}\frac{1}{R + sL} = \frac{E/L}{s(s + R/L)} \qquad (6\text{-}97)$$

A comparison of this equation with Eq. (6-87) shows them to be identical. This happens here because in both cases initial conditions are zero. It is interesting to note in Eq. (6-97) that the presence of s in the denominator is associated with the step forcing function, and this in turn is responsible for generating the steady-state solution. This fact is borne out by Eq. (6-88) and (6-92). On the other hand, the operational impedance $(R + sL)$ is associated with the generation of the transient term, as also revealed by Eqs. (6-88) and (6-92).

In many instances in electrical engineering, particularly in control systems, it is useful to formulate the ratio of the output response† (in this case, current)

---

† The output quantity need not be restricted to current. For example, it could be taken as the voltage across R or L, in which case a different transfer function results.

to the input forcing function (voltage) of the Laplace-transformed function for zero initial conditions. This ratio is called the *transfer function* of the circuit. In the case of the *RL* circuit the transfer function can be written as

$$\frac{I(s)}{E(s)} = \frac{1}{Z(s)} = \frac{1}{R + sL} \tag{6-98}$$

Note that it is unnecessary to specify the specific form of the forcing function. As the right side of Eq. (6-98) indicates, the transfer function indirectly conveys information about the transient response of the circuit. The transfer function involves just the circuit parameters.

EXAMPLE 6-5 In the circuit depicted in Fig. 6-7, the battery has been applied for a long time. For the values of the parameters indicated find the complete expression for the current after closing the switch which removes R_1 from the circuit.

Solution: The governing differential equation which applies when the switch is closed is

$$Eu(t) = R_2 i + L\frac{di}{dt} \tag{6-99}$$

Laplace transforming yields

$$\frac{E}{s} = R_2 I(s) + sLI(s) - Li(0^-) \tag{6-100}$$

$$\frac{E}{s} + Li(0^-) = I(s)[R_2 + sL] \tag{6-101}$$

Equation (6-101) is written with the forcing function and initial conditions on the left side and the transformed response and the operational impedance on the right side. This formulation indicates that the operational impedance may be singled out even in the presence of nonzero initial conditions.

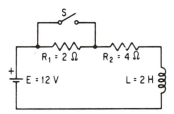

Fig. 6-7 Circuit arrangement for Example 6-5.

From Eq. (6-101) the solution in the *s* domain may accordingly be written as

$$I(s) = \frac{E + sLi(0^-)}{s(R_2 + sL)} = \frac{E/L + si(0^-)}{s(s + R_2/L)} \tag{6-102}$$

A comparison of Eq. (6-102) with Eq. (6-97) shows how an initial-condition term enters into the formulation of the solution. It is combined with the forcing function to give an altered magnitude.

By means of a partial-fraction expansion Eq. (6-102) becomes

$$I(s) = \frac{K_0}{s} + \frac{K_1}{s + R_2/L} \tag{6-103}$$

Then

$$K_0 = \left[\frac{E/L + si(0^-)}{s + R_2/L}\right]_{s=0} = \frac{E}{R_2} = \frac{12}{4} = 3 \text{ A} \tag{6-104}$$

and

$$K_1 = \left[\frac{E/L + si(0^-)}{s}\right]_{s=-R_2/L} = -\frac{E}{R_2} + i(0^-) \tag{6-105}$$

The current just before switching is given by

$$i(0^-) = \frac{E}{R_1 + R_2} = \frac{12}{2 + 4} = 2 \text{ A} \tag{6-106}$$

Inserting this result into Eq. (6-105) yields

$$K_1 = -\frac{E}{R_2} + i(0^-) = -\frac{12}{4} + 2 = -1 \tag{6-107}$$

Accordingly, the expression for $I(s)$ from Eq. (6-103) becomes

$$I(s) = \frac{3}{s} - \frac{1}{s + R_2/L} = \frac{3}{s} - \frac{1}{s + 2} \tag{6-108}$$

Performing the inverse Laplace transformation by means of items 1 and 2 of Table 6-1 furnishes the desired response.

$$i(t) = 3 - \varepsilon^{-2t} \tag{6-109}$$

As a check note that at $t = 0^+$ the value of $i(t)$ is 2 A and upon full decay of the transient it assumes a value of 3 A.

The time constant for this problem is

$$T = \frac{L}{R} = \frac{2\text{H}}{4\,\Omega} = \frac{1}{2} s \tag{6-110}$$

Thus after 2.5 s the transient term has a value less than 1% of its initial value of 1 A.

6-9 STEP RESPONSE OF AN *RC* CIRCUIT

Appearing in Fig. 6-8 is the circuit arrangement of a resistor in series with a capacitor. The capacitor is assumed to have an initial charge of value q_0 owing to the flow of a previously applied current. Let us now find the complete expression for the current after the switch S is closed. The governing differential equation is obtained upon applying Kirchhoff's voltage law to the closed circuit. Thus

$$Eu(t) = Ri + \frac{1}{C}\int i \, dt \tag{6-111}$$

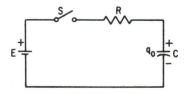

Fig. 6-8 Series *RC* circuit. The capacitor has an initial charge q_0.

Laplace transforming each term yields

$$E(s) = RI(s) + \frac{1}{C}\left[\frac{I(s)}{s} + \frac{i^{-1}(0)}{s}\right]$$

(6-112)

where

$$i^{-1}(0) = \int_{-\infty}^{0} i \, dt = q_0$$

(6-113)

and represents the accumulation of charge on the capacitor brought about by the flow of current due to a previously applied action. This previous action is stressed in Eq. (6-113) by taking the limits of the integral from $-\infty$ to 0 where 0 refers to the instant of the closing of the switch in Fig. 6-7. Equation (6-112) may be rewritten as

$$E(s) = RI(s) + \frac{1}{C}\left[\frac{I(s)}{s} + \frac{q_0}{s}\right]$$

(6-114a)

or

$$E(s) - \frac{q_0}{Cs} = I(s)\left[R + \frac{1}{sC}\right]$$

(6-114b)

On the left side appear the source function and the effect of the initial condition. Appearing in brackets on the right side is the operational impedance of the RC circuit.

For a step-voltage forcing function $E(s) = E/s$, so that Eq. (6-114b) becomes

$$\frac{1}{s}\left(E - \frac{q_0}{C}\right) = I(s)\left[R + \frac{1}{sC}\right]$$

(6-115)

The quantity in parentheses in Eq. (6-115) represents the effective voltage acting in the circuit to produce current. The transformed solution for the current is therefore

$$I(s) = \frac{E - q_0/C}{s(R + 1/sC)} = \frac{E - q_0/C}{R(s + 1/RC)}$$

(6-116)

The corresponding total solution in the time domain readily follows from item 2 of Table 6-1. Thus

$$i(t) = \left(\frac{E}{R} - \frac{q_0}{RC}\right)\varepsilon^{-t/RC}$$

(6-117)

A plot of this equation as a function of time is depicted in Fig. 6-9. Note that

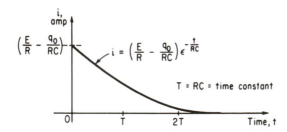

Fig. 6-9 Exponential decay of the current in an RC circuit.

at $t = 0^+$ there occurs a step change in current in the circuit. Its magnitude is equal to the difference between the applied voltage E and the initial voltage across the capacitor q_0/C divided by the resistance of the circuit. As time elapses this current gradually decays to zero.

An inspection of the exponential term of Eq. (6-117) reveals that the time constant of an RC circuit is given by

$$\boxed{T = RC} \quad \text{s} \tag{6-118}$$

The complete current solution in the case of the simple RC circuit consists merely of the transient term because the capacitor presents an open circuit to the battery source at steady state. Hence the forced solution is zero in this instance.

EXAMPLE 6-6 In the circuit of Fig. 6-10 the capacitor has an initial charge of such value that the voltage appearing across its plates is 40 V. For the values of the parameters indicated find the expression for the current which flows when the switch is closed. Also, find the total energy dissipated in the resistor during the transient state and compare it with the energy initially stored in the capacitor.

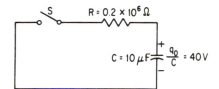

Fig. 6-10 Circuit configuration for Example 6-6.

Solution: Taking voltage drops in a clockwise direction as positive, Kirchhoff's voltage law applied to the circuit with S closed yields

$$0 = Ri + \frac{1}{C} \int i \, dt \tag{6-119}$$

Upon Laplace transforming there results

$$0 = RI(s) + \frac{1}{C} \left[\frac{I(s)}{s} + \frac{q_0}{s} \right] \tag{6-120}$$

or

$$-\frac{q_0}{Cs} = I(s)\left(R + \frac{1}{sC} \right) \tag{6-121}$$

By the statement of the problem $q_0/C = 40$ V. Hence

$$I(s) = \frac{-40}{s(R + 1/sC)} = -\frac{40}{R}\left(\frac{1}{s + 1/RC} \right) \tag{6-122}$$

From Table 6-1 the corresponding time solution is found to be

$$i(t) = -\frac{40}{R}\varepsilon^{-t/RC} = -2 \times 10^{-4}\varepsilon^{-t/2} \text{ A} \tag{6-123}$$

The negative sign indicates that the actual current flow is counterclockwise rather than clockwise as initally assumed.

The amount of energy dissipated in the resistor is readily found from

$$W_{diss} = \int_0^\infty i^2(t)R\,dt = \int_0^\infty (4 \times 10^{-8})2 \times 10^5 \varepsilon^{-t}\,dt$$

$$= -8 \times 10^{-3}[\varepsilon^{-t}]_0^\infty = 8 \times 10^{-3}\,J \tag{6-124}$$

The amount of energy initially stored in the electric field of the capacitor is given by

$$W_e = \frac{1}{2}CV_0^2 = \frac{1}{2}10^{-5}(40)^2 = 8 \times 10^{-3}\,J \tag{6-125}$$

A comparison of Eqs. (6-124) and (6-125) shows that all of the capacitor energy is dissipated as heat after the switch is closed.

6-10 THE IMPULSE RESPONSE

A direct solution of the impulse response of the series RC circuit is obtainable from a Laplace transform solution of Eq. (5-119). Thus, applying the Laplace transform to each term of Eq. (5-119) allows us to write

$$A = RI(s) + \frac{1}{C}\left[\frac{I(s)}{s} + \frac{\int_{-\infty}^{0^-} i\,dt}{s}\right] \tag{6-126}$$

But the initial condition on the capacitor at time 0^- is zero. Accordingly, the last expression becomes simply

$$A = \left(R + \frac{1}{sC}\right)I(s) = \frac{R}{s}\left(s + \frac{1}{RC}\right)I(s) \tag{6-127}$$

or

$$I(s) = \frac{A}{R}\left(\frac{s}{s + 1/RC}\right) = \frac{A}{R}\left[1 - \frac{1/RC}{s + 1/RC}\right] \tag{6-128}$$

The bracketed version of the transformed solution is in the form of a proper fraction. Hence the inverse Laplace transform may be found by use of Table 6-1. Thus

$$i(t) = \mathcal{L}^{-1}I(s) = \mathcal{L}^{-1}\frac{A}{R} - \mathcal{L}^{-1}\frac{A}{RT}\frac{1}{s + 1/RC} \tag{6-129}$$

which by transform pairs 17 and 2 yields the time expression for the current response of

$$i(t) = \frac{A}{R}\delta(t) - \frac{A}{RT}\varepsilon^{-(t/T)} \tag{6-130}$$

Then, by Eq. (5-123), this may be rewritten as

$$i(t) = Q\delta(t) - \frac{A}{RT}\varepsilon^{-(t/T)} \tag{6-131}$$

Examination of Eq. (6-131) discloses that the total current response consists of a current impulse of magnitude $Q = A/R$ occurring in the time interval from 0^- to 0^+, and a second component, $-(A/RT)\varepsilon^{-(t/T)}$, which exists for all time from 0^+ to ∞. It is instructive to note the consistency of Eq. (6-131) with Eqs. (5-122) and (5-123).

6-11 STEP RESPONSE OF SECOND-ORDER SYSTEM (*RLC* CIRCUIT)

Let it be desired to find the complete current response in the circuit of Fig. 6-11 after the switch is closed. Assume that the capacitor has an initial charge of q_0. The governing differential equation for the circuit is found upon applying Kirchhoff's voltage law. Thus

$$Eu(t) = Ri + L\frac{di}{dt} + \frac{1}{C}\int i \, dt \qquad (6\text{-}132)$$

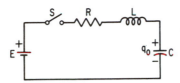

Fig. 6-11 Series *RLC* circuit with an initial charge on *C*.

Equation (6-132) is a second-order nonhomogeneous differential equation because it involves a derivative as well as an integral term. Laplace transforming each term of the equation yields

$$\frac{E}{s} = RI(s) + sLI(s) + \frac{1}{C}\left[\frac{I(s)}{s} + \frac{q_0}{s}\right] \qquad (6\text{-}133)$$

where

$$q_0 = i^{-1}(0^-) = \int_{-\infty}^{0^-} i \, dt$$

and represents the accumulation of charge on the capacitor owing to a previous current flow. Rearranging, we get

$$\left(\frac{E}{s} - \frac{q_0}{sC}\right) = I(s)\left[R + sL + \frac{1}{sC}\right] \qquad (6\text{-}134)$$

The function in brackets on the right side is the operational impedance of the series *RLC* circuit. The expression for the current solution in the *s* domain then readily becomes

$$I(s) = \frac{\dfrac{E}{s} - \dfrac{q_0}{sC}}{\dfrac{L}{s}\left(s^2 + \dfrac{R}{L}s + \dfrac{1}{LC}\right)} = \frac{\dfrac{E}{L} - \dfrac{q_0}{LC}}{\left(s^2 + \dfrac{R}{L}s + \dfrac{1}{LC}\right)} \qquad (6\text{-}135)$$

A glance at Eq. (6-135) shows that the denominator expression does not include an s factor standing alone as was the case with Eq. (6-87) for the *RL* circuit. This means that a constant term in the steady-state solution does not exist. A glance at the circuit configuration verifies this conclusion because in the steady state the capacitor presents an open circuit to the battery so that the particular solution must be zero.

It is significant to note that the quadratic expression in the denominator of Eq. (6-135) results from an algebraic manipulation of the operational impedance. Accordingly, it involves each of the three circuit parameters. Moreover, the expression is quadratic because of the presence of two energy-storing elements. From the experience we have gained so far in evaluating the inverse Laplace transform, we know that the first step in the procedure is to identify the specific root factors of the poles of the transformed solution $I(s)$. This then permits a partial-fraction expansion to be made, each root factor being readily identifiable in Table 6-1. For each of the cases treated so far there was no need to find the root factors of the denominator of $I(s)$ because they automatically appeared in the desired form, as reference to Eqs. (6-87) and (6-116) indicates. This simplicity was a consequence of dealing with first-order circuits. Before we can proceed further in the solution procedure for $i(t)$ in the situation now under consideration, we must first find the specific roots of the quadratic expression. That is, we must determine those values of s which satisfy the equation

$$s^2 + \frac{R}{L}s + \frac{1}{LC} = 0 \qquad (6\text{-}136)$$

This expression is called the *characteristic equation* of the second-order system. Keep in mind that it results from setting the operational impedance equal to zero and then performing an algebraic manipulation which serves to isolate the s^2 term. The roots of this equation, which are also the poles of $I(s)$ in this instance, depend solely upon the circuit parameters. In turn these roots determine entirely the nature of the transient response. To understand this better, consider that s_1 and s_2 are the roots of the characteristic equation. That is,

$$(s - s_1)(s - s_2) = s^2 + \frac{R}{L}s + \frac{1}{LC} \qquad (6\text{-}137)$$

Then Eq. (6-135) can be rewritten as

$$I(s) = \frac{\dfrac{E}{L} - \dfrac{q_0}{LC}}{s^2 + \dfrac{R}{L}s + \dfrac{1}{LC}} = \frac{\dfrac{E}{L} - \dfrac{q_0}{LC}}{(s - s_1)(s - s_2)} = \frac{K_1}{s - s_1} + \frac{K_2}{s - s_2} \qquad (6\text{-}138)$$

The formal expression for the corresponding time solution then becomes

$$i(t) = K_1 \varepsilon^{s_1 t} + K_2 \varepsilon^{s_2 t} \qquad (6\text{-}139)$$

Thus the characteristic manner in which the transient terms decay is solely dependent on the roots s_1 and s_2.

By using the results appearing in Eq. (5-143) for s_1 and s_2 it is possible to write the characteristic equation entirely in terms of ζ and ω_n for the *underdamped* case. Thus

$$s^2 + \frac{R}{L}s + \frac{1}{LC} = s^2 + 2\zeta\omega_n s + \omega_n^2 = (s + \zeta\omega_n)^2 + \omega_d^2 \qquad (6\text{-}140)$$

Therefore, returning to Eq. (6-135), the expression for the solution in the s domain may be written

$$I(s) = \frac{\dfrac{E}{L} - \dfrac{q_0}{LC}}{s^2 + 2\zeta\omega_n s + \omega_n^2} = \frac{\dfrac{E}{L} - \dfrac{q_0}{LC}}{(s + \zeta\omega_n)^2 + \omega_d^2} \qquad (6\text{-}141)$$

To put this equation in a form which is readily identifiable in Table 6-1, it is necessary to multiply the right side of Eq. (6-141) by ω_d/ω_d. Then

$$I(s) = \left(\frac{E - q_0/C}{\omega_d L}\right)\left[\frac{\omega_d}{(s + \zeta\omega_n)^2 + \omega_d^2}\right] \qquad (6\text{-}142)$$

The quantity in parentheses is merely a constant, while that in brackets is seen to be the Laplace transform of an exponentially damped sine function as indicated by item 9 in Table 6-1. The corresponding solution in the time domain is therefore

$$i(t) = \frac{E - q_0/C}{\omega_d L}\varepsilon^{-\zeta\omega_n t}\sin\omega_d t \qquad (6\text{-}143)$$

This expression is the specific version of the formal solution represented by Eq. (6-139). The total solution in this case involves only a transient term because, as already pointed out, the forced solution is zero.

EXAMPLE 6-7   Refer to the circuit of Fig. 6-12. Assume it is initially deenergized.

(a) When this circuit is disturbed by the application of the source function $e(t)$, determine whether or not the ensuing dynamic response will be a damped oscillation.

(b) If the response in part (a) is found to be oscillatory, compute the time it takes for the transient response to decay to within 1 per cent of the initial value. (This is the setting time of the circuit.)

Solution: (a) The first step in arriving at a solution is to find the characteristic equation. Subsequent examination will then disclose the answers to the question. The characteristic equation can be found by obtaining the transformed solution of *any* of the circuit variables.

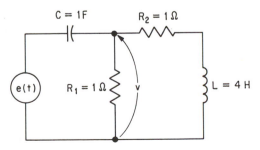

Fig. 6-12   Circuit of Example 6-7. Parameter values are chosen for convenience.

The characteristic equation is the corresponding denominator expression set equal to zero.

To illustrate, let the nodal voltage $V(s)$ in Fig. 6-12 be the circuit variable for which the transformed solution is to be found. Kirchhoff's current law at the node allows us to write

$$\frac{V(s) - E(s)}{1/sC} + \frac{V(s)}{R_1} + \frac{V(s)}{R_2 + sL} = 0 \tag{6-144}$$

Inserting the specified parameter values and rearranging terms yields

$$V(s)\left[s + 1 + \frac{4}{1 + 4s} \right] = sE(s) \tag{6-145}$$

from which the transformed solution for the nodal voltage becomes

$$V(s) = \frac{s(s + \frac{1}{4})}{s^2 + \frac{5}{4}s + \frac{5}{4}} E(s) \tag{6-146}$$

It is instructive to note here that the identity of the Laplace transform of the source function is unnecessary, because the character of the dynamic response is determined by the circuit parameters and the configuration in which they are placed and not at all by the form of the source function. The characteristic equation is now found by setting the denominator expression of $V(s)$ equal to zero. Thus

$$s^2 + \frac{5}{4}s + \frac{5}{4} = 0 \tag{6-147}$$

Comparing this equation with the standard formulation for a second-order system, that is,

$$s^2 + 2\zeta\omega_n s + \omega_n^2 = 0 \tag{6-148}$$

leads us to

$$\omega_n = \sqrt{\frac{5}{4}} = \frac{\sqrt{5}}{2} = 1.12 \tag{6-149}$$

and

$$\zeta\omega_n = \frac{1}{2}\left(\frac{5}{4}\right) = \frac{5}{8} \tag{6-150}$$

or

$$\zeta = \frac{\frac{5}{8}}{\omega_n} = \frac{\frac{5}{8}}{1.12} = 0.56 \tag{6-151}$$

Since the damping ratio ζ is less than unity, it follows that the ensuing response is a damped oscillation. The maximum overshoot is found from Fig. 5-25 to be 12%.

It is instructive to note here that if we had chosen to solve for any other variable, for example, the source current, the resulting transformed solution would lead to the same expression for the characteristic equation. This should not be a surprising result, for the reaction of the circuit to a disturbance anywhere in the circuit is the same.

(b) The settling time in this case corresponds to five time constants of the second-order system. Thus, from Eq. (5-152) we have

$$t_s = 5T = \frac{5}{\zeta\omega_n} = \frac{5}{\frac{5}{8}} = 8 \text{ s} \tag{6-152}$$

Eight seconds after this circuit is disturbed—whether by the application of a source function, or the discharge of the capacitor, or whatever—the ensuing transients will disappear.

6-12 COMPLETE RESPONSE OF *RL* CIRCUIT TO SINUSOIDAL INPUT

In this chapter up to this point attention has been confined to just one type of deterministic input, the step forcing function. In this section we now determine the total response of a series *RL* circuit when a second type of deterministic input, the sinusoidal function, is applied to the circuit terminals. The circuit diagram is depicted in Fig. 6-13.

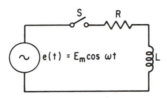

Fig. 6-13 Series *RL* circuit with sinusoidal forcing function.

   Since our concern is with finding the total solution, it follows that a steady-state solution as well as a transient solution must be determined. A point worthy of note here is that we shall be finding the steady-state solution to a sinusoidal forcing function in a linear *RL* series circuit in spite of the fact that we have not yet discussed the sinusoidal steady-state theory of electrical networks.† That we can do this in a rather simple and direct fashion is attributable to the use of Laplace transforms as a means of solving linear nonhomogeneous differential equations. This is just another of the advantages the Laplace transform has to offer as a solution procedure.

   The voltage source in the circuit of Fig. 6-13 is assumed to be varying in a cosinusoidal fashion. When the switch is closed the governing differential equation for the circuit becomes

$$E_m u(t) \cos \omega t = Ri + L\frac{di}{dt} \tag{6-153}$$

Since the switch is initially open, the initial value of the current through the inductor is obviously zero. With the use of Laplace-transform pair 4 of Table 6-1, the *s*-domain form of Eq. (6-153) is

$$\frac{E_m s}{s^2 + \omega^2} = I(s)[R + sL] \tag{6-154}$$

Hence the solution for the response in the *s* domain readily becomes available as

$$I(s) = \frac{E_m}{L}\left(\frac{s}{s^2 + \omega^2}\right)\frac{1}{s + R/L} \tag{6-155}$$

It is always impressive to see how easily the transformed solution is obtained with the Laplace transform. Also, note that the poles (or root factors of the

† This is the subject matter of Chapter 7.

denominator) of $I(s)$ are directly available. Therefore, a partial-fraction expansion can be written forthwith.

$$I(s) = \frac{s(E_m/L)}{(s^2 + \omega^2)(s + R/L)} = \frac{K_1}{s - j\omega} + \frac{K_1^*}{s + j\omega} + \frac{K_2}{s + R/L} \qquad (6\text{-}156)$$

The cosinusoidal function is responsible for the presence of the purely imaginary complex conjugate poles. The associated coefficients for these poles in the partial-fraction expansion must themselves in general be complex conjugate numbers. To stress this point the symbol K_1^* is used to denote that this quantity is the complex conjugate of K_1. The evaluation of any coefficient in the partial-fraction expansion is accomplished by following the usual procedure as called for by Eq. (6-63) or Eq. (6-68). Thus for K_1 we have

$$
\begin{aligned}
K_1 &= \left[(s - j\omega) \frac{s(E_m/L)}{(s - j\omega)(s + j\omega)(s + R/L)} \right]_{s = +j\omega} \\
&= \left[\frac{E_m}{L} \frac{j\omega}{j2\omega(j\omega + R/L)} \right] = \frac{E_m}{2} \frac{1}{R + j\omega L} \qquad (6\text{-}157) \\
&= \left(\frac{E_m}{2} \right) \frac{R - j\omega L}{R^2 + \omega^2 L^2}
\end{aligned}
$$

Since K_1^* must be the complex conjugate we can write forthwith

$$K_1^* = \frac{E_m}{2} \frac{R + j\omega L}{R^2 + \omega^2 L^2} \qquad (6\text{-}158)$$

Recall that the rule for finding the complex conjugate is to replace all the j terms with opposite signs. For K_2 the determining expression is

$$
\begin{aligned}
K_2 &= \left[\left(s + \frac{R}{L} \right) \frac{s(E_m/L)}{(s^2 + \omega^2)(s + R/L)} \right]_{s = -R/L} \\
&= \left[\frac{E_m}{L} \frac{s}{s^2 + \omega^2} \right]_{s = -R/L} = -\frac{E_m R}{R^2 + \omega^2 L^2} \qquad (6\text{-}159) \\
&= \frac{E_m}{\sqrt{R^2 + \omega^2 L^2}} \times \frac{R}{\sqrt{R^2 + \omega^2 L^2}} = -\frac{E_m}{\sqrt{R^2 + \omega^2 L^2}} \cos \theta
\end{aligned}
$$

where θ is defined by the right triangle shown in Fig. 6-14, i.e., $\theta = \tan^{-1}(\omega L/R)$. The introduction here of θ and of the algebraic manipulation which leads to its definition has the purpose of putting the final solution for the response in a form which allows easy comparison with similar results found by other means in Chapter 7. Actually, to reach a correct solution for the problem at hand it would suffice to work with that form of K_2 which does not involve the radical.

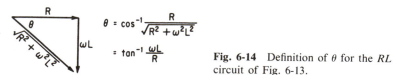

$$\theta = \cos^{-1} \frac{R}{\sqrt{R^2 + \omega^2 L^2}}$$

$$= \tan^{-1} \frac{\omega L}{R}$$

Fig. 6-14 Definition of θ for the RL circuit of Fig. 6-13.

Inserting Eqs. (6-157), (6-158), and (6-159) into Eq. (6-156) yields an expression for $I(s)$ which is ready to be inverse Laplace transformed. Hence

$$I(s) = \frac{E_m}{2(R^2 + \omega^2 L^2)}\left[(R - j\omega L)\frac{1}{s - j\omega} + (R + j\omega L)\frac{1}{s - j\omega}\right]$$
$$- \frac{E_m}{\sqrt{R^2 + \omega^2 L^2}}\cos\theta\left(\frac{1}{s + R/L}\right) \tag{6-160}$$

The corresponding time solution is then

$$i(t) = \frac{E_m/2}{R^2 + \omega^2 L^2}[(R - j\omega L)\varepsilon^{j\omega t} + (R + j\omega L)\varepsilon^{-j\omega t}]$$
$$- \frac{E_m}{\sqrt{R^2 + \omega^2 L^2}}\cos\theta[\varepsilon^{-(R/L)t}] \tag{6-161}$$

Collecting terms yields

$$i(t) = \frac{E_m/2}{R^2 + \omega^2 L^2}[R(\varepsilon^{j\omega t} + \varepsilon^{-j\omega t}) + j\omega L(\varepsilon^{-j\omega t} - \varepsilon^{+j\omega t})]$$
$$- \frac{E_m}{\sqrt{R^2 + \omega^2 L^2}}\cos\theta[\varepsilon^{-(R/L)t}] \tag{6-162}$$

Recalling that

$$\varepsilon^{j\omega t} + \varepsilon^{-j\omega t} = 2\cos\omega t \quad \text{and} \quad (\varepsilon^{-j\omega t} - \varepsilon^{j\omega t}) = -2j\sin\omega t \tag{6-163}$$

simplifies Eq. (6-162) to

$$i(t) = \frac{E_m}{R^2 + \omega^2 L^2}[R\cos\omega t + \omega L\sin\omega t] - \frac{E_m}{\sqrt{R^2 + \omega^2 L^2}}\cos\theta[\varepsilon^{-(R/L)t}] \tag{6-164}$$

Now since the cosine function leads the sine function by 90°, the quantity in brackets in the last equation may be found with the aid of the right triangle in Fig. 6-14. If the horizontal line is assumed to represent the cosine having magnitude R, then the vertical downward line, which is 90° displaced from the cosine line, represents the sine function having magnitude ωL. Clearly, then, the addition of these two quadrature terms is the hypotenuse of the right triangle the amplitude of which is $\sqrt{R^2 + \omega^2 L^2}$ and the position of which is θ degrees *behind* the horizontal line which denotes the cosine function. On the basis of this reasoning it follows that

$$R\cos\omega t + \omega L\sin\omega t = \sqrt{R^2 + \omega^2 L^2}\cos(\omega t - \theta) \tag{6-165}$$

where θ is defined in Fig. 6-14. The minus sign is used with θ to denote that the hypotenuse is in a position below the horizontal line.

Substituting Eq. (6-165) into Eq. (6-164) yields the final form of the total solution for the current which flows in the RL circuit when a sinusoidal forcing function is applied. Thus

$$i(t) = \frac{E_m}{\sqrt{R^2 + \omega^2 L^2}}[\cos(\omega t - \theta) - \varepsilon^{-(R/L)t}\cos\theta] \tag{6-166}$$

A plot of the solution appears in Fig. 6-15.

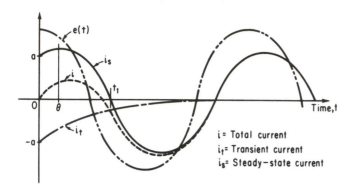

i = Total current
i_t= Transient current
i_s= Steady-state current

Fig. 6-15 Current response of an *RL* circuit to a sinusoidal forcing function.

An inspection of Eq. (6-166) and Fig. 6-15 reveals that the total solution is composed of a forced solution as well as a transient solution. The former is represented by the first term in brackets. That is,

$$i_s = \text{steady-state solution} = \frac{E_m}{\sqrt{R^2 + \omega^2 L^2}} \cos (\omega t - \theta) \qquad (6\text{-}167)$$

Because we are dealing with linear circuits, the forced solution has the same form as the forcing function, i.e., cosinusoidal. However, the response differs in two respects. Its amplitude is modified by the factor $1/\sqrt{R^2 + \omega^2 L^2}$, and its argument is altered by the angle $-\theta$. Of course the transient solution behaves as expected. Initially it provides a magnitude of current which is equal and opposite to the instantaneous value of the steady-state current at the instant of switching. In Fig. 6-15 it is assumed that the switch is closed at the instant when the voltage has its positive maximum value. Note that the corresponding steady-state current is not zero at this instant. In Fig. 6-15 it is shown to have a value $0a$. Since the boundary condition demands that the current at this instant be zero, it is necessary for the transient term to have a value equal and opposite. This is depicted as $-0a$ in Fig. 6-15. As time elapses after switching, the transient term decays to zero at a rate determined by the circuit time constant, and the total current in Fig. 6-15 then becomes identical to the steady-state current.

Summary review questions

1. Describe briefly the role of a mathematical transform. What is a transformed quantity?
2. Identify the special quality in the definition of the Laplace transform of a time function that accounts for its usefulness in the solution of linear differential equations. How is this usefulness manifested?
3. State the restriction that is placed on the transformed variable s in the Laplace transform of a function and describe its importance.
4. How does the Laplace transform method of solving integrodifferential equations treat the initial conditions associated with energy-storing elements? Contrast this procedure with that of the classical method studied in Chapter 5 and identify the advantages and disadvantages of the two schemes.

5. The Laplace transform method is said to provide a systematic formulation of the solution process of linear differential equations. Explain and illustrate the meaning of this statement.

6. Describe and illustrate how the Laplace transform of a deterministic function is found. Does this result need to be found again? Explain.

7. Describe the importance of a fairly complete table of Laplace-transform pairs. Illustrate.

8. State the time-displacement theorem of Laplace-transform theory and explain its importance especially in electrical engineering.

9. State the final-value theorem of Laplace transform theory and tell why it is useful.

10. Explain what is meant by the statement "performing the inverse Laplace transformation via tables." Illustrate.

11. Once a solution of a problem in circuits is found in the transformed domain (i.e., the s domain), how do you recognize the steady-state component of the solution in the resulting algebraic expression for the transformed solution?

12. Write the expression in the transformed domain for the current that flows in a series RL circuit that has an initial current $i(0)$ and is driven by a unit step voltage. Identify the poles of the transformed current solution that are associated with the steady-state and transient components of the solution.

13. What is the operational impedance of an RC circuit? Describe its usefulness.

14. Compare the manner of finding the total solution of the current response in a series RLC circuit driven by a step voltage by the classical and Laplace-transform methods. Assume the presence of initial energy for both energy-storing elements.

15. Explain the meaning of the statement "the characteristic equation can be found by obtaining the transformed solution of *any* of the dependent circuit variables."

Problems

GROUP I

6-1. Obtain the Laplace transform of the following differential equations. Assume zero initial conditions except where noted.

(a) $\dfrac{d^2m(t)}{dt^2} + \dfrac{dm(t)}{dt} = A \sin 2\omega t.$

(b) $\dfrac{d^3m(t)}{dt^2} + a_2\dfrac{d^2m(t)}{dt^2} + a_1\dfrac{dm(t)}{dt} + a_0m(t) + a_{-1}\displaystyle\int m(t)\,dt = t\varepsilon^{-\alpha t}.$

(c) $\dfrac{d^2c(t)}{dt^2} + 3\dfrac{dc(t)}{dt} + 2c(t) = t\left(\text{initial conditions: } \dfrac{dc}{dt} = 2, c = -1\right).$

6-2. Find the Laplace transform of the following functions:

(a) $f(t) = \dfrac{\varepsilon^{-\alpha t} + \alpha t - 1}{\alpha^2}.$

(b) $f(t) = \sin(\omega t + \theta).$

(c) $f(t) = \varepsilon^{-\alpha t}\sin(\omega t + \theta).$

6-3. Find the Laplace transform of the time function depicted in Fig. P6-3.

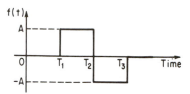

Fig. P6-3

6-4. Determine the Laplace transform of the time function depicted in Fig. P6-4.

6-5. Determine the Laplace transform of the time function shown in Fig. P6-5.

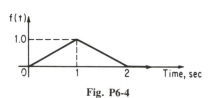

Fig. P6-4

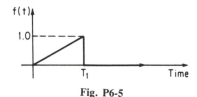

Fig. P6-5

6-6. Find the inverse Laplace transform of $F(s) = 1/(s + 2)^3$.

6-7. The behavior of an engineering device is described by the differential equation

$$\frac{d^2x}{dt^2} + 14\frac{dx}{dt} + 40x = 5$$

It is desired to find the solution for x subject to the initial conditions $dx/dt = 2$ and $x = 1$.

(a) By using the Laplace transform find the s-domain solution for x.

(b) Determine the corresponding time solution for the result found in part (a).

6-8. Find the initial value (at $t = 0^+$) of the time function specified by

$$\mathscr{L}f(t) = F(s) = \frac{(2s + 5)(s + 4)}{(s + 6)^2(s + 2)}$$

6-9. Determine the time function corresponding to

$$F(s) = \frac{s + 3}{(s + 1)(s + 2)^2}$$

6-10. Find the general form of the differential equation which leads to the $F(s)$ function specified in Prob. 6-9.

6-11. The Laplace transform of a time function is

$$F(s) = \frac{10}{s^2 + 4s + 8}$$

Find the time function.

6-12. Determine the time function whose Laplace transform is given by

$$F(s) = \frac{10}{s^2 + 6s + 8}$$

6-13. Obtain the time function whose Laplace transform is given by

$$F(s) = \frac{10}{s(s^2 + 6s + 8)}$$

6-14. Compute the initial value of the time function found in Prob. 6-13. Check this value by applying the initial-value theorem to the $F(s)$ function.

6-15. An engineering device contains two independent energy-storing elements. When subjected to a step source function, the governing differential equation becomes

$$\frac{dx}{dt} + 12x + 100 \int x\, dt = 300u(t)$$

Just prior to the application of the forcing function the initial value of x is known to be 10 and the initial value of the integrating device is known to be 6.4, i.e., $\int_{-\infty}^{0} x_0\, dt = 6.4$.

(a) By means of the Laplace transform, determine the response in the transformed domain.

(b) Does the response have a steady-state value different from zero in the time domain? Why?

(c) Find the expression for the complete solution in the time domain.

6-16. The Laplace-transformed solution for the current in a circuit that contains two energy-storing elements is given by

$$I(s) = \frac{105}{s(s^2 + 10s + 21)}$$

The initial conditions are known to be zero.

(a) What type and what magnitude of forcing function was used?

(b) What are the characteristic modes of the dynamic response?

(c) What is the final value of the current?

6-17. Find the expression for the Laplace transform of $v(t)$ depicted in Fig. P6-17.

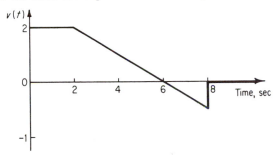

Fig. P6-17

6-18. Repeat Prob. 5-7 using the Laplace-transform method.

6-19. Repeat Prob. 5-8 using the Laplace-transform method.

6-20. Repeat Prob. 5-11 using the Laplace-transform method.

6-21. Repeat Prob. 5-13 using the Laplace-transform method.

6-22. Repeat Prob. 5-15 using the Laplace-transform method.

6-23. Repeat Prob. 5-33 using the Laplace-transform method.

GROUP II

6-24. The Laplace transform function is given by

$$F(s) = \frac{3s + 8}{s^2 + 6s + 25}$$

What is the corresponding $f(t)$?

6-25. The flow of current in an electric circuit is described by the differential equation

$$\frac{d^3i}{dt^3} + 14.1\frac{d^2i}{dt^2} + 41.4\frac{di}{dt} + 4i = 10$$

It is desired to find the solution for i subject to the initial condition that $i(0) = 0.725$ ampere.

(a) Through use of the Laplace transform, obtain the s-domain solution for the current.

(b) Determine the corresponding time solution for the result found in part (a).

6-26. Repeat Prob. 5-34 using the Laplace-transform method.

6-27. Repeat Prob. 5-38 using the Laplace-transform method.

6-28. Repeat Prob. 5-45 using the Laplace-transform method.

chapter seven

Sinusoidal Steady-State Response of Circuits

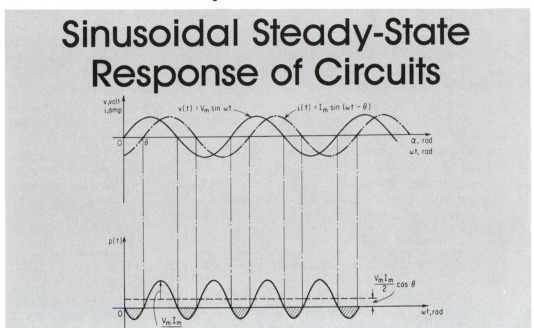

The theory of the sinusoidal steady-state response of circuits occupies a position of preeminence in electric-circuit theory. The analysis of many circuits and devices throughout all branches of electrical engineering is accomplished by the techniques embodied in the sinusoidal theory. Particularly impressive in this regard is the fact that the sinusoidal circuit theory is applicable not only in situations involving sinusoidal forcing functions but equally as well in those situations where the forcing functions are of a nonsinusoidal character.

It is not an accident that the bulk of the electric power generated in power plants throughout the world and distributed to the consumer appears in the form of sinusoidal variations of voltage and current. There are many technical and economical advantages associated with the use of sinusoidal voltages and currents. A significant appreciation of this statement will be gained upon the completion of the study of this book. In Chapter 18, for example, it will be learned that the use of sinusoidal voltages applied to appropriately designed coils results in a revolving magnetic field which has the capacity to do work. As a matter of fact it is this principle which underlies the operation of almost all the electric motors found in home appliances and about 90% of all electric motors found in commercial and industrial applications. Although other waveforms can be used in such devices, none leads to an operation which is as efficient and economical as that achieved through the use of sinusoidal functions.

In addition to these practical aspects, however, the sinusoidal function offers some very important and significant advantages in a mathematical sense.

Recall that by Euler's theorem the sine as well as the cosine function can be represented quite simply by the exponential function. Thus $\varepsilon^{j\omega t} = \cos \omega t + j \sin \omega t$. Since as described in Chapter 5, the equations which govern the behavior of electric circuits frequently involve derivative and integral terms, this exponential character of the sinusoid is of prime importance. The reason is that it permits a simplification of mathematical analysis which cannot be achieved by any other function. The exponential function is the only mathematical function, the original form of which is preserved even though such operations as differentiation and integration are performed on it. Consequently, as described in this chapter, it is possible to treat sinusoidal time functions entirely in terms of complex numbers.

Knowledge of the sinusoidal theory has another notable advantage in the mathematical sense. By means of the Fourier series it is possible to represent *any* periodic function of whatever form in terms of an infinite series of sinusoids. Accordingly, the steady-state response of electric circuits to nonsinusoidal forcing functions can be obtained through a repeated application of the sinusoidal theory followed by a summation of the individual sinusoidal response.

7-1 SINUSOIDAL FUNCTIONS—TERMINOLOGY

In dealing with sinusoidal functions we must become familiar with the nomenclature before proceeding with the sinusoidal steady-state analysis of circuits. This makes it easier to describe and to interpret the results. Appearing in Fig. 7-1 are two sinusoids—one denoting voltage and the other current. The voltage sinusoid is a sine function which is represented mathematically as

$$v = V_m \sin \alpha = V_m \sin \omega t \qquad \text{V} \tag{7-1}$$

where

$$\alpha = \omega t \qquad \text{radians} \tag{7-2}$$

and is called the argument of the sine function. The *amplitude* of the sine function is V_m and it denotes the *maximum* value of the sine function. Equation (7-1) is often referred to as the expression for the *instantaneous voltage* because by insertion of any particular value for t the corresponding value of v is determined.

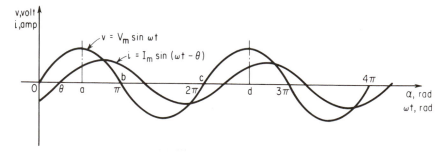

Fig. 7-1 Sinusoidal voltage and current waves of the same frequency. The current sinusoid has a phase lag of θ relative to the voltage wave.

Note that as we substitute into Eq. (7-1) values of α that lie between 0 and 2π radians, different and distinct values of the function result. However, as α is made to take on values between 2π and 4π there occurs a repetition of the first set of values obtained. Any complete set of positive and negative values of the function is called a *cycle*. Thus portions *oabc* and *abcd* of the voltage wave of Fig. 7-1 qualify as cycles of the time function.

Another characteristic of the sinusoid is its *frequency,* which is defined as the number of cycles of the function which is traversed in 1 s. Accordingly, the frequency is measured in units of *cycles per second,* which has recently come to be called the hertz and abbreviated Hz. In radio and radar transmission work where high frequencies are involved it is customary to use units of kilohertz, megahertz, and gigahertz. The time duration of one cycle is called the *period* of the function. It follows then that if f denotes the frequency of the periodic function, the period is related to the frequency by

$$T = \frac{1}{f} \quad \text{s} \tag{7-3}$$

Moreover, a glance at Fig. 7-1 indicates that each cycle spans 2π radians. Hence, if this quantity is divided by the period, there results the *angular velocity* (or angular frequency) of the sine function. It is denoted by ω and has units of rad/second.

$$\omega = \frac{2\pi}{T} = 2\pi f \tag{7-4}$$

The second form of ω is obtained from Eq. (7-3).

The equation to be used to describe the second sinusoid appearing in Fig. 7-1, i.e., the current wave, obviously cannot be identical to that used for the voltage wave. The reason is that the two sinusoids are displaced from one another by the angle θ and so cannot have the same instantaneous value. For example, at $t = 0$, $v = 0$, whereas the current has a finite negative value. A comparison of the two sine waves shows that the zero value of the current wave for positive slope occurs *later* in time by an amount $\alpha = \theta$. Accordingly, the current wave is said to be *lagging* behind the voltage wave by the relative phase angle θ. *Relative phase* is the term used to denote the angular displacement between two sinusoids of the same frequency. Thus in Fig. 7-1 the current sinusoid can be described as having a *phase lag* of θ degrees relative to the voltage sinusoid. Alternatively we can say that the voltage wave has a *phase lead* of θ degrees relative to the current wave.

The equation which describes the instantaneous value of the current wave must include the angular displacement (or phase angle) existing between the two sinusoids. A little thought reveals that the current expression is

$$i = I_m \sin (\omega t - \theta) \tag{7-5}$$

Note that the phase angle is included as part of the argument of the sine function. A minus sign is used because there is a phase lag between the voltage and current sinusoids. This formulation assures that at $t = 0$ a finite, negative value for the current exists as called for by the plot of Fig. 7-1.

A word of caution is appropriate at this point concerning the units of the argument $(\omega t - \theta)$. The form of Eq. (7-5) demands that the unit for this total quantity be radians. However, in engineering usage of this equation it is customary to express ωt in radians and θ in degrees. The reason lies in the fact that in engineering calculations it is the phase angle which is important, not the total angular displacement. Therefore, when Eq. (7-5) is occasionally seen written as $i = I_m \sin (\omega t - 45°)$, the inconsistency in the units of the argument should be accepted in light of the foregoing comment.

There exist alternate forms with which to express sinusoids. These originate with Euler's identity, i.e.,

$$\varepsilon^{j\omega t} = \cos \omega t + j \sin \omega t \qquad (7\text{-}6)$$

It should be apparent from this expression that a cosine and a sine function can be expressed in terms of the exponential notation as follows:

$$\cos \omega t = \text{Re}[\varepsilon^{j\omega t}] \qquad (7\text{-}7)$$

$$\sin \omega t = \text{Im}[\varepsilon^{j\omega t}] \qquad (7\text{-}8)$$

where Re[ ] denotes the real part of the expression in brackets and Im[ ] the imaginary part. This representation is used on various occasions throughout the book when it is convenient to do so. The sinusoids may also be expressed by the equations

$$\cos \omega t = \frac{\varepsilon^{j\omega t} + \varepsilon^{-j\omega t}}{2} \qquad (7\text{-}9)$$

$$\sin \omega t = \frac{\varepsilon^{j\omega t} - \varepsilon^{-j\omega t}}{2j} \qquad (7\text{-}10)$$

The insertion of Eq. (7-6) into the right side of either of the last two equations bears out the validity of the equivalence.

7-2 AVERAGE AND EFFECTIVE VALUES OF PERIODIC FUNCTIONS

The energy sources throughout the treatment of Chapter 3 are all of the nonvarying, constant type—often called the *direct* kind. Thus when a voltage source of fixed magnitude is applied to a network, the forced solution is found to be a constant quantity too. This constant character of the response makes it a simple matter to identify the number of amperes flowing in the circuit and thereby to describe the energy-transferring capability of the circuit. Moreover, the computation of the power absorbed by each circuit element is accomplished in a direct manner through the use of Eq. (3-5) by inserting the constant values of voltage and current. The situation, however, is quite different when the voltages and currents associated with a circuit element are varying functions of time such as those appearing in Fig. 7-1. Note that the sinusoidal current is an *alternating* current, i.e., one which has positive and negative values. In such a case the manner of describing the energy-transferring capability of the current is not at all obvious, as it is when direct sources are used; in the latter case the average current flow is identical to the direct (or constant) value.

In view of the fact that the average current serves as a useful criterion in determining the energy transfer in circuits involving direct sources, let us investigate its usefulness in situations involving periodic driving functions. The term periodic is used rather than sinusoidal in order that the treatment be general. The sinusoid is only one example of a periodic function. Any function whose cycle is repeated continuously irrespective of waveform is called a periodic function.

A general definition of the average value of any function $f(t)$ over the specified interval between t_1 and t_2 is expressed mathematically as

$$F_{av} \equiv \frac{1}{t_2 - t_1} \int_{t_1}^{t_2} f(t)\, dt \qquad\qquad (7\text{-}11)$$

In the instance when $f(t)$ is a periodic function having a period of T sec, Eq. (7-11) becomes

$$F_{av} = \frac{1}{T} \int_0^T f(t)\, dt \qquad\qquad (7\text{-}12)$$

If the time function is expressed in radians through the use of Eq. (7-2), then an alternative form for the average value of the function results. Thus

$$F_{av} = \frac{1}{2\pi} \int_0^{2\pi} f(\omega t)\, d(\omega t) = \frac{1}{2\pi} \int_0^{2\pi} f(\alpha)\, d\alpha \qquad\qquad (7\text{-}13)$$

Average Value of a Sinusoid. By means of Eq. (7-13) let us now find the average value of the sinusoidal current variation of Fig. 7-1. Hence

$$I_{av} = \frac{1}{2\pi} \int_0^{2\pi} I_m \sin(\omega t - \theta)\, d(\omega t) = \frac{1}{2\pi} \int_0^{2\pi} I_m \sin(\alpha - \theta)\, d\alpha \qquad (7\text{-}14)$$

where $\alpha = \omega t$. Integrating and inserting limits yields

$$I_{av} = \frac{I_m}{2\pi} \left[-\cos(\alpha - \theta)\right]_{\alpha=0}^{\alpha=2\pi} = \frac{I_m}{2\pi} \left[-\cos(2\pi - \theta) + \cos(-\theta)\right] \equiv 0 \quad (7\text{-}15)$$

Therefore, the average value of a sinusoid over *one complete cycle* is identically equal to zero. A study of the plot of the sinusoid makes this conclusion obvious because for one cycle there is as much area above the abscissa axis as there is below. Accordingly, the net area is zero.

It is interesting to note that a finite average value can be found for the sinusoid for the *positive or negative half-cycle*. Because of the usefulness of this result in subsequent work, we determine the general expression for this quantity here. Hence

$$I_{av-1/2\,cycle} = \frac{1}{\pi} \int_\theta^{\pi+\theta} I_m \sin(\alpha - \theta)\, d\alpha \qquad\qquad (7\text{-}16)$$

The sinusoid being used is the one depicted in Fig. 7-1 and described by Eq.

(7-5). The lower and upper limits of the integral are selected to include only the area above the abscissa axis. Upon integrating and substituting limits, we obtain

$$I_{\text{av}-1/2 \text{ cycle}} = \frac{I_m}{\pi}[-\cos(\alpha - \theta)]_{\alpha=\theta}^{\alpha=\pi+\theta} = \frac{I_m}{\pi}[-\cos(\pi) + \cos 0] = \frac{2}{\pi}I_m \quad (7\text{-}17)$$

or

$$\boxed{I_{\text{av}-1/2 \text{ cycle}} = \frac{2}{\pi}I_m = 0.636I_m} \qquad (7\text{-}18)$$

Thus the average value of either the positive or negative half of a sine function can be found simply by multiplying the amplitude of the wave by 0.636. When taken over a full cycle the equal and opposite average values cancel out.

On the basis of the foregoing results it should be clear that, although the criterion of the average value of current works well in describing the energy-transferring capacity for direct sources, it is a meaningless criterion for symmetrical† periodic functions because its value is always equal to zero. Therefore we must search for a more suitable criterion to measure the effectiveness of a periodic function. Preferably the chosen standard should in some way be related to the energy or power capability associated with the periodic function. Herein lies a clue about the procedure to follow in establishing such a criterion. Consider that the sinusoidal current of Eq. (7-5) is made to flow through a resistor having R ohms. Then by Joule's law the instantaneous power absorbed by the resistor and converted to heat is

$$p = i^2(t)R \qquad \text{W} \qquad (7\text{-}19)$$

Since the current in this expression varies usually from a positive maximum through zero to a negative maximum, the insertion of any one value of the current is of little usefulness in determining the actual power absorbed by the resistor. Rather a much more meaningful approach is to sum the instantaneous power consumption over one full cycle. Such a formulation cannot lead to a zero result because the power dissipation is real whether the current flows clockwise (positive) or counterclockwise (negative) in this circuit. Therefore a nonzero value must result. This conclusion is also borne out by the presence of the second power of the current in Eq. (7-19). Thus even if the current is negative the contribution to the power expression is positive. Proceeding on this basis, we find that the expression for the average power absorbed by the resistor becomes

$$P_{\text{av}} = \frac{1}{T}\int_0^T i^2(t)\,R\,dt = \left[\frac{1}{T}\int_0^T i^2(t)\,dt\right]R \qquad (7\text{-}20)$$

A study of the quantity in brackets reveals some interesting points. Note first that this quantity is the average value of the function $f(t) = i^2(t)$. A comparison with Eq. (7-12) makes this obvious. We are still dealing with average effects—

† The term *symmetrical* is used to emphasize that we are here concerned only with those periodic functions with equal positive and negative areas.

but with one important difference: instead of working with the periodic function $i(t)$ directly we are dealing with the *squared* function. This not only gives significance to the formulation in terms of a description involving the power capability of the current, but it also eliminates the possibility of dealing with a zero average value because the square of a periodic function always yields a positive contribution. Another point of interest is that the unit of the bracketed quantity is amperes squared. Accordingly we can interpret this quantity as the square of an *effective current*, which, when multiplied by R, yields the average power. Expressing this mathematically, we can write

$$I_{\text{eff}}^2 \equiv \frac{1}{T} \int_0^T i^2(t)\, dt = \text{average } i^2(t) \qquad (7\text{-}21)$$

From this it follows that the effective current is the *root mean square* value. Thus

$$I_{\text{eff}} = I_{\text{rms}} = \sqrt{\text{average } i^2(t)} = \sqrt{\frac{1}{T} \int_0^T i^2(t)\, dt} \qquad (7\text{-}22)$$

Although the effective current may be denoted by either of the subscript notations used in the last equation, it is frequently written without a subscript. The understanding is that reference is always to the effective value of a periodic function as defined in Eq. (7-22) unless otherwise specified.

Finally, as a third point of interest, let us make an analogy with the direct-current case. A direct voltage source applied to the resistor R causes an average power dissipation of I^2R to take place. Here I denotes the direct or average current. The flow of a periodic function of current through the same resistor yields an average power dissipation of $I_{\text{eff}}^2 R$. Comparing this result with that of the direct-current case yields another interpretation of effective current: it is that current which produces the same heating effect as the direct current.

Effective Value of a Sinusoid. Let us now assume that the periodic function $i(t)$, referred to in the foregoing discussion, takes on the specific form of the sinusoid as depicted in Fig. 7-2. Then by Eq. (7-21) we have

$$I_{\text{eff}}^2 = \frac{1}{T} \int_0^T i^2(t)\, dt = \frac{1}{T} \int_0^T I_m^2 \sin^2 \omega t\, dt \qquad (7\text{-}23)$$

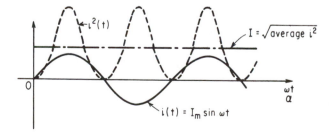

Fig. 7-2 Illustrating the computation of the effective value of a sinusoidal current.

Introducing the trigonometric identity

$$\sin^2 \omega t = \tfrac{1}{2} - \tfrac{1}{2}\cos 2\omega t \tag{7-24}$$

permits Eq. (7-23) to be written as

$$I_{\text{eff}}^2 = \frac{I_m^2}{2T}\int_0^T (1 - \cos 2\omega t)\, dt \tag{7-25}$$

$$= \frac{I_m^2}{2T}[T] - \frac{I_m^2}{4\omega T}[\sin 2\omega t]_0^T = \frac{I_m^2}{2}$$

because $T = 2\pi$.

Therefore,

$$\boxed{I = I_{\text{eff}} = \frac{I_m}{\sqrt{2}} = 0.7071 I_m} \tag{7-26}$$

This is an important result which is used often in the study of sinusoidal steady-state theory. It is important to remember that it is valid only for sinusoidal functions.

EXAMPLE 7-1 Find the average value of the periodic function depicted in Fig. 7-3.

Solution: A study of the waveshape shows that all the information about the curve is contained in the portion which lies from 0 to $\pi/2$. In this region two equations are needed to describe the function. Thus

$$v(t) = \begin{cases} \dfrac{V_m}{\pi/3}(\omega t) = \dfrac{V_m}{\pi/3}\alpha & \text{for } 0 \leqslant \alpha \leqslant \dfrac{\pi}{3} \tag{7-27} \\[4mm] V_m & \text{for } \dfrac{\pi}{3} \leqslant \alpha \leqslant \dfrac{\pi}{2} \tag{7-28} \end{cases}$$

A direct application of Eq. (7-13) over the region from $\alpha = 0$ to $\alpha = \pi/2$ yields the average value. Thus

$$V_{\text{av}} = \frac{1}{\pi/2}\left\{ \int_0^{\pi/3} \frac{V_m}{\pi/3}\alpha\, d\alpha + \int_{\pi/3}^{\pi/2} V_m\, d\alpha \right\}$$

$$= \frac{1}{\pi/2}\left\{ \frac{V_m}{\pi/3}\left[\frac{\alpha^2}{2}\right]_0^{\pi/3} + V_m[\alpha]_{\pi/3}^{\pi/2} \right\} \tag{7-29}$$

$$= \frac{V_m}{\pi/2}\left\{ \frac{\pi}{6} + \frac{\pi}{6} \right\} = \frac{2}{3}V_m$$

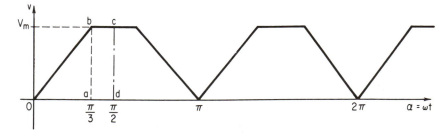

Fig. 7-3 Trapezoidal waveform for Example 7-1.

We can check this result by finding the total area beneath the $v(t)$ curve over the portion from 0 to $\pi/2$ and dividing the result by $\pi/2$. Thus

$$\text{area under } v(t) \text{ curve } = \text{area } 0ab + \text{area } abcd$$

$$= \frac{1}{2} V_m \left(\frac{\pi}{3} \right) + V_m \left(\frac{\pi}{6} \right) = \frac{\pi}{3} V_m \tag{7-30}$$

$$\therefore V_{av} = \frac{\text{area}}{\pi/2} = \frac{(\pi/3)V_m}{\pi/2} = \frac{2}{3} V_m \tag{7-31}$$

7-3 INSTANTANEOUS AND AVERAGE POWER. POWER FACTOR

Our interest in this section is to develop a general expression for the average power associated with a voltage and current in an a-c* circuit. The restrictive limitation of confining the treatment to resistive circuits is now dropped. In this connection then let $v(t) = V_m \sin \omega t$ represent the potential difference appearing across the branch terminals of a given circuit and let $i(t) = I_m \sin (\omega t - \theta)$ denote the corresponding current flowing through that branch. The relative phase angle is given by θ. The voltage and current sinusoids are shown in Fig. 7-4. It

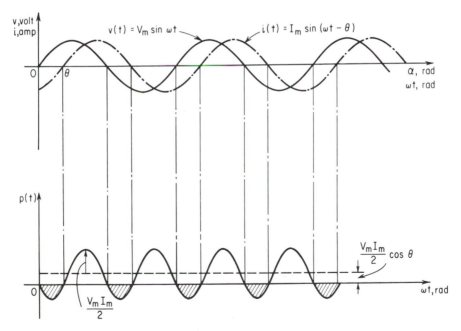

Fig. 7-4 Plot of instantaneous power as determined from the given voltage and current sinusoids.

* This notation is often used to denote circuits subjected to sinusoidal (i.e., alternating) forcing functions.

follows then that the expression for the instantaneous power is

$$p(t) = v(t)i(t) = V_m I_m \sin \omega t \sin (\omega t - \theta) \tag{7-32}$$

To put this in a more suitable form insert the identity

$$\sin (\omega t - \theta) = \sin \omega t \cos \theta - \cos \omega t \sin \theta \tag{7-33}$$

into Eq. (7-32) to yield

$$p(t) = V_m I_m (\sin^2 \omega t \cos \theta - \sin \omega t \cos \omega t \sin \theta) \tag{7-34}$$

To effect a further reduction introduce

$$\sin^2 \omega t = \tfrac{1}{2} - \tfrac{1}{2} \cos 2\omega t \tag{7-35}$$

and

$$\sin \omega t \cos \omega t = \tfrac{1}{2} \sin 2\omega t \tag{7-36}$$

into Eq. (7-34). Thus

$$p(t) = \frac{V_m I_m}{2} \cos \theta - \frac{V_m I_m}{2} (\cos \theta \cos 2\omega t + \sin 2\omega t \sin \theta) \tag{7-37}$$

Finally, by substituting the relation

$$\cos (2\omega t - \theta) = \cos 2\omega t \cos \theta + \sin 2\omega t \sin \theta \tag{7-38}$$

into Eq. (7-37) the desired form of the expression for the instantaneous power is obtained. Thus

$$\boxed{p(t) = \frac{V_m I_m}{2} \cos \theta - \frac{V_m I_m}{2} \cos (2\omega t - \theta)} \tag{7-39}$$

A plot of Eq. (7-39) appears in Fig. 7-4. Note that for a fixed θ the instantaneous power consists of two components—a constant part and a time-varying part. Note, too, that the varying part has a frequency which is twice that of the voltage and current sinusoids. The shaded portions of the plot of $p(t)$ refer to those time intervals when the power is negative. In effect this means that the circuit is returning power to the source during these intervals. It should be apparent, then, that the branch circuit under consideration contains at least one energy-storing element. A glance at Fig. 7-4 shows that the instantaneous power is negative whenever the voltage and current are of opposite sign. However, for the case plotted in Fig. 7-4 notice that the positive area under the $p(t)$ curve exceeds the negative area. Therefore, the average power is positive and finite, and specifically it is equal to the constant term of Eq. (7-39). As θ is made smaller, i.e., as i is brought more nearly in phase with v, the negative areas of the $p(t)$ curve of Fig. 7-4 become smaller and so the average power increases. This is equivalent to raising the $p(t)$ curve higher above the abscissa axis. When $\theta = 0$ the current and voltage are in phase. There are no negative areas associated with the $p(t)$ curve; hence all the power is consumed between the circuit branch terminals. The circuit may then be called purely resistive. On the other hand, when θ is increased, the negative areas become larger and so less power is consumed between the terminals and more returned to the source. At the extreme value of θ, i.e., $\theta = \pi/2$, the $p(t)$ curve is dropped to that position which makes the negative and positive areas equal. In this instance there is no power consumed between the circuit terminals.

The relative phase angle θ is determined by the values of the circuit parameters appearing between the circuit branch terminals across which $v(t)$ is assumed to exist. Because of the passive nature of these circuit parameters the value of θ is restricted to lie in the range expressed by $-\pi/2 \leq \theta \leq \pi/2$.

The general expression for the instantaneous power in an a-c circuit is described by Eq. (7-39). The really useful quantity in terms of the capability of the circuit to do work is the average value of the power over one cycle. Since each cycle is continuously repeated, whatever is done for one cycle applies equally as well for each succeeding cycle. It has already been stated in connection with the discussion of Fig. 7-4 that this average power is given by the constant term of Eq. (7-39). A mathematical verification now follows. From the general definition of the average value over one cycle, we have

$$P_{av} = \frac{1}{T} \int_0^T p(t)\, dt \tag{7-40}$$

Inserting Eq. (7-39) for $p(t)$ yields

$$P_{av} = \frac{1}{T} \left[\int_0^T \frac{V_m I_m}{2} \cos \theta\, dt - \int_0^T \frac{V_m I_m}{2} \cos (2\omega t - \theta)\, dt \right] \tag{7-41}$$

Since the second term on the right side involves the integration of a simple sine function over a time interval equal to two complete periods of the double frequency sine function, the value is always identically equal to zero. This leaves just the first term and, since θ is independent of t, it follows that the average power is

$$P_{av} = \frac{V_m I_m}{2} \cos \theta \qquad \text{W} \tag{7-42}$$

It is helpful at this point to rewrite Eq. (7-42) as

$$P_{av} = \frac{V_m}{\sqrt{2}} \left(\frac{I_m}{\sqrt{2}} \right) \cos \theta \tag{7-43}$$

Recalling that the effective value of a sinusoidal quantity is the amplitude divided by $\sqrt{2}$, Eq. (7-43) can be expressed more significantly in terms of the corresponding effective values of the voltage and current sinusoids. Therefore,

$$\boxed{P_{av} = V_{eff} I_{eff} \cos \theta = VI \cos \theta \qquad \text{W}} \tag{7-44}$$

Although it is customary to drop the subscripts entirely when writing this equation, one of our reasons for leaving them here is to emphasize better that *average power* is determined in terms of the *effective voltage and current* values. Equation (7-44) points out in a general and significant fashion the usefulness of the root mean square value of a periodic function as a criterion to measure its effectiveness. The effective values of voltage and current play key roles in measuring the ability of a circuit to do work.

In the interest of introducing another term of electric-circuit theory rewrite Eq. (7-44) as follows:

$$\cos \theta = \frac{P}{VI} \tag{7-45}$$

The quantity P is the average power and is expressed in watts—a unit which conveys the capability to do work. However, note that the denominator of Eq. (7-45) involves a quantity whose units are represented by the product of volts by amperes. When we are dealing with direct sources this product is called watts, because it is real power which can be entirely converted to work. The same is not true when sinusoidal quantities are involved. For example, in Fig. 7-4, corresponding to $\theta = \pi/2$ there is no useful (or work-producing) power in the circuit in spite of the large values which V and I may have. For this reason the product VI is called *apparent power*. This power is not always realizable in the circuit for doing work. The useful part depends upon the value of $\cos \theta$, and because of this, $\cos \theta$, is called the *power factor* (abbreviated pf) of the circuit. Thus

$$\boxed{\text{pf} = \cos \theta = \frac{\text{average power}}{\text{apparent power}} = \frac{P}{VI}} \tag{7-46}$$

where V and I are effective values.

EXAMPLE 7-2 A voltage $v(t) = 170 \sin (377t + 10°)$ is applied to a circuit. It causes a steady-state current to flow which is described by $i(t) = 14.14 \sin (377t - 20°)$. Determine the power factor and the average power delivered to the circuit.

Solution: A comparison of the expressions for $v(t)$ and $i(t)$ reveals that the relative phase angle is 30°. Hence

$$\text{pf} = \cos \theta = \cos 30° = 0.866 \tag{7-47}$$

Also,

$$V = \frac{V_m}{\sqrt{2}} = \frac{170}{\sqrt{2}} = 120 \text{ V} \tag{7-48}$$

and

$$I = \frac{I_m}{\sqrt{2}} = \frac{14.14}{\sqrt{2}} = 10 \text{ A} \tag{7-49}$$

Therefore,

$$P = P_{\text{av}} = VI \cos \theta = 120(10)(0.866) = 1040 \text{ W} \tag{7-50}$$

7-4 PHASOR REPRESENTATION OF SINUSOIDS

Often in determining the sinusoidal steady-state response of circuits it is necessary to perform algebraic operations such as addition, subtraction, multiplication, and division on two or more sinusoidal quantities of the same frequency. Usually the sinusoids differ in amplitude and phase. Specifically consider the matter of adding two sinusoidal currents whose equations are

$$i_1 = I_{m1} \sin \omega t \qquad (7\text{-}51)$$

$$i_2 = I_{m2} \sin (\omega t + \theta_2) \qquad (7\text{-}52)$$

The current i_2 leads i_1 by the relative phase angle θ_2. The resultant current i_3 can obviously be written as

$$i_3 = I_{m1} \sin \omega t + I_{m2} \sin (\omega t + \theta_2) \qquad (7\text{-}53)$$

Since the addition of two sinusoids of the same frequency always results in another sinusoid, it is desirable to express Eq. (7-53) in terms of a resultant appropriate amplitude and phase. Keep in mind that any sinusoid at a given frequency is completely specified once its amplitude and phase are known. Of course one obvious way of obtaining the result called for in Eq. (7-53) is to plot each sinusoid and then make a point-by-point summation of the two sine waves. The amplitude and phase of the resultant sinusoid can then be measured, thus allowing i_3 to be written in the more useful form

$$i_3 = I_{m3} \sin (\omega t + \theta_3) \qquad (7\text{-}54)$$

where θ_3 is the phase angle measured with respect to the same reference point used for θ_2. Needless to say, such a procedure is laborious and time-consuming.

An alternative to the graphical solution is an analytical one which simplifies Eq. (7-53) to the form of Eq. (7-54) through the use of trigonometric identities. Accordingly, by introducing

$$\sin (\omega t + \theta_2) = \sin \omega t \cos \theta_2 + \cos \omega t \sin \theta_2 \qquad (7\text{-}55)$$

into Eq. (7-53), it becomes

$$i_3 = (I_{m1} + I_{m2} \cos \theta_2) \sin \omega t + (I_{m2} \sin \theta_2) \cos \omega t \qquad (7\text{-}56)$$

Moreover, the use of Eq. (7-55) in Eq. (7-54) allows Eq. (7-54) to be expressed alternatively as

$$i_3 = (I_{m3} \cos \theta_3) \sin \omega t + (I_{m3} \sin \theta_3) \cos \omega t \qquad (7\text{-}57)$$

Since the last two equations for i_3 are to be identical, it follows that

$$I_{m3} \cos \theta_3 = I_{m1} + I_{m2} \cos \theta_2 \qquad (7\text{-}58)$$

$$I_{m3} \sin \theta_3 = I_{m2} \sin \theta_2 \qquad (7\text{-}59)$$

Equations (7-58) and (7-59) are obtained by equating the coefficients of like terms in Eqs. (7-56) and (7-57). By means of the last two equations the amplitude (I_{m3}) and the phase (θ_3) of the resultant sinusoid can be found in terms of the amplitudes and phase angles of i_1 and i_2, which are known.

Although this analytical procedure requires less effort than the graphical one, the method is still too cumbersome to be practical. This is particularly so when situations arise where more than two sinusoidal quantities must be summed. Furthermore, multiplication and division present additional complications. Clearly, then, we need a simpler and more direct method of treating sinusoidal quantities. In 1893 such a method was introduced, when Charles P. Steinmetz advanced the idea of using a constant-amplitude line rotating at a frequency ω to represent a sinusoid. Let us now see how such an idea is effective in simplifying the algebraic operations involving sinusoidal quantities.

Attention is first directed to the expression for i_1 as given by Eq. (7-51). By employing the notation of Eq. (7-8) we can write

$$i_1 = I_{m1} \sin \omega t = \text{Im}[I_{m1}\varepsilon^{j\omega t}] \qquad (7\text{-}60)$$

Keep in mind that the exponential function $\varepsilon^{j\omega t}$ may be treated as a rotational operator. Its amplitude is always unity, but the cosine and sine components vary as time progresses. This is illustrated in Fig. 7-5(a). As ωt moves through one full period of 2π radians (i.e., one complete cycle) the line OA makes one complete traversal of the circle in a counterclockwise direction. Line OA is fixed in value to the amplitude of the sine function it represents. Note that the vertical component of line OA is the sine function. As a matter of fact this is the meaning of the notation $\text{Im}[\ \ ]$—it refers to the values generated by taking the projections of a rotating line on a pre-established reference line (the vertical in this case). Accordingly, if we plot the vertical components of OA as it makes one complete revolution, the sine function shown in Fig. 7-5(b) is generated. When $\omega t = 0$ the position of OA is on the horizontal axis directed towards the right. Its vertical component at this instant is zero, as it should be for the sine function.

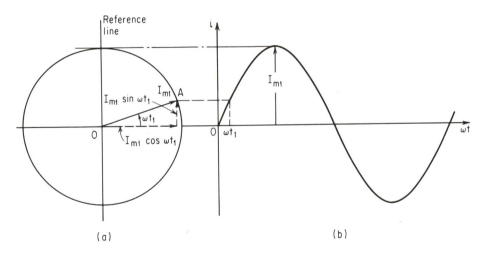

(a)   (b)

Fig. 7-5 Generating the sine function from the vertical component of the rotating line $I_{m1}\varepsilon^{j\omega t} = OA$.

The current i_2 as given by Eq. (7-52) can be represented in a similar fashion. Thus in terms of the exponential notation, we have

$$i_2 = I_{m2} \sin (\omega t + \theta_2) = \text{Im}[I_{m2}\varepsilon^{j(\omega t + \theta_2)}] = \text{Im}[I_{m2}\varepsilon^{j\theta_2}\varepsilon^{j\omega t}] \qquad (7\text{-}61)$$

To simplify the notation we next define

$$\boxed{\bar{I}_{m2} \equiv I_{m2}\varepsilon^{j\theta_2} = I_{m2}(\cos \theta_2 + j \sin \theta_2)} \qquad (7\text{-}62)$$

and denote it as line OB in Fig. 7-6(a). It is this quantity—$\bar{I}_{m2}$—which is called the *phasor* of the sinusoidal function of Eq. (7-61). In general the phasor can be represented by a *complex number* which is a result of locating a line in a plane. Thus the phasor OB can be located by specifying its magnitude, I_{m2}, and its displacement from the horizontal axis θ_2. The reader certainly recognizes these quantities as the polar coordinates of line OB. However, note that OB can also be located in terms of a horizontal (i.e., real-axis) component and a vertical (i.e., imaginary or j-axis) component as indicated by the second form of $\bar{I}_{m2}$ in Eq. (7-62). For a given θ_2 the real part of the complex number is $I_{m2} \cos \theta_2$ and the imaginary or j part is $I_{m2} \sin \theta_2$. It is important to understand that the position of OB in Fig. 7-6(a) corresponds to time $t = 0$ in Eq. (7-61). Hence the corresponding value of i_2 at this instant must be the projection of OB on the vertical reference line, which clearly is $I_{m2} \sin \theta_2$.

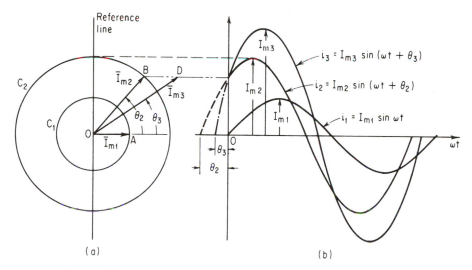

Fig. 7-6 Illustrating the phasor representation and addition of two sinusoids, $i_3 = i_1 + i_2$: (a) position of the phasors for $t = 0$; (b) sinusoidal function for increasing time.

The phasor of current i_1 is $\bar{I}_{m1} = I_{m1}\varepsilon^{j0°} = I_{m1}$. Hence it has no vertical (or j) component and so initially (i.e., at $t = 0$) must lie along the horizontal axis as shown by line OA in Fig. 7-6(a). Note too in this figure that the phasors $\bar{I}_{m1}$ and $\bar{I}_{m2}$ are displaced from one another by θ_2. This is entirely consistent with Eqs. (7-51) and (7-52). Now, as time elapses, both phasors $\bar{I}_{m1}$ and $\bar{I}_{m2}$ revolve at a constant frequency of ω radians per second, and a continuous plot of the vertical components of both phasors results in the sinusoids shown in Fig. 7-6(b). In this way a sinusoid can be represented by a line fixed at one end and rotating at a frequency equal to the angular velocity of the sinusoid. The magnitude of the line must be equal to the amplitude of the sinusoid.

How does the phasor representation of sinusoids simplify the procedure for adding two sinusoidal quantities? To understand this, let us find $i_3 = i_1 + i_2$ by rewriting Eq. (7-53) in exponential form. Thus

$$i_3 = i_1 + i_2 = \text{Im}[\bar{I}_{m1}\varepsilon^{j\omega t}] + \text{Im}[\bar{I}_{m2}\varepsilon^{j\omega t}] \tag{7-63}$$

Keep in mind that attaching the factor $\varepsilon^{j\omega t}$ to the phasors $\bar{I}_{m1}$ and $\bar{I}_{m2}$ has the effect of "animating" the phasors, i.e., it causes them to revolve at the angular frequency ω. From our knowledge of complex algebra we can rewrite Eq. (7-63) as

$$i_3 = \text{Im}[(\bar{I}_{m1} + \bar{I}_{m2})\varepsilon^{j\omega t}] \tag{7-64}$$

By factoring out $\varepsilon^{j\omega t}$ in this last expression we are recognizing that both phasors are revolving at the same frequency ω. Therefore the only thing that needs to be done here is to perform the operation called for in parentheses—which, significantly, in no way involves ωt. Expressed in words, Eq. (7-64) states that *to add two sinusoidal quantities it is necessary merely to add the corresponding phasors.* Accordingly, if we call the phasor quantity of the resultant $\bar{I}_{m3} = I_{m3}\varepsilon^{j\theta_3}$, we may write

$$\bar{I}_{m3} = \bar{I}_{m1} + \bar{I}_{m2} \tag{7-65}$$

or

$$I_{m3}\varepsilon^{j\theta_3} = I_{m1} + I_{m2}\varepsilon^{j\theta_2} = (I_{m1} + I_{m2}\cos\theta_2) + jI_{m2}\sin\theta_2 \tag{7-66}$$

The right side of the last equation involves all known quantities. Moreover, since it is a complex number having a horizontal and vertical component, it follows that the magnitude of the component is

$$I_{m3} = \sqrt{(I_{m1} + I_{m2}\cos\theta_2)^2 + (I_{m2}\sin\theta_2)^2} \tag{7-67}$$

and the angle is

$$\theta_3 = \tan^{-1}\frac{I_{m2}\sin\theta_2}{I_{m1} + I_{m2}\cos\theta_2} \tag{7-68}$$

The resultant phasor $\bar{I}_{m3}$ is depicted in Fig. 7-6(a) as line OD. Note that phasors are added in the same manner as vectors in mechanics. The sinusoid i_3 in Fig. 7-6(b) can then be considered as having been generated by the rotation of phasor $\bar{I}_{m3}$ in Fig. 7-6(a).

When the value of $\bar{I}_{m3}$ has been established, the corresponding time expression for i_3 readily follows from Eq. (7-64). Hence

$$i_3 = \text{Im}[\bar{I}_{m3}\varepsilon^{j\omega t}] = \text{Im}[I_{m3}\varepsilon^{j\theta_3}\varepsilon^{j\omega t}] = \text{Im}[I_{m3}\varepsilon^{j(\omega t + \theta_3)}]$$
$$= \text{Im}[I_{m3}\cos(\omega t + \theta_3) + jI_{m3}\sin(\omega t + \theta_3)] \tag{7-69}$$

or

$$i_3 = I_{m3}\sin(\omega t + \theta_3) \tag{7-70}$$

EXAMPLE 7-3   Two sinusoidal currents are described as follows:
$$i_1 = 10\sqrt{2}\sin\omega t \quad \text{and} \quad i_2 = 20\sqrt{2}\sin(\omega t + 60°)$$
Find the expression for the sum of these currents.

Solution: The solution is found by performing phasor addition in the manner indicated by Eq. (7-65). In passing, note that i_1 and i_2 have effective values of 10 and 20 A, respectively. The phasor quantities are

$$\bar{I}_{m1} = I_{m1} = 10\sqrt{2}$$

$$\bar{I}_{m2} = 20\sqrt{2}\,(\cos 60° + j \sin 60°)$$

$$= 20\sqrt{2}\left(\frac{1}{2} + j\frac{\sqrt{3}}{2}\right) = 10\sqrt{2} + j10\sqrt{6}$$

Hence

$$\bar{I}_{m3} = \bar{I}_{m1} + \bar{I}_{m2} = (10\sqrt{2} + 10\sqrt{2}) + j10\sqrt{6} \qquad (7\text{-}71)$$

$$= 20\sqrt{2} + j10\sqrt{6}$$

So that

$$I_{m3} = \sqrt{800 + 600} = 37.4$$

$$\theta_3 = \tan^{-1}\frac{10\sqrt{6}}{20\sqrt{2}} = \tan^{-1}\frac{\sqrt{3}}{2} = 41°$$

Therefore,

$$\bar{I}_{m3} = 37.4\varepsilon^{j41°}$$

and

$$i_3 = 37.4 \sin(\omega t + 41°)$$

It is pointed out in Secs. 7-2 and 7-3 that the most important and useful quantity of a sinusoidal function is its effective value. Although knowledge of the peak value of a sinusoid can be useful, it is not nearly so useful as the effective value. In Example 7-3 it is much more significant to speak in terms of the effective current than in terms of the peak value, because the former conveys information about its energy-transferring capability per cycle whereas the latter does not. Therefore in finding the sum of two currents the engineer is often really interested in the resultant effective current. Of course, from Eq. (7-26) we know that the effective current can readily be obtained by dividing the peak value by $\sqrt{2}$. In view of the far greater importance of effective values it is therefore customary to use a phasor diagram in which the phasors are expressed as effective values rather than maximum values. Accordingly, this procedure shall be followed in the remainder of the book.

To express the addition of two sinusoids in terms of the effective values of the phasors, it follows readily from Eq. (7-66) upon dividing each term in the expression by $\sqrt{2}$. Thus

$$\frac{\bar{I}_{m3}}{\sqrt{2}} = \frac{\bar{I}_{m1}}{\sqrt{2}} + \frac{\bar{I}_{m2}}{\sqrt{2}} \qquad (7\text{-}72)$$

or, more simply,

$$\bar{I}_3 = \bar{I}_1 + \bar{I}_2 \qquad (7\text{-}73)$$

We take note of one other convention which is used in phasor diagrams and the algebraic manipulations associated with them. In the interest of simplicity it is customary to replace $\varepsilon^{j\theta}$ by $\underline{/\theta}$. Accordingly the phasor quantity for a current such as i_2 of Eq. (7-52) can be expressed in terms of its effective value as

$$\bar{I}_2 = I_2\varepsilon^{j\theta_{22}} = I_2\underline{/\theta_2} \tag{7-74}$$

where I_2 denotes the effective current and $\underline{/\theta_2}$ denotes "at an angle θ_2 degrees counterclockwise." The angle θ_2 is always measured relative to the horizontal axis. For negative values of the angle the direction is taken as clockwise. Also keep in mind that

$$\underline{/\theta} = \varepsilon^{j\theta} = \cos\theta + j\sin\theta \tag{7-75}$$

Example 7-3 is now repeated in order to illustrate the use of the foregoing modifications.

EXAMPLE 7-4 A sinusoidal current having an effective value of 10 $\underline{/0°}$ A is added to another sinusoidal current of effective value 20 $\underline{/60°}$. Find the effective value of the resultant current.

Solution: By Eq. (7-73) we have

$$\bar{I}_3 = \bar{I}_1 + \bar{I}_2 = 10\underline{/0°} + 20\underline{/60°} = 10 + (10 + j10\sqrt{3}) \tag{7-76}$$

$$= 20 + j10\sqrt{3} = 26.4\underline{/41°}\text{ A}$$

To obtain the complete time expression we write

$$i_3 = \sqrt{2}\,(26.4)\sin(\omega t + 41°) = 37.4\sin(\omega t + 41°) \tag{7-77}$$

which checks with the result of Example 7-3.

Multiplication and Division of Complex Quantities. Up to now attention has been directed exclusively to the problem of adding sinusoidal quantities which are out of phase with one another. The use of phasor representation of the sinusoids simplified the procedure considerably. The method of solution was reduced to one of dealing with complex numbers as illustrated in Examples 7-3 and 7-4. In dealing with the sinusoidal steady-state response of electric circuits the need frequently arises to multiply and divide complex numbers. The complex numbers, however, do not always represent sinusoidal functions. In the interest of illustrating how the product of two complex numbers is obtained, consider the following two complex numbers. One is denoted by the phasor $\bar{I} = I\varepsilon^{j\theta}$; the other is represented by the operator† $\bar{Z} = Z\varepsilon^{j\phi}$. The product is desired. As the first step in the procedure, formulate the product in terms of the exponential form. The exponential form is preferred initially because as a legitimate part of the language of mathematics we are familiar with the rules governing its manipulation. Accordingly, we can write

$$\bar{I}\bar{Z} = I\varepsilon^{j\theta}Z\varepsilon^{j\phi} = IZ\varepsilon^{j(\theta+\phi)} \tag{7-78}$$

When the notation of Eq. (7-75) is inserted this expression becomes

$$\bar{I}\bar{Z} = IZ\underline{/\theta + \phi} \tag{7-79}$$

Therefore the product of two complex numbers is found by taking the product of their magnitudes and the sum of their angles.

---

† This term will be better understood after a study of Sec. 7-5.

The division of one complex quantity by another is treated in a similar fashion. To illustrate, let it be required to divide the complex quantity $\overline{Z} = Z\varepsilon^{j\phi}$ into the phasor $\overline{V} = V\varepsilon^{j\theta}$, which represents the sinusoid $v = \sqrt{2}\,V \sin(\omega t + \theta)$. We shall call the quotient $\overline{I}$. Thus

$$\overline{I} = \frac{\overline{V}}{\overline{Z}} = \frac{V\varepsilon^{j\theta}}{Z\varepsilon^{j\phi}} = \frac{V}{Z}\,\varepsilon^{j(\theta-\phi)} \tag{7-80}$$

Expressed in terms of the shorthand notation of Eq. (7-75), we have

$$\overline{I} = \frac{V\angle\theta}{Z\angle\phi} = \frac{V}{Z}\angle\theta - \phi \tag{7-81}$$

Therefore the division of one complex number by another involves the division of their magnitudes to yield the magnitude of the quotient and the difference of their phase angles to yield the phase of the quotient. For the sake of completeness we give the corresponding time expression for the phasor of Eq. (7-81), which is

$$i = \sqrt{2}\,\frac{V}{Z}\sin(\omega t + \theta - \phi) \tag{7-82}$$

The angular frequency ω is the same as that for v.

Powers and Roots of Complex Numbers. Occasionally, it becomes necessary to find the square or cube of a complex quantity. The manner of treatment again is revealed by using the exponential form. Hence to find the nth power of the complex quantity $\overline{Z} = Z\varepsilon^{j\phi}$ we proceed as follows.

$$\overline{Z}^n = (Z\varepsilon^{j\phi})^n = Z^n\varepsilon^{jn\phi} = Z^n\angle n\phi \tag{7-83}$$

Therefore, the nth power of a complex number is a complex number whose magnitude is the nth power of the magnitude of the original complex number and whose angle is n times as large as that of the original complex number.

The root of a complex number can be found by making n a proper fraction in Eq. (7-83). However, one additional modification is necessary: the angle of the original complex number must be increased by $2k\pi$ (where k is an integer) in order to bring into evidence all those root values which satisfy Eq. (7-83). Thus to find the fourth power of $\overline{Z} = Z\varepsilon^{j\phi}$ we proceed as follows. First replace ϕ by $\phi + 2k\pi$. Note that this in no way alters the value of the original complex number. Assign the value 1/4 to n in Eq. (7-83). Then

$$\overline{Z}^{1/4} = [Z\varepsilon^{j(\phi+2k\pi)}]^{1/4} = Z^{1/4}\angle\frac{\phi}{4} + \frac{k\pi}{2} \tag{7-84}$$

Therefore, the four different and distinct values which satisfy Eq. (7-84) are

$$\overline{Z}_1^{1/4} = Z^{1/4}\angle\frac{\phi}{4} \qquad \text{for } k = 0$$

$$\overline{Z}_2^{1/4} = Z^{1/4}\angle\frac{\phi}{4} + \frac{\pi}{2} \qquad \text{for } k = 1$$

$$\tag{7-85}$$

$$\overline{Z}_3^{1/4} = Z^{1/4} \bigg/ \underline{\dfrac{\phi}{4} + \pi} \qquad \text{for } k = 2$$

$$\overline{Z}_4^{1/4} = Z^{1/4} \bigg/ \underline{\dfrac{\phi}{4} + \dfrac{3\pi}{2}} \qquad \text{for } k = 3$$

Any further values assigned to k will yield results which are repetitions of those already listed in Eqs. (7-85).

EXAMPLE 7-5 The following three sinusoidal currents flow into a junction: $i_1 = 3\sqrt{2} \sin \omega t$, $i_2 = 5\sqrt{2} \sin (\omega t + 30°)$, and $i_3 = 6\sqrt{2} \sin (\omega t - 120°)$. Find the time expression for the resultant sinusoidal current which leaves the junction.

Solution: The corresponding phasor expressions are

$$\overline{I}_1 = 3 \underline{/0°}$$

$$\overline{I}_2 = 5 \underline{/30°} = 5(\cos 30° + j \sin 30°) = 2.5\sqrt{3} + j2.5$$

$$\overline{I}_3 = 6 \underline{/-120°} = 6[\cos(-120) + j \sin(-120)] = -3 - j3\sqrt{3}$$

Note that since we are concerned with addition it is the *rectangular form* of the complex number which is used. The resultant phasor quantity is therefore

$$\overline{I} = \overline{I}_1 + \overline{I}_2 + \overline{I}_3 = 2.5\sqrt{3} - j(3\sqrt{3} - 2.5) \tag{7-86}$$

The phasor diagram representing the addition called for in the last equation is depicted in Fig. 7-7.

$$\overline{I} = 2.5\sqrt{3} - j2.7 = 5.1 \underline{/-32°} \tag{7-87}$$

The time expression for the resultant current is

$$i = \sqrt{2}(5.1)\sin(\omega t - 32°) = 7.22 \sin(\omega t - 32°) \tag{7-88}$$

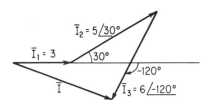

Fig. 7-7 Phasor diagram for Example 7-5 illustrating the addition of three sinusoidal currents by the use of phasors.

EXAMPLE 7-6 Find the quotient of $\overline{V} \, \overline{Z}_1/\overline{Z}_2$ where

$$\overline{V} = 45\sqrt{3} - j45$$

$$\overline{Z}_1 = 2.5\sqrt{2} + j2.5\sqrt{2} \tag{7-89}$$

$$\overline{Z}_2 = 7.5 + j7.5\sqrt{3}$$

Solution: Since we are not interested in addition but rather multiplication and division, it is simpler to work with the given complex quantities in polar form. Accordingly,

$$\overline{V} = \sqrt{(45\sqrt{3})^2 + 45^2} \bigg/ \underline{\tan^{-1} \dfrac{-45}{45\sqrt{3}}} = 90 \underline{/-30°} \tag{7-90}$$

$$\overline{Z}_1 = \sqrt{(2.5\sqrt{2})^2 + (2.5\sqrt{2})^2} \underline{/\tan^{-1}1} = 5\underline{/45°} \tag{7-91}$$

$$\overline{Z}_2 = \sqrt{7.5^2 + (7.5\sqrt{3})^2} \underline{/\tan^{-1}\sqrt{3}} = 15\underline{/60°} \tag{7-92}$$

Hence

$$\frac{\overline{V}\overline{Z}_1}{\overline{Z}_1} = \frac{90\angle{-30°} \times 5\angle{45°}}{15\angle{60°}} = \frac{450}{15}\angle{-30 + 45 - 60} = 30\angle{-45°} \qquad (7\text{-}93)$$

7-5 SINUSOIDAL STEADY-STATE RESPONSE OF SINGLE ELEMENTS—R,L,C

The response of each circuit parameter to a sustained sinusoidal source function is found individually—for two reasons. First, it provides an opportunity to illustrate the manner in which the response can be found easily and directly by the use of the phasor representation of sinusoids. Second, it affords the opportunity to establish once and for all the phase-angle relationships existing between the current and voltage for each circuit parameter. It is seen that these relationships are fixed and must always be satisfied irrespective of whether a given circuit element is in a series or parallel arrangement with other circuit elements. Let us start by treating the simplest of the three parameters—resistance.

Resistive Circuit. The circuit of interest is depicted in Fig. 7-8. Assume a sinusoidal forcing function which can be described by

$$v = V_m \sin \omega t = \text{Im}[V_m \varepsilon^{j\omega t}] \qquad (7\text{-}94)$$

Because the circuit is linear we know that the steady-state response must also be a sinusoid having the same frequency as the source function. However, in general, the response sinusoid (i.e., the current) differs in two respects—*amplitude and phase*. Therefore, on this basis we can say that the form of the response must be

$$i = I_m \sin (\omega t + \theta) = \text{Im}[I_m \varepsilon^{j\theta} \omega^{j\omega t}] \qquad (7\text{-}95)$$

where I_m and θ are respectively the amplitude and phase of the response which must be determined to achieve a solution of the problem.

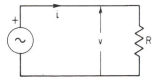

Fig. 7-8 Series resistive circuit with sinusoidal source.

The equation which makes available the amplitude and phase information in this instance is Kirchhoff's voltage law applied to the circuit of Fig. 7-8. Thus

$$v = Ri \qquad (7\text{-}96)$$

Note that this expression is nothing more than Ohm's law with the modification that sinusoidal quantities are involved for the variables v and i. Inserting Eqs. (7-94) and (7-95) into Eq. (7-96) yields

$$\text{Im}[V_m \varepsilon^{j\omega t}] = R \times \text{Im}[I_m \varepsilon^{j\theta} \varepsilon^{j\omega t}] \qquad (7\text{-}97)$$

The exponential form is preferred because it leads to the solution in a direct and easy fashion as is apparent from the material which follows. But first let us rewrite the last equation in a simpler form by dropping the notation "imaginary part of . . . ." *This is permissible because such equations reduce to an identity for all values of time whenever the coefficients of $\varepsilon^{j\omega t}$ on one side equal the coefficients of $\varepsilon^{j\omega t}$ on the other side.* There is a precaution to be observed, however, when using such a procedure. After completion of the algebraic manipulations which lead to the solution, it must be remembered that the actual solution is found by taking only the j part of the resulting exponential solution.

On this basis Eq. (7-97) can be written as

$$RI_m \varepsilon^{j\theta} \varepsilon^{j\omega t} = V_m \varepsilon^{j\omega t} \tag{7-98}$$

Since the time factor $\varepsilon^{j\omega t}$ is common to both sides of the equation, it may be suppressed, thus leading to

$$I_m \varepsilon^{j\theta} = \frac{V_m}{R} = \frac{V_m}{R} \varepsilon^{j0°} \tag{7-99}$$

On the left side of this equation appear the two unknown quantities I_m and θ, while on the right side appear the known quantities. Although in general both sides of this equation represent complex numbers, it is clear from the right side that in this case we have just a real number. This is characteristic of purely resistive circuits. A comparison of amplitudes and angles of the left and right sides of Eq. (7-99) leads to

$$\boxed{I_m = \frac{V_m}{R} \quad \text{and} \quad \theta = 0°} \tag{7-100}$$

Substituting this information into Eq. (7-95) yields the final expression for the solution. Hence

$$i = \frac{V_m}{R} \sin \omega t \tag{7-101}$$

Comparing Eq. (7-101) with Eq. (7-94) reveals that for a resistive circuit the current and voltage are in time phase—i.e., the peak values and the zero values occur at the same time instants. This is illustrated in Fig. 7-9.

As noted in the discussion of phasor representation of sinusoids, it is more frequently desirable to deal in terms of the effective value of voltages and currents rather than their maximum values. Accordingly, to express the solution as rep-

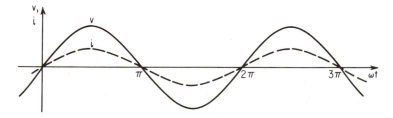

Fig. 7-9 Illustrating that i and v are in phase for the circuit of Fig. 7-8.

resented by Eqs. (7-100) in terms of effective values, use is made of Eq. (7-26). Thus

$$I_m = \sqrt{2}I = \frac{\sqrt{2}V}{R} \tag{7-102}$$

where I and V denote effective values. Therefore, one can very simply describe the current response caused by a sinusoidal voltage of effective value V applied to a circuit of resistance R as

$$\boxed{\bar{I} = \frac{V}{R} \angle 0°} \tag{7-103}$$

After experience has been gained in solving problems involving sinusoidal sources, it is usually unnecessary to go beyond this point in the solution. This is because the corresponding interpretation of Eq. (7-103) in the time domain is so well understood as to be self-evident. Recall from Sec. 7-4 that the procedure is to multiply the effective value of the response by $\sqrt{2}$ and add the computed phase angle to ωt to get the total argument. Applying this procedure to Eq. (7-103) yields

$$i = \frac{\sqrt{2}V}{R} \sin(\omega t + 0°) = \frac{V_m}{R} \sin \omega t \tag{7-104}$$

which obviously is identical to Eq. (7-101).

Appearing in Fig. 7-10 is the phasor diagram for the resistive circuit. The reference phasor is arbitrarily taken to be the applied voltage and is placed along the horizontal. By Eq. (7-103) the current phasor has an angle of 0° which means that it is in phase with the voltage. Hence it, too, is located on the horizontal axis. It is customary to show only effective values of voltage and current in the phasor diagrams. Note that the relationship depicted in Fig. 7-10 is consistent with that shown in Fig. 7-9, where time appears explicitly.

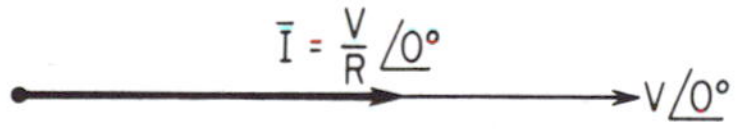

$$\bar{I} = \frac{V}{R} \angle 0°$$
$$\longrightarrow V\angle 0°$$

Fig. 7-10 Phasor diagram for resistive circuit.

Inductive Circuit. The same method of analysis is used to find the response of a purely inductive circuit to a sinusoidal source function in the steady state. Figure 7-11 shows the circuit. Assume that the potential difference appearing across the inductor terminals is given by Eq. (7-94). Because the circuit is assumed linear, the response can again be represented by Eq. (7-95). However,

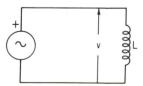

Fig. 7-11 Sinusoidal voltage applied to a pure inductor.

for the inductive circuit the defining equation is

$$v = L\frac{di}{dt} \tag{7-105}$$

Substituting Eqs. (7-94) and (7-95) gives†

$$\text{Im}[V_m \varepsilon^{j\omega t}] = L\frac{d}{dt}\,\text{Im}[I_m \varepsilon^{j\theta}\varepsilon^{j\omega t}] \tag{7-106}$$

where I_m and θ in general will be different from those values found for the resistive circuit. This is to be expected, since the defining equation involves a derivative of the current. Performing the differentiation and suppressing $\varepsilon^{j\omega t}$ leads to

$$I_m \varepsilon^{j\theta} = \frac{V_m}{j\omega L} \tag{7-107}$$

Again note that the term involving the unknown amplitude and phase angle is isolated to the left side of the equation. The quantities appearing on the right side are known. Examination of the right side in this instance shows that it is a number which is located along the vertical axis. In mathematical language this is described as the imaginary number. An alternative way of writing the right side follows upon recalling that the factor j is a rotational operator defined as

$$j \equiv \varepsilon^{j(\pi/2)} = 1\underline{/90°} \tag{7-108}$$

Any real number which is multiplied by j changes its position from the horizontal axis to the vertical axis. Thus $j10$ means that a line of 10 units (normally measured

---

† To perform the proper operations on Eq. (7-106) so that it leads to Eq. (7-107), a lemma is needed which shows that the derivative of the imaginary part of a rotating phasor ($\overline{A}\varepsilon^{j\omega t}$) is the same as the imaginary part of the derivative of the rotating phasor. Expressed mathematically, we need to show that

$$\frac{d}{dt}[\text{Im}(\overline{A}\varepsilon^{j\omega t})] = \text{Im}\left[\frac{d}{dt}(\overline{A}\varepsilon^{j\omega t})\right]$$

where $\overline{A}$ denotes the phasor.

Equivalency is demonstrated by manipulating the left side of the foregoing equation into the form represented by the right side. Thus

$$\frac{d}{dt}[\text{Im}(\overline{A}\varepsilon^{j\omega t})] = \frac{d}{dt}[\text{Im}(\overline{A}\cos\omega t + j\overline{A}\sin\omega t)]$$

$$= \frac{d}{dt}[\overline{A}\sin\omega t] = \omega\overline{A}\cos\omega t$$

Now this last form may also be expressed in terms of the notation Im (for "imaginary part of . . ."). This merely requires attaching the j factor and then introducing Im. Thus

$$\omega\overline{A}\cos\omega t = \text{Im}[j\omega\overline{A}\cos\omega t]$$

In turn, the right side of this last expression can also be written as

$$\omega\overline{A}\cos\omega t = \text{Im}\left[\frac{d}{dt}(\overline{A}\varepsilon^{j\omega t})\right]$$

which proves the identity of the two forms.

on the horizontal axis) is now measured on the vertical axis. Accordingly, Eq. (7-108) can be rewritten

$$I_m \varepsilon^{j\theta} = \frac{V_m}{\omega L \,\underline{/90°}} = \frac{V_m}{\omega L} \underline{/-90°} \qquad (7\text{-}109)$$

Therefore,

$$I_m = \frac{V_m}{\omega L} \quad \text{and} \quad \theta = -90° \qquad (7\text{-}110)$$

Expressed in terms of effective values, the solution for the response can be found conveniently by writing Eq. (7-109) as

$$\bar{I} = \frac{\bar{V}}{j\omega L} \qquad (7\text{-}111)$$

or

$$I \,\underline{/\theta} = \frac{V \,\underline{/0°}}{\omega L \,\underline{/90°}} = \frac{V}{\omega L} \underline{/-90°} \qquad (7\text{-}112)$$

Therefore,

$$I = \frac{V}{\omega L} = \frac{V}{X_L} \quad \text{and} \quad \theta = -90° \qquad (7\text{-}113)$$

where

$$X_L \equiv \omega L \qquad (7\text{-}114)$$

and is called the *inductive reactance* of the coil. The term *reactance* is used to distinguish it from resistance. A resistive circuit element causes no phase shift between v and i. However, an inductive element causes i to lag behind v by 90°. Any circuit element which exhibits this property in the sinusoidal steady state is said to have inductive reactance.

The phasor diagram for the purely inductive circuit is illustrated in Fig. 7-12(a). The location of the current phasor is drawn consistent with the results of Eq. (7-113), which calls for the current phasor to *lag* behind the voltage phasor by $\theta = -90°$. It is always true that for sinusoidal source functions *the current flowing through an inductor always lags behind the potential difference across the inductor terminals by 90°*. In the interest of completeness keep in mind the fact that the phasors of Fig. 7-12(a) are actually rotating counterclockwise (the assumed positive direction) at the angular frequency ω radians/second.

Equation (7-113) contains all the useful information needed for the solution of the problem. However, if the complete *time* solution for the current response is desired, it readily follows from Eq. (7-113) by recalling the meaning of phasor quantities. This leads to

$$i = \sqrt{2}\,\frac{V}{X_L} \sin(\omega t + \theta) = \frac{V_m}{X_L} \sin(\omega t - 90°) \qquad (7\text{-}115)$$

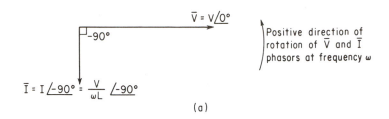

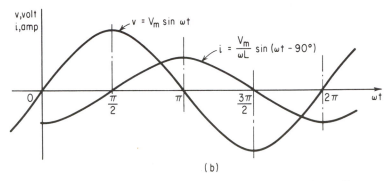

(b)

Fig. 7-12 Inductive circuit response to sinusoidal source: (a) phasor diagram; (b) time diagram.

A plot of Eq. (7-115) appears in Fig. 7-12(b). Note that the response sinusoid in the case of an inductive circuit is displaced 90° behind the potential difference across the inductor. Note also that just as Eq. (7-113) is a simpler and more direct way of representing the solution of Eq. (7-115) so too is Fig. 7-12(a) a simpler and more convenient way of expressing the results shown in Fig. 7-12(b).

Capacitive Circuit. In the circuit depicted in Fig. 7-13 the current which flows through the capacitor is related to the potential difference v by

$$i = C\frac{dv}{dt} \tag{7-116}$$

For an assumed sinusoidal source function of $v = V_m \sin \omega t$ the general form of the current response can be represented by $i_m = I_m \sin (\omega t + \theta)$. Inserting the exponential forms of v and i into the last expression yields

$$I_m \varepsilon^{j\omega t}\varepsilon^{j\theta} = C\frac{d}{dt}[V_m \varepsilon^{j\omega t}] \tag{7-117}$$

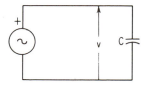

Fig. 7-13 Capacitive circuit with sinusoidal source function.

Performing the differentiation leads to

$$I_m \varepsilon^{j\theta} = j\omega C V_m \qquad (7\text{-}118)$$

or

$$I_m \angle \theta = \frac{V_m}{1/\omega C} \angle 90° \qquad (7\text{-}119)$$

Equating the magnitudes and angles of the right and left sides then yields

$$I_m = \frac{V_m}{1/\omega C} \qquad \text{and} \qquad \theta = +90° \qquad (7\text{-}120)$$

Once Eq. (7-119) is obtained by using the exponential forms of v and i in the governing differential equation, it can then be rewritten in terms of effective values. It is helpful at this point to employ effective values, because from here on the solution of capacitive circuits can be found by applying Ohm's law in the modified form dictated by Eq. (7-119). Accordingly, we can write

$$\bar{I} = \frac{\bar{V}}{1/j\omega C} \qquad (7\text{-}121)$$

This states that the phasor current response $\bar{I}$ is equal to the phasor potential difference $\bar{V}$ across the capacitor divided by the quantity $1/j\omega C$. It is important to understand that the quantity $1/j\omega C$ arises from the exponential formulation as revealed by Eq. (7-118). Moreover, the units of $1/j\omega C$ are volts/ampere, which is ohms. This is evident from Eq. (7-121). Therefore, $1/j\omega C$ may be looked upon as a kind of resistance, but it is not called this because it involves the rotational operator j. For this reason it is called a *reactance*; more specifically, it is called a *capacitive reactance* because a capacitor is involved. Equation (7-121) often appears in the form

$$\bar{I} = \frac{\bar{V}}{X_c \angle -90°} = \frac{\bar{V}}{X_c} \angle 90° \qquad (7\text{-}122)$$

where

$$X_c \equiv \frac{1}{\omega C} = \text{capacitive reactance in ohms} \qquad (7\text{-}123)$$

Also,

$$I \angle \theta = \frac{V}{X_c} \angle 90° \qquad (7\text{-}124)$$

Therefore, for the capacitive circuit

$$I = \frac{V}{X_c} = V\omega C \qquad \text{and} \qquad \theta = 90° \qquad (7\text{-}125)$$

By the information contained in Eq. (7-125) the phasor diagram of a purely capacitive circuit is readily drawn. Figure 7-14(a) depicts the result. The cor-

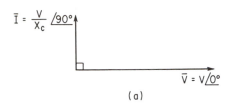

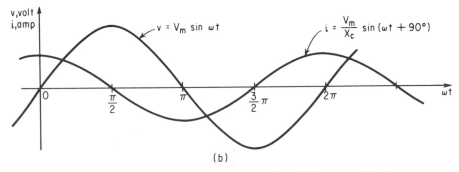

(b)

Fig. 7-14 Capacitive circuit response to sinusoidal source: (a) phasor diagram; (b) time diagram.

responding time-domain description of the current response is illustrated in Fig. 7-14(b). The time-domain equation for the response is obtained in the usual way and found to be

$$i = \sqrt{2}\,\frac{V}{X_c}\sin(\omega t + \theta) = \frac{V_m}{X_c}\sin(\omega t + 90°) \qquad (7\text{-}126)$$

A study of Eq. (7-125) as well as Fig. 7-14 reveals an important characteristic of the capacitive circuit: *the current flowing through a capacitor in the sinusoidal steady state always leads the potential difference across the capacitor by 90°.*

7-6 THE SERIES *RL* CIRCUIT

In this section we direct attention to the sinusoidal steady-state analysis of a circuit configuration which involves the series connection of a resistive and an inductive element. This leads to the concepts of complex impedance and reactive power.

Appearing in Fig. 7-15 is a diagram of the series *RL* circuit. Our objective

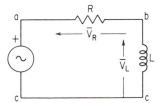

Fig. 7-15 Series *RL* circuit.

is to find the steady-state current response assuming that R, L, and the forcing function are known. The differential equation for the circuit is

$$v = iR + L\frac{di}{dt} \tag{7-127}$$

If the applied forcing function is assumed to be $v = V_m \sin \omega t = \text{Im}[V_m \varepsilon^{j\omega t}]$, then the current response must be of the form $i = I_m \sin(\omega t + \theta)$ where I_m and θ are to be determined. Introducing these quantities into Eq. (7-127), we have

$$V_m \varepsilon^{j\omega t} = RI_m \varepsilon^{j\theta} \varepsilon^{j\omega t} + j\omega L I_m \varepsilon^{j\theta} \varepsilon^{j\omega t} \tag{7-128}$$

Suppressing $\varepsilon^{j\omega t}$ and collecting terms yields

$$V_m = I_m \varepsilon^{j\theta}(R + j\omega L) \tag{7-129}$$

This expression can be written in terms of effective values by dividing both sides by $\sqrt{2}$. Accordingly,

$$\overline{V} = \overline{I}(R + j\omega L) \tag{7-130}$$

Equation (7-130) is Kirchhoff's voltage law for the RL circuit written in terms of phasor quantities. This expression can be written directly without going through the preceding steps. It is obtained by summing the voltage drops across each circuit element. However, it is important to attach the rotational operator j to the inductive reactance and $1/j$ to the capacitive reactance. This will be the procedure followed in the remainder of the book whenever dealing with such circuits. The analysis of the RC circuit in the next section starts at this point.

Upon transposing Eq. (7-130) the solution for the current becomes

$$\overline{I} = \frac{\overline{V}}{R + j\omega L} = \frac{\overline{V}}{\overline{Z}} \tag{7-131}$$

where

$$\overline{Z} \equiv R + j\omega L = Z \left/ \tan^{-1}\frac{\omega L}{R} \right. \tag{7-132}$$

This quantity $\overline{Z}$ is called the *complex impedance* of the RL circuit, and can always be represented by a right triangle as shown in Fig. 7-16. An inspection of Eq. (7-131) indicates that effectively $\overline{Z}$ may be considered as a complex operator which when divided into the voltage phasor yields the desired current phasor. Information about the magnitude and relative phase angle of the response

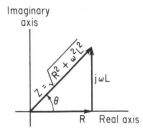

Fig. 7-16   Complex impedance triangle.

then readily follows when Eq. (7-131) is rewritten as

$$I\angle\theta = \frac{V\angle 0°}{Z\angle\tan^{-1}(\omega L/R)} = \frac{V}{Z}\left/-\tan^{-1}\frac{\omega L}{R}\right. \tag{7-133}$$

where

$$Z = \sqrt{R^2 + \omega^2 L^2} \tag{7-134}$$

Therefore,

$$\boxed{\begin{aligned} I &= \frac{V}{Z} = \frac{V}{\sqrt{R^2 + \omega^2 L^2}} \\ \theta &= -\tan^{-1}\frac{\omega L}{R} \end{aligned}} \tag{7-135}$$

and

Equation (7-135) furnishes all the information that is needed to identify the current response. Of course if the complete time expression is desired, it is found in the usual manner. In this case the result is

$$i = \frac{\sqrt{2}V}{Z}\sin(\omega t + \theta) = \frac{V_m}{\sqrt{R^2 + \omega^2 L^2}}\sin\left(\omega t - \tan^{-1}\frac{\omega L}{R}\right) \tag{7-136}$$

Comparing this equation with Eq. (6-163) shows the two to be equivalent. The difference in form lies in the fact that here the source function is assumed to be sinusoidal rather than cosinusoidal.

Examination of the solution for the relative phase angle reveals that for finite resistance the angle must be less than 90°. Note too that it is a *lag* angle, which means the current sinusoid is behind the voltage in time. These facts are represented graphically in Fig. 7-17. Appearing in Fig. 7-17(a) is a graphical representation of Eq. (7-130). The potential difference across the resistive element is clearly

$$\overline{V}_R \equiv \overline{I}R \tag{7-137}$$

Because there is no phase shift associated with a resistor, it follows that the voltage drop $\overline{I}R$ must be in phase with $\overline{I}$. Hence $\overline{V}_R$ is drawn on the same line as $\overline{I}$. Equation (7-137) is represented by ab in the phasor diagram. Similarly, the potential difference across the inductive element is shown by the second term of the right side of Eq. (7-130) to be expressed as

$$\overline{V}_L \equiv j\overline{I}X_L \tag{7-138}$$

The presence of the rotational operator j conveys the meaning that the quantity $\overline{I}X_L$ is to be rotated in the positive (or counterclockwise) direction through 90°. Accordingly, line bc in Fig. 7-17(a) represents Eq. (7-138). Of course the sum of $\overline{V}_R$ and $\overline{V}_L$ must add up to the line voltage ac. This too is borne out in the phasor diagram.

EXAMPLE 7-7   A 60-Hz sinusoidal voltage $v = 141\sin\omega t$ is applied to a series RL circuit. The value of resistance is 3 Ω and the inductance value is 0.0106 H.

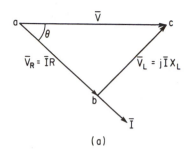

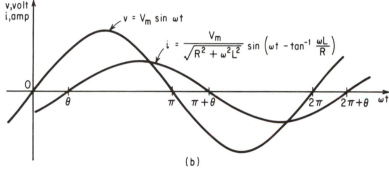

Fig. 7-17 Sinusoidal response of RL circuit: (a) phasor diagram; (b) time diagram.

(a) Compute the effective value of the steady-state current as well as the relative phase angle.

(b) Write the expression for the instantaneous current.

(c) Compute the effective magnitude and phase of the voltage drops appearing across each circuit element.

(d) Find the average power dissipated by the circuit.

(e) Calculate the power factor.

Solution: (a) The applied source function expressed in terms of its effective value is

$$\bar{V} = V\underline{/0°} = \frac{V_m}{\sqrt{2}}\underline{/0°} = 100 + j0 = 100 \text{ V} \tag{7-139}$$

To find the current we must first compute the complex impedance. Thus by Eqs. (7-132) and (7-134) we have

$$\bar{Z} = R + j\omega L = 3 + j2\pi f L = 3 + j377(0.0106) = 3 + j4$$
$$= \sqrt{3^2 + 4^2}\underline{/\tan^{-1}\tfrac{4}{3}} = 5\underline{/53.1°} \tag{7-140}$$

Therefore,

$$\bar{I} = \frac{\bar{V}}{\bar{Z}} = \frac{100\underline{/0°}}{5\underline{/53.1°}} = 20\underline{/-53.1°} \tag{7-141}$$

The effective value of the steady-state current is 20 amperes and it lags behind the voltage by 53.1°.

(b) $i = 20\sqrt{2} \sin(\omega t - 53.1°)$ \hfill (7-142)

(c) The voltage drop across the resistor is

$$\overline{V}_R = \overline{I}R = (20\angle -53.1)3 = 60\angle -53.1° \qquad (7\text{-}143)$$

In Fig. 7-17(a) this quantity is denoted by line ab. Note that $\overline{V}_R$ is drawn in a lagging position of $-53.1°$ with respect to the horizontal reference axis along which $\overline{V}$ is located. The voltage drop across the coil is

$$\overline{V}_L = j\overline{I}X_L = j(20\angle -53.1)4 = 1\angle 90°(20\angle -53.1)4 \qquad (7\text{-}144)$$
$$= 80\angle 90 - 53.1 = 80\angle 36.9°$$

This quantity is represented by line bc in Fig. 7-17(a). Note that $\overline{V}_L$ makes an angle of $36.9°$ relative to the horizontal axis.

It is also worthwhile to note that the phasor sum of $\overline{V}_R$ and $\overline{V}_L$ yields the applied voltage V. Thus

$$\overline{V} = \overline{V}_R + \overline{V}_L = 60\angle -53.1° + 80\angle 36.9° = 100 + j0 \qquad (7\text{-}145)$$

(d) From Eq. (7-44) we have

$$P_{av} = VI \cos \theta = 100(20) \cos 53.1° = 200(0.6) = 1200 \text{ W} \qquad (7\text{-}146)$$

Because the resistive element is the only element which dissipates power, the average power may also be found from

$$P_{av} = I^2R = (20)^2 3 = 1200 \text{ W} \qquad (7\text{-}147)$$

(e) Power factor can be found by either of two methods:

$$\text{pf} = \frac{\text{average power}}{\text{apparent power}} = \frac{P_{av}}{VI} = \frac{1200}{100(20)} = 0.6 \qquad (7\text{-}148)$$

or

$$\text{pf} = \cos \theta = \cos 53.1° = 0.6 \qquad (7\text{-}149)$$

Reactive Power. From the expression for the average power dissipated in a circuit it should be apparent that power is dependent upon the in-phase component of the current. This point can be stressed by writing the expression for average power as

$$P_{av} = V[I \cos \theta] \qquad (7\text{-}150)$$

and then noting that the quantity in brackets is the projection of the current phasor $\overline{I}$ onto the voltage phasor $\overline{V}$. Of course, θ must be the angle between $\overline{V}$ and $\overline{I}$ as depicted in Fig. 7-17(a). When conditions in the circuit are such that the projection of the current phasor $\overline{I}$ onto $\overline{V}$ is zero, then no real power can be delivered or dissipated in the circuit. This condition is represented in Fig. 7-4 when θ equals $90°$. An alternative way of expressing Eq. (7-150) is to introduce the following equivalences:

$$\cos \theta = \frac{R}{Z} \qquad (7\text{-}151)$$

and

$$V = IZ \qquad (7\text{-}152)$$

Equation (7-151) immediately follows from Fig. 7-16 and Eq. (7-152) states that the magnitude of the effective voltage is equal to the magnitude of the effective current multiplied by the magnitude of the complex impedance. Thus Eq. (7-150) becomes

$$P_{av} = V(I \cos \theta) = (IZ)I\frac{R}{Z} = I^2R \quad \text{W} \tag{7-153}$$

This equation states that the average power is equal to the power dissipated in the resistive portion of the circuit—a result already familiar to us.

A logical question to ask at this point is: What meaning, if any, can be given to the analogous equation involving the inductive reactance? That is, does the expression

$$P_X = I^2X_L = I^2\omega L \tag{7-154}$$

have any meaning? To seek out an answer, let us return to first principles—the expression for the instantaneous "power" across the inductance as expressed by the product of its current and potential difference. Thus

$$p_X = iv_L = iL\frac{di}{dt} \tag{7-155}$$

Inserting $i = I_m \sin (\omega t + \theta)$, we get

$$p_X = I_m^2\omega L \sin (\omega t + \theta) \cos (\omega t + \theta) \tag{7-156}$$

Substituting the trigonometric identity

$$\sin (\omega t + \theta) \cos (\omega t + \theta) = \tfrac{1}{2} \sin 2(\omega t + \theta) \tag{7-157}$$

into the last equation yields

$$p_X = \frac{I_m^2}{2}(\omega L) \sin 2(\omega t + \theta) = I^2\omega L \sin 2(\omega t + \theta) \tag{7-158}$$

Equation (7-158) reveals that the product of the current and voltage of an inductor is a sinusoid having double the frequency of either the current or voltage sinusoids. It is important here to note that, unlike the case for resistance, the average value of p_X over one cycle yields a result which is identically equal to zero. This means that over part of the cycle energy is delivered to the inductance where it is stored in the magnetic field, but in the next half-cycle it is returned to the source. The net transfer of energy in a pure inductance is thereby zero.

A comparison of the amplitude of the double-frequency sinusoid of Eq. (7-158) with Eq. (7-154) reveals the meaning of the latter quantity: it denotes the amplitude of the energy which is interchanged between the source and the energy-storing inductive element. Although the units of Eq. (7-154) are voltamperes, the units of P_X are not described in terms of watts. To distinguish P_X from real power, it is called *reactive power* expressed in units of *reactive voltamperes* (abbreviated *var*).

The reactive power P_X can also be expressed in terms of the effective values of current and voltage as well as the relative phase angle. From the impedance triangle of Fig. 7-16 we have

$$X_L = Z \sin \theta \tag{7-159}$$

Inserting this expression into Eq. (7-154) gives

$$P_X = I^2 X_L = I^2 Z \sin \theta = (IZ)I \sin \theta = VI \sin \theta \qquad (7\text{-}160)$$

Accordingly, reactive power in a circuit may also be found by taking the product of V and the component of the current which is in quadrature with $\overline{V}$.

There remains one final item in our discussion of the RL circuit. It is the treatment of the sinusoidal response of the parallel combination of resistance and inductance. The circuit arrangement is shown in Fig. 7-18. Because the same effective value of the sinusoidal voltage appears across each circuit element, the effective current through the resistor is

$$\overline{I}_R = \frac{\overline{V}}{R} \qquad (7\text{-}161)$$

and that through the inductor is

$$\overline{I}_L = \frac{\overline{V}}{jX_L} = \frac{\overline{V}}{j\omega L} \qquad (7\text{-}162)$$

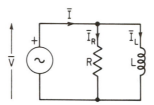

Fig. 7-18 Parallel RL circuit; $G = 1/R$ and $B = 1/\omega L$.

Accordingly, by Kirchhoff's current law the total current is

$$\overline{I} = \overline{I}_1 + \overline{I}_2 = \overline{V}\left(\frac{1}{R} + \frac{1}{j\omega L}\right) = \overline{V}\,\overline{Y} \qquad (7\text{-}163)$$

where $\overline{Y}$ is the *complex admittance*, and for the RL case is obviously defined as

$$\overline{Y} \equiv \frac{1}{R} + \frac{1}{j\omega L} = \frac{1}{R} - j\frac{1}{\omega L} \qquad (7\text{-}164)$$

The admittance too may be interpreted in terms of the role of an operator. It is that quantity which when multiplied by the voltage phasor yields the current phasor. In general, $\overline{Y}$ is a complex number which often is denoted as

$$\overline{Y} \equiv G + jB \qquad (7\text{-}165)$$

where G is the real part of the admittance and is called the *conductance*, and B is the quadrature component and is called *susceptance*. Upon comparing Eq. (7-165) with Eq. (7-164) it follows that for the parallel RL case

$$G = \frac{1}{R} \quad \text{and} \quad B = -\frac{1}{\omega L} \qquad (7\text{-}166)$$

Furthermore, the form of Eq. (7-164) makes it clear that the total current supplied by the source in steady state *lags* the voltage by the impedance angle, which for the parallel combination is

$$\theta = \tan^{-1}\frac{-1/\omega L}{1/R} = \tan^{-1}\frac{-R}{\omega L} \qquad (7\text{-}167)$$

Therefore the same general result applies for the parallel arrangement of R and L as it does for the series combination: the resultant current *lags* the voltage.

In situations involving the sinusoidal steady-state analysis of circuits it is often helpful to replace a series combination of resistance and inductance by a corresponding parallel equivalent circuit which is expressed in terms of conductance and susceptance. The equivalent arrangement is depicted in Fig. 7-19. We consider this subject matter for two reasons. First, it is a useful technique to employ when three or more complex impedances appear in parallel because it is easier to add admittances than to deal with the reciprocal of impedances. Second, it is important to understand the distinction between the equivalent admittance (i.e., parallel equivalence) of a series RL circuit and the admittance of an actual arrangement of R and L such as appears in Fig. 7-18. The admittance of the parallel R and L, of course, is represented by Eqs. (7-165) and (7-166).

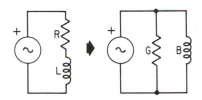

Fig. 7-19 Parallel equivalent of a series RL circuit. $G = R/(R^2 + X_L^2)$ and $B = -X_L/(R^2 + X_L^2)$.

Since the impedance of a series RL circuit is $\overline{Z} = R + j\omega L$, the corresponding equivalent admittance is expressed as the reciprocal of this quantity. Thus

$$\overline{Y} \equiv \frac{1}{\overline{Z}} = \frac{1}{R + j\omega L} \qquad (7\text{-}168)$$

To express this in terms of an equivalent conductance and susceptance Eq. (7-168) is rationalized as follows:

$$\overline{Y} = \frac{1}{R + j\omega L} \times \frac{R - j\omega L}{R - j\omega L} = \frac{R}{R^2 + \omega^2 L^2} - j\frac{\omega L}{R^2 + \omega^2 L^2} \qquad (7\text{-}169)$$

Recalling that $\omega L = X_L$ and that $\overline{Y} = G + jB$, we may rewrite Eq. (7-169) as

$$G + jB = \frac{R}{R^2 + X_L^2} - j\frac{X_L}{R^2 + X_L^2} \qquad (7\text{-}170)$$

Equating the real and j parts then yields

$$G = \frac{R}{R^2 + X_L^2} \quad \text{and} \quad B = \frac{-X_L}{R^2 + X_L^2} \qquad (7\text{-}171)$$

where G is the conductance and B is the susceptance of the equivalent parallel combination of the series RL configuration. A comparison of Eqs. (7-166) and (7-171) draws the distinction respectively of the circuit arrangements of Figs. 7-18 and 7-19.

7-7 THE SERIES *RC* CIRCUIT

The principles underlying the method of solution are presented in detail in the two preceding sections. Accordingly, in the interest of brevity as well as to illustrate the method to be employed from here on in solving such problems, we proceed directly with Kirchhoff's voltage law for the circuit of Fig. 7-20 by using effective quantities throughout. Thus

$$\overline{V} = \overline{I}R + \overline{I}\frac{1}{j\omega C} \tag{7-172}$$

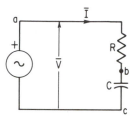

Fig. 7-20 The series *RC* circuit.

The effective potential difference appearing across the capacitor is expressed always in terms of the effective current passing through the capacitor multiplied by its capacitive reactance, which is written in conjunction with the rotational operator *j* as called for by Eq. (7-121). Rewriting Eq. (7-172) gives

$$\overline{V} = \overline{I}\left(R + \frac{1}{j\omega C}\right) = \overline{I}\left(R - j\frac{1}{\omega C}\right) = \overline{I}\overline{Z} \tag{7-173}$$

where $\overline{Z}$ denotes the complex impedance for the *RC* circuit and specifically is defined as

$$\overline{Z} \equiv R - j\frac{1}{\omega C} = \sqrt{R^2 + \frac{1}{\omega^2 C^2}}\Big/ -\tan^{-1}\frac{1}{\omega RC} \tag{7-174}$$

Therefore, the expression for the current response is

$$I\angle\theta = \frac{\overline{V}}{\overline{Z}} = \frac{V\angle 0°}{\sqrt{R^2 + 1/\omega^2 C^2}}\Big/ \tan^{-1}\frac{1}{\omega RC} \tag{7-175}$$

Thus

$$I = \frac{V}{\sqrt{R^2 + 1/\omega^2 C^2}} \tag{7-176}$$

and

$$\theta = \tan^{-1}\frac{1}{\omega RC} \tag{7-177}$$

The corresponding time solution is then

$$i = I_m \sin(\omega t + \theta) = \frac{V_m}{\sqrt{R^2 + 1/\omega^2 C^2}}\sin\left(\omega t + \tan^{-1}\frac{1}{\omega RC}\right) \tag{7-178}$$

The phasor diagram for the series *RC* circuit is depicted in Fig. 7-21(a). The current now *leads* the voltage in time—a result which is consistent with Eq. (7-177) showing a positive value for the phase angle. Again note that the effective potential difference across the resistor is in phase with the current, while that across the capacitor is 90° *behind* the position of the current phasor in Fig. 7-21(a). This $-90°$ rotation is brought about in the diagram by the attachment of $-j = \angle -90°$ to the quantity $\bar{I}X$. The corresponding time diagram of the current response is illustrated in Fig. 7-21(b).

The analysis of the parallel combination of resistance and capacitance is left as an exercise for the reader.

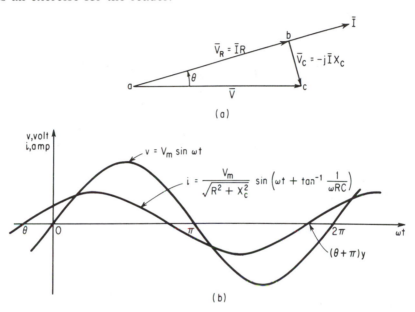

Fig. 7-21 (a) Phasor diagram for the *RC* circuit of Fig. 7-20; (b) time diagram for the *RC* circuit of Fig. 7-20.

7-8 THE *RLC* CIRCUIT

The circuit configuration for the series combination of the three parameters is shown in Fig. 7-22; the terminals are lettered in order to facilitate the discussion of the phasor diagram. Writing Kirchhoff's voltage law expressed in terms of effective values we have

$$\bar{V} = \bar{I}\left(\frac{1}{j\omega C}\right) + \bar{I}R + \bar{I}(j\omega L) = \bar{I}\left(R + j\omega L + \frac{1}{j\omega C}\right) \qquad (7\text{-}179)$$

Calling the quantity in parentheses the complex impedance and denoting it as

$$\bar{Z} \equiv R + j\left(\omega L - \frac{1}{\omega C}\right) = \sqrt{R^2 + (\omega L - 1/\omega C)^2} \, \Big/ \tan^{-1}\frac{\omega L - 1/\omega C}{R} \qquad (7\text{-}180)$$

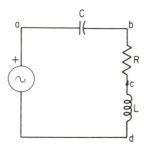

Fig. 7-22 The series *RLC* circuit.

enables us to write the expression for the current response

$$I\angle\theta = \frac{\overline{V}}{\overline{Z}} = \frac{V\angle 0°}{\sqrt{R^2 + (\omega L - 1/\omega C)^2}} \Big/ -\tan^{-1}\frac{\omega L - 1/\omega C}{R} \quad (7\text{-}181)$$

Therefore, the magnitude and phase of the effective current are

$$I = \frac{V}{\sqrt{R^2 + (\omega L - 1/\omega C)^2}} \quad (7\text{-}182)$$

and

$$\theta = -\tan^{-1}\frac{\omega L - 1/\omega C}{R} \quad (7\text{-}183)$$

The corresponding time solution is

$$i = \sqrt{2}\,I \sin(\omega t + \theta) = \frac{V_m}{\sqrt{R^2 + (\omega L - 1/\omega C)^2}} \sin(\omega t + \theta) \quad (7\text{-}184)$$

A study of Eq. (7-183) reveals that the current phasor may either lead or lag the voltage phasor, depending upon the relative values of the inductive and capacitive reactance ωL and $1/\omega C$. Whenever $\omega L > 1/\omega C$ the *RLC* circuit essentially behaves as an inductive circuit insofar as the current is concerned. It is interesting to note that this condition can be satisfied either by having a large inductance or else by operating at a high frequency. On the other hand, whenever $\omega L < 1/\omega C$ the current leads the voltage, thereby indicating that the *RLC* circuit behaves as a capacitive circuit as far as the current is concerned. However, there are some differences; these are discussed presently.

Appearing in Fig. 7-23 is the phasor diagram for the circuit of Fig. 7-22.

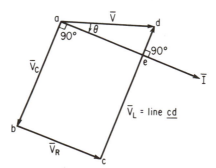

Fig. 7-23 Phasor diagram of the *RLC* circuit for $\omega L > 1/\omega C$.

It is drawn for the case where $\omega L > 1/\omega C$. Hence the current phasor must lag the voltage phasor. The component values of the effective potential difference appearing across each circuit element are also depicted. The voltage across the capacitor terminals $\overline{V}_c$ is represented by line ab. Note that the $\overline{I}$ leads $\overline{V}_c$ by 90° as it always must for sinusoidal forcing functions. The voltage drop across the resistor terminals $\overline{V}_R$ must be in phase with $\overline{I}$. It is represented by line bc, which is parallel to phasor $\overline{I}$. In phasor diagrams any line drawn parallel to another line means that the quantities represented by the two lines are in phase. Finally, note that line cd represents the effective potential difference appearing across the inductor terminals, $\overline{V}_L$. The current $\overline{I}$ lags $\overline{V}_L$ by 90° as expected.

Although the phasor diagram depicted in Fig. 7-23 is for an *RLC* circuit in which the inductive reactance predominates, note that the circuit does behave differently than the straight *RL* circuit in two respects. One, the potential difference across the inductor contains a component (line ce) which is equal and opposite to the total voltage across the capacitor terminals. This leaves a net reactive voltage, as "seen" by the source, of amount ed. Two, the voltage across the inductor can be several times greater *in magnitude* than the source voltage $\overline{V}$. This cannot occur in the simple *RL* circuit. The large value of $\overline{V}_L$ should not be disturbing, however, because Kirchhoff's voltage law continues to be satisfied. Keep in mind that with a-c circuits it is the *phasor* sum that is important. An algebraic sum is meaningless except in those instances where only elements of the same type appear in the circuit.

The net reactive power which is interchanged between the source and the circuit continues to be given validly by Eq. (7-160), namely

$$P_X = VI \sin \theta = I(V \sin \theta) \qquad (7\text{-}185)$$

Here the quantity in parentheses is equivalent to line ed in Fig. 7-23, which is the net reactive voltage of the circuit.

7-9 APPLICATION OF NETWORK THEOREMS TO COMPLEX IMPEDANCES

In Chapter 3 the elementary network theory was illustrated by circuit configurations which included only resistive elements. This restriction was introduced in the interest of simplicity and also because we did not then know how to treat the voltage across a capacitor or an inductor for a time-varying current such as the sinusoid. We learned in Secs. 7-5 to 7-7 that for sinusoidal forcing functions the inductance and capacitance parameters exhibit an impediment to the flow current which is called reactance and which is characterized chiefly by the time phase displacement it causes between the response and the forcing function in the steady state. In the material that follows, the principal network theorems are applied to a-c circuits, i.e., to networks which include the three circuit parameters in various combinations. The methods of analysis in arriving at the desired solutions are the same as those used in Chapter 3. The difference lies merely in working with complex impedances rather than just resistances. Since the impedances as well as the phasor currents and voltages are expressed by complex numbers, the algebraic manipulations are more extensive. The application of the theory is best illustrated by means of examples.

EXAMPLE 7-8 *Series-parallel reduction of complex impedance.* In the circuit of Fig. 7-24 find the equivalent impedance which appears between points a and c.

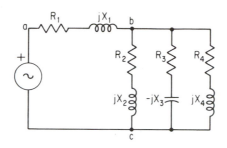

Fig. 7-24 Network used to illustrate series-parallel reduction of complex impedances.

Solution: Assume that this circuit is subjected to a sinusoidal forcing function of frequency ω radians per second. At this frequency the impedances have the following values:

$$\bar{Z}_1 = R_1 + jX_1 = 10 + j10 = 14.14\underline{/45°}$$

$$\bar{Z}_2 = R_2 + jX_2 = 15 + j20 = 25\underline{/53.1°}$$

$$\bar{Z}_3 = R_3 - jX_3 = 3 - j4 = 5\underline{/-53.1°}$$

$$\bar{Z}_4 = R_4 + jX_4 = 8 + j6 = 10\underline{/36.9°}$$

Because there are more than two impedances in parallel between terminals b and c it is simpler to work first in terms of the admittance of the parallel combination. Thus

$$\bar{Y}_{bc} = \frac{1}{\bar{Z}_{bc}} = \frac{1}{\bar{Z}_2} + \frac{1}{\bar{Z}_3} + \frac{1}{\bar{Z}_4} = \frac{1}{25\underline{/53.1}} + \frac{1}{5\underline{/-53.1}} + \frac{1}{10\underline{/36.9°}}$$

$$= 0.04\underline{/-53.1} + 0.2\underline{/53.1°} + 0.1\underline{/-36.9} \tag{7-186}$$

$$= 0.024 - j0.032 + 0.12 + j0.16 + 0.08 - j0.06$$

Note that a quantity such as $\bar{Y}_2 = 1/\bar{Z}_2 = 0.024 - j0.032$ is the equivalent admittance of the impedance $\bar{Z}_2$. In fact these values can be found also by using Eq. (7-171). Thus

$$\bar{Y}_2 = G_2 + jB_2 = \frac{15}{15^2 + 20^2} - j\frac{20}{15^2 + 20^2}$$

$$= \frac{15}{625} - j\frac{20}{625} = 0.024 - j0.032$$

A similar procedure may be employed in treating $\bar{Z}_3$ and $\bar{Z}_4$. Summing the real and j terms in Eq. (7-186) yields

$$\bar{Y}_{bc} = 0.224 + j0.068 = 0.234\underline{/16.9°} \tag{7-187}$$

$$\therefore \bar{Z}_{bc} = \frac{1}{\bar{Y}_{bc}} = \frac{1}{0.234\underline{/16.9}} = 4.27\underline{/-16.9} = 4.09 - j1.24$$

Thus, by Eq. (7-187) the series equivalent impedance between b and c is reduced to an equivalent resistance of 4.09 Ω and an equivalent capacitive reactance of 1.24 Ω. The total impedance between a and c is then given by

$$\overline{Z}_{ac} = \overline{Z}_1 + \overline{Z}_{bc} = 10 + j10 + 4.09 - j1.24$$
$$= 14.09 + j8.76$$

Therefore,
$$\overline{Z}_{ac} = 16.2\underline{/31.85°}\ \Omega$$

EXAMPLE 7-9 *Illustrating the mesh method of analysis.* Appearing in Fig. 7-25 is a two-mesh network with a sinusoidal source voltage appearing in each mesh. At a given frequency the impedance and phasor voltages are as follows:

$$\overline{Z}_1 = R_1 + jX_1 = 1 + j = \sqrt{2}\underline{/45°}$$
$$\overline{Z}_2 = R_2 - jX_2 = 1 - j = \sqrt{2}\underline{/-45°}$$
$$\overline{Z}_3 = R_3 + jX_3 = 1 + j2 = \sqrt{5}\underline{/63.5°}$$
$$\overline{V}_1 = 10\underline{/0°} \text{ and } \overline{V}_2 = 10\underline{/-60°}$$

The phasor voltage $\overline{V}_1$ is taken as the reference. On this basis $\overline{V}_2$ lags $\overline{V}_1$ in time by $\omega t = 60°$. Find the phasor expression for the current which flows through branch $\overline{Z}_3$.

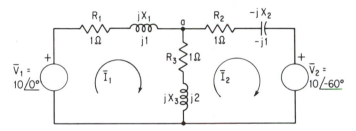

Fig. 7-25 Two-mesh network with complex impedances.

Solution: The first step is to write the mesh equations in terms of phasor quantities. Thus for mesh 1 we have
$$\overline{I}_1\overline{Z}_1 + \overline{I}_1\overline{Z}_3 - \overline{I}_2\overline{Z}_3 - \overline{V}_1 = 0$$
or
$$\overline{I}_1(\overline{Z}_1 + \overline{Z}_3) - \overline{I}_2\overline{Z}_3 = \overline{V}_1 \tag{7-188}$$
For the second mesh Kirchhoff's voltage law yields
$$\overline{I}_2\overline{Z}_2 + \overline{I}_2\overline{Z}_3 - \overline{I}_1\overline{Z}_3 + \overline{V}_2 = 0$$
or
$$\tag{7-189}$$
$$-\overline{I}_1\overline{Z}_3 + \overline{I}_2(\overline{Z}_2 + \overline{Z}_3) = -\overline{V}_2$$
Equations (7-188) and (7-189) can be written more formally as
$$\overline{I}_1\overline{Z}_{11} - \overline{I}_2\overline{Z}_{12} = \overline{V}_1 \tag{7-190}$$
$$-\overline{I}_1\overline{Z}_{21} + \overline{I}_2\overline{Z}_{22} = -\overline{V}_2 \tag{7-191}$$
where
$$\overline{Z}_{11} = (R_1 + R_3) + j(X_1 + X_3) = 2 + j3$$
$$\overline{Z}_{12} = \overline{Z}_{21} = \overline{Z}_3 = 1 + j2$$
$$\overline{Z}_{22} = (R_2 + R_3) + j(X_3 - X_2) = 2 + j1$$

Applying determinants to the pair of equations (7-190) and (7-191) allows the expression for $\bar{I}_1$ to be written as

$$\bar{I}_1 = \frac{\begin{vmatrix} \bar{V}_1 & -\bar{Z}_{12} \\ -\bar{V}_2 & \bar{Z}_{22} \end{vmatrix}}{\begin{vmatrix} \bar{Z}_{11} & -\bar{Z}_{12} \\ -\bar{Z}_{21} & \bar{Z}_{22} \end{vmatrix}} = \frac{\bar{V}_1\bar{Z}_{22} - \bar{V}_2\bar{Z}_{12}}{\bar{Z}_{11}\bar{Z}_{22} - \bar{Z}_{12}^2}$$

$$= \frac{10(2 + j) - 10\angle-60(1 + j2)}{(2 + j3)(2 + j1) - (1 + j2)^2} = \frac{20 + j10 - (5 - j5\sqrt{3})(1 + j2)}{1 + j8 + 3 - j4} \qquad (7\text{-}192)$$

$$= \frac{20 + j10 - 22.32 - j1.34}{4 + j4} = \frac{8.97\angle105°}{5.66\angle45°}$$

$$= 1.58\angle60° = 0.79 + j1.37$$

Similarly,

$$\bar{I}_2 = \frac{\begin{vmatrix} \bar{Z}_{11} & \bar{V}_1 \\ -\bar{Z}_{21} & -\bar{V}_2 \end{vmatrix}}{\bar{Z}_{11}\bar{Z}_{22} - \bar{Z}_{12}^2} = \frac{-\bar{V}_2\bar{Z}_{11} + \bar{Z}_{21}\bar{V}_1}{5.66\angle45°}$$

$$= \frac{-10\angle-60(2 + j3) + (1 + j2)10}{5.66\angle45°} = \frac{34.1\angle139.3°}{5.66\angle45°} \qquad (7\text{-}193)$$

$$= 6.03\angle94.3° = -0.46 + j6.01$$

Therefore, the current in branch $\bar{Z}_3$ is

$$\bar{I}_{\text{branch 3}} = \bar{I}_1 - \bar{I}_2 = (0.79 + j1.37) - (0.46 + j6.01)$$

$$= 1.25 - j4.64 = 4.82\angle-75° \text{ A}$$

Hence the effective value of the current is 4.82 A and in the steady state this current lags behind the voltage $\bar{V}_1$ by 75°.

EXAMPLE 7-10 *Illustrating the nodal method.* Find the current in the $\bar{Z}_3$ branch by employing the nodal method of solution.

Solution: Inspection of the circuit reveals that there is only one independent node which is marked terminal a in Fig. 7-25. Upon assuming that all currents flow away from the node, the describing equation which results from Kirchhoff's current law is

$$\frac{\bar{V}_a - \bar{V}_1}{\bar{Z}_1} + \frac{\bar{V}_a}{\bar{Z}_3} + \frac{\bar{V}_a - \bar{V}_2}{\bar{Z}_2} = 0 \qquad (7\text{-}194)$$

or

$$\bar{V}_a\left(\frac{1}{\bar{Z}_1} + \frac{1}{\bar{Z}_2} + \frac{1}{\bar{Z}_3}\right) = \frac{\bar{V}_1}{\bar{Z}_1} + \frac{\bar{V}_2}{\bar{Z}_2} \qquad (7\text{-}195)$$

Let

$$\bar{Y} = \frac{1}{\bar{Z}_1} + \frac{1}{\bar{Z}_2} + \frac{1}{\bar{Z}_3} = \left(\frac{1}{2} - j\frac{1}{2}\right) + \left(\frac{1}{2} + j\frac{1}{2}\right) + \left(\frac{1}{5} - j\frac{2}{3}\right) \qquad (7\text{-}196)$$

$$= 1.2 - j0.4 = 1.263\angle-18.4°$$

Also,

$$\frac{\bar{V}_1}{\bar{Z}_1} + \frac{\bar{V}_2}{\bar{Z}_2} = \frac{10\angle0°}{\sqrt{2}\angle45°} + \frac{10\angle-60°}{\sqrt{2}\angle-45°} = 5 - j5 + 6.81 - j1.83$$

$$= 11.81 - j6.83 = 13.65\angle-30° \qquad (7\text{-}197)$$

Inserting Eqs. (7-196) and (7-197) into Eq. (7-195) yields the phasor expression for the nodal voltage:

$$\overline{V}_a = \frac{13.65\,\underline{/-30°}}{1.263\,\underline{/-18.4°}} = 10.8\,\underline{/-11.6°} \text{ V}$$

Therefore, the branch current is simply

$$\overline{I}_{\text{branch 3}} = \frac{\overline{V}_a}{\overline{Z}_3} = \frac{10.8\,\underline{/-11.6°}}{\sqrt{5}\,\underline{/63.5°}} = 4.82\,\underline{/-75°}$$

which checks with the result found by using the mesh method. Note, however, that considerably less effort is required to arrive at the solution.

EXAMPLE 7-11 *Solution by Thévenin's theorem.* Figure 7-25 is redrawn in Fig. 7-26 to separate branch $\overline{Z}_3$ from the remainder of the circuit. By using Thévenin's theorem find the current which flows through $\overline{Z}_3$.

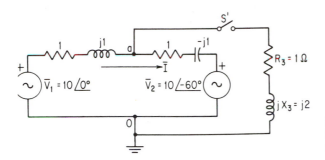

Fig. 7-26 Circuit of Fig. 7-25 redrawn in preparation for applying Thévenin's theorem.

Solution: The open-circuit voltage of the Thévenin equivalent circuit is equal to the potential difference appearing between points 0 and a in Fig. 7-26 when the switch is open. To determine this quantity let us first find the current $\overline{I}$ that flows in the circuit between the two sources when S is open. Kirchhoff's voltage law yields

$$-\overline{V}_1 + \overline{I}\overline{Z}_1 + \overline{I}\overline{Z}_2 + \overline{V}_2 = 0 \tag{7-198}$$

or

$$\overline{I} = \frac{\overline{V}_1 - \overline{V}_2}{\overline{Z}_1 + \overline{Z}_2} = \frac{10 - 10\,\underline{/-60}}{1 + j + 1 - j} = \frac{10 - 5 + j8.66}{2} = 5\,\underline{/60°}$$

Hence the voltage between points 0 and a is

$$\overline{V}_{0a} = \overline{V}_{\text{open-circuit}} = \overline{V}_1 - \overline{I}\overline{Z}_1 = 10\,\underline{/0°} - 5\,\underline{/60°}\,\sqrt{2}\,\underline{/45°} \tag{7-199}$$
$$= 10 + 1.83 - j6.82 = 13.65\,\underline{/-30°}$$

Note that the open-circuit voltage has a larger magnitude than either source voltage. The equivalent impedance looking into the circuit at points 0 and a is

$$\overline{Z}_i = \frac{\overline{Z}_1\overline{Z}_2}{\overline{Z}_1 + \overline{Z}_2} = \frac{\sqrt{2}\,\underline{/45°}\sqrt{2}\,\underline{/-45}}{1 + j + 1 - j} = 1\,\underline{/0°}\ \Omega \tag{7-200}$$

Therefore, the branch current in $\overline{Z}_3$ is given by

$$\overline{I}_{\text{branch}} = \frac{\overline{V}_{\text{oc}}}{\overline{Z}_i + \overline{Z}_3} = \frac{13.65\,\underline{/-30°}}{1 + 1 + j2} = 4.82\,\underline{/-75°} \text{ A}$$

Again note the exact correspondence of this result with those obtained by the other methods.

The use of Norton's theorem in solving this problem is left as an exercise for the reader.

7-10 RESONANCE

Resonance is identified with engineering situations which involve energy-storing elements subjected to a forcing function of varying frequency. Specifically, resonance is the term used to describe the steady-state operation of a circuit or system at that frequency for which the resultant response is in time phase with the source function despite the presence of energy-storing elements. Resonance cannot take place when only one type of energy-storing element is present, e.g., inductance or mass. There must exist two types of independent energy-storing elements capable of interchanging energy between one another—for example, inductance and capacitance or mass and spring. Although attention here is confined to electric circuits, resonance is a phenomenon found in any system involving two independent energy-storing elements be they mechanical, hydraulic, pneumatic, or whatever. In the material which follows, the parallel as well as the series arrangement of the energy-storing elements—L and C—are treated as functions of frequency, with particular emphasis focused on the performance at resonance.

Series Resonance. The series arrangement of L and C along with resistance R is shown in Fig. 7-22. The expression for the effective current flow caused by a sinusoidal forcing function is given by Eq. (7-179) and is repeated for convenience. Thus

$$\bar{I} = \frac{\bar{V}}{R + j(\omega L - 1/\omega C)} = \frac{\bar{V}}{\bar{Z}} \tag{7-201}$$

What is the effect on $\bar{I}$ of increasing the frequency ω of the source function from zero to infinity? A glance at Eq. (7-201) indicates that a change in frequency means a change in the magnitude and phase angle of the complex impedance. Note that for ω close to zero the inductive reactance is almost zero but the capacitive reactance approaches infinity. Hence the current magnitude approaches zero. As ω increases, the reactance part of $\bar{Z}$ decreases, thus causing an increase in current. As ω continues to increase, a point is reached where the reactance term is zero. Calling this frequency ω_0 we have

$$\omega_0 L - \frac{1}{\omega_0 C} = 0 \tag{7-202}$$

or

$$\omega_0^2 = \frac{1}{LC} \tag{7-203}$$

or

$$\omega_0 = \frac{1}{\sqrt{LC}} \qquad (7\text{-}204)$$

The frequency ω_0 is called the *resonant frequency* of the circuit. Its value is specified entirely in terms of the parameters of the two energy-storing elements of the circuit in the manner called for by Eq. (7-204).

At resonance the impedance of the circuit is a minimum, and specifically it is equal to R. Consequently, when a series RLC circuit is at resonance, the current is a maximum and is also in time phase with the voltage. The power factor is unity. The complete expression for the current phasor is then

$$\bar{I}_0 = \frac{\bar{V}}{\bar{Z}} = \frac{V\angle 0°}{R\angle 0°} = \frac{V}{R}\angle 0° \qquad (7\text{-}205)$$

where $\bar{I}_0$ denotes the current at resonance.

For operation at a frequency below ω_0 the resultant j part of $\bar{Z}$ is capacitive so that the current leads the voltage. When $\omega > \omega_0$ the inductive reactance prevails, so that the current then lags the voltage. Depicted in Fig. 7-27 is the plot of the impedance in the complex plane. The magnitude of $\bar{Z}$ at a frequency ω_1 is obtained as the length of the line drawn from the origin to the point ω_1 on the heavy vertical line. This heavy vertical line denotes the value of the reactance portion of $\bar{Z}$. Note that at $\omega = \omega_0$ it has a zero value. An alternative way of representing the information of Fig. 7-27 is illustrated in Fig. 7-28. A plot of the variation of the magnitude and phase of the current with frequency for two values of the ratio $\omega_0 L/R$ is depicted in Fig. 7-29. Note that when the ratio of inductive reactance to resistance in a series RLC circuit is high, a very rapid rise of current to the maximum value occurs in the vicinity of the resonant frequency. A characteristic of this type is especially useful in radio and other communication applications.

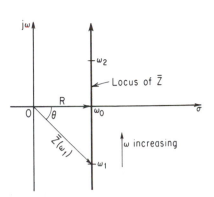

Fig. 7-27 Locus of the complex impedance $\bar{Z}$ as a function of frequency ω; $\bar{Z} = R + j(\omega L - 1/\omega C)$.

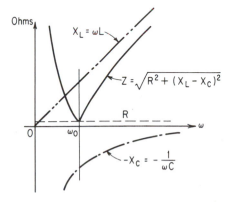

Fig. 7-28 Variation of impedance, inductive reactance, and capacitive reactance as a function of frequency.

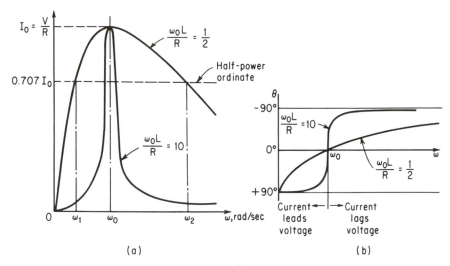

Fig. 7-29 Variation of current in a series RLC circuit with frequency: (a) magnitude; (b) phase angle.

Bandwidth. To describe the width of the resonance curve the term *bandwidth* is used. For the series RLC circuit bandwidth is defined as the range of frequency for which the power delivered to R is greater than or equal to $P_0/2$, where P_0 is the power delivered to R at resonance. From the shape of the resonance curve it should be clear that there are two frequencies for which the power delivered to R is half the power at resonance. For this reason these frequencies are referred to as those corresponding to the half-power points. The magnitude of the current at each half-power point is the same. Hence we can write

$$I_1^2 R = \tfrac{1}{2} I_0^2 R = I_2^2 R \qquad (7\text{-}206)$$

where subscript 1 denotes the lower half-power point and subscript 2 the higher half-power point. It follows then that

$$I_1 = I_2 = \frac{I_0}{\sqrt{2}} = 0.707 I_0 \qquad (7\text{-}207)$$

Accordingly, the bandwidth may be identified on the resonance curve as that range of frequency over which the magnitude of the current is equal to or greater than 0.707 of the current at resonance. In Fig. 7-29(a) for $\omega_0 L/R = \tfrac{1}{2}$, the frequency at the lower half-power point is denoted ω_1 and that of the upper half-power point is denoted ω_2. Hence the bandwidth is $\omega_2 - \omega_1$.

In view of the fact that the current at the half-power points is $I_0/\sqrt{2}$, it follows that the *magnitude* of the impedance must be equal to $\sqrt{2}\,R$ to yield this current. This information can now be used to obtain an expression for the bandwidth in terms of the parameters of the series circuit. Calling the reactance at the lower half-power frequency X_1, we have

$$X_1 = \omega_1 L - \frac{1}{\omega_1 C} = -R \qquad (7\text{-}208)$$

The minus sign appears on the right side of the equation because below resonance the capacitive reactance exceeds the inductive reactance. Rearranging Eq. (7-208) leads to

$$\omega_1^2 + \frac{R}{L}\omega_1 - \frac{1}{LC} = 0 \qquad (7\text{-}209)$$

The roots are therefore

$$(\omega_1)_{1,2} = -\frac{R}{2L} \pm \sqrt{\left(\frac{R}{2L}\right)^2 + \frac{1}{LC}} = -\alpha \pm \sqrt{a^2 + \omega_0^2} \qquad (7\text{-}210)$$

where

$$\alpha \equiv \frac{R}{2L} \qquad (7\text{-}211)$$

Although Eq. (7-210) provides two solutions to Eq. (7-209), only one of these is physically realizable. The negative frequency is meaningless and so may be discarded. Hence the expression for the lower half-power frequency is

$$\omega_1 = -\alpha + \sqrt{\alpha^2 + \omega_0^2} \qquad (7\text{-}212)$$

The upper half-power frequency is found in a similar fashion. In this instance we have

$$X_2 = \omega_2 L - \frac{1}{\omega_2 C} = +R \qquad (7\text{-}213)$$

which leads to

$$\omega_2^2 - \frac{R}{L}\omega_2 - \frac{1}{LC} = 0 \qquad (7\text{-}214)$$

The expression for the useful ω_2 is then

$$\omega_2 = \alpha + \sqrt{\alpha^2 + \omega_0^2} \qquad (7\text{-}215)$$

Therefore, the expression for the bandwidth becomes, from Eqs. (7-212) and (7-215),

$$\omega_{\text{bw}} \equiv \omega_2 - \omega_1 = 2\alpha = \frac{R}{L} \qquad (7\text{-}216)$$

This expression is significant because it reveals that the bandwidth of the series RLC circuit depends solely upon the R/L ratio. Note that it is not the individual values of R or L but rather their ratio that is important. Note too that bandwidth depends not at all upon the capacitance parameter C.

Quality Factor Q_0 of the RLC Circuit.   By forming the ratio of the resonant frequency to the bandwidth we obtain a factor which is a measure of the *selectivity* or *sharpness of tuning* of the series *RLC* circuit. This quantity is called the *quality factor* of the circuit and is denoted by Q_0. Thus

$$Q_0 \equiv \frac{\omega_0}{\omega_{\text{bw}}} \qquad (7\text{-}217)$$

Moreover, when Eq. (7-216) is inserted into the last equation an alternative expression results:

$$Q_0 = \frac{\omega_0}{R/L} = \frac{\omega_0 L}{R} = \frac{X_{L0}}{R} \qquad (7\text{-}218)$$

A glance at the curves appearing in Fig. 7-29(a) should make it clear as to why the quality factor Q_0 is used as a measure of the selectivity or sharpness of tuning. Note how much narrower the resonance curve is for $Q_0 = 10$ than for $Q_0 = \frac{1}{2}$. Equation (7-216) shows that the value of C in no way influences the bandwidth but it does alter the value of ω_0. Hence if C is changed, ω_0 may be made to occur at a different position along the frequency scale in Fig. 7-29(a) without in any way affecting the sharpness of tuning. This feature of the high-Q circuit is used often in communication networks. For example, in a radio the antenna may be considered in terms of an equivalent RLC circuit where C is adjusted by means of the dial tuning knob. If the dial is turned to the position which tunes in the frequency (ω_0) of a given radio transmitting station, the sharpness of tuning (i.e., small bandwidth) allows only signals from that station to produce large resonant current signals. The signals from other broadcasting stations, although present in equal strength at the antenna, produce little or no signal strength in the circuit because the dial is tuned to a frequency considerably off resonance relative to their broadcasting frequencies. Values of Q_0 of the order of 100 are typical in radio circuits.

By making use of the knowledge that Q_0 is very large in many resonant circuits, we can express the lower and upper half-power points in terms of the resonant frequency and the bandwidth. Returning to Eq. (7-212), we can rewrite it as

$$\omega_1 = -\frac{R}{2L} \pm \sqrt{\left(\frac{R}{2L}\right)^2 + \omega_0^2} = -\frac{R}{2L} + \sqrt{\left(\frac{R}{2L}\right)^2 + \frac{R^2}{L^2}\left(\frac{\omega_0^2 L^2}{R^2}\right)}$$

$$= -\frac{R}{2L} + \sqrt{\left(\frac{R}{2L}\right)^2 + Q_0^2 \frac{R^2}{L^2}} \qquad (7\text{-}219)$$

For values of Q_0 which are 5 or greater very little error is made by writing

$$\omega_1 = -\frac{R}{2L} + \frac{R}{L}Q_0 = -\frac{R}{2L} + \omega_0 \qquad (7\text{-}220)$$

Inserting Eq. (7-216) into Eq. (7-220) yields

$$\omega_1 = \omega_0 - \tfrac{1}{2}\omega_{bw} \qquad (7\text{-}221)$$

Similarly, it can be shown that

$$\omega_2 = \omega_0 + \tfrac{1}{2}\omega_{bw} \qquad (7\text{-}222)$$

Another characteristic of a series RLC circuit at resonance is worth noting. It has to do with the magnitude of the voltage drop appearing across L and C. In terms of a phasor formulation the phasor voltage drop across L at resonance is given by

$$\overline{V}_{L0} = \overline{I}_0(jX_L) = \frac{V}{R}\omega_0 L\underline{/90°} = Q_0 V\underline{/90°} \tag{7-223}$$

Therefore the magnitude of the voltage across the terminals of the inductor is Q_0 times the rms value of the applied voltage. For high-Q circuits this represents a considerable amplification of the source voltage. By proceeding in a similar fashion for the capacitor we find

$$\overline{V}_{C0} = \overline{I}_0\left(\frac{1}{j\omega C}\right) = \frac{V}{R}\left(\frac{1}{\omega_0 C}\right)\underline{/-90°}$$

$$= \frac{V}{R}(\omega_0 L)\underline{/-90°} = Q_0 V\underline{/-90°} \tag{7-224}$$

Again note the resonant voltage-rise effect across the capacitor terminals. The phasor voltage diagram at resonance is depicted in Fig. 7-30.

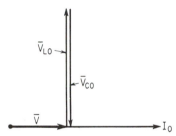

Fig. 7-30 Voltage phasor diagram of the series RLC circuit at resonance.

On the basis of Eqs. (7-223) and (7-224) another definition of the quality factor Q_0 is possible. It represents the extent to which the voltage across L or C rises at resonance expressed as a multiple of the applied voltage. Stated mathematically, we have

$$Q_0 = \frac{V_{L0}}{V} = \frac{V_{C0}}{V} \tag{7-225}$$

where V_{L0} and V_{C0} are both measured at resonance.

The definitions of Q_0 given by Eqs. (7-218) and (7-225) are restrictive in the sense that they apply specifically to the series RLC circuit. The definition of Eq. (7-217) is more generally applicable because it is based on the frequency-selectivity characteristic of the circuit. However, the most universally applicable definition is one that is expressed in terms of energy. In this connection, then, let us assume that the instantaneous expression for the forced solution of current at resonance is

$$i_0 = I_{m0} \cos \omega_0 t \tag{7-226}$$

The total energy stored by the circuit in both L and C can be expressed as

$$W = \tfrac{1}{2}Li_0^2 + \tfrac{1}{2}Cv_{c0}^2 \tag{7-227}$$

Inserting the appropriate expressions for i_0 and v_{c0} yields

$$W = \frac{1}{2}LI_{m0}^2 \cos^2 \omega_0 t + \frac{1}{2}C\left(\frac{1}{C}\int_0^t I_{m0}^2 \cos \omega_0 t\right)^2$$

$$= \frac{1}{2}LI_{m0}^2 \cos^2 \omega_0 t + \frac{I_{m0}^2}{2\omega_0^2 C} \sin^2 \omega_0 t$$

$$= \frac{1}{2}LI_{m0}^2 \cos^2 \omega_0 t + \frac{1}{2}LI_{m0}^2 \sin^2 \omega_0 t \qquad (7\text{-}228)$$

$$= \frac{1}{2}LI_{m0}^2 = LI_0^2$$

where I_0 is the rms current at resonance. It is interesting to note that at resonance this energy is a constant quantity.

Returning to Eq. (7-218) and multiplying both numerator and denominator by I_0^2 we can write for Q_0

$$Q_0 = \frac{\omega_0 LI_0^2}{RI_0^2} = \frac{2\pi f_0 LI_0^2}{RI_0^2} = 2\pi \frac{I_0^2 L}{I_0^2 R/f_0} \qquad (7\text{-}229)$$

Now since $I_0^2 L$ represents the total stored energy at resonance and $I_0^2 R/f_0$ is the energy dissipated per cycle, Eq. (7-229) may be written more generally as

$$\boxed{Q_0 = 2\pi \frac{\text{total stored energy}}{\text{energy dissipated per cycle}}} \qquad (7\text{-}230)$$

Equation (7-230) is applicable to any resonant system regardless of its composition. As long as the quantities on the right side can be determined the quality factor can, in turn, be found.

Parallel Resonance. The circuit configuration for the study of parallel resonance appears in Fig. 7-31. The expression for the total current is

$$\bar{I} = \bar{V}\bar{Y} = \bar{V}(G + jB) \qquad (7\text{-}231)$$

where $\bar{I}$ is a fixed, known quantity and $\bar{V}$ is the nodal voltage, which varies as $\bar{Y}$ varies with frequency. For the circuit of Fig. 7-31 the expression for the admittance is

$$\bar{Y} = G + jB = \frac{1}{R} + j\omega C + \frac{1}{j\omega L} = \frac{1}{R} + j\left(\omega C - \frac{1}{\omega L}\right) \qquad (7\text{-}232)$$

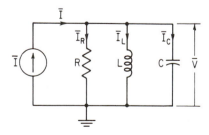

Fig. 7-31 Circuit configuration for studying parallel resonance.

Hence

$$\bar{I} = \bar{V}\left[\frac{1}{R} + j\left(\omega C - \frac{1}{\omega L}\right)\right] \qquad (7\text{-}233)$$

A study of this equation reveals that for small values of ω the susceptance B will be large. Then to keep the product of the right side constant and equal to $\bar{I}$, the nodal voltage $\bar{V}$ must be correspondingly small. As ω increases, B decreases, and so $\bar{V}$ increases. When B equals zero, $\bar{V}$ has its maximum value. At this point the output nodal voltage is in time phase with the current source, and the circuit is said to be in parallel resonance. The frequency at which resonance occurs is found from

$$B = \omega_0 C - \frac{1}{\omega_0 L} = 0 \qquad (7\text{-}234)$$

or

$$\omega_0 = \frac{1}{\sqrt{LC}} \qquad (7\text{-}235)$$

which is identical to the expression which applies to the series RLC circuit. As ω increases beyond ω_0, the susceptance gets ever larger, thus causing the nodal voltage to diminish toward zero. A sketch of $\bar{V}$ as a function of frequency is depicted in Fig. 7-32. The bandwidth is identified as the frequency range lying between ω_1 and ω_2. For the parallel case, the lower frequency is found from

$$\omega_1 C - \frac{1}{\omega_1 L} = -\frac{1}{R} \qquad (7\text{-}236)$$

which leads to

$$\omega_1^2 + \frac{1}{RC}\omega_1 - \frac{1}{LC} = 0 \qquad (7\text{-}237)$$

Ignoring the negative solution, the desired expression is

$$\omega_1 = -\frac{1}{2RC} + \sqrt{\left(\frac{1}{2RC}\right)^2 + \frac{1}{LC}} = -\alpha + \sqrt{\alpha^2 + \omega_0^2} \qquad (7\text{-}238)$$

where

$$\alpha = \frac{1}{2RC} \qquad (7\text{-}239)$$

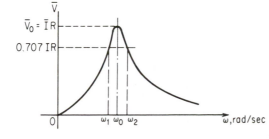

Fig. 7-32 Variation of the nodal voltage $\bar{V}$ of Fig. 7-31 as a function of frequency.

Similarly for the upper half-power frequency point

$$\omega_2 C - \frac{1}{\omega_2 L} = \frac{1}{R} \tag{7-240}$$

This leads to

$$\omega_2 = \frac{1}{2RC} + \sqrt{\left(\frac{1}{2RC}\right)^2 + \frac{1}{LC}} = \alpha + \sqrt{\alpha^2 + \omega_0^2} \tag{7-241}$$

Therefore, the expression for the bandwidth is

$$\boxed{\omega_{bw} = \omega_2 - \omega_1 = 2\alpha = \frac{1}{RC}} \tag{7-242}$$

Employing the definition of the quality factor given by Eq. (7-217) leads to

$$Q_0 \equiv \frac{\omega_0}{\omega_{bw}} = \frac{\omega_0}{1/RC} = \omega_0 RC \tag{7-243}$$

It can be shown that Eq. (7-230) also applies with equal validity in determining Q_0.

Just as there occurs a resonant rise in voltage associated with L and C in the series-resonant case, so too there occurs a resonant rise in current in the L and C elements when the circuit is at parallel resonance. This is readily demonstrated. Recall that the value of the nodal voltage $\overline{V}_0$ at resonance is

$$\overline{V}_0 = IR\underline{/0°} \tag{7-244}$$

Since this voltage appears across L we can write

$$\overline{I}_{L0} = \frac{\overline{V}_0}{j\omega_0 L} = \frac{IR}{\omega_0 L}\underline{/-90°} = I\omega_0 RC\underline{/-90°} = IQ_0\underline{/-90°} \tag{7-245}$$

For the capacitor

$$\overline{I}_{C0} = \frac{\overline{V}_0}{1/j\omega_0 C} = jIR\omega_0 C = Q_0 I\underline{/90°} \tag{7-246}$$

The subscript 0 denotes the resonant condition. An inspection of each of the last two expressions shows that the magnitude of the current through L or C is Q_0 times the source current I. It follows then that another way of expressing Q_0 in the parallel-resonance case is to write

$$Q_0 = \frac{I_{L0}}{I} = \frac{I_{C0}}{I} \tag{7-247}$$

Equation (7-247) is analogous to Eq. (7-225) for the series-resonant case.

Appearing in Fig. 7-33 is the phasor diagram for the currents using the nodal voltage at resonance, $\overline{V}_0$, as the reference phasor. Note that I_{L0} lags and I_{C0} leads $\overline{V}_0$ by 90°, as called for by Eqs. (7-245) and (7-246). Of course, the resultant of the three currents is the source current $\overline{I}$.

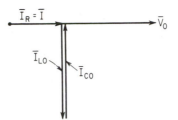

Fig. 7-33 Current phasor diagram for parallel resonance.

7-11 BALANCED THREE-PHASE CIRCUITS

Almost all the electric power used in this country is generated and transmitted in the form of balanced three-phase voltage systems. The single-phase voltage sources referred to in the preceding sections originate as part of the three-phase system. The manner in which a three-phase voltage is generated is described in detail in Sec. 19-1. A balanced three-phase voltage system is composed of three single-phase voltages having the same amplitude and frequency of variation but time-displaced from one another by 120°. A schematic representation is depicted in Fig. 7-34(a). The three single-phase voltages appear in a Y configuration: a Δ configuration is also possible. These single-phase voltages are generated by a

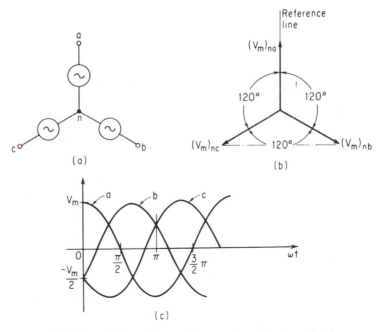

Fig. 7-34 Balanced three-phase voltage system: (a) schematic diagram; (b) phasor diagram in terms of the maximum value of each phasor; (c) time diagram.

common rotating flux field in three identical windings which are separated from each other by 120° inside the housing of the electric generator. When we join one end of each winding together to form terminal n, the Y connection results. Appearing in Fig. 7-34(b) is the phasor diagram of the three-phase voltage system. The corresponding time diagram for each single-phase voltage is shown in Fig. 7-34(c). For the time instant depicted in Fig. 7-34(b) [i.e., $(V_m)_{na}$ in phase with the vertical reference line] phase a is at its maximum value while simultaneously phases b and c have values of minus one-half the maximum value. Recall that in the phasor diagram instantaneous values are obtained by projecting the maximum-value phasor onto the reference line. Hence for $(V_m)_{nb}$ and $(V_m)_{nc}$ these projections are negative and equal to one-half V_m. Note that these instantaneous values exactly correspond to those obtained from Fig. 7-34(c) for $\omega t = 0$.

As the phasors in Fig. 7-34(b) revolve at the angular frequency ω with respect to the reference line in the counterclockwise (positive) direction, the complete time diagram of Fig. 7-34(c) is generated. The positive maximum value first occurs for phase a and then in succession for phases b and c. For this reason the three-phase voltage of Fig. 7-34 is said to have the *phase order abc*. The terms *phase sequence* and *phase rotation* are used synonymously. Phase order is important in certain applications. For example, in three-phase induction motors it determines whether the motor turns clockwise or counterclockwise.

Current and Voltage Relations for the Y Connection. Every balanced three-phase system can be analyzed in terms of the procedures which apply to a single phase. Hence all the techniques developed in the preceding sections are applicable to symmetrical three-phase systems. The arrangement of the three single-phase voltages into a Y or Δ configuration calls for some modifications in dealing with total effects. For example the voltages appearing between any pair of the line terminals a, b, and c bear a different relationship in magnitude and phase to the voltages appearing between any one line terminal and the common terminal n called the *neutral*. The former set of voltages—$\overline{V}_{ab}$, $\overline{V}_{bc}$, and $\overline{V}_{ca}$—are called the *line voltages*, and the latter set of voltages—$\overline{V}_{na}$, $\overline{V}_{nb}$, and $\overline{V}_{nc}$—are called the *phase voltages*. The relationships existing between the phase and line voltages are readily determined from an analysis of the phasor diagram in conjunction with Kirchhoff's voltage law. In this connection the phasor diagram of Fig. 7-34(b) is redrawn in Fig. 7-35 in terms of effective values rather than peak values. Moreover, V_{na} is made to coincide with the horizontal axis of the complex plane. The use of the double subscript notation is a convenience in treating three-phase circuits. If $\overline{V}_{na}$ is made to denote the voltage rise in going through the source from terminal n to a, then the quantity $-\overline{V}_{na} = \overline{V}_{an}$ represents the voltage drop in going from a to n.

The effective values of the phase voltages are shown in Fig. 7-35 as V_{na}, V_{nb}, and V_{nc}. Each has the same magnitude and each is displaced 120° from the other two phasors. To obtain the magnitude and phase angle of the line voltage from a to b, i.e, $\overline{V}_{ab}$, we apply Kirchhoff's voltage law. Thus in Fig. 7-34(a)

$$\overline{V}_{ab} = \overline{V}_{an} + \overline{V}_{nb} \tag{7-248}$$

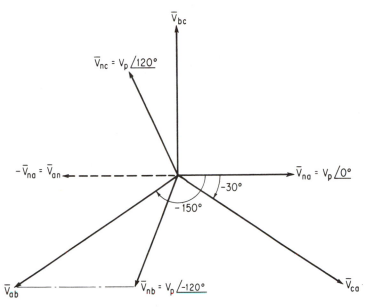

Fig. 7-35 Illustrating the phase and magnitude relations between the phase and line voltages of a Y connection.

This equation states that the voltage existing from a to b is equal to the voltage from a to n (i.e., $-V_{na}$) plus the voltage from n to b. Equation (7-248) can be rewritten as

$$\overline{V}_{ab} = -\overline{V}_{na} + \overline{V}_{nb} \tag{7-249}$$

Since for a balanced system each phase voltage has the same magnitude, let us set

$$V_{na} = V_{nb} = V_{nc} = V_p \tag{7-250}$$

where V_p denotes the effective magnitude of the phase voltage. Accordingly, we may write

$$\overline{V}_{na} = V_p\angle 0° \tag{7-251}$$

$$\overline{V}_{nb} = V_p\angle -120° \tag{7-252}$$

$$\overline{V}_{nc} = V_p\angle -240° = V_p\angle 120° \tag{7-253}$$

Inserting Eqs. (7-251) and (7-252) into Eq. (7-249) yields

$$\overline{V}_{ab} = -V_p + V_p\angle -120°$$

$$= -V_p + V_p(\cos 120° - j\sin 120°) \tag{7-254}$$

$$= V_p\left(-\frac{3}{2} - j\frac{\sqrt{3}}{2}\right) = \sqrt{3}\,V_p\angle -150°$$

Similarly, we obtain

$$\overline{V}_{bc} = \overline{V}_{bn} + \overline{V}_{nc} = \sqrt{3}\,V_p\angle -270° \tag{7-255}$$

and

$$\overline{V}_{ca} = \overline{V}_{cn} + \overline{V}_{na} = \sqrt{3}\, V_p \angle -30° \qquad (7\text{-}256)$$

A study of the expressions for the line voltages shows that they constitute a balanced three-phase voltage system whose magnitudes are greater than the phase voltages by the $\sqrt{3}$. That is, the magnitude of the line voltage V_L is given by

$$\boxed{V_L = \sqrt{3}\, V_p} \quad \text{valid for Y connection} \qquad (7\text{-}257)$$

Any current that flows out of line terminal a (or b or c) must be the same as that which flows through the phase source voltage appearing between terminals n and a (or n and b, or n and c). Therefore, in the Y-connected three-phase generator the line current equals the phase current. That is,

$$\boxed{I_L = I_p} \qquad (7\text{-}258)$$

where I_L denotes the effective value of line current and I_p denotes the effective value of the phase current.

Current and Voltage Relationships in a Δ-*Connected Three-Phase System.* Assume next that the three single-phase sources of Fig. 7-34(a) are rearranged to form the Δ connection shown in Fig. 7-36. A three-phase load is shown connected to the three-phase source for the purpose of allowing line currents to flow. From inspection of the circuit diagram it should be obvious that the line and phase voltages have the same magnitude. Hence

$$\boxed{V_L = V_p} \quad \text{for } \Delta \text{ connection} \qquad (7\text{-}259)$$

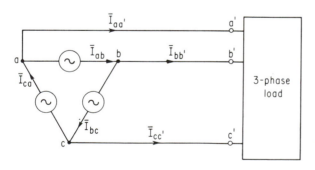

Fig. 7-36 Δ-connected three-phase source.

The phasor diagram appears in Fig. 7-37. It should also be obvious from Fig. 7-36 that the phase and line currents are not identical. Applying Kirchhoff's current law at one of the line terminals, e.g., a, emphasizes this point. Thus

$$\overline{I}_{aa'} = \overline{I}_{ca} + \overline{I}_{ba} = \overline{I}_{ca} - \overline{I}_{ab} \qquad (7\text{-}260)$$

where $\overline{I}_{aa'}$ denotes the current which flows in the line joining a to a'. To illustrate the relationship existing between line and phase currents, let us consider that

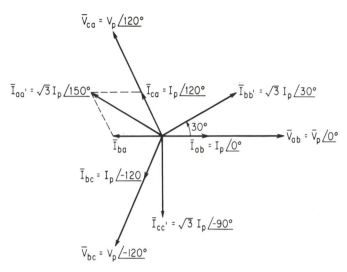

Fig. 7-37 Illustrating relation between phase and line currents in a Δ connection.

the nature of the load circuit is such that it causes the three phase currents to be expressed as

$$\bar{I}_{ab} = I_p \angle 0° \tag{7-261}$$

$$\bar{I}_{bc} = I_p \angle -120° \tag{7-262}$$

$$\bar{I}_{ca} = I_p \angle 120° \tag{7-263}$$

As Fig. 7-37 indicates, these phase currents are assumed to be in phase with their respective phase voltages. Inserting Eqs. (7-261) and (7-263) into Eq. (7-260) then yields

$$\bar{I}_{aa'} = I_p \angle 120° - I_p = I_p \left(-\frac{3}{2} + j\frac{\sqrt{3}}{2} \right) = \sqrt{3}\, I_p \angle 150° \tag{7-264}$$

In a similar fashion

$$\bar{I}_{bb'} = \bar{I}_{ab} + \bar{I}_{cb} = \sqrt{3}\, I_p \angle 30° \tag{7-265}$$

and

$$\bar{I}_{cc'} = \bar{I}_{bc} + \bar{I}_{ac} = \sqrt{3}\, I_p \angle -90° \tag{7-266}$$

Note that a set of balanced phase currents yields a corresponding set of balanced line currents which differ in magnitude by the factor $\sqrt{3}$. Thus

$$\boxed{I_L = \sqrt{3}\, I_p} \quad \text{for } \Delta \text{ connection} \tag{7-267}$$

where I_L denotes the magnitude of any one of the three line currents.

Power in Three-Phase Systems. It is shown in Sec. 7-3 and in Fig. 7-4 that the instantaneous power for a single-phase sinusoidal source varies itself sinusoidally at twice the frequency of the source. By Eq. (7-39) the expression for the instantaneous power is

$$p(t) = \frac{V_m I_m}{2} \cos \theta - \frac{V_m I_m}{2} \cos (2\omega t - \theta) \qquad (7\text{-}268)$$

In terms of effective values this becomes

$$p(t) = VI \cos \theta - VI \cos (2\omega t - \theta) \qquad (7\text{-}269)$$

Equation (7-269) can now be applied to each phase of the three-phase system. The only modification needed is introducing the 120° displacement that exists between phases. Accordingly, we may write

$$p_a(t) = V_p I_p \cos \theta - V_p I_p \cos (2\omega t - \theta) \qquad (7\text{-}270)$$

$$p_b(t) = V_p I_p \cos \theta - V_p I_p \cos (2\omega t - \theta - 120°) \qquad (7\text{-}271)$$

$$p_c(t) = V_p I_p \cos \theta - V_p I_p \cos (2\omega t - \theta - 240°) \qquad (7\text{-}272)$$

where phase a is taken as the reference phase, V_p and I_p denote the effective values of the phase voltage and phase current, and θ denotes the phase impedance angle of the balanced three-phase load. As a result of the 120° displacement between p_a, p_b, and p_c, the total instantaneous power in the three-phase system is a constant equal to three times the average power per phase. This situation is depicted in Fig. 7-38. Expressed mathematically the total average three-phase power is

$$P = \frac{1}{T} \int_0^T [p_a(t) + p_b(t) + p_c(t)] \, dt \qquad (7\text{-}273)$$

or

$$\boxed{P = 3V_p I_p \cos \theta} \qquad (7\text{-}274)$$

For a Y-connected system the insertion of Eqs. (7-257) and (7-258) permits rewriting Eq. (7-274) as

$$P = 3\frac{V_L}{\sqrt{3}}I_L \cos \theta = \sqrt{3}\, V_L I_L \cos \theta \qquad (7\text{-}275)$$

For the Δ connection Eq. (7-274) becomes

$$P = 3V_L\frac{I_L}{\sqrt{3}} \cos \theta = \sqrt{3}\, V_L I_L \cos \theta \qquad (7\text{-}276)$$

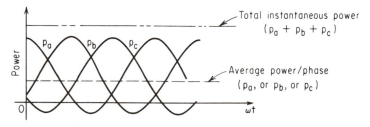

Fig. 7-38 Total instantaneous power in a three-phase system.

Sinusoidal Steady-State Response of Circuits　Chap. 7

A comparison of the last two expressions thus indicates that the equation for the power in a three-phase system is the same either for a Y or a Δ connection when the power is expressed in terms of *line* quantities. However, it is important to remember that θ denotes the angle of the load impedance per phase and not the angle between V_L and I_L.

Before leaving this discussion on three-phase power, let us emphasize the fact that the total instantaneous power for the three-phase system is a constant as illustrated by Fig. 7-38. This stands in sharp contrast to the single-phase case where the single-phase power pulsates at twice the line frequency. Herein, then, lies another significant advantage of the three-phase system over the single-phase system. Wherever large loads must be driven mechanically in commercial and industrial applications, the three-phase motor rather than the single-phase motor is used.

Three-Phase Load Circuits.   Appearing in each leg of the Y connection shown in Fig. 7-39 are identical impedances having the same values of resistance and inductive reactances per phase. Such an arrangement is called a balanced three-phase load circuit. In the interest of demonstrating how to carry out the computation of line currents and power for such circuits, let us assume that a balanced three-phase voltage is applied to the load having an effective line voltage of 173.2 V. At this point it is not necessary for us to know whether the generator is Y or Δ connected. Moreover, assume that each phase impedance is properly specified by

$$\bar{Z}_a = \bar{Z}_b = \bar{Z}_c = 7.07 + j7.07 = 10\underline{/45°} \qquad (7\text{-}277)$$

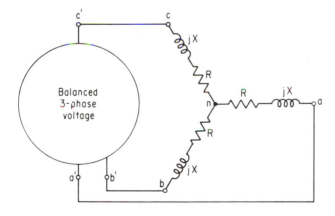

Fig. 7-39 Balanced three-phase system.

First find the current in line $a'a$. Because of the Y connection this line current is the same as the phase current $\bar{I}_{an}$. To find $\bar{I}_{an}$, however, we must first determine $\bar{V}_{an}$. Since the line voltage is given as 173.2 V, it follows that the magnitude of the phase voltage V_p is

$$V_p = \frac{V_L}{\sqrt{3}} = \frac{173.2}{\sqrt{3}} = 100 \text{ V} \qquad (7\text{-}278)$$

We can now select the voltage drop from a to n, i.e., $\overline{V}_{an}$, as our reference phasor and write it as

$$\overline{V}_{an} = V_p\angle 0° = 100\angle 0° \tag{7-279}$$

Then because of the balanced nature of the system we can further write

$$\overline{V}_{bn} = V_p\angle -120° = 100\angle -120° \tag{7-280}$$

and

$$\overline{V}_{cn} = V_p\angle 120° = 100\angle 120° \tag{7-281}$$

Each of these line-to-neutral voltages is depicted in Fig. 7-40. The required line currents can then be specified as follows:

$$\overline{I}_{a'a} = \overline{I}_{an} = \frac{\overline{V}_{an}}{\overline{Z}_a} = \frac{100\angle 0°}{10\angle 45°} = 10\angle -45° \tag{7-282}$$

$$\overline{I}_{b'b} = \overline{I}_{bn} = \frac{\overline{V}_{bn}}{\overline{Z}_b} = \frac{100\angle -120°}{10\angle 45°} = 10\angle -165° \tag{7-283}$$

$$\overline{I}_{c'c} = \overline{I}_n = \frac{\overline{V}_{cn}}{\overline{Z}_c} = \frac{100\angle 120°}{10\angle 45°} = 10\angle 75° \tag{7-284}$$

These phasor quantities also appear in Fig. 7-40.

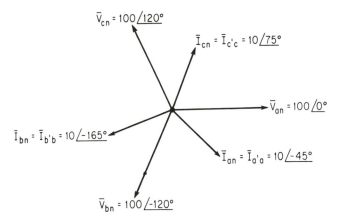

Fig. 7-40 Phasor diagram for the system of Fig. 7-39.

The power consumed in phase a is given by

$$P_a = V_p I_p \cos \theta \tag{7-285}$$

where

$$\theta = \tan^{-1}\frac{X}{R} = \tan^{-1}\frac{7.07}{7.07} = \tan^{-1} 1 = 45° \tag{7-286}$$

Hence

$$P_a = V_p I_p \cos 45° = (100)(10)(0.707) = 707 \text{ W} \tag{7-287}$$

This power can also be computed from

$$P_a = I_{an}^2 R = (10)^2(7.07) = 707 \text{ W} \tag{7-288}$$

Since the power consumed in each phase is the same, it follows that the total power delivered to the three-phase load by the source is

$$P_T = 3P_a = 2121 \text{ W} \tag{7-289}$$

The total power can also be calculated from a knowledge of the line quantities. Thus, by Eq. (7-275)

$$P_T = \sqrt{3}\ V_L I_L \cos \theta = \sqrt{3}\ (173.2)(10)(0.707) = 2121 \text{ W} \tag{7-290}$$

Note that the θ used in Eq. (7-290) is the very same one which is used in Eq. (7-287).

Let us now rearrange the given phase impedances $\overline{Z}_a$, $\overline{Z}_b$, and $\overline{Z}_c$ into a Δ connection as shown in Fig. 7-41(a). The source voltage is assumed to be the same. Since in a Δ connection the line and phase voltages are the same, we arbitrarily select $\overline{V}_{ab}$ as the reference voltage and thereby draw the phasor diagram for the voltages as illustrated in Fig. 7-41(b). The expressions for the load phase

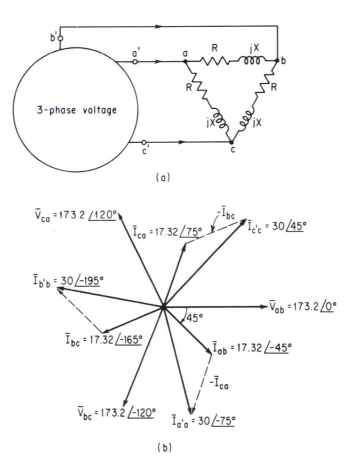

(a)

(b)

Fig. 7-41 Balanced three-phase Δ-connected load: (a) wiring diagram; (b) phasor diagram.

currents become

$$\bar{I}_{ab} = \frac{\bar{V}_{ab}}{\bar{Z}_a} = \frac{173.2 \underline{/0°}}{10 \underline{/45°}} = 17.32 \underline{/-45°} \tag{7-291}$$

$$\bar{I}_{bc} = \frac{\bar{V}_{bc}}{\bar{Z}_b} = \frac{173.2 \underline{/-120°}}{10 \underline{/45}} = 17.32 \underline{/-165°} \tag{7-292}$$

$$\bar{I}_{ca} = \frac{\bar{V}_{ca}}{\bar{Z}_c} = \frac{173.2 \underline{/120°}}{10 \underline{/45°}} = 17.32 \underline{/75°} \tag{7-293}$$

The line current $\bar{I}_{a'a}$ is found upon applying the current law at terminal a. Thus

$$\bar{I}_{a'a} = \bar{I}_{ab} + \bar{I}_{ac} = \bar{I}_{ab} - \bar{I}_{ca} = 17.32(0.448 - j1.673)$$
$$= 30 \underline{/-75°} \tag{7-294a}$$

Similarly

$$\bar{I}_{b'b} = 30 \underline{/-195} \text{ and } \bar{I}_{cc} = 30 \underline{/45°} \tag{7-294b}$$

A comparison of the line current for a Δ-connected arrangement of three fixed impedances with a Y-connected arrangement shows that the Δ connection draws *three* times as much current. Compare Eq. (7-282) with Eq. (7-294). The total power delivered is also three times greater.

7-12 FOURIER SERIES

Much is said in the preceding sections about the response of electric circuits to a periodic forcing function having a sinusoidal variation. This attention is merited, since the bulk of the electric power in the world is generated, transmitted, and consumed as a sinusoidally varying quantity at constant frequency. At this point, however, it is appropriate to ask: How can the response of electric circuits be obtained when the forcing function is not sinusoidally varying but is periodic? There are numerous applications in engineering where nonsinusoidal forcing functions such as rectangular, triangular, trapezoidal, or other waveshapes are used. Moreover, in communication engineering situations exist which involve the summation of sinusoidal signals of many frequencies, the resultant waveshape of which is extremely nonsinusoidal albeit periodic. This condition is illustrated in Fig. 7-42, which shows how it is possible to represent three entirely different waveforms by means of two sinusoidal components. Note that in each case one of the components is assumed to have a frequency three times greater than the other. For the condition depicted in Fig. 7-42(a) the resultant waveform may be expressed mathematically as

$$f(t) = f_1(t) + f_3(t) = F_1 \sin \omega t + F_3 \sin 3\omega t \tag{7-295}$$

where $f_1(t)$ is the *fundamental* sinusoidal component of frequency ω, and $f_3(t)$ is the *third harmonic* component of frequency 3ω. Both the fundamental and the third harmonic terms are called *periodic functions* because they satisfy the condition

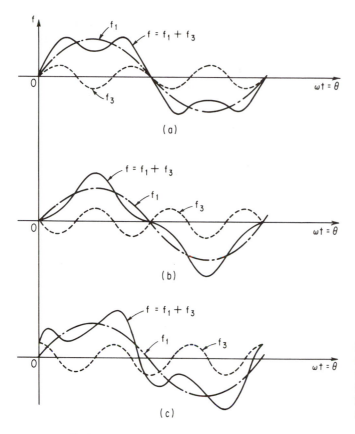

Fig. 7-42 (a) Addition of fundamental sine term plus positive third harmonic; (b) addition of fundamental plus negative third harmonic; (c) addition of fundamental sine term and a third harmonic cosine term.

that

$$f(t) = f(t + T) \tag{7-296}$$

where T denotes the period of the fundamental wave. Observe that the period of the third harmonic is one-third that of the fundamental and that the resultant wave has the same period as the fundamental.

The equation which describes the waveform of Fig. 7-42(b) is

$$f(t) = f_1(t) - f_3(t) = F_1 \sin \omega t - F_3 \sin 3\omega t \tag{7-297}$$

By merely reversing the third harmonic term the resultant waveshape changes from one which is essentially flat-topped (Fig. 7-42a) to one which is peaked.

The resultant waveshape of Fig. 7-42(c) is obtained by shifting the third harmonic term of Fig. 7-42(a) 90° in the lead direction. The corresponding mathematical formulation is

$$f(t) = F_1 \sin \omega t + F_3 \sin \left(\omega t + \frac{\pi}{2} \right) \tag{7-298}$$

A little thought should make it apparent that even with only two sinusoidal components manipulated in the manner just described it is possible to represent a great number of different waveshapes.

A general formulation of this situation is provided by the Fourier series theorem,† which is stated here without proof. *A periodic function f(t) with period T can be represented by the sum of a constant term, a fundamental of period T, and its harmonics.* Expressed mathematically, we have

$$f(\theta) = \left(\frac{a_0}{2}\right) + a_1 \cos \theta + a_2 \cos 2\theta + a_3 \cos 3\theta + \cdots$$
$$+ b_1 \sin \theta + b_2 \sin 2\theta + b_3 \sin 3\theta + \cdots$$

(7-299)

where

$$\theta \equiv \omega t$$

and

$$\omega = \frac{2\pi}{T}$$

where T is the period of the function $f(\theta)$. Thus Eq. (7-299) states that any periodic function of whatever shape can be replaced by the sum of a constant term $(a_0/2)$ and sine and/or cosine terms involving odd and/or even harmonics. The a- and the b-coefficients are evaluated from the shape of the $f(\theta)$ function. General expressions can readily be derived for evaluating these coefficients, as shown below. In this connection it is useful to keep in mind the following identities.

$$\sin^2 \theta = \tfrac{1}{2} - \tfrac{1}{2} \cos 2\theta \tag{7-300a}$$

$$\cos^2 \theta = \tfrac{1}{2} + \tfrac{1}{2} \cos 2\theta \tag{7-300b}$$

$$\int_0^{2\pi} \cos m\theta \sin n\theta \, d\theta = 0 \tag{7-301}$$

$$\int_0^{2\pi} \cos m\theta \cos n\theta \, d\theta = 0 \text{ for } m \neq n \tag{7-302}$$

$$\int_0^{2\pi} \sin m\theta \sin n\theta \, d\theta = 0 \text{ for } m \neq n \tag{7-303}$$

Consider first the evaluation of the constant term of Eq. (7-299). The waveform of $f(\theta)$ is assumed to be known for all values of θ and to be periodic. To evaluate a_0 it is necessary merely to integrate both sides of Eq. (7-299) over one period. Thus

$$\int_0^{2\pi} f(\theta) \, d\theta = \int_0^{2\pi} \frac{a_0}{2} \, d\theta + \int_0^{2\pi} a_1 \cos \theta \, d\theta + \cdots + \int_0^{2\pi} b_1 \sin \theta \, d\theta + \cdots \tag{7-304}$$

Since the average value over one period of sine and cosine functions is zero, all terms on the right side of the last equation are zero with the exception of the first one. Hence we have

$$\int_0^{2\pi} f(\theta) \, d\theta = \frac{a_0}{2}(2\pi) \tag{7-305}$$

† Published in 1822.

or

$$a_0 = \frac{1}{\pi} \int_0^{2\pi} f(\theta) \, d\theta = \frac{2}{T} \int_0^T f(t) \, dt \qquad (7\text{-}306)$$

The strategy to be employed in evaluating a_1 is to manipulate Eq. (7-299) in such a way as to cause all terms to drop out with the exception of a_1. A little thought reveals that this is readily achieved by multiplying each term of Eq. (7-299) by $\cos \theta$ and then integrating over one period. Thus

$$\int_0^{2\pi} f(\theta) \cos \theta \, d\theta = \int_0^{2\pi} \frac{a_0}{2} \cos \theta \, d\theta + \int_0^{2\pi} a_1 \cos^2 \theta \, d\theta$$

$$+ \int_0^{2\pi} b_1 \sin \theta \cos \theta + \int_0^{2\pi} a_2 \cos \theta \cos 2\theta \, d\theta \qquad (7\text{-}307)$$

$$+ \int_0^{2\pi} b_2 \cos \theta \sin 2\theta \, d\theta + \cdots$$

All terms on the right side drop out but the one involving $\cos^2 \theta$. Hence

$$\int_0^{2\pi} f(\theta) \cos \theta \, d\theta = \int_0^{2\pi} a_1 \left(\frac{1}{2} + \frac{1}{2} \cos 2\theta \right) d\theta \qquad (7\text{-}308)$$

$$= a_1 \pi$$

Therefore,

$$a_1 = \frac{1}{\pi} \int_0^{2\pi} f(\theta) \cos \theta \, d\theta = \frac{2}{T} \int_0^T f(t) \cos \omega t \, dt \qquad (7\text{-}309)$$

To evaluate b_1 a similar procedure is used, but now since we are dealing with the coefficient of the sine term in the Fourier series it is necessary to multiply each term of Eq. (7-299) by $\sin \theta$ and then to integrate over T. Thus

$$\int_0^{2\pi} f(\theta) \sin \theta \, d\theta = \int_0^{2\pi} \frac{a_0}{2} \sin \theta \, d\theta + \int_0^{2\pi} a_1 \sin \theta \cos \theta \, d\theta$$

$$+ \int_0^{2\pi} b_1 \sin^2 \theta \, d\theta + \int_0^{2\pi} a_2 \sin \theta \cos 2\theta \, d\theta \qquad (7\text{-}310)$$

$$+ \int_0^{2\pi} b_2 \sin \theta \sin 2\theta \, d\theta + \cdots$$

Performing the integration leads to

$$\int_0^{2\pi} f(\theta) \sin \theta \, d\theta = b_1 \pi \qquad (7\text{-}311)$$

or

$$b_1 = \frac{1}{\pi} \int_0^{2\pi} f(\theta) \sin \theta \, d\theta = \frac{2}{T} \int_0^T f(t) \sin \omega t \, dt \qquad (7\text{-}312)$$

By following this procedure for the coefficients of each of the higher harmonic terms it can be shown that the general expressions for the a- and b-coefficients for the nth harmonic are given by

$$a_n = \frac{1}{\pi} \int_0^{2\pi} f(\theta) \cos n\theta \, d\theta = \frac{2}{T} \int_0^T f(t) \cos n\omega t \, dt \qquad (7\text{-}313)$$

$$n = 0, 1, 2, 3, \cdots$$

and

$$b_n = \frac{1}{\pi} \int_0^{2\pi} f(\theta) \sin n\theta \, d\theta = \frac{2}{T} \int_0^T f(t) \sin n\omega t \, dt \qquad (7\text{-}314)$$

$$n \quad 1, 2, 3, \cdots$$

It is interesting to note that by writing the constant term in Eq. (7-299) as $a_0/2$ it is permissible to use Eq. (7-313) to evaluate a_0 as well as the coefficients of the cosine components in the wave.

Symmetry Properties of Periodic Waves.   Before proceeding with the solution of a problem illustrating the use of Eqs. (7-313) and (7-314), let us turn attention briefly to the simplification of these equations brought about by the presence of certain symmetry conditions which the $f(\theta)$ function may possess.

A periodic function is said to have *half-wave* symmetry when it satisfies the relation

$$f\left(\theta + \frac{T}{2}\right) = -f(\theta) \qquad (7\text{-}315)$$

Each of the waveforms shown in Fig. 7-43 exhibits this property. Moreover, all periodic functions which obey Eq. (7-315) contain *no even harmonics*.

Sometimes periodic functions are encountered which contain either only sine functions or only cosine functions. For a periodic function to contain only sine terms it must satisfy the relation

$$f(-\theta) = -f(\theta) \qquad (7\text{-}316)$$

The waveform depicted in Fig. 7-43(b) is an example of such a function. Because this waveform satisfies both Eqs. (7-315) and (7-316), its Fourier series representation includes only sine terms and in particular the fundamental sine function and the odd harmonics. For a periodic function to contain only cosine terms it must satisfy the equation

$$f(-\theta) = f(\theta) \qquad (7\text{-}317)$$

The waveform of Fig. 7-43(a) meets this condition. Such functions are sometimes referred to as *even* functions because the value of the function is the same for a positive or negative value of the argument. Note that the waveshape of Fig. 7-43(c) satisfies neither Eq. (7-315) nor Eq. (7-316), hence it contains cosine as well as sine terms. However, it does not contain even harmonics.

EXAMPLE 7-12   Find the Fourier series representation of the waveform depicted in Fig. 7-44.

Solution: A check of the symmetry conditions indicates that none is satisfied. Hence we can expect both even and odd harmonics to exist as well as sine and cosine terms. The average value of the function is found by evaluating Eq. (7-313) for $n = 0$. The periodic

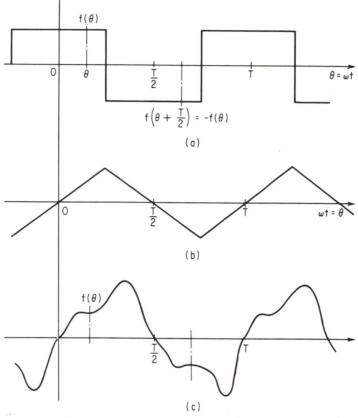

(a)

(b)

(c)

Fig. 7-43 Typical periodic waveforms illustrating symmetry properties:
(a) rectangular wave; (b) triangular wave; (c) arbitrary wave.

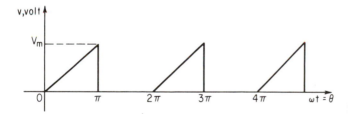

Fig. 7-44 Waveform for Example 7-12.

function in the first period may be expressed as

$$v(\theta) = \frac{V_m}{\pi}\theta, \qquad 0 \leqslant \theta \leqslant \pi$$
$$v(\theta) = 0, \qquad \pi \leqslant \theta \leqslant 2\pi$$

(7-318)

Accordingly we get

$$a_0 = \frac{1}{\pi}\int_0^{2\pi} f(\theta)\,d\theta = \frac{1}{\pi}\int_0^{\pi}\frac{V_m}{\pi}\theta\,d\theta = \frac{V_m}{2\pi^2}[\theta^2]_0^{\pi} = \frac{V_m}{2}$$

(7-319)

$$\therefore \left(\frac{a_0}{2}\right) = \frac{V_m}{4} = \text{average value}$$

(7-320)

This average value of the function can also be found in this case by inspection. How?

The general expression for the coefficient a_n of the cosine terms is found as follows

$$a_n = \frac{1}{\pi} \int_0^\pi \frac{V_m}{\pi} \theta \cos n\theta \, d\theta = \frac{V_m}{\pi^2} \left[-\theta \frac{\sin n\theta}{n} + \frac{\cos n\theta}{n^2} \right]_0^\pi = \frac{V_m}{\pi^2 n^2}(\cos n\pi - 1) \quad (7\text{-}321)$$

The limits of the integral for a_n are taken as zero to π rather than as zero and 2π because the contribution of the function from π to 2π is obviously zero. Inserting consecutive integer values for n yields

$$a_1 = \frac{V_m}{\pi^2}(-1 - 1) = -\frac{2V_m}{\pi^2}$$

$$a_2 = \frac{V}{4\pi^2}(1 - 1) = 0 = a_4 = a_6 = a_8 = \cdots$$

$$a_3 = \frac{V_m}{9\pi^2}(-1 - 1) = -\frac{2V_m}{9\pi^2}$$

$$a_5 = -\frac{2V_m}{25\pi^2}$$

$$\begin{array}{cc} \vdots & \vdots \\ \vdots & \vdots \end{array}$$

The evaluation of the coefficients of the sine terms follows in a similar fashion. Thus

$$b_n = \frac{1}{\pi} \int_0^\pi \frac{V_m}{\pi} \theta \sin n\theta \, d\theta = \frac{V_m}{\pi^2} \left[\frac{-\theta \cos n\theta}{n} + \frac{\sin n\theta}{n^2} \right]_0^\pi = -\frac{V_m}{\pi} \frac{\cos n\pi}{n} \quad (7\text{-}322)$$

Therefore,

$$b_1 = -\frac{V_m}{\pi}(-1) = \frac{V_m}{\pi}$$

$$b_2 = -\frac{V_m}{2\pi}$$

$$b_3 = +\frac{V_m}{3\pi}$$

$$b_4 = -\frac{V_m}{4\pi}$$

$$\begin{array}{cc} \vdots & \vdots \\ \vdots & \vdots \end{array}$$

Hence the Fourier series representation of the waveform is

$$v(\theta) = v(\omega t) = \frac{V_m}{4} + \frac{V_m}{\pi} \left[\sin \theta - \frac{1}{2} \sin 2\theta + \frac{1}{3} \sin 3\theta - (-1)^n \frac{1}{n} \sin n\theta \right]$$

$$- \frac{2V_m}{\pi^2} \left[\cos \theta + \frac{1}{9} \cos 3\theta + \frac{1}{25} \cos 25\theta + \frac{1}{k^2} \cos k\theta \right] \quad (7\text{-}323)$$

where $n = 1, 2, 3, 4, \ldots$, all integers,

$k = 1, 3, 5, 7, \ldots$, odd integers

Equation (7-323) shows that a nonsinusoidal periodic function can be expressed entirely in terms of sines and cosines. It should be apparent then that if this waveform is used as the forcing function in a series RL circuit, the corresponding current response can be found by a systematic application of the sinusoidal steady-state theory developed in the preceding sections. The current response at each frequency is found in the accustomed manner for each sine and cosine term as well as the constant term in $v(\theta)$ and then summed. The waveform of the current response in such a case, however, will be different than that of $v(\theta)$ because at each frequency the amount of phase lag caused by the inductance is different.

The Fourier series representation of a periodic time function has one other useful property which is worth noting. From the treatment so far it should be clear that an infinite number of terms in the series is required to get an exact representation of the original function. If we use only the first N terms of the series, then instead of an exact representation we have an approximation of the original time function yielding an error which can be expressed as

$$e(\theta) = f(\theta) - f_N(\theta) \tag{7-324}$$

where $e(\theta)$ denotes the error, $f(\theta)$ the original periodic function, and $f_N(\theta)$ the approximate function using the first N terms of the series. It is possible to identify an average value for this error by writing

$$E_{av} = \frac{1}{2\pi} \int_0^{2\pi} e(\theta)d\theta \tag{7-325}$$

However, this is not a very useful criterion of how well $f_N(\theta)$ represents $f(\theta)$ because large positive and negative errors may exist which could cancel each other, thereby giving a misleading result. A way of avoiding this difficulty is to use as a criterion the average value of the squared error. This is called the *mean squared error* (MSE). Thus

$$\text{MSE} = \frac{1}{2\pi} \int_0^{2\pi} e^2(\theta)d\theta \tag{7-326}$$

Here is the interesting aspect of this result; it can be shown that by using the coefficients of the Fourier series (i.e., a_n and b_n) for any given N the MSE is minimized. In other words, any other choice of coefficients results in a larger mean squared error.

Summary review questions

1. Define the following terms as they relate to sinusoidal functions: argument, amplitude, cycle, frequency, period, instantaneous value, angular velocity, phase.
2. Distinguish between a periodic time function and the sinusoidal time function.
3. Define mathematically the average value of a periodic function. What is the average value of one period of a sine function having amplitude A? What is the average value of the positive half-cycle of a sine wave having amplitude A?

4. How is the effective value of a periodic function defined? Why is such a definition consistent with a description of power-transfer capability of the associated variable?

5. When the periodic function is a sinusoid of amplitude A, what is the effective value of the sinusoid? Over what interval of time is the effective value valid?

6. Do all periodic functions possess an effective value that is different than zero? Explain.

7. Do all periodic functions possess an average value that is different than zero? Explain and illustrate.

8. In a sinusoidal steady-state circuit the *average* power is expressed in terms of the *effective* values of the current and its associated voltage. Explain why average power is dependent upon *effective* values of voltage and current.

9. Define power factor and explain why it arises in sinusoidal a-c circuits.

10. Distinguish between instantaneous power and average power in a sinusoidal a-c circuit.

11. What is apparent power? How is it related to average power? What practical significance does it have?

12. Describe several ways of adding two sinusoids of the same frequency.

13. How is the *phasor* of a sinusoidal quantity defined? Mention specifically the information that is conveyed by the phasor about the corresponding sinusoidal function.

14. It is said that the phasor representation of sinusoids is a mathematical transform. In light of the background of Chapter 6, explain this statement and identify specifically the transform variable.

15. Why is knowledge of the effective value of a sinusoid more useful than its maximum value?

16. When a circuit is driven by a sinusoidal function, state the phasor relationship that exists between the voltage across a resistor and the current that flows through it.

17. Describe and illustrate the phasor relationship between the voltage that appears across the terminals of a pure inductor and the current that flows through it in steady state when the inductor is excited by a sinusoidal source. What is the power factor of such a circuit?

18. What is inductive reactance, and how does it arise in a-c circuits?

19. What is a phasor diagram? Why is it useful in the study of sinusoidal steady-state analysis of circuits?

20. What is capacitive reactance, and how does it arise in a-c circuits?

21. Draw the phasor diagram of the purely capacitive circuit. Are the electrical quantities used in your phasor diagram effective or peak values? Does it matter? Explain.

22. What is meant by the complex impedance of an RL circuit? How does it arise in the analysis of a-c circuits.

23. Complex impedance is often described as an *operator* that serves to convert a voltage source to a current response. Explain and illustrate the meaning of this statement.

24. What is reactive power? Why is such a term not encountered when d-c sources are used in an electric circuit?

25. Distinguish among admittance, susceptance, and conductance. What is complex admittance?

26. Draw the phasor diagram of a series RLC circuit using the source voltage as the reference phasor and assuming that the capacitive reactance exceeds the inductive reactance. Show the corresponding time diagram of the source voltage and response current.

27. Can the phenomenon of resonance occur in engineering situations when a single energy-storing element is present? Explain.

28. In a series RLC circuit, how is the resonant frequency related to the natural frequency? What is the character of the impedance at resonance for this circuit? How is the current response related to the voltage source at the resonant frequency?

29. In the series RLC circuit, identify the character of the complex impedance when the circuit is driven at a frequency that exceeds the resonant frequency.

30. How is the bandwidth of a series RLC circuit defined? Upon which circuit parameters does the bandwidth depend? What role do the remaining circuit parameters play is this consideration?

31. How is the quality factor of a series RLC circuit defined? Of what practical significance is this number? Illustrate.

32. Is it possible for the voltage that appears across the L or C elements in a series RLC circuit to exceed the voltage applied to the circuit? Explain.

33. At the resonant frequency in a series RLC circuit, how is the voltage across the inductor coil related to the applied source voltage in phase and magnitude?

34. Identify the salient features of a balanced three-phase voltage system of the Y and Δ types.

35. If the phase voltage of one phase of a Y-connected three-phase source is know to be $V_{na} = 120\underline{/30°}$, what are the expressions for the other two phase voltages? What is the expression for the line voltage $\overline{V}_{ab}$?

36. In balanced three-phase systems, explain what gives rise to the presence of the factor $\sqrt{3}$ in expressions for power and between line and phase quantities as well.

37. The instantaneous power per phase in a single-phase circuit contains a double-frequency sinusoidal component. Is a similar component present in the total instantaneous power of a three-phase system? Explain.

38. Three fixed and balanced load impedances are placed across a balanced three-phase voltage source. What is the relationship between the resulting line currents when the load impedances are first connected in a Δ arrangement and then in a Y arrangement?

39. State the Fourier series theorem and comment on its importance.

40. How does one take advantage of the symmetry properties of periodic waves? Illustrate.

Problems

GROUP I

7-1. Find the average and effective values of each waveshape depicted in Fig. P7-1.

7-2. In each of the sketches of Fig. P7-2 the curves are sine waves or parts thereof. Find the average and effective value for each waveshape.

7-3. Compute the average and effective values for each waveshape shown in Fig. P7-3.

7-4. A sinusoidal source of $e(t) = 170 \sin 377t$ is applied to an RL circuit. It is found that the circuit absorbs 720 W when an effective current of 12 A flows.
   (a) Find the power factor of the circuit.

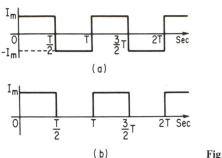

(a)

(b)

Fig. P7-1

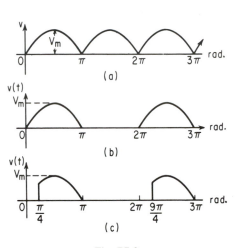

(a)

(b)

(c)

Fig. P7-2

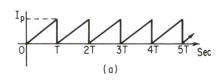

(a)

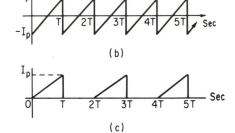

(b)

(c)

Fig. P7-3

(b) Compute the value of the impedance.

(c) Calculate the inductance of the circuit in henrys.

7-5. A voltage source of $e(t) = 141 \sin 377t$ is applied to two parallel branches. The time expression for the current in the first branch is

$$i_1(t) = 7.07 \sin\left(\omega t - \frac{\pi}{3}\right)$$

In the second branch it is

$$i_2(t) = 10 \sin\left(\omega t + \frac{\pi}{6}\right)$$

Compute the total power supplied by the source.

7-6. Refer to Prob. 7-5 and answer the following questions:

(a) Find the resultant current delivered by the source expressed in effective amperes.

(b) Write the expression for the instantaneous value of the resultant current.

(c) Compute the apparent power of the complete circuit.

(d) Find the resultant power factor of the circuit.

(e) Draw the phasor diagram showing the voltage and all current phasors.

7-7. A voltage of

$$e(t) = 141 \sin \left(377t + \frac{\pi}{3} \right)$$

is applied to a 20-Ω resistor.
(a) Find the effective value of the current in amperes.
(b) Write the expression for the instantaneous current.
(c) Compute the power supplied by the source.

7-8. A sinusoidal voltage

$$e(t) = 170 \sin \left(377t + \frac{\pi}{3} \right)$$

is applied to a 0.1-H inductor.
(a) Find the effective value of the steady-state current in amperes.
(b) Write the expression for the instantaneous current.
(c) Draw a properly labeled phasor diagram. Use the rms value of the voltage phasor as reference.

7-9. A sinusoidal voltage

$$e(t) = 170 \sin \left(377t + \frac{\pi}{3} \right)$$

is applied to a 100μF capacitor.
(a) Find the effective value of the steady-state current in amperes.
(b) Write the expression for the instantaneous current.
(c) Draw the phasor diagram. Use the rms value of the voltage phasor as reference.

7-10. A circuit is composed of a resistance of 9 Ω and a series-connected inductive reactance of 12 Ω. When a voltage $e(t)$ is applied to the circuit the resulting steady-state current is found to be $i(t) = 28.3 \sin 377t$.
(a) What is the value of the complex impedance?
(b) Determine the time expression for the applied voltage.
(c) Find the value of the inductance in henrys.

7-11. When a sinusoidal voltage of 120 V rms is applied to a series RL circuit, it is found that there occurs a power dissipation of 1200 W and a current flow given by $i(t) = 28.3 \sin (377t - \theta)$.
(a) Find the circuit resistance in ohms.
(b) Find the circuit inductance in henrys.

7-12. In an RC series circuit to which is applied a voltage of 170 sin ωt it is found that a steady-state current flows which leads the voltage by 30°. Find the effective voltage drops across the resistive and reactive elements.

7-13. A circuit is composed of a resistance of 6 Ω and a series capacitive reactance of 8 ohms. A voltage $e(t) = 141 \sin 377t$ is applied to the circuit.
(a) Find the complex impedance.
(b) Determine the effective and instantaneous values of the current.
(c) Compute the power delivered to the circuit.
(d) Find the value of the capacitance in farads.

7-14. A sinusoidal voltage is applied to three parallel branches yielding branch currents as follows:

$$i_1 = 14.14 \sin (\omega t - 45°), \ i_2 = 28.3 \cos (\omega t - 60°), \ i_3 = 7.07 \sin (\omega t + 60°)$$

(a) Find the complete time expression for the source current.

(b) Draw the phasor diagram in terms of effective values. Use the voltage as reference.

7-15. A sinusoidal voltage $V_m \sin \omega t$ is applied to three parallel branches. Two of the branch currents are given by

$$i_1 = 14.14 \sin (\omega t - 37°)$$

$$i_2 = 28.28 \cos (\omega t - 143°)$$

The source current is found to be

$$i = 63.8 \sin (\omega t + 12.8°)$$

(a) Find the effective value of the current in the third branch.

(b) Write the complete time expression for the instantaneous value of the current in part (a).

(c) Draw the phasor diagram showing the source current and the three branch currents. Use voltage as the reference phasor.

7-16. A voltage wave $e(t) = 170 \sin 120t$ produces a net current of $i(t) = 14.14 \sin 120t + 7.07 \cos (120t + 30°)$.

(a) Express the effective value of the current as a single phasor quantity.

(b) Draw the phasor diagram of part (a). Show the components of the current as well as the resultant.

(c) Find the power delivered by the voltage source of part (a).

7-17. A voltage wave has the variation shown in Fig. P7-17.

(a) Find the average and the effective value of the voltage.

(b) If the voltage of part (a) is applied to a 10-Ω resistance, find the dissipated power in watts.

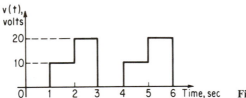

Fig. P7-17

7-18. A voltage of $e(t) = 150 \sin 1000t$ is applied across a series RLC circuit where $R = 40\ \Omega$, $L = 0.13$ H, and $C = 10\ \mu$F.

(a) Compute the rms value of the steady-state current.

(b) Find the expression for the instantaneous voltage appearing across the capacitor terminals.

(c) Determine the expression for the instantaneous voltage appearing across the inductor terminals.

(d) Compare the rms value of the voltages appearing across L and C with that of the applied voltage and comment.

(e) Draw the complete phasor diagram for the solution of this problem showing all voltage components.

7-19. In the circuit of Prob. 7-18 determine:

(a) The power supplied by the source.

(b) The reactive power supplied by the source.

(c) The reactive power of the capacitor.

(d) The reactive power of the inductor.

(e) The power factor of the circuit.

7-20. In the circuit shown in Fig. P7-20 the applied forcing function is given by $e(t) =$ 141 sin ωt.

(a) Express the voltage drops across terminals ab and bc in terms of phasor notation.

(b) Draw a phasor diagram showing $\overline{V}_{ab} + \overline{V}_{bc}$ and using current as the reference phasor.

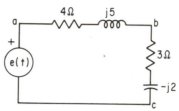

Fig. P7-20

7-21. An rms voltage of $100\underline{/0°}$ is applied to the series combination of $\overline{Z}_1$ and $\overline{Z}_2$ where $\overline{Z}_1 = 20\underline{/30°}$. The effective voltage drops across $\overline{Z}_1$ is known to be $40\underline{/-30°}$ V. Find the reactive component of $\overline{Z}_2$.

7-22. An effective voltage of 100 V is applied to the parallel combination of two impedances $\overline{Z}_1 = R_1 + jX_1$ and $\overline{Z}_2 = R_2 + jX_2$. Assuming that $R_1 = 3$ Ω and $R_2 = 4$ Ω and that the magnitudes of the two branch currents are the same, determine the values of X_1, X_2 and the resultant source current.

7-23. In the network configuration shown in Fig. P7-23 find the current which flows through the $\overline{Z}_3$ branch. Use the nodal method.

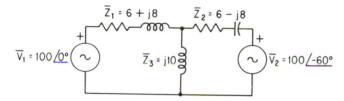

Fig. P7-23

7-24. Solve Prob. 7-23 by using Thévenin's equivalent circuit.

7-25. In the circuit shown in Fig. 7-25 find the Thévenin and Norton equivalent circuits looking in at terminals a-b.

7-26. A circuit has the configuration depicted in Fig. P7-26.

(a) Find the equivalent impedance appearing to the right of points ab.

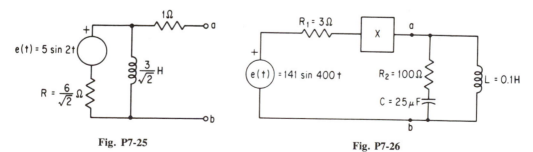

Fig. P7-25

Fig. P7-26

(b) Determine the value of the reactance X which makes the source current in phase with the source voltage.

(c) Should the reactance X of part (b) be inductive or capacitive? Find the required value of L or C.

(d) Compute the effective value of the source current for the condition described in part (b).

7-27. A sinusoidal forcing function having a frequency of 2 rad/s is applied to the circuit of Fig. P7-27.

(a) For the parameter values specified find the value of the impedance appearing at the input terminals.

(b) Identify the type, location, and value of the circuit element which must be used with this circuit in order that the impedance presented to the source be entirely resistive.

(c) For the condition of part (b) find the phasor expression for the voltage across the 0.1-F capacitance when the applied voltage is $e(t) = 14.14 \sin 2t$.

7-28. The inductive reactance in series with $\overline{Z}$ in the circuit of Fig. P7-28 has a value of 25 Ω. A voltmeter placed across $\overline{Z}$ registers a reading of 179 V when 4 A flows through Z. The power dissipated in the circuit is known to be 320 W.

(a) Find the power factor of the circuit.

(b) What is the circuit resistance?

(c) Is it possible for the magnitude of the voltage across $\overline{Z}$ to be greater than the source voltage? Assume $\overline{Z}$ contains an inductive reactance. Explain fully.

(d) Compute the reactive component of Z.

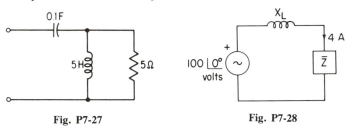

Fig. P7-27   Fig. P7-28

7-29. Refer to the circuit of Fig. P7-29.

(a) Find the equivalent reactance for the parallel branch.

(b) Find the rms line current.

(c) Determine the rms voltage across the parallel branch. Comment on this result.

(d) What is the power dissipated?

(e) What is the power factor?

7-30. Refer to the circuit depicted in Fig. P7-30.

(a) What is the power factor?

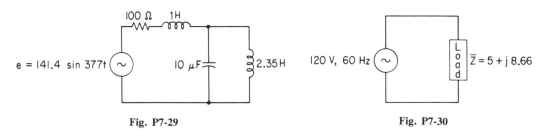

Fig. P7-29   Fig. P7-30

(b) Find the power dissipated.

(c) Find that value of capacitance which when placed across the load will make the overall power factor unity.

7-31. The circuit shown in Fig. P7-31 is in the sinusoidal steady state.
   (a) Determine the phasor voltage $\overline{V}_{ab}$.
   (b) To what value of radian frequency must the source voltage be set in order that $v_{cb}(t)$ and $i(t)$ be in phase?

7-32. In the circuit of Fig. P7-32 the following relationships hold:

$$v_{ab}(t) = 4\sqrt{2}\ \sin\ (\omega t + 135°)$$

$$v_{bc}(t) = -4\sqrt{3}\ \sin\ (\omega t + 60°)$$

$$v_{cd}(t) = 4\ \cos\ (\omega t - 150°)$$

   (a) Draw a clearly labeled phasor diagram for the voltages $v_{ab}(t)$, $v_{bv}(t)$, and $v_{cd}(t)$.
   (b) Find the phasor voltage $\overline{V}_{ad}$.
   (c) Write the expression for $v_{ad}(t)$.
   (d) If $i(t) = 2\ \sin\ (\omega t + 165°)$, find the average power delivered by the current source.

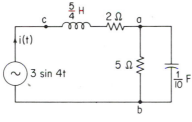

Fig. P7-31

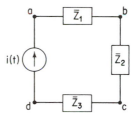

Fig. P7-32

7-33. A series RLC circuit has the following parameter values: $R = 10\ \Omega$, $L = 0.01\ H$, $C = 100\ \mu F$.
   (a) Compute the resonant frequency in radians per second.
   (b) Calculate the quality factor of the circuit.
   (c) What is the value of the bandwidth?
   (d) Compute the lower and upper frequency points of the bandwidth.
   (e) If a signal of $e(t) = 1\ \sin\ 1000t$ is applied to this series RLC circuit, calculate the maximum value of the voltage appearing across the capacitor terminals.

7-34. Repeat Prob. 7-33 for circuit elements having the following values: $R = 10'\ \Omega$, $L = 1\ H$, and $C = 1\ \mu F$.

7-35. A current source is applied to the parallel arrangement of R, L, and C where $R = 10\ \Omega$, $L = 1\ H$, and $C = 1\ \mu F$.
   (a) Compute the resonant frequency.
   (b) Find the quality factor.
   (c) Calculate the value of the bandwidth.
   (d) Compute the lower and upper frequency points of the bandwidth.
   (e) If a signal of $i(t) = 1\ \sin\ 1000t$ is applied to this parallel RLC circuit, calculate the maximum value of voltage appearing across the capacitor terminals.
   (f) What is the capacitor current in part (e)?

7-36. In the circuit of Prob. 7-14 compute the value of the current which flows when the input signal is (a) $e(t) = 1\ \sin\ 300t$ V, (b) $e(t) = 1\ \sin\ 1800t$ V.

7-37. In the circuit of Prob. 7-35 compute the value of the voltage appearing across the parallel elements when the input signal is (a) $i(t) = 1 \sin 300t$ A, (b) $i(t) = 1 \sin 1800t$ A.

7-38. A voltage of $e(t) = 10 \sin \omega t$ is applied to a series RLC circuit. At the resonant frequency of the circuit the maximum voltage across the capacitor is found to be 500 V. Moreover, the bandwidth is known to be 400 rad/sec and the impedance at resonance is 100 Ω.
(a) Find the resonant frequency.
(b) Compute the upper and lower limits of the bandwidth.
(c) Determine the value of L and C for this circuit.

7-39. A 220-V, three-phase voltage is applied to a balanced delta-connected three-phase load in the manner illustrated in Fig. P7-39. The rms value of the phase current measured between points a and b is $\bar{I}_{ab} = 10\angle -30°$.
(a) Find the magnitude and phase of the line current $\bar{I}$. Draw the phasor diagram showing clearly the line voltages, phase currents, and line currents.
(b) Compute the total power received by the three-phase load.
(c) Find the value of the resistive portion of the phase impedance.

7-40. The delta-connected load of Fig. P7-39 consists of phase impedances each equal to $15 + j20$.
(a) Find the phasor current in each line.
(b) What is the power consumed per phase?
(c) What is the phasor sum of the three line currents? Why does it have this value?

7-41. A three-phase, 208-V generator supplies a total of 1800 W at a line current of 10 A when three identical impedances are arranged in a wye connection across the line terminals of the generator. Compute the resistive and reactive components of each phase impedance.

7-42. A balanced three-phase wye-connected load has an impedance of $4\angle 60°$ Ω from line to neutral. Moreover, from line a to neutral n the voltage in $\bar{V}_{an} = 20\angle 30°$.
(a) What is the current in phases b and c?
(b) What is the voltage from line b to neutral?
(c) What is the phasor expression for the voltage from line a to line c, i.e., $\bar{V}_{ac}$?

7-43. Determine the Fourier series for the waveforms shown in Fig. P7-1.

7-44. Find the Fourier series for the waveform of Fig. P7-47(a).

7-45. Determine the Fourier series of the rectified waveform shown in Fig. P7-2(a).

7-46. Find the Fourier series of the sawtooth waveform of Fig. P7-3(a).

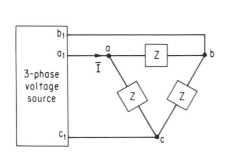

Fig. P7-39

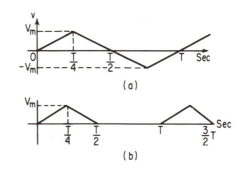

Fig. P7-47

GROUP II

7-47. For each waveshape shown in Fig. P7-47 find the average and effective values.

7-48. Compute the average and effective values of each of the waveshapes shown in Fig. P7-48.

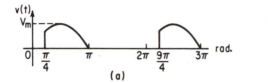
(a)

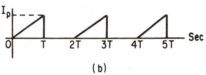
(b)

Fig. P7-48

7-49. A sinusoidal voltage $e(t) = 141 \sin \omega t$ is applied to a series RL circuit. In the steady state the effective value of the voltage measured across the terminals of the inductor is 60 volts. What is the voltage appearing across the resistor terminals?

7-50. In the circuit shown in Fig. P7-50 the reactance of capacitor C_1 is 4 Ω, the reactance of C_2 is 8 Ω, and the reactance of L is 8 Ω. A sinusoidal voltage having an effective value of 120 V is applied to the circuit.
(a) Find the effective value of the current delivered by the source.
(b) Write the expression for the instantaneous value of the current found in part (a).
(c) Draw a carefully labeled phasor diagram showing the source voltage, the source current, and voltages $\overline{V}_{ab}$ and $\overline{V}_{bc}$.

7-51. For the circuit shown in Fig. P7-51:
(a) Find $\overline{I}_1$, $\overline{I}_2$, $\overline{I}_3$, $\overline{I}_4$, and $\overline{I}_5$.
(b) Compute $\overline{V}_{bc}$ and $\overline{V}_{cd}$.
(c) Draw the phasor diagram showing all currents and voltages.
(d) Compute the power supplied by the source.
(e) Find the line power factor.

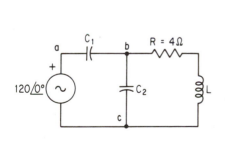

Fig. P7-50

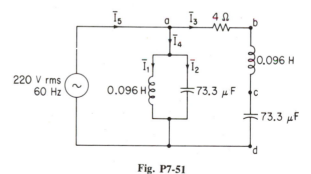

Fig. P7-51

7-52. The AM radio dial spreads over a frequency range 570 to 1560 kHz.
(a) Calculate the range of capacitance that is needed in series with a 20-μH inductance to tune over the entire frequency band.
(b) If the quality factor for a radio station operating at 570 kHz is 100, what is the bandwidth of the tuning circuit?

(c) What is the resistance of the tuning circuit?

(d) Determine the value of the quality factor at the upper end of the radio band.

7-53. A 208-V, three-phase generator supplies power to both a delta- and a wye-connected load in the manner shown in Fig. P7-53. All the phase impedances are identical and specifically equal to $5 + j8.66$. Compute the total generator current which flows in line a.

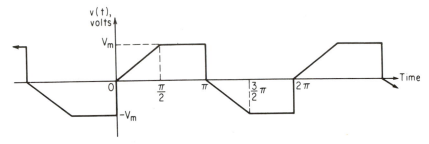

Fig. P7-53

7-54. A voltage wave has the form depicted in Fig. P7-54. Obtain the Fourier series representation of this periodic function.

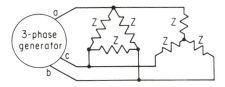

Fig. P7-54

chapter eight

Electron Control Devices: Semiconductor Types

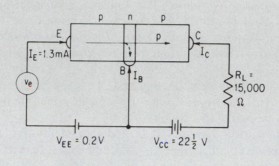

In Chapter 1 it is pointed out that electricity involves the flow of electrons no matter whether this flow takes place in a vacuum tube, in a semiconductor, or in a network consisting of circuit elements in whatever combination. In a broad sense, therefore, one may describe this electron flow as electronics. However, in common usage this term has a more restrictive meaning.

In accordance with the definition of the term published by the Institute of Electrical and Electronic Engineers "*electronics* is that branch of science and engineering which deals with electron devices and their utilization." Moreover, the IEEE goes on to define an *electron device* as "a device in which conduction by electrons takes place in a vacuum, gas, or semiconductor." Taken together, these two statements identify electronics not only in terms of the theory and design of electron devices such as the vacuum tube and the transistor, but also in terms of the vast variety of the circuitry built around them to do countless tasks in such fields as communication systems, computers, control systems, instrumentation, and telemetry.

Activity in electronics has been increasing at a phenomenal rate ever since World War II. Needless to say, national defense needs and space projects are important factors in this growth. An indication of the extent of this activity may be had by referring to some of the fields of interest of the organized professional technical groups of the IEEE as listed below:

Audio. The technology of communication at audio frequencies including acoustics and recordings as well as the reproductions from recordings.

Broadcasting. The design and use of broadcast equipment.

Nuclear Science. The application of electronic techniques and devices to the nuclear field.

Broadcast and Television Receivers. The design and manufacture of broadcast and television receivers and components, and related activities.

Space Electronics and Telemetry. The measurement and recording of data from remote points by electromagnetic media.

Navigational Electronics. The application of electronics to the operation and traffic control of aircraft and to the navigation of all craft.

Industrial Electronics and Control Instrumentation. Electronics pertaining to applied control, treatment, and measurement specifically directed to industrial processes.

Electronic Computers. Design and operation of electronic computers.

Biomedical Electronics. The use of electronic theory and techniques in problems of medicine and biology.

Communication Systems. Radio and wire telephony, telegraph and facsimile in marine, aeronautical, radio-relay, coaxial cable, and fixed-station services.

Automatic Control. The theory and application of automatic control techniques including feedback control.

Geoscience Electronics. Research and development in electronic instrumentation for geophysics and geochemistry especially regarding gravity, measurements, seismic measurements, space exploration, meteorology, and oceanography.

Aerospace. Theory and application of electronics and electricity for aerospace instrumentation and vehicle electric systems including electric propulsion and system application of electrically operated subsystems.

Although the foregoing list is only partial, the widespread activity in electronics is certainly apparent. This list is included to furnish the reader with a better perspective of that which electronics entails. The objective of Part II of this book is surely not to provide the detailed background needed to make useful contributions in any one of these fields of interest. Rather the goal is to furnish a general background, directing attention especially to a study of the theory and characteristics of the more important electron devices and their utilization in standard electronic circuitry. In this chapter and the next one, therefore, a study is made of solid-state electron devices, emphasizing their external characteristic and control capabilities. The next two chapters offer a detailed study of the operation, analysis, and performance description of the electronic circuitry commonly found in the various fields of interest in electronics.

8-1 THE BOLTZMANN RELATION AND DIFFUSION CURRENT IN SEMICONDUCTORS

The availability as well as the flow of electrons in semiconductor materials such as germanium and silicon† depend upon processes which are distinctly different from those which occur in vacuum devices and in metallic conductors. As we study this distinction in the material which follows, we shall conclude with a description of the external characteristics of the semiconductor diode. This is basic to the understanding of semiconductor electronics. Once the theory of operation of the semiconductor diode is understood, other related devices such as the transistor can be readily treated because they are modifications of the fundamental device. For example, as is pointed out later, the transistor may be looked upon as consisting of two diodes connected back to back.

The properties of metal are quite different from those of semiconductor materials. For example, in a metal the valence electrons (i.e., those in the outermost orbit) are completely free and roam about from atom to atom, whereas in a semiconductor the motion of the valence electrons of one atom is coordinated with the motion of those of an adjacent atom, so that a covalent bond is imposed between them. The result is that under ordinary circumstances there are few, if any, free electrons in a pure semiconductor element. Accordingly, the conductivity property of the two elements is very different. Thus the application of just a fraction of a volt across the ends of a copper conductor can produce a large current flow, whereas the same potential difference appearing across a germanium material having the same dimensions results in negligible current flow.‡

Electronics is concerned with the control of electron flow in engineering devices. A primary consideration of any electron device, then, must be the ability and ease with which electrons can be made available and controlled. In a metal, electrons are available by virtue of the free-electron nature of the space lattice structure of the material, so that again the application of an electric field in the metal exerts control over the flow of the electrons. It should be apparent by now that since the pure semiconductor element lacks a significant source of electrons it can be only of limited use as an electronic device. To be useful the pure semiconductor material must be altered so as to significantly increase its capability to transport charge carriers. Fortunately, this result can be achieved by the addition of small amounts of impurities to the composition of the pure semiconductor materials. Thus the addition of a small amount of pentavalent arsenic to a melt of pure tetravalent silicon makes one free electron per arsenic atom available for flow, and the impure silicon-arsenic mixture becomes a semiconductor. Because of the limited number of free electrons made available by this process, the conductivity of the semiconductor is less than that of a metal;

† Refer to column IV of the periodic table of the elements in Appendix B.

‡ A minute current flow can result because at room temperature the associated thermal energy can free some electrons from their covalent bonds, thus providing conduction.

however, the important point here is that there now exists a means of making available charge carriers for control purposes. It merely requires introducing impurities into a material characterized by covalent bonds. The higher the impurity content, the greater the conductivity of the semiconductor.

Essentially, there are two types of impurities that can be added to a tetravalent element such as germanium or silicon—elements found in column IV of the periodic table. When the impurity has five valence electrons (pentavalent), it is referred to as a *donor* or *n-type impurity*. Depicted in Fig. 8-1 is a schematic

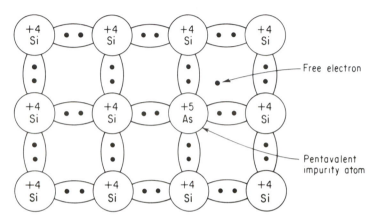

Fig. 8-1 Diagrammatic representation of the filled covalent bonds of the valence electrons. Note the free electron donated by the pentavalent impurity atom, arsenic.

representation of the covalent bonds existing between the valence electrons of the silicon atoms. Although silicon has a total of 14 orbital electrons, only the four in the outermost orbit are shown; for consistency the nucleus is shown carrying a corresponding charge of $+4$. Of course the total charge on the nucleus is $+14$. Note too that in the case of the arsenic† impurity atom four of its five valence electrons have coordinated motion with electrons of adjacent germanium atoms, thus forming the covalent bonds. The fifth electron is unbonded and at room temperature has sufficient thermal energy so that it is not even electrostatically attracted to the positive charge at the nucleus of the arsenic atom. Hence this electron is free to roam through the material. It is for this reason that the arsenic atom is called a donor impurity: it donates one charge carrier in the form of an electron. The symbolism to be used to depict this situation is illustrated in Fig. 8-2. The circle around the plus sign emphasizes that the positive nucleus is an immobile charge and thus cannot be part of a current flow. On the other hand the free electron can.

A second type of impurity semiconductor results when a trivalent element (such as indium, boron, gallium, or aluminum) is added to pure silicon. Since indium has only three valence electrons in the outer shell, the schematic rep-

† Other pentavalent impurities include phosphorus and antimony.

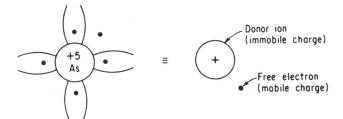

Fig. 8-2 Symbolism used to represent the donor impurity atom.

resentation of the covalent bonds between electrons of adjacent atoms appears as shown in Fig. 8-3. Note that only three covalent bonds can be filled. The vacancy that exists in the fourth bond is called a *hole*. The presence of this hole means that conditions are favorable for this impurity atom to *accept* an electron, thus filling the vacancy. For this reason this kind of impurity is called an *acceptor* or *p-type impurity*. Note that when an electron is received by the impurity atom it takes on a net negative charge that is immobile. The hole, which now no longer exists at this place in the material, may be considered to have moved elsewhere—for example, to the place from which the electron came. With such terminology one can look upon the hole as being a mobile charge carrier. This situation is depicted in Fig. 8-4, where the circle around the minus sign indicates

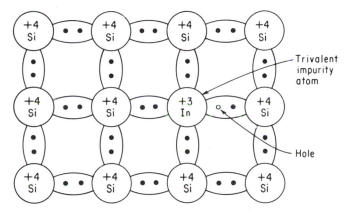

Fig. 8-3 Depicting the three filled covalent bonds of the trivalent impurity and the vacancy existing in the fourth bond.

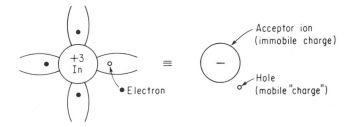

Fig. 8-4 Symbolism used to represent a trivalent impurity upon accepting an electron. When the electron fills the vacancy, the hole appears to go elsewhere. Hence the hole may be treated as a positive charge carrier.

Sec. 8-1   The Boltzmann Relation and Diffusion Current in Semiconductors   **337** that the trivalent impurity atom has accepted an electon, leaving the hole to appear elsewhere.

It is useful at this point to consider the energy-band description of materials. It can be shown by quantum theory that in crystalline materials there are bands of allowed energy levels for the atoms separated by bands of forbidden energy levels. Although for a single atom the energy bands are discrete, it is found that the proximity of other atoms modifies the discrete energy levels of the individual atoms to the point where the energy distributions take on the appearances of bands. The energy-band description of a metallic conductor is illustrated in Fig. 8-5. The partially filled band refers to the free or valence electrons possessing the highest energy. Note that if small additional energy is added to the crystal these electrons are raised to higher energy levels without prohibition. This statement cannot be made for those electrons with energy levels that place them in the filled band. Moreover, if energy is applied to this material in the form of an externally applied electric field, conduction is found to occur. For this reason the uppermost band is also called the *conduction band*.

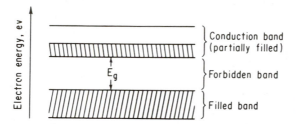

Fig. 8-5 Energy-band description of a metallic conductor.

In a semiconductor material without impurities the energy-band description appears as shown in Fig. 8-6. The significant factor here is that the forbidden energy gap separating the filled valence band and the empty conduction band is about one electron-volt. Accordingly, as the temperature is increased, electrons whose energy levels are raised beyond the gap level will be freed of their covalent bonds and become available for conduction. At room temperature approximately one out of 10^{10} atoms has electrons so liberated.

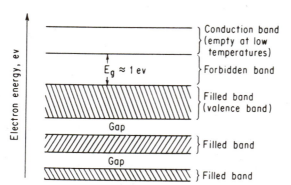

Fig. 8-6 Energy-band structure of a semiconductor.

Figure 8-7 depicts the situation which prevails for insulators. In this case the gap energy is considerably larger than it is for semiconductors, with the result that at room temperature no covalent bonds are broken to provide free electrons in the conduction band. Even the application of extremely large electric fields fails to rupture any appreciable number of covalent bonds. Thus the conductivity of such materials is practically nil.

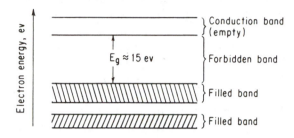

Fig. 8-7 Energy-band structure of an insulator.

Semiconductor Types. We are now ready to consider the impurity-type semiconductors in terms of the energy-band concept. But first let us consider the *intrinsic silicon semiconductor*, which contains no impurities. The thermal energy associated with room temperature is sufficient to dislodge an electron from a covalent bond. As already mentioned, this happens to about one out of 10^{10} atoms. Figure 8-8(a) depicts this situation in terms of the lattice arrangement, and Fig. 8-8(b) does it by means of the energy-band diagram. Note the presence

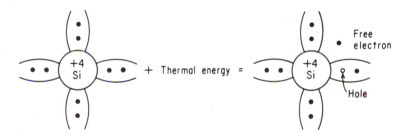

Fig. 8-8 (a) Creation of electron-hole pair through thermal agitation.

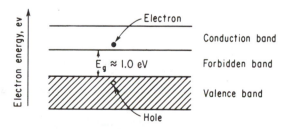

Fig. 8-8 (b) Intrinsic semiconductor at room temperature.

of the electron in the conduction band. As thermal energy increases, more electrons become available for conduction, which explains why a semiconductor material becomes a better conductor with rising temperature. In sharp contrast is the conductivity of metals, which decreases with temperature because of the greater impediment to the free flow of the electrons presented by the more violently vibrating atoms in the space lattice. Note too that current flow is now possible in two ways—the flow of electrons as well as the displacement of holes.

The band-structure description of the *n-type impurity semiconductor* is depicted in Fig. 8-9. Impurities are normally added in the ratio of about one part in 10^8. Accordingly, the energy diagram for the impurities remains discrete because of the relatively wide spacing between the impurity atoms. This situation is indicated in Fig. 8-9 by showing the individual impurity atoms. Furthermore, it is found that there are additional energy levels associated with pentavalent impurity atoms which are located just slightly below the lower level of the conduction band by approximately 0.01 eV.† Because of this small difference in energy levels the thermal energy of room temperature supplies enough energy to each excess electron (i.e., the fifth valence electron) of the impurity atom to raise it into the conduction region. The impurity atoms are thus said to be fully ionized and are so depicted in Fig. 8-9 by the circles around the positive charges. Keep in mind, too, that the normal thermal agitation of the pure silicon atoms continues to make electron-hole pairs available. However, the number of electrons furnished from this source is small compared to that provided by the impurity atoms. A glance at Fig. 8-9 reveals that, like the intrinsic semiconductor, the *n*-type impurity semiconductor brings about current flow through the action of electrons and holes. But in the *n*-type impurity semiconductor the number of electrons (negative carriers) far exceeds the number of holes, which is why it is called "*n*-type." The holes constitute the minority carriers. In a pure semiconductor the number of holes is always equal to the number of electrons.

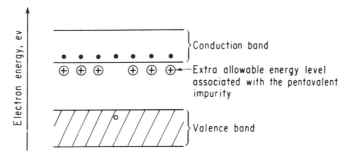

Fig. 8-9 Band structure of an *n*-type impurity semiconductor. Electrons are majority carriers; holes are minority carriers.

An analogous condition is found to occur when a trivalent impurity is added to germanium or silicon. The presence of the impurity allows an extra unfilled energy level to exist, which significantly is located a few hundredths of an

† The electron-volt (eV) is a unit of energy equal to 1.6×10^{-19} J.

electron-volt above the filled valence band as illustrated in Fig. 8-10. Thus enough electrons in the covalent bonds acquire the energy to be elevated to the additional energy level so that at room temperature all the holes of the impurity atoms are filled; the impurity is then said to be ionized 100%. This ionization process, of course, leaves holes in the valence band which are free to move about to other locations. Thermal agitation of the pure silicon atoms also generates electron-hole pairs, thus adding to the total number of holes present in the valence band. Note that when a trivalent impurity is used the minority carriers are the electrons whereas the holes are the majority charge carriers. For this reason such a semiconductor is referred to as a *p-type impurity semiconductor*. The *p* stands for "positive" charge carrier. Clearly, even where holes are involved current flow always occurs because of the movement of electrons, but the effect is that of a hole moving in the opposite direction. Therefore, electrically, one can say that the movement of a hole is equivalent to the movement of a positive ion having a mobility comparable to that of the electron. For convenience this description is used in all further work dealing with *p*-type devices.

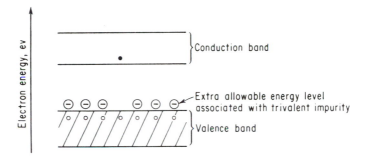

Fig. 8-10 Band structure of *p*-type impurity semiconductor. Holes are the majority carriers; electrons are minority carriers.

The p-n Junction Diode with Forward Bias.   The foregoing material was treated here for the purpose of providing the background for understanding the diffusion process which underlies the operation of the semiconductor diode. A *p-n* junction diode can be formed by growing a single crystal of semiconductor material in which the acceptor impurity is made to predominate in one part as the crystal is drawn out of a suitable melt and the donor impurity is made to predominate in the other part. A schematic diagram of the *p-n* junction diode appears in Fig. 8-11. Keeping in mind that the *n* material has a relatively high density of electrons while the *p* material has a high density of holes (or electron vacancies), it is reasonable to expect that there will be a tendency for the electrons to *diffuse* over to the *p*-side and vice versa. As a matter of fact this is precisely

Fig. 8-11 A grown *p-n* junction.

what takes place initially, but the process does not continue unhindered. As electrons move across the junction from the *n*-side to the *p*-side, positive immobile charges are uncovered on the *n*-side and negative immobile charges are formed upon acceptance of an electron by the acceptor impurity on the *p*-side. When a sufficient number of these charges are uncovered, a potential energy barrier V_0 is created and it prevents any further diffusion from occurring. This situation is illustrated in some detail in Fig. 8-12. Note that the figure depicts only the impurity atoms and one electron-hole pair resulting from the thermal agitation of the pure germanium. As a result of the absence of mobile charge carriers on either side of the junction, this region is called the *depletion region* and is of the order of 10^{-4} cm wide.

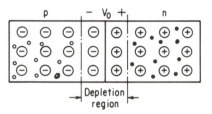

Fig. 8-12 The *p-n* junction diode showing the creation of the potential energy barrier V_0 through diffusion of mobile charge carriers across the junction.

The potential energy barrier V_0 plays a role analogous to that of the potential energy barrier of metals. The magnitude of V_0 is of the order of a few tenths of a volt, and in any given situation its value may be found from the Boltzmann relation and a knowledge of the charge densities in the *p* and *n* regions. As an illustration, refer to Fig. 8-13 which depicts on a logarithmic scale the charge densities of the majority and minority carriers in the *p* as well as the *n* materials. Consider the equilibrium condition—switch S open. Note that the *p* material is more heavily doped with impurity than is the *n* material. However, it is a law of the semiconductor, impure or otherwise, that the product of holes and electrons in the *p* or *n* material is equal to a constant determined by the temperature. The Boltzmann relation is an equation which relates the density of particles in one region to those in an adjacent region when the densities involved are relatively sparse. The Boltzmann relation therefore is particularly useful, for example, in atmospheric studies. However, to a good approximation it is also applicable to the *p-n* junction diode because the density of holes in the valence band and the density of electrons in the conduction band for semiconductor materials is quite low. Expressed mathematically, the Boltzmann relation is

$$N_1 = N_2 \varepsilon^{V_{21}/E_T} \tag{8-1}$$

where N_1 = density of particles in region 1

N_2 = density of particles in region 2

E_T = energy equivalent of temperature = $T/11{,}600$ eV

V_{21} = the potential energy of region 2 with respect to region 1

Upon applying this equation to the *p-n* junction diode of Fig. 8-13, we have for the holes in the two regions

$$p_p = p_n \varepsilon^{V_0/E_T} \tag{8-2}$$

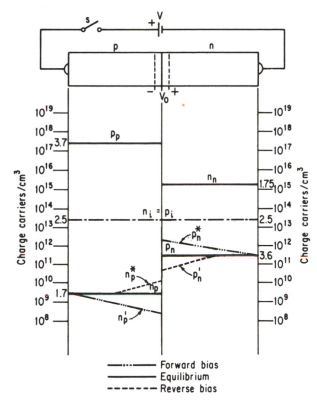

Fig. 8-13 Carrier densities in a *p-n* junction diode:

p_p = holes, or majority carriers in *p* material;

n_p = electrons, or minority carriers in *p* material;

n_n = electrons, or majority carriers in *n* material;

p_n = holes, or minority carriers in *n* material.

Inserting the values of p_n and p_p (appearing in Fig. 8-13) as well as $E_T = 0.026$ V for room temperature yields

$$\varepsilon^{V_0/0.026} = \frac{3.7 \times 10^{17}}{3.6 \times 10^{11}} = 1.027 \times 10^6 \tag{8-3}$$

and leads to

$$V_0 = 0.36 \text{ V}$$

We are now at the point where we can begin to establish the volt-ampere characteristic of the semiconductor diode. Let us consider first the behavior of the diode when a forward bias voltage is applied. This merely requires closing switch S in Fig. 8-13. Note that a forward bias is applied by connecting the positive side of the battery to the *p* material and the negative side of the battery to the *n* material. Moreover, assume that the battery voltage V is less than the potential energy barrier V_0. Observe that as a forward bias voltage, V is opposed to V_0. Consequently, the density of holes in the *n* region at the junction must be altered in accordance with the modified Boltzmann relation, namely,

$$p_p = p_n^* \varepsilon^{(V_0 - V)/E_T} \tag{8-4}$$

where p_n^* is the new value of hole density in the *n* material. Upon rewriting Eq. (8-4), we get

$$p_p = p_n^* \varepsilon^{V_0/E_T} \varepsilon^{-V/E_T} \tag{8-5}$$

Inserting Eq. (8-2) into Eq. (8-5) then yields

$$p_n^* = p_n \varepsilon^{V/E_T}$$ (8-6)

This is one of the most important relationships in junction theory. This equation states that upon applying a small forward bias there is an exponential increase in the minority charge carriers. Thus for $V = 0.102$ V the new value of hole density in the n material at the junction is

$$p_n^* = p_n \varepsilon^{0.102/0.026} \approx 50 p_n$$ (8-7)

Thus there occurs a fiftyfold increase over the equilibrium value of the hole density.

Where does this large increase of holes come from? The presence of the forward bias voltage V decreases the effective potential energy barrier from V_0 to $(V_0 - V)$, thereby allowing holes to move across the junction into the n region. It is interesting to note that although the hole density at the junction in the n material increases from 3.6×10^{11} to 180×10^{11}, this quantity is only a very small fraction of the hole density in the p material. The hole density in the p material is 3.7×10^{17}. Furthermore, in accordance with the *law of charge neutrality*, every time a hole leaves the p material, it is replenished by one which enters the p material from the positive side of the battery.

What happens to this large increase in hole density after it is injected into the n material across the junction? Since we are dealing here with two contiguous regions having different particle densities there is a natural action of *diffusion* taking place. As the holes cross the junction into the n material the preponderance of negative charge carriers throughout the n material causes swift recombination to take place. Thus the charge density rapidly decreases as the distance from the junction increases. Again, by the law of charge neutrality, for every electron in the n material which combines with a hole, there is another electron entering the n region from the negative side of the battery. It is important to understand that this change of charge density with distance in the semiconductor involves a transport of charge and thereby constitutes a current flow. But this is not a current flow in the usual sense that we have come to associate with metallic conductors. Rather in semiconductor diodes current flow takes place by *diffusion*. In fact, the hole current may be generally written as

$$I_{pn} = -AeD_p \frac{dp_n^*}{dx} \quad \text{A}$$ (8-8)

where D_p = diffusion constant for holes, m²/s
A = cross-sectional area of p-n junction, m²
e = electron charge, coulombs/electron
I_{pn} = *hole* current in the n region

Equation (8-8) emphasizes the fact that the diffusion current is dependent upon the rate of change of charge density with distance. Diffusion current does not obey Ohm's law.

In the foregoing treatment attention was focused on the difference in the hole densities in the p and n regions. A similar treatment holds for the electron densities in the two regions. Accordingly, by the Boltzmann relation we can describe a simultaneous increase in the minority carriers of the p region expressed as

$$n_p^* = n_p \varepsilon^{V/E_T}$$

(8-9)

As these electrons diffuse through the p region they quickly recombine with holes, which are present in great numbers. Again there is a transport of charge with distance, leading to a second component of diffusion current but this time attributable to the electrons in the p region. Hence we can write

$$I_{np} = AeD_n \frac{dn_p^*}{dx} \quad \text{A}$$

(8-10)

where D_n = diffusion constant for electrons m^2/s

A = cross-sectional area, m^2

e = electron charge

I_{np} = *electron* diffusion current in the p region

The total diffusion current is given by the sum of Eqs. (8-8) and (8-10) evaluated at the junction, i.e., $x = 0$. Thus

$$I = I_{pn}(0) + I_{np}(0)$$

(8-11)

It follows from Eqs. (8-6) and (8-8) as well as Eqs. (8-9) and (8-10) that the p-n junction diode diffusion current is exponentially related to the forward bias voltage. This is a consequence of the Boltzmann relation. Hence the external volt-ampere characteristic of the p-n junction diode appears as shown in Fig. 8-14 for positive values of V.

What happens to the p-n junction diode current when a reverse bias is applied (i.e., the positive side of the battery is connected to the n region and the negative side to the p region)? The Boltzmann relation reveals that under these

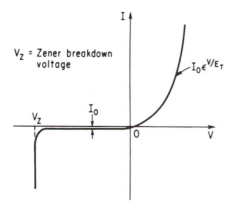

Fig. 8-14 Volt-ampere characteristic of the p-n junction diode.

conditions there is a marked decrease in holes at the junction of the n material. Thus we have $p'_n = p_n \varepsilon^{-V/E_T}$ where p'_n is the hole density for reverse bias. This means that holes diffuse from the n region, where they are indeed scarce, into the p region. Similarly, the reverse bias causes a decrease of electrons at the junction in the p region. Effectively then electrons diffuse from the p material, where they are relatively very few, into the n region. Because of the small numbers of charge carriers involved, the current flow in the reverse direction is correspondingly very small too. It is frequently of the order of 10^{-5} A and is called the *reverse saturation current*, denoted by I_0.

Accounting for both forward and reverse bias voltages, it is possible to write a single expression for the semiconductor diode current which is valid for positive as well as negative voltages up to V_z—the Zener breakdown voltage. Thus

$$\boxed{I = I_0(\varepsilon^{V/E_T} - 1)} \tag{8-12}$$

where I_0 is the reverse saturation current. Note that for even small positive values of V the exponential term predominates so that the current variation is exponential, which is consistent with the Boltzmann relation. At $V = 0$ the current, of course, should be zero, and the form of Eq. (8-12) certainly bears this out. Finally, for negative V the exponential term approaches zero rapidly so that the only significant current flow is the reverse saturation current, $-I_0$.

It is important to keep in mind when using Eq. (8-12) that the exponential character of the variation for positive V is valid as long as V is less than the potential energy barrier V_0. When V exceeds V_0, ohmic drops within the material must be accounted for.

A glance at Fig. 8-14 indicates that when a large negative voltage is applied to the diode, it is accompanied by a tremendous increase in current. Essentially, this condition is brought about by the complete disruption of all the covalent bonds, which releases huge numbers of electrons for conduction. In this state, the semiconductor behaves as a metallic conductor.

8-2 THE SEMICONDUCTOR DIODE

The *p-n* semiconductor junction diode with a forward bias is shown in Fig. 8-15. Recall that the p region is heavily doped with a trivalent impurity and the n region is doped with a pentavalent impurity. The application of a forward bias V, which is assumed smaller than the potential energy barrier V_0, causes an

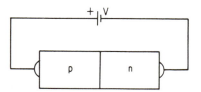

Fig. 8-15 The *p-n* junction diode with forward bias.

increase in the hole density in the n region at the junction. This increase is expressed by Eq. (8-6) and is repeated here. Thus

$$p_n^* = p_n \varepsilon^{V/E_T}$$

where p_n^* = new value of hole density in the n region

p_n = equilibrium value of hole density

E_T = energy equivalent of temperature

At the same time there is a similar action taking place which involves the electron density in the p region. The same forward bias causes an increase in the electron density of the p region which at the junction is described by Eq. (8-9). For convenience this equation too is repeated here:

$$n_p^* = n_p \varepsilon^{V/E_T}$$

where n_p^* = new value of electron density in the p region

n_p = equilibrium value

The increased concentration of the hole density at the junction of the n region is followed by a diffusion of the charge carriers through the n material. Recombination with electrons then takes place rapidly because of the very high concentration of negative charge carriers in the n region. This action gives rise to a *diffusion current* which depends upon the change of charge density with distance from the junction. This hole (p) current in the n region, I_{pn}, is depicted in Fig. 8-16(a). Two things are worth noting here. One is the exponential decay in I_{pn} that occurs with increasing displacements from the junction. The other is the presence of a *recombining electron current* which is required by the law of charge neutrality. For every electron that recombines with a hole in the n region

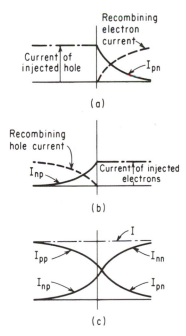

(a)

(b)

(c)

Fig. 8-16 Identifying the total diffusion current I in terms of the hole and electron current components.

there enters into this region from the negative side of the battery another electron which preserves the charge neutrality of the n region.

A similar situation takes place in the p region. The increased electron density n_p^* brought about by the forward bias causes a diffusion of electrons away from the junction into the p region. However, because of the very high concentration of holes in this region, electrons rapidly recombine with holes. Since the decay in charge density from the junction can be shown to be an exponential decay, it follows that the electron (n) current in the p region, I_{np}, undergoes a similar decay as depicted in Fig. 8-16(b). In this instance, however, note the presence of a *recombining hole current*. Thus for every hole that recombines with an electron in the p region the positive side of the battery supplies an additional hole in order to preserve charge neutrality in the p region.

The total diffusion current is found by adding the effects depicted in Figs. 8-16(a) and (b). The total hole (p) current in the p region, I_{pp}, is composed of the recombining hole current of Fig. 8-16(b) plus the value of I_{pn} as *it exists at the junction*. It is important to understand that the increased hole density p_n^* originates from the positive side of the battery. The forward bias causes holes to be injected from the positive terminal through the entire p region and thence across the junction into the n region where diffusion occurs. Little or no recombination of the injected holes with electrons occurs in the p region because of the great scarcity of electrons there compared to holes. Accordingly, I_{pp} assumes the variation shown in Fig. 8-16(c). By a similar line of reasoning the total electron (n) current in the n region, I_{nn}, can be shown to vary as indicated.

Further examination of Fig. 8-16(c) reveals that the total diffusion current I is a constant throughout the body of the junction diode even though I_{nn} and I_{pp} vary with the distance from the junction. A study of Fig. 8-16(c) makes it apparent that the diffusion current can be expressed in any one of three ways. Thus

$$I = I_{pp} + I_{np} \tag{8-13a}$$

or

$$I = I_{nn} + I_{pn} \tag{8-13b}$$

or

$$I = I_{pn}(0) + I_{np}(0) \tag{8-13c}$$

In this last equation $I_{pn}(0)$ refers to the hole current in the n region evaluated at the junction.

In dealing with the diffusion current it is important to understand the distinction between the two exponential functions with which it is related. An increase in forward bias voltage V causes an exponential increase in the diffusion current as described by Eq. (8-12). That is, the values of I_{np} and I_{pn} at the junction ($x = 0$) are increased exponentially with V. On the other hand, the values of these same currents as a function of the distance away from the junction decrease exponentially with x.

Semiconductor Diode Resistance. The volt-ampere characteristic of the semiconductor junction diode takes on the exponential character depicted in Fig.

8-17. If the forward bias is fixed at the value V_Q, there is associated with the diode an apparent forward resistance given by

$$R_f \equiv \frac{V_Q}{I_Q} \quad \Omega \qquad (8\text{-}14)$$

When we are interested in the resistance of the diode as it appears to a signal source causing small changes in the bias voltage about the quiescent point Q, its value is determined as the reciprocal of the slope of the curve at point Q. This resistance is called the *incremental resistance*. An expression for it readily follows from Eq. (8-12), which mathematically describes Fig. 8-17. Thus

$$I = I_0(\varepsilon^{V/E_T} - 1) \qquad (8\text{-}15)$$

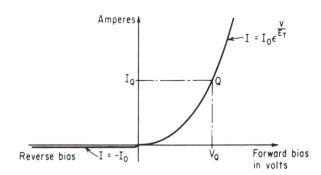

Fig. 8-17 Volt-ampere characteristic of *p-n* junction diode.

Differentiating yields

$$dI = \frac{I_0}{E_T}\varepsilon^{V/E_T}\, dV \qquad (8\text{-}16)$$

Therefore, the forward incremental resistance is

$$r_f \equiv \frac{dV}{dI} = \frac{E_T}{I_0\varepsilon^{V/E_T}} = \frac{E_T}{I + I_0} \qquad (8\text{-}17)$$

At room temperature the value of E_T is 0.026 V. Moreover, for a forward bias, i.e. V positive, Eq. (8-17) may be simplified to

$$r_f = \frac{E_T}{I} = \frac{0.026}{I} \qquad (8\text{-}18)$$

Finally, if I is assumed to be expressed in milliamperes (mA), as is frequently the case, then Eq. (8-18) becomes

$$\boxed{r_f = \frac{26}{I(\text{ma})}} \quad \Omega \qquad (8\text{-}19)$$

This result indicates that the incremental forward resistance of the *p-n* junction diode is quite low. For example, when a current of 1 mA flows, this resistance has a value of 26 Ω.

   The incremental backward resistance (i.e., the resistance with reverse bias) is ideally infinite because of the zero slope of the characteristic in this region.

This conclusion also follows from Eq. (8-17) upon inserting $I = -I_0$ for negative values of bias voltage. Thus

$$r_b = \frac{E_T}{I + I_0} = \frac{0.026}{-I_0 + I_0} = \infty \qquad (8\text{-}20)$$

However, the apparent backward resistance is not infinite because for any value of reverse bias there does exist a small current I_0. For germanium and silicon junction diodes this resistance is large compared to the apparent forward resistance. In germanium diodes the ratio of apparent backward to forward resistance is 400,000:1, while for silicon this ratio is 1,000,000:1.

The volt-ampere characteristic of an ideal semiconductor diode is depicted by Fig. 8-18, and the circuit symbol is shown in Fig. 8-19.

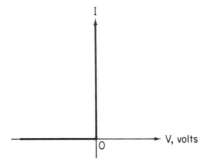

Fig. 8-18 Volt-ampere characteristic of an ideal diode, i.e., $r_f = 0$.

Fig. 8-19 Circuit symbol for the ideal diode. Conduction takes place from p to n only.

Diffusion Capacitance.   Appearing in Fig. 8-20 is the variation of the hole density injected into the n material of a p-n junction diode shown as a function of the distance from the junction for two values of forward bias voltage. At the junction the variation of the hole density with forward bias is an increasing exponential function. Thus for a forward bias voltage V_1 the hole density at the junction is

$$p_{n1} = p_n \varepsilon^{V_1/E_T} \qquad (8\text{-}21)$$

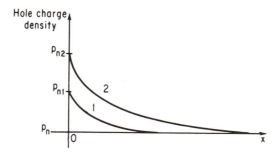

Fig. 8-20 Variation of hole charge density as a function of distance from the junction for two values of forward-biased voltage.

and for $V_2 > V_1$, it is

$$p_{n2} = p_n \varepsilon^{V_2/E_T} \qquad (8\text{-}22)$$

where in each case p_n denotes the equilibrium value of the hole density at the junction in the n material. When a forward bias is applied, however, as one moves away from the junction the hole charge density diminishes because of the recombination which takes place. It can be shown that this decrease proceeds in accordance with an exponential decay ε^{-x/L_p} where L_p denotes the mean diffusion length for holes and x is the distance into the n material measured from the junction.

It is important to note that the area beneath either curve in Fig. 8-20 represents the charge stored at the junction per unit cross-sectional area for each value of forward bias voltage. As the forward bias is changed from V_1 to V_2 operation goes from curve 1 to curve 2 in Fig. 8-20. Because the area beneath the charge-density curve increases with an increase in forward bias, it follows that the charge at the junction increases. Accordingly, there occurs at the junction a change in charge, ΔQ, per change in forward bias voltage, $\Delta V = V_2 - V_1$, so that the ratio of the two, $\Delta Q / \Delta V$, can be interpreted in terms of an incremental capacitance. A similar analysis applies to the electron densities in the p region. The cumulative effect of the change in stored charge of the hole and electron densities at the junction corresponding to a change in forward bias voltage is called the *diffusion capacitance* and is denoted by C_D. Typical values of this quantity fall in the range of several thousand picofarads.

8-3 THE TRANSISTOR (OR SEMICONDUCTOR TRIODE)

The addition of a third doped element to a semiconductor diode results in a triode. Figure 8-21(a) depicts one possible configuration of the semiconductor triode. It is a *p-n-p* type and is composed of two sections of silicon, heavily doped with a trivalent impurity, separated by a very thin section (of the order

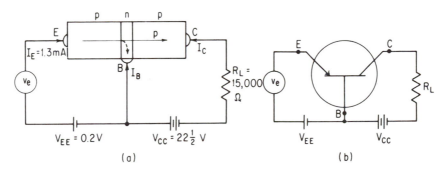

Fig. 8-21 Semiconductor triode in a common-base configuration: (a) circuit arrangement showing typical operating values for a *p-n-p* triode; (b) schematic representation.

of 1 mil thickness) of germanium which is heavily doped with a pentavalent impurity. The semiconductor triode can also be formed by two n sections separated by a p section. Appearing in Fig. 8-21(b) is the symbol used to represent the semiconductor triode.

How are control and amplification achieved in the semiconductor triode? This can be understood by investigating the manner in which the circuit of Fig. 8-21(a) operates. The thin center section of the semiconductor triode is called the *base* and the terminal connection associated with it is denoted by B. The input circuit consists of that part of the configuration associated with terminals E and B. Note that the base voltage V_{EE} is connected in the input circuit in a way that ensures *forward bias*. Thus when the input circuit is closed, V_{EE} causes holes to be injected into the p material. For this reason the second terminal of the semiconductor triode is denoted by E. The voltage V_{EE} is a fixed quantity which serves to establish an operating point about which changes may be effected by the action of the varying component of the signal v_e. The output circuit consists of that part of the configuration associated with terminals C and B. This section of the semiconductor triode circuit is *reverse-biased*, as a glance at Fig. 8-21(a) reveals. The third terminal of the triode is denoted by C in order to refer to the collector action that takes place at this terminal. Because the base terminal is common to both the input and output sections, the circuit configuration of Fig. 8-21(a) is called the *common-base* mode. In the representation of the semiconductor triode shown in Fig. 8-21(b), note that the emitter terminal has an arrowhead associated with it. When the arrowhead points towards the base terminal the triode is of the *p-n-p* type; when the arrowhead points away from the base terminal, it means that an *n-p-n* type is being used. Of course whenever the latter type is used, the voltages V_{EE} and V_{CC} must be connected in the reverse sense from that shown in Fig. 8-21.

Let us now analyze the circuit of Fig. 8-21 in terms of some typical operating figures. Assume that when the input circuit is quiescent (i.e., $v_e = 0$), the bias voltage V_{EE} causes an emitter current I_E to flow of value 1.3 mA. In other words a sufficient number of charge carriers (holes in this instance) are being injected from the forward bias battery into the p material to cause a current of 1.3 mA to flow. These injected holes then appear at the emitter-base junction, beyond which point a diffusion process should take place. However, the base region was made intentionally small with the expressed purpose of preventing the injected holes from recombining with the excess electrons existing in the n region. This result is achieved by making the thickness of the base region only a small fraction of the average diffusion length of a hole. Thus if the average diffusion length of a hole is L_p and the thickness of the base section is of the order of $0.05L_p$, then the chances of a hole recombining with an electron in the base region are small indeed. As a result the vast majority of the holes injected across the emitter-base junction never recombine but rather move on across the base-collector junction into the p material of the collector. Once the injected holes appear in this region, the effect of the voltage V_{CC} accelerates them toward the collector terminal and then through the load resistor R_L, through V_{CC}, and finally back to V_{EE}, thus completing the circuit. The remaining fraction of the injected holes, of course, take a path through the base terminal to the negative side of V_{EE}. In most semiconductor triodes operating in the common-base mode approximately 95 to 98% of the injected holes take a path to the collector terminal and thence through the output circuit.

A figure of particular importance in semiconductor triodes is the ratio of the change of current appearing at the collector output terminal to a given current change appearing at the emitter input terminal. It is called the *current transfer ratio* and is denoted by α. This quantity is also called the *current amplification factor*.

Expressing this mathematically, we can write

$$\alpha = \text{current transfer ratio} \equiv -\left(\frac{\partial i_c}{\partial i_e}\right)_{V_{CB} \text{ constant}} \tag{8-23}$$

This definition is valid for the common-base configuration. Moreover, α has a value which is always less than unity but frequently greater than 0.9. In this sense, then, it appears that use of the word "amplification" is misleading. However, it is used in order to keep clear the distinction between current amplification factor and current gain. The former is a property of a given semiconductor triode; the latter is influenced by the external circuitry connected to the triode.

In the interest of simplicity let us for the moment assume that α equals unity in the circuit of Fig. 8-21. Hence the entire emitter current of 1.3 mA flows to the collector and on through the 15,000-Ω resistance. Upon passing through this resistor, there occurs a voltage drop of $(1.3 \times 10^{-3})(15 \times 10^3) = 19.5$ V. Consequently, the voltage from collector to base becomes

$$V_{CB} = +19.5 - 22.5 = -3 \text{ V} \tag{8-24}$$

The existence of this negative voltage from collector to base assures that the output circuit is in fact reverse biased.

Consider next the situation where a change in the emitter voltage is introduced by letting v_e assume a value of 1 mV; that is,

$$v_e = 0.001 \text{ V} \tag{8-25}$$

To determine the corresponding change in the emitter current caused by v_e, it is necessary first to obtain the forward resistance of the p-n section of the triode appearing in the input circuit. This readily follows from Eq. (8-19). Thus

$$r_e = \frac{26}{I_E(\text{mA})} = \frac{26}{1.3} = 20 \ \Omega \tag{8-26}$$

The subscript e indicates that the input change occurs at the emitter terminal. Therefore the change in emitter current is

$$\Delta I_E = \frac{v_e}{r_e} = \frac{0.001}{20} = 0.05 \text{ mA} \tag{8-27}$$

Continuing with the assumption that α is 1 yields a change in collector current of

$$\Delta I_C \approx \Delta I_E = 0.05 \text{ mA}$$

Accordingly, the change in output voltage, which is obtained across the 15,000-Ω resistor, is

$$v_c = \Delta I_C R_L = (0.05 \times 10^{-3})(15 \times 10^3) = 0.75 \text{ V} \qquad (8\text{-}28)$$

Here v_c denotes the change in voltage appearing at the collector terminal.

A comparison of Eq. (8-28) with Eq. (8-25) shows that there occurs a voltage gain A_v of 750. Thus

$$A_v \equiv \frac{v_c}{v_e} = \frac{0.75}{0.001} = 750 \qquad (8\text{-}29)$$

The corresponding gain in power, G, is

$$G \equiv \frac{v_c \Delta I_C}{v_e \Delta I_E} = A_v \alpha = 750 \qquad (8\text{-}30)$$

A study of the results shown in the last two equations makes it clear that the semiconductor triode may be used to amplify small voltage variations as well as power. It is important at this stage to avoid erroneous impressions regarding the power-amplifying capability of the circuit of Fig. 8-19. Although it is correct to say that one unit of power in the input circuit controls 750 times as many units in the output circuit, one must guard against thinking that the circuit is capable of generating any level of power called for by the input circuit. This is not so. Any power that is delivered to R_L must come from the source V_{CC}. Because of the nature of the circuit of Fig. 8-19, it is possible for a small signal in the input circuit to command V_{CC} to deliver large changes of power in the output circuit. In other words, the varying signals of the input circuit serve as valves controlling the output derived from V_{CC} at much higher voltage and power levels. The overall efficiency of this circuit, however, is considerably less than 100%.

The key factor which is responsible for the amplifying capability of the semiconductor triode is the arrangement which causes the emitter input current to flow almost entirely through the output circuit. Thus, easy control of the emitter current is obtained because the emitter-base junction is forward biased, and yet most of this current is made to flow through the output circuit because the average diffusion length of the charge carriers is many times greater than the thickness of the base region. By this scheme, therefore, there occurs a *transfer* of the input signal current from a low-*resistor* circuit (i.e., the forward-biased input circuit) to a high-resistor circuit (namely the output circuit containing R_L). By combination of parts of the two key words which describe the operation of Fig. 8-19, there is coined the new word *transistor*. Accordingly, the transistor is a semiconductor device which has a forward-biased input section and a reverse-biased output section. Moreover, because of the low-resistance forward-biased input section, the transistor is essentially a current-sensitive device.

A general formulation of the amplifying capabilities of the common-base transistor further bears out this transfer aspect from a low- to a high-resistance circuit. Corresponding to a change v_e in emitter voltage, we can write

$$v_e = \Delta I_E r_e \qquad (8\text{-}31)$$

Moreover, the change in collector current is related to ΔI_E by

$$\Delta I_C = \alpha \Delta I_E \tag{8-32}$$

Furthermore, the change in collector output voltage is described by

$$v_c = (\Delta I_C)R_L = \alpha \Delta I_E R_L \tag{8-33}$$

Inserting the expression for ΔI_E from Eq. (8-31) leads to

$$v_c = \alpha \frac{R_L}{r_e} v_e \tag{8-34}$$

Equation (8-34) indicates that any change in emitter input voltage v_e appears as a change in the collector output circuit amplified by $\alpha(R_L/r_e)$. Since α is very close to unity and R_L is very large compared to r_e, it follows that the input signal change is greatly amplified at the output terminal. Returning to the values used in the circuit of Fig. 8-19 we get

$$v_c = (1)\frac{15,000}{20}(0.001) = 0.75 \text{ V} \tag{8-35}$$

which agrees with Eq. (8-28).

Volt-Ampere Characteristic of the Common-Base Mode. The volt-ampere characteristic of the common-base transistor involves a plot of the variation of the output circuit current I_C versus the collector-to-base voltage V_{CB}. A family of curves results when these plots are made for various values of fixed emitter currents. Note that for the transistor the input current is the parameter. As long as the reverse bias voltage across terminals C and B exceeds several tenths of a volt, Eq. (8-32) is valid, so that practically all the emitter current drifts through the base region to the collector terminal. Depicted in Fig. 8-22 is a set of typical common-base output characteristics for a *p-n-p* junction transistor. Note that the curves are almost straight horizontal lines. This is consistent with the theory that the emitter current flows almost entirely to the collector terminal.

By convention a transistor current is said to be positive when it *enters* a terminal. Thus the positive direction of I_C is into terminal C in Fig. 8-19. However,

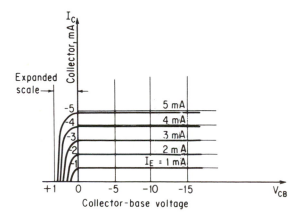

Fig. 8-22 Common-base output characteristics.

if I_E is assumed to flow into terminal E, it then follows that almost all of this current flows out of terminal C. Because this is opposite to the assumed positive direction of I_C, the collector current is made to carry a minus sign as shown in Fig. 8-22.

Appearing in Fig. 8-23 are the corresponding input characteristics of the same junction transistor whose output characteristics appear in Fig. 8-22. Note that the exponential character of these curves is consistent with Eq. (8-12).

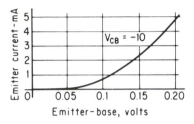

Fig. 8-23 Common-base input characteristics.

Transistor Parameters. Certain useful parameters may be identified for the junction transistor. Two of the parameters have already been identified. Equation (8-23) defines the current transfer ratio α, and Eq. (8-19) determines the forward-biased resistance r_e of the input circuit. The third useful parameter is the intrinsic transconductance, which is defined as the change in output collector current per unit change in emitter voltage. Expressing this mathematically we have

$$g_m \equiv \left(\frac{\partial i_c}{\partial v_e}\right)_{V_{CB}\ \text{constant}} \qquad \text{S} \qquad (8\text{-}36)$$

Under the assumption that α is essentially constant, this expression may be rewritten as

$$g_m = \alpha\left(\frac{\partial i_e}{\partial v_e}\right)_{V_{CB}\ \text{constant}} \qquad (8\text{-}37)$$

A little thought reveals that the quantity in parentheses is the slope of the input characteristics of Fig. 8-23, which is the reciprocal of the forward-biased resistance r_e. Accordingly, Eq. (8-37) may be rewritten as

$$g_m = \alpha\left(\frac{1}{r_e}\right) \qquad (8\text{-}38)$$

or

$$\boxed{\alpha = g_m r_e} \qquad (8\text{-}39)$$

8-4 THE JUNCTION FIELD-EFFECT TRANSISTOR (JFET)

The conventional transistor described in the preceding section is characterized chiefly by its current sensitivity. The changes that are registered at the output terminals of the transistor are a direct consequence of changes that occur in the input current. The signal source that is applied to the input terminals must accordingly be capable of bringing about the required range of input current change if the transistor is to serve as a meaningful amplifier. There are many situations in electronics, however, where the signal source is capable of effecting an appreciable change in voltage level but not an appreciable change in current. This happens, for example, whenever the source has a high internal impedance. In such instances it is convenient to have available an amplifying device that possesses a very high input impedance. Such a device is the *field-effect transistor*, and it essentially provides features which, in the era of vacuum-tube electronics, were furnished by such popular devices as the vacuum triodes.

The field-effect transistor has been a well-established laboratory device since 1952. But it was not until a decade later that it began to receive wide acceptance. It had to await the perfection of thin-film and related technology to achieve with repeatable reliability configurations of thin, lightly doped layers of semiconductor material between more heavily doped layers of opposite type.

The composition of the junction field-effect transistor is best described by referring to Fig. 8-24, which depicts an *n*-channel JFET. The silicon body is lightly doped with a pentavalent impurity thereby forming the *n*-channel. At the ends of the channel are placed two terminals—the drain D and the source S. On either side of the channel there is deposited a much more heavily doped *p*-type impurity in the manner illustrated in Fig. 8-24. A terminal connection is provided for each section and these are called the gate terminals and denoted by G_1 and G_2. In use both gate terminals are wired together.

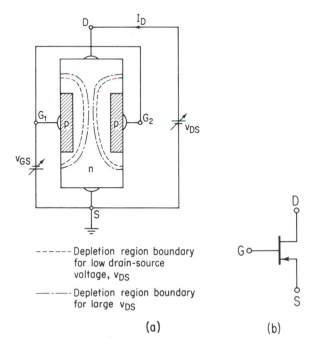

- - - - - - Depletion region boundary
for low drain-source
voltage, v_{DS}

— - — - Depletion region boundary
for large v_{DS}

(a)　　　　(b)

Fig. 8-24　(a) JFET, showing associated source and supply voltages; (b) symbol.

For the n-channel JFET the positive terminal of the supply voltage v_{DS} must be connected to the drain terminal of the JFET. The negative side of the drain source is connected to S. As a result the supply voltage which appears entirely across the n-channel is capable of producing a current flow from drain to source, I_D. At low values of v_{DS} this current is limited essentially by the magnitude of v_{DS} and by the resistance of the n material. The latter in turn depends upon the length and cross-sectional area of the channel and the conductivity.

The success of the JFET depends upon the establishment of a substantial depletion region the width of which is readily and conveniently controlled. A glance at Fig. 8-24 discloses that a reverse-biased voltage is made to appear between each p section and the n-channel section about the drain terminal. The size of the voltage varies with the distance along the n-channel because of the effect of the ohmic drop in the n-channel. For points closer to the drain terminal the reverse bias is larger than it is for points closer to the source terminal. This results in the creation of a depletion region surrounding each p section which is nonuniformly wide. It is also important here to understand that the depletion region width that spreads into the n-channel is many times greater than that appearing in the p material simply because the n material is lightly doped whereas the p material is much more heavily doped. This situation is emphasized in the diagram of Fig. 8-24 by showing that the spread of the depletion region takes place only in the n-channel. At moderate values of reverse-biased voltage the depletion region about the two p sections is denoted by the dashed lines. At high values of reversed-biased voltage, brought about by increased values for v_{DS} and/or v_{GS}, the depletion region spreads to cover an area enclosed by the dash-dot lines. As the depletion region changes from one of less width to one of greater width, an effective control of the drain current, I_D, results. It is instructive here to recall that there exist only uncovered immobile charges in the depletion region so that the conductivity of this region is virtually negligible. Consequently, as the depletion region is made to increase in width, the path of the drain current in the n-channel becomes more narrow. Accordingly, the path resistance increases and the drain current diminishes.

The sensitivity of the JFET to variations in the conductivity of the channel that joins the D and S terminals of Fig. 8-24(a) and which is subject to changes in the *negative* potential of the gate terminal has caused this electronic device to become known as a *depletion-mode* JFET. Schematically, it is represented by the symbol shown in Fig. 8-24(b).

Volt-Ampere Characteristic of the JFET. A typical set of drain characteristics for the 2N4222 JFET is depicted in Fig. 8-25. This set of curves very strongly resembles those of the conventional transistor in the common-base mode. Of course, the parameter here is the gate-source voltage, not the emitter-base current.

Let us direct attention first to the curve corresponding to $v_{GS} = 0$ in the configuration of Fig. 8-25. For positive values of the drain source voltage v_{DS} in the vicinity of zero the current is limited solely by the ohmic resistance of the

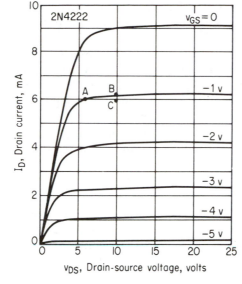

Fig. 8-25 Drain characteristics of the 2N4222 JFET.

n-channel. By Ohm's law then we have

$$i_D = \frac{v_{DS}}{l/\sigma A} \tag{8-40}$$

where *l* denotes the length of the *n*-channel, *A* is the cross-sectional area, and σ denotes the conductivity. As v_{DS} increases, a reverse-biased voltage begins to appear, thus developing a depletion region which acts to decrease the effective cross-sectional area of the path of the drain current. Now as v_{DS} increases further the rate of increase of path resistance proceeds faster than the rate of increase in drain current owing to the increases in drain source voltage. Consequently, the current curve begins to level off. Finally, a value of v_{DS} is reached beyond which no further significant increase in drain current occurs even for large changes in v_{DS}. In other words the current saturates. The value of drain-source voltage at which current saturation begins is called the *pinch-off* voltage. This nomenclature is used to describe the squeezing effect on the drain current brought about by the two spreading depletion regions in the channel of the JFET. At the pinch-off voltage and in the vicinity of the drain terminal the depletion regions have spread as far as they can go. It is not possible for the spread of the depletion regions to close off the channel completely. If this were to happen, the ohmic drop that is responsible for the nonuniform distribution of the depletion region would disappear and the drain current would be reestablished. Apparently, then, in the region beyond the pinch-off voltage, the JFET might best be described as being in a condition of dynamic balance.

What is the effect of varying the gate-source voltage when the drain-source voltage is kept fixed beyond the pinch-off value? It is helpful to keep in mind that the narrowest part of the drain-source current path occurs nearest the drain terminal for $v_{GS} = 0$. As v_{GS} is increased in the negative direction at fixed drain-source voltage, the reverse-biased voltage increases. But now the effect is to change the configuration of the depletion region about the source terminal S where the depletion region is much less wide than it is near the drain terminal beyond pinch-off. The effect is an appreciable drop in drain current as a glance at the drain characteristics of Fig. 8-25 readily discloses.

We have here a situation where output current control is made possible by variation of the gate-source voltage. As a result, the JFET takes on the character of being a *voltage sensitive* device with a very high input resistance. The input resistance is that which appears between the gate and source terminals. In view of the fact that these are reverse-biased terminals, negligible current flows. This explains why the value of the input resistance is of the order of 10^8 Ω for such devices.

The JFET is aptly named. As the gate-source voltage is varied to increase (or decrease) the reverse-biased voltage, especially as it appears between each p section and that part of the n-channel about the source terminal, there occurs a further increase (or decrease) in the spread of the depletion region. Hence, more (or less) of the n-channel becomes a region of uncovered charges and thereby reduced (or increased) conductivity. The drain current then drops (or rises). Throughout this action it is the *field effect* of the uncovered charges that influences the conductivity of the n-channel and therefore the level of permissible drain current.

Although the treatment so far has made reference solely to an n-channel JFET, p-channel units are equally valid. The use of the latter requires reversing the polarity of the drain-source and gate-source voltages.

The useful portion of the volt-ampere characteristic of the JFET in amplifier applications is the region beyond the pinch-off voltages. Note incidentally that the pinch-off voltage diminishes with increasing values of the gate-source voltage. An expression that gives a good approximation of the manner in which the gate-source voltage determines the drain current is

$$I_{DS} = I_{DSS} \left(1 - \frac{V_{GS}}{V_{po}} \right)^2 \qquad (8\text{-}41)$$

where I_{DSS} denotes the drain saturation current with the gate short circuited and V_{po} is the value of the pinch-off voltage when v_{GS} is zero. Equation (8-41) describes the common-source transfer characteristic at constant drain-source voltage. A plot of this curve for a typical JFET is shown in Fig. 8-26.

The JFET Parameters. At this point we continue the policy established early in the chapter of identifying appropriate parameters for each new device as it is studied. This facilitates the analysis of those electronic circuits in which the device appears as a circuit element.

On the basis of the foregoing theory it is certainly correct to state that generally the drain-source current is a function of both the gate-source voltage and the drain-source voltage. Expressed mathematically we have

$$i_D = f(v_{GS}, v_{DS}) \qquad (8\text{-}42)$$

This equation involves the three variables i_D, v_{GS}, and v_{DS}. Accordingly, it is

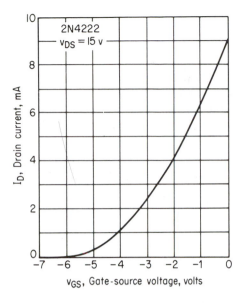

Fig. 8-26. Common-source transfer characteristics.

possible to identify three parameters. We need merely allow any two variables to undergo change subject to the condition that the third variable be kept constant. To illustrate, consider the JFET to be operating initially at point A in the drain characteristics depicted in Fig. 8-25. At this operating point the drain current is 6 mA, the drain-source voltage is approximately 6 V and the gate-source voltage equals -1 V. In the interest of measuring the relative influence of a change in the drain-source voltage and the gate-source voltage on the drain current, consider first that with v_{GS} held momentarily fixed at -1 V the drain-source voltage is increased to 10 V. This changes the operating point from A to B in Fig. 8-25. Note that the 4-V increase in v_{DS} causes a slight increase in drain current. This increase may now be neutralized by allowing the gate-source voltage to increase slightly in the negative direction. It can be estimated from Fig. 8-25 that the new operating point moves from B to C by allowing a negative increase in the gate-source voltage of 0.1 V. These figures thus point out that the change in the gate-source voltage is forty times more effective than a change in the drain-source voltage. This ratio is called the *voltage amplification factor* of the JFET. It is readily computed for incremental changes from

$$\mu = -\left(\frac{\Delta v_{DS}}{\Delta v_{GS}}\right)_{i_D \text{ constant}} \tag{8-43}$$

A little thought about this calculation procedure should make it plain that the specific value of μ depends not only upon the size of the changes used but also upon the particular initial operating point. For example, if A in Fig. 8-25 were selected initially at the ordinate line corresponding to a drain-source voltage of 15 V, the value of the amplification factor would be found to be substantially greater than 40. In this connection therefore a more precise definition of the amplification factor is one that employs differentials rather than increments. Thus

$$\mu \equiv -\left(\frac{\partial v_{DS}}{\partial v_{GS}}\right)_{i_D \text{ constant}}$$ (8-44)

The minus sign appears because the change in the quantities involved must move in opposite directions to maintain constant drain current. Since the JFET is used in the flat portions of the drain characteristics, high amplification factors of the order of several hundred are not uncommon.

A second parameter of the JFET can be identified by assuming operation to take place at constant gate-source voltage. The variables are then v_{DS} and i_D. Starting at the same operating point A of Fig. 8-25, it is clear that an incremental change of $+4$ V in drain-source voltage brings about an increase of approximately 0.20 mA. Accordingly, we can identify a resistance parameter defined in incremental terms by

$$r_d = \left(\frac{\Delta v_{DS}}{\Delta i_D}\right)_{v_{GS} \text{ constant}}$$ (8-45)

This quantity is called the *incremental drain (or output) resistance*. For line AB of the drain characteristic of Fig. 8-25, this quantity has a value of approximately 20 kΩ. However, if operation were taking place beyond B, the quantity could easily take on values of several hundred thousand ohms. Typical values range anywhere from 0.1 to 1.0 MΩ.

Again, a more precise definition is obtained by expressing Eq. (8-25) in terms of differentials. Thus

$$r_d \equiv \left(\frac{\partial v_{DS}}{\partial i_D}\right)_{v_{GS} \text{ constant}}$$ (8-46)

Note that this quantity at any operating point is conveniently identified in Fig. 8-25 as the reciprocal of the slope of the drain characteristic at fixed values of v_{GS}.

The third parameter of the JFET relates the drain current to the gate-source voltage at fixed drain-source voltage. Returning to Fig. 8-25, assume the operating point now to be B corresponding to which $V_{DS} = 10$ V, $I_{DS} = 6.2$ mA, and $V_{DS} = -1$ V. Let the gate-source voltage increase negatively to -2 V at a constant V_{DS} of 10 V. The new drain current decreases to 4.2 mA. Accordingly, there occurs a change of 2 mA/gate-source volts. Since this quantity has the units of inverse ohms and since it relates an output quantity (drain current) to an input quantity (gate-source voltage), it is called the *mutual or transfer conductance* or *transconductance* for short. It is also frequently identified in manufacturers' literature as the *forward admittance*. In terms of incremental quantities we can write

$$g_m = \left(\frac{\Delta i_D}{\Delta v_{GS}}\right)_{v_{DS} \text{ constant}}$$ (8-47)

Typical values range anywhere from 0.1 mA/V to 10 mA/V depending upon the operating point and the particular JFET used.

The differential form of Eq. (8-47) is expressed by

$$g_m \equiv \left(\frac{\partial i_D}{\partial v_{GS}}\right)_{v_{DS} \text{ constant}} \tag{8-48}$$

A glance at Fig. 8-26 reveals that this last expression is simply the slope of the transfer characteristic.

The foregoing three parameters of the JFET defined in terms of the partial derivatives of Eqs. (8-44), (8-46), and (8-48) are not independent because they derive from a common expression, i.e. Eq. (8-42). The functional relationship existing between these parameters readily follows by writing the expression for the total differential of Eq. (8-42). Thus

$$\Delta i_D = \frac{\partial i_D}{\partial v_{GS}} \Delta v_{GS} + \frac{\partial i_D}{\partial v_{DS}} \Delta v_{DS} \tag{8-49}$$

Inserting Eqs. (8-46) and (8-48) yields

$$\Delta i_D = g_m \Delta v_{GS} + \frac{1}{r_d} \Delta v_{DS} \tag{8-50}$$

For convenience, if the condition is now imposed that Δv_{GS} and Δv_{DS} are so manipulated as to yield a Δi_D equal to zero, Eq. (8-50) may then be written as

$$0 = g_m \Delta v_{GS} + \frac{1}{r_d} \Delta v_{DS} \tag{8-51}$$

or

$$g_m r_d = -\left(\frac{\Delta v_{DS}}{\Delta v_{GS}}\right)_{i_D \text{ constant}} \tag{8-52}$$

Introducing Eq. (8-44) then leads to the simple result

$$\mu = g_m r_d \tag{8-53}$$

This last expression states that for any operating point on the drain characteristics the product of the transconductance and the dynamic drain (or output) resistance is equal to the voltage amplification factor. It is instructive to note the similarity of Eq. (8-53), which applies to a voltage sensitive device such as the JFET, with Eq. (8-39), which applies to a current sensitive device such as the conventional transistor.

8-5 THE INSULATED-GATE FET (OR MOSFET)

Recently a new version of the JFET has been developed which shows greater promise of achieving much wider commercial acceptance. A much larger input resistance (about 100 times greater) and bigger transconductances are two of the reasons for this promise. A schematic diagram illustrating the composition of this device is depicted in Fig. 8-27. It is called the *insulated-gate field-effect transistor* or IGFET.

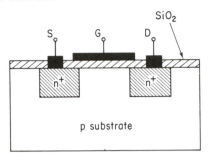

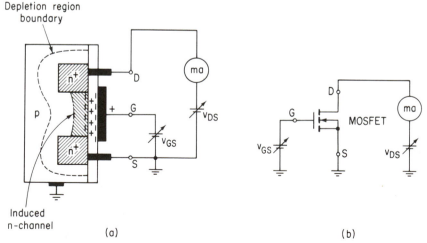

Fig. 8-27 Cross-sectional view of an insulated-gate FET illustrating composition.

The body of the IGFET consists of a lightly doped p substrate. Next, by means of masking and etching techniques, two highly doped isolated n regions are formed. These are called n^+ regions to emphasize the very large n impurity content. This combination is then capped with a layer of silicon dioxide insulation by placing the substrate in a hot oven supplied with an oxygen environment. Finally, aluminum is deposited in the places indicated to establish the three terminals—D (drain), S (source), and G (gate). It is helpful to note that by this construction the metal gate terminal is insulated from the p and n^+ regions of the semiconductor by the silicon dioxide. For this reason this device is alternatively referred to as a metal-oxide-semiconductor FET or, in terms of its acronym, the MOSFET.

Figure 8-28(a) depicts the manner in which the MOSFET is wired in order to obtain the drain characteristics that appear in Fig. P8-13. The symbolic representation of the MOSFET appears in Fig. 8-28(b). It is important to note that

Fig. 8-28 (a) Diagram showing how the MOSFET is wired to obtain the drain characteristics depicted in Fig. P8-13; (b) symbolic representation.

unlike the JFET the MOSFET is operated at positive gate-source voltages. Because of the lightly doped character of the substrate and the heavy dosage of n impurities in the n^+ sections, a widely spread depletion region is created that completely surrounds the n^+ sections, as shown in Fig. 8-28(a). For the moment, assume that the gate-source voltage is shorted and let us recall what is taking place inside the depletion region. Besides the uncovered immobile charges that exist, electron-hole pairs are constantly being created as a result of thermal agitation of the pure silicon material of the substrate. Consider what happens next as the gate-source voltage is made to assume a positive potential. Immediately, this positive potential on the gate metal brings about a polarization of charges within the silicon dioxide insulator as illustrated in Fig. 8-28(a). The positive charges take on a position of alignment along the inner surface of the insulator. Consequently, they attract the electrons in the depletion region to the area of the insulator, thereby creating an induced n-channel that bridges the two n^+ sections of the MOSFET. If a positive drain voltage is now applied between the D and S terminals, n-channel electrons provide a conducting path between the two n^+ sections. This situation now resembles that of an n-channel JFET.

The magnitude of the gate-source voltage essentially determines how many thermally generated electrons from the depletion region will participate in the current flow from D to S. Hence drain current increases rapidly at first with increasing drain-source voltage as depicted in Fig. P8-13. But a point of saturation is soon reached beyond which only small increases of drain current accompany large increases in drain-source voltage. To effect further appreciable increases in drain current, it becomes necessary to increase the gate-source voltage in order that additional electrons be drawn from the depletion region into the induced channel to partake in the current flow from drain to source.

The process of increasing the participating charge carriers in the induced channel between the two highly doped n^+ sections is called *enhancement*. In this connection the dashed line that is used in Fig. 8-28(b) to join the drain terminal D to the source terminal S is for the purpose of denoting the normally *off* state of the enhancement-mode FET. It is only on application of a positive voltage to the gate terminal that induced charges are made to exist and thereby to enhance the current flow between the D and S terminals.

The parameters of the MOSFET can be identified and obtained in exactly the same manner as was done for the JFET. Hence Eqs. (8-44), (8-46), (8-48), and (8-56) apply with equal validity.

The DE MOSFET. In a modified version of the MOSFET of Fig. 8-27 a channel is diffused between the source and the drain. Moreover, it is diffused with an impurity which matches that of the drain and source terminals but of lesser density. The situation is illustrated in Fig. 8-29(a). The presence of such a channel causes an increased drain-source current to flow for a zero gate-source voltage compared to that obtained without the diffused channel. This current is identified as I_{DSS} in Fig. 8-30.

A depletion-mode operation of this field-effect transistor is realized by continuing the variation of the gate-source voltage to the negative region. Because

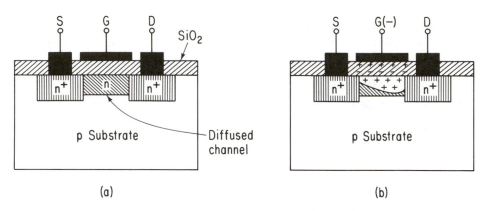

Fig. 8-29 Fabrication features of the DE MOSFET: (a) structure; (b) charge distribution for operation as a depletion-mode MOSFET corresponding to negative gate-source voltages.

of the capacitive effect created by the insulating silicon dioxide layer that separates the gate terminal from the n regions, charges are induced in the manner illustrated in Fig. 8-29(b) whenever the gate-source voltage goes negative. The resultant uncovered positive charges that occur in the diffused n-channel between the drain and source sections reduce the conductivity of the n-channel and thereby reduce the drain-source current. In other words, this increased redistribution of the positive charges in the n-channel brings about a *depletion* of the majority mobile charges that contribute to the flow of current from the drain to the source terminals.

An enhancement-mode operation of this transistor occurs whenever the gate-source voltage is allowed to vary in the positive direction. At positive potentials the gate terminal induces negative charges in the diffused n-channel and thereby increases the conductivity of the channel. This in turn produces an increase in the drain-source current beyond I_{DSS} as depicted in Fig. 8-30.

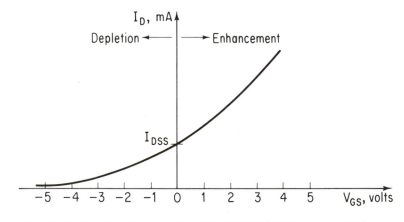

Fig. 8-30 Transfer characteristic of the DE MOSFET at a constant V_{DS}.

Electron Control Devices: Semiconductor Types   Chap. 8

The inherent capability possessed by the structure of Fig. 8-29 to operate effectively at both positive and negative excursions of the gate-source voltage is an advantage that it has over the junction field effect transistor as well as the MOSFET that is fabricated without a diffused n-channel. In an effort to distinguish between these FET types, the unit shown in Fig. 8-29 is called the depletion-enhancement MOSFET or simply the DE MOSFET.

Complementary MOS (CMOS).   A big advantage in the reduction of power dissipation can be achieved by fabricating MOSFETs of the enhancement type in a complementary orientation of the n substrate and the p substrate. The situation is illustrated in Fig. 8-31. Observe that the diffusion pattern is such that the drain of the p MOSFET (D_1) is permanently joined by deposition to the drain of the n MOSFET (D_2). Such CMOS devices are very useful in digital systems for building inverters and logic gates in great numbers (see Sec. 11-3).

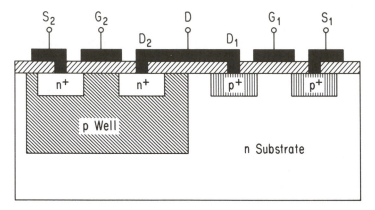

Fig. 8-31   Illustrating the fabrication features of the complementary MOS. Observe that the drain terminals of the MOSFET that operates with a p-induced channel is permanently joined to the complementary MOSFET that operates with an n-induced channel.

8-6 THE SILICON-CONTROLLED RECTIFIER

This treatment begins with a discussion of the four-layered diode. A schematic diagram of the arrangement is shown in Fig. 8-32. The upper terminal is the anode and represents the terminal to which positive potential is applied; the lower terminal is the cathode. When a positive voltage is applied between these terminals, junctions j_1 and j_3 are forward biased and junction j_2 is reverse biased. Consequently, the major part of the anode-cathode voltage (V_{AK}) appears across junction j_2. For small values of V_{AK} very small currents flow so that the resistance of the four-layered diode between terminals A and K is found to be very high (of the order of 100 MΩ). As V_{AK} is increased, the current increases slowly until a so-called *breakover* voltage, V_{BO}, is reached. Once this voltage is exceeded,

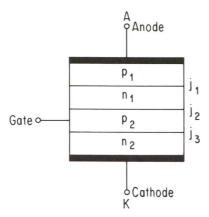

Fig. 8-32 Four-layered diode arranged to accommodate a gate terminal.

the size of the resulting diode current is limited essentially by the parameters of the circuit in which the four-layered diode finds itself.

The abrupt change that occurs in the diode current gives it the characteristic of a switch not unlike that of the conventional two-layered diode. However, what makes the four-layered diode significantly different is the capability to control the level of the anode-cathode voltage at which breakover occurs. This is achieved by introducing a cathode gate voltage between terminals G and K in Fig. 8-32.

The manner in which the increasing forward-biased currents flowing across junction j_3 influence the breakover voltage is depicted in Fig. 8-33. Note that the larger these gate currents are, the smaller are the corresponding breakover voltages. Hence the *firing* of the diode can be made to take place over a wide range of anode-cathode voltages. Once firing occurs, the four-layered diode

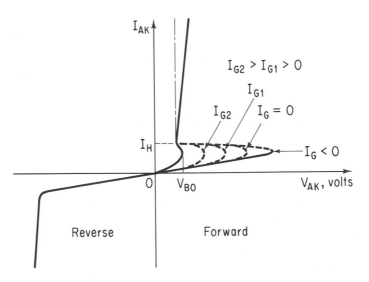

Fig. 8-33 The *i-v* characteristics of the SCR, illustrating the effect of the gate-cathode current, I_G, on the breakover voltage.

remains in the conductive state unless the conditions in the external circuit are such that they command a current flow below I_H, the holding current. In that case the diode switch returns to its *off* state.

When the four-layered *pnpn* diode is equipped with a gate terminal, it is popularly known as a *silicon-controlled rectifier* or SCR for short. The SCR is schematically represented by the symbol shown in Fig. 8-34. It is called a rectifier because it allows current to flow only in one direction—from anode to cathode. It is called silicon because the foregoing control characteristics are found to be peculiar to silicon-type diodes. For example, the control features that are illustrated in Fig. 8-33 are not achievable with germanium-type semiconductors. The SCR finds many useful industrial applications especially in the electronic speed control of a-c and d-c motors.

The *triac* is a modification of the SCR that is arranged to allow the diode switch to be triggered with either positive or negative gate pulses in conformity with whether or not the anode is positive or negative. As a result the triac finds wide use in circuits where the voltage source is a-c. In such circumstances the triac offers the decided advantage of permitting conduction to take place on both the positive and negative halves of the a-c cycle. Figure 8-35 shows the symbol used for this effective, controllable electronic switch.

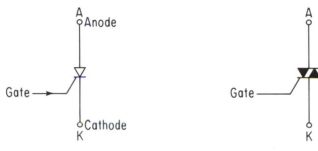

Fig. 8-34 Symbol for the SCR.   Fig. 8-35 Symbol for the triac.

Summary review questions

1. How is electronics defined?
2. Describe in some detail how charge carriers are made available for controlled conduction in a semiconductor device.
3. How do the electrical properties of metal differ from those of a semiconductor?
4. What is a donor impurity? What is an acceptor impurity?
5. Describe the meaning of the following terms: hole, trivalent impurity, pentavalent impurity, conduction band, donor ion, forbidden band, covalent band, acceptor ion.
6. Contrast the effect of temperature on the conducting property of metals and semiconductors.
7. A pentavalent impurity semiconductor is said to be fully ionized at room temperature. Explain what the statement means and why it is so.

8. What are the majority and minority carriers in a pentavalent impurity semiconductor?

9. What are the majority and minority carriers in a trivalent impurity semiconductor?

10. Describe the diffusion process that takes place at the junction of a semiconductor diode and explain the presence of a depletion region.

11. What is diffusion current as it relates to a semiconductor diode, and how is it functionally related to the applied forward-biased voltage of the diode?

12. What is the Boltzmann equation? Describe its role in establishing the ampere-voltage characteristic of the semiconductor diode.

13. What is the potential energy barrier of a *p-n* junction diode? How does it arise, and what is its order of magnitude?

14. Define the law of charge neutrality and outline its role in semiconductor diode theory.

15. What effect does the application of a forward-biased voltage V ($<V_0$, the potential energy barrier) have on the hole density in the *n*-section of a *p-n* junction diode? Moreover, describe the effect that it has simultaneously on the electron density in the *p*-section of the *p-n* junction diode.

16. Explain the role of the recombining electron current as well as the recombining hole current in the operation of the *p-n* junction diode.

17. Sketch the *i-v* characteristic of the *p-n* junction diode for forward bias voltages. Distinguish between the incremental resistance and the apparent resistance of the diode.

18. What is the diffusion capacitance of a semiconductor diode? Explain how it arises.

19. Describe the essential fabrication details of the semiconductor triode. Show the biasing voltages and indicate which sections of the triode are forward biased and which are reverse biased.

20. Sketch a circuit configuration that permits the semiconductor triode to perform as an electronic amplifier. Identify the construction feature of the triode that is responsible for the amplifying characteristic.

21. Define the current transfer ratio of the semiconductor triode as used in the common-base circuit configuration.

22. What is a transistor? Why is this electronic device aptly named?

23. The transistor of Fig. 8-21 is considered to be a current-sensitive electronic device. Explain what this means.

24. Why is the current transfer ratio of the semiconductor triode always less than unity?

25. The voltage gain of the common-base transistor amplifier of Fig. 8-21 involves the incremental resistance of the forward-biased input section of the semiconductor triode. Why?

26. Define the intrinsic transconductance of the transistor. How is this parameter related to the current transfer ratio?

27. How does the field-effect transistor differ from the junction-type version?

28. Describe the mechanism that comes into play in the junction field-effect transistor that permits the output current to respond to changes in the input voltage appearing at the gate-source terminals.

29. The field effect transistor is called a voltage-sensitive electronic control device. Explain why this is the case.

30. Why is the term *depletion mode* an apt description of the operation of the JFET?

31. Define the pinch-off voltage of the JFET. Of what importance is this voltage?

32. Explain why the input resistance of the field-effect transistor is so very high.

33. Name and define the circuit parameters of the JFET. How are they related to each other?

34. Describe the salient construction features of the metal-oxide semiconductor field-effect transistor (MOSFET) and point out how it differs from the JFET.

35. What is the source of the charge carriers that is found in the induced channel of the MOSFET?

36. Describe the kind of operation that takes place in the enhancement-mode MOSFET? How does this differ from the depletion-mode type?

37. How does the DE MOSFET differ in construction from the conventional enhancement-mode or depletion-mode MOSFET? What advantage does it offer?

38. What is a complementary metal-oxide semiconductor (CMOS) integrated circuit? What is its chief advantage?

39. What is a silicon-controlled rectifier (SCR)? How is the breakover voltage of the SCR defined?

40. What is meant by the expression: firing of the SCR?

41. Describe how the triac differs from the SCR.

Problems

GROUP I

8-1. Indicate whether the following statements are true or false. When the statement is false, supply a reason to justify it:

(a) An example of an intrinsic semiconductor is pure germanium at $0°$ K.

(b) An n-type impurity semiconductor is one in which the electron energy level of the impurity is selected close to the valence band of the intrinsic material.

(c) In a p-type impurity semiconductor, minority carriers exist by virtue of the presence of the impurity.

(d) A transistor is equipped with a filament heater supply voltage as is the vacuum-tube device.

8-2. A p-n junction diode has an equilibrium-value charge density in the n region of 2×10^{11} holes/cm^3. The electron density at equilibrium in the p region is 3×10^9 electrons/cm^3. The energy equivalent of ambient temperature is 0.026 V.

(a) Find the density of holes injected into the n region when a forward bias of 0.2 V is applied to the junction diode.

(b) What is the density of injected electrons in part (a)?

(c) Sketch the variation of the injected charge densities in the p and n regions. Show also the corresponding variation of the diffusion currents.

8-3. Explain why Eq. (8-13a) and Eq. (8-13b) correctly represent the total diffusion current of the p-n junction diode.

8-4. A p-n junction diode has a reverse saturation current of 10^{-5} A. The diode is operated in an ambient temperature of $600°$ K. A forward bias of 0.2 V is applied.

(a) Find the apparent diode resistance.

(b) Find the incremental resistance.

8-5. The volt-ampere characteristic of a junction diode is described by the equation $I = 10^{-5} (\varepsilon^{38.5V} - 1)$. The diode is placed in series with an ohmic resistance of 5 Ω and a 1-V d-c source. Determine the diode current.

8-6. A junction diode at room temperature has a reverse saturation current of 5 μA. Find the maximum and minimum currents that flow in this diode when a voltage wave of $0.1 \sin \omega t$ is superimposed on a 0.2-V d-c source and applied to the diode.

8-7. The junction diode of Prob. 8-5 is placed in series with a resistor and a 1-V battery. The resulting current is found to be 100 mA. What is the value of the series resistance?

8-8. A 2N104 transistor is used in the common-base configuration. The average collector characteristics can be found in any RCA semiconductor handbook.
  (a) The emitter currnt is increased from 15 mA to 20 mA at fixed collector-to-base voltage. The corresponding increase in collector current is from 14.6 mA to 19.4 mA. What is the value of the current amplification factor?
  (b) The transistor is operating at $V_{CB} = -5$ V and $I_E = 10$ mA. Find the value of the collector current when the emitter current is decreased by 5 mA and the collector-to-base voltage is doubled.

8-9. A 2N104 transistor is used in the circuitry of Fig. 8-21(a). Also $V_{CC} = 12$ V, $R_L = 400$ Ω, $\alpha = 0.97$, and $I_E = 10$ mA.
  (a) Determine the value of the collector-to-base voltage.
  (b) Find the change in the output voltage developed across R_L when the input voltage changes by 2.6 mV.
  (c) What is the voltage gain in part (b)?

8-10. A 2N104 transistor is used in the circuitry of Fig. 8-21(a) where $R_L = 400$ Ω and $\alpha = 0.97$. For a specific change in v_e the voltage gain is found to be 125. What is the level of the collector current during this point of operation of the transistor?

8-11. Indicate whether the following statements are true or false. When the statement is found to be false, supply a reason to justify it.
  (a) The field-effect transistor is so named because it produces an electric field that completely surrounds the body of the transistor.
  (b) A p-channel JFET is used with the drain terminal connected to the positive side of the drain-source voltage supply.
  (c) The channel of the JFET is more heavily doped than the regions immediately around the gate terminals.
  (d) In a properly biased JFET the depletion region spreads over a larger region as the magnitude of either the drain-source or the gate-source voltage is increased.
  (e) At the pinch-off voltage the drain-source current is cut off.
  (f) Changes in the gate-source voltage primarily influence the depletion region in the vicinity of the drain terminal.
  (g) The JFET is a current sensitive device.
  (h) The ohmic drop in the n-channel of a properly biased JFET in no way influences the boundaries of the depletion regions.
  (i) The conductivity of the depletion region is appreciable.
  (j) Changes in the drain-source supply voltage do not affect the spread of the depletion regions.

8-12. The drain and transfer characteristics of the 2N4222 JFET are depicted in Figs. 8-25 and 8-26. This JFET is operated at a drain-source voltage of 15 V and a gate-source voltage of -2 V. The incremental drain resistance is known to be 120 kΩ.

Find the value of the forward admittance and amplification factor as they apply to the specified operating point.

8-13. The 2N3796 MOSFET shown in Fig. P8-13 is operated at a drain-source voltage of 16 V and a gate source voltage of +2 V.
  (a) Find the drain-source current in milliamperes.
  (b) Determine the incremental drain resistance.
  (c) What is the value of transconductance at the specified operating point?
  (d) Find the value of the amplification factor.

8-14. A MOSFET 2N3797 depicted in Fig. P8-14 is operated at a drain-source voltage of 10 V. Compare the value of the forward admittance prevailing at a gate-source voltage of −3 V with that prevailing at +3 V and comment.

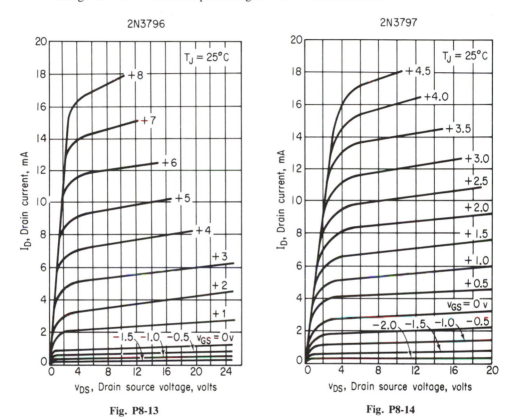

Fig. P8-13   Fig. P8-14

GROUP II

8-15. A *p-n* junction diode has an equilibrium hole density of 2×10^{11} holes/cm^3. The average diffusion length of the holes is 0.1 cm and the length of the *n* section is 0.5 cm. The cross-sectional area at the junction is 0.001 cm^2. The diode is used at room temperature. Find the value of the diffusion capacitance due to holes associated with a change in forward bias from 0.1 V to 0.2 V.

8-16. Devise an appropriate circuit model for a zener diode. Assume that breakdown occurs at $-V_Z$ volts and that the slope of the v-i curve associated with the avalanche breakdown is R_Z.

8-17. A silicon diode has the i-v characteristic shown in Fig. P8-17 and is used in series with a load resistor of 100 Ω. A supply voltage of $v(t) = 208 \sin 377t$ is applied to the series combination.

    (a) Draw the variation of the voltage across the load resistor for one full period of the supply voltage.

    (b) Find the average current that flows through the load resistor.

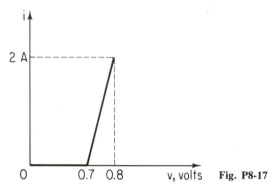

Fig. P8-17

8-18. In the arrangement of Fig. 8-28(a), assume that a 47-kΩ resistor is connected in series with the drain-source supply voltage of 12 V and the combination is placed at the D and S terminals. When the gate-source voltage undergoes a change of 1.5 V, the corresponding change in the drain-source current is found to be 0.2 mA. Find (a) the voltage gain and (b) the current gain.

chapter nine

Semiconductor Electronic Circuits

In this chapter we study the composition, behavior, and performance of the main variety of circuits that use semiconductor diodes and triodes for control and amplifying purposes. These circuits are commonly referred to as *electronic circuits*. Although several chapters could be used to treat the subject matter dealt with here, this procedure is avoided in the interest of stressing that each section heading is an appropriate topic of electronic circuitry.

9-1 GRAPHICAL ANALYSIS OF TRANSISTOR AMPLIFIERS

Common-Emitter Configuration. Since a brief description of the transistor amplifier in the common-base mode is given in Sec. 8-3, in the interest of expanding our knowledge we shall now direct our attention to the *common-emitter* arrangement which is depicted in Fig. 9-1. A glance at the diagram shows that the emitter section of the junction transistor is common to both the input and output circuits. This mode of operation is very frequently used because it offers an appreciable current gain as well as a sizable voltage gain.

How is a current gain exceeding unity achievable with the common-emitter circuit? To provide the answer to this question, let us first write Kirchhoff's current law for the transistor. In dealing with the transistor it is customary to consider the currents as positive when they flow into the terminals, as illustrated in Fig. 9-1. Thus

$$I_E + I_B + I_C = 0 \qquad (9\text{-}1)$$

This equation applies irrespective of the circuit configuration used. Moreover, for the *n-p-n* transistor shown the collector current at a quiescent condition may be written as

$$I_C = I_{CO} - \alpha I_E \qquad (9\text{-}2)$$

where I_{CO} is the reverse saturation current which exists in the reverse-biased section of the transistor and α is the common-base current transfer ratio. The minus sign is used for the I_E term because in this case the conventional current flow in the input forward-biased section is *out* of the emitter terminal and not *into* the terminal, which is the assumed positive direction. When a number is introduced for I_E in Eq. (9-2), it must be negative, thus yielding the algebraic sum of the two terms involved. Equation (9-2) states that of the total number of electrons injected into the transistor at the emitter terminal constituting the current, $-I_E$, the portion arriving at the collector terminal is $-\alpha I_E$. Keep in mind that α lies between 0.9 and 1.0.

The current transfer ratio for a common-base transistor can be found directly from Eq. (9-2) because it involves explicitly the input current, which is I_E, and the current in the output circuit I_C. However, note that for the common-emitter circuit under study, the input current is now I_B. Therefore, to determine the relationship existing between the output and input currents for the common-emitter mode, it is necessary first to eliminate I_E from Eq. (9-2) in terms of I_B and I_C. This is readily accomplished by means of Eq. (9-1). Thus Eq. (9-2) becomes

$$I_C = I_{CO} + \alpha(I_B + I_C) \qquad (9\text{-}3)$$

Solving for I_C yields

$$I_C = \frac{I_{CO}}{1 - \alpha} + \frac{\alpha}{1 - \alpha} I_B \qquad (9\text{-}4)$$

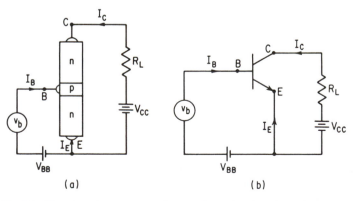

Fig. 9-1 Transistor in common-emitter configuration: (a) schematic diagram; (b) circuit diagram.

Upon considering I_{CO} fixed and subjecting the input current to an incremental change ΔI_B, the corresponding change in the output current is found to be

$$\Delta I_C = \frac{\alpha}{1 - \alpha} \Delta I_B \tag{9-5}$$

Consequently, the *common-emitter current amplification factor, β*, becomes

$$\boxed{\beta \equiv \frac{\Delta I_C}{\Delta I_B}\bigg|_{V_{CE} \text{ constant}} = \frac{\alpha}{1 - \alpha}} \tag{9-6}$$

where α denotes the common-base current transfer ratio. Typical values of β range from 20 to 80 depending upon α. Thus, if α equals 0.98, it follows that β is 49.

The common-emitter characteristics of the 2N241 transistor are depicted in Fig. 9-2. Unlike the common-base characteristics of Fig. 8-22, these curves are not horizontal but rather exhibit a positive slope. This comes about because the slight changes which occur in α with changes of the collector-to-emitter voltage appear in β with much greater sensitivity. Inspection of Eq. (9-6) makes this apparent since the denominator is a number appreciably less than unity.

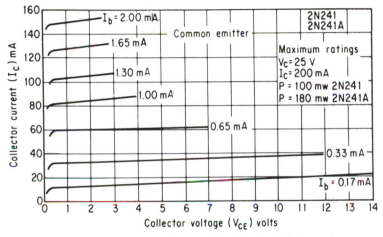

Fig. 9-2 Common-emitter characteristics for the 2N241 transistor.

Before proceeding with the analysis of the performance, it is worthwhile first to become familiar with the symbols to be used in connection with electronic circuits involving transistors. These are given in Table 9-1 for ready reference.

The circuitry for a common-emitter amplifier is depicted in Fig. 9-3. Use is made of an *n-p-n* 2N377 transistor.

How is a circuit such as the one depicted in Fig. 9-3 analyzed? Or, more specifically, how does one find the collector current i_c corresponding to a specified value of base current I_B? One step towards arriving at a solution would be to write Kirchhoff's voltage law for the output circuit. Thus

$$V_{CC} = v_C + i_C R_L \tag{9-7}$$

TABLE 9-1 TRANSISTOR SYMBOLS

	Base terminal	Emitter terminal	Collector terminal
Current quantities:			
Instantaneous total value	i_B	i_E	i_C
Quiescent value	I_B	I_E	I_C
Instantaneous value of varying component	i_b	i_e	i_c
Effective value of varying component	I_b	I_e	I_c
Voltage quantities:			
Instantaneous total value	v_B	v_E	v_C
Quiescent value[a]	V_B	V_E	V_C
Instantaneous value of varying component	v_b	v_e	v_c
Effective value of varying component	V_b	V_e	V_c
Supply voltage	V_{BB}	V_{EE}	V_{CC}

[a] Measured with respect to a ground reference.

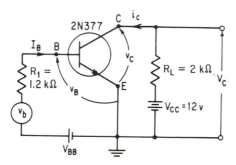

Fig. 9-3 Common-emitter transistor amplifier. Appearing at the output terminals is the varying component V_c of the collector-to-emitter voltage.

From this equation we then get an expression for the desired collector current,

$$i_C = \frac{V_{CC}}{R_L} - \frac{1}{R_L} v_C \qquad (9\text{-}8)$$

Inspection of Eq. (9-8) quickly reveals that the collector-to-emitter voltage v_C is also unknown. We thus have a situation of a single equation involving two unknown quantities: i_C and v_C. Therefore, additional information is needed, and in particular it must relate i_C to v_C. Fortunately, once the semiconductor is specified, the required relationship is available graphically by means of the appropriate common-emitter characteristics which for the 2N377 appears in Fig. 9-4(a).

A convenient and direct solution to the problem can now be obtained by plotting Eq. (9-8) on the same set of axes as are used for the common-emitter characteristics. Clearly, Eq. (9-8) is the equation of a straight line having a vertical intercept of V_{CC}/R_L and a slope of $-1/R_L$. This straight line is called the *d-c*

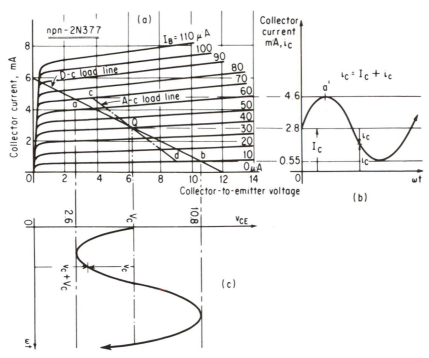

Fig. 9-4 Graphical analysis of transistor amplifier (these curves are drawn for an input signal of $V_b = 36 \times 10^{-3} \sin \omega t$):
(a) common emitter characteristics; (b) variation of collector current;
(c) variation of collector-to-emitter voltage.

load line. The point of intersection of the d-c load line with that curve of the common-emitter characteristics corresponding to the specified value of base current provides the solution for the collector current.

Let us illustrate this solution procedure for the circuit of Fig. 9.3. The manufacturers' published characteristics for the specified transistor appear in Fig. 9-4(a) drawn to scale. The d-c load line is readily plotted by identifying the endpoints of the straight line. Thus from Eq. (9-8) it follows that when i_C is zero, v_C equals 12 V. This locates one point on the load line—the one on the horizontal axis. Next, when v_C is set equal to zero, the corresponding value of collector current becomes $V_{CC}/R_L = 12/2000 = 6$ mA. This identifies the second point—the one on the vertical axis. The line drawn between these two points yields the d-c load line. Note that the slope is negative and has a value equal to $1/R_L$, all of which is in keeping with Eq. (9-8). If it is now desired to find the value of the collector current for a base current of 40 μA, a glance at Fig. 9-4(a) reveals the answer to be 2.8 mA. This point is identified as Q.

The base current of 40 μA is assumed to be produced by the voltage V_{BB} acting on the resistor R_1 of 1200 Ω in the input circuit of Fig. 9-3. The forward-bias resistance of the *p-n* section of the input circuit is neglected because generally R_1 is many times larger. The signal voltage v_b is assumed to be zero. Consequently, because of the absence of a changing input signal, point Q is called the *quiescent operating* point of the circuit of Fig. 9-3. This is the operating point about which changes occur in the output circuit in response to changes in v_b. It is customary to identify the coordinates of the quiescent point as (I_C, V_C). In our example these quantities take on the values (2.8 mA, 6.2 V). Thus, at the quiescent point it can be said for the circuit of Fig. 9-3 that the potential of the collector terminal is 6.2 corresponding to a collector current flow of 2.8 mA. Finally, note that the location of the quiescent point Q can be moved along the load line by changing the value of V_{BB}.

We next consider how the circuit of Fig. 9-3 behaves in the presence of a sinusoidal variation of $v_b = \sqrt{2}(25.4) \sin \omega t = 36 \sin \omega t$ where the amplitude is expressed in millivolts. The quantity $V_b = 25.4$ mV is the effective value of the specified sinusoidal input voltage. The corresponding change in base current produced by this voltage is plainly

$$i_b = \frac{36}{1.2} \sin \omega t = 30 \sin \omega t \qquad \mu A \qquad (9\text{-}9)$$

where the amplitude here is expressed in microamperes (μA). The collector current and voltage for any value of i_b as called for by Eq. (9-9) can be found in the same manner as was done for the quiescent point. Thus if ωt is assumed to increase from 0 to $\pi/2$, i_B changes from 40 μA to 70 μA, which is consistent with

$$i_B = I_B + i_b \qquad (9\text{-}10)$$

Equation (9-10) describes the total instantaneous base current (i_B) in terms of the quiescent value (I_B) plus the instantaneous values of the varying component (i_b). In the plot of Fig. 9-4(a) this variation means that the *instantaneous operating point* moves along the load line from Q to a. The instantaneous operating point is confined to move along the load line because at every instant Kirchhoff's voltage law in the collector circuit which contains R_L must be satisfied. The value of the collector current corresponding to $i_B = 70 \mu A$, which is point a in Fig. 9-4(a), is read off the ordinate axis as 4.6 mA. A plot of the collector currents corresponding to the instantaneous operating points between Q and a on the d-c load line results in the variation shown in Fig. 9-4(b) between the points marked 2.8 (I_B) and a'. A corresponding plot of the collector voltage is depicted in Fig. 9-4(c). Note that the variation of the varying component of the collector voltage about the quiescent value is a negative sine wave. A comparison of this result with a plot of Eq. (9-9) brings out one of the distinguishing characteristics of an electronic amplifier: *phase inversion*. This means that the sign of varying collector voltage v_c is opposite to that of the varying input base current i_b. Finally, note that as i_b continues its variation through a complete sine wave, corresponding variations take place in the plots of i_c and v_c.

Does it follow that a sinusoidal variation of i_b results in a corresponding sinusoidal variation of the collector current and voltage? The answer depends on how the Q-point is selected and on how large a variation about the Q-point is allowed. If the Q-point is located in that region of the common-emitter characteristics where equal increments in base current are accompanied by equal spacing of the curves of the common-emitter characteristics, then the answer is yes. The selection of Q in Fig. 9-4(a) was made with this in mind. Otherwise, the answer is no.

As indicated in Fig. 9-4(b) the instantaneous value of the varying component of the collector current is denoted by i_c. The amplitude, I_{cm}, of this current is readily found from the relationship

$$I_{cm} = \frac{I_{cmax} - I_{cmin}}{2} \tag{9-11}$$

where I_{cmax} denotes the maximum value of the collector current corresponding to the positive maximum value of i_b, and I_{cmin} denotes the minimum current corresponding to the negative maximum value of i_b. For the example illustrated in Fig. 9-4 this becomes

$$I_{cm} = \frac{4.6 - 0.55}{2} = 2.03 \text{ mA} \tag{9-12}$$

The rms or effective value then follows from

$$I_c = \frac{I_{cm}}{\sqrt{2}} = \frac{2.03}{\sqrt{2}} = 1.43 \text{ mA} \tag{9-13}$$

Inserting Eq. (9-11) into Eq. (9-13) leads to an expression for I_c related directly to the minimum and maximum values of collector current. Thus

$$\boxed{I_c = \frac{I_{cmax} - I_{cmin}}{2\sqrt{2}}} \tag{9-14}$$

In view of the sinusoidal character of the varying component of the collector voltage, a similar analysis leads directly to the following expression for the effective value of the varying component of the collector voltage:

$$\boxed{V_c = \frac{V_{cmax} - V_{cmin}}{2\sqrt{2}}} \tag{9-15}$$

Inserting the values found from Fig. 9-4 yields

$$V_c = \frac{10.8 - 2.6}{2\sqrt{2}} = 2.9 \text{ V rms} \tag{9-16}$$

It is important to understand that this is the effective value of voltage developed across the load resistor in response to the varying component i_b in the input circuit. It should be apparent from Fig. 9-4 that if i_b were zero, V_c would also be zero.

We are now at a point in the analysis where we can compute the voltage gain of the amplifier. The *voltage gain* is defined as the ratio of the rms value of the varying component of the collector voltage to the rms value of the varying component of the input voltage. Expressed mathematically,

$$A_v \equiv \frac{V_c}{V_b}$$

(9-17)

where A_v = voltage gain
V_c = rms value of v_c
V_b = rms value of v_b
For our example,

$$A_v = \frac{V_c}{V_b} = -\frac{2.9}{0.0254} = -114$$

(9-18)

The minus sign signifies the phase reversal between input and output signals.

We can proceed in a similar fashion to find the current gain. The effective value of the varying component of the base current is found to be

$$I_b = \frac{I_{b\max}}{\sqrt{2}} = \frac{30}{\sqrt{2}} = 21.2 \ \mu A$$

(9-19)

The *current gain* is then obtained as

$$A_c = \frac{I_c}{I_b} = \frac{1.43 \times 10^{-3}}{21.2 \times 10^{-6}} = 68.5$$

(9-20)

The listed current amplification factor β for this transistor is 70. In the presence of a finite load resistor the current gain is always less than the current amplification factor.

The average value of a-c power that is delivered to the load resistor as a result of the varying signal at the input circuit is readily found to be

$$P_{\text{a-c}} = V_c I_c = 2.9(1.43)10^{-3}$$

$$= 0.0041 \ W = 4.1 \ mW$$

(9-21)

The power gain G of the transistor amplifier is the ratio of the average a-c power delivered to the load resistor to the average a-c power supplied by the signal source. Thus,

$$G = \frac{V_c I_c}{V_b I_b} = 114(68.5) = 7809$$

(9-22)

Note that the power gain for the common-emitter configuration is considerably greater than for the common-base mode.

Since the transistor is a device that transfers a signal current from a low-resistance circuit to a high-resistance circuit and thereby brings about a voltage (and power) amplification, the voltage gain can also be computed by using an expression analogous to Eq. (7-34). A modification is necessary. The current gain A_c for the common-emitter circuit must replace the common-base current transfer ratio α. Accordingly, the alternative expression for the voltage gain becomes

$$A_v = -\frac{R_L}{R_1} A_c$$

(9-23)

where R_1 refers to the total resistance of the input circuit. Inserting the numbers that apply for the circuit of Fig. 9-3 yields

$$A_v = -\frac{2000}{1200}(68.5) = -114 \qquad (9\text{-}24)$$

which agrees with Eq. (9-18). It should be noted, however, that if care is not exercised in using the graphical results leading to the determination of A_c and A_v discrepancies will arise in these alternative computations of A_v.

Practical Voltage-Amplifier Configuration: A-C Load Line. Practical considerations dictate the modification of the circuit of Fig. 9-3 to that shown in Fig. 9-5. An important change in this circuit is the insertion of the capacitor C_C between the collector load resistor at the collector and the input to a succeeding stage. The resistor R_2 represents the input resistance of a succeeding stage of amplification. By the proper selection of C_C it can be made to pass the varying component of the collector voltage v_c practically undiminished while completely blocking out the d-c component. To appreciate the importance of blocking out the d-c component, keep in mind that the a-c output of the circuit of Fig. 9-5 often serves as the input signal to a succeeding stage of amplification. If C_C were not used and if it were desired to operate the succeeding stage at an advantageous Q-point, it would be necessary to use a V_{BB} supply for the following stage which would be large as well as inconvenient. Another reason for including C_C is that it prevents changes in the collector supply voltage V_{CC} from being interpreted by the succeeding stage as variations brought about by an input signal change. Usually, these changes occur very slowly so that C_C effectively blocks them out entirely. It is for this reason that the capacitor-coupled amplifier is said to be more stable than the direct coupled amplifier (i.e., C_C omitted).

The presence of the coupling capacitor changes the a-c amplifier performance somewhat. As the varying component i_c of the collector current i_C leaves the emitter and flows toward terminal n, it no longer sees the single path through R_L in Fig. 9-5 but rather sees two paths. One is through R_L to C; the second is through R_2 and C_C to C. Of course the d-c component (i.e. the quiescent current I_C) continues to flow through R_L to C because C_C presents an open circuit to

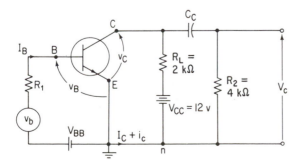

Fig. 9-5 Common-emitter transistor amplifier with capacitor coupling to a succeeding stage.

direct current. In view of the presence of the two paths for i_c, it follows that the effective load resistance seen by i_c changes from R_L to R_L in parallel with R_2. Accordingly, in the amplifier circuit of Fig. 9-5 the equivalent load resistance is

$$R'_L = \frac{R_L R_2}{R_L + R_2} \tag{9-25}$$

When $R_2 \gg R_L$ very little effect on a-c amplifier performance results. Often, however, R_2 and R_L are comparable so that adjustments become necessary. To illustrate, let us compute the a-c performance of the amplifier of Fig. 9-5 for the value of R_2 indicated. By means of Eq. (9-25) the equivalent load resistance becomes

$$R'_L = \frac{R_L R_2}{R_L + R_2} = \frac{2000(4000)}{6000} = 1333 \ \Omega \tag{9-26}$$

This value is sufficiently different from 2000 Ω to require the construction of a new load line. Since the quiescent point has not been changed, the new load line must also pass through point Q in Fig. 9-4(a). A convenient way of plotting the new load line is to write

$$R'_L = \frac{\Delta v_C}{\Delta i_C} \tag{9-27}$$

Then arbitrarily choose Δv_C as some convenient value measured from the Q-point. For example, choose $\Delta v_C = 1.8$ V taken to the right of the Q-point. Equation (9-27) then yields the corresponding change in collector current of

$$\Delta i_C = \frac{\Delta v_C}{R'_L} = \frac{1.8}{1.333} 10^{-3} = 1.35 \ \text{mA} \tag{9-28}$$

Thus by moving from the Q-point by 1.8 V to the right along the horizontal and then moving downward along the vertical by an amount equal to 1.35 mA a second point on the new load line is established. The line drawn between this point and the Q-point locates the new load line which is called the *a-c load line* for obvious reasons. In Fig. 9-4(a) the a-c load line is marked cQd. This line is sometimes called the *dynamic load line*.

As the base current varies in the manner described by Eq. (9-9), the instantaneous operating point for the circuit of Fig. 9-5 is now confined to the a-c load line. This leads to altered values of maximum and minimum collector voltages and current, thus resulting in a change in a-c performance. From Fig. 9-4(a) we have

At point c: $I_{cmax} = 4.7$ mA   and $V_{cmin} = 3.6$ V

and

At point d: $I_{cmin} = 0.5$ mA   and $V_{cmax} = 9$ V

Then by Eq. (9-14) the new value of the rms collector current is

$$I_c = \frac{4.7 - 0.5}{2\sqrt{2}} = 1.48 \ \text{mA} \tag{9-29}$$

Hence the new current gain becomes

$$A_c = \frac{1.48 \times 10^{-3}}{21.2 \times 10^{-6}} = 70.2 \qquad (9\text{-}30)$$

A comparison with Eq. (9-20) shows little difference occurs in this quantity. This is largely attributable to the horizontal nature of the collector volt-ampere characteristics. The corresponding rms value of the collector voltage is similarly found to be

$$V_c = \frac{9 - 3.6}{2\sqrt{2}} = 1.92 \text{ V} \qquad (9\text{-}31)$$

This leads to a voltage gain of

$$A_v = \frac{V_c}{V_b} = \frac{-1.92}{0.0254} = -75.6 \qquad (9\text{-}32)$$

The minus sign denotes a phase reversal between input and output signals. Clearly, the voltage gain undergoes an appreciable reduction brought about by the shunting action of R_2. The related power gain then becomes

$$G = \left|\frac{V_c}{V_b}\right|\left|\frac{I_c}{I_b}\right| = 75.6(70.2) = 5307 \qquad (9\text{-}33)$$

As expected, this quantity also diminishes by a good deal by operation on the a-c load line. However, we must not overlook the fact that this still represents a very healthy increase in signal power level.

JFET Amplifier Configuration. In Chapter 8 it is pointed out that the field-effect transistor is a voltage-sensitive device. Accordingly, control of the output current is achieved by variations that take place in the *gate-source voltage*. In the interest of comparing the performance of an amplifier circuit employing a JFET to those already considered in this section, we focus attention on the simple circuit appearing in Fig. 9-6. This amplifier employs the silicon *n*-channel field-effect transistor 2N4220 whose drain characteristics appear in Fig. 9-7. The resistor R_g provides a path for maintaining a negative gate-source voltage even when the varying component of the gate-source voltage is removed.

For the specified value of drain load resistor R_d the load line assumes the location depicted in Fig. 9-7. The *Q*-point is determined by the intersection of

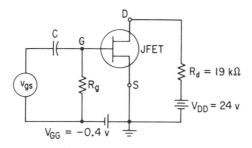

Fig. 9-6 Common-source amplifier employing the JFET 2N4220.

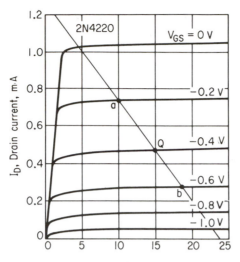

Fig. 9-7 Drain characteristic for the JFET amplifier of Fig. 9-7.

v_{DS}, Drain-source voltage, volts

this load line with the drain characteristic corresponding to $V_{GS} = V_{GG} = -0.4$ volt. Assume next that the input signal voltage varies in accordance with $v_{gs} = 0.2 \sin \omega t$. The total instantaneous gate-source voltage can then be written as

$$v_{GS} = V_{GG} + v_{gs} = -0.4 + 0.2 \sin \omega t \qquad (9\text{-}34)$$

Equation (9-34) states that as the varying component of the gate-source voltage swings through a complete cycle, operation on the load line goes from Q to a, back to Q, then to b, and back to Q again. From Fig. 9-7 it follows that the collector voltage and current at point a are

$$V_{d\min} = 10 \text{ V} \qquad \text{and} \qquad I_{d\max} = 0.74 \text{ mA}$$

while at point b there results

$$V_{d\max} = 18.8 \text{ V} \qquad \text{and} \qquad I_{d\min} = 0.27 \text{ mA}$$

Accordingly, we can compute the rms value of the varying component of the drain voltage as

$$V_d = \frac{V_{d\max} - V_{d\min}}{2\sqrt{2}} = \frac{18.8 - 10}{2\sqrt{2}} = 3.11 \text{ V} \qquad (9\text{-}35)$$

Similarly, the effective value of the varying component of the drain current is found to be

$$I_d = \frac{I_{d\max} - I_{d\min}}{2\sqrt{2}} = \frac{0.74 - 0.27}{2\sqrt{2}} = 0.166 \text{ mA} \qquad (9\text{-}36)$$

The rms value of the varying component of the gate-source voltage is known to be

$$V_{gs} = \frac{0.2}{\sqrt{2}} = \frac{\sqrt{2}}{10} \qquad (9\text{-}37)$$

Therefore, the voltage gain is

$$A_v = \frac{V_d}{V_{gs}} = \frac{-3.11}{\frac{\sqrt{2}}{10}} = -22 \qquad (9\text{-}38)$$

and the corresponding a-c power is

$$P_{\text{a-c}} = V_d I_d = 3.11(0.166) = 0.516\,\text{mW} \qquad (9\text{-}39)$$

It is instructive to note that the voltage gain of the JFET amplifier is lower than those encountered for the common-emitter amplifier configurations of the conventional transistors. This is typical of such amplifiers. Moreover, the field-effect transistor amplifiers also produce more distortion than the conventional transistor amplifiers. An examination of the characteristics of Fig. 9-7 makes this readily apparent. Note that equal increments in gate-source voltage do not produce equal increments in drain voltage and current. Or, stated graphically, distance Qa is not equal to Qb. For this reason the JFET and MOSFET amplifiers are frequently used in situations where the input signal variations produce a small excursion about the quiescent point. However, one should not lose sight of the fact that this amplifier type also has several outstanding advantages. Because the input circuit is reversed-biased, the JFET amplifier draws negligibly small current. The input resistance is thereby extremely high. It can control power in the output circuit with negligibly small input power, thus yielding extremely high power gains.

9-2 LINEAR EQUIVALENT CIRCUITS

It is often advantageous, for reasons of accuracy as well as greater ease of analysis, to replace an engineering device by an equivalent circuit. Once this is done, all the tools and techniques of circuit theory can then be used in the analysis, design, and computation of performance of the device and the circuit or system of which it may be a part. This procedure is applied generally throughout the study of electrical engineering. In this section we apply it to the transistor.

Although the performance of a transistor amplifier can be conveniently determined by a graphical analysis for large variations of the input signal, a little thought should make it apparent that this approach leads to appreciable inaccuracies when the input signal varies over a very small range. In the transistor amplifier of Fig. 9-5 a variation of a few microamperes in base current cannot be treated graphically with any acceptable degree of accuracy. It is precisely in such instances that the equivalent-circuit representation of the device is useful. In essence the equivalent circuit is a means of replacing the transistor by a suitable arrangement of circuit parameters and voltage and/or current sources which are consistent with the characteristics of the original device. Moreover, since we are concerned with calculating the changes which occur about the operating point, the parameters which are selected for the equivalent circuit are those which apply for the given operating point. For small variations of the input signal insignificant changes take place in these parameters, so that even greater confidence may be put in the calculated results. In other words, as the input signal variations become smaller the results computed graphically become less accurate while those computed by the equivalent circuit become more accurate.

Bipolar Junction Transistor (BJT) Equivalent Circuit in Terms of h Parameters. There are several types of equivalent circuits that may be used to represent the transistor. However, we treat here only that equivalent circuit which today has received the widest acceptance: the *h*-parameter equivalent circuit. The reasons for this preference are twofold: One, a relatively simple circuit results which has the very desirable feature that the input and output circuits are completely isolated for the normal frequency range. Two, each of the parameters appearing in the equivalent circuit is readily found either from the volt-ampere characteristic of the input circuit or the volt-ampere characteristic of the output circuit. A departure for this preference occurs only when the transistor amplifier is analyzed at very high frequencies. Under this condition the transition and diffusion capacitances of the transistor became important and the easiest way to handle these quantities is to employ the hybrid-π equivalent circuit. More is said about this matter in Sec. 9-5.

Because of the vast importance of the common-emitter configuration, the equivalent circuit is derived specifically for this mode of operation. The procedure holds with equal validity, however, for the other operating modes—common collector and common base. It is important to keep in mind that we are exclusively concerned in this equivalent circuit derivation with a description of changes that take place about the quiescent point. There are four variables involved in describing these changes. Two on the input side, v_B and i_B; and two on the output side, v_C and i_C. Since the transistor is a current sensitive device on the input side, it is reasonable to choose i_B as one of the independent variables. Moreover, on the output side, it is really the voltage developed across the load resistor that is most important. Hence, v_C is arbitrarily chosen as the second independent variable. The remaining two variables—v_B and i_C—are then considered to be the dependent variables. On this basis we may accordingly write the following functional relationships

$$v_B = f_1(i_B, v_C) \tag{9-40}$$

and

$$i_C = f_2(i_B, v_C) \tag{9-41}$$

From calculus the corresponding expressions for the total differentials become

$$\Delta v_B = \left(\frac{\partial v_B}{\partial i_B}\right)_{V_C} \Delta i_B + \left(\frac{\partial v_B}{\partial v_C}\right)_{I_B} \Delta v_C \tag{9-42}$$

$$\Delta i_C = \left(\frac{\partial i_C}{\partial i_B}\right)_{V_C} \Delta i_B + \left(\frac{\partial i_C}{\partial v_C}\right)_{I_B} \Delta v_C \tag{9-43}$$

where the subscripts of the partial derivatives employing uppercase letters denotes that these quantities are held at fixed values. Now recalling that the changes in total instantaneous quantities are the same as the changes in the varying component of these quantities we can write as equivalencies

$$\Delta v_B = v_b \tag{9-44}$$

$$\Delta i_B = i_b \tag{9-45}$$

$$\Delta v_C = v_c \tag{9-46}$$

$$\Delta i_C = i_c \tag{9-47}$$

Inserting the last four equations into Eqs. (9-42) and (9-43) allows the latter to be written as

$$v_b = \left(\frac{\partial v_B}{\partial i_B}\right)_{V_C} i_b + \left(\frac{\partial v_B}{\partial v_C}\right)_{I_B} v_c \tag{9-48}$$

$$i_c = \left(\frac{\partial i_C}{\partial i_B}\right)_{V_C} i_b + \left(\frac{\partial i_C}{\partial v_C}\right)_{I_B} v_c \tag{9-49}$$

To further simplify these equations, let us take a closer look at each of the partial derivatives.

The coefficient of i_b in Eq. (9-48) clearly has the unit of resistance because it is the ratio of the base voltage and the base current. More simply, it is the slope of the input characteristic. See Fig. 8-23. It is denoted by h_{ie}. Thus

$$h_{ie} \equiv \left(\frac{\partial v_B}{\partial i_B}\right)_{V_C} = \begin{array}{l}\text{input resistance in common-emitter} \\ \text{mode for constant collector voltage}\end{array} \tag{9-50}$$

Typical values for amplifier applications range from a few hundred ohms to several kilohms.

The coefficient of v_c in Eq. (9-48) is a numeric. It represents that fraction of the collector (or output) voltage appearing at the input when the input is open circuited. Essentially it is a "reverse amplification factor." An appreciation of how this quantity comes about and its order of magnitude can be had by referring to Fig. 9-1. In normal operation the p-n section from base to emitter is forward-biased and hence of very low resistance. The n-p section from collector to base is reverse-biased and so is characterized by extremely high resistance. Accordingly, for a given collector-to-emitter voltage there will be a portion appearing across the forward-biased p-n section between base and emitter terminals but it will be only a very small fraction. Typical values are of the order of 10^{-4}. In most computations involving amplifier performance little error results from neglecting this term entirely. However, it is included here for completeness and is denoted by h_{re}. Thus,

$$h_{re} \equiv \left(\frac{\partial v_B}{\partial v_c}\right)_{I_B} = \text{reverse amplification factor} \tag{9-51}$$

In Eq. (9-49) the coefficient of i_b is also a numeric. Now the ratio involves changes in current. Essentially, this coefficient describes the change that occurs in the collector current per unit change in base current at constant collector-emitter voltage. But this is precisely the definition of the common-emitter current amplification factor β. Refer to Eq. (9-6). In terms of the generally accepted hybrid notation this ratio is denoted as

$$h_{fe} \equiv \left(\frac{\partial i_c}{\partial i_b} \right)_{V_C} = \beta = \text{forward current transfer ratio} \qquad (9\text{-}52)$$

Typical values range from 20 to 80.

The coefficient of v_c in Eq. (9-49) carries the unit of inverse ohms or conductance. This quantity relates to the output circuit since it represents the change in collector current produced by a unit change in collector voltage at constant base current. The condition of fixed base current is equivalent to stating that the input terminals are short circuited. Because of the reversed-biased condition of the transistor in the collector output circuit, very little change in collector current is caused by a given change in collector voltage. Accordingly, this quantity assumes relatively small values and frequently can be neglected in computing amplifier performance without noticeably deteriorating accuracy. Expressed mathematically we have

$$h_{oe} \equiv \left(\frac{\partial i_c}{\partial v_c} \right)_{I_B} = \text{output conductance with input short circuited} \qquad (9\text{-}53)$$

Equations (9-50) to (9-53) identify the four parameters that can be used to completely describe the operation of the transistor in small-signal applications. Two parameters are numerics, one has the unit of resistance, and the other has the unit of conductance. In short, the parameters have *mixed* units. It is for this reason that these quantities are called the *hybrid* parameters or more simply the *h* parameters. The *e* subscript carried by each of these parameters merely serves to remind us that we are dealing with the common-emitter configuration.

The equations that describe the behavior of the transistor can now be written in terms of the *h* parameters. Thus,

$$v_b = h_{ie} i_b + h_{re} v_c \qquad (9\text{-}54)$$

$$i_c = h_{fe} i_b + h_{oe} v_c \qquad (9\text{-}55)$$

The linear equivalent circuit of the transistor immediately follows from a circuit interpretation of Eqs. (9-54) and (9-55). The first equation is a Kirchhoff voltage equation. It states that the varying component of the base voltage is made up of a resistance drop involving the input resistance and the varying component of the base current plus a fractional portion of the varying component of the collector voltage. The circuit interpretation appears as the left side of the diagram of Fig. 9-8. The second equation is a Kirchhoff current equation. It states that

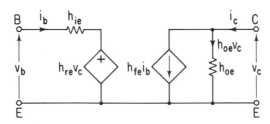

Fig. 9-8 The *h*-parameter equivalent circuit of the transistor in the common-emitter mode.

Semiconductor Electronic Circuits   Chap. 9

the varying component of the collector current is an amplified version of the varying component of the base current plus a current component associated with the varying component of the collector voltage appearing across the output resistance $(1/h_{oe})$. Recall that the positive direction of collector current is defined as entering the collector terminal. Hence the right side of the circuit of Fig. 9-8 depicts i_c entering the C terminal and then splitting into two paths as called for by Eq. (9-55).

An equivalent circuit is a convenient way of describing the internal operation of a device as seen from the externally available terminals. For the transistor these terminals are B, C, and E. Figure 9-8 is a means of representing this operation in terms of familiar elements so that performance can be readily determined by the powerful methods of circuit analysis studied in Chapter 3. Note that no d-c quantities such as V_{BB} and V_{CC} appear in the equivalent circuit because the emphasis is on the *changes* that take place about the Q-point.

Depicted in Fig. 9-9 is an approximate form of the linear model of Fig. 9-8. Here both h_{re} and h_{oe} are set equal to zero. In many applications entirely satisfactory results are obtained by the use of the approximate equivalent circuit.

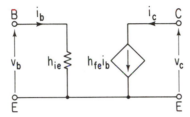

Fig. 9-9 The h-parameter approximate equivalent circuit of the common-emitter transistor.

Determination of h-Parameters. The four h-parameters described in the foregoing can be conveniently found once a set of input and output characteristics are available. These may be obtained either from the manufacturer or experimentally. A typical set of these curves is shown in Fig. 9-10.

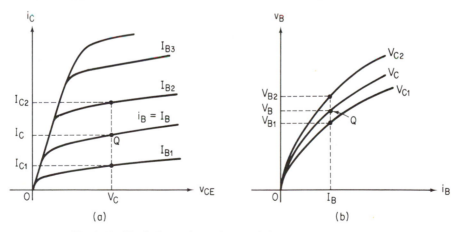

Fig. 9-10 Typical transistor characteristic curves:
(a) the output characteristics from which h_{fe} and h_{oe} are obtained;
(b) the input characteristics from which h_{ie} and h_{re} are obtained.

The output characteristics can be made to yield values for the parameters h_{oe} and h_{fe}. The output conductance h_{oe} is merely the slope of the curve at the quiescent point in Fig. 9-10(a) for fixed base current I_B. A typical variation with collector current is illustrated in Fig. 9-11(b). The current amplification factor h_{fe} can be found at constant collector voltage by measuring the change in collector current $(I_{C2} - I_{C1})$ corresponding to a change in base current from I_{B1} to I_{B2}. Thus

$$h_{fe} = \frac{I_{C2} - I_{C1}}{I_{B2} - I_{B1}} \tag{9-56}$$

It is important in making this calculation that the spreads $(I_{C2} - I_{C1})$ and $(I_{B2} - I_{B1})$ be as small as is consistent with reliable measurements. The parameter h_{fe} varies also with the location of the Q-point. Figure 9-11(a) depicts a typical variation.

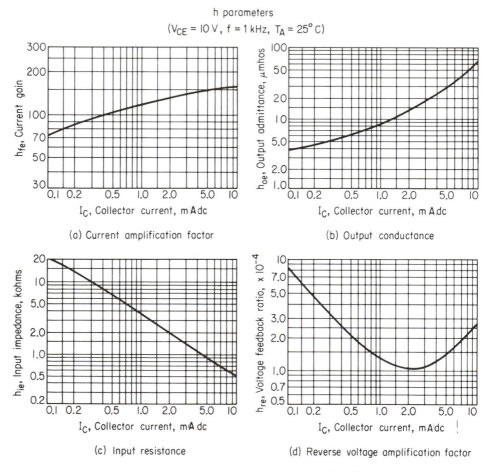

Fig. 9-11 Variations of the four transistor parameters with collector current at fixed collector-emitter voltage. Transistor 2N3903.

Semiconductor Electronic Circuits   Chap. 9

Information about h_{ie} and h_{re} results from the input characteristics of Fig. 9-10(b). The slope of the curve at the Q-point at fixed V_C yields the value of h_{ie} directly. The manner in which this parameter varies with the collector current is shown in Fig. 9-11(c). The reverse voltage amplification factor can be determined from

$$h_{re} = \frac{V_{B2} - V_{B1}}{V_{C2} - V_{C1}} \qquad (9\text{-}57)$$

where the quantities on the right side follow from Fig. 9-10(b). The corresponding variation with the Q-point at fixed collector-emitter voltage appears in Fig. 9-11(d).

Linear Model of the FET. The small-signal equivalent circuit of the field-effect transistor as it appears between its three external terminals is developed next. Since the FET is a voltage sensitive device which draws negligible current on the input side, there are only three variables involved—the total instantaneous drain current i_D, the total instantaneous gate-source voltage v_{GS}, and the total instantaneous drain-source voltage v_D. Consequently, the derivation of the equivalent circuit is based on a single functional relationship, namely,

$$i_D = f(v_{GS}, v_D) \qquad (9\text{-}58)$$

In deriving the equivalent circuit our objective is to replace the FET between terminals G, D, and S by a suitable arrangement of circuit elements. Although our interest at this point is solely with the FET, it is helpful in understanding the development to keep in mind the manner in which the FET is used in a typical amplifier circuit such as the one depicted in Fig. 9-6. The equivalent circuit is useful in analyzing the behavior of such an electronic circuit whenever the input signal undergoes small variations. Again, because of this emphasis on the changes about the quiescent point, all d-c quantities are omitted from the equivalent circuit. It is the a-c performance which is important. Consistent with this theme, therefore, let us assume that a change Δv_{GS} occurs in the gate-source circuit. In general this causes a change in drain voltage Δv_D and in drain current Δi_D. Since these variables are related by Eq. (9-58), we may write

$$\Delta i_D = \left(\frac{\partial i_D}{\partial v_{GS}}\right)_{V_D} \Delta v_{GS} + \left(\frac{\partial i_D}{\partial v_D}\right)_{V_{GS}} \Delta v_D \qquad (9\text{-}59)$$

As a convenience the changes in the total instantaneous quantities can be replaced by the varying components of the same quantities. Accordingly,

$$\Delta i_D \equiv i_d \qquad \Delta v_{GS} = v_{gs} \qquad \Delta v_D = v_d \qquad (9\text{-}60)$$

Inserting Eq. (9-60) into Eq. (9-59) yields

$$i_d = \left(\frac{\partial i_D}{\partial v_{GS}}\right)_{V_D} v_{gs} + \left(\frac{\partial i_D}{\partial v_D}\right)_{V_{GS}} v_d \qquad (9\text{-}61)$$

A further simplification results upon inserting Eqs. (8-46) and (8-48) into Eq. (9-61). This then brings into play two of the three parameters previously defined for the FET. Thus,

$$i_d = g_m v_{gs} + \frac{v_d}{r_d} \qquad (9\text{-}62)$$

Equation (9-62) is meaningful because it describes the functional relationship between the FET parameters and the changing quantities that take place between the gate-source and drain-source terminals. In view of the fact that Eq. (9-62) is an equation relating currents, a circuit interpretation naturally leads to the Norton equivalent circuit. This is shown in Fig. 9-12. Note that i_d is drawn flowing into the D terminal in accordance with the conventional positive direction established for it. The input circuit merely shows the varying component of voltage appearing between the gate and source terminals. No closed path is provided for current flow on the input side since there is none.

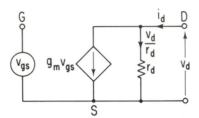

Fig. 9-12   Linear equivalent circuit of the FET, current-source version.

A linear model that emphasizes a voltage source and voltage drops rather than a current source and current paths is also possible. The expression for current of Eq. (9-62) is readily converted to an expression relating voltages by multiplying through by r_d. Thus

$$i_d r_d = g_m r_d v_{gs} + v_d \qquad (9\text{-}63)$$

or

$$i_d r_d = \mu v_{gs} + v_d \qquad (9\text{-}64)$$

where μ is the voltage amplification factor described by Eq. (8-53). Upon rearranging terms, we have

$$v_d = i_d r_d - \mu v_{gs} \qquad (9\text{-}65)$$

The circuit interpretation of this expression is shown in Fig. 9-13. A comparison

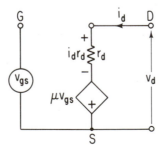

Fig. 9-13   Linear equivalent circuit of the FET, voltage-source version.

with the circuit of Fig. 9-12 should make it evident that the voltage source equivalent circuit could be obtained directly from Fig. 9-12 by applying Thévenin's theorem. Both i_d and v_d in these figures are shown in their defined positive directions.

Equivalent Circuit of FET at Very High Frequencies. When the gate-source signal variations occur at very high frequencies, the equivalent circuit must be modified to take into account the capacitances that exist between the FET terminals. Keep in mind that at each pair of terminals voltages and charges appear. These in turn identify corresponding capacitances. Thus a capacitance C_{gd} exists between the gate and drain terminals, another exists between the gate and source terminals denoted by C_{gs}, and a third, C_{ds}, exists between the drain and source terminals. Usually the drain-source capacitance is neglected entirely. The gate-drain capacitance has a value of the order of C_{ds} but it is not neglected because its value is amplified for a reason that is pointed out in Sec. 9-5. Values of C_{gs} and C_{gd} are measured in picofarads (pF). Accordingly, it is only at very high frequencies (well above the audio range) that the associated reactances $(1/\omega C)$ become small enough to be reckoned with.

The complete versions of the FET equivalent circuits depicting the foregoing capacitances appear in Fig. 9-14. The presence of capacitor C_{gd} indicates that a path exists which connects the output circuit at D to the input circuit at G. This direct connection can cause oscillation difficulties if appropriate precautions are not taken to limit the amount of feedback that occurs between the output and input circuits. These capacitances also result in a decrease in voltage gain when t' ⋅ FET is used as an amplifier.

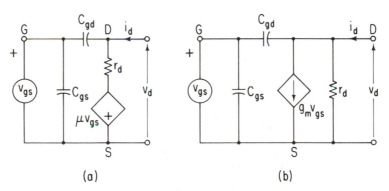

Fig. 9-14 Complete FET equivalent circuit including interterminal capacitances: (a) voltage-source equivalent; (b) current-source equivalent.

9-3 BIASING METHODS FOR TRANSISTOR AMPLIFIERS

In the transistor amplifier circuit of Fig. 9-3 the quiescent operating point was established with the aid of a battery source which was separate and distinct from the battery used in the output circuit. In the interest of simplicity and economy it is desirable that the amplifier circuit have a single source of supply—the one in the output circuit. We focus attention, therefore, in this section on the more commonly used methods of obtaining a fixed bias voltage or current in the input circuit without the use of a second power source.

Biasing Bipolar Junction Transistor Amplifiers. The treatment here is confined to the common-emitter configuration for two reasons. One, it is the most commonly used mode of transistor amplifier, and two, the high current amplification which occurs in this mode leads to a stability problem concerning the Q-point which must be reckoned with when selecting a bias method.

The simplest scheme for fixing the Q-point is to use the circuit shown in Fig. 9-15. A single source voltage V_{CC} is used. The quiescent base current is determined by the proper selection of the base resistor R_B. The closed circuit for this current is N-A-B-E-N. Applying Kirchhoff's voltage law to this circuit yields

$$V_{CC} - I_B R_B - V_{BE} = 0 \tag{9-66}$$

where V_{BE} denotes the voltage drop between the base and emitter terminals. Since V_{BE} is at most several tenths of a volt, it can often be neglected with little error. It then follows from Eq. (9-66) that the required value of base resistance needed to yield the base current which for a specified load resistance establishes the Q-point is given by

$$R_B = \frac{V_{CC}}{I_B} \tag{9-67}$$

Keep in mind that V_{CC} is a known fixed quantity and I_B is chosen at some suitable value. Hence R_B can always be computed directly, and for this reason this procedure is called the *fixed-bias* method.

The circuit shown in Fig. 9-15(b) is a simplified form of Fig. 9-15(a). It is customary in drawing circuits of this kind to indicate only that side of V_{CC} which is connected to the load resistor R_L. It is assumed that the other side of the battery is connected to the ground (chassis) or reference point. The capacitor

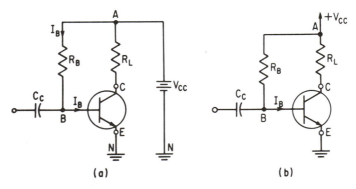

(a) (b)

Fig. 9-15 Common-emitter transistor amplifier with fixed bias resistor R_B: (a) circuit diagram; (b) simplified form.

which appears in series with the base terminal is inserted for the purpose of blocking any d-c component in the signal from passing on through the base terminal.

On the basis of a superficial examination it would appear that the circuit of Fig. 9-15 should perform satisfactorily. However, experience shows that this is not so. The trouble lies in the inability of this circuit to maintain a fixed Q-point in the absence of signal inputs. Basically the reason for this behavior is that the reverse saturation current between the collector and emitter, I_{CO}, is greatly influenced by the temperature. In fact it can be shown that the value of I_{CO} is proportional to the cube of temperature. Thus

$$I_{CO} = f(T^3) \tag{9-68}$$

When this fact is coupled with the current-amplifying feature of the common-emitter mode as reflected in Eq. (9-4), it is easy to understand why there occurs an instability of the Q-point. For convenience Eq. (9-4) is repeated here:

$$I_C = \frac{I_{CO}}{1 - \alpha} + \frac{\alpha}{1 - \alpha} I_B \tag{9-69}$$

As the collector current flows in response to the base current in accordance with Eq. (9-69) heat is produced at the collector terminal, causing the temperature to rise. As the temperature rises, there occurs a rapid increase in I_{CO} as called for by Eq. (9-68). Then, as revealed by Eq. (9-69), this causes a greatly magnified increase in collector current because I_{CO} is divided by a very small number $(1 - \alpha)$. In turn the increased collector current causes a further increase in temperature which further increases the collector current. This behavior continues to that point where I_C is limited by the load resistor. This means that the Q-point has moved higher up along the load line to a position almost on the ordinate axis of the common-emitter characteristics. Any subsequent application of a sinusoidal input signal leads to an output signal that is greatly distorted. This instability of the Q-point is referred to as *thermal runaway* for obvious reasons.

A useful figure of merit which can be used to compare the relative instability of the various bias methods found in transistor amplifiers is the stability factor. It is simply defined as

$$S \equiv \frac{\partial I_C}{\partial I_{CO}} \tag{9-70}$$

Applying Eq. (9-70) to Eq. (9-69) reveals that the stability factor for the fixed bias circuit of Fig. (9-15) is

$$S = \frac{\partial I_C}{\partial I_{CO}} = \frac{1}{1 - \alpha} \tag{9-71}$$

Accordingly, a value of $\alpha = 0.98$ yields a value of $S = 50$. This states that a per-unit change in the reverse saturation current results in a fiftyfold increase in collector current. Experience shows that values of S exceeding 25 result in unsatisfactory performance.

An improvement over the fixed-bias scheme of Fig. 9-15 is the use of the *collector-to-base bias* method depicted in Fig. 9-16. In this arrangement note

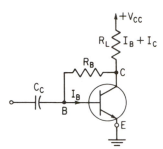

Fig. 9-16 Collector-to-base method of biasing a transistor amplifier.

that the quiescent base current is determined not by V_{CC} but by the collector-to-base voltage. Consequently, any increase in I_C is partially offset by a decreased collector-to-base voltage swing due to the increased voltage drop in the load resistance R_L. The stability factor for this bias circuit is found to be

$$S = \frac{\partial I_C}{\partial I_{CO}} = \frac{1}{1 - \alpha + \alpha\left(\dfrac{R_L}{R_L + R_B}\right)} \tag{9-72}$$

Since the denominator term is increased the improvement in the stability of the Q-point is apparent.

The most frequently used bias method is the one illustrated in Fig. 9-17. It is called the *self-bias* or *emitter-bias* circuit, and it has a stability factor of the order of 10 which yields very satisfactory results. To understand how this circuit brings about an effective stabilization of the Q-point, recall that the desired quiescent collector current I_C is determined by the voltage between the base-emitter terminals V_{BE}, which in turn determines the value of I_B in Eq. (9-69). This equation may be rewritten as

$$I_C = \frac{I_{CO}}{1 - \alpha} + \frac{\alpha}{1 - \alpha} I_0 \varepsilon^{V_{BE}/E_T} \tag{9-73}$$

where I_O is the reverse saturation current of the forward-biased input section. Note too that V_{BE} is obtained as the difference of the voltages appearing across R_2 and R_E. That is

$$V_{BE} = V_{R2} - V_{RE} = V_{R2} - (I_B - I_C)R_E \tag{9-74}$$

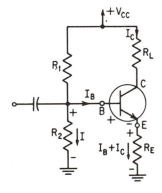

Fig. 9-17 Emitter-bias method of biasing an *n-p-n* transistor amplifier.

With this arrangement any increase in I_C due to a temperature rise causes the voltage drop across the emitter resistor R_E to increase. Since the voltage drop across R_2 is independent of I_C, it follows that V_{BE} diminishes [see Eq. (9-74)] which in turn causes I_B to decrease. The reduced value of I_B tends to restore I_C to its original value [see Eq. (9-69)].

A mathematical description of this behavior is given by the derivation of the stability factor. In the interest of simplifying the circuit analysis the d-c circuit to the left of the base terminal is replaced by its Thévenin equivalent circuit as depicted in Fig. 9-18. As far as the base terminal is concerned, looking towards the left, it sees a voltage source of magnitude

$$V = \frac{R_2}{R_1 + R_2} V_{CC}$$

and an internal resistance R_B composed of the parallel combination of R_1 and R_2. Accordingly, in the base-emitter circuit of Fig. 9-18 Kirchhoff's voltage law yields

$$V = I_B R_B + (I_B + I_C)R_E = I_B(R_B + R_E) + I_C R_E \qquad (9\text{-}75)$$

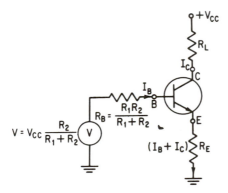

Fig. 9-18 Replacing the bias portion of the circuit of Fig. 9-17 by its Thévenin equivalent.

Hence the expression for the base current becomes

$$I_B = \frac{V}{R_B + R_E} - \frac{R_E}{R_B + R_E} I_C \qquad (9\text{-}76)$$

which upon insertion into Eq. (9-69) leads to an expression for the collector current given by

$$I_C\left(1 + \frac{\alpha}{1 - \alpha}\frac{R_E}{R_E + R_B}\right) = \frac{I_{CO}}{1 - \alpha} + \frac{\alpha}{1 - \alpha}\frac{V}{R_B + R_E} \qquad (9\text{-}77)$$

Multiplying through on both sides by $(1 - \alpha)$ yields

$$I_C\left(1 - \alpha + \alpha\frac{R_E}{R_E + R_B}\right) = I_{CO} + \alpha\frac{V}{R_B + R_E} \qquad (9\text{-}78)$$

Equation (9-78) is now in a form which leads directly to the stability factor.

Thus

$$S = \frac{\partial I_C}{\partial I_{CO}} = \frac{1}{1 - \alpha + \alpha \dfrac{R_E}{R_E + R_B}} \tag{9-79}$$

Examination of this expression indicates that there are two ways in which the stability may be increased: (1) by decreasing R_B and (2) by increasing R_E. However, there is a limit on how much R_B can be decreased. Low values of R_B are obtained by making R_2 very small. In turn this has two drawbacks. It causes a shunting effect on the input signal, and it causes a heavy current drain on V_{CC}. Similarly, there is an upper limit on the value of R_E. For a fixed source V_{CC}, the larger R_E is made, the smaller the allowable input signal must be to prevent distortion. On the other hand, if it is desirable to maintain a suitably centered Q-point relative to the common-emitter characteristics, then the larger R_E, the greater V_{CC} must be. Typical values for the stability factor of the circuit of Fig. 9-17 are around ten.

EXAMPLE 9-1   The purpose of this example is to discuss the manner in which the resistors R_1, R_2, and R_E are selected in the circuit of Fig. 9-17, and then to consider how the actual Q-point can be found graphically.

Assume that the following information is known:

transistor type (see Fig. 9-19)

$R_L = 1000 \ \Omega$

$V_{CC} = 12 \ \text{V}$

Solution: Without R_E the d-c load line would be determined by R_L so that on the ordinate axis it would pass through the 12-ma point and on the abscissa axis through the 12-V point. Refer to Fig. 9-19. With R_E the 12-V point remains fixed but the ordinate-axis intercept is lowered. Since this lowering of the load line should not be excessive, a good

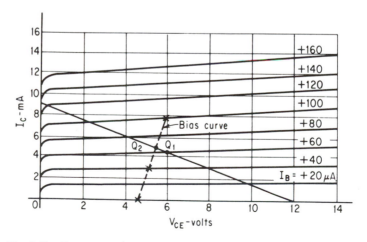

Fig. 9-19   Common-emitter characteristics for transistor of Example 9-1.

rule of thumb is to choose R_E so that the new coordinate-axis intercept is about three-fourths of the value without R_E. Thus we may write

$$I_{C\text{-axis intercept}} = \frac{V_{CC}}{R_L + R_E} = \frac{3}{4}\left(\frac{V_{CC}}{R_L}\right) \qquad (9\text{-}80)$$

or

$$R_E = \frac{R_L}{3} = 333 \quad \Omega \qquad (9\text{-}81)$$

Next construct the d-c load line and select a suitable Q-point at about midrange on the load line. This choice is identified as Q_1 in Fig. 9-19. Corresponding to this point read off the following pertinent quantities:

$$I_B = 60\ \mu A,\ V_{CE} = 6\ V,\ I_C = 4.4\ mA \qquad (9\text{-}82)$$

To find R_2 use is made of Eq. (9-74) as it applies to Fig. 9-17. Thus

$$IR_2 = V_{BE} + (I_B + I_C)R_E \qquad (9\text{-}83)$$

Inserting a typical value for V_{BE} of 0.2 volt as well as I_B and I_C from Eq. (9-82) yields

$$IR_2 = 0.2 + 4.46 \times 10^{-3} \times 333 = 1.69\ V \qquad (9\text{-}84)$$

In this expression neither I nor R_2 is known, so that we may arbitrarily select one and compute the other. In the interest of limiting the drain on the voltage source, I is generally selected as a fraction of I_C (often one-half to one-tenth of I_C). In this case upon choosing $I \approx 0.5$ mA we get from (9-84),

$$R_2 = \frac{1.69}{I} = \frac{1.69}{0.5}10^3 = 3.38\ k\Omega \qquad (9\text{-}85)$$

To find R_1 write Kirchhoff's voltage law for the loop containing V_{CC}, R_1, and R_2. Thus from Fig. 9-17 we have

$$V_{CC} = (I + I_B)R_1 + IR_2 \qquad (9\text{-}86)$$

or

$$R_1 = \frac{V_{CC} - IR_2}{I + I_B} \qquad (9\text{-}87)$$

Inserting the values for the quantities on the right yields

$$R_1 = \frac{12 - 1.69}{(0.5 + 0.06)10^{-3}} = 18.4\ k\Omega \qquad (9\text{-}88)$$

Now that R_1, R_2, and R_E have specific values, the actual Q-point can be found so that it may be compared with the arbitrarily selected one. If sufficient disagreement occurs, a refinement on R_1 and R_2 will be introduced.

Kirchhoff's voltage law for the collector-emitter circuit of Fig. 9-18 leads to

$$V_{CC} = I_C R_L + V_{CE} + (I_C + I_B)R_E \qquad (9\text{-}89)$$

Next insert the values for R_L and R_E expressed in kilohms. This permits I_B and I_C to be expressed in milliamperes. Thus

$$V_{CC} = 12 = V_{CE} + 1.333I_C + 0.333I_B \qquad (9\text{-}90)$$

Similarly for the base-emitter loop we write

$$V_{CC}\frac{R_2}{R_1 + R_2} = I_B\frac{R_1 R_2}{R_1 + R_2} + (I_B + I_C)R_E \qquad (9\text{-}91)$$

Inserting the values of R_1, R_2, and R_E just computed yields

$$1.86 = 0.333I_C + 3.193I_B \qquad (9\text{-}92)$$

It is helpful at this point to eliminate I_C from Eqs. (9-90) and (9-92). This is readily done by multiplying Eq. (9-92) by 4 and then subtracting it from Eq. (9-90). The result is

$$V_{CE} = 4.56 + 12.44I_B \qquad (9\text{-}93)$$

where I_B must be expressed in ma.

Equation (9-93) is called the *bias equation* and involves two unknowns. The point of intersection of this equation with the d-c load line identifies the actual Q-point for the specified values of R_1, R_2, and R_E. Since Eq. (9-93) is a straight line, only two points are needed to plot it. Thus for

$$\begin{aligned} I_B &= 0 & V_{CE} &= 4.56 \text{ V} \\ I_B &= 0.100 \text{ mA} & V_{CE} &= 5.8 \text{ V} \end{aligned} \qquad (9\text{-}94)$$

This leads to the *bias curve* shown in Fig. 9-19. Note that it intersects the d-c load line at Q_2.

Since Q_2 is away from the assumed position of the Q-point, an adjustment is in order for R_1 and R_2. A little thought indicates that R_2 should be selected somewhat higher. Hence choose

$$R_2 = 4 \text{ k}\Omega \qquad (9\text{-}95)$$

The adjusted value of I then becomes

$$I = \frac{1.69}{4} = 0.423 \text{ mA} \qquad (9\text{-}96)$$

The new value of R_1 can still be found from Eq. (9-87), but the value of I_B corresponding to Q_2 must be used. This value is 0.065 mA. Hence

$$R_1 = \frac{10.31}{0.423 + 0.065} = 21.2 \text{ k}\Omega \qquad (9\text{-}97)$$

We must now find the new bias curve so that the Q-point corresponding to the adjusted values of R_1 and R_2 can be found and compared to Q_1. Proceeding as before, the new bias equation becomes

$$V_{CE} = 4.4 + 13.48I_B \qquad (9\text{-}98)$$

A plot of this equation yields an intersection point with the load line which is almost identical to Q_1.

Therefore the values of the resistors in the final design are:

$$R_E = 333 \ \Omega \qquad R_1 = 4000 \ \Omega \qquad R_2 = 21{,}000 \ \Omega \qquad (9\text{-}99)$$

Although the introduction of R_E in the emitter leg helps to provide satisfactory stability of the quiescent point, the circuit configuration shown in Fig. 9-17 will cause a drop in gain unless suitably modified. This may be illustrated by assuming that a positive increment in base current occurs, which in turn causes a corresponding amplified increase in collector current. However, since this increased current causes the voltage drop across R_E to increase, thereby reducing the base-to-collector voltage, there occurs a subsequent drop in collector current. The result is that the overall gain is less with R_E present than without it. This diminution in gain can be avoided by providing an alternate path from the emitter terminal through which the *changing* component of the emitter current may pass with little or no hindrance. The insertion of a low-impedance capacitor in parallel with R_E effectively accomplishes this purpose. See Fig. 9-20. The size of C_E is chosen so that at the frequency of the input signal the reactance $1/\omega C_E$ is small compared to R_E. Of course the presence of C_E does not upset the quiescent condition because the capacitor is an open circuit to d-c.

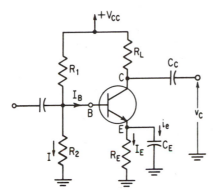

Fig. 9-20 Complete common-emitter amplifier circuit. The varying component of emitter current passes through C_E.

Biasing the FET. Reference to Fig. 9-6 shows that for a specified load resistance and junction field-effect transistor the quiescent operating point is determined by the battery source V_{GG} in the gate-source circuit. As was true with the transistor amplifier it is desirable to eliminate this battery by using an appropriate biasing circuit. The most common procedure is to put a resistor R_s in the source leg as illustrated in Fig. 9-21. Then at the quiescent condition the d-c drain current I_{DQ} flows through R_s thus establishing a negative gate-source voltage given by

$$V_{GQ} = -I_{DQ}R_s \tag{9-100}$$

The minus sign appears because the voltage drop across R_s has the negative end on the gate side. Equation (9-100) is the bias equation and is analogous to Eq. (9-98) for the transistor case. The corresponding bias curve is found by plotting Eq. (9-100) on the drain volt-ampere characteristics for various assumed values of quiescent drain current I_{DQ}. The point of intersection of this bias curve with the load line then locates the actual quiescent point.

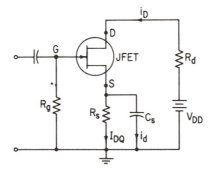

Fig. 9-21 A JFET voltage amplifier with self-biasing source resistor R_s.

The source by-pass capacitor C_s is also included in the amplifier circuitry for the FET in order to prevent the drop in gain that results when the varying component of the drain current is made to flow through the self-bias source resistor. If C_s is selected sufficiently large so that its reactance at the lowest signal frequency is small compared to R_s, the voltage gain remains virtually intact.

EXAMPLE 9-2   A 2N4223 JFET is used in the amplifier circuit of Fig. 9-21. The drain supply voltage is 24 V. It is desired that the quiescent point be located at $I_{DQ} = 3.4$ mA, $V_{DQ} = 15$ V, and $V_{GQ} = 1$ V. Find the values of self-bias source resistance R_s and drain load resistance R_d to assure this Q-point.

Solution: The self-bias source resistance is found directly from Eq. (9-100). Thus

$$V_{GQ} = -I_{DQ}R_s$$

$$-1 = -3.4 \times 10^{-3} R_s$$

or

$$R_s = 293 \; \Omega$$

To find R_d write Kirchhoff's voltage law for the output circuit of the quiescent point. Thus

$$V_{DD} = I_{DQ}R_d + V_{DQ} + I_{DQ}R_s \qquad (9\text{-}101)$$

and

$$R_d = \frac{V_{DD} - V_{DQ}}{I_{DQ}} - R_s = \frac{24 - 15}{3.4 \times 10^{-3}} - 293$$

$$= 2645 - 293 = 2352 \; \Omega$$

Biasing the MOSFET.   The self-bias circuit configuration that is employed for the BJT amplifier (see Fig. 9-20) is also applicable to the enhancement MOSFET and the DE MOSFET as well. If in the circuit of Fig. 9-20 R_E is replaced with R_S and the bipolar junction transistor is replaced with the MOSFET, the expression for the gate-source voltage can be written as

$$V_{GS} = \frac{R_2}{R_1 + R_2} V_{DD} - I_D R_S \qquad (9\text{-}102)$$

By properly selecting the circuit values to be used in this equation, V_{GS} may be made to assume either a positive or negative value to make it consistent with the particular mode of operation that is desired.

9-4 COMPUTING AMPLIFIER PERFORMANCE

In this section attention is directed to the manner in which single-stage and multistage amplifier performance can be computed. In situations where the input signal variations are small the amplifier performance is determined through use of the equivalent circuit. Typical parameters are used for the transistors as obtained from the manufacturers' handbooks. The effects of the necessary biasing circuits on the overall amplifier performance are also treated. In those instances where the input signal goes through a wide excursion of values about a quiescent point, the performance is found more correctly by a graphical analysis. This is often necessary in the last stages of a multistage amplifier.

General Expressions for Current and Voltage Gains. Appearing in Fig. 9-22 is the small-signal equivalent circuit of a transistor amplifier that includes the appropriate transistor parameters, the base bias equivalent resistor R_B and the collector load resistor R_L. The source signal V_s and its internal resistance r_s are also shown. All symbols for voltages and currents are those that denote the effective values of the varying components of the quantities involved. Moreover, I_b, I_c, I_L, V_c, and V_b are shown in directions consistent with their assumed positive values. If actual current or voltage polarities are opposite to these, then the quantities are considered negative.

There are several current gains that can be identified with the configuration of Fig. 9-22. The first is the current amplification factor of the transistor itself— h_{fe}. This is an ideal current gain. It is realizable at the load resistor only when h_{oe} approaches zero and R_B approaches infinity. The second current gain is that which relates the collector (or load) current to the base current of the transistor. This quantity can be called the *transistor current gain* and can be expressed in terms of the load resistor and the transistor parameters. In the output circuit the expression for load current becomes

$$I_L = -I_c = \frac{-1/h_{oe}}{(1/h_{oe}) + R_L} h_{fe} I_b = \frac{-h_{fe}}{1 + h_{oe} R_L} I_b \qquad (9\text{-}103)$$

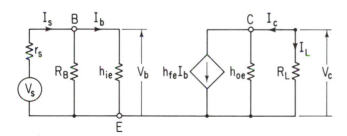

Fig. 9-22 Equivalent circuit of a transistor amplifier in the common-emitter mode.

The transistor current gain results upon formulating the ratio of output current to base current. Thus

$$A_I \equiv \frac{I_L}{I_b} = \frac{-I_c}{I_b} = -\frac{h_{fe}}{1 + h_{oe} R_L} \qquad (9\text{-}104)$$

The third current gain relates the load current to the source current. Accordingly, this is the current gain of the complete stage of the transistor amplifier. It is the most useful of the three current gains for obvious reasons. An asterisk notation is used to emphasize this fact. We have then

$$A_I^* \equiv \frac{I_L}{I_s} = \frac{I_L}{I_b} \times \frac{I_b}{I_s} \qquad (9\text{-}105)$$

The first part of the right side of Eq. (9-105) is available from the expression for the transistor current gain Eq. (9-104). The second part is readily obtained by using the current division principle as it applies to the parallel resistors R_B and h_{ie}. Thus

$$\frac{I_b}{I_s} = \frac{R_B}{R_B + h_{ie}} = \frac{1}{1 + h_{ie}/R_b}$$

(9-106)

Inserting Eq. (9-104) and the last expression into Eq. (9-105) then yields the desired result.

$$A_I^* = \frac{-h_{fe}}{(1 + h_{oe}R_L)(1 + h_{ie}/R_B)}$$

(9-107)

It is instructive to note here that the current gain of the complete amplifier stage becomes the current amplification factor only when $h_{oe} \to 0$ and $R_B \to \infty$. The more h_{oe} and R_B deviate from these ideal values, the greater will be the deterioration of the stage current gain from the ideal value of h_{fe}.

An analogous set of equations can be developed to express the voltage gain of the transistor amplifier. By definition the voltage gain of the transistor (but not the complete stage) is

$$A_V \equiv \frac{V_c}{V_b} = \frac{-I_c R_L}{V_b}$$

(9-108a)

Here $-I_c$ can be replaced by its equivalent expression, $A_I I_b$ [see Eq. (9-104)]. Hence

$$A_V = \frac{A_I I_b R_L}{V_b} = \frac{A_I R_L}{V_b/I_b} = A_I \frac{R_L}{h_{ie}}$$

(9-108b)

It is significant to note here that the transistor voltage gain is determined by its current gain multiplied by the ratio of load resistor in the output circuit to the forward-biased resistor of the input circuit. This behavior was first encountered in Sec. 8-3. Refer to Eq. (8-34). Upon inserting Eq. (9-103) into Eq. (9-108b), the transistor voltage gain may be written as

$$A_V = \frac{-h_{fe}}{1 + h_{oe}R_L} \frac{R_L}{h_{ie}}$$

(9-109)

The role played by each of the transistor parameters and the load resistor in determining voltage gain is made clearly evident by this formulation.

The voltage gain of the entire stage is defined as

$$A_V^* \equiv \frac{V_c}{V_s} = \frac{V_c}{V_b} \times \frac{V_b}{V_s} = A_V \frac{V_b}{V_s}$$

(9-110)

The first ratio on the right side is given by Eq. (9-109). The second ratio follows from an analysis of the input circuit of Fig. 9-22. Let the parallel arrangement of R_B and h_{ie} be represented by R_i, where the subscript stresses that this quantity is the input resistance of the amplifier. Thus

$$R_i = \frac{R_B h_{ie}}{R_B + h_{ie}} = \frac{h_{ie}}{1 + h_{ie}/R_B}$$

(9-111)

Then by voltage division

$$\frac{V_b}{V_s} = \frac{R_i}{R_i + r_s} = \frac{1}{1 + r_s/R_i} \tag{9-112}$$

The final expression for the overall voltage gain thus becomes

$$A_V^* = \frac{-h_{fe}}{1 + h_{oe}R_L}\left(\frac{R_L}{h_{ie}}\right)\frac{1}{1 + r_s/R_i} \tag{9-113}$$

Single-Stage Transistor Amplifier. As stated in the introduction to this section our interest next is to illustrate all of the computation features involved in computing the performance of a single-stage as well as a multistage amplifier. In this effort a step-by-step analysis is made of a transistorized intercommunication system composed of four amplifying stages. Each stage is considered to be appropriately equipped with biasing circuits, bypass capacitors, and coupling capacitors. A secondary objective of the analysis is to show how a succeeding amplifier stage "loads down" a preceding stage, thereby causing an overall reduction in performance.

Although the general expressions developed above could be applied directly to obtain the current and voltage gains, this procedure will not be used. Rather, in the interest of emphasizing first principles, each equivalent circuit as it arises is solved as a separate problem in circuitry. Besides offering an opportunity of exploiting our knowledge of circuit theory, such a procedure also makes us less dependent upon memorizing formulas.

As the first step in the development of the intercommunication system, consider the single-stage transistor amplifier depicted in Fig. 9-23. Assume the input signal to be a sinusoidal voltage having an effective value of 25 mV and an internal resistance of 1000 Ω. Let us find the performance of this amplifier-circuit for the 2N104 transistor and the values of the circuit parameters shown in Fig. 9-23. Keep in mind that finding the performance of the amplifier generally means computing current gain, voltage gain, power gain, the input resistance as seen by the signal source, and the output resistance appearing at the output terminals.

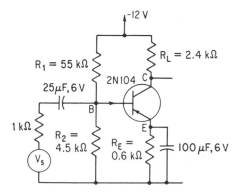

Fig. 9-23 Single-stage transistor amplifier. The input signal is a generator having an rms voltage V_b and internal resistance of 1 kΩ.

In view of the small magnitude of the input signal the performance is best determined by using an equivalent-circuit analysis. Also, because we are dealing here with an audio amplifier, the diffusion and transition capacitors in the complete equivalent circuit of the transistor may be neglected. These capacitors are important only at very high signal frequencies. Accordingly the equivalent circuit of Fig. 9-8 may be employed to replace the transistor in the circuit of Fig. 9-23, which in turn leads to the equivalent circuit of the transistor amplifier shown in Fig. 9-24. Here h_{re} is assumed equal to zero. Recall that R_B is the parallel combination of R_1 and R_2 of the base bias circuit which must be considered in the analysis. Depicted in Fig. 9-24 is the equivalent circuit showing the transistor manufacturer's published values of the equivalent circuit parameters.

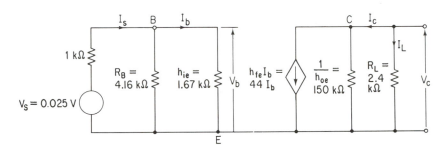

Fig. 9-24 The h-parameter equivalent circuit of the common-emitter amplifier of Fig. 9-23.

The amplifier performance is now readily calculated by applying the techniques of circuit analysis to the circuit of Fig. 9-24. To find the signal source current I_s, it is necessary first to find the parallel combination of the two resistances between terminals B and E. Thus

$$R_p = R_i = \frac{4.16(1.67)}{4.16 + 1.67}\text{k}\Omega = 1.19 \text{ k}\Omega \tag{9-114}$$

Then the effective value of the source current is

$$I_s = \frac{V_s}{1000 + R_p} = \frac{0.025}{2.19 \times 10^3} = 11.4\mu\text{A} \tag{9-115}$$

By the current-divider rule the effective value of the varying component of the base current I_b is found to be

$$I_b = \frac{4.16}{5.825} I_s = 8.15\mu\text{A} \tag{9-116}$$

The current source generator in the output section of the equivalent circuit has the value

$$h_{fe}I_b = 44(8.15) \ \mu\text{A} = 0.358 \text{ mA} \tag{9-117}$$

This leads to an effective value of collector current I_c of

$$I_c = \frac{150}{152.4}(0.358) = 0.352 \text{ mA} \tag{9-118}$$

The current gain of the *complete amplifier stage* is then

$$A_I^* \equiv \frac{I_L}{I_s} = \frac{-I_c}{I_s} = -\frac{0.352 \times 10^{-3}}{11.4 \times 10^{-6}} = -30.9 \tag{9-119}$$

This result is readily verified by direct application of Eq. (9-106). Thus

$$A_I^* = \frac{-44}{(1 + 2.4/150)(1 + 1.67/4.16)} = -30.9 \tag{9-120}$$

The transistor current gain which is the ratio of the collector current to the base current is given either by

$$A_I = \frac{I_L}{I_b} = \frac{-I_c}{I_b} = -\frac{352}{8.15} = -43.2 \tag{9-121}$$

or computed directly from Eq. (9-104) as

$$A_I = \frac{-h_{fe}}{1 + h_{oe}R_L} = \frac{-44}{1 + 2.4/150} = -43.2 \tag{9-122}$$

The voltage gain for the complete amplifier stage is found to be

$$A_V^* \equiv \frac{V_L}{V_s} = \frac{V_c}{V_s} = \frac{-I_c R_L}{V_s} = -\frac{0.352(2.4)}{0.025} = -33.8 \tag{9-123}$$

A direct application of Eq. (9-113) provides

$$A_V^* = \frac{-h_{fe}}{1 + h_{oe}R_L}\left(\frac{R_L}{h_{ie}}\right)\frac{1}{1 + r_s/R_i} \tag{9-124}$$

$$= \frac{-44}{1.016}\left(\frac{2.4}{1.67}\right)\frac{1}{1 + 1/1.19} = -33.8$$

The corresponding value of the power gain is

$$G = |A_V^*| \, |A_I^*| = 33.8(30.9) = 1042 \tag{9-125}$$

The input resistance of the amplifier circuit is the resistance seen by the signal source between terminals B and E in Fig. 9-24. Accordingly,

$$R_i = R_p = 1.19 \text{ k}\Omega \tag{9-126}$$

The input resistance is of importance because it provides a measure of the extent to which the internal resistance of the source is significant in making available a signal variation to the transistor.

The output resistance R_0 of the amplifier circuit is defined here as the resistance appearing between the output terminals C and E in the equivalent circuit of Fig. 9-24 with R_L connected. The value is

$$R_0 = \frac{2.4(150) \text{ k}\Omega}{152.4 \text{ k}\Omega} = 2.36 \text{ k}\Omega \tag{9-127}$$

Information about the output resistance of an amplifier stage is useful in determining matching characteristics when the first stage is followed by a second stage.

Two-Stage Transistor Amplifier. Attention is next directed to the two-stage transistor amplifier depicted in Fig. 9-25. Note that the first stage is identical to

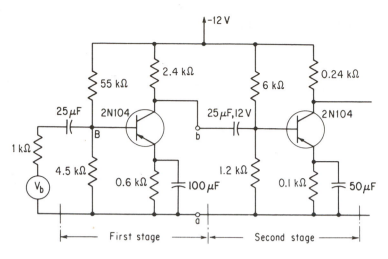

Fig. 9-25 Schematic diagram of a two-stage BJT amplifier.

that shown in Fig. 9-23. The input terminals to the second stage are identified as *ab*. As far as the first stage is concerned, it is helpful to note that it may be replaced by an equivalent current source having a value of $44I_b = 358\ \mu A$ and shunted by an equivalent resistance which is the output resistance of the first stage. This current source serves as the input signal to the second stage, the equivalent circuit of which is depicted in Fig. 9-26. Since the second stage also uses the 2N104 transistor, the same equivalent-circuit parameters are used as appear in Fig. 9-24. However, note that the equivalent resistance R_{B2} associated with the bias circuit is now equal to 1 kΩ, which is the parallel combination of the 1.2-kΩ and 6-kΩ resistors of the second stage.

The signal input current to the second stage is the current, I_{s2}, flowing into terminal *a* in Fig. 9-26. Its value is readily found by applying the current-divider rule. But first it is helpful to obtain the equivalent resistance of the parallel combination of the 1-kΩ and 1.67-kΩ resistances. Thus

$$R_{p2} = \frac{1(1.67)}{1 + 1.67} = 0.625\ \text{k}\Omega \tag{9-128}$$

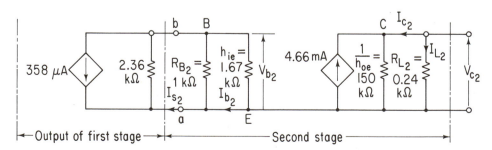

Fig. 9-26 The *h*-parameter equivalent circuit of the two-stage amplifier of Fig. 9-25.

Semiconductor Electronic Circuits   Chap. 9

Then

$$I_{s2} = \frac{2.36}{2.36 + 0.625}(358) = 283\mu A \tag{9-129}$$

In order to obtain the current amplification as it appears in the output circuit of the second stage, we need the base current to the transistor in the second stage. Again by the current-divider rule we have

$$I_{b2} = \frac{1}{1 + 1.67} I_{s2} = \frac{1}{2.67}(283) = 106\mu A \tag{9-130}$$

It is worthwhile at this point in the computations to check on the validity of proceeding with an equivalent-circuit approach for finding the performance. Because I_{b2} in Eq. (9-130) represents the effective value of the base current variation, it follows that the maximum input current swing is $\sqrt{2}(106) = 150$ μA. A glance at the average collector characteristics for the 2N104 transistor reveals that this swing is sufficiently small to allow computations to be made by means of the equivalent circuit rather than by a graphical analysis. However, we can well expect that for a third stage the swing will be so large that the equivalent-circuit representation is no longer valid. A graphical analysis will then be necessary; this matter is considered in the next phase of our study.

The current source appearing in the output section of the second stage is found to be

$$h_{fe}I_{b2} = 44(106)\mu A \tag{9-131}$$
$$= 4.66mA$$

The output resistance of the second stage as seen between terminals C and E in Fig. 9-26 is for all practical purposes equal to 240 Ω. It also follows from this approximation that the collector current at the second stage is equal to the source current. Thus

$$I_{c2} \approx h_{fe}I_{b2} = -4.66mA \tag{9-132}$$

The minus sign indicates flow is opposite to the assumed positive direction.

We are now in a position to make an interesting comparison of the current gain of the first stage both with and without the second stage. The useful version of the current gain of the one-stage amplifier was found in Eq. (9-119) to be

$$A_I = -\frac{I_{c1}}{I_{s1}} = -\frac{352\mu A}{11.4\mu A} = -30.9 \tag{9-133}$$

When a second stage is made to follow the first stage, a loading effect occurs which serves to reduce the current gain. This comes about because the input resistance of the second stage is finite and of a value comparable to the output resistance of the first stage. A glance at the input section of Fig. 9-26 makes this apparent. Accordingly, a more significant expression for the magnitude of current gain of the first stage of a two-stage amplifier is defined as

$$A_I' = \frac{I_{s2}}{I_{s1}} = -\frac{283\ \mu A}{11.4\ \mu A} = -24.8 \tag{9-134}$$

Note the substantial reduction that takes place from the unloaded current gain of Eq. (9-133).

The current gain for the second stage may be identified as

$$A_{l2}^* = \frac{I_{L2}}{I_{s2}} = -\frac{I_{c2}}{I_{s2}} = \frac{4.66 \text{ mA}}{-0.283 \text{ mA}} = -16.5 \tag{9-135}$$

Keep in mind that this value is an unloaded current gain too. Therefore, in situations where the second stage is followed by a third stage, the same adjustment must be made for the second-stage gain as was done above for the first stage.

The total current gain of the complete two-stage amplifier (but with the second stage unloaded, i.e., as it appears in Fig. 9-25), is given by

$$\frac{I_{L2}}{I_{s1}} = -\frac{I_{c2}}{I_{s1}} = \frac{4.66 \text{ mA}}{0.0114 \text{ mA}} = 408 \tag{9-136}$$

A check on this computation may be made by taking the product of the first-stage current gain, Eq. (9-134), and the second-stage current gain, Eq. (9-135). Thus

$$\frac{I_{L2}}{I_{s1}} = -\frac{I_{c2}}{I_{s1}} = \left(\frac{I_{s2}}{I_{s1}}\right)\left(-\frac{I_{c2}}{I_{s2}}\right) = 24.8(16.5) = 408 \tag{9-137}$$

Continuing with the description of the performance of the two-stage amplifier we next compute the voltage gain. Clearly, this is defined as the ratio of the a-c output voltage at the second stage divided by the input signal voltage. Thus

$$\frac{V_{c2}}{V_s} = -\frac{I_{c2}R_{L2}}{V_s} = \frac{4.66(0.24)}{0.025} = 44.74 \tag{9-138}$$

The corresponding power gain for both stages then becomes

$$G = 408(44.74) = 18,254 \tag{9-139}$$

Finally, the input resistance to the second stage is

$$R_{i2} = \frac{1(1.67)}{1 + 1.67} = 0.625 \text{ k}\Omega = R_{p2} \tag{9-140}$$

and the output resistance of the second stage is

$$R_{o2} = \frac{(150 \text{ k}\Omega)(0.24 \text{ k}\Omega)}{150.24 \text{ k}\Omega} \approx 0.24 \text{ k}\Omega \tag{9-141}$$

Three-Stage Transistor Amplifier. The circuit configuration of a three-stage transistor amplifier appears in Fig. 9-27. It is formed by adding an appropriate third stage to the two-stage amplifier of Fig. 9-25. Before proceeding with the performance analysis of the three-stage amplifier, it is important first to determine how big a swing of the input current to the third stage occurs. Figure 9-28 shows the input section of the equivalent circuit of the third stage, which can be used to check on this swing. The input impedance as seen by the current source looking into terminals c and d of the third stage is

$$R_{i3} = \frac{R_{B3}h_{ie}}{R_{B3} + h_{ie}} \tag{9-142}$$

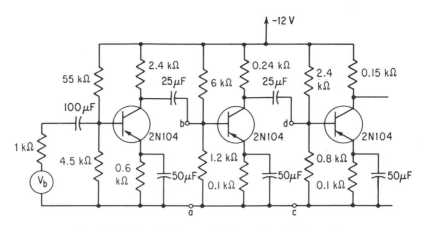

Fig. 9-27 Circuit diagram of a three-stage BJT amplifier.

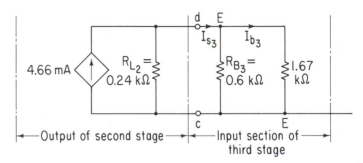

Fig. 9-28 Portion of input section of the equivalent circuit of the third stage which is used to check the maximum swing of the input signal.

where

$$R_{B3} = \frac{(0.8 \text{ k}\Omega)(2.4 \text{ k}\Omega)}{0.8 \text{ k}\Omega + 2.4 \text{ k}\Omega} = 0.6 \text{ k}\Omega \tag{9-143}$$

Hence

$$R_{i3} = \frac{0.6(1.67)}{0.6 + 1.67} = 0.441 \text{ k}\Omega \tag{9-144}$$

Therefore,

$$I_{s3} = \frac{R_{i2}}{R_{i2} + R_{i3}} 4.66 = \frac{0.24}{0.681}(4.66) = 1.64 \text{ mA} \tag{9-145}$$

Accordingly, the effective value of the varying component of the base current to the transistor in the third stage is

$$I_{b3} = \frac{0.6}{0.6 + 1.67} I_{s3} = \frac{0.6}{2.27}(1.64) = 0.435 \text{ mA} \tag{9-146}$$

The maximum input current swing is thus $\sqrt{2}$ (435) μA or 615 μA. A glance at the average collector characteristics of the 2N104 depicted in Fig. 9-29 indicates

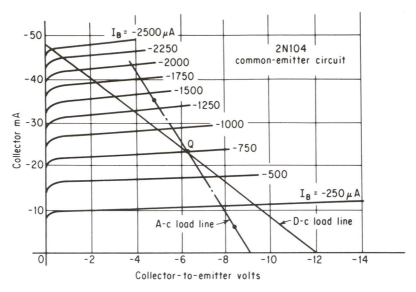

Fig. 9-29   Average collector characteristics of the 2N104 transistor.

that for a properly located quiescent point the base-current swing is so large as to require a graphical analysis.

To find the performance of the third stage graphically, it is necessary first to establish the quiescent point on the collector characteristics of Fig. 9-29. By following the procedure outlined in Sec. 9-3 for the transistor amplifier, the bias curve can be drawn and the Q-point located. For the third stage of the amplifier depicted in Fig. 9-27 the quiescent point is found to be

$$I_B = 750 \ \mu A \qquad I_C = 28 \ mA \qquad V_{CE} = -6.3 \ V \qquad (9\text{-}147)$$

Therefore the maximum value of the base current in the third stage becomes

$$I_{b3max} = 750 + 615 = 1365 \ \mu A \qquad (9\text{-}148)$$

Since capacitor C_E is a virtual short circuit to varying input signals, changes about the Q-point must be found on the a-c load line, which here has a slope equal to the reciprocal of the 150-Ω load resistance. The corresponding values of maximum collector current and collector-to-emitter voltage as obtained on the a-c load line in Fig. 9-29 are then

$$I_{cmax} = 35 \ mA \qquad and \qquad V_{CEmin} = 4.8 \ V \qquad (9\text{-}149)$$

Similarly, the minimum value of the base current is

$$I_{bmin} = 750 - 615 = 135 \ \mu A \qquad (9\text{-}150)$$

which in turn yields

$$I_{cmin} = 6 \ mA \qquad and \qquad V_{CEmax} = 8.4 \ V \qquad (9\text{-}151)$$

The effective value of the varying component of the collector current then becomes

$$I_{c3} = \frac{I_{cmax} - I_{cmin}}{2\sqrt{2}} = \frac{35 - 6}{2\sqrt{2}} = 10.2 \ mA \qquad (9\text{-}152)$$

This leads to a current gain for the three-stage amplifier of

$$\frac{I_{L3}}{I_{s1}} = -\frac{I_{c2}}{I_{s1}} = -\frac{10.2 \text{ mA}}{0.0114 \text{ mA}} = -898 \tag{9-153}$$

The effective value of the varying component of the collector voltage is given by

$$V_{c3} = \frac{8.4 - 4.8}{2\sqrt{2}} = \frac{3.6}{2\sqrt{2}} = 1.28V \tag{9-154}$$

The corresponding overall voltage gain is then

$$\frac{V_{c3}}{V_s} = -\frac{1.28}{0.025} = -51.2 \tag{9-155}$$

Finally, the overall power gain becomes

$$G = 898 \mid 51.2 \mid = 45,978 \tag{9-156}$$

Four-Stage Transistor Amplifier. The addition of still another stage to the output terminals of the circuit of Fig. 9-27 leads to the fourstage amplifier appearing in Fig. 9-30. The selection of the transistor for the fourth stage must be such that it can accommodate properly the wide swing in signal appearing at the output terminals of the third stage. The 2N301 transistor meets this condition.

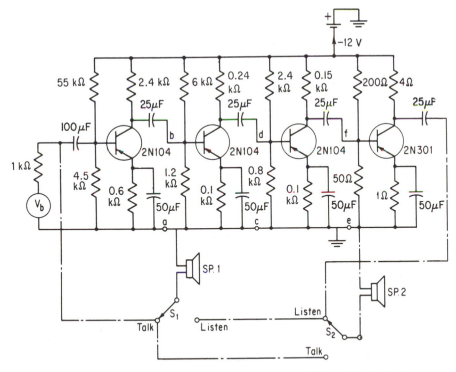

Fig. 9-30 Transistorized four-stage amplifier. The addition of the dashed-line connections to the two speakers yields an intercommunication system.

In order to compute the output at the fourth stage, it is necessary to find the maximum and minimum values of the input base current to this stage. This is readily accomplished by representing the output of the third stage by the equivalent current source, 10.2 ma, and its shunt output resistance, $R_{L3} = 150$ Ω, and combining it with the input resistance of the fourth stage. Refer to Fig. 9-31. The manufacturer's transistor manual specifies the resistance between the base and emitter terminals for the 2N301 as 23 Ω. Hence the input resistance of the fourth stage, taking into account the effect of the bias circuit, becomes

$$R_{p4} = \frac{R_{B4}h_{ie}}{R_{B4} + h_{ie}} = \frac{40(23)}{40 + 23} = 14.6 \ \Omega \tag{9-157}$$

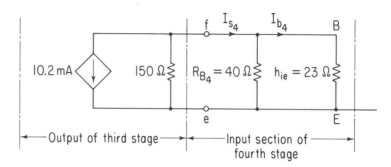

Fig. 9-31 Circuit used to obtain maximum swing of base current into four-stage transistor.

Again by current division we find

$$I_{s4} = \frac{-150}{150 + 14.6} 10.2 = -9.12 \ \text{mA} \tag{9-158}$$

and

$$I_{b4} = \frac{-40}{40 + 23} 9.12 = -5.79 \ \text{mA} \tag{9-159}$$

Negative signs again indicate current flow is opposite to assumed positive directions. Thus the peak value of the varying component of the base current for the fourth stage becomes $\sqrt{2} (5.79) = 8.2$ mA.

Construction of the bias curve for the circuit configuration of the fourth stage locates the quiescent point at $I_B = 20$ mA, $I_C = 1.4$ A, and $V_{CE} = -5$ V. This is identified as Q in Fig. 9-32. Moreover, the a-c load line is drawn corresponding to the 4-Ω load resistance. The maximum value of the base current in this case then becomes

$$I_{b\text{max}} = 20 + 8.2 = 28.2 \ \text{mA} \tag{9-160}$$

and is marked as point $a\dagger$ on the a-c load line of Fig. 9-32. The corresponding

† This point as well as b were found on a large version of Fig. 9-32 for greater accuracy.

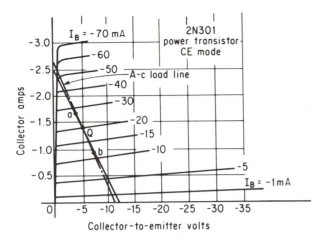

Fig. 9-32 Average collector characteristics of the 2N301 power transistor in the common-emitter mode.

values of the collector current and collector-to-emitter voltage are

$$I_{c\text{max}} = 1.7 \text{ A} \quad \text{and} \quad V_{CE\text{min}} = 3.5 \text{ V} \quad (9\text{-}161)$$

For the minimum value of base current, $I_{b\text{min}} = 20 - 8.2 = 11.8$ mA, the collector current and voltage values are

$$I_{c\text{min}} = 0.85 \text{ A} \quad \text{and} \quad V_{CE\text{max}} = 7.5 \text{ V} \quad (9\text{-}162)$$

This point is identified as b on the a-c load line. It is worthwhile to note here that for the same swing above and below the Q-point of 8.2 mA in base current, the corresponding changes brought about in the collector current are quite different. The nonlinear nature of the spacing of collector characteristics is responsible for this effect.

The effective value of the magnitude of the varying component of the collector current is

$$| I_{c4} | = \frac{1.7 - 0.85}{2\sqrt{2}} = 0.3 \text{ A} \quad (9\text{-}163)$$

This leads to an overall current gain of the four-stage amplifier of

$$\frac{I_{L4}}{I_{s1}} = -\frac{I_{c4}}{I_{s1}} = \frac{-(-0.3)}{11.4 \times 10^{-6}} = 26,300 \quad (9\text{-}164)$$

The value of I_{c4} is negative when expressed relative to I_s. Proceeding in a similar fashion we find the rms value of the a-c component of the collector voltage to be

$$V_{c4} = \frac{7.5 - 3.5}{2\sqrt{2}} = \frac{2}{\sqrt{2}} = 1.4 \text{ V} \quad (9\text{-}165)$$

and thereby yielding a four-stage voltage gain of

$$\frac{V_{c4}}{V_s} = \frac{1.4}{0.025} = 56 \quad (9\text{-}166)$$

The corresponding power gain is

$$G = 26.3 \times 10^3 \times 56 = 1.47 \times 10^6 \qquad (9\text{-}167)$$

This completes the performance analysis of the four-stage amplifier.

As a matter of application interest it is useful to note that by the addition of two speakers wired in the manner shown by the dashed-line connections depicted in Fig. 9-30, a transistorized intercommunication system results.

Single-Stage FET Amplifier. The manner of computing the performance of a FET amplifier by the equivalent circuit can be illustrated by considering the FET amplifier configuration of Fig. 9-6. It is assumed that the varying input signal v_{gs} undergoes very small changes and that R_g is so large that it may be neglected from further consideration. Accordingly, the linear model for this amplifier may be drawn as shown in Fig. 9-33. We direct attention first to the current-source version. The effective value of the varying component of the drain voltage can be expressed as

$$V_d = I_L R_L = -I_d R_L \qquad (9\text{-}168)$$

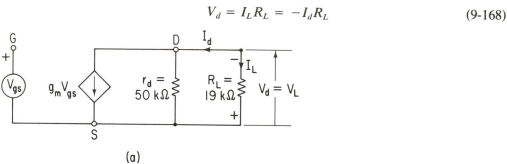

(a)

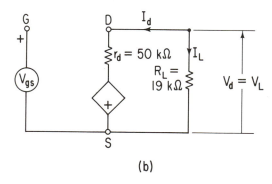

(b)

Fig. 9-33 Equivalent circuit of FET amplifier of Fig. 9-6: (a) current-source version; (b) voltage-source version.

Here too the assumed positive directions for the load current and the drain current are opposite. This linear model of the amplifier states that the effective value of the voltage V_{gs} in the input circuit makes itself felt in the output circuit in the form of a current source of value $g_m V_{gs}$. This current then divides itself between r_d and R_L. By the current divider principle the drain current can be expressed as

$$I_d = \frac{r_d}{R_L + r_d} g_m V_{gs} = \frac{g_m}{1 + R_L/r_d} V_{gs} \qquad (9\text{-}169)$$

The rms value of the varying drain voltage is thus related to the rms value of the varying gate-source voltage by

$$V_d = \frac{-g_m R_L}{1 + R_L/r_d} V_{gs} \tag{9-170}$$

The voltage gain of the FET amplifier readily follows from Eq. (9-170) upon formulating the ratio of drain to gate-source voltage. Thus

$$\boxed{A_V \equiv \frac{V_L}{V_{gs}} = \frac{V_d}{V_{gs}} = \frac{-g_m R_L}{1 + R_L/r_d}} \tag{9-171}$$

where g_m is the transconductance of the FET. The presence of the minus sign in this expression attests to the phase reversal that takes place in this amplifier. That is, instantaneously as the varying gate-source signal moves in a positive direction from the Q-point, the varying component of the drain-source voltage moves in a negative direction from the Q-point.

The 2N4220 FET used in the circuit of Fig. 9-6 has the following parameters at the specified Q-point:

$$g_m = 1200 \ \mu S \qquad r_d = 50 \ k\Omega \qquad \mu = 60$$

Accordingly, by Eq. (9-171) the specific value of the voltage gain for this amplifier is

$$A_V = \frac{-1.2 \times 10^{-3}(19 \times 10^3)}{1 + \dfrac{19}{50}} = \frac{22.8}{1.38} = -16.5 \tag{9-172}$$

An alternative expression for computing the voltage gain follows from an analysis of Fig. 9-33(b). Here the voltage appearing across the load is found by applying the voltage division principle. Thus

$$V_d = V_L = \frac{-R_L}{R_L + r_d} \mu V_{gs} \tag{9-173}$$

Rearranging terms leads to

$$\boxed{A_V = \frac{V_d}{V_{gs}} = \frac{-R_L}{R_L + r_d} \mu = \frac{-\mu}{1 + r_d/R_L}} \tag{9-174}$$

Inserting the specified values yields

$$A_V = \frac{-60}{1 + 50/19} = \frac{-60}{1 + 2.63} = -16.5 \tag{9-175}$$

As expected this result is identical to that of Eq. (9-172).

It is instructive to note that Eq. (9-174) could be derived from Eq. (9-171) by merely multiplying the right side of the latter by unity in the form of r_d/r_d and recalling that $\mu = g_m r_d$.

9-5 FREQUENCY RESPONSE OF *RC* COUPLED TRANSISTOR AMPLIFIERS

A glance at the amplifier system depicted in Fig. 9-30 reveals that the coupling from one stage to the next is achieved by a coupling capacitor followed by a connection to a shunt resistor. Because of this arrangement such amplifiers are called *resistance-capacitance coupled amplifiers*. Since amplifiers are often designed to provide essentially uniform gain over some specified frequency range, it is important to know the factors which place upper and lower limits on this range. In this way appropriate steps may be taken in the amplifier design to assure satisfactory performance over the specified frequency range. For example, in the case of an audio amplifier, which is used to amplify speech and music sounds as detected by a microphone, it is essential that all the frequencies in the sound spectrum (15 to 15,000 Hz) be uniformly amplified. Otherwise the amplified sound, emanating from the loudspeaker, will be a distorted version of that picked up by a quality microphone.

There are certain factors in *RC* coupled amplifiers which will cause the amplifier not to provide the same gain at very low and very high frequencies. It is the purpose of this section to study what these factors are and how they may be controlled to yield a bandwidth which furnishes the desired fidelity of the output signals. Attention is directed first to the common-emitter transistor *RC* coupled amplifier.

Bipolar Junction Transistor RC Coupled Amplifiers. If an amplifier configuration contained no time-sensitive elements (such as capacitors), there would be no limit on the range of signal frequencies which could be amplified uniformly. That is, the bandwidth would be infinite. This situation, however, is not achievable in practice for two reasons. One, a coupling capacitor is generally employed between stages in order to eliminate transfer of the high d-c voltage levels to the input section of the following stage. This puts a limitation on the low-frequency transmission. Two, the amplifying devices themselves inherently possess capacitance, which although subject to diminution cannot be entirely eliminated. In transistors we must reckon with diffusion and transition capacitances, while in the vacuum tube it is the interelectrode capacitance which is important. These factors limit the high-frequency capability of the amplifier.

Three regions of the frequency spectrum are of concern to us in this study. One is the *midband frequency range*. This is defined as that frequency range over which the effects of all capacitors (coupling, diffusion, and transition) are negligible. The linear model of the common-emitter transistor amplifier as it applies to the midband frequency range takes on the form depicted in Fig. 9-34, which clearly is the same one used in Sec. 9-4 to compute amplifier performance. For our purposes here I_2 in Fig. 9-34 is considered to be the useful output current of the first stage (and therefore the input current of the second stage). The resistance R_2 denotes the net input resistance appearing at the base terminal of the second transistor. In this instance it consists of the bias circuit resistance, R_{B2}, of the second stage in parallel with the h_{ie} parameter of the second transistor.

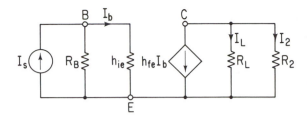

Fig. 9-34 Equivalent circuit of the common-emitter transistor amplifier as it applies to the midband frequency range.

The relationship of the current I_2 to the base current is found by applying the current division principle to the output section of Fig. 9-34. Thus

$$I_2 = \frac{-R_L}{R_L + R_2} h_{fe} I_b = \frac{-h_{fe}}{1 + R_2/R_L} I_b \qquad (9\text{-}176)$$

The minus sign simply recognizes that the actual current flow in this circuit is opposite to the assumed positive direction. Moreover, from the input circuit section we can write further that

$$I_b = \frac{R_B}{R_B + h_{ie}} I_s = \frac{1}{1 + h_{ie}/R_B} I_s \qquad (9\text{-}177)$$

Inserting Eq. (9-177) into Eq. (9-176) yields

$$I_2 = \frac{-h_{fe}}{(1 + R_2/R_L)(1 + h_{ie}/R_B)} I_s \qquad (9\text{-}178)$$

From Eq. (9-178) the current gain of the complete stage including the loading effect of the next stage becomes simply

$$A_{Im} \equiv \frac{I_2}{I_s} = \frac{-h_{fe}}{(1 + R_2/R_L)(1 + h_{ie}/R_B)} \qquad (9\text{-}179)$$

This expression is the literal equivalent of Eq. (9-134) of the preceding section. The subscript m is included to denote reference to the midband frequency range. In view of the absence of any frequency sensitive terms in Eq. (9-179), it follows that the gain in the midband frequency range remains invariant.

The second region of interest is the *low-frequency range*. How should the equivalent circuit of Fig. 9-34 be modified so that it becomes applicable in this range? In this connection it is helpful to recall that the diffusion and transition capacitances (C_D and C_T) of the transistor are measured in picofarads while the coupling capacitor C_C is measured in microfarads. Hence the reactances of the former two parameters are of the order of a million times larger than that of C_C. Recalling further that reactance is $1/\omega C$ and that ω is small at low frequencies, it follows that the reactance values of the diffusion and transition capacitors become so high as to appear as virtual open circuits. Hence C_D and C_T may be omitted from the low-frequency model. The capacitor C_C, however, cannot be removed, for if it were, the transmission to the second stage would be zero. In fact, C_C is intentionally chosen large in order to ensure satisfactory transmission even at low frequencies. Clearly, the larger the value of C_C, the smaller will be the frequency of the signal that can be successfully passed to the next stage. Of course, it is impossible to transmit a zero frequency (d-c) signal because the capacitor then behaves as an open circuit. The applicable equivalent circuit at low frequencies is thus that shown in Fig. 9-35.

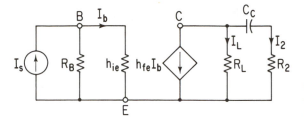

Fig. 9-35 Linear model of a common-emitter BJT amplifier that is applicable at low frequencies.

In this case current division applied to the output section of the equivalent circuit yields

$$I_2 = \frac{-R_L}{R_L + R_2 + 1/j\omega C_C} h_{fe}I_b = \frac{-R_L}{R_L + R_2}\left[\frac{1}{1 + \dfrac{1}{j\omega C_C(R_L + R_2)}}\right] \tag{9-180}$$

Inserting Eq. (9-177) for I_b and rearranging terms gives

$$I_2 = \frac{-h_{fe}}{(1 + R_2/R_L)(1 + h_{ie}/R_B)}\left[\frac{1}{1 - j\dfrac{1}{\omega(R_L + R_2)C_C}}\right]I_s \tag{9-181}$$

The quantity in brackets is a complex number, the value of which depends upon the frequency ω. Of particular interest is the value of frequency that makes the magnitude of the resulting complex number equal to $1/\sqrt{2}$ or 0.707. Examination of the bracketed term reveals that this magnitude prevails whenever the frequency has such a value as to make the coefficient of the j term unity. Hence, if we call this frequency ω_l, we have

$$\frac{1}{\omega_l(R_L + R_2)C_C} = 1$$

or

$$\omega_l = \frac{1}{(R_L + R_2)C_C} \qquad \text{rad/s} \tag{9-182}$$

Expressed in hertz this frequency becomes

$$\boxed{f_l = \frac{1}{2\pi(R_L + R_2)C_C}} \qquad \text{Hz} \tag{9-183}$$

Upon inserting Eqs. (9-182) and (9-179) into Eq. (9-181) and formulating the ratio of currents, the current gain at low frequency for the entire stage under loaded conditions is obtained. Thus

$$\boxed{A_{Il} = A_{Im}\frac{1}{1 - j(\omega_l/\omega)}} \tag{9-184}$$

This expression shows that as the input signal frequency ω gets smaller the magnitude of the low-frequency gain gets correspondingly smaller by the factor

$$\frac{1}{\sqrt{1 + (\omega_l/\omega)^2}}$$

Of course at $\omega = \omega_l$ there occurs about a 30% decrease in gain from the midband value. The frequency ω_l identifies the lower end of the useful bandwidth of the amplifier.

The third and final region of interest is the *high-frequency range*. By definition the high-frequency range is that range of frequencies for which the reactance value of the diffusion capacitance† C_D and the transition capacitance‡ C_T are low enough to command inclusion in the equivalent circuit. However, to locate these parameters, appropriate pairs of terminals must be identified across which to place C_D and C_T. In view of the fact that the diffusion capacitance is a property of the forward-biased section of the transistor, it would seem reasonable at first to place C_D between the base (B) and emitter (E) terminals. However, a more detailed study of the internal workings of the transistor discloses that the diffusion process of current flow begins at some internal point (B') of the base section after the current undergoes an *ohmic* drop. Accordingly, a more refined representation of the h_{ie} parameter is one which treats it in two parts: a base-spreading resistor $r_{BB'}$ that accounts for the ohmic drop, and the forward-biased resistor $r_{B'E}$ that relates the increase in diffusion current I_b to the corresponding increment of forward-biased applied voltage. This distinction is shown in the model appearing in Fig. 9-36. On this basis the correct location for the diffusion capacitance is across terminals B' and E. The value of base-spreading resistance lies in the vicinity of 20 per cent of h_{ie} and so needs to be reckoned with. The transition capacitance C_T is placed between terminals B' and C. This too is a correct location because C_T arises from a reversed-bias state in the transistor existing between the collector terminal and the base section. It is important to note now that the input and output sections are no longer isolated. In fact, this bridging effect of C_T creates a π form which is the reason why this version of the equivalent circuit is called the *hybrid-π* model. However, by means of an appropriate

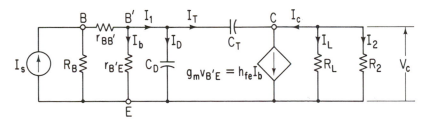

Fig. 9-36 Hybrid-π equivalent circuit for computing high-frequency performance.

† See p. 350.

‡ C_T is associated with the charge distribution across the junction in the depletion region and the potential energy barrier V_0.

analytical procedure it is possible to redraw the circuit of Fig. 9-36 so as once again to bring into evidence isolated input and output sections. Let us do this before proceeding with computing the high-frequency behavior.

Applying Kirchhoff's current law at the common junction point of C_D and C_T, we have

$$I_1 = I_D + I_T \tag{9-185}$$

The currents, I_D and I_T, may be expressed in terms of the product of the admittance of the circuit element carrying the current and the voltage appearing across its terminals. Thus

$$I_D = j\omega C_D v_{B'E} \tag{9-186}$$

and

$$I_T = j\omega C_T (v_{B'E} - V_c) \tag{9-187}$$

where V_c is the rms value of the collector voltage. Also, by the current law as it applies to terminal C, we can write

$$I_T - g_m v_{B'E} + I_c = 0 \tag{9-188}$$

where I_c is the current flowing through R_p, which is the equivalent resistance of the parallel combination of R_L and R_2.

The varying component of the collector-to-emitter voltage is given by

$$V_c = -I_c R_p \tag{9-189}$$

where

$$\frac{1}{R_p} = \frac{1}{R_L} + \frac{1}{R_2} \tag{9-190}$$

By Eq. (9-188) we can then write

$$V_c = -g_m v_{B'E} R_p + I_T R_p \tag{9-191}$$

However, in practice I_T is very small compared to $g_m v_{B'E}$ so that the last equation may be written

$$V_c \approx -g_m v_{B'E} R_p \tag{9-192}$$

Introducing this equation into Eq. (9-187) then allows I_T to be expressed as

$$I_T = j\omega C_T (1 + g_m R_p) v_{B'E} \tag{9-193}$$

Finally, by substituting Eqs. (9-193) and (9-186) into Eq. (9-185), we get

$$I_1 = j\omega [C_D + C_T (1 + g_m R_p)] v_{B'E} = j\omega C v_{B'E} \tag{9-194}$$

where C is an equivalent capacitance as seen between terminals B' and E, and is defined by

$$C = C_D + C_T (1 + g_m R_p) \tag{9-195}$$

It is interesting to note here that the effect of C_T is amplified by the factor $(1 + g_m R_p)$. This behavior is readily explained as follows. When terminal B' (which is the left side of C_T) is driven positive in potential about the Q point simultaneously through the action of the current generator, $g_m v_{B'E}$, terminal C (the right side of C_T) is driven negative, thereby accounting for a larger potential difference across the capacitor terminals and hence a larger current. This effect is called the Miller effect.

By means of Eq. (9-194) the current I_1 in Fig. 9-36 may be looked upon as flowing into a terminal capacitor C. In this way the circuit of Fig. 9-36 may be replaced by the arrangement shown in Fig. 9-37(a). The high-frequency performance of the transistor amplifier may now be found by working with the equivalent circuit of Fig. 9-37(a).

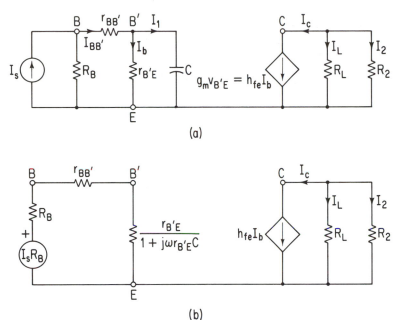

Fig. 9-37 The high-frequency equivalent circuit:
(a) representation in terms of an isolated input and output section;
(b) one-loop equivalent of the input section of (a).

The input current to the second stage I_2 is related to the base diffusion current I_b by

$$I_2 = \frac{-R_L}{R_L + R_2} h_{fe} I_b = \frac{-h_{fe}}{1 + R_2/R_L} I_b \qquad (9\text{-}196)$$

The parallel combination of $r_{B'E}$ and $1/j\omega C$ is expressed as $r_{B'E}/(1 + j\omega r_{B'E} C)$. Moreover, the input section of Fig. 9-37(a) can be replaced by the single-loop voltage source equivalent circuit shown in Fig. 9-37(b). By the voltage division principle we can then write

$$
\begin{aligned}
v_{B'E} &= \frac{\dfrac{r_{B'E}}{1 + j\omega r_{B'E} C}}{R_B + r_{BB'} + \dfrac{r_{B'E}}{1 + j\omega r_{B'E} C}} R_B I_s \\[2mm]
&= \frac{r_{B'E} R_B I_s}{R_B + h_{ie} + j\omega r_{B'E}(R_B + r_{BB'})C}
\end{aligned}
\qquad (9\text{-}197)
$$

where $h_{ie} = r_{BB'} + r_{B'E}$. The diffusion base current thus becomes

$$I_b = \frac{v_{B'E}}{r_{B'E}} = \frac{1}{1 + h_{ie}/R_B} \frac{1}{1 + j(\omega/\omega_h)} I_s \qquad (9\text{-}198)$$

where

$$\omega_h \equiv \frac{R_B + h_{ie}}{r_{B'E}(R_B + r_{BB'})C} \quad \text{rad/s} \qquad (9\text{-}199)$$

Substituting Eq. (9-198) into Eq. (9-196) and formulating the ratio of I_2 to I_s yields the desired expression for the complete-stage current gain under loaded conditions for high-band frequencies. Thus

$$A_{Ih} = \frac{I_2}{I_s} = \frac{-h_{fe}}{1 + R_2/R_L} \frac{1}{1 + h_{ie}/R_B} \frac{1}{1 + j(\omega/\omega_h)}$$

or

$$\boxed{A_{Ih} = A_{Im} \frac{1}{1 + j(\omega/\omega_h)}} \qquad (9\text{-}200)$$

Inspection of this result reveals that when the frequency of the input signal ω is allowed to become considerably larger than ω_h there is a sharp falloff of gain. For this reason the useful upper frequency limit of the amplifier is taken to be ω_h. It has already been pointed out that the useful lower frequency limit is ω_l. Accordingly, the useful range of frequencies over which the transistor amplifier performs effectively spreads from ω_l to ω_h. This range is called the *bandwidth*.

A plot of the current gain as a function of frequency on semilogarithmic paper appears as shown in Fig. 9-38. The behavior of the curve at low frequencies is dictated by Eq. (9-184). At the high frequencies the curve is described by Eq. (9-200). At both ends there is a serious dropoff of gain. Therefore, the useful portion lies between ω_l and ω_h.

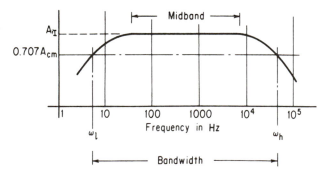

Fig. 9-38 Frequency response curve of an *RC* coupled amplifier.

EXAMPLE 9-3   Refer to Fig. 9-30. The 2N104 transistor has the following parameters:

$$r_{BB'} = 0.29 \text{ k}\Omega \qquad g_m = 0.032 \text{ S} \qquad h_{fe} = 44$$

$$r_{B'E} = 1.375 \text{ k}\Omega \qquad C_T = 40 \text{ pF}$$

$$\frac{1}{h_{oe}} = 150 \text{ k}\Omega \qquad C_D = 6900 \text{ pF}$$

(a) Compute the midband frequency current gain of the first stage of the circuit of Fig. 9-30. The first stage is to be taken from the base terminal of the first transistor to the base terminal of the second transistor.

(b) Find the bandwidth of the first-stage amplifier.

Solution: Refer to the equivalent circuit shown in Fig. 9-39. In the midband frequency range C is removed and C_C is replaced with a short circuit. The resulting equivalent circuit then becomes the same as that of Fig. 9-34 upon recalling that $h_{ie} = r_{BB'} + r_{B'E}$. At low frequencies C is removed and C_C is retained. At high frequencies C is retained and C_C is replaced by a short circuit.

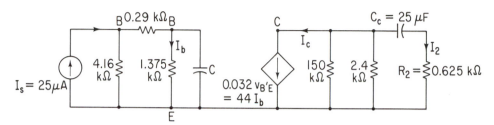

Fig. 9-39 Equivalent circuit for Example 9-3.

(a) The midband frequency current gain for the complete stage follows directly from Eq. (9-179):

$$A_{Im} = \frac{-h_{fe}}{(1 + 0.625/2.36)(1 + 1.67/4.16)} = \frac{-44}{1.265(1.402)} = 24.8 \qquad (9\text{-}201)$$

Here the quantity 2.36 kΩ is an equivalent load resistor which is the parallel resistance of R_L and $1/h_{oe}$. Note that this result is identical to that found in Eq. (9-134).

(b) To find the bandwith it is necessary to compute ω_l and ω_h. From Eq. (9-183) we get

$$f_l = \frac{1}{2\pi} \frac{1}{(R'_L + R_2)C_C} = \frac{1}{2\pi} \frac{1}{(2360 + 625)25 \times 10^{-6}} = 2.14 \text{ Hz} \qquad (9\text{-}202)$$

Note that R'_L is used in place of R_L to denote the parallel resistance of R_L with $1/h_{oe}$. Similarly, from Eq. (9-199),

$$\omega_h = \frac{R_B + h_{ie}}{r_{B'E}(R_B + r_{BB'})C} = \frac{4.16 + 1.67}{1.375(4160 + 290)C} \qquad (9\text{-}203)$$

where

$$C = C_D + C_T(1 + g_m R_p) = 6900 \times 10^{-12} + 40 \times 10^{-12}(1 + 0.032R_p) \qquad (9\text{-}204)$$

Now

$$R_p = \frac{R'_L R_2}{R'_L + R_2} = \frac{2.36(0.625)}{2.985} = 0.49 \text{ k}\Omega \qquad (9\text{-}205)$$

Thus

$$C = [6900 + 40(1 + 15.8)]10^{-12} = 7572 \times 10^{-12} \qquad (9\text{-}206)$$

Inserting Eq. (9-206) into Eq. (9-203) yields

$$\omega_h = 125,836 \text{ rad/s} \quad \text{or} \quad f_h = 20,027 \text{ Hz} \qquad (9\text{-}207)$$

Thus, the bandwidth ranges from approximately 2 Hz to 20,000 Hz, which gives excellent coverage of the audio range.

RC Coupled FET Amplifiers. The equivalent circuit of the FET showing the coupling capacitance and the FET interterminal capacitances appears in Fig. 9-40. For the midband frequencies C_{gs} and C_{gd} are removed and C_C is replaced by a short circuit. At the low frequencies C_C is inserted while C_{gs} and C_{gd} continue to be left out. At the high frequencies the equivalent circuit is solved with C_{gs} and C_{gd} present and C_C again replaced with a short circuit. The expression for the complete-stage voltage gain (gate of one stage to gate of next stage) at midfrequencies is

$$A_{Vm} = \frac{-g_m}{\dfrac{1}{r_d} + \dfrac{1}{R_L} + \dfrac{1}{R_g}} = \frac{-\mu}{1 + \dfrac{r_d}{R_L} + \dfrac{r_d}{R_g}} \tag{9-208}$$

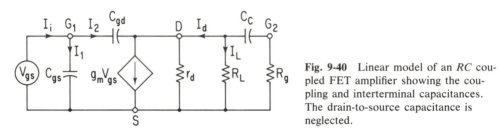

Fig. **9-40** Linear model of an *RC* coupled FET amplifier showing the coupling and interterminal capacitances. The drain-to-source capacitance is neglected.

At the low end of the bandwidth this equation is modified to

$$A_{Vl} = A_{Vm} \frac{1}{1 - j(\omega_l/\omega)} \tag{9-209}$$

where

$$\omega_l = \frac{1}{R_l C_c} \quad \text{and} \quad R_l = R_g + \frac{r_d R_L}{r_d + R_L} \tag{9-210}$$

At high frequencies the expression becomes

$$A_{Vh} = A_{Vm} \frac{1}{1 + j(\omega/\omega_h)} \tag{9-211}$$

where

$$\omega_h = \frac{1}{R_h C_i}, \quad \frac{1}{R_h} = \frac{1}{r_d} + \frac{1}{R_L} + \frac{1}{R_g} \quad \text{and} \quad C_i = C_{gs} + (1 + A_{Vm})C_{gd} \tag{9-212}$$

The bandwidth of the FET amplifier is equal to the range of frequencies from ω_l to ω_h. Expressed mathematically, we can write

$$\omega_{bw} = \omega_h - \omega_l = \frac{1}{R_h C_i} - \frac{1}{R_l R_C} \tag{9-213}$$

Examination of Eq. (9-213) discloses that C_i plays the key role in establishing the upper frequency limit while the coupling capacitor C_C exerts the key influence in fixing the lower limit of the bandwidth.

9-6 INTEGRATED CIRCUITS

In constructing the electronic amplifier circuits treated in the foregoing sections use is made of discrete components such as resistors, capacitors, diodes, and transistors which are assumed to be suitably interconnected with appropriate wiring. In recent years, however, highly sophisticated technological advances have been achieved that allow many *complete* amplifier circuits to be formed simultaneously including all the interconnections on small silicon wafers. These circuits are called *integrated circuits*. The aim of this section is to give a brief description of the fabrication procedure involved in producing integrated components and then to show how a complete amplifier can be so formed.

The current great popularity of integrated circuits, especially in the area of digital computers is attributable to several outstanding advantages. One is the tremendous saving in space. Many complete amplifier circuits can be formed on a wafer measuring only a small fraction of a square inch. Another is the considerable improvement in frequency response. The close spacing of the various diffused components and the necessarily short interconnections result in smaller capacitances which in turn extend the frequency range over which the gain can remain constant. These short connections have also made possible appreciable improvements in the speeds of digital computers. A third noteworthy advantage is the increased reliability of such circuits. This attribute stems primarily from the fact that *all* circuit components (including the interconnections) are made at the same time. Finally, the cost of integrated circuits is lower than for the more conventional types.

There are a few shortcomings too. The current state of the art of fabrication limits production to essentially low-power amplifiers. Moreover, the diffusion processes and other related procedures in the fabrication process are not good enough to permit a precise control of the parameter values for the circuit elements. However, control of the ratios is at a sufficiently acceptable level.

Fabrication Procedure. Attention is directed first to the forming of two diffused transistors on a single crystal silicon chip. The process begins by using a *p* substrate as a base having a thickness of approximately 6 mils. An epitaxial layer of about 25 mils is then applied to the *p*-substrate by allowing a thin film of silicon in the gas phase to grow in such a manner as to trap *n*-type impurity atoms. This done, the chip is next placed in an oven with an oxygen atmosphere and heated to 1000°C, thereby permitting a silicon dioxide (SiO_2) layer of one-half micron to be placed over the entire top surface of the *n*-type epitaxial layer. The chip has now been prepared for the creation of isolated *n*-regions. As will be seen presently, these *n*-regions will serve as the collector sections of the diffused transistors. The silicon dioxide layer is useful because it forbids impurities to diffuse through it.

Let it now be desired to create *n*-regions in the center of each half of the silicon chip. To achieve this result the following steps are necessary. First a photosensitive emulsion is placed on the chip covering the SiO_2 layer. This is followed by a mask (or stencil) with transparent windows appearing where the *n*-regions are to be located. The remainder of the mask is opaque. Refer to Fig. 9-41 for a pictorial description of the chip at this stage. Finally, the chip is

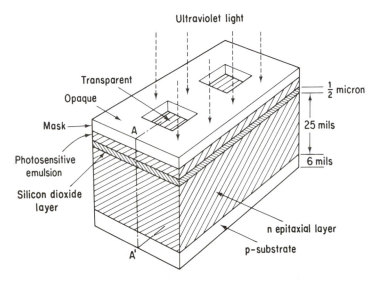

Fig. 9-41 Depicting the initial stages in the development of two diffused transistors for integrated circuits.

subjected to ultraviolet light, which has the effect of polymerizing the photosensitive emulsion that appears beneath the transparent areas but not that hidden beneath the opaque sections. The mask is then removed and the chip is developed, which means that it is washed in a suitable chemical thus allowing the unexposed portions of the chip to wash away. At this point the wafer then has a *p*-type impurity applied to it with the result that what was originally an *n*-type epitaxial layer is now converted to a *p*-type layer *with the sole exception of those regions that remained protected by the layers of SiO_2 and photosensitive emulsion.* At this stage of the process, if a cross-sectional view of the chip were taken at a section such as AA' in Fig. 9-41, the result would appear as depicted in Fig. 9-42. Here the isolated *n*-regions are clearly evident. It is instructive to note too that the *p*-regions between the isolation islands are doped more heavily than the

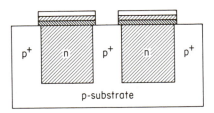

Fig. 9-42 Showing cross-sectional view through AA' of Fig. 9-40. Formation of isolated *n*-regions and p^+ doping.

p-substrate. In any integrated circuit in which the chip is used the *p*-substrate is always connected to the most negative part of the circuit in order to ensure that the *p-n* sections are all reversed-biased. If the region between the *n*-sections in Fig. 9-42 were not heavily doped, there would exist the danger that the depletion region of the reversed-biased *p-n* sections might spread sufficiently to touch, thereby connecting the two *n*-regions. The heavier *p*-doping guards against this and is represented by the *p*[+] notation.

Keep in mind that our aim is to develop two diffused transistors of the *n-p-n* type. Portions of the *n*-regions already formed are to serve as the collector sections. The next step in the fabrication process is to develop successive diffusions for the purpose of creating the base and emitter sections. Before proceeding with the creation of the base section, it will be necessary to remove the remaining parts of the SiO_2 and mask as shown in Fig. 9-42. This is readily accomplished by chemical solvents and mechanical abrasions. At this point another layer of SiO_2 is formed over the entire top surface of the chip. This is then followed by suitable masking in conjunction with the photolithographic process to create new openings in accordance with the configuration of Fig. 9-43. The wafer is subjected to *p*-doping to create the *p*-sections shown within the *n*-regions of this diagram. Because a portion of the newly created *p*-sections will serve as the base of the diffused transistor, this part of the process is often referred to as the *base diffusion* process.

The foregoing process is now repeated once again for the purpose of creating isolated *n*-sections in the newly formed isolated *p*-regions. Accordingly, by proper application of the masking and photolithographic process the openings depicted in Fig. 9-44 result. Then an *n*-type impurity is applied, thus creating the isolated *n*[+]-regions which serve as the emitter sections of the transistors. Here again the + notation is used to denote a heavy amount of doping. It is common practice to use an especially heavy doping at those places of the isolated regions where metal contacts are to be inserted. This helps to reduce contact resistance.

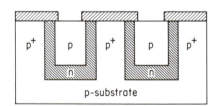

Fig. 9-43 Cross-sectional view illustrating openings for base diffusion within the isolated *n*-regions.

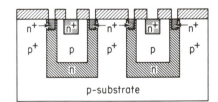

Fig. 9-44 Cross-sectional view illustrating openings for emitter diffusion within the isolated *p*-regions. Also illustrates *n*[+] diffusion for metal contacts for collector sections.

The final stage of the fabrication process involves depositing aluminum contacts by employing the masking and etching techniques previously applied. The final result is depicted in Fig. 9-45. Typical dimensions of this double diffused transistor wafer might measure 10 by 15 by 30 mils.

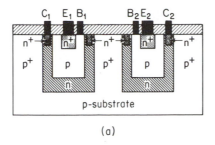

(a)

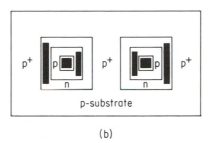

(b)

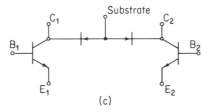

(c)

Fig. 9-45 Complete chip for two diffused transistors showing aluminum contacts (heavy shading): (a) cross-sectional view; (b) top view; (c) schematic representation.

Diffused Circuit Elements. The diffusion process described in the foregoing pages to produce transistors has also been applied to produce microresistors, microcapacitors, and diodes. Resistive elements can be had by making use of the ohmic property of the *p*- or *n*-material. In this connection refer to Fig. 9-43. It is instructive here to note that if metallic contacts were placed at the extreme ends of the opening inside the isolated *p*-regions a resistance would appear between the terminals which is dependent upon the dimensions of the *p*-region as well as its conductivity. Present technology permits values of resistance from 100 to 30,000 Ω to be realized by this scheme. The limits are dictated by practical factors associated with the fabrication procedure.

A typical microresistor is depicted in Fig. 9-46. Aluminum contacts are embedded in the *p*-material strip, which is designed for a specific length *l* and cross-sectional area as represented by a thickness dimension *t* and width ω. Accordingly, for a net effective length *l* and cross-sectional area $t\omega$ the resistance of the microresistor is given by $R = \rho(l/t\omega)$, where ρ denotes the resistivity of the *p* material. In practice, lasers are used to trim the resistor strip in order to obtain a fairly precise resistor value.

The microcapacitor can be formed by employing a diffusion procedure that leads to the configuration appearing in Fig. 9-47(a). There are two junctions in

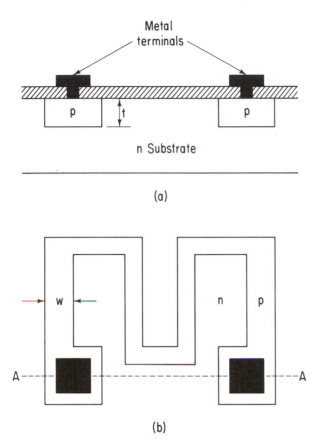

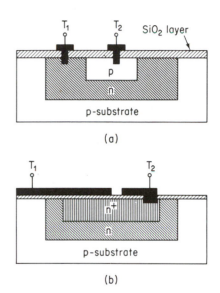

Fig. 9-46 Illustrating the construction details of a microresistor: (a) side view along cross section AA; (b) top view.

Fig. 9-47 The microcapacitor: (a) *p-n* junction type that must be kept reverse-biased; (b) nonpolarized MOS type.

this arrangement. The junction between the *p*-substrate and the isolated *n*-region serves to provide isolation. The other junction between the *n*-region and the isolated *p*-region furnishes the desired capacitance provided that this *p-n* junction is reversed-biased. The need for a reverse-bias represents a notable shortcoming of the *p-n* junction microcapacitor. Another disadvantage is the small size of capacitance that can be realized by this scheme owing to the limitations on the size of the microcircuit in which it appears. Typical values of capacitance are of the order of tens of picofarads.

An improvement over the *p-n* junction microcapacitor is the MOS microcapacitor which is illustrated in Fig. 9-47(b). Here the construction closely resembles that of the conventional capacitor. Terminal T_1 serves as one plate while the heavily doped isolated n^+-region serves as the other plate. The SiO_2 fulfills the role of a thin film dielectric. This arrangement offers the advantage of a nonpolarized capacitor and provides a slightly higher range of capacitance.

A diffused diode is obtained in integrated circuits from the transistor geometry. The two methods most preferred are either to use the base and emitter terminals of the diffused transistor or to connect the collector and base terminals to form the anode of the diode and then to use the emitter terminal as the cathode.

The Integrated Transistor Amplifier. We are now in a position to illustrate how a complete transistor amplifier circuit including bias resistors can be formed on a silicon wafer employing the masking and photolithographic procedures previously described. The difference now is that the masking pattern must be layed out to include the appropriate interconnections as well. Appearing in Fig. 9-48(b) and (c) is the complete integrated circuit layout of the transistor amplifier shown in Fig. 9-48(a). The five circuit elements—one capacitor, three resistors,

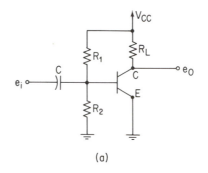

(a)

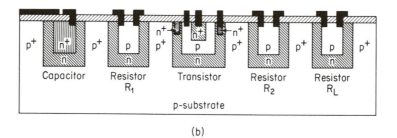

(b)

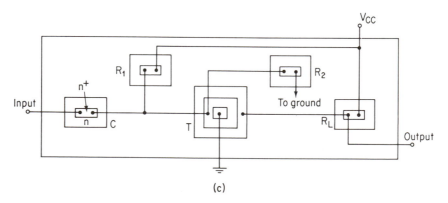

(c)

Fig. 9-48 Integrated transistor amplifier: (a) schematic diagram; (b) cross-sectional view showing each element; (c) top view showing the interconnections.

and one transistor—and all the interconnections are created by the same masking, etching, and diffusion process. In an actual integrated circuit the circuit elements would not appear in the tandem arrangement shown in Fig. 9-48(b) and (c). Rather the elements would be so placed as to make optimum use of the available space and to reduce the length of the interconnections to the minimum possible. The tandem arrangement is used here merely for convenience and clarification. The total area on the chip covered by this amplifier is a very small fraction of a square inch.

Summary review questions

1. Name the advantages that the common-emitter transistor configuration has over the common-base configuration.
2. Define the current amplification factor of the common-emitter configuration. How does this figure compare with that of the common-base circuit?
3. Describe how the d-c load line of a transistor amplifier in the common-emitter mode is constructed and explain its role in amplifier behavior.
4. How is the quiescent operating point of a common-emitter transistor amplifier determined?
5. Describe how the voltage gain of a common-emitter transistor amplifier is determined. The voltage gain involves a sign reversal. Explain why this is so.
6. Distinguish between the current gain and the current amplification factor of the common-emitter transistor amplifier. Which has the larger value? Why?
7. What is meant by the power gain of a common-emitter transistor amplifier? What is the source of this power gain?
8. State the reasons for the use of a coupling capacitor in joining one transistor amplifier stage to another.
9. How does the a-c load line differ from the d-c load line, and when is it necessary to use the a-c load line?
10. What is responsible for the distortion that can take place in the amplified output signal of a transistor amplifier? What can be done to reduce the degree of distortion?
11. When is it appropriate to do amplifier analysis by linear models?
12. Define mathematically and in words the four hybrid parameters that can be used to model a common-emitter transistor amplifier for small variations about the quiescent operating point.
13. Explain why the reverse amplification factor of the common-emitter transistor amplifier has so small a value.
14. Draw the linear equivalent circuit of the common-emitter transistor amplifier and identify the meaning of each circuit parameter and each circuit variable. Why are no d-c sources present in the linear model?
15. Draw an approximate equivalent circuit of the common-emitter transistor amplifier and state what assumptions are made to arrive at this configuration.
16. How are the *h*-parameters of the CE transistor amplifier found?
17. Are the *h*-parameters of the CE transistor amplifier constants over a small operating range? Over a large operating range?

18. Draw the small-signal equivalent circuit of an FET amplifier. Explain the meaning of each parameter and circuit variable shown.

19. Why is it necessary to modify the linear equivalent circuit of the FET amplifier when it is used at very high frequencies? Draw an equivalent circuit that shows the changes needed and explain the meaning of each additional parameter.

20. Why are biasing circuits used in transistor amplifiers?

21. Describe *thermal runaway* as it applies to the biasing of BJT amplifiers.

22. Describe the self-bias method of biasing a BJT amplifier and explain why this leads to satisfactory values of the stability factor.

23. What is the purpose of the shunt capacitor that is placed across the emitter resistor when the self-bias method of biasing BJT amplifiers is used? How is the value of this capacitance determined?

24. Show how the linear equivalent circuit of the BJT amplifier is modified by the presence of a self-bias resistor as well as a load resistor. Assume a CE configuration.

25. For the linear model of Question 24 distinguish between the transistor current gain, the current gain of the BJT amplifier and the current amplification factor.

26. What useful purpose does knowledge of the input and output resistances of a given amplifier stage of a multistage amplifier serve?

27. List at least two factors that cause a deterioration of the frequency response of a transistor amplifier. Explain how each factor affects the performance of the amplifier and the portion of the frequency range where it is effective.

28. Draw the low-frequency model of the BJT amplifier and comment on how it differs from the midfrequency model.

29. How is the current gain of the model drawn in Question 28 related to the current gain that prevails at midfrequency? How is the value of the lower-frequency bound of the bandwidth determined?

30. Draw the equivalent circuit of the BJT amplifier as it applies to the very high frequencies. Identify each circuit element and give a reason for its presence in the linear model.

31. How is the current gain of the BJT amplifier at high frequencies related to the value at midfrequencies? Identify the specific factor(s) responsible for the falloff of gain as the frequency of the input signal is allowed to assume large values.

32. Define the *bandwidth* of the BJT amplifier.

33. What is the Miller effect, and why is it important?

34. List some of the advantages of monolithic integrated electronic circuits over the discrete version of such circuits.

35. Outline briefly the fabrication process that yields (a) a diffused resistor, (b) a diffused capacitor, (c) a diffused diode, and (d) a diffused transistor.

Problems

GROUP I

9-1. A transistor amplifier has the configuration shown in Fig. 9-5. The circuit elements and power sources are specified as follows:

$$\text{2N109 transistor type} \qquad V_{CC} = 12 \text{ V}$$

$$R_L = 300 \ \Omega \qquad\qquad V_{BB} = 0.2 \text{ V}$$

$$R_2 = 900 \ \Omega \qquad\qquad R_1 \ = 1000 \ \Omega$$

Neglect reactance of C_C.

(a) What is the value of the collector-to-emitter voltage at the quiescent state?

(b) Assume R_1 represents the total resistance of the forward-biased input section. If $v_b = 0.2 \cos \omega t$ volt, find the current gain.

(c) Determine the voltage gain in part (b).

(d) Compute the power gain in part (b).

9-2. A transistor amplifier has the circuit shown in Fig. P9-2.

(a) As v_b is allowed to increase in amplitude, will cutoff (i.e., zero collector current) be reached before saturation occurs? Explain.

(b) What is the value of v_b beyond which saturation occurs?

9-3. Repeat Prob. 9-2 for the case where the base bias battery V_{BB} is changed to 0.7 V.

9-4. In the circuit of Fig. P9-4 the quiescent base current is 30 μA.

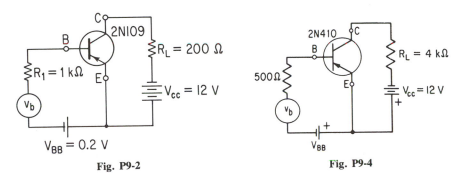

Fig. P9-2   **Fig. P9-4**

(a) Find the change in base input voltage needed to cause a 2-V drop in the collector-to-emitter voltage.

(b) Compute the ratio of the change in collector-to-emitter voltage to the change in input voltage for part (a).

(c) What is the value of the current gain?

(d) Compute the value of the voltage gain by using Eq. (9-23) and compare the result with that of part (b).

9-5. The 2N175 transistor whose average collector characteristics appear in Appendix E draws a quiescent current of 0.85 mA when the collector supply voltage is -6 V and the base current is -10 μA.

(a) Find the value of a load resistance.

(b) Assume the output of the amplifier of part (a) is connected to a succeeding stage whose input resistance is 5100 Ω. Compute the total swing in collector current for a swing in base current of -5 to -15 μA.

9-6. In the JFET amplifier circuit of Fig. 9-6 the input signal is given by $v_{gs} = 0.4 \sin \omega t$.

(a) Compute the voltage gain and compare with the result given by Eq. (9-38). Comment.

(b) Make a rough sketch of the variation of the drain-source current as a function of time, and comment.

9-7. The voltage across the load resistor R_d in Fig. P9-7 is found to be 14 V at the quiescent operating point. The average drain characteristics for this JFET appear in Fig. 9-7.

(a) Find the quiescent drain-source current.

(b) Compute the value of R_d.

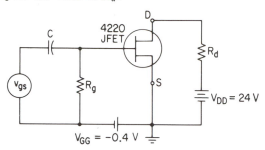

Fig. P9-7

9-8. The equivalent circuit parameters of the 2N139 transistor as they apply to the quiescent point $v_{CE} = -9$ V and $I_C = -1.0$ mA are as follows:

$$h_{ie} = 1325 \ \Omega$$

$$h_{fe} = 48.3$$

$$h_{oe} = 8.6 \times 10^{-6} \ \text{S}$$

This transistor is used in the common-emitter amplifier configuration having a load resistance of 2000 Ω. Neglect bias circuit effects.

Determine the change in base current which causes a 0.4-V change across the terminals of the load resistor.

9-9. The fixed-bias method illustrated in Fig. 9-15 is to be used for the 2N109 transistor operating with a supply voltage of $V_{CC} = 12$ V. The manufacturer's handbook lists the d-c base-to-emitter voltage as 0.15 V. Find the value of base bias resistance which for a load resistance of 200 Ω yields a Q-point lying midway between cutoff and saturation.

9-10. In the circuit of Fig. P9-10 the quiescent point is to be at $V_{CE} = 8.8$ V, $I_c = 3$ mA, and $I_B = 200 \ \mu\text{A}$. Find R_B and R_E if $R_L = 3$ kΩ.

9-11. In the circuit of Fig. 9-17 the following data apply:

$$V_{CC} = 12 \ \text{V} \qquad R_1 = 6000 \ \Omega$$

$$R_L = 240 \ \Omega \qquad R_2 = 1200 \ \Omega$$

$$R_E = 100 \ \Omega \qquad \text{2N104 transistor type}$$

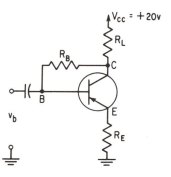

Fig. P9-10

(a) Find the bias equation for this amplifier.

(b) Determine the Q-point.

9-12. A current source of 2 μA rms value is applied as the input signal of Fig. P9-12. The equivalent circuit parameters of the 2N175 transistor are listed in the statement of Prob. 9-25.

(a) Draw a completely labeled equivalent-circuit diagram of the transistor amplifier.

(b) What is the rms value of the base signal current?

(c) What is the rms value of the varying component of the collector current?

(d) Compute the transistor amplifier current gain (i.e., from signal current to collector current).

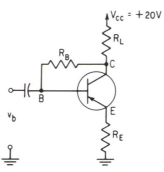

Fig. P9-12

9-13. In the circuit of Prob. 9-12 a signal source having an rms voltage of 6 mV and an internal resistance of 3 kΩ is applied to the input terminals.

(a) Find the stage amplifier current gain.

(b) Determine the value of voltage gain.

(c) Calculate the value of power gain.

(d) What is the value of the input resistance of the amplifier circuit?

(e) What is the value of output resistance?

9-14. A transistor amplifier has the circuit configuration shown in Fig. P9-12. However, the values of the circuit elements are as follows:

$$R_1 = 42 \text{ k}\Omega \qquad R_L = 3.2 \text{ k}\Omega$$

$$R_2 = 6.5 \text{ k}\Omega \qquad R_E = 0.8 \text{ k}\Omega$$

Also, the transistor used is the 2N139, the equivalent-circuit parameters of which are given in Prob. 9-8.

Compute the rms value of a signal-source current which causes the varying component of the collector voltage to change by 0.8 V rms. Assume the reactance of C_E to be negligibly small.

9-15. A 2N410 transistor is used in the amplifier configuration shown in Fig. P9-15. The parameters for this transistor are:

$$h_{ie} = 1325 \ \Omega$$

$$h_{fe} = 48.3$$

$$h_{oe} = 8.6 \times 10^{-6} \text{ S}$$

(a) Draw a clearly labeled equivalent circuit for this transistor amplifier.

(b) Compute the rms value of the output voltage for the specified input signal.

(c) Find the overall current, voltage, and power gains.

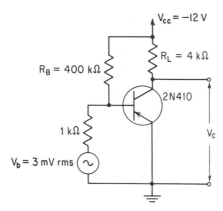

Fig. P9-15

9-16. A two-stage transistor amplifier has the configuration shown in Fig. P9-16. The input signal is a sinusoid of 20-mV rms value. Assume the reactance of the coupling capacitors and the base-to-emitter voltages of the transistors to be negligibly small. The manufacturer's semiconductor handbook lists the following information:

$$2N104 \text{ transistor: } h_{ie} = 1665 \ \Omega \ h_{fe} = 44$$

$$h_{oe} = 6.6 \times 10^{-6} \text{ S}$$

2N270 transistor: input resistance between base and emitter

terminals is 400 Ω

The average collector characteristics are available in Appendix E.

(a) Find the rms value of the current through R_{L1} when the second stage is disconnected.

(b) With the second stage connected, find the rms value of the base current in the 2N270 transistor.

(c) Compute the rms value of the varying components of the collector voltage and current at the output terminals of the second stage.

(d) Calculate the two-stage current gain.

(e) What is the total voltage gain?

(f) Find the overall power gain.

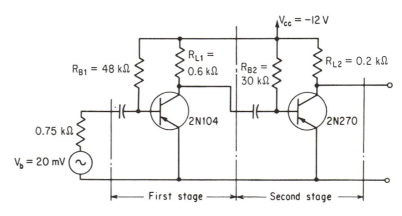

Fig. P9-16

Semiconductor Electronic Circuits   Chap. 9

9-17. Answer the questions of Prob. 9-16 for the two-stage transistor amplifier circuit shown in Fig. P9-17. Assume the reactance of all capacitors negligible. Note that the rms value of the input signal is 30 mV.

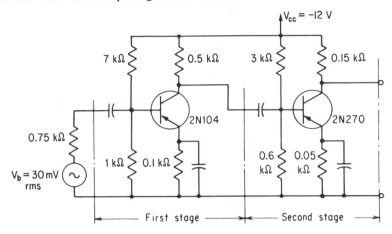

Fig. P9-17

9-18. The transistor amplifier of Fig. P9-15 is coupled by a 25-μF capacitor to a second-stage amplifier, the input resistance of which is 1200 Ω. Moreover, the diffusion capacitance of the 2N410 transistor is 1100 pF and the transition capacitance is 9.5 pF.

(a) Compute the midband-frequency current gain of the first stage. Consider the first stage as appearing between the base terminal of the first transistor and the base terminal of the second transistor.

(b) Find the lower limit of the bandwidth in Hz. Assume $r_{B'E} = 1250$ Ω, $r_{B'B} = 75$ Ω.

(c) Calculate the upper limit of the bandwidth in Hz. Assume that $g_m = 0.0386$.

(d) What is the magnitude and phase of the current which flows into the second stage? Assume that the 3-mV source has a midband frequency.

9-19. In the circuit of Fig. P9-16 the coupling capacitor between the first and second stages has a capacitance of 25 μF. The diffusion capacitance of the 2N104 is 6900 pF and the transition capacitance is 40 pF. The input resistance of the 2N270 between base and emitter terminals is 400 Ω.

(a) Draw the complete equivalent circuit of the first stage showing all parameters clearly labeled.

(b) What is the midband-frequency current gain of the first stage?

(c) Find the bandwidth in hertz.

(d) What is the effect of decreasing the value of load resistance on the upper limit of the bandwidth?

(e) Repeat part (d) for the lower limit of the bandwidth.

9-20. Repeat Prob. 9-19 for the circuit shown in Fig. P9-17.

9-21. A solid-state amplifier employs the junction field-effect transistor whose parameters are as follows:

$$g_m = 3 \times 10^{-3}\,S \qquad C_{gs} = 6\,pF$$

$$r_d = 70\,k\Omega \qquad C_{gd} = 2\,pF$$

In use this amplifier is coupled to a succeeding stage by a 25-μF capacitor feeding into an input resistor of $R_g = 0.1$ MΩ. The amplifier load resistor has a value of 10 kΩ.

(a) Determine the midfrequency voltage gain.

(b) Find the frequency at which the amplifier circuitry causes the output signal to lead the input signal by 45°.

(c) Find the frequency at which the amplifier circuitry causes the output signal to lag behind the input signal by 45°.

(d) What is the bandwidth of this amplifier?

9-22. For the JFET amplifier of Prob. 9-21 compute the value of input capacitance as seen by the source voltage and explain why this value is larger than the sum of the gate-source and gate-drain capacitances.

9-23. Design the layout for the integrated-circuit version of the amplifier depicted in Fig. P9-23. Show both a cross-sectional and a top view, and avoid overlapping connections.

9-24. Repeat Prob. 9-23 for the amplifier circuit of Fig. P9-24.

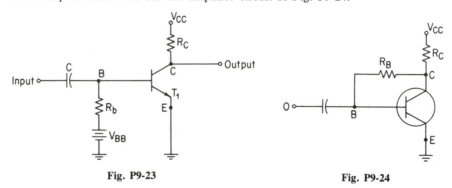

Fig. P9-23   Fig. P9-24

GROUP II

9-25. A manufacturer's handbook lists the following parameters for a 2N175 transistor as they apply to the quiescent point $v_{CE} = -4$ and $I_C = 0.5$ mA.

$$h_{ie} = 3570 \ \Omega$$

$$h_{fe} = 65$$

$$h_{oe} = 6.6 \times 10^{-6} \ \text{S}$$

(a) Draw the equivalent circuit showing all parameter values.

(b) A load resistor of 10 kΩ is placed in the collector-to-emitter circuit of this transistor. If the base current is increased about its quiescent value by 5 μA, compute the change in collector current assuming a conventional common-emitter amplifier configuration. Neglect the effect of the bias circuit.

(c) Find the current gain in part (b).

(d) Repeat parts (b) and (c) for a load resistor of 1 kΩ.

(e) Repeat parts (b) and (c) for a 100-Ω load resistor.

9-26. Prove the validity of Eq. (9-72).

9-27. The *n-p-n* germanium transistor shown in Fig. P9-27 has $\alpha = 0.98$, $I_{CO} = 4$ μA, and $V_{BE} \approx 0$.

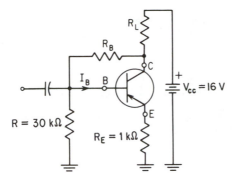

Fig. P9-27

(a) Calculate R_B and R_L so that the quiescent point is placed at $I_E = 2$ mA and $V_{CE} = 6$ V.

(b) Give a physical explanation of how this circuit provides d-c bias stabilization and compare it with other bias methods.

9-28. An amplifier employing the 2N104 transistor has the configuration shown in Fig. 9-17. Also, $R_L = 150 \ \Omega$, $R_E = 100 \ \Omega$, and $V_{CC} = 12$ V. Find the values of R_1 and R_2 which yield a quiescent point located at approximately the center of the load line.

9-29. In the circuit of Fig. P9-16 design an appropriate third stage. Select a suitable transistor and load resistor which can be properly driven by the preceding stage. Compute the a-c performance of the third stage.

9-30. Design a suitable transistor amplifier stage which can be made to follow the two-stage amplifier of Fig. P9-17. Be sure to specify an appropriate transistor, a load resistor, and suitable bias-circuit resistors.

chapter ten

Special Topics and Applications

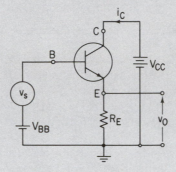

The subject matter that can be appropriately included in a chapter of this kind is extensive. In the interest of saving space, however, the selection of topics is restricted to those which are commonly found in electronic circuits and systems and at the same time are simple enough to be included in a book at this level.

10-1 ELECTRONIC-CIRCUIT APPLICATIONS INVOLVING DIODES

The diodes referred to in this section are of the semiconductor type. Moreover, the diode is assumed to be ideal, which means it has zero forward resistance and infinite backward resistance. Thus the symbolism used to denote the diode is the same as that shown in Fig. 8-19.

The Half-Wave Rectifier. The conversion of an alternating quantity to a unidirectional quantity is achieved by the circuitry depicted in Fig. 10-1. The alternating source e is connected across a load resistor R_L through an ideal diode D. Recalling that the diode conducts whenever its anode side (the arrowhead) is positive with respect to its cathode side (the quadrature line), it follows that during the positive half-cycle of the voltage the entire value appears across R_L. When the voltage is in its negative half-cycle, the diode is open so that no voltage appears across the load. The resulting waveshape of the voltage across the load

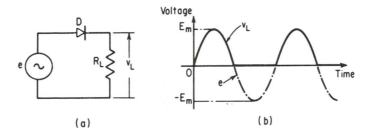

Fig. 10-1 (a) Half-wave rectifier circuit; (b) Unidirectional output voltage waveshape.

resistor, v_L, is shown in Fig. 10-1(b) as the solid curve. Expressed mathematically we can write

$$v_L = e = E_m \sin \omega t \qquad \text{for } e \text{ positive}$$
$$v_L = 0 \qquad \text{for } e \text{ negative}$$

(10-1)

Note the unidirectional aspect of the load voltage, which by virtue of the diode leads to an average different than zero. Specifically since the average value of a positive half sine wave is $(2/\pi)E_m$ and since there is zero contribution in the second half-cycle, the average value of the voltage appearing across R_L is clearly E_m/π.

The Peak Rectifier. If the output voltage of a half-wave rectifier is developed across a capacitor as shown in Fig. 10-2(a) rather than across a resistor, a nonpulsating unidirectional output voltage is obtained. Moreover, the magnitude of this voltage will be E_m rather than E_m/π. To understand how this is accomplished refer to Fig. 10-2 and assume that the source voltage is increasing from zero in the positive direction with the capacitor initially de-energized. As e increases positively, diode D conducts, thus permitting a charging current to flow through the capacitor. This builds up a voltage across the capacitor plates. At time t_1 in Fig. 10-2(b) the capacitor is charged to the peak value of the source voltage E_m. As time increases slightly beyond t_1, the potential of the anode side of the diode drops below E_m, but the potential of the cathode side remains at E_m because of the charge on the capacitor. Hence the diode opens. In fact the diode remains open for all time thereafter provided that there is no charge loss on the capacitor through leakage. Consequently the output voltage available at the capacitor terminals remains constant at E_m.

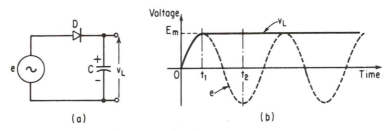

Fig. 10-2 (a) Peak rectifier circuit; (b) resulting output waveshape.

All diodes have a limit on the maximum reverse voltage they can reasonably be expected to withstand. For convenience this is called the *peak inverse voltage* (*PIV*) and must be checked when applying diodes so as to be sure not to exceed the value recommended by the manufacturer. The value of the PIV is readily found for the circuit of Fig. 10-2(a). Note that the peak negative voltage appearing across the diode terminals occurs when the cathode side is at the potential $+E_m$ and the plate side is at the potential $-E_m$. This occurs at time t_2 in Fig. 10-2(b). Thus the PIV is equal to $2E_m$. Accordingly in this application one must first compute the PIV and then select a diode capable of withstanding it. Otherwise, the peak value of the source voltage must be correspondingly reduced.

In practical applications of the peak rectifier the load connected across the capacitor terminals is very often resistive. The circuit therefore takes on the configuration illustrated in Fig. 10-3(a). The presence of the load resistor across the capacitor terminals provides a path for the capacitor to discharge during that part of the cycle when the source voltage has a magnitude less than the capacitor voltage or when it is negative. As a result the output voltage starts to diminish and the rate of decrease is dependent upon the time constant $R_L C$. Usually this time constant is very large compared to the period of the a-c source so that little loss of voltage occurs. The waveshape of the output voltage is shown in Fig. 10-3(b). Note that it is no longer nonpulsating. Also, in the interval between t_3 and t_4, the diode once again conducts, thus allowing the source to reestablish the output voltage to the E_m level at time t_4. This is repeated once in each cycle.

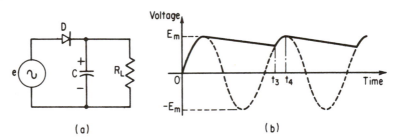

(a)

(b)

Fig. 10-3 (a) Peak rectifier circuit with connected load R_L; (b) resulting waveshape.

The Diode Clamper.   An interesting variation of the peak rectifier circuit is obtained by taking the output across the diode terminals rather than the capacitor terminals in the manner depicted in Fig. 10-4. Assume that the voltage source has been applied for some time so that the magnitude of the voltage appearing across the capacitor terminals is E_m. Recalling that in this configuration the diode remains open for all time after the initial charging quarter-cycle,† the expression for the voltage appearing across the diode terminals is

$$e_d = e - E_m \tag{10-2}$$

where e_d denotes the output voltage across the diode.

† Assuming the charging process starts at the beginning of the positive half-cycle.

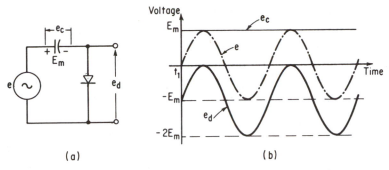

(a)　　　　　　　　　　　　(b)

Fig. 10-4　(a) Diode clamper circuit; (b) resulting waveshapes for $t_1 \gg 0$.

The quantity e is the instantaneous value of the source voltage having a peak E_m. A plot of Eq. (10-2) yields the variation shown in Fig. 10-4(b). Note that the only difference between the input waveshape e and the output waveshape e_d is a downward shift of the latter relative to the former by the amount $-E_m$, which is the fixed voltage appearing across the capacitor. This circuit is said to *clamp* the positive peak value of the source voltage at zero volts at the output terminals. The diode clamper finds wide application in television circuits because of the need to provide fixed peak voltages at various points throughout the circuit.

The Voltage Doubler.　If the diode clamper circuit is combined with the peak rectifier circuit in the fashion shown in Fig. 10-5, a voltage-doubler circuit results. This arrangement is commonly found in the power-supply sections of electronic circuits such as tape recorders, radios, and so on.

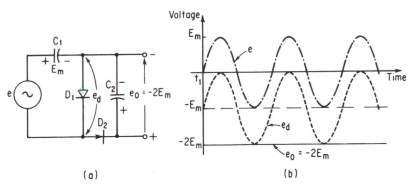

(a)　　　　　　　　　　　　(b)

Fig. 10-5　(a) Voltage-doubler circuit; (b) resulting waveshapes for $t_1 \gg 0$.

The basic idea of the voltage doubler is to take the output of the diode clamper and apply it to a peak rectifier circuit. Since the output voltage at the $D1$ diode terminals is a sinusoidal voltage with a peak negative value of $-2E_m$, application of this voltage to the peak rectifier circuit results in voltage appearing on C_2 equal to $2E_m$. The output waveshape is shown in Fig. 10-5(b).

The Diode Limiter.　Diodes are frequently used in electronic circuits to generate nonlinear functions for simulation purposes. One example is the generation of the limit-stop action often found in mechanical systems. For example, in an aircraft the elevator, aileron, and rudder control surfaces can be deflected only through a restricted region, beyond which no further movement of the control surfaces is possible in spite of increasing input signals. This behavior is described as a *limiting* action and may be simulated by the electronic circuit shown in Fig. 10-6.

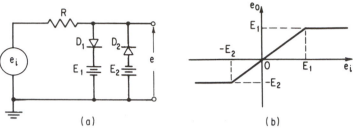

Fig. 10-6   (a) Diode limiter circuit; (b) transfer characteristic relating output and input voltages.

The voltage E_1 is used to simulate the upper limit of the output quantity corresponding to positive inputs. Similarly the voltage E_2 is used to simulate the lower limit corresponding to negative inputs. By making E_1 and E_2 different, we can simulate limits of different values. To understand the operation of this circuit, first consider that the input signal e_i is increasing in the positive sense and is less in magnitude than E_1. For this condition then diode D_1 is open because the anode side (arrowhead) is at potential $e_i < E_1$ and the cathode side is at the potential E_1. Moreover, diode D_2 is also open because any positive value of e (with respect to ground) makes the voltage across D_2 more negative. Consequently the input voltage e_i passes directly to the output terminals so that the output and input voltages are the same. Since there is no current flow in this instance, there is no loss of voltage across R. A straight line having unity slope graphically represents this situation.

Consider next that the input voltage e_i exceeds E_1. Now the voltage across D_1 is positive and so it conducts. When this happens the voltage E_1 appears directly at the output terminals. Hence the output voltage remains fixed at this value irrespective of any further increase in e_i. The difference between e_i ($> E_1$) and E_1 appears across R. The greater this difference, the larger the voltage drop across R. A graphical description of the output for $e_i > E_1$ is thus a straight horizontal line as depicted in Fig. 10-6(b).

By similar reasoning it is found that for negative inputs, diode D_1 is always open while diode D_2 is open only so long as the magnitude of this negative voltage is smaller than E_2.

10-2 THE EMITTER-FOLLOWER AMPLIFIER

An amplifier configuration which has some very useful properties is depicted in Fig. 10-7. It differs from the circuitry of the conventional amplifier by the absence

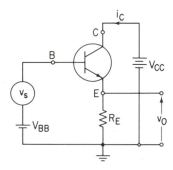

Fig. 10-7 Emitter-follower amplifier.

of a load resistor in the collector circuit as well as the absence of a bypass capacitor for the emitter resistor. Another point of interest is that with respect to a-c signals the collector is at ground potential. For this reason this configuration is sometimes called a common-collector amplifier.

As the input signal v_s increases, the a-c current i_c also increases, and as i_c passes through the emitter resistor it produces an increase in the a-c output voltage v_o. Similarly, as v_s decreases, there occurs a corresponding decrease in v_o. Since v_o is measured between the emitter and ground and because v_o follows the variations in v_s, this circuit is widely known as the *emitter-follower* amplifier.

In the interest of pointing out some of the useful features of this amplifier, it is helpful to analyze the emitter follower in terms of its equivalent circuit. As is customary the transistor is replaced by its appropriate equivalent representation in the manner shown in Fig. 10-8. The simplified version of the equivalent circuit is used here and of course all d-c quantities are omitted. One notable characteristic of the emitter follower is that it possesses a high input resistance—considerably higher in fact than the common-emitter or common-base modes of transistor amplifiers. This fact can be readily demonstrated by redrawing the equivalent circuit of Fig. 10-8(b) in the form shown in Fig. 10-9(a). Note that the shunt combination of R_E and the current source $h_{fe}i_b$ has been replaced by the Thévenin equivalent of a voltage source, $h_{fe}R_E i_b$, and series resistor R_E. In turn this voltage

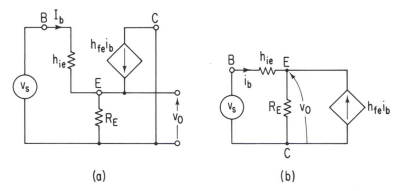

Fig. 10-8 Equivalent circuit of the emitter-follower amplifier:
(a) direct derivation from Fig. 10-7 with all d-c quantities omitted;
(b) rearrangement of the circuit elements in (a).

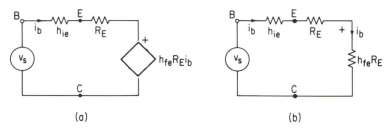

Fig. 10-9 (a) Replacing the current source of Fig. 10-8(b) by the Thévenin equivalent voltage source; (b) replacing the Thévenin voltage source by an equivalent resistance drop.

source may be replaced by the current i_b flowing through a resistor $h_{fe}R_E$ as illustrated in Fig. 10-9(b). It then follows from this circuit that the input resistance of the emitter follower is

$$R_i = \frac{v_s}{i_b} = h_{ie} + (1 + h_{fe})R_E \tag{10-3}$$

Since h_{fe} is a large quantity (50 to 80 typically) and R_E is greater than h_{ie}, an appreciable increase in the level of input resistance occurs with the common-collector configuration.

In view of the high input resistance it would be useful to know the value of the output resistance. This quantity is readily found by formulating the ratio of the open-circuit voltage to the short-circuit current as they appear at the emitter-collector terminals. The open-circuit voltage is found by removing R_E in Fig. 10-8(a) and measuring $v_o = v_{oc}$ for variations in v_s. A study of this circuit discloses that for the open-circuit condition $i_b = 0$ and so $v_{oc} = v_s$. The short-circuit current is found by replacing R_E with a short circuit in Fig. 10-8(b) and measuring the current. An examination of the circuit reveals that both i_b and $h_{fe}i_b$ flow through the short circuit, thus yielding a total current of $(i_b + h_{fe}i_b)$. Accordingly, the output resistance becomes

$$R_o = \frac{v_{oc}}{i_{sc}} = \frac{v_s}{i_b(1 + h_{fe})} \tag{10-4}$$

But for the short-circuit condition

$$\frac{v_s}{i_b} = h_{ie} \tag{10-5}$$

Therefore, we have finally,

$$R_o = \frac{h_{ie}}{1 + h_{fe}} \tag{10-6}$$

Recall that h_{ie} is the forward-biased input resistance of the common-emitter mode and as such has a value of resistance that is relatively low. In the emitter-follower configuration, however, this quantity is even further reduced by a substantial factor, thus leading to a very low output resistance.

The combination of a very high input resistance and a very low output resistance makes the emitter-follower a very popular circuit for providing *resistance transformation.* Very often in electronic circuits the loads (i.e., the places where the amplified signals finally do their useful work) are characterized by low resistance, while signals to be amplified are associated with circuits having high resistance levels. Without the use of an appropriate transformation device, such as the emitter-follower, these amplified signals could not succeed in bringing about substantial changes in energy at the loads because most of the signal would be consumed as an internal resistance drop. But with the emitter-follower providing the coupling to the load, the load assumes the role of prime importance in the output circuit.

What is the voltage gain of the emitter-follower amplifier? The answer can be had by relating v_o to v_s in the configuration of Fig. 10-8. Writing Kirchhoff's current law at the emitter junction E in Fig. 10-8(b) yields

$$i_b - \frac{v_o}{R_E} + h_{fe}i_b = 0 \qquad (10\text{-}7)$$

or

$$v_o = i_b(1 + h_{fe})R_E \qquad (10\text{-}8)$$

Inserting the expression for i_b from Eq. (10-3) provides the relationship involving v_o and v_s. Thus

$$v_o = (1 + h_{fe})R_E\left[\frac{v_s}{h_{ie} + (1 + h_{fe})R_E}\right] \qquad (10\text{-}9)$$

Formulating the ratio of output voltage v_o to input signal v_s and manipulating for compactness leads to

$$\frac{v_o}{v_s} = \frac{1}{1 + \dfrac{h_{ie}}{(1 + h_{fe})R_E}} \qquad (10\text{-}10)$$

A study of this last expression makes it quite clear that the voltage "gain" of an emitter-follower amplifier is never unity but comes extremely close. Values of 0.99 and 0.995 are typical. This is a significant conclusion because it means that the high level of signal voltage which is often applied to the input of the emitter-follower amplifier appears at the output circuit almost undiminished. The important difference is that in the output circuit this voltage level is associated with a greatly amplified collector current. Hence, although there is no voltage gain, there is a very substantial gain in power delivered to the load as represented by the emitter resistor R_E.

10-3 THE PUSH-PULL AMPLIFIER

The push-pull amplifier is a power amplifier, and it is frequently found in the output stages of electronic circuits. It is used whenever high output power at high efficiency and little distortion is required. The push-pull amplifier has these advantages to offer over other arrangements such as the parallel combination of two power amplifiers to provide the same output. Figure 10-10 shows the circuitry

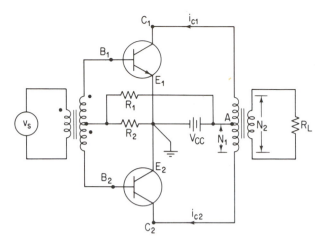

Fig. 10-10 Push-pull amplifier employing transistors.

of a typical push-pull amplifier. Resistors R_1 and R_2 are used to furnish the proper bias for the circuit. The advantages of the push-pull arrangement, however, do not come without a price. Note the need in the base circuit of each transistor of signals of equal magnitude and opposite signs with respect to ground. Accordingly, it becomes necessary to apply the input signal to the push-pull amplifier through a transformer as shown in Fig. 10-10 or else to include a driver stage to generate these signals.

In the push-pull circuit the actual load resistor R_L is connected to the transistors by means of a transformer which is equipped with a center-tapped primary winding (point A in Fig. 10-10). The power supply voltage is connected between the emitter terminals and this center tap. The quiescent current of each transistor flows through each half of the primary winding in opposite directions so that no saturation of the magnetic core occurs.

When the base current of one transistor is being driven positive with respect to the Q-point, the collector current increases, thus causing a decrease in collector potential relative to ground. At the same time, however, a reverse action is taking place in the base circuit of the second transistor. Its base current is decreasing, this causing a drop in the collector current with a consequent rise in collector potential with respect to ground. This means that the a-c current flowing through the transformer primary winding is in the *same* direction. As i_{c1} in Fig. 10-10 increases (i.e. pulls), the current i_{c2} decreases (i.e. pushes). Hence the name push-pull amplifier.

In view of the fact that the load current in the secondary of the transformer is proportional to the *difference* of the collector currents, it follows that the harmonic content will be less than it is in either of the collector currents because of cancellation effects. Consequently, the push-pull arrangement yields much less distortion in the output.

The use of transformers in the push-pull amplifier of Fig. 10-10 makes it bulky and expensive, especially in this day of the integrated circuit. Needless to say, an integrated circuit version of the push-pull amplifier is also available and the circuitry is depicted in Fig. 10-11. Here, by choosing the bias resistor

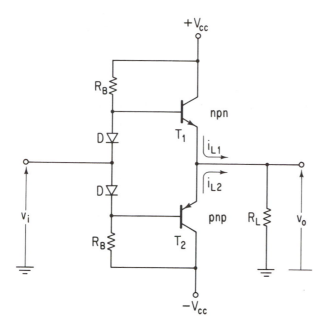

Fig. 10-11 An IC version of the push-pull amplifier.

R_B to bias transistor T_1 to cutoff, V_o is caused to follow V_i during the positive excursion of the input signal V_i and to be zero during the negative half-cycle. This corresponds to Class B amplifier operation. In a similar fashion, during the negative excursion of the input signal it is transistor T_2 which supplies the current i_{L2} to the load. Note that the diodes provide the base-emitter offset voltages. Therefore, with this IC version of the push-pull amplifier, Class B operation, which is characterized by low distortion of the input signal and high efficiency, is achieved at low cost and minimal space requirements.

It is worthwhile to note at this point that the more common practice today in audio frequency solid state circuitry is to utilize designs that do not require transformers.

10-4 MODULATION

The term *modulation* is used to describe the process by which some characteristic of a carrier is varied in accordance with a modulating wave. A simple example of modulation can be illustrated by referring to the circuit of Fig. 10-12. If the switch is opened and closed, say in a periodic fashion, the voltage (or power) delivered to resistor R is altered. Thus when the switch is closed the d-c carrier

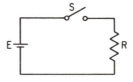

Fig. 10-12 A periodic opening and closing of the switch illustrates the modulation process.

is applied to the load; when the switch is opened the d-c carrier is reduced to zero. The opening and closing of the switch represents the modulating wave. Although in this example the carrier is a d-c quantity, in general it may be d-c, sinusoidal, or a series of pulses.

One very important application of modulation occurs in radio transmission. To transmit radio signals successfully and in an efficient manner, it is necessary to radiate electric power. At d-c it is impossible to radiate electrical energy. At audio frequencies (15 to 15,000 Hz) radiation is not practicable because of the huge antenna sizes required; in addition, the efficiency of radiation is very poor. Radiation of electrical energy is practicable only at high frequencies—for example, above 20 kHz. It should be apparent then that if speech is to be transmitted properly, some means must be devised which will permit transmission to occur at high frequencies while it simultaneously allows carrying the intelligence contained in the audio-frequency signals. The solution lies in modulating the high-frequency signals (called the *carrier wave* and denoted by ω_c) by the audio-frequency signals (called the *modulating wave* and denoted by ω_m).

Although there are many modulation procedures possible, our attention is confined here to the two most important ones. Since it often happens that the wave to be modulated is sinusoidal, we may write for the unmodulated signal

$$i = A_c \sin (\omega_c t + \phi) \tag{10-11}$$

where A_c denotes the amplitude of the carrier signal and ω_c is its frequency in radians per second. A study of Eq. (10-11) reveals that the carrier wave may have either its amplitude A_c or its argument $\omega_c t$ altered. When the amplitude is modified in accordance with a much lower-frequency modulating wave, the alteration process is called *amplitude modulation* and abbreviated AM. When the argument is modified by changing the frequency ω_c, the process is described as *frequency modulation* and abbreviated FM. We shall discuss here only amplitude modulation.

It is helpful at this point to identify some of the distinguishing characteristics of an amplitude-modulated carrier wave. For such a wave the amplitude becomes time-dependent and may be expressed as

$$A_c(t) = A_c + A_m \sin \omega_m t \tag{10-12}$$

Introducing

$$m = \frac{A_m}{A_c} \tag{10-13}$$

Eq. (10-12) may be rewritten as

$$A_c(t) = A_c(1 + m \sin \omega_m t) \tag{10-14}$$

The quantity m is called the *modulation index* and is restricted to values between zero and unity. The last expression indicates that the amplitude of the carrier signal varies sinusoidally at the modulating frequency ω_m between the limits $A_c(1 + m)$ and $A_c(1 - m)$. The complete time expression for the amplitude-modulated wave then becomes

$$i = A_c(1 + m \sin \omega_m t) \sin \omega_c t \tag{10-15}$$

A normalized plot of Eq. (10-15) appears in Fig. 10-13(a). Note the maximum and minimum values of the amplitude. It is the variation in the amplitude of the carrier wave which contains the intelligence to be transmitted.

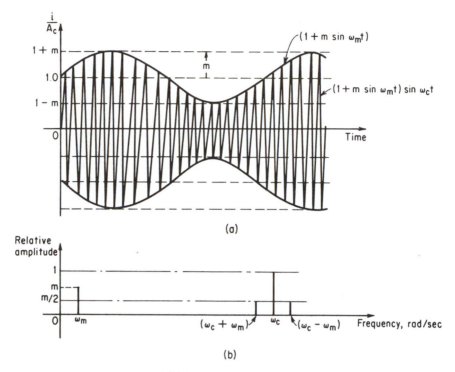

(a)

(b)

Fig. 10-13 (a) Amplitude-modulated sine wave with $m < 1$; (b) frequency spectrum of a sinusoidally modulated wave.

An alternative expression may be used to represent the modulated carrier wave which reveals additional information about the nature of amplitude modulation. From Eq. (10-15) we can write

$$\frac{i}{A_c} = \sin \omega_c t + m \sin \omega_m t \sin \omega_c t \qquad (10\text{-}16)$$

Inserting

$$\sin \omega_m t \sin \omega_c t = \tfrac{1}{2} \cos (\omega_c - \omega_m)t - \tfrac{1}{2} \cos (\omega_c + \omega_m)t \qquad (10\text{-}17)$$

into Eq. (10-16) yields

$$\frac{i}{A_c} = \sin \omega_c t + \frac{m}{2} \cos (\omega_c - \omega_m)t - \frac{m}{2} \cos (\omega_c + \omega_m)t \qquad (10\text{-}18)$$

Inspection of Eq. (10-18) reveals that the amplitude-modulated wave is equivalent to the summation of three sinusoids: one having a unity amplitude and frequency ω_c, the second having amplitude $m/2$ and frequency $(\omega_c - \omega_m)$, and the third having amplitude $m/2$ and frequency $(\omega_c + \omega_m)$. In practical radio transmission ω_c may be many times greater than ω_m. Hence the frequency of the second and third terms on the right side of Eq. (10-18) is generally close to the carrier frequency. Figure 10-13(b) represents this situation graphically on the frequency-spectrum plot. The frequency components contained in the amplitude-modulated wave are shown by vertical lines appropriately located along the frequency axis. The height of each vertical line is drawn proportional to its amplitude. The lower frequency component $(\omega_c - \omega_m)$ is called the *lower side frequency*. The upper frequency component $(\omega_c + \omega_m)$ is called the *upper side frequency*. Any amplitude-modulated wave must contain at least these frequencies.

How is the amplitude-modulation process implemented in electronic circuitry? Usually implementation is achieved by one of two methods: varying the quiescent operating point or using a square-law device. In either case the objective is to modify the single-frequency carrier wave so as to create new frequencies—the sideband frequencies. Appearing in Fig. 10-14 is an amplitude-modulated amplifier which achieves modulation by base injection. That is, the modulating signal develops a voltage across R_1, which in turn alters the quiescent point in the base circuit. Thus as the bias current is increased the signal at the collector terminal is correspondingly increased. Similarly, as the bias current is decreased during the reversed part of the cycle of the modulating signal, the output at the collector is also decreased. Consequently, a plot of the collector output will be found to yield a variation similar to that shown in Fig. 10-13. The resistor R_E in Fig. 10-14 is the self-bias emitter resistor and C_E is its bypass capacitor. Also, R_2 is a dropping resistor which limits the d-c voltage drop across R_1, which in turn establishes the initial quiescent base current. Capacitor C_2 is a by-pass capacitor for the carrier signal in the input circuit. In understanding the operation of this circuit it is helpful to keep in mind the fact that ω_m is very small compared to ω_c. Hence for any given cycle of the carrier frequency the bias in the base-to-emitter circuit appears to be fixed.

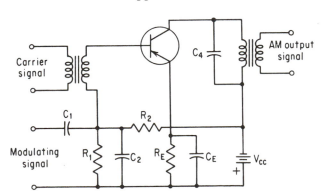

Fig. 10-14 Amplitude-modulated amplifier with base injection of the modulating signal.

Another arrangement which permits the introduction of a time-varying element to create new frequencies is shown in Fig. 10-15. Here the modulating signal is injected in the collector circuit. Accordingly, the total supply voltage in the collector circuit is

$$\text{total collector-supply voltage} = V_{CC} + kV_m \sin \omega_m t$$
$$= V_{CC}(1 + m \sin \omega_m t) \tag{10-19}$$

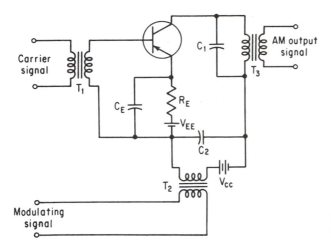

Fig. 10-15 Amplitude-modulated amplifier with collector injection.

where

$$m = \frac{kV_m}{V_{CC}} \qquad (10\text{-}20)$$

and k denotes a proportionality factor of transformer $T2$. For a given load impedance as the collector-supply voltage varies in accordance with Eq. (10-19), the load line moves parallel to itself up and down the transistor output characteristics, thereby changing the quiescent operating point. As the collector-supply voltage takes on its peak value of $V_{CC}(1 + m)$, the a-c current increases. Similarly as the collector-supply voltage diminishes to its minimum of $V_{CC}(1 - m)$, there occurs a corresponding decrease in a-c collector current. Once again the output appears as depicted in Fig. 10-13. Capacitor C_2 is used in Fig. 10-15 to by-pass the carrier signal in the collector circuit and C_1 is selected to provide series resonance with the primary winding of transformer $T3$.

The use of a square-law device to produce an amplitude-modulated signal can be described by the circuitry depicted in Fig. 10-16. Here it is assumed that the current is describable by two terms of the Taylor series expansion

$$i = A_1 e_s + A_2 e_s^2 \qquad (10\text{-}21)$$

where

$$e_s = E_c \sin \omega_c t + E_m \sin \omega_m t \qquad (10\text{-}22)$$

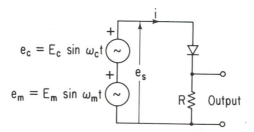

Fig. 10-16 Square-law diode modulator.

Upon inserting the expression for e_s into Eq. (10-21), we obtain

$$i = A_1 E_c \sin \omega_c t + A_1 E_m \sin \omega_m t \tag{10-23}$$
$$+ A_2 [E_c^2 \sin^2 \omega_c t + E_c E_m \sin \omega_c t \sin \omega_m t + E_m^2 \sin^2 \omega_m t]$$

The quantity in brackets derives from the square term in Eq. (10-21). A little thought reveals that it is this term which gives rise to the product of the carrier and modulating sinusoids, which in turn yields the upper and lower frequency sidebands thereby permitting the amplitude-modulated signal to be identified. By introducing the trigonometric identity of Eq. (10-17) and $\sin^2 \omega_c t = \frac{1}{2} - \cos 2\omega_c t$, Eq. (10-23) may be rearranged to read as follows:

$$i = \frac{A_2}{2}(E_c^2 + E_m^2) + A_1 E_m \sin \omega_m t - \frac{A_2}{2} E_c^2 \cos 2\omega_c t - \frac{A_2}{2} E_m^2 \cos 2\omega_m t$$

$$+ A_1 E_c \sin \omega_c t + \frac{A_2 E_c E_m}{2} \cos (\omega_c - \omega_m)t - \frac{A_2 E_c E_m}{2} \cos (\omega_c + \omega_m)t \tag{10-24}$$

The last three terms of Eq. (10-24) are identical in form to those appearing in Eq. (10-18), so that together they yield in the output resistor the desired amplitude-modulated signal. The remaining four terms are of no use; they can be filtered out by appropriate band-pass filters which will pass ω_c and the two sideband frequencies but reject all else.

Frequency modulation is obtained in electronic circuits by varying the frequency of the carrier signal in a way which makes it dependent upon the instantaneous value of the modulating signal. Expressed mathematically, the expression for ω_c in Eq. (10-11) becomes

$$\omega_c(t) = \omega_c + k_f E_m \sin \omega_m t \tag{10-25}$$

A frequency-modulated output is sometimes obtained by using the modulating signal to vary the gain of an oscillator, thereby changing its frequency of oscillation in accordance with Eq. (10-25). For further treatment of this subject matter the reader is referred to the bibliography at the end of the book.

10-5 AMPLITUDE DEMODULATION OR DETECTION

At the transmitting end of a radio station it is easier to radiate electrical energy at high frequencies (above 500 kHz). For this reason the intelligence contained in an audio signal (15 Hz to 15 kHz) is used to modify the amplitude of the high-frequency carrier wave. However, once the transmission has been achieved by being picked up by a local radio antenna and the AM signal suitably amplified, the radio receiver must also be equipped with electronic circuitry which will allow the recovery of the information appearing in the form of the amplitude variations of the carrier wave. The process of recovering the modulating wave from the carrier is called *amplitude demodulation* or, more simply, *detection*. Appearing in Fig. 10-17 is a circuit which is commonly used to accomplish this result. It is called a *diode detector*. The principle of operation is simple, since

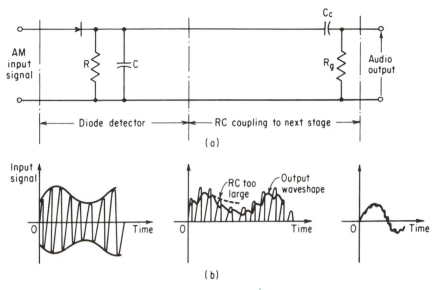

Fig. 10-17 Illustrating the operation of a diode detector: (a) diode detector and coupling circuit; (b) waveshapes at various points in the circuit of (a).

it essentially requires rectifying the carrier wave. As a matter of fact some form of rectification is involved in all demodulation circuits.

When the amplitude-modulated wave is applied to the input terminals of the diode detector, the diode conducts when the carrier signal is positive and cuts off when it is negative. Consequently, each positive half-cycle of the carrier wave appears across the parallel combination of R and C in the manner depicted in Fig. 10-17(b). The capacitor C charges to nearly the peak value of the positive half-cycle of the carrier wave. (A small voltage drop occurs across the diode.) During the negative half-cycle of the carrier the capacitor C discharges through the resistor R at a time constant RC. In choosing R and C care must be taken not to select too small a value for RC, otherwise the charge will be dissipated and the voltage reduced to zero during the negative half-cycle. This condition can be avoided by choosing RC large compared to the time it takes for the carrier to pass through one cycle. On the other hand, it is important that RC not be chosen excessively large, for then the capacitor will discharge so slowly that in the decreasing portion of the rectified output *diagonal clipping* occurs as indicated in Fig. 10-17(b). The best value for RC is that which causes the rate of discharge of the capacitor to follow the variations in the modulating signal.

The output signal of the diode detector has the somewhat jagged variation depicted in Fig. 10-17(b). It has a waveshape which very closely resembles that of the modulating signal, with the exception that it contains a d-c level. This, however, is readily removed by using capacitance-resistance coupling to the next stage as shown in Fig. 10-17(a). The output of this coupling network is a signal which very closely matches the original modulating signal. Since this signal may be thought to represent one of the many similar signals found in speech sounds, the combined process of modulation and demodulation thus performs a key role in the successful transmission of the audio frequencies generated by the human voice.

It is possible to achieve amplitude demodulation by using a three-element electronic device. A typical example of such a configuration is illustrated in Fig. 10-18. This circuit is called a *transistor detector*. In addition to detection of the modulated carrier input signal, it also furnishes amplification of the modulating signal. Recalling that for the transistor shown in Fig. 10-18 conduction occurs only from emitter to base because of the diode action associated with these elements, it is easy to see that the amplitude-modulated wave is rectified in the emitter-base circuit in precisely the same manner as occurs in the diode detector of Fig. 10-17(a). In fact resistor R_1 and capacitor C_1 serve the same function as do R and C in Fig. 10-17(a). Recall, too, that the voltage developed across R_1 in Fig. 10-18 varies in accordance with the modulating signal rate. Since this variation exists in the transistor input circuit, it causes a corresponding amplified variation to occur in the collector circuit—specifically across R_2. Capacitor C_2 is selected to by-pass the carrier-frequency signal and C_3 serves to block the d-c voltage level from passing to the next stage. As a result, the voltage appearing across the output terminals of this circuit is an amplified form of the modulating portion of the input signal.

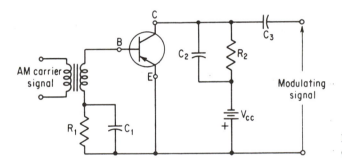

Fig. 10-18 Simplified transistor detector circuit.

Summary review questions

1. Draw a half-wave rectifier circuit and sketch the voltage waveshape that appears across the load resistor.
2. By means of a schematic circuit diagram illustrate the peak rectifier. If the supply voltage is $v(t) = V_m \sin \omega t$, what is the voltage across the load resistor?
3. Explain and illustrate the meaning of the peak inverse voltage. Why is this important to know in rectifier circuits?
4. Draw and explain the operation of the voltage doubler.
5. Draw and explain the operation of the diode limiter circuit. State the usefulness of this circuit.

6. Name the notable features of the emitter-follower amplifier. How can these features be used to advantage in electronic circuits? Illustrate.

7. What is a push-pull amplifier? What special properties does it offer?

8. Explain what is meant by the term *modulation* as it applies to electronic circuits.

9. Describe and illustrate amplitude modulation.

10. Explain the meaning of the following terms: modulation index, lower-side frequency, upper-side frequency, sideband frequencies.

11. Describe how a square-law electronic device can be used to provide an amplitude-modulated signal.

12. What is meant by *amplitude demodulation*? What is another name for this process?

13. Explain the operation of a diode detector. What care must be taken in choosing the time constant of the diode detector?

Problems

GROUP I

10-1. In the circuit of Fig. 10-1(a) assume that the sinusoidal source is replaced by a periodic symmetrical rectangular wave having a positive amplitude of 10 V and a negative amplitude of 5 V. Determine
   (a) The waveform of the voltage developed across R_L.
   (b) The average value of the load voltage.
   (c) The rms value of the load voltage.

10-2. A voltage of $100 \sin \omega t$ is applied to the circuit of Fig. P10-2.
   (a) Sketch the variation of current as a function of time. Clearly indicate all points of interest.
   (b) Find the average value of the current.

10-3. In the circuit of Fig. P10-3 each diode is assumed to have a resistance of 10 Ω. With respect to ground the voltages e_1 and e_2 are equal and opposite. Specifically, $e_1 = 100 \sin \omega t$.

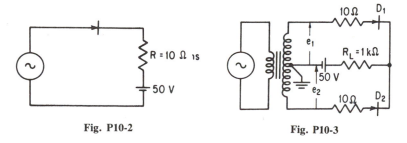

Fig. P10-2   Fig. P10-3

   (a) Sketch the variation of the battery current as a function of time.
   (b) Find the average value of this current.
   (c) What advantage does this circuit have over that depicted in Fig. P10-2?

10-4. The sinusoidal voltage $e = E_m \sin \omega t$ is applied to the circuitry shown in Fig. P10-4.

(a) Sketch the variation of the voltage across R for several cycles.

(b) If E_m is 100 V, what is the average value of the voltage across R?

10-5. If a sinusoidal source voltage e having an amplitude of E_m is applied to the circuit illustrated in Fig. P10-5, find the value of the output voltage in terms of E_m.

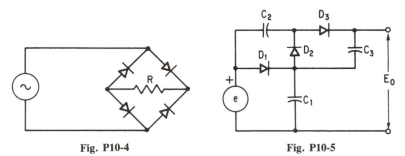

Fig. P10-4   Fig. P10-5

10-6. For the circuit shown in Fig. P10-6 plot the output voltage e_0 as a function of the input signal voltage e_i. Assume ideal diodes.

10-7. In the circuit of Fig. P10-7 determine the magnitude of the voltage appearing across R if the voltage source is $e = E_m \sin \omega t$. Assume that the resistance R has a very high value.

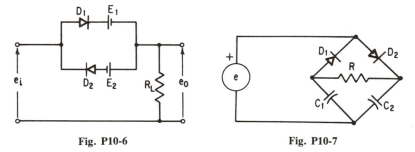

Fig. P10-6   Fig. P10-7

10-8. The transistor in the circuit shown in Fig. P10-8 has $h_{ie} = 3570\ \Omega$ and $h_{fe} = 65$.

(a) Obtain the expression for the voltage gain.

(b) Find the value of the voltage gain when $R_L = 10\ \text{k}\Omega$ and $R_E = 1$.

(c) Suggest a way of increasing the a-c output voltage while maintaining the same quiescent point, and compute the new value of gain.

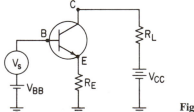

Fig. P10-8

10-9. Refer to the circuit of Fig. P10-8. Consider $h_{ie} = 3570\ \Omega$, $h_{fe} = 65$, and let $R_E = 10\ \text{k}\Omega$ and $R_L = 0$.

(a) Find the value of the input resistance and compare with the case where $R_L = 10$ kΩ and $R_E = 0$.

(b) Compute the value of output resistance.

(c) Compare the result of part (b) with that commonly encountered in the conventional common-emitter configuration, and comment.

(d) Determine the value of the output voltage appearing across R_E when the signal voltage is sinusoidal and has an rms value of 5 V.

10-10. If an emitter-follower circuit is to be found in a transistor radio circuit, where is it likely to be located—immediately following the antenna pickup circuitry or at the other end where the loudspeaker is located? Explain.

10-11. A high-level signal source has an internal resistance of 50 kΩ.

(a) Design a circuit employing a transistor having $h_{ie} = 1.7$ kΩ and $h_{fe} = 44$ which makes the source signal feed into a total input resistance of 150 kΩ.

(b) Find the percentage of the output voltage that appears across the load in the designed circuit of part (a).

10-12. A signal source having an internal resistance of 50 kΩ must feed into a 2-kΩ load resistor.

(a) Design a circuit employing a transistor that will permit the successful transmission of the signal to the load without excessive attenuation.

(b) Find the portion of the signal voltage that appears at the base-emitter terminals of the transistor.

(c) Compute the portion of the signal voltage that appears across the load resistor terminals.

10-13. An amplitude modulated wave is described by the equation

$$e_s = 10(1 + 0.4 \cos 1000t + 0.2 \cos 1800t) \cos 10^6 t$$

(a) List all the frequencies which exist in this wave.

(b) What is the modulation index of each frequency?

(c) Plot the frequency spectrum for this wave.

10-14. An amplitude modulated wave $e_s = (1 + 0.8 \sin 1000t) \sin 10^6 t$ is received by a square-law detector described by $i = 0.01e_s^2$. If the upper sideband frequency is eliminated from the transmitted wave, find the frequency components in the detector circuit. What are their relative amplitudes?

10-15. Repeat Prob. 10-14 for the case where the lower sideband frequency is removed from the transmitted wave.

GROUP II

10-16. A sinusoidal voltage of amplitude E_m is applied to the circuit of Fig. P10-16.

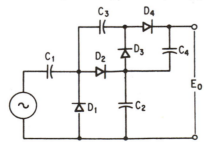

Fig. P10-16

(a) Identify the polarity and magnitude of voltage on each capacitor.

(b) What is the value of the output voltage?

10-17. An amplitude-modulated wave having the form

$$e = E_m[1 + m \sin \omega_m t] \sin \omega_c t$$

is transmitted to a receiving station where the signal is demodulated by a square-law detector. If the volt-ampere characteristic of the detector is described by $i = ke^2$, find the frequencies which exist in the detector current. What are their relative amplitudes?

chapter eleven

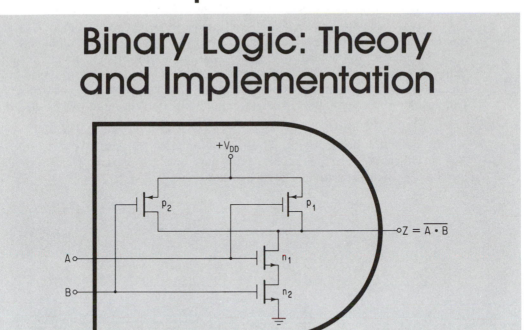

Binary Logic: Theory and Implementation

$$Z = \overline{A \cdot B}$$

Digital circuits are pervading every aspect of life. These devices are to be found increasingly not only in engineering devices and systems, but in many consumer-oriented devices as well. Such products as personal computers, microwave ovens, automobiles, airline reservation systems, bank teller machines, calculators, digital watches, and many others, all involve digital circuits for the purpose of processing information and/or providing desired control functions. The digital circuits themselves are composed of a few basic units which are easy to build in great numbers with reliable repeatability at low cost employing the techniques of integrated circuit design. By ingeniously interconnecting large numbers of these basic units a digital system can be constructed that is capable of performing many impressive tasks. It is the goal of this chapter to describe these basic digital circuits.

11-1 BINARY LOGIC

Binary logic presupposes two distinguishing characteristics: *two-valued variables* and appropriate *logical operations*. Unlike the situation of ordinary numbers, the values of the variables in binary logic can be only two in number. In fact, they do not even need to be numbers. They should perhaps more properly be called states. One pair of terms that can be used to identify the two states that binary variables can assume is: *open* and *closed*. Another suitable pair is: *high* and *low*. Two others are: *hot* and *cold,* and *true* and *false*. In each case the binary variable can take on at any given time only one of two values. However, on the basis of logic, these two variables should be mutually exclusive. The examples cited obviously possess this feature. However, the preferred choice for the two values to be assigned to the binary variables are: *0* and *1*. It is also customary to represent the binary variables by the letters of the alphabet.

There are three logical operations associated with binary logic. These are called AND, OR, and NOT. In the interest of facilitating the description of these logical operations we introduce the two binary variables A and B, each of which can assume the value 0 or 1. The AND function is represented by the equation

$$Z = A \cdot B \tag{11-1}$$

and by definition of the logical AND operation $Z = 1$ if and only if $A = 1$ *and* $B = 1$. Otherwise, the result of the AND logical operation is zero. This information can be represented in tabular form as follows:

A	B	Z
0	0	0
0	1	0
1	0	0
1	1	1

Such a table is called a *truth table*. The dot symbol is used to identify the AND function and an inspection of the truth table reveals this choice to be an appropriate one if we permit ourselves for the moment to treat the dot as multiplication. Accordingly, in the first row of the truth table we have $0 \cdot 0 = 0$. Similarly, in the subsequent rows we observe $0 \cdot 1 = 0$, $1 \cdot 0 = 0$, and $1 \cdot 1 = 1$. Clearly, the logical AND operation resembles multiplication as we know it for ordinary numbers.

The OR function is represented by the equation

$$Z = A + B \tag{11-2}$$

By definition of the logical OR operation $Z = 1$ if $A = 1$ *or* $B = 1$; otherwise, $Z = 0$. The truth table for the OR function is given below.

A	B	Z
0	0	0
0	1	1
1	0	1
1	1	1

A study of the truth table in this case reveals that the logical OR operation resembles addition. Observe that on successive rows we have: $0 + 0 = 0$, $0 + 1 = 1$, $1 + 0 = 1$, and $1 + 1 = 1$. Use of the $+$ symbol in Eq. (11-2) is therefore entirely appropriate. Of course, to say that the OR function resembles addition is not to say that it is addition in the sense that we use the term in ordinary arithmetic. This is obviously illustrated by the last row of the truth table. The value of the variable that represents the output of the OR function is still confined to 1, even when both A and B are at value 1.

The NOT function is represented by the equation

$$Z = \overline{A} \qquad (11\text{-}3)$$

The NOT function is the logical operation of *negation*. By definition, if $A = 0$, then $Z = 1$. Similarly, if $A = 1$, then $Z = 0$. The NOT function acts as an *inversion* that provides the opposite state of A whatever it may be. This process is also described as one that provides the *complement* of A. Thus Eq. (11-3) is read as: Z equals the complement of A. It is interesting to note that the NOT operation is not part of ordinary arithmetic or algebra.

We are now in a position to illustrate how the foregoing logical operations can be used to make logical decisions in practical situations centered around electrical circuits. Keep in mind that the two-valuedness of binary logic makes it a natural for implementation by circuits where the binary variables are switches which are either in the open (i.e., OFF) state or in the closed (i.e., ON) state. In this connection consider the design of a switching circuit that causes a buzzer to sound whenever the key is inserted into the ignition slot and the driver of a car has neglected to fasten his/her seat belt. This problem involves two binary variables: the ignition key and the seat belt. Let K denote the state of the ignition key and let's assign a 0 when the key is out of the key slot and a 1 when it is in the slot. Let B denote the seat belt variable and set the value of the fastened state as 0 and the unfastened state as 1. Because it is desirable to have the buzzer activated when the key is in the slot ($K = 1$) *AND* the seat belt is unfastened ($B = 1$), the AND function is obviously to be invoked. The digital circuit that represents this condition is depicted in Fig. 11-1. Observe that the AND logical operation requires the use of a series arrangement of the two switches that represent the binary input information. Switches K and B are in the closed state only when the key is inserted *and* the seat belt is unfastened.

To represent the OR logical operation also in terms of an appropriate switching circuit, let it be required that a circuit be designed that causes a buzzer to sound whenever either the car door is open or the seat belt is unfastened

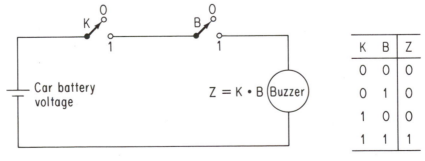

K	B	Z
0	0	0
0	1	0
1	0	0
1	1	1

Fig. 11-1 Switching circuit to represent the AND logical operation. A and B are binary variables represented by switches.

after the key is inserted into the ignition slot. This problem now involves three binary variables. They are the key K (in, 1 or out, 0), the door D (closed, 0 or open, 1) and the belt B (unfastened, 1 or fastened, 0). Of course, it is desirable that the buzzer not sound under any circumstance when the ignition key is out. Hence the switch that represents the key should be placed in series with the buzzer. But when the key is in place, if either the car door is open or the seat belt is unfastened, the buzzer should be made to sound. This either/or condition on the door and the belt clearly suggests a *parallel* arrangement of the switches. The complete circuit together with its truth table is displayed in Fig. 11-2. The function of the parallel switches is represented by the logical statement $(D + B)$. The complete logical statement for the buzzer then becomes $Z = (B + D) \cdot K$. The last four rows of the truth table are the really pertinent ones (corresponding to the ignition key in place) and a glance at the binary values for B, D, and Z readily reveals the pattern for the OR function cited previously.

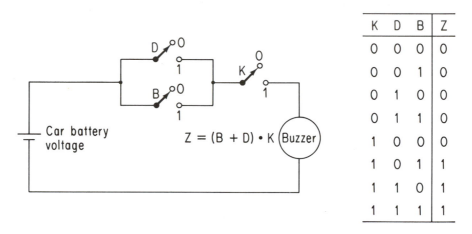

K	D	B	Z
0	0	0	0
0	0	1	0
0	1	0	0
0	1	1	0
1	0	0	0
1	0	1	1
1	1	0	1
1	1	1	1

Fig. 11-2 Switching circuit to represent the OR function. B, D, and K are binary variables.

It is worthwhile to note at this point that the rules described in the foregoing material for binary logic are entirely consistent with the formal mathematical development of such rules known as *Boolean† algebra*. This is important because we will invoke the postulates and theorems of this algebra when it suits our purposes. For now we simply recognize that both the associative and distributive rules of ordinary algebra also apply to Boolean algebra. Moreover, because Boolean algebra does not have an additive or multiplicative inverse, subtraction and division are not part of this algebra.

† G. Boole, *Investigation of the Laws of Thought* [New York: Dover Publications, Inc. (reprint), 1954].

11-2 LOGIC GATES

Very simply put, a logic gate is a hardware implementation of a logical operation. There can be N inputs and a two-level (0, 1) output. Figure 11-3 illustrates the situation. The circuitry contained within the box, identified as a gate, performs the logical operation AND when the output level undergoes a transfer to logic 1 in response to *all* inputs being placed at logic 1. The output goes to logic 0 when any one input is at logic 0.

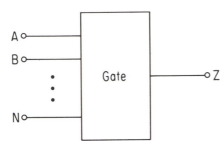

Fig. 11-3 General representation of a logic gate showing N inputs and a single two-level output, Z.

Similarly, the gate is said to perform the OR function when the output transfers to logic 1 in response to the change of *any one* input signal to logic 1 in the manner called for by its truth table.

The AND, OR, and NOT logical operations are not the only ones that are possible, however. The theory of Boolean algebra shows that there are as many as 16 functions for two binary variables. Of this number two are either 0 or 1, four involve the unary operations of identity or complement, two others are undesirable because they are not commutative or associative operations and one is simply a buffer operation. The remaining seven operations are the useful ones including the three already introduced. All seven logical operations are tabulated in Table 11-1 for the convenience of reference. Each logical operation is assigned its own symbol together with the expression representing the logical operation it performs. Also included are the corresponding truth tables. Up to this point emphasis was placed on the AND, OR, and NOT functions chiefly because they are easier to deal with as illustrated by the design of the digital circuits for use in automobiles. But this is not necessarily true about the electronic circuitry that is used to build these gates. In fact, the most extensively used gates are not the AND and OR gates but rather their complemented versions NAND and NOR. This is because such circuits commonly involve transistors where inversion occurs automatically.

The exclusive-OR gate (XOR) is so named because by definition $Z = 1$ if $A = 1$ or $B = 1$ but not if A and B are both 1. In other words, this logic circuit excludes the output from being 1 when both inputs are 1. The hardware implementation of the XOR gate is difficult to achieve for more than two inputs even though this is allowed because the logical operation does satisfy the commutative and associative laws. The last item in Table 11-1 is the exclusive-NOR gate,

TABLE 11-1 DIGITAL LOGIC GATES AND ASSOCIATED LOGICAL OPERATIONS FOR BINARY VARIABLES

Name	Symbol	Logical operation	Truth table

AND $Z = A \cdot B$

A	B	Z
0	0	0
0	1	0
1	0	0
1	1	1

OR $Z = A + B$

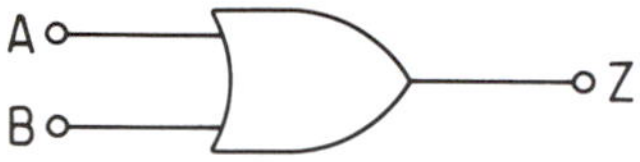

A	B	Z
0	0	0
0	1	1
1	0	1
1	1	1

NOT $Z = \overline{A}$

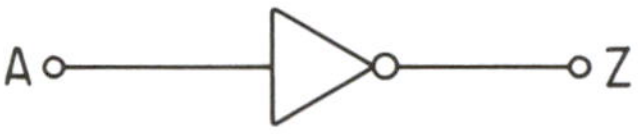

A	Z
0	1
1	0

NAND $Z = \overline{A \cdot B}$

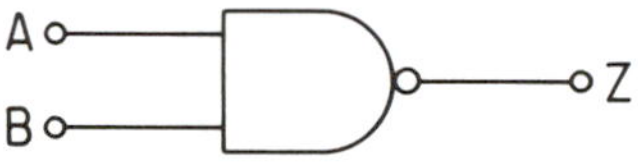

A	B	Z
0	0	1
0	1	1
1	0	1
1	1	0

NOR $Z = \overline{(A + B)}$

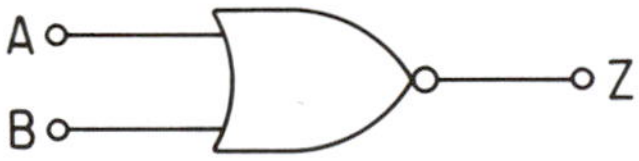

A	B	Z
0	0	1
0	1	0
1	0	0
1	1	0

XOR $Z = A \oplus B$

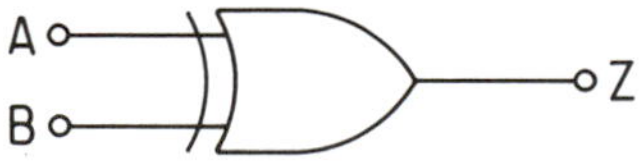

A	B	Z
0	0	0
0	1	1
1	0	1
1	1	0

XNOR $Z = A \odot B$

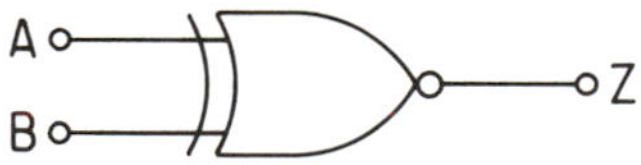

A	B	Z
0	0	1
0	1	0
1	0	0
1	1	1

which is simply the complement of the exclusive-OR gate. This circuit has the same limitations as the XOR gate. However, both logical functions have specialized applications and are found to be particularly useful in performing arithmetic operations.

The input signals to the logic gates listed in Table 11-1 are voltages that take on the binary values of high or low. The digital circuitry that provides these signals can be of several types. In those cases where *positive logic* is used logic 1 is assigned to the high voltage and logic 0 to the low voltage. The low voltage band is often identified as between 0 and 0.8 V, whereas the high voltage band might have a range of 2 to 5 V. When a digital signal is introduced at the input terminal of a gate, it means that the input is considered to be a logic 0 if the voltage value is less than 0.8 V and a logic 1 if it lies between 2 and 5 V. Moreover, when the logic gate is based on transistors, the output of the gate switches very rapidly to the corresponding logic level called for by its logical function. *Negative logic* refers to those cases where the high voltage is called logic 0 and the low voltage is assigned logic 1. Often, logic 1 corresponds to the negative voltage and logic 0 to the voltage about zero. *Mixed* logic is also used.

It is important to become familiar with the logic gates listed in Table 11-1 because they serve as the basic building blocks of all digital systems however simple or complex. Logical operations can readily be represented in terms of these basic logic gates. As an illustration of the procedure, let us draw a logic diagram of the logical function for the digital circuit of Fig. 11-2. Recall that the logical operation is described by

$$Z = (B + D) \cdot K$$

The process begins by applying the signals B and D to an OR gate to obtain an output represented by $(B + D)$. This signal is then combined with K at an AND gate in the fashion illustrated in Fig. 11-4 to obtain Z.

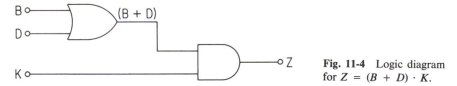

Fig. 11-4 Logic diagram for $Z = (B + D) \cdot K$.

Occasions often arise in drawing logic diagrams where inversion of a signal is required. Although Table 11-1 lists an appropriate symbol to denote this operation, nevertheless it is customary to replace this symbol with an open circle whenever the inverter is used in conjunction with other gates. Figure 11-5 illustrates the situation.

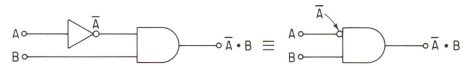

Fig. 11-5 When used with another gate, the inverter is usually replaced with an open circle in the connection line.

11-3 IMPLEMENTATION OF DIGITAL LOGIC GATES

Our attention is confined in this section to a description of the major types of digital logic gates that are fabricated by integrated circuit procedures. These include transistor-transistor logic gates (TTL), emitter-coupled logic gates (ECL), metal-oxide-semiconductor logic gates (MOS), and complementary metal-oxide-semiconductor logic gates (CMOS). The acronyms refer either to the particular arrangement of circuit elements or to the mode of fabrication employed. In each case the negation feature is manifested at the output as a consequence of the use of transistors as the switching elements in the circuits used. Hence the natural operations of these circuits are NAND, NOR, and INVERSION. Moreover, four groups of gates are treated because each group has specific advantages which makes it superior to the others in certain circumstances. These features are pointed out as each type is discussed. However, first it is helpful to examine in some detail the manner in which the basic transistor amplifier circuit is made to behave like a switch.

Refer to Fig. 11-6. The digital input signal is assumed to be a square wave where the low voltage level is zero and represents logic 0. The high voltage level is 5 V and it represents logic 1 [see Fig. 11-6(a)]. The transistor amplifier circuitry is shown in Fig. 11-6(b) and appearing in Fig. 11-6(c) are the v-i characteristics corresponding to various base currents. The transistor supply voltage, V_{cc}, is assumed to be 5 V and the d-c load line is also indicated. The resistors R_L and R_B are selected so that when the input voltage goes high the operating point will be at point S, which corresponds to the saturation region of the transistor.

Let's examine how this circuit performs as the input is allowed to traverse a cycle of low and high voltages. In the time interval $0 < t < t_1$ when the input to the transistor amplifier is zero, the associated zero base current puts the operating point at Q in Fig. 11-6(c). This is called the *cutoff* point. At this operating point the output voltage, which is taken between the collector terminal and ground reference, is very close in value to the supply voltage as is apparent from Fig. 11-6(c). This high output, of course, denotes logic 1. In the subsequent time interval $t_1 < t < t_2$ the digital input signal goes high (logic 1) and quickly moves the operating point from Q to S in the saturated region. Now, as the abscissa axis of Fig. 11-6(c) readily reveals, the output voltage drops abruptly to approximately 0.2 V. This is the saturated value of the collector-to-emitter voltage and is essentially the same for all silicon transistors. The high state of the input (logic 1) thus causes the output to switch to the low state (logic 0). We see therefore that inversion occurs automatically when transistors are used. Moreover, although this characteristic has been demonstrated for the case of positive logic, the same result prevails with negative logic as well.

The time it takes for the transistor to switch from the cutoff point Q to the saturated point S is dependent upon a number of factors. Chiefly, it is determined by the effective capacitance and resistance that appears at the input and output terminals. As a rule these switching times are of the order of 10 to 30 ns.

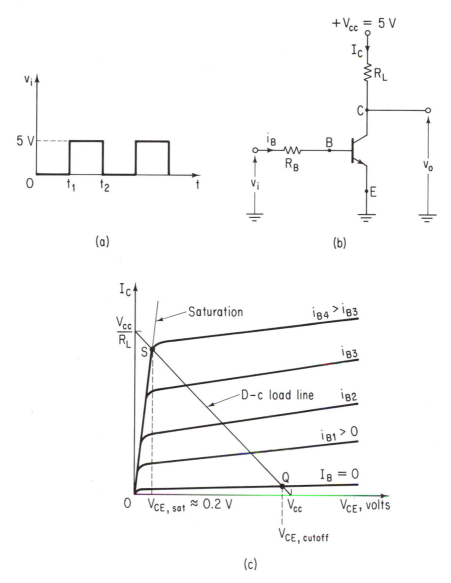

Fig. 11-6 Operation of the transistor amplifier as an electronic switch that performs the logical operation of inversion.

Transistor-Transistor Logic (TTL).   The inversion operation that is available from the transistor amplifier can be used to implement the NAND and NOR logical functions. For reasons of improved density, simplicity, and versatility, as well as other factors, the preferred way of implementing the NAND logical function is to use the transistor-transistor logic gate that is depicted in Fig. 11-7. This is one of the most popular logic gates in use today.

To aid in the understanding of the operation of this circuit, several points

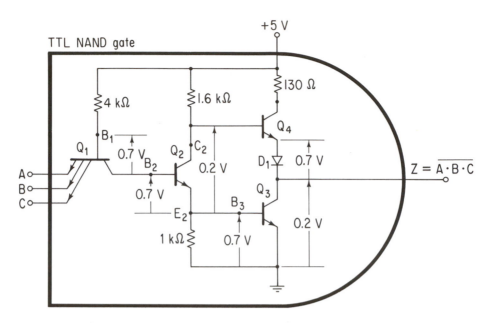

Fig. 11-7 Circuit diagram for a TTL NAND gate. Indicated voltages apply to the case where Q2 and Q3 conduct and Q4 is cut off. The output is low (logic 0) and *all* inputs are high (logic 1).

are worth noting at the outset. The basic function of transistor Q_1 is to operate as a diode on two counts. It is arranged to provide diode action from its base terminal to each of the emitter terminals at which the digital logic signals are applied. But it is also expected to furnish diode action from its base-to-collector terminals. Of course, the latter is possible provided that the potential of the base terminal exceeds that of the collector terminal by an appropriate amount. The circuitry is designed to ensure this whenever all inputs are high. In the output line of the TTL circuit, transistor Q_4 together with the 130-Ω resistor are designed to serve as a load resistor for the output transistor Q_3. An active load resistor is preferred because much lower values of resistance can be used, and this in turn yields the decided advantage of faster switching times. Moreover, it is helpful to keep in mind that the voltage drop for silicon transistors from base to emitter terminals is in general approximately equal to 0.7 V while in the active region of operation, and about 0.8 V when in a saturated state. Also useful to have at hand is the information that while in or near its saturated state the collector-to-emitter voltage of the transistor is approximately 0.2 V. Such figures are important because they help to reveal whether or not a particular transistor in the TTL circuitry is ON (saturated) or OFF (cutoff).

We begin the explanation by assuming that the output is at its low voltage level (logic 0). This is possible if Q_2 and Q_3 are both ON *and* Q_1 is acting as a diode from base-to-collector terminal. With Q_1 so biased a path is established to allow current to flow from the 5-V supply to the base terminals of transistors Q_2 and Q_3 to put them in the active conducting state, which in turn puts the collector of Q_3 at low voltage (≈ 0.2 V). The voltage values that appear at the various transistors in Fig. 11-7 correspond to this low-voltage state of the output. When a semiconductor device is forward biased and conducting, it is assumed that the associated voltage drop between base and emitter terminals is 0.7 V. This is also true of diode D_1 that appears between Q_4 and Q_3.

It is instructive to note in this TTL logic circuit that the potential of the base terminal of Q_1 with respect to ground has a value of $V_{B1} \approx 3(0.7) = 2.1$ V. In other words, in order for Q_2 and Q_3 to be in a conducting state (ON), it is necessary for the potential of the base terminal of Q_1 to be high. In turn, this can be ensured only if all three input signals A, B, and C are themselves high (logic 1). Accordingly, this circuit is said to be performing the NAND logical operation.

What is the state of transistor Q_4 under the conditions just described? The answer readily follows from knowledge of its base-to-emitter voltage. Remember that for silicon transistors to begin conduction this voltage must be about 0.5 V. Figure 11-7 shows that the potential of the base terminal of Q_4 relative to ground is $0.2 + 0.7 = 0.9$ V and its emitter terminal is also 0.9 V relative to ground. Hence the net voltage across the base-to-emitter terminals is zero, which means cutoff. Diode D_1 was placed into the circuit in order to ensure that Q_4 is off when Q_2 and Q_3 are on.

In order for the circuit of Fig. 11-7 to act as a proper NAND gate, it is necessary next to show that the output terminal switches to the high level (logic 1) when *any one* of the input signals goes low (logic 0). For the sake of illustration let us assume that the input signal at A goes low. The ensuing diode action between base terminal B_1 and emitter terminal A serves to drop the potential of B_1 to 0.7 V above ground rather than the required 2.1 V which is necessary to keep transistors Q_2 and Q_3 on. Consequently, Q_2 and Q_3 are cut off. With Q_3 off, the base of Q_4 goes to supply potential which turns Q_4 on. The output circuit now consists of the low resistance represented by the 130-Ω resistor and the low collector-to-emitter resistance of the conducting Q_4 in series with the very high resistance represented by the cutoff transistor Q_3. This has the effect of driving the output voltage high (3.6 V). Thus the given TTL circuit behaves as a NAND gate in all respects.

The switching time that is generally achievable with the circuit of Fig. 11-7 is of the order of 10 ns. This figure can be improved by a factor of 3 or so by making certain changes in the circuitry which involve specially designed transistors. The modified circuit is called the *Schottky TTL gate*. It is important to note, however, that the increased switching speed is obtained at the expense of increased power dissipation. The "goodness criterion" of these circuits is best described by a speed-power product. On this basis the Schottky TTL gate exhibits a figure which is about half that of the standard TTL gate. But still in those cases where faster switching is indispensable, the Schottky TTL gate is preferred.

Emitter-Coupled Logic (ECL). The fastest switching time of all the integrated circuit families of logic gates is offered by the emitter-coupled logic gate. Switching occurs in approximately 1 ns and is achieved because the transistors are operated so that they never enter a saturated state. As a result it is easier to charge and discharge the capacitance associated with the logic circuitry. This avoidance of saturation as one of the two operating states of the transistors does exact its costs. One important cost is reduced immunity to noise. In the ECL logic gate the separation between low voltage (logic 0) and high negative voltage (logic 1) can be as low as 0.2 V, whereas in the TTL logic circuit it is 1.5 V. Hence there is less room for error in drawing a distinction between logic 0 and logic 1. It is also implied by the foregoing remark that the ECL gate employs negative logic. Besides the reduced tolerance to noise, the ECL gate is also the one that has the largest power dissipation of all the gate families. It is approximately thirty times greater than the TTL gate. Accordingly, the ECL gate is used only in special-purpose applications where the fast switching time is indispensable, such as in high-speed computers equipped with fast-speed counters.

The essential features of the circuit configuration of the ECL gate are displayed in Fig. 11-8(a). Before an explanation of this circuit is undertaken, several points are worth noting. There is a section of the gate which is designed to maintain the potential of the base of transistor Q_3 at a value of -1.3 V. This figure corresponds to the midvoltage range of the digital input signals, which are assumed to have the variation depicted in Fig. 11-8(b). The supply voltage for the entire circuit is fixed at -5.2 V. Also, while the transistors are in the active (or conducting) state, the voltage drop from base to emitter is about 0.8 V. This figure is typical of ECL transistors. Furthermore, in order for conduction to start in an ECL transistor, V_{BE} must be at least 0.6 V. The voltage values that appear at the transistors in Fig. 11-8 apply for the case where at least one input is at logic high (-0.8 V).

The parameters of this gate circuit are so selected that when a logic input, say A, goes high, the transistor to which it is connected turns ON. In this case Q_1 conducts. In so doing it causes the potential of the *coupled* emitters to be put at -1.6 V. This figure is obtained by adding to the potential of A (-0.8) the drop across the base-to-emitter terminals of Q_1 (another -0.8 V). Consequently, at transistor Q_3, the base-to-emitter voltage becomes

$$V_{BEQ3} = -1.3 - (-1.6) = 0.3 \text{ V}$$

Since this value is well below the 0.6 V that is needed to initiate conduction in these ECL transistors, Q_3 cuts off. Transistors Q_4 and Q_5 are introduced as emitter followers in order to furnish signal levels at their output terminals that match those of the digital input levels, namely, -0.8 V and -1.8 V. When Q_3 cuts off, only a very small current remains to flow through R_{L3}, thus producing a negligible voltage drop across this resistor. But since Q_4 is conductive the drop across its base-to-emitter terminals is 0.8 V, which serves to put the potential of Z at -0.8 V relative to ground. Since this value represents the high voltage level and is produced when A or B goes high, this makes available at the Z terminal the logical OR function.

At resistor R_{L1}, which is in the collector circuit of the conducting transistor, the much greater current flow produces a voltage drop of 1 V. Combining this

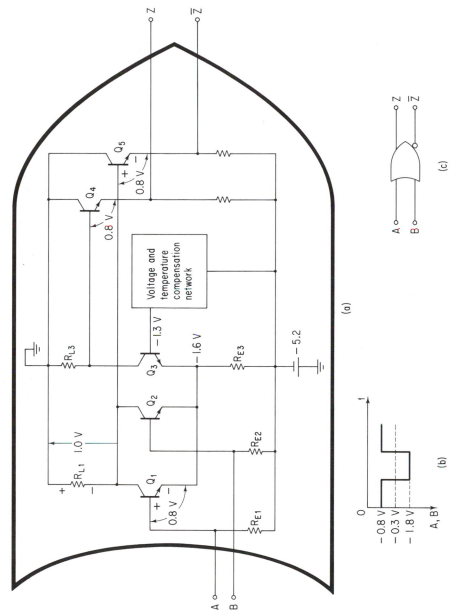

Fig. 11-8 ECL logic gate: (a) circuit details; (b) illustrating high and low voltage levels of digital input signals; (c) symbol to represent the OR/NOR feature.

477

voltage with the base-to-emitter voltage of Q_5 yields a total potential at terminal $\bar{Z}$ relative to ground of -1.8 V. Because this is the low voltage level and is associated with A high, the logical operation appearing at the $\bar{Z}$ terminal is clearly the NOR function. Thus the circuit of Fig. 11-7 has the advantage of offering complementary outputs. Figure 11-8(c) shows the symbolic model of such a gate.

Metal-Oxide-Semiconductor Logic (MOS). The enhancement-type metal-oxide semiconductor offers as salient features relatively low power dissipation and high fabrication density. It is possible to get many more MOS gates on a given size of semiconductor chip than in the case of the junction-field-effect transistor, for example. Unfortunately the gate-to-drain and gate-to-source capacitances are sufficiently large to cause this gate to have a switching time that ranks it among the highest of the digital integrated-circuit family of gates. However, in contrast, the MOS gate does offer the additional feature of permitting operation either as a transistor or as an active load resistor. The latter mode of operation is achieved usually by tying the gate terminal to the drain in order to furnish a constant bias voltage for conduction.

The symbolic representation of the enhancement n-channel and p-channel MOS transistors are illustrated in Fig. 11-9. However, in the interest of simplifying the circuit diagrams in which these devices appear, the alternative symbols for these units, which are also shown in Fig. 11-9, are used in all subsequent diagrams where they are present. A typical circuit configuration in which the enhancement-mode MOS device is used as a transistor appears in Fig. 11-10(a). When the load resistance line is constructed on the i-v characteristic of the MOS unit, a plot of output voltage (appearing at the drain terminal) as a function of the input voltage, which represents the gate-to-source signal, takes on the variations displayed in Fig. 11-10(b). Especially notable is the presence of a threshold value of input voltage up to which no change in output voltage appears. This threshold value generally has a value slightly below 3 V, during which the output voltage is relatively constant at the supply value. As the input voltage increases further, there is a moderately fast change in output voltage to slightly below 1 V. It follows then from this description that if a square-type digital signal is applied to the input terminal, the low-voltage state (logic 0) produces a high-voltage

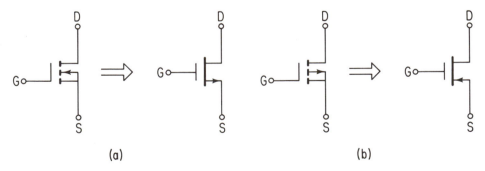

Fig. 11-9 Symbolic representations of the enhancement-mode MOS transistors: (a) n-channel type; (b) p-channel type.

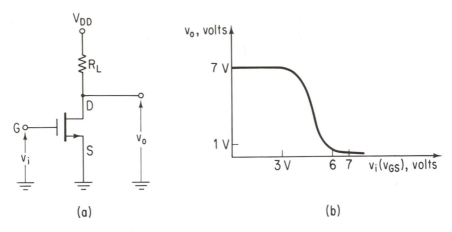

Fig. 11-10 (a) Amplifier configuration using the *n*-channel MOS transistor; (b) voltage transfer characteristic.

output (logic 1). Similarly, a high-voltage input (say, 7 V) produces a low-voltage output (<1 V). Clearly, such performance is nothing more than the logical operation of inversion.

The circuitry of Fig. 11-10(a), however, is not used for generating a NOT function because the fabrication of the load resistor is such that it requires many times more space than is used for the MOS transistor. In such cases, therefore, it is customary to use in place of this resistor another enhancement mode transistor which is biased to cause it to act as a resistor. In this role the MOS unit is described as an *active* resistor. We will not direct further attention to MOS gates because as a practical matter they have been virtually displaced in applications by their very close cousin the complementary-symmetry MOS or CMOS.

Complementary Metal-Oxide-Semiconductor Logic (CMOS).   As shown in Fig. 7-31, the distinguishing characteristic of the complementary-symmetry MOS gate is that both the *n*-channel and *p*-channel transistors are fabricated on the same substrate. In fact, the drain terminals of each channel are joined together on the chip itself. As a result the use of one MOS transistor as an active load is conveniently available. The formation of logic circuits is thereby enhanced. These circuits are particularly notable for their very low power dissipation. They also exhibit a fairly good immunity to noise, a high fabrication packing density, and use with a fairly wide range of supply voltages. Unfortunately, the disadvantage of large switching time (about 25 ns) still remains. Despite this shortcoming, the CMOS today is replacing the TTL gate in many applications save those that require high speeds and some other special features not available from the CMOS gates.

Before undertaking an explanation of the manner in which the CMOS transistors are used in generating the logical functions, it is helpful to keep the following points in mind. Logic 0 corresponds to a low voltage in the vicinity of zero. Logic 1 corresponds to a voltage level near the drain-source supply voltage V_{DD} and this can vary from 3 to 10 V in most typical CMOS gates.

Moreover, because the complementary-symmetry MOS has a *p*-channel as well as an *n*-channel transistor, the behavior of the gate circuits are easier to comprehend by remembering that the PMOS unit conducts when its gate-source voltage goes negative ($V_{GS} < 0$), the NMOS unit conducts when $V_{GS} > 0$ and neither transistor conducts if $V_{GS} = 0$.

The circuitry that implements the logical operation of inversion is depicted in Fig. 11-11. One striking feature of this circuit is clearly its simplicity. It requires merely the one pair of transistors that constitutes the complementary-symmetry MOS device, which is fabricated on a common substrate taking up very little space on the integrated chip. When the input digital signal *A* is at its low level (logic 0), the gate-source voltage of the *n*-channel MOS (or NMOS) is zero so that it cuts off. Also, the potential of D_2 relative to the gate potential of the PMOS is positive, which makes the gate-source voltage of the PMOS transistor negative. In turn, this puts the PMOS transistor into a conductive state. Since the nonconducting NMOS exhibits a high resistance between its drain-source terminals while the conducting PMOS presents a much smaller resistance, it follows by the voltage-divider principle that the output terminal assumes a voltage level that is very close to V_{DD}, which denotes logic 1. Accordingly, inversion of the input signal takes place.

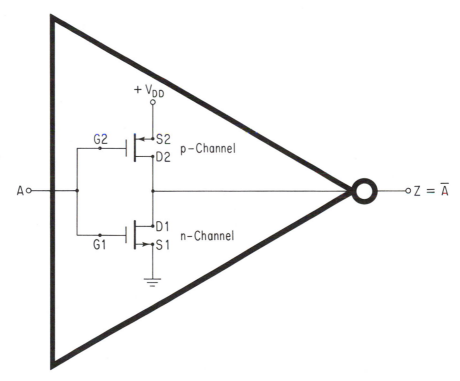

Fig. 11-11 Implementation of the logical operation inversion by the use of a CMOS gate.

When the input signal goes high (logic 1), the gate-source voltage of the NMOS transistor goes positive, thus causing it to conduct. Simultaneously, the gate-source voltage of the PMOS unit goes positive, thereby cutting it off from conduction. Now the low resistance state occurs between the output terminal and ground, while the high resistance appears between the output terminal and the supply terminal. The result is a low-level voltage at the output terminal, which is logic 0. Inversion is achieved again. Note that in either state one transistor is on and the other is off. It is this situation that is responsible for the low power dissipation of CMOS gates.

The implementation of the NAND logical operation calls for the use of a double pair of complementary-symmetry MOS transistors arranged in the manner illustrated in Fig. 11-12. The operation of this circuit can be explained as follows. When A and B are both low, n_1 and n_2 turn off while p_1 and p_2 turn on. The low effective resistance associated with the conducting p units put the potential of the output terminal near supply value. Hence Z takes on the logic 1 state. Next assume that just A goes high. It causes n_1 to turn on and p_1 to turn off. However, because p_2 is still on and n_2 continues off, the new states of p_1 and n_1 are irrelevant in establishing the state of the output terminal. The reason, of course, is that when a high resistance (p_1 off) is in parallel with a low resistance (p_2 on), the low resistance dominates. Similarly, when a high resistance (n_2 off) is in series with a low resistance (n_1 on), the high resistance exerts the controlling influence. Hence Z stays at logic 1. Finally, if B is allowed to join A in going high, then n_1 and n_2 both go to the on state and p_1 and p_2 turn off. The output terminal in response moves slow in voltage to logic 0. For operation of the NOR gate employing CMOS technology, see Prob. 11-28.

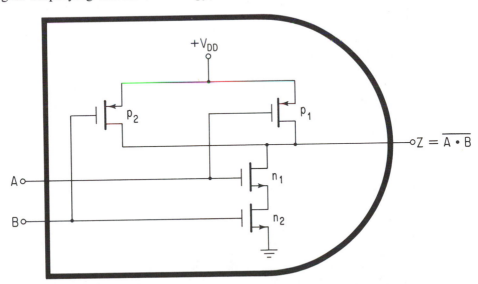

Fig. 11-12 Implementation of a NAND logical operation using a double pair of complementary-symmetry MOS transistors.

11-4 BINARY AND OTHER NUMBER SYSTEMS

It is the purpose of the remaining sections of this chapter to show that the logic gate circuits of the preceding section, besides fulfilling the important functions of performing logical operations, play a crucial role too in the processing of data be it in numerical or alphabetical form. The sole requirement for the manipulation of such data by the gates is that the information be supplied in binary form, i.e., in the form of low voltages (logic 0) and high voltages (logic 1), for example. Accordingly, it becomes necessary for us to represent data such as numbers as well as the arithmetical operations of addition, subtraction, multiplication, and division in the binary system. Moreover, appropriate codes can be introduced so that even the letters of the alphabet and many other useful symbols such as the dollar sign can be identified in terms of a proper sequence of 0s and 1s. We direct attention first to the binary system of numbers.

The treatment of a binary system of numbers can perhaps best be understood by first reviewing the characteristics of the decimal system, then resorting to a general formulation that applies to all number systems, and finally applying the general form to the specific case of a binary system. In the decimal system the number 5432 has a meaning with which we are all familiar. The placement of the digits in the sequenced order from right to left is understood to carry a specific meaning. For the case at hand the total value of the number is obtained by adding 2 unit values, 3 tens, 4 hundreds, and 5 thousands. Expressed more formally, we can write

$$5432. = 5 \cdot 10^3 + 4 \cdot 10^2 + 3 \cdot 10^1 + 2 \cdot 10^0$$

The figure 10 plays the role of a *base* or *radix*, hence the name *decimal number system*. The powers of 10 are incremented from zero by one as the position is moved successively toward the left from the decimal point. Furthermore, the coefficients of the powers of 10 are allowed to take on values from 0 to 9 (i.e., $r - 1 = 10 - 1 = 9$). In our use of numbers in the decimal system it is our custom just to use these coefficients because we all understand which powers of 10 are implied by their *ordered placement*. Clearly, the decimal system is just one of many number systems.

The general expression for a number that belongs to a system with a base or radix value r can be written as

$$
\begin{aligned}
N &= \sum_{i=-m}^{n-1} a_i r^i \\
&= a_{n-1} r^{n-1} + a_{n-2} r^{n-2} + \cdots + a_1 r + a_0 r^0 + a_{-1} r^{-1} \\
&\quad + \cdots + a_{-m} r^{-m}
\end{aligned}
\tag{11-4}
$$

Here a_{n-1} denotes the most significant coefficient and a_{-m} is the least significant coefficient. In addition, the integer part of the number corresponds to nonnegative powers of r provided that $r > 1$. The portion that contains the negative powers of the radix represents the fractional part of the number. For a radix r the allowable values of the coefficients are 0, 1, 2, . . ., $r - 1$. Thus for an octal number system where $r = 8$, the a coefficients in Eq. (11-4) are 0, 1, 2, 3, 4, 5, 6, and 7. In some digital systems representation in terms of a radix of 16 is not uncommon. The name given to the corresponding number system is *hexadecimal* because the range given to the digits is now described by the following group: 0, 1, 2, 3, 4, 5, 6, 7, 8, 9, A, B, C, D, E, and F. Observe that in addition to the 10 decimal numbers the first six letters of the alphabet are also used to yield a total of 16 digits. The letter A obviously is serving as a 10 and F as 15.

In view of the fact that digital circuits can only respond to two-valued input signals, the number system that is indispensable for these circuits, if they are to process data, is the binary system. This means the use of a radix of 2 and coefficients which can only be 0 or 1. Equation (11-4) accordingly reduces to the following form for binary numbers:

$$N = B_{n-1} \cdot 2^{n-1} + B_{n-2} \cdot 2^{n-2} + \cdots + B_1 \cdot 2 + B_0 \cdot 2^0$$
$$+ B_{-1} \cdot 2^{-1} + \cdots + B_{-m} \cdot 2^{-m} \tag{11-5}$$

Since the B coefficients are restricted to values of 0 and 1, they are called *binary digits* or *bits* for short. It follows, then, that B_{n-1} is the *most significant bit* (MSB) and B_{-m} denotes the *least significant bit* (LSB). When a binary number has no fractional part and $n = 4$, the result is a four-digit binary number which is referred to as a *nibble*. In the case where $n = 8$, the resulting eight-digit number is called a *byte*. It is useful to mention, however, that, although our discussion is at the moment focused on numbers, these terms have a general applicability to all information that has a representation in terms of 4-bit and 8-bit ordered sequences.

EXAMPLE 11-1   (a) Determine the decimal number that is represented by the binary ordered sequence $N_2 = B_2 B_1 B_0 = 101$.
 (b) Repeat (a) for $N_2 = B_4 B_3 B_2 B_1 B_0 = 11011$.

Solution: (a) Applying Eq. (11-4), we can write
$$N = 1 \cdot 2^2 + 0 \cdot 2^1 + 1 \cdot 2^0 = 1(4) + 0(2) + 1(1) = 5 \tag{11-6}$$
Hence $(101)_2 = 5_{10}$. Note that the quantities in parentheses in Eq. (11-6) are the weighting factors associated with the various powers of the base 2.
 (b) Here we have

16	8	4	2	1	as the weighting factors
1	1	0	1	1	as the associated coefficients

Hence we must add 16, 8, 2, and 1, for a total of 27. Thus
$$(11011)_2 = (27)_{10}$$
where the subscript denotes the base of the number.

EXAMPLE 11-2   (a) Determine the decimal equivalent of the octal number which is expressed by $N_8 = 432_8$.
 (b) Find the decimal equivalent of the hexadecimal number C4F.

Solution: (a)

64	8	1	are the weighting factors
4	3	2	are the associated digits

Hence

$$N_{10} = 4(64) + 3(8) + 2 = 282$$

Thus

$$(432)_8 = (282)_{10}$$

(b) Here the weighting factors are: 256   16   1
and the associated digits are: C   4   F

Hence

$$N_{10} = 12(256) + 4(16) + 1(15) = (3151)_{10}$$

or

$$(C4F)_{16} = (3151)_{10}$$

Observe that the hexadecimal system requires only a three-digit number in this instance.

Table 11-2 displays the ordered sequence of digits that are associated with the binary, octal, and hexadecimal number systems for the first 16 decimal numbers. In each system note how the digits move to the 10 (one-zero) position after moving through the first $r - 1$ coefficients. Observe too that in the binary system four digits are needed to yield a total of 16 numbers. This corresponds to 2^n, where n denotes the number of digits (or bits). Of course, the hexadecimal number system needs a single digit for the same number of values.

TABLE 11-2   FOUR POPULAR NUMBER SYSTEMS

Decimal $(r = 10)$	Binary $(r = 2)$	Octal $(r = 8)$	Hexadecimal $(r = 16)$
0	0000	0	0
1	0001	1	1
2	0010	2	2
3	0011	3	3
4	0100	4	4
5	0101	5	5
6	0110	6	6
7	0111	7	7
8	1000	10	8
9	1001	11	9
10	1010	12	A
11	1011	13	B
12	1100	14	C
13	1101	15	D
14	1110	16	E
15	1111	17	F

11-5 CONVERSIONS BETWEEN NUMBER SYSTEMS

Our chief concern in this section is to devise a scheme that allows an easy conversion from the decimal number system (and others as well) to the binary system because our interest invariably is to establish communication with the digital system which is inherently binary. Moreover, it is necessary to have procedures which can deal both with the integer part of a decimal number and the fractional part as well. Attention is directed first to the integer part. Let us begin by undertaking the task of finding the binary ordered sequence representation of the decimal number 23. Since the total of numbers represented by an n-digit binary sequence is 2^n, the determination of n readily follows from

$$n \geq \frac{\log N_r}{\log 2} \tag{11-7}$$

which for $N_r = 23_{10}$ yields $n = 4.524$.† Therefore, a five-digit binary number is needed to obtain a representation for 23_{10}. In other words, the binary number may be expressed by the ordered sequence

$$N_2 = B_4 B_3 B_2 B_1 B_0$$

and the problem then reduces to one of determining which of these coefficients is to be assigned a 1 and which a 0. For a number as small as 23 this is not difficult to do if one examines the weighting factors for a few seconds. However, our aim here is to devise a general algorithm that works just as easily for large numbers as for the smaller ones.

By invoking Eq. (11-4) for $n = 5$ and dealing just with the integer part, we can write

$$(23)_{10} = B_4 \cdot 2^4 + B_3 \cdot 2^3 + B_2 \cdot 2^2 + B_1 \cdot 2 + B_0 \cdot 2^0 \tag{11-8}$$

It is now useful to divide both sides of this last expression by the base value 2. Thus

$$11 + \frac{1}{2} = B_4 \cdot 2^3 + B_3 \cdot 2^2 + B_2 \cdot 2 + B_1 + \frac{B_0}{2} \tag{11-9}$$

The importance of this operation is that it immediately reveals the value of B_0 as equal to 1. Moreover, the integer, 11, on the left side is now represented by those terms on the right side with the exception of the fraction, which we know to be $\frac{1}{2}$. Accordingly, the reduced equation becomes

$$11 = B_4 \cdot 2^3 + B_3 \cdot 2^2 + B_2 \cdot 2 + B_1 \tag{11-10}$$

In the interest of uncovering the binary value to be assigned to B_1, we again perform division by 2 but this time it is applied to the reduced expression, Eq. (11-10). This yields

$$5 + \frac{1}{2} = B_4 \cdot 2^2 + B_3 \cdot 2^1 + B_2 + \frac{B_1}{2} \tag{11-11}$$

Again a comparison of the fractional terms tells us that B_1 must be assigned the digit value 1 and the integer 5 is now represented by

$$5 = B_4 \cdot 2^2 + B_3 \cdot 2 + B_2 \tag{11-12}$$

---

† Whenever n comes out to be a whole number, the required number of bits is increased by one. This accommodates the borderline cases.

Dividing by 2 once more yields

$$2 + \frac{1}{2} = B_4 \cdot 2 + B_3 + \frac{B_2}{2} \tag{11-13}$$

Hence $B_2 = 1$ and the new integer 2 has the representation

$$2 = B_4 \cdot 2 + B_3 \tag{11-14}$$

Repeating the process again gives

$$1 + \frac{0}{2} = B_4 + \frac{B_3}{2} \tag{11-15}$$

Therefore, $B_3 = 0$. This last equation indicates that when the division-by-2 process is applied to an even number, the remainder is treated as zero. Keeping in mind that the evaluation of the binary digits is done in conjunction with a fractional relationship, we can also divide the last reduced equation by 2. Thus

$$0 + \frac{1}{2} = \frac{B_4}{2}$$

which yields the last digit to be determined. Consequently, the binary equivalent of decimal 23 is found to be

$$N_2 = B_4 B_3 B_2 B_1 B_0 = 10111 \tag{11-16}$$

A simple check on the validity of this result can be accomplished by following the procedure employed in Example 11-1. Thus

weighting factors: 16   8   4   2   1

digits: 1   0   1   1   1

Hence

$$(10111)_2 = (23)_{10}$$

The foregoing procedure for obtaining the digit values of the ordered binary sequence can be neatly executed in a simple tabular form as illustrated below.

	Integer part	Remainder
	23	
These are the	11	$1 = B_0$
quotients resulting	5	$1 = B_1$
from successive	2	$1 = B_2$
division of 23 by 2	1	$0 = B_3$
	0	$1 = B_4$

On the basis of the explanation already given, these entries should be self-evident.

How does one treat the fractional part of a decimal number? In the interest of simplicity we now consider just a fractional number without an integer part. Again from the general expression for a number given by Eq. (11-5), we can write

$$(0.7)_{10} = B_{-1} \cdot 2^{-1} + B_{-2} \cdot 2^{-2} + B_{-3} \cdot 2^{-3} + B_{-4} \cdot 2^{-4} \tag{11-17}$$

where 0.7 is the decimal fraction to be converted to binary form to four places (chosen arbitrarily). It should be apparent that where fractional numbers are involved, division by 2 is not fruitful because of the negative powers of the radix. A quick study of Eq. (11-17), however, indicates that multiplication by 2 can be helpful because this process serves to isolate B_{-1} in preparation for its subsequent evaluation. Multiplying Eq. (11-17) by 2 on both sides gives

$$1.4 = B_{-1} + B_{-2} \cdot 2^{-1} + B_{-3} \cdot 2^{-2} + B_{-4} \cdot 2^{-3} \qquad (11\text{-}18)$$

Inspection of the right side of this expression reveals that the fractional part must be represented by the three rightmost terms since they involve negative powers of the base 2. In turn, it follows that the integer part must be represented by B_{-1}. Hence $B_{-1} = 1$. The new fractional part is expressed by

$$0.4 = B_{-2} \cdot 2^{-1} + B_{-3} \cdot 2^{-2} + B_{-4} \cdot 2^{-3} \qquad (11\text{-}19)$$

Multiplication by 2 changes this expression to

$$0.8 = B_{-2} + B_{-3} \cdot 2^{-1} + B_{-4} \cdot 2^{-2} \qquad (11\text{-}20)$$

The absence of an integer part on the left side of Eq. (11-20) means that $B_{-2} = 0$ and this equation can be rewritten as

$$0.8 = B_{-3} \cdot 2^{-1} + B_{-4} \cdot 2^{-2} \qquad (11\text{-}21)$$

Upon multiplying once more by 2, we get

$$1.6 = B_{-3} + B_{-4} \cdot 2^{-1}$$

This expression yields $B_{-3} = 1$ and

$$0.6 = B_{-4} \cdot 2^{-1}$$

With a final multiplication by 2 this last expression becomes

$$1.2 = B_{-4}$$

so that $B_{-4} = 1$. The remaining 0.2 represents the part of the binary number that is associated with all the digits beyond the fourth one. As a result we can expect an error between the original value and the four-digit binary representation that we arbitrarily chose at the outset. The extent of this error can be calculated by finding the decimal number that is actually represented by the just-calculated binary digits. Thus

$$
\begin{aligned}
N &= 1(2^{-1}) + 0(2^{-2}) + 1(2^{-3}) \quad + 1(2^{-4}) \\
&= 1(0.5) + 0 \quad\quad + 1(0.125) + 1(0.0625) \\
&= 0.6875
\end{aligned}
$$

The difference between the original fractional number 0.7 and this figure is 0.0125 (i.e., 0.2×2^{-4}) and is called the *round-off error*. If we had extended the number of digits to eight rather than four, clearly the error would be much smaller. In fact, the greater the number of digits the better the accuracy of the conversion.

A tabular format of the foregoing algorithm is also possible for the fractional part of a decimal number. This is shown as follows.

Value of binary digit	Remaining fractional part	
	$\times 2$ ⟶ 0.7	Original decimal fraction
$B_{-1} = 1 \quad .4$	$\times 2$ ⟶ 0.4	
$B_{-2} = 0 \quad .8$	$\times 2$ ⟶ 0.8	Fractional part of number after multiplication by 2
$B_{-3} = 1 \quad .6$	$\times 2$ ⟶ 0.6	
$B_{-4} = 1 \quad .2$	⟶ 0.2	

Further manipulation of the last fractional remainder (0.2) in the manner indicated makes it a routine matter to find subsequent binary digits of the decimal number. See Prob. 11-18.

EXAMPLE 11-3   The procedures described in this section are applicable to all number systems for achieving conversions. The differences lie in using the appropriate radix value. To illustrate the point, find the binary and octal equivalents of the decimal number 247.

Solution: The tabular procedure is employed in each case. For the binary equivalent we apply successive division by the base value 2 to get

247	
123	$1 = B_0$
61	$1 = B_1$
30	$1 = B_2$
15	$0 = B_3$
7	$1 = B_4$
3	$1 = B_5$
1	$1 = B_6$
0	$1 = B_7$

Thus

$$(247)_{10} = (11110111)_2$$

The octal equivalence is found by performing repeated division by the base value 8, which leads to the following tabulation.

247	
30	$7 = B_0$
3	$6 = B_1$
0	$3 = B_2$

Thus

$$(247)_{10} = (367)_8$$

It is instructive to note in Example 11-3 the existence of a simple strategy that permits the binary result to be expressed in octal form virtually by inspection. This is achieved by performing the operations as indicated below.

$$\underbrace{0 \ \ 1 \ \ 1}_{3} \quad \underbrace{1 \ \ 1 \ \ 0}_{6} \quad \underbrace{1 \ \ 1 \ \ 1}_{7} \quad \begin{array}{l} \text{binary form} \\ \\ \text{octal form} \end{array}$$

Starting immediately to the left of the binary point, identify successive groups of three binary digits. Then, using weighting factors of 4, 2, and 1, determine the octal equivalent of each group. In the example at hand the group on the right has the value: $4 + 2 + 1 = 7$. The middle group carries a value of $4 + 2 = 6$ and the group on the left has the value $2 + 1 = 3$.

A similar procedure can be used to represent the binary number in terms of an equivalent hexadecimal number. However, since the number of digits in the hexadecimal system is 2^4, it is necessary to use groupings of 4 and an associated weighting-factor sequence of 8, 4, 2, 1. Accordingly, we can write

$$\underbrace{1 \ \ 1 \ \ 1 \ \ 1}_{F} \quad \underbrace{0 \ \ 1 \ \ 1 \ \ 1}_{7} \quad \begin{array}{l} \text{binary} \\ \\ \text{hexadecimal} \end{array}$$

Summarizing, we can state that

$$(247)_{10} = (11110111)_2 = (367)_8 = (F7)_{16}$$

This example illustrates a particularly striking feature of the hexadecimal form, namely, simplicity, especially when compared to the binary version of the data. It is this feature that makes communication with such digital systems as digital computers and robots a much less tiring chore.

11-6 APPLICATIONS OF GATE CIRCUITS FOR DATA PROCESSING

In addition to the important logical operations that can be performed by digital gate circuits they are also capable of achieving the equally important task of data processing. Attention is directed here to just two examples to illustrate the point. The first deals with the binary-coded-decimal (BCD) decoder, which serves the function of converting the binary version of the decimal digits from 0 to 9 to its decimal form for communication with human beings. The second involves binary arithmetic (in this instance, addition) to show how the use of these same logic circuits permits easy data manipulation. In each case the standard logic gates are used.

The BCD-to-Decimal Decoder. The first ten entries for the decimal and binary columns of Table 11-2 describe the bit ordered sequence of a binary number and its corresponding decimal value. This is the ordinary relationship that is dictated by Eq. (11-4) for conversion. A binary-coded-decimal number, however, differs from the result obtained by the conversion equations whenever the number exceeds 9. For example, the binary equivalent of the number 25 is 1 1 0 0 1. In contrast, the binary-coded-decimal equivalent of 25 is $\underbrace{0010}_{2}\ \underbrace{0101}_{5}$. Observe that when the BCD code is used, *each* decimal digit is represented by four digits with which are attached the usual weights of 8, 4, 2, and 1. Of course, with this code there exist bit patterns representing 10 through 15 which are not used, but this is part of the cost of having a simple code. Many other codes are possible and in use which offer additional features.

Decoding is said to occur whenever digital circuitry is used to convert the four-digit BCD number to its corresponding decimal equivalent. The manner in which the NAND logic gates can be used to effect this conversion is depicted in Fig. 11-13. The inputs are the four binary signals B_3, B_2, B_1, and B_0, where, as customary, the lowest-numbered subscript denotes the least significant bit and the highest-numbered subscript represents the most significant bit. The complements of these signals are also available through use of the four inverters that are supplied by the decoder chip manufacturer. These are indicated by the open circles drawn on the left side of the chip. Also shown are the 10 output lines—one for each decimal digit. However, for illustration purposes, only one of the 10 NAND gates is shown, namely, the one that makes available information about decimal digit 4. The BCD representation of this number is 0100. Expressed as a logic statement we can say that the output of the decoder will indicate decimal 4(D_4) when the binary inputs B_3, B_1, and B_0 are at logic 0 and B_2 is at logic 1. Expressed in terms of Boolean algebra, we have

$$D_4 = \overline{B}_3 \cdot B_2 \cdot \overline{B}_1 \cdot \overline{B}_0$$

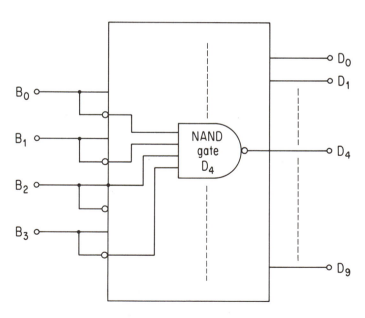

Fig. 11-13 Illustrating the conversion of the BCD number 0100 to decimal 4. Only one of the ten gates is shown.

Observe in Fig. 11-13 that when line B_0 is at logic 0 the output at its inverter is at logic 1, and it is this latter signal that is connected to the D_4 NAND gate. The situation is the same for lines B_1 and B_3. In the case of line B_2 the connection is made directly to the gate. Thus only when the inputs at NAND gate D_4 are 0100 will the output at terminal D_4 go low (logic 0), because the inputs are now all high assuming positive logic. Simultaneously, all the other output terminals will be high (logic 1), which represent the invalid codes for this number.

Binary Addition. Other examples of the application of digital logic circuits in the processing of data are offered by all aspects of binary arithmetic. This topic receives a good deal more attention in the next chapter, but for our purposes here we focus solely on the matter of the addition of two binary numbers. At the outset it is important to adhere to the same rules for binary arithmetical operations that are used for the decimal numbers. Keep in mind that the two number systems differ in their base values as well as the consequences that accrue from this difference. Therefore, in doing addition in the binary system, care must be taken to use only the $(r - 1)$ allowed digits. Since $r = 2$ for the binary system, a carry operation must be executed whenever we exceed 1. This corresponds to a carry in the decimal system whenever addition produces a sum that exceeds 9 (i.e., $r - 1 = 10 - 1 = 9$). Similar precautions apply for the borrowing process involved in subtraction. More is said about the latter in Chapter 12.

The process of binary addition is readily illustrated by assuming that we want to add decimal 10 to decimal 12 in binary form. This leads to the following:

$$
\begin{array}{rrcccc}
(10)_{10} = & & 1 & 0 & 1 & 0 \\
+ \quad\quad\quad + & & & & & \\
\hline
(12)_{10} = & & 1 & 1 & 0 & 0 \\
\hline
(22)_{10} = & 1 & 0 & 1 & 1 & 0 \quad \text{sum} \\
\end{array}
$$

Performing the addition in the accustomed manner for the first three columns starting from the right is routine and self-evident. At the fourth column, however, binary $1 + 1$ must be equated to 10 as indicated. This has the effect of placing the binary 1 in the next adjacent position to the left, corresponding to which is attached the next higher weighting factor. In this illustration the leftmost binary 1 carries a weight of 16, which, upon addition to the remaining contributions of 4 and 2, yields the required total of 22.

The numbers 10 and 12 were selected for the example because they bring into evidence the four combinations that can be encountered in binary addition, namely, $0 + 0 = 0$, $1 + 0 = 1$, $0 + 1 = 1$, and $1 + 1 = 10$. It is important to exert care that the binary addition $1 + 1 = 10$ not be confused with the logical-AND operation $1 + 1 = 1$. In order to implement the binary addition process with appropriate digital circuitry, it is necessary to arrange it so that it provides two outputs, one representing the sum S and the other the carry information C.

When the binary addition of two numbers is performed in an arithmetic device, the addition is performed on each pair of bits at a specific dedicated circuit. Hence a truth table for the four possible combinations that can arise on each bit line together with the sum and carry information can be expressed as follows:

B_i	B_i'	S	C
0	0	0	0
0	1	1	0
1	0	1	0
1	1	0	1

Here B_i denotes the ith bit of the first binary number and B_i' is the ith bit of the second binary number. A glance at the results of the carry column quickly reveals the results associated with an AND gate, whereas the column for the sum readily identifies the output of the exclusive-OR gate. The implementation of binary addition, therefore, involves an AND gate and an XOR gate in the fashion illustrated in Fig. 11-14. This arrangement is called a *half-adder* because, as will be explained in the next chapter, it lacks a provision for accommodating the carry signal from the preceding summations of the binary digits B_{i-1} and B_{i-1}'.

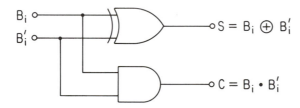

Fig. 11-14 Illustrating binary addition using digital logic gates.

Summary review questions

1. Name the two distinguishing characteristics of binary logic.
2. Name the three logical operations associated with binary logic.
3. Describe and illustrate with a truth table the logical AND function. Explain why this resembles multiplication.
4. Describe and illustrate with a truth table the logical OR function. Explain why this resembles addition.
5. Describe and illustrate the NOT function.
6. Describe a practical situation that calls for use of the logical AND operation.
7. Describe a practical situation that calls for use of the logical OR operation.
8. Define a logic gate. Indicate the number of input variables allowed and the number of logical output variables.
9. Why are the NAND and NOR gates commonly used in place of the AND or OR gates? Illustrate.

10. Distinguish the exclusive-OR gate from the OR gate.

11. What is positive logic? What is negative logic? Give illustrations. How are the logic variables, 0 and 1, assigned in each case?

12. Draw the logic diagram for the logical function $Z = (A + B + C) \cdot D$.

13. Describe and illustrate how the transistor amplifier can be used as a fast-acting electronic switch. In particular, point out the inversion property.

14. What is the advantage of using an active load resistor in the output circuit of the TTL NAND logic gate?

15. Explain the behavior of the TTL NAND gate that is depicted in Fig. 11-7 when input terminal B is low.

16. What is meant by the speed-power product of transistor logic gates? Why is this figure important?

17. What feature of the ECL logic gate makes it the fastest of the electronic switching families? What are the attendant disadvantages?

18. Explain the operation of the ECL logic gate of Fig. 11-8 when the input terminal A goes low.

19. Compare the relative merits of TTL, ECL, and CMOS gates regarding switching time, power dissipation, and fabrication density.

20. Name the serious shortcoming of the enhancement-mode MOS gate. How is this matter resolved?

21. List the advantages and disadvantages of CMOS gates.

22. Describe the operation of the inversion function of a CMOS inverter gate.

23. The complementary symmetry of the CMOS gate makes it easy to describe the behavior of such gates. Explain why this is so and illustrate.

24. Describe the behavior of the CMOS NAND gate of Fig. 11-12 in your own words.

25. What is meant by the radix of a number system? What is the radix value of the decimal system? Of the octal system? Of the hexadecimal system? Of the binary number system?

26. What is the range of coefficients that is possible in the octal number system? In the tertiary system? In the binary system? In the hexadecimal system?

27. Define the following terms: bit, byte, most significant bit, least significant bit, nibble, ordered sequence.

28. How many binary digits are needed to represent decimal 524?

29. Why is it necessary to have conversion strategies that permit conversion from one number system to the binary number system?

30. Describe the basic notion involved in converting an integer member of the number system with radix r into a binary form.

31. Describe the basic notion involved in converting a fractional member of the number system with radix r into a binary form.

32. What is round-off error as it applies to the conversion of numbers from one system into another? How can it be controlled?

33. Describe a simple strategy that can be used to express a binary number in octal form; in hexadecimal form.

34. What is the advantage of using the hexadecimal equivalents of binary numbers when communicating with digital systems? Illustrate.

35. Logic gates can be used to perform logical operations as well as data processing. Give examples of each application.

Problems

GROUP I

11-1. Modify the logic circuit of Fig. 11-2 so that it is made to include a seat pressure switch S which gives an indication that the seat is occupied. What difference in behavior does this new circuit exhibit over that of Fig. 11-2?

11-2. Devise a logic circuit that serves to sound a buzzer B whenever the ignition key is in place and either the seat belt of the driver, D, or the passenger, P, is unfastened. Write the logical function and draw the truth table.

11-3. Modify the switching circuit of Prob. 11-2 to include a seat pressure switch S in order to provide information about seat occupancy. Write the logical expression for the devised switching circuit.

11-4. Devise a logic gate circuit that produces $Z = 1$ when the logical variables take on the values $A = 0$, $B = 1$, $C = 1$, or $A = 1$, $B = 0$, $C = 0$.

11-5. Devise a logic gate circuit that produces $Z = 0$ when the logical variables take on the values $A = 1$, $B = 1$, $C = 1$, or $A = 0$, $B = 0$, $C = 0$.

11-6. A light bulb L is placed in series with a voltage source and two switches. The switch arrangement is to be such that either switch can be used to put the light on or off. Devise a switching circuit to perform this task. Write the truth table and the associated logical function.

11-7. If a process is governed by three logical variables, how many rows are there in the associated truth table? What is the result for four logical variables? Illustrate. What is the general expression for the number of rows of the truth table for n logical variables?

11-8. Express the switching circuit depicted in Fig. P11-8 in binary logic notation.

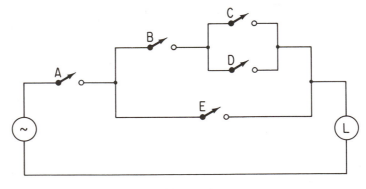

Fig. P11-8

11-9. A process involves four logical variables A, B, C, and D. Draw the logic gate circuitry that represents the logical expression $Z = A \cdot (B + C \cdot D)$.

11-10. Repeat Prob. 11-9 for the case where $Z = (A + B) \cdot (C + B \cdot D) \cdot C$.

11-11. Repeat Prob. 11-9 for the case where $Z = \overline{A \cdot B} + \overline{B \cdot C} + \overline{D} + C$.

11-12. Appearing in Fig. P11-12 is a TTL NAND gate that employs three transistors with Q_1 designed to operate as a multiple-emitter transistor in a manner similar to Q_1 in Fig. 11-7.

   (a) Explain the behavior of this circuit when A is low and B and C are at logical highs.

   (b) Repeat part (a) for the case where all three logical inputs are high.

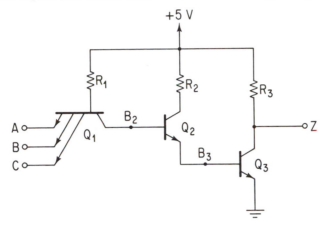

Fig. P11-12

11-13. Find the decimal equivalents of the following binary ordered sequences: (a) $N_2 = 1001101$, (b) $N_2 = 11100011$, and (c) $N_2 = 10101010$.

11-14. Determine the binary equivalents of the following decimal numbers: (a) 13, (b) 35, (c) 45, and (d) 875.

11-15. Find the decimal equivalents of the following octal numbers: (a) $N_8 = 704$, (b) $N_8 = 2456$, and (c) $N_8 = 74621$.

11-16. Is $N_8 = 257682$ a proper octal number? Explain.

11-17. Determine the decimal equivalents of the following hexadecimal numbers: (a) $N_{16} = $ F4F, (b) $N_{16} = $ FED, and (c) $N_{16} = $ FEED.

11-18. Find the binary representation of the decimal fraction 0.65 to 8 bits. Is there a round-off error? How large is it?

11-19. Convert the decimal number 466 to binary, octal, hexadecimal, and BCD forms.

11-20. Convert the binary number $N_2 = 101100110110$ to octal, hexadecimal, decimal, and BCD forms.

11-21. Find the binary equivalent of the decimal number 2.74 to six digits on the right of the binary point.

11-22. Write the first 16 digits of the number system with a radix of 4.

11-23. Convert the decimal number 17.6 to base 7.

11-24. Convert the following numbers to decimal: (a) 3401_5 and (b) 1202_3.

11-25. Convert the decimal number 17.6 to base 3.

GROUP II

11-26. Devise a logic gate circuit that produces an output logic signal $Z = 1$ when the input logical variables are $A = 1$, $B = 1$, $C = 1$, $D = 0$ or $A = 1$, $B = 0$, $C = 0$, $D = 1$ or $A = 0$, $B = 0$, $C = 1$, $D = 0$.

11-27. Refer to the ECL gate depicted in Fig. 11-8. Explain the operation of this circuit when A and B are both at the low voltage level.

11-28. The circuit illustrated in Fig. P11-28 employs CMOS gates. Determine the logical function that is represented by this configuration.

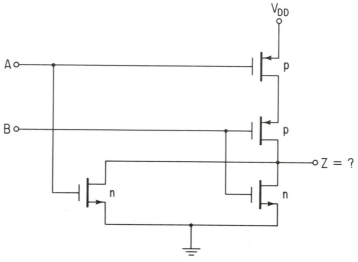

Fig. P11-28

11-29. The BCD code uses weighting factors 8, 4, 2, and 1 for the ordered sequence of bits to represent each decimal number in its ordered position. However, other binary-decimal codes are possible which can be made to possess certain useful features. One such code is that which carries the weighting factors 2, 4, 2, and 1. Observe that the 8 in the BCD code is replaced with a 2.
(a) Write the decimal number 83 using this new code.
(b) Write the decimal number 16 using the new code. (*Note:* Give preference to the 2 on the left when representing decimal 6.)
(c) Compare the results in parts (a) and (b) and identify the special feature associated with this code. When can such a feature be useful?

11-30. Codes can be written with error-detecting capabilities. The biquinary code is such a code. The assigned weighting factors in this 7-bit code are: 5, 0, 4, 3, 2, 1, and 0.
(a) To uncover the special feature of this code, write the decimal numbers 0 through 9 using these weighting factors but subject to the condition that each number must contain five 0s and two 1s.
(b) How can such representation of data be used to detect errors that might be incurred during the transmission and manipulation of digital data?

chapter twelve

Simplifying Logical Functions

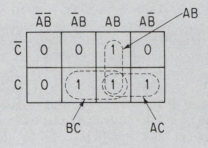

Logical operations serve two important functions in digital systems. This was described and illustrated at some length in the preceding chapter. First, logical functions contain the information by which a logical sequence of events can be executed to achieve a desired goal. Second, logical functions that contain the three important operations of AND, OR, and NOT can be used effectively to process data in a prescribed fashion.

The specific logical functions that meet the needs and objectives of a problem can always be written from an examination of the pertinent truth table. In general, the direct implementation of such functions by the various families of gates that are described in Chapter 11 leads to a more complex, more expensive, and usually slower-acting digital circuit than is actually necessary. The reason is that, in general, the logical functions that derive from the truth table contain redundancies which, if eliminated, readily simplifies the design of the logic circuitry. Methods are available to achieve this goal. Two methods are treated in this chapter. The first is an algebraic procedure based on the postulates and theorems of Boolean algebra. The second method is one that exploits the truly simplifying features of a geometrical representation of the truth table which is called a Karnaugh map. Alternative solutions of the simplification problem are discussed as well through use of the duality property of logical functions.

497

12-1 BOOLEAN ALGEBRA: POSTULATES AND THEOREMS

The notion of Boolean algebra was briefly referred to in the previous chapter after a two-valued binary variable set (0, 1) and the logical operations of AND, OR, and NOT were introduced in connection with switching circuits. As it happens, these are precisely the conditions that are needed to define a two-valued Boolean algebra for which a systematic and formal set of postulates and theorems are available. It is our purpose in this section to state and describe the more important aspects of this mathematical formulation as it relates to digital switching circuits. By becoming familiar with the algebraic rules of two-valued Boolean algebra, it will be possible in many cases to manipulate logical expressions, derived from truth tables, into their simplest forms. Such a procedure offers the distinct advantage of reducing the complexity of the logic diagrams that are required for the implementation of the corresponding truth tables. Inevitably, this in turn leads to cheaper fabrication costs for the digital circuits.

A group of five postulates of Boolean algebra, which is sufficient to meet our purposes, is listed in Table 12-1. These postulates are axioms which reveal self-evident features of the Boolean algebra. However, this self-evidency can be quickly demonstrated as illustrated in Fig. 12-1 by the corresponding truth table or the associated switching circuit, which is nothing more than an analog representation of the truth table for the AND and OR logical operations. Observe that in each case the output terminal exactly duplicates the logic level of A. Thus when A in Fig. 12-1 is at logic 0, so is the output since there is no transmission with switch A open. When A is at logic 1, the switches are in the closed positions and transmission occurs exactly in step with A.

It is useful and instructive at this point to study with some care the first two pairs of logical expressions listed in Table 12-1. Upon so doing, one begins to see a certain symmetry emerge. Because of the inherent symmetry, the second entry is called the *dual* of the first and the fourth is the dual of the third entry. A dual version of a logical expression involves the interchange of the logical

TABLE 12-1 POSTULATES OF BOOLEAN ALGEBRA

1. $A \cdot 1 = A$	1 is the identity element for AND
2. $A + 0 = A$	0 is the identity element for OR
3. $A \cdot \overline{A} = 0$ 4. $A + \overline{A} = 1$	Complement law
5. $AB = BA$ 6. $A + B = B + A$	Commutative law
7. $A \cdot (B + C) = (AB) + (AC)$	Distributive law
8. $A \cdot (B \cdot C) = (AB) C$ 9. $A + (B + C) = (A + B) + C$	Associative law

Simplifying Logical Functions   Chap. 12

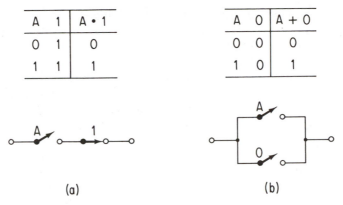

A	1	A · 1
0	1	0
1	1	1

A	0	A + 0
0	0	0
1	0	1

(a)   (b)

Fig. 12-1 Illustrating the identity operation for AND and OR logical operations: (a) $A \cdot 1 = A$; (b) $A + 0 = A$.

binary operation and the identity elements (0, 1) as well. Thus the fourth entry is obtained from the third by replacing the AND operation with the OR function and also replacing the 0 with logic 1. The *principle of duality* applies to the AND and OR functions generally. This is easily verified by replacing all the 0s with 1s in the truth table of the AND operation for two logical variables. The result becomes the truth table for the OR function between these variables. Presently, use is made of this principle to provide an expanded table of theorems for Boolean algebra.

The truth tables and their associated switching models for the logical operations of ANDing and ORing of a variable and its complement (i.e., entries 3 and 4) are displayed in Fig. 12-2. Clearly, for the complement rule of the AND operation no transmission can take place because the path is always open whether A is at logic 1 or logic 0. On the other hand, the parallel orientation of the switches that model the OR function ensures transmission for either value of the logical variable.

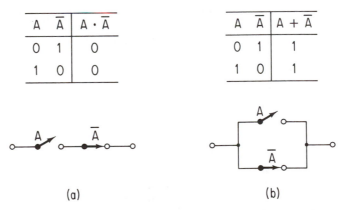

A	$\overline{A}$	$A \cdot \overline{A}$
0	1	0
1	0	0

A	$\overline{A}$	$A + \overline{A}$
0	1	1
1	0	1

(a)   (b)

Fig. 12-2 Illustrating the complement rule for the logical operations AND and OR for a logical variable A: (a) AND function; (b) OR function.

The laws of commutation, distribution, and association are valid for all logical operations in Boolean algebra. Examples are illustrated in postulates 5 through 9 in Table 12-1. These manipulations are easily verified by means of the respective truth tables.

We turn our attention next to some of the more useful theorems of Boolean algebra. The listing appears in Table 12-2 in two columns in order to stress the dual character of logical expressions. Also included are the analog models of these operations; they should enhance visualization especially in cases where the logical equivalence appears strange (e.g., theorem 6). For convenience the dot symbol for the AND operation is omitted. Thus $A \cdot B$ becomes AB. This policy will be in force from here on except in cases where ambiguity may arise. The switching model for theorem 3 makes it obvious that transmission of information from the input to output terminal depends solely upon A. The logical variable B in these dual logical expressions is irrelevant. This elimination of unnecessary variables is identified as the *absorption rule* of Boolean algebra. Theorem 4 illustrates another example of the absorption rule. The operation in either case is independent of the state of the B variable. Transmission takes place only when A is at logic 1. The models make this fact self-evident, but a simple manipulation of the dual version of this entry together with postulate 4 offers the same conclusion. Thus

$$AB + A\overline{B} = A(B + \overline{B}) = A \cdot 1 = A$$

Theorem 6 is deserving of a further remark. Surely, at first sight the equivalency appears strange. However, by redrawing the switching model to an equivalent form as indicated in the first column of Table 12-2, the reasonableness of the equivalent logical expression is apparent. When strange logical equations such as this one arise, it is helpful to draw a model for both sides of the equation as was done for theorem 6.

De Morgan's Theorems. The last entry, pair (7), in Table 12-2 is singled out for special attention because of its importance in the design of digital circuits and the analysis of logical expressions. This pair provides a specific relationship between NAND and OR as well as NOR and AND logical expressions and thereby provides the means of developing logic diagrams entirely in terms of NAND gates or NOR gates. The employment of such a strategy leads to cheaper digital circuit design. It also attaches to these gates the distinction of serving as *universal* gates. All logical functions can be expressed fully in terms of either NAND or NOR operations.

De Morgan's theorem states that the inverse of any logical function is found by complementing all input variables and replacing all AND operations with OR and all OR operations with AND.

It is instructive to apply this theorem to our two basic logical operations. The complement of the logical expression AB is found by complementing the inputs to $\overline{A}$ and $\overline{B}$ and then replacing the AND operation with the OR operation. Expressed mathematically, we have

$$\overline{AB} = \overline{A} + \overline{B} \tag{12-1}$$

A diagrammatic representation of this equation using the appropriate symbols of the gates involved is shown in Fig. 12-3(a). The left side of Eq. (12-1) is modeled by the NAND gate and the right side of Eq. (12-1) is represented by an OR gate with inverted inputs. It is this depicted equivalency that is used to replace OR gates with NAND gates in logic diagrams. To find how to express an AND gate by this technique, it is necessary first to complement both sides of Eq. (12-1). This yields

$$AB = \overline{\overline{A} + \overline{B}} \tag{12-2}$$

The diagrammatic representation of this expression appears in Fig. 12-3(b), where it should be clear that an AND gate is equivalent to a NOR gate with inverted inputs. Note that Fig. 12-3(b) derives from Fig. 12-3(a) upon complementing the outputs.

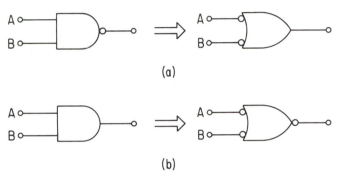

(a)

(b)

Fig. 12-3 (a) Illustrating the equivalency of a NAND gate to an OR gate with inverted inputs; (b) the AND gate as equivalent to a NOR gate with inverted inputs.

The application of De Morgan's theorem to the OR logical operation permits us to write

$$\overline{A + B} = \overline{A} \cdot \overline{B} \tag{12-3}$$

Figure 12-4(a) depicts the model for this equation and clearly indicates that a NOR gate is equivalent to an AND gate with inverted inputs. The equivalency

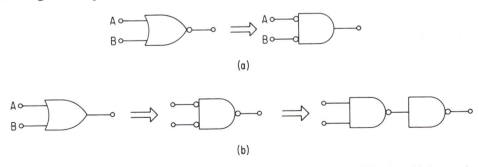

(a)

(b)

Fig. 12-4 (a) Illustrating the equivalency between a NOR gate and an AND gate with inverted inputs; (b) the OR gate as equivalent to the NAND gate with inverted inputs or two NAND gates without inverted inputs.

TABLE 12-2 SOME THEOREMS OF BOOLEAN ALGEBRA

Model

1.

2.

3.

4.

5.

6.

7.

for the OR circuit can be found by complementing both sides of Eq. (12-3). The result is

$$A + B = \overline{\overline{A} \cdot \overline{B}} \tag{12-4}$$

The corresponding model appears in Fig. 12-4(b). Observe that this diagram readily follows from the one in Fig. 12-4(a) by merely complementing both outputs. As a matter of fact, any one of the four diagrams in Figs. 12-3 and

(a) Logical expression	(b) Dual version	Model
$A \cdot 0 = 0$	$A + 1 = 1$	
$A \cdot A = A$	$A + A = A$	
$A(A + B) = A$	$A + AB = A$	
$(A + B)(A + \overline{B}) = A$	$AB + A\overline{B} = A$	
$A(\overline{A} + B) = AB$	$A + \overline{A}B = A + B$	
$A + BC = (A + B)(A + C)$	$A(B + C) = AB + AC$	
$\overline{AB} = \overline{A} + \overline{B}$	$\overline{A + B} = \overline{A} \cdot \overline{B}$	

12-4 can be derived from one another by performing the appropriate complementing operations.

A final remark is in order regarding the sufficiency aspect of the NAND and NOR gates. Although every logic diagram can be drawn using exclusively either one or the other type gate circuit, it is important to keep in mind that the transformation of the final logic diagram into either NAND or NOR gates is performed as the last step in the design of the digital system. The design process itself is executed in terms of the AND, OR, and NOT operations. The reason for this preference lies simply in the fact that the associative rule does not apply to the NAND and NOR operations, whereas it is entirely valid for AND and OR.

EXAMPLE 12-1   Using the postulates and theorems of Boolean algebra, demonstrate the validity of the logical statement: $A \cdot 0 = 0$.

Solution: Begin by replacing logic zero on the left side by its logical equivalent as represented by postulate 3 of Table 12-1. The successive steps then proceed as follows:

$$A \cdot 0 = 0$$
$$A \cdot (A\overline{A}) = 0 \qquad \text{by postulate 3, Table 12-1}$$
$$(A \cdot A)\,\overline{A} = 0 \qquad \text{by associative law}$$
$$A \cdot \overline{A} = 0 \qquad \text{by theorem 2, Table 12-2}$$
$$0 = 0 \qquad \text{by postulate 3, Table 12-1}$$

Of course, it then follows by the duality principle that

$$A + 1 = 1$$

EXAMPLE 12-2   Prove theorem 5(b) of Table 12-2.

Solution:

$$A + \overline{A}B = A + B$$
$$= A + B \cdot 1 \qquad \text{by postulate 1, Table 12-1}$$
$$= A + B\,(A + \overline{A}) \qquad \text{postulate 4, Table 12-1}$$
$$= A + BA + B\overline{A} \qquad \text{distributive law}$$
$$= A + AB + \overline{A}B \qquad \text{commutative law}$$
$$= A + \overline{A}B \qquad \text{theorem 3(b), Table 12-2}$$

12-2 ALGEBRAIC SIMPLIFICATION OF LOGICAL FUNCTIONS

There is a definite need to examine the logical expressions that describe a particular device for the elimination of redundancy if the complexity and thereby the cost of the digital circuitry is to be kept to a minimum. In applications of the design of digital structures the engineer has information concerning the logical variables and the associated truth table that defines the desired behavior of the device. A proper logical expression that describes this behavior can then be written. However, invariably these expressions are not in the form that command the fewest number of gates and inverters. Fortunately, strategies exist that do permit the designers to reduce the logical expressions to a minimum form which serves effectively to eliminate all redundancies. Two important advantages are thereby gained. One is circuit simplicity, which is the obvious result. The other is a reduction in propagation time in moving the digital signal through the digital logic circuitry.

Two methods for simplifying the logical expressions of combinational circuits are studied in this section. The first method uses the theorems of Boolean algebra in a proper algebraically manipulative fashion to remove the redundant terms that lie hidden in the original logical expression. The second method employs an analog mapping strategy that has the attribute of revealing redundancies by inspection. These maps are called *Karnaugh maps*.

To illustrate the situation, let us suppose that a process is described by the three logical variables A, B, and C, which in accordance with the truth table are related by the logical expression

$$Z = \overline{(A + \overline{B})} \, B(A + \overline{C}) \tag{12-5}$$

A digital circuit that generates the logic in this equation involves one NOR gate (for $A + \overline{B}$), one OR gate (for $A + \overline{C}$), two inverters (for $\overline{B}$, and $\overline{C}$), and one AND gate where the output of the NOR and OR gates are combined with variable B to yield the logical output variable Z. The logic circuit diagram is shown in Fig. 12-5(a). Observe that the signals B and C must propagate through three digital elements before arriving at the output terminal Z. If this number can be reduced, the digital circuit will be faster and this is always a desirable objective. We now demonstrate the simplification of Eq. (12-5) by the algebraic manipulations associated with the theorems of Boolean algebra.

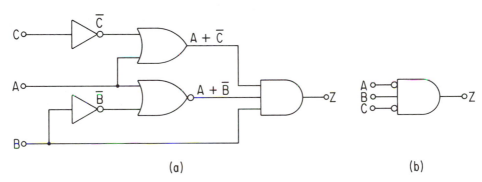

(a)   (b)

Fig. 12-5 (a) Circuit implementations of $Z = \overline{(A + \overline{B})} \, B \, (A + \overline{C})$; (b) the circuit after algebraic reduction by Boolean theorems.

Simplification by the Theorems of Boolean Algebra.   The procedure is best illustrated by the following step-by-step manipulation of Eq. (12-5).

$$
\begin{aligned}
Z &= \overline{(A + \overline{B})} B(A + \overline{C}) & \\
&= (\overline{A} \cdot B) B(A + \overline{C}) & \text{by De Morgan's theorem} \\
&= \overline{A}\,(B \cdot B)(A + \overline{C}) & \text{associative law} \\
&= \overline{A}B\,(A + \overline{C}) & \text{theorem 2(a), Table 12-2} \\
&= \overline{A}BA + \overline{A}B\overline{C} & \text{distributive law} \\
&= B(A \cdot \overline{A}) + \overline{A}B\overline{C} & \text{associative law} \\
&= 0 + \overline{A}B\overline{C} & \text{postulate 3, Table 12-1} \\
&= \overline{A}B\overline{C} & \text{postulate 2, Table 12-1}
\end{aligned}
\tag{12-6}
$$

Thus we see that the logical information that is contained in Eq. (12-5) is reduced to the much simpler expression represented by Eq. (12-6). It is at step 6 in the foregoing development that the algebraic manipulation uncovers a redundancy. The presence of a term containing $A \cdot \overline{A}$ means that such a switching path is always in the open state (i.e., nontransmissive). The gate circuit implementation of the reduced logical expression is depicted in Fig. 12-5(b). Now only a single AND gate is needed in addition to the two inverters. Moreover, propagation delay is determined by passage through two digital elements (an inverter and the AND gate) rather than three.

EXAMPLE 12-3   (The purpose of this example is to provide a more complete description of how the logical function arises and the effort required to reduce it to a minimal formulation that eliminates redundancy. The drudgery associated with the algebraic manipulations will also serve as a motivation for seeking out an easier procedure to arrive at a minimized equivalent logical expression.)

An agreement exists between three persons (A, B, C) that calls for all business policy decisions to be decided by a majority vote. But in the interest of avoiding animosity among them the voting is to be accomplished secretly through a voting machine.

(a) Develop an appropriate truth table and then design a digital logic circuit that provides an output of logic 1 whenever the majority vote is yes. Be sure to furnish a minimum implementation.

(b) Modify the logic diagram of the voting machine of part (a) so that only NAND gates are used.

Solution: (a) There are three logical variables. Hence there is a total of $2^3 = 8$ combinations of votes. If we let logic 1 denote a *yes* vote and logic 0 a *no* vote, the truth table becomes as follows:

A	B	C	Z
0	0	0	0
0	0	1	0
0	1	0	0
0	1	1	1
1	0	0	0
1	0	1	1
1	1	0	1
1	1	1	1

Here Z takes on the value of logic 1 if at least two persons vote yes. Because there are four places in the truth table where the vote is yes, the corresponding logical expression can be written as

$$Z = \overline{A}BC + A\overline{B}C + AB\overline{C} + ABC \qquad (12\text{-}7)$$

The right side of Eq. (12-7) contains a logical expression for each 1 in the column for Z. Thus the first 1 in the Z column corresponds to NOT A (or $\overline{A}$), which is a no vote, and a yes vote for B and C. Consistent expressions are written for the remaining 1s. Although we can design a digital logic circuit to implement Eq. (12-7) directly, our previous experience indicates that an excess number of gates and connections are very likely. Accordingly, in the interest of simplification, we now bring to bear the algebraic manipulations permitted by Boolean algebra. The result is as follows.

$$Z = \overline{A}BC + A\overline{B}C + AB\overline{C} + ABC$$

$$= \overline{A}BC + A\overline{B}C + AB\overline{C} + ABC + ABC \qquad \text{Theorem 4(b)}$$

$$= \overline{A}BC + A\overline{B}C + AB\overline{C} + ABC + ABC + ABC \qquad \text{Theorem 4(b)}$$

$$= (\overline{A} + A)BC + AC\overline{B} + ACB + AB(\overline{C} + C) \qquad \begin{array}{l}\text{Associative and} \\ \text{commutative laws}\end{array} \qquad (12\text{-}8)$$

$$= (\overline{A} + A)BC + AC(\overline{B} + B) + AB(\overline{C} + C) \qquad \text{Associative law}$$

$$= BC + AC + AB \qquad \text{Postulate 4}$$

This expression provides the minimal form of the logical expression for this problem. A little thought quickly uncovers the reasonableness of this result. A combination of any two of the three votes provides a majority and the possibilities can only be, of course, AB, or BC, or CA. In this simple case we could have said at the outset that logical expression (12-7) had to reduce to Eq. (12-8). However, our focus here is on the instructive aspect of the procedure. Accordingly, two points are particularly noteworthy: (1) the logical expression that is obtained from the truth table in general is in need of simplification; and (2) even for moderately complex problems (which this is not) the algebraic manipulations can become burdensome.

A study of Eq. (12-8) readily discloses that implementation requires three AND gates and one OR gate arranged in the manner depicted in Fig. 12-6(a).

(b) A convenient way to convert the logic diagram of Fig. 12-6(a) to an all-NAND gate implementation is to place inversion at the output of each AND circuit as well as at the input to the OR gate. Since the double inversion is present in each line connecting the AND and OR gates, the logical expression remains intact. The result is shown in

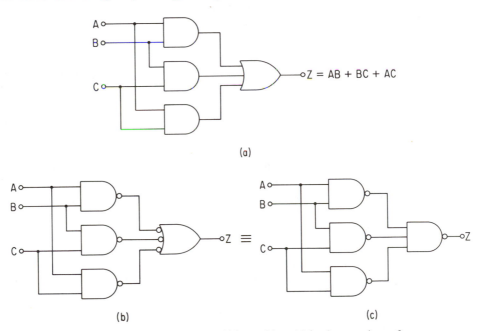

(a)

(b)   (c)

Fig. 12-6 A three-person voting machine: (a) implementations of Eq. (12-8); (b) introducing inversions at outputs of AND gates and inputs of OR gate; (c) replacing the OR gate with inverted inputs by a NAND gate.

Fig. 12-6(b). Now we have an OR gate with inverted inputs which, as shown in Fig. 12-3(a), is equivalent to a NAND gate. The all-NAND gate implementation appears in Fig. 12-6(c).

12-3 SIMPLIFICATION VIA KARNAUGH MAPS

The truth table is a first step in the design of a digital logic circuit because it describes what the circuit must do for all possible combinations of the input logical variables. Once the truth table is developed, a logical expression, similar to Eq. (12-7) for the simple voting machine, can be written in routine fashion. Boolean expressions that originate in this manner possess the special feature that each term of the ensuing logical expression involves *all* the logical variables either in the normal or complemented state. Moreover, if the Boolean expression is written for coincidence with the logical 1s, then there results what is called the *canonical* form of the *sum* of the *products* of the logical variables. To illustrate this statement, return for the moment to Eq. (12-7). This expression was written to correspond to those rows in the truth table which make Z equal to logical 1. Therefore, each term on the right side, such as $\overline{A}BC$, implies the AND operation and involves each of the logical variables. When so formulated, each such product term is called a *minterm*. Minterms can be written for each row of the truth table but our interest is only on those corresponding to Z equal to logical 1. Thus the canonical form of the Boolean expression becomes a *sum of minterms*. The sum merely takes note of the fact that an OR operation appears in the logical expression for Z, whenever the truth table shows more than one logical 1.

Unfortunately, it is the nature of such canonical Boolean expressions to contain redundancies which must be removed to simplify the design of logic circuits. The Karnaugh map provides a systematic procedure for achieving just such a goal without entailing the awkwardness associated with the application of Boolean algebra. Experience with the algebraic procedure, however, is useful for it lends understanding to the method of the Karnaugh (K) map. It is also important to keep in mind that the development of the K map is based on the canonical form of the sum of minterms of the Boolean expression.

The Karnaugh map is essentially a geometrical pattern of squares that carry the labels of the logical variables in both normal and complemented forms. The unique feature of the K map is that the arrangement is oriented in a manner that reveals the presence of redundancy by inspection. As a result it is a routine matter to write the minimal logical expression that is needed for the digital circuit design. The number of squares contained in the K map is dependent upon the number of logical variables. For n variables there are 2^n squares arranged in a specific order. The K map for the two variable case is shown in Fig. 12-7(a). Since $n = 2$, the K map contains four squares. The logical values of the A variable are depicted along the top of the square and those of the B variable are along the left side. An alternative representation is shown in Fig. 12-7(b), where $\overline{A}$ replaces logic 0 and A replaces logic 1. The 0s and 1s within the squares are

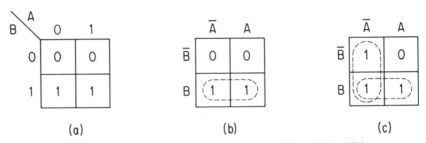

Fig. 12-7 Karnaugh map for two logical variables: (a) identifying the individual squares with the logic values 0, 1 along the top and side; (b) alternative representations using $\overline{A}$, A and $\overline{B}$, B; (c) a different case corresponding to the presence of three 1s in the truth table. The 0s and 1s originate from the truth table.

assumed to be the values of the Boolean expression Z for the given variable values. Thus in this case the canonical expression of the sum of minterms is

$$Z = \overline{A}B + AB \tag{12-9}$$

Of course, in this simple case algebraic manipulation readily indicates that the minimal form of Eq. (12-9) is simply

$$Z = B(\overline{A} + A) = B \qquad \text{by postulate 4} \tag{12-10}$$

In other words, for the situation illustrated by the K map of Fig. 12-7(a) and (b) the logical expression for Z is merely B. The K map reveals this algebraic absorption of variable A by means of the rounded dashed rectangle that encloses the two logical 1s on the B line. Whenever a rectangle (or rounded square in other cases) can be drawn enclosing the 1s and the rectangle is seen to span both the normal and complemented state of the same variable, that variable is absorbed and does not appear in the minimal logic expression. Thus in Fig. 12-7(b) the curved rectangle is found to expand both boxes $\overline{A}$ and A. Hence variable A is absorbed and the logical expression reduces to $Z = B$.

Diagrammed in Fig. 12-7(c) is a different situation, one where the truth table reveals the presence of three logical 1s. The canonical form of the sum of minterms is accordingly

$$Z = \overline{A}\,\overline{B} + \overline{A}B + AB \tag{12-11}$$

The minimized version of this expression will now be found from the K map rather than through algebraic manipulation. Observe that two rounded rectangles can be drawn. Overlapping rectangles are entirely permissible. The overlapping in this case takes place in the square corresponding to minterm $\overline{A}B$. The net logical expression associated with the horizontally oriented rectangle is B since as in the case of Fig. 12-7(b) the variable A is absorbed. In the case of the vertical rectangle it is the B variable which is absorbed. Consequently, the minimum sum of minterms becomes simply

$$Z = B + \overline{A} \tag{12-12}$$

Because Eqs. (12-11) and (12-12) are equivalent, it would be a useful exercise for the reader to verify the equivalency by applying Boolean algebra.

When the number of logical variables is three, the K map contains $2^3 =$ 8 squares, one for each minterm. In this case the K map is drawn as shown in Fig. 12-8. Depicted in Fig. 12-8(a) is the K map expressed in terms of the 0,1 nomenclature to identify the individual squares. Note that for the three-variable case, the first two variables, A and B, are placed along the top of the K map and the third along the side. It is important to observe that when two variables are placed along a line, the logical values must be so arranged that *only one bit change occurs* as one moves to an adjacent box. It is this feature which permits the absorption of a variable to take place (whenever 1s appear in adjacent boxes) and thereby leads to simplification without the cumbersome algebraic manipulations. Figure 12-8(b) presents an alternative presentation of the identification of the squares in terms of the complemented and uncomplemented forms of the logical variables. This is the preferred drawing of the K map in view of the fact that the originating Boolean expression for Z must be of the canonical form, involving each of the logical variables of the design.

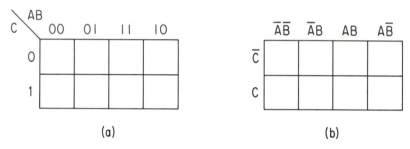

Fig. 12-8 Orientation of a Karnaugh map for three logical variables: (a) K maps with 0,1 notation; (b) K map in terms of the normal and complemented variables.

The procedure for writing the minimal logical expression by inspection of an appropriately drawn K map is illustrated by the series of diagrams that appear in Fig. 12-9. In Fig. 12-9(a) it is assumed that the truth table indicates the presence of 1s in each of the upper squares of the map. The canonical sum of minterms is accordingly described by

$$Z = \overline{A}\,\overline{B}\,\overline{C} + \overline{A}B\overline{C} + AB\overline{C} + A\overline{B}\,\overline{C} \tag{12-13}$$

In this case the rounded rectangle is drawn to enclose all four 1s. Along the horizontal portion of this rectangle both the complemented and uncomplemented states of A and B are encountered. Hence both variables are absorbed and so the logical expression of Eq. (12-13) reduces very simply to

$$Z = \overline{C} \tag{12-14}$$

Figure 12-9(b) represents a different orientation of 1s for the truth table. In this case a rounded square is allowed to be drawn encircling the logical 1s. Along the upper portion of this rounded square we encounter B and $\overline{B}$, and so the B variable is absorbed, but variable A remains. Along the side of the rounded square both states of the C variable are encountered, so that the C variable is

Simplifying Logical Functions   Chap. 12

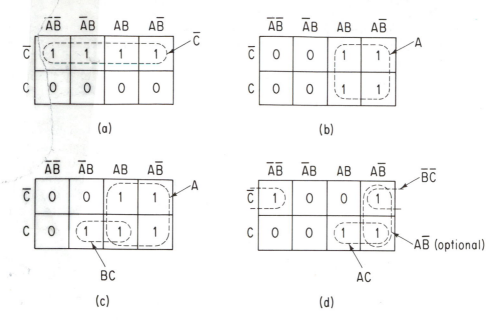

Fig. 12-9 Illustrating the minimization associated with various minterms with logic 1 values for the case of three logical variables.

absorbed. Consequently, the canonical sum of minterms associated with Fig. 12-9(b) and described by

$$Z = AB\overline{C} + A\overline{B}\overline{C} + ABC + A\overline{B}C \qquad (12\text{-}15)$$

actually reduces to

$$Z = A \qquad (12\text{-}16)$$

The decided advantage offered by the K map in effecting a minimization of the canonical logical expression should be quite apparent at this stage.

The case illustrated in Fig. 12-9(c) has five minterms at logical 1 in its truth table. Here an overlapping rounded rectangle in addition to the rounded square of the previous case can be drawn as indicated. The canonical sum of minterms includes those that are displayed in Eq. (12-15) plus $\overline{A}BC$. Thus

$$Z = AB\overline{C} + A\overline{B}\overline{C} + ABC + A\overline{B}C + \overline{A}BC \qquad (12\text{-}17)$$

The contribution of the rounded rectangle to the minimal logical sum of products is BC because the upper part of the rectangle has associated with it A and $\overline{A}$, so this variable is absorbed. When this result is combined with the contribution of the rounded square, the final expression for the reduced logical function reads as

$$Z = A + BC$$

In dealing with a situation such as the one shown in Fig. 12-9(d), it is important to remember that the left edge of the K map is adjacent to the right edge. The K map must be treated as a rolled-out cylinder. Hence the 1 in the upper leftmost square has to be treated as adjacent to the 1 appearing in the upper rightmost square. The variable that is absorbed by this rectangle is A which means the contribution to the minimal sum of products is $\overline{B}\,\overline{C}$. Because the vertical rectangle has both of its 1s as parts of other rectangles, its contribution may be ignored. The remaining rectangle offers AC. The final minimal sum of products becomes

$$Z = AC + \overline{B}\,\overline{C}$$

The labeling that is used for a K map of four logical variables is shown in Fig. 12-10. Once the logical 1s are properly placed in the squares in a manner that is consistent with the truth table, the procedure described for the three-variable case is equally applicable. However, care must be taken now whenever 1s appear in the top and bottom rows of squares in the same column to treat them as adjacent. The notion of a rolled-out cylinder applies to the top and bottom edges as well.

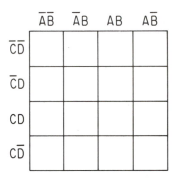

Fig. 12-10 Orientation of the squares in a four-variable K map.

EXAMPLE 12-4 Draw the K map for the problem of Example 12-3 and obtain the minimal sum of products by inspection of the diagram.

Solution: The total of three variables calls for a K map with $2^3 = 8$ squares, one for each minterm. Moreover, the canonical sum of minterms (i.e., those that have a value of logical 1) is given by Eq. (12-7). This expression furnishes the instructions as to which boxes in the K map receive a 1. The result is shown in Fig. 12-11. Also depicted are the appropriate rectangles, of which there are three. The reduced product terms for each rectangle are indicated as well. Thus the minimal sum of products becomes simply

$$Z = AB + BC + AC$$

A comparison of the effort needed to obtain this result through use of the K map with the awkward manipulations executed in Example 12-3 is striking indeed. It becomes more so for the four-variable case.

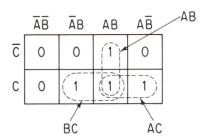

Fig. 12-11 K map for Example 12-4.

EXAMPLE 12-5 The K map of a problem that involves four logical variables is shown in Fig. 12-12. Indicated too are the minterms that are at logical 1.

(a) Find the canonical sum of minterms.

(b) Determine the minimal sum of products.

Solution: (a) In the top row the leftmost 1 is the minterm that corresponds to the coincidence of $\overline{A}B$ with $\overline{C}D$ to yield $\overline{A}B\overline{C}D$. Continuing this procedure to include all minterms at logic 1, the result is

$$Z = \overline{A}B\overline{C}D + AB\overline{C}D + AB\overline{C}D + ABCD + \overline{A}B\overline{C}\overline{D} + AB\overline{C}\overline{D}$$

(b) By inspection of the K map for the rounded rectangle and square we get

$$Z = AB + B\overline{D}$$

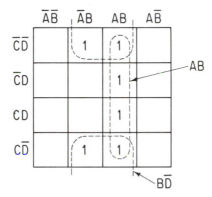

Fig. 12-12 K map for Example 12-5.

12-4 AN ALTERNATIVE SCHEME: PRODUCT OF SUMS

It was demonstrated in connection with Fig. 12-6 that a logical expression that is formulated as a sum of products [refer to Eq. (12-7)] leads rather naturally to a NAND logic diagram. Moreover, in our treatment of Karnaugh maps, we identified expressions such as Eq. (12-7) as a canonical sum of minterms whenever the minterms are those products of the logical variables, in either state, for which the Boolean expression Z is 1. At this point an obvious question suggests itself: What meaning can we attach to those remaining minterms for which the logical expression has the value 0? To explore this query let us return to the truth table of Example 12-3 on p. 506. A glance at the first row shows that $Z = 0$ when $A = 0$, $B = 0$, and $C = 0$. The second row shows that $Z = 0$ when $A = 0$, $B = 0$, and $C = 1$. Similar statements can be made for the remaining two 0s at rows 3 and 5. It is useful to note that an alternative description of the first row is to say that $\overline{Z} = 1$ when $\overline{A} = 1, \overline{B} = 1$, and $\overline{C} = 1$. This is simply the complement rule. At row 2 we can state alternatively that $\overline{Z} = 1$ when $\overline{A} = 1$, $\overline{B} = 1$, $C = 1$. The complemented form of the rows where the logical function has a logical 0 value now allows us to express these 0s as a standard sum of products. Thus for the four rows of this truth table we can state

$$\overline{Z} = \overline{A}\,\overline{B}\,\overline{C} + \overline{A}\,B\,C + \overline{A}B\overline{C} + A\overline{B}\,\overline{C} \tag{12-18}$$

Keep in mind that $\overline{Z}$ has value 1. However, we can return to the 0 state, which is really the focus of our investigation, by complementing Eq. (12-18) through the application of De Morgan's theorem. We get

$$Z = \overline{\overline{Z}} = (A + B + C)(A + B + \overline{C})(A + \overline{B} + C)(\overline{A} + B + C) \qquad (12\text{-}19)$$

This result is significant because it identifies those rows in the truth table that correspond to logical 0 as a *canonical product of sums*. It is called canonical because all the logical variables are present in each parentheses, where the variables are related by the OR operation. Moreover, Eq. (12-19) is also described as a canonical product of *maxterms*. The name *maxterm* is a general term that refers to the logical sum of each variable of the truth table and is formulated so that where a 0 occurs the literal representation of the variable is uncomplemented and where a 1 occurs it is complemented. By continuing the policy used for minterms, a *canonical* product of maxterms refers only to those maxterms for which $Z = 0$ in the truth table. In the case at hand these are the four that appear in Eq. (12-19). These matters can be conveniently summarized by repeating the truth table of Example 12-3 but modified to include two additional columns which are captioned to be self-explanatory.

A	B	C	Z	Minterms for canonical sum of *products*	Maxterms for canonical product of *sums*
0	0	0	0		$A + B + C$
0	0	1	0		$A + B + \overline{C}$
0	1	0	0		$A + \overline{B} + C$
0	1	1	1	$\overline{A}BC$	
1	0	0	0		$\overline{A} + B + C$
1	0	1	1	$A\overline{B}C$	
1	1	0	1	$AB\overline{C}$	
1	1	1	1	ABC	

Equation (12-19) makes it quite clear that when implementing circuits from a product of maxterms the first logic level would involve four OR gates to take care of the four parenthesized maxterms followed by a second level that accommodates the product of the sums at a four-input AND gate. We have every reason to suspect, however, that if we were to implement the logic circuit by Eq. (12-19), we would not have a minimal solution. At this stage a second question arises: Can we employ the Karnaugh map strategy to obtain a minimal product of sums, and how is this achieved?

The response is affirmative and the procedure is similar to that used in connection with the sum of minterms. An excellent starting point is to work with the *equivalent sum of products* that is obtained from the logical expression, Z, at the 0s in the truth table. In our case this means dealing with Eq. (12-18) rather than Eq. (12-19). The advantage of this approach is that $\overline{Z}$ is now at logical 1 and we can thereby treat all the terms on the right side of Eq. (12-18) just as we did for the sum of minterms in the K map. However, it is necessary to make one important change in the K map. The complementing operation that changes $Z = 0$ to $\overline{Z} = 1$ makes it necessary to change the orientation of 0s and 1s (which applied to the sum of minterms) to the complemented state. This means the 0s are changed to 1s, and vice versa. Of course, as a practical matter, this procedure is unnecessary if we remember to draw our rectangles and squares around the 0s whenever we are implementing a solution to a logical problem by a strategy that revolves about a product of maxterms.

Displayed in Fig. 12-13 is the K map with the 0s and 1s placed in those squares that was dictated by a standard sum-of-minterm solution (compare with Fig. 12-11). However, the rounded rectangles now are drawn about the 0s because we really want a product-of-maxterms solution. We get the solution by applying the technique associated with a sum-of-products approach provided that we continue to deal with the complemented expression $\overline{Z}$. By inspection the K map reveals the *equivalent* minimal sum of products related to $\overline{Z}$ to be

$$\overline{Z} = \overline{A}\,\overline{C} + \overline{B}\,\overline{C} + \overline{A}\,\overline{B} \qquad (12\text{-}20)$$

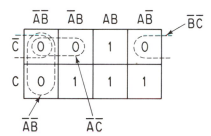

Fig. 12-13 K map of the voting machine problem for a solution by the product of maxterms (sums).

The solution of the logical problem from the viewpoint of the product of maxterms can be ensured if Eq. (12-20) is complemented because by so doing we return to the 0 state values of the logical expression. The application of De Morgan's theorem readily achieves this objective. Thus, by complementing each logical variable and replacing every AND operation with OR, and vice versa, the result becomes

$$\overline{\overline{Z}} = Z = (A + C)(B + C)(A + B) \qquad (12\text{-}21)$$

This equation represents the minimal product-of-sums solution to the voting machine problem. A comparison of this solution with Eq. (12-8), which denotes

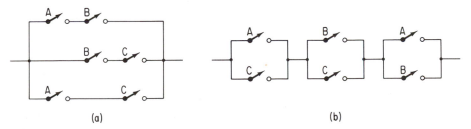

Fig. 12-14 Switching-circuit representation of the voting machine problem of Example 12-3: (a) the sum-of-products solution $Z = AB + BC + AC$; (b) the product-of-sums solution $Z = (A + B)(B + C)(A + C)$.

the sum-of-minterm solution to the same problem, discloses the one to be the *dual* of the other.

From the point of view of human beings, however, the solution to the logical problem of the voting machine is easier to understand in terms of the sum-of-products solution, where $Z = AB + BC + AC$. It is obvious that a majority vote of two occurs when A and B vote yes, or, B and C vote yes, or,

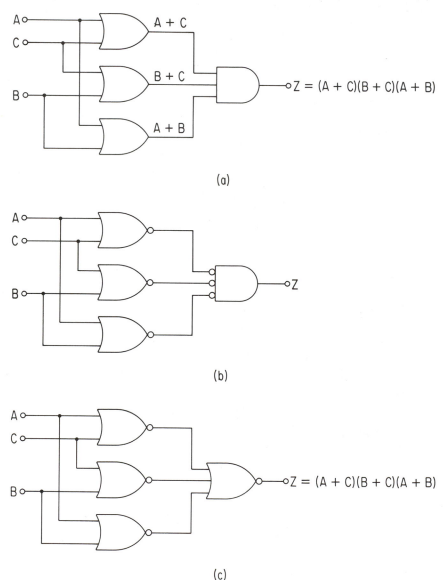

(a)

(b)

(c)

Fig. 12-15 Illustrating the development of an all-NOR gate solution of the voting machine: (a) the product-of-sums implementation; (b) introducing inversion in the lines that connect the OR gates to the AND gate; (c) replacing the AND gate with inverted inputs by its equivalent NOR gate.

A and *C* vote yes. It is not nearly as obvious to draw this conclusion from the equivalent description of the solution given by Eq. (12-21). The use of a switching-circuit model for each solution, however, uncovers the equivalency rather handily. The diagrams are depicted in Fig. 12-14. The sum-of-products solution is illustrated in Fig. 12-14(a) and needs no further comment. The product-of-sums solution is shown in Fig. 12-14(b). In essence this is a series arrangement (i.e., the AND or product operation) of parallel (i.e., the OR or sum operation) switches taken two at a time. To point out the equivalency observe that with *A* and *B* closed in Fig. 12-14(a) the upper path is closed, so a yes vote is registered. In Fig. 12-14(b) with *A* and *B* at logical 1 all three upper switches are closed. Here this circuit also records a yes vote. An affirmative vote by *B* and *C* closes the middle path of Fig. 12-14(a). In Fig. 12-14(b) a closed path also occurs but it takes a zigzag path. In fact, it is this aspect of the product-of-sums solution to a logical problem that makes it less easy to visualize directly from the resulting minimal expression.

Often, the designer of digital circuits will choose to obtain a product-of-sums solution to a logical problem in order to obtain a logic diagram that uses only NOR gates. This is a natural consequence of the product-of-sums formulation. The situation is readily illustrated by continuing with the example of the voting machine. To implement the logical expression represented by Eq. (12-21), the first logic level calls for three OR gates where the sums $A + C$, $B + C$, and $A + B$ are generated. This is then followed by a second logic level where the outputs of the three OR gates undergo the AND operation (i.e., the product of the sums). The resulting logic diagram is depicted in Fig. 12-15(a). At this stage in the design modifications can be made to put the emphasis on the use of NOR gates rather than OR gates. Figure 12-15(b) illustrates the transition step. Observe that this diagram differs from Fig. 12-15(a) solely by the inclusion of inversion at both ends of the connections from the OR gates to the AND gate. The final logic diagram of Fig. 12-15(c) results upon replacing an AND gate with inverted inputs by its equivalent, the NOR gate.

Summary review questions

1. Name two methods that can be used to simplify the logical functions that derive from the truth table.
2. List the requirements for a two-valued Boolean algebra.
3. Identify and illustrate the identity element for the AND logical function.
4. Identify and illustrate the identity element for the OR operation.
5. State and illustrate the complement rule of Boolean algebra.
6. State and illustrate the commutative law of Boolean algebra.
7. State and illustrate the associative law of Boolean algebra.
8. State and illustrate the distributive law of Boolean algebra.
9. Describe the dual property that is present in the postulates of Boolean algebra.

10. State the rule that applies in obtaining the dual of a logical expression for a two-valued Boolean algebra.

11. State and illustrate the absorption rule of Boolean algebra.

12. State and illustrate De Morgan's theorem. What important role does the application of this theorem play in the design of digital circuits?

13. What is a universal gate? Identify them.

14. Use De Morgan's theorem to derive the NOR equivalent of an AND gate. Assume the availability of inverters.

15. Use De Morgan's theorem to derive the OR equivalent of a NAND gate. Assume that inverters are available.

16. Use De Morgan's theorem to derive the AND gate version of a NOR gate. Assume that inverters are available.

17. Use De Morgan's theorem to derive the NAND gate version of the OR gate. Assume that inverters are available.

18. List two important reasons for simplifying the logical expressions that originate from the truth table that describes a specific process or problem.

19. Comment in a general way about the strategy for simplifying logical expressions by algebraic manipulations.

20. What specific property can you identify for the logical expressions that originate from the truth table of a process or problem?

21. What is meant by a logical expression written in the canonical form of the sum of products? How are the terms in this logical expression recognized in the truth table?

22. Define a minterm. Illustrate.

23. Describe a Karnaugh map and identify its salient features making reference in particular to the role provided by some of the postulates of Boolean algebra.

24. How is the number of squares to be used in a Karnaugh map determined? How is the orientation of the squares established for two logical variables? For three logical variables? For four logical variables?

25. Describe the meaning of the phrase: canonical product of sums.

26. Distinguish between the canonical sum of products and the canonical product of sums.

27. Define a maxterm. Illustrate.

28. Distinguish between minterms and maxterms. How are these terms identified?

29. The product-of-sums formulation of a logical problem is said to lead naturally to a hardware implementation that uses the NOR circuit as the universal gate. Explain why this is so.

Problems

GROUP I

12-1. Using the postulates of Boolean algebra show that $A \cdot A = A$.

12-2. Applying the techniques of algebraic manipulation show that $A + AB = A$.

12-3. Demonstrate the validity of theorem 6(b) in Table 12-2.

12-4. Use Boolean algebra to prove the following identities.

(a) $(A + B)(A + \bar{B}) = A$

(b) $A + BC = (A + B)(A + C)$

(c) $AB + \bar{A}C = (A + C)(\bar{A} + B)$

(d) $(A + B)(\bar{A} + C) = AC + \bar{A}B$

12-5. Determine the truth table for the logical function $Z = AB + AC + \bar{A}\bar{B}$.

12-6. Simplify the following Boolean logical functions to a minimum number of literals using Boolean algebra.

(a) $Z = ABC + \bar{A}B + AB\bar{C}$

(b) $Z = AB + BCD + B\bar{C}D$

12-7. A Boolean logical expression is described by $Z = AB + BCD + B\bar{C}D$.

(a) Express Z in terms of a minimum number of literals.

(b) Implement the logic circuit with AND, OR, and NOT gates.

(c) Implement the logic circuit with the universal NAND gate.

(d) Implement the logic circuit with the universal NOR gate.

12-8. Repeat Prob. 12-7 for a logical expression which is given by the following canonical sum of minterms:

$$Z = \bar{A}\bar{B}C + \bar{A}B\bar{C} + A\bar{B}C + AB\bar{C}$$

12-9. Repeat Prob. 12-7 for the case where the logical function is given by

$$Z = \bar{A}\bar{B}\bar{C} + AB\bar{C} + ABC$$

12-10. Find the truth table that corresponds to the minimal logical expression $Z = B\bar{C} + \bar{B}C$. The process involves the three logical variables A, B, and C.

12-11. Refer to the logic circuit that is depicted in Fig. P12-11.

(a) Determine the logical expression for Z.

(b) Find the truth table that corresponds to this logical function.

(c) Redesign the given logic circuit so that it uses the universal NAND gate exclusively.

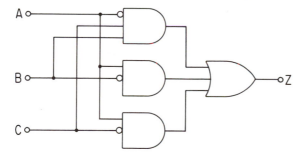

Fig. P12-11

12-12. Refer to the logic circuit drawn in Fig. P12-12.

(a) Find the logical expression for Z.

(b) Is the result found in part (a) in canonical form? Explain.

(c) Minimize the logical expression found in part (a).

(d) Draw the logic circuit for the expression found in part (c) and compare with that given in Fig. P12-12 and make appropriate comments.

(e) Redesign the circuit of part (d) to involve only the NAND gate.

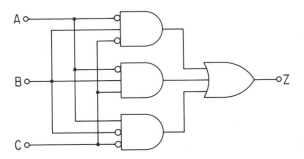

Fig. P12-12

12-13. Use the Karnaugh map to minimize the logical expression given in Prob. 12-8.

12-14. Use the Karnaugh map to minimize the logical expression given in Prob. 12-9.

12-15. A process involves three logical variables (A, B, C) and two logical outputs (Z_1, Z_2). The associated truth table is as follows:

A	B	C	Z_1	Z_2
0	0	0	0	1
0	0	1	1	1
0	1	0	0	1
0	1	1	1	0
1	0	0	0	0
1	0	1	1	0
1	1	0	0	1
1	1	1	1	0

(a) Write expressions for Z_1 and Z_2 that are sums of minterms.
(b) Write expressions for Z_1 and Z_2 that are products of maxterms.
(c) Using Karnaugh maps, find the minimized logical expressions in part (a).
(d) Repeat (c) for part (b).

12-16. A process is described by three logical variables (A, B, C). The sum of minterms is given by

$$Z = AB\overline{C} + A\overline{B}\,\overline{C} + \overline{A}BC + ABC$$

(a) Find the simplified logical expression using the Karnaugh map.
(b) Develop an all-NAND logic circuit that implements the result found in part (a).

12-17. Refer to the process described in Prob. 12-16.
(a) Find the logical expression for the output that is expressed as a product of maxterms.
(b) Determine the minimized logical expression using the K map.
(c) Develop an all-NOR logic circuit that implements the result found in part (b).

12-18. A process is described by the following truth table:

A	B	C	Z
0	0	0	0
0	0	1	1
0	1	0	1
0	1	1	0
1	0	0	0
1	0	1	0
1	1	0	1
1	1	1	0

(a) Write the logical expression for Z as a canonical product of sums.

(b) Reduce the expression found in part (a) to its simplest form using the Karnaugh map.

12-19. A process is represented by the logical expression $Z = ABC + AC + A\bar{B}C$. Find the expression for the minimal sum of products.

GROUP II

12-20. Show that

$$\overline{A[B + C(D + \bar{A})]} = \bar{A} + B\bar{C} + B\bar{D}$$

12-21. Use Boolean algebra to prove the following identity:

$$AB + B\bar{C} + CD + B\bar{D} + BC = B + CD$$

12-22. Using algebraic manipulations simplify the following algebric expression to a minimum number of literals:

$$Z = B(A + CD) + \bar{D}(BC + D) + A(\bar{B} + D) + \bar{A}D$$

12-23. A process is described by four logical variables (A, B, C, D). The sum of minterms is given by

$$Z = AB\bar{C}\bar{D} + A\bar{B}\bar{C}\bar{D} + \bar{A}BCD + \bar{A}\bar{B}\bar{C}D + A\bar{B}\bar{C}D + A\bar{B}CD$$

(a) Find the minimal form of this logical expression using Karnaugh maps.

(b) Implement the result of part (a) with AND, OR, and NOT gates.

12-24. The sum of the minterms in a given logical problem is given by

$$Z = \bar{A}\bar{B}C\bar{D} + \bar{A}B\bar{C}\bar{D} + A\bar{B}C\bar{D} + \bar{A}BCD + AB\bar{C}D + \bar{A}BC\bar{D}$$

Find the product of maxterms.

chapter thirteen

Components of Digital Systems

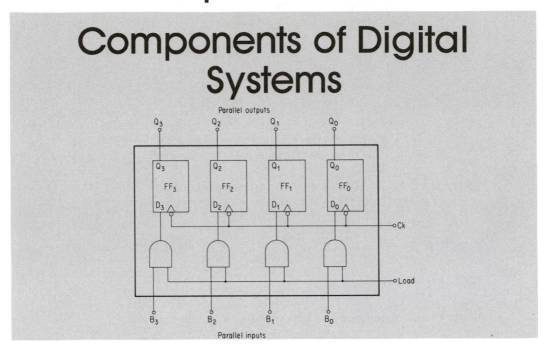

In this chapter we undertake the task of describing the salient features of the most commonly used components of digital systems. These components are divided into two categories: combinational logic circuits and sequential logic circuits. It is useful to point out here that the background of the preceding two chapters should serve well in making the study of these components a satisfying experience. We have many occasions here to illustrate the application aspects of the foregoing gate circuits and theory.

Combinational logic circuits

Combinational circuits are composed of gate structures which are driven by input signals that are completely independent of the state of the output signals. The gates themselves are arranged in a manner that is consistent with the instructions that derive from a truth table or a desired logical function. In some combinational logic circuits the designer may use chips such as the ones that are depicted in Fig. 13-1 in order to generate the required Boolean function. These generally contain a few gates and are produced by the techniques of small-scale integration (SSI). In other instances the designer may resort to specially designed packaged IC chips on which appears the complete logical function containing all the gates together with interconnections. Circuits of this type are produced by medium-

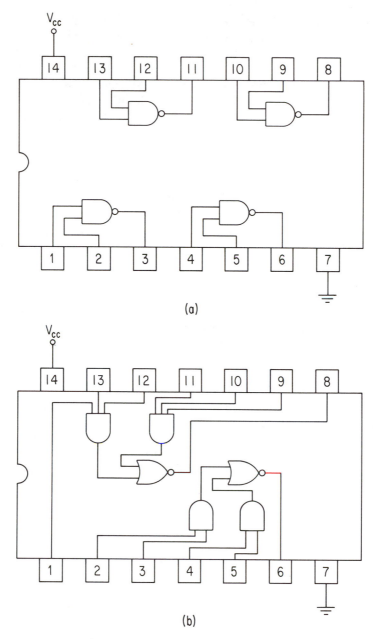

Fig. 13-1 Illustrating the gates available on integrated circuit chips: (a) a quad two-input NAND gate; (b) a dual two-input AND-NOR gate.

scale integration (MSI) and large-scale integration (LSI). The MSI chips contain up to 100 gates, while the number on the LSI devices reaches a thousand or more. If a digital circuit is needed, for example, to do binary addition, the judicious approach is to purchase an MSI packaged unit rather than to build the circuit from the gates available from SSI chips. The latter approach exacts a very costly penalty that is associated with the required interconnections between the gate chips. Often, this can be the more expensive part of the design and can be effective in reducing reliability as well.

In this first part of the chapter attention is directed to the design and description of some of the more popular combinational circuits that have found wide applications in digital computers and digital systems generally. In most cases the devices are available as MSI and/or LSI packaged units.

13-1 ENCODERS

We treat the encoder first because this is one of the few digital devices that is constructed from gate chips such as those illustrated in Fig. 13-1.

Situations arise in digital systems where it is necessary to provide a code for the unique signal that appears on each input line of an encoder. Consequently, if the code is to be expressed with three binary bits, the number of input lines that can be *encoded* is eight. In general, 2^n input lines can be encoded by an n-bit code. It is important to understand, too, that at any one time only *one* of the input lines can be activated (i.e., at logic 1) while the encoding takes place. The process of building an encoder can be demonstrated by working with a truth table and applying the principles of Secs. 12-3 and 12-4.

As an illustration let us design an encoder which translates (encodes) the decimal numbers 0 to 7 into binary form. Since there are eight input lines (one for each number), the encoder needs only three output bits. The corresponding truth table thus becomes

Inputs								Outputs		
D_0	D_1	D_2	D_3	D_4	D_5	D_6	D_7	X	Y	Z
1	0	0	0	0	0	0	0	0	0	0
0	1	0	0	0	0	0	0	0	0	1
0	0	1	0	0	0	0	0	0	1	0
0	0	0	1	0	0	0	0	0	1	1
0	0	0	0	1	0	0	0	1	0	0
0	0	0	0	0	1	0	0	1	0	1
0	0	0	0	0	0	1	0	1	1	0
0	0	0	0	0	0	0	1	1	1	1

The D-quantities serve as the input signals to the encoder. Note that only one line can be at logic 1 at any given time in order to ensure a unique code for that line at the output. Moreover, because the output is coded with three bits, it follows that a gate is needed for each of the three output bits, X, Y and Z, where Z is the least significant bit. The output code for any given input line is determined by the simultaneous state of each output line. Hence if input line D_5 is at logic 1, then correspondingly $X = 1$, $Y = 0$, and $Z = 1$. Further study of the output section of the truth table indicates that Z is at logic 1 whenever input lines D_1, D_3, D_5, or D_7 are activated. Expressed as a logical function, we can write

$$Z = D_1 + D_3 + D_5 + D_7 \qquad (13\text{-}1)$$

Clearly, the implementation of this logical expression calls for an OR gate. The truth table also shows that the logical expressions for Y and X are

$$Y = D_2 + D_3 + D_6 + D_7 \qquad (13\text{-}2a)$$

and

$$X = D_4 + D_5 + D_6 + D_7 \qquad (13\text{-}2b)$$

Each of these equations also calls for an OR gate. Therefore, the encoder that achieves the task specified here can be built with three OR gates arranged as depicted in Fig. 13-2.

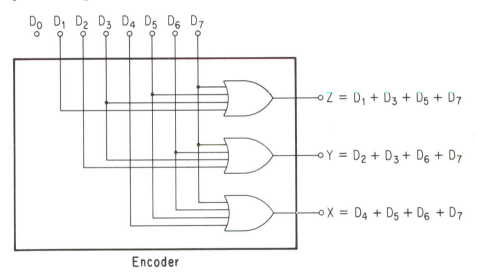

Fig. 13-2   Octal-to-binary encoder.

13-2 ADDERS

The addition of two binary numbers is carried out in each column by moving from right to left. The addition of the least significant bits involves the simplest operation in the process because there is no need to accommodate a carry bit. The combinational logic circuit that deals with binary addition without carry is called a *half-adder*. This name is used for the reason that two such units can

Inputs			Outputs	
A	B	C_i	C_o	S_o
0	0	0	0	0
0	0	1	0	1
0	1	0	0	1
0	1	1	1	0
1	0	0	0	1
1	0	1	1	0
1	1	0	1	0
1	1	1	1	1

A study of this table discloses the following features of the full-adder. The ouput sum S_o is at logic 1 whenever only one of the three inputs is also at logic 1. Whenever two of the three inputs are at logic 1, the sum S_o is at logic 0 *and* the carry output C_o is at logic 1. Finally, when all three inputs are at logic 1, then both S_o and C_o must be at logic 1. Observe that the full-adder must have the capability of handling a number as high as 3. This occurs when the addend and augend are at logic 1 and the carry input is also at logic 1. In this case we can represent the situation as follows: $A_k = 1$, $B_k = 1$, and $C_i = 1$, where k denotes the kth binary digit. Hence

$$A_k = \quad 1 \quad \text{and} \quad A_k + B_k = \quad 10$$
$$\text{plus} \quad + \qquad\qquad\qquad \text{plus} \quad +$$
$$B_k = \quad \underline{1} \qquad\qquad\qquad C_i = \quad \underline{1}$$
$$\qquad\quad 10 \qquad\qquad\qquad\qquad\quad 11$$

where 11 is the binary code for number 3.

One design for the logic circuit that performs full addition of two binary numbers can be obtained by expressing the output quantities in terms of a sum of minterms as these are obtained from the truth table. Thus, by noting the rows in the truth table where S_o and C_o are at logical 1, we can write

$$S_o = \overline{A}\overline{B}C_i + \overline{A}B\overline{C}_i + A\overline{B}\overline{C}_i + ABC_i \qquad (13\text{-}5)$$

and

$$C_o = \overline{A}BC_i + A\overline{B}C_i + AB\overline{C}_i + ABC_i \qquad (13\text{-}6)$$

Because these expressions often include redundancy, an examination for simplification is called for by invoking the minimizing techniques associated with Karnaugh maps. The K map for S_o is displayed in Fig. 13-4(a) and that for C_o is shown in Fig. 13-4(b). Since there are no adjacent placements of 1s for S_o, it follows that Eq. (13-5) is already in its simplest formulation. On the other hand, the adjacency feature associated with the 1s of C_o permits Eq. (13-6) to be reduced to the following simpler logical expression:

$$C_o = AB + AC_i + BC_i \qquad (13\text{-}7)$$

The design of the logic circuit for the full-adder readily follows from a gate-circuit interpretation of the logical expressions given by Eqs. (13-5) and (13-7).

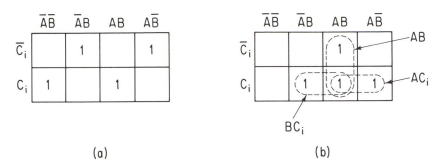

Fig. 13-4 Karnaugh maps for the full-adder: (a) K map for output sum S_o; (b) K map for output carry C_o.

To implement the former expression, four AND gates for each minterm and one OR gate for the OR'ing process are needed. Similarly, the implementation of Eq. (13-7) requires three AND gates and one OR gate. The complete logic diagram is depicted in Fig. 13-5.

An alternative implementation of the full-adder can be found by an appropriate algebraic manipulation of Eqs. (13-5) and (13-6). The steps of the manipulation are indicated below together with the algebraic operations involved. But first it is helpful to keep in mind that the exclusive-OR operation between two variables can be represented by

$$A \oplus B = \overline{A}B + A\overline{B} \qquad (13\text{-}8)$$

This identity follows directly as the sum of minterms in the truth table of the exclusive-OR operation. For Eq. (13-5) we have

$$
\begin{aligned}
S_o &= \overline{A}\,\overline{B}C_i + \overline{A}B\overline{C}_i + A\overline{B}\,\overline{C}_i + ABC_i \\
&= \overline{A}B\overline{C}_i + A\overline{B}\,\overline{C}_i + \overline{A}\,\overline{B}C_i + ABC_i \qquad && \text{commutative law} \\
&= \overline{C}_i(\overline{A}B + A\overline{B}) + C_i(\overline{A}\,\overline{B} + AB) \qquad && \text{associative law} \\
&= \overline{C}_i(\overline{A}B + A\overline{B}) + C_i(\overline{A}A + \overline{A}\,\overline{B} + AB + B\overline{B}) \qquad && \text{identity: } \overline{A}A = 0; \ B\overline{B} = 0 \\
&= \overline{C}_i(\overline{A}B + A\overline{B}) + C_i(\overline{A} + B)(A + \overline{B}) \qquad && \text{factoring} \\
&= \overline{C}_i(\overline{A}B + A\overline{B}) + C_i(\overline{\overline{A}B + A\overline{B}}) \qquad && \text{De Morgan's theorem}
\end{aligned}
\qquad (13\text{-}9)
$$

Now for simplicity introduce

$$D \equiv \overline{A}B + A\overline{B} = A \oplus B \qquad (13\text{-}10)$$

Equation (13-9) can then be written as

$$S_o = \overline{C}_i D + C_i \overline{D} = C_i \oplus D \qquad (13\text{-}11)$$

The final expression for the output at the sum terminal of the full-adder thus becomes simply

$$S_o = C_i \oplus (A \oplus B) \qquad (13\text{-}12)$$

A circuit interpretation of this expression discloses the need for two exclusive-OR gates. At the first one the inputs are A and B. At the following X-OR gate the inputs are C_i and the ouput of the first X-OR gate.

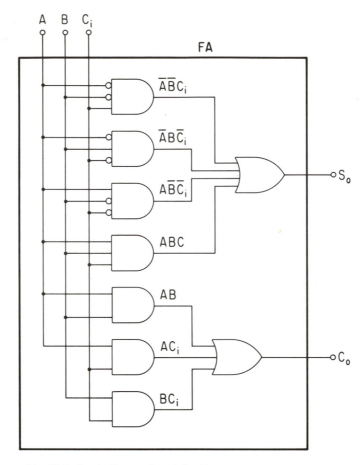

Fig. 13-5 Logic diagram for a full-adder based on sum of minterms.

A similar procedure is applied to Eq. (13-6). We get

$$C_o = \overline{A}BC_i + A\overline{B}C_i + AB\overline{C}_i + ABC_i$$
$$= C_i(\overline{A}B + A\overline{B}) + AB(C_i + \overline{C}_i) \qquad (13\text{-}13)$$
$$= C_i(A \oplus B) + AB$$

The instructions contained in this last expression for a logic circuit representation invoke the use of two AND circuits and a summing OR gate. The inputs to one AND gate are simply A and B. The inputs to the second AND gate are C_i and the output of the first X-OR gate used in connection with S_o. The final gate configuration is depicted in Fig. 13-6.

A close inspection of the gate layout of this version of the full-adder quickly discloses the presence of two half-adders: hence the name.

The 4-Bit Parallel Adder. The integrated circuit implementation of a 4-bit adder is shown in Fig. 13-7. Note that as many adders are needed as there are

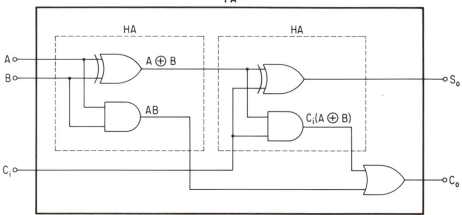

Fig. 13-6 Alternative implementation of a full-adder. Observe the presence of two half-adders.

bits in the binary numbers. The adder for the least significant bit can be a half-adder since it is not necessary to accommodate a carry input. However, the adders for each of the remaining bits must be full-adders. In the operation of this adder it is assumed that all bits are available simultaneously and applied to each adder at the same time: hence the reason for calling this device a parallel adder. Moreover, because of the need for ready access of the binary numbers which are to be added, they frequently are stored in *storage registers* before being applied to the adder.

Although the parallel nature of the adder favors fast addition, there will be delay encountered because of the time that is required to propagate the carry information through the various levels of the gate logic. This is called the *carry ripple*. Practical versions of this 4-bit adder have been modified in design to diminish this ripple effect. It is achieved by endowing the design with a look-ahead carry feature. Such units are capable of adding two 8-bit numbers in about 25ns.

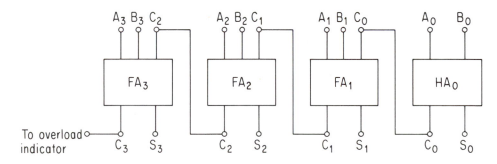

Fig. 13-7 Four-bit parallel adder. The binary numbers are $A_3A_2A_1A_0$ and $B_3B_2B_1B_0$.

13-3 SUBTRACTORS

The design of logic circuitry for subtracting two numbers can proceed in the manner that was used for the addition of two numbers. We can begin by developing an appropriate truth table replacing the sum column with one entitled *difference* and the carry column with one called the *borrow* column. An appropriate sum of minterms or product of maxterms can subsequently be written from which the logic circuit readily follows (see Prob. 13-8). We do not describe the details of this procedure here because digital systems such as computers can perform subtraction rather easily by a process of addition. The sole modification that is required in the process is that the negative number be treated as the addend in the form of the so-called *2s complement* of the binary number and *added* to the addend. In this way an adder logic circuitry can do subtraction and thereby eliminate the need to provide a special device for this purpose. Accordingly, our chief concern in this section reduces to a description of the use of the complements of numbers in performing subtraction.

When two numbers are to be subtracted by human beings, we find it easy to do so by working successively with each column of the minuend and the subtrahend (i.e., the number to be subtracted). When a given digit of the subtrahend exceeds the corresponding digit of the minuend, a borrow from the next highest digit is made and the procedure continues. Although human beings find this process a routine matter to execute, the digital computer does not. Rather, the digital computer finds it much more convenient to deal with the 2s complement of the subtrahend, principally because it can generate this version of the number by a simple operation that involves not much more than changing all 1s to 0s and all 0s to 1s. More is said about this presently. Now it serves our purpose to show how the use of the complement of a number allows subtraction to be accomplished through a process of addition even though this is not the method normally used when human beings do subtraction.

We begin with the definition of the complement of a number, but at the outset it is important to distinguish this from the complement of a logical variable, which we have defined and used many times in the two preceding chapters. For a number system with radix r the so-called r's complement of the number N_r in that system is defined by

$$\overline{N}_r \equiv r^n - N_r \tag{13-14}$$

where n denotes the number of digits in the integer part of N_r and $\overline{N}_r$ is the r's complement of the number N_r. To illustrate the meaning of this equation, consider first the decimal number system for which $r = 10$. If $N_r = N_{10} = 2652$, then since $n = 4$ and $r = 10$, we can say that the 10s complement of 2652 is

$$\overline{N}_r = 10^4 - 2652 = 10,000 - 2652 = 7348 \tag{13-15}$$

As another example, assume that $N_r = N_{10} = 0.795$. Now $n = 0$, so that

$$\overline{N}_r = 10^0 - 0.795 = 1 - 0.795 = 0.205$$

If $N_r = N_{10} = 263.4$, the 10s complement becomes

$$\overline{N}_r = 10^3 - 263.4 = 1000 - 263.4 = 736.6$$

Next we illustrate how the 10s complement is used to perform subtraction by addition. Initially, attention continues to be focused on the decimal system because the process and the results are easier to understand. Let it be required that the subtrahend, $N_{10} = 2652$, be subtracted from the minuend, $M_{10} = 7234$. By the difference-and-borrow method we can determine the result directly as

$$M_{10} - N_{10} = 7234 - 2652 = 4582 \qquad (13\text{-}16)$$

However, by the 10s-complement method, the procedure calls for the *addition* of the minuend ($M_{10} = 7234$) and the 10s complement of the subtrahend [$\overline{N}_{10} = 7348$ by Eq. (13-15)]. Thus we can write

$$
\begin{array}{rl}
M_{10} = & 7234 \\
\text{plus} & \text{plus} \\
\overline{N}_{10} = & 7348 \\
\hline
\text{sum} = & 1|4582
\end{array}
$$

Whenever the minuend is greater than the subtrahend, the result of subtraction must be a number that also has n digits. In this case the result must be a four-digit number. The presence of the leftmost digit 1 in the foregoing computation is associated with a carry into a fifth place when the complement of the number is computed by Eq. (13-14). For example, in this case $r^n = r^4 = 10,000$. The 1 in this number is superfluous as regards the four-digit minuend and must be ignored. Only the first $n = 4$ digits (reading from right to left) have meaning. Therefore, we see that the answer by the method of complements is 4582, which exactly agrees with the result displayed in Eq. (13-16).

Consider next the case where the subtrahend is larger in magnitude than the minuend. Specifically, for $M_{10} = 235$ and $N_{10} = 481$, we want to find $M_{10} - N_{10}$. By straight subtraction we get a difference of -246, which obviously is negative. By the method of the 10s complement we get

$$
\begin{array}{rll}
M_{10} = & 235 \\
\text{plus} & \text{plus} \\
\overline{N}_{10} = & 519 & \text{(10s complement of 481)} \\
\hline
& 754
\end{array}
$$

Several points are worth noting about the result, 754. First, note the absence of a leftmost end carry digit. In other words, when the subtrahend exceeds the minuend, the number of digits in the result remains fixed at n (in this case 3). Second, because the number to be subtracted is the larger of the two and we know it to be in a complemented state, so too must the result of the addition be in a complemented state. Therefore, to get the true difference, the number 754 must be complemented and a negative sign attached as well. Hence the answer is $-(1000 - 754) = -246$. The nature of these operations should make it apparent that these modifications must be easy to do in digital systems; otherwise, it is difficult to justify such a convoluted procedure.

Proceeding with this background, we turn attention next to the matter of subtracting two *binary* numbers by addition of the minuend and the 2s complement of the subtrahend. The 2s complement of a binary number is found from a direct application of Eq. (13-14) where $r = 2$. Thus to find the 2s complement of decimal 10 in the binary number system, we first identify the binary form of the number and then apply Eq. (13-14). Hence

$$(10)_{10} = N_2 = 1010_2$$

$$\overline{N}_2 = (2^n)_{10} - N_2 = 2^4 - 1010$$

Here $2^4 = 16$, which in the binary number system is 10000. Accordingly, the 2s complement of 10 is

$$\overline{N}_2 = 10000 - 1010 = 0110 \tag{13-17}$$

This result is obtained by applying the normal rules of subtraction, which for binary numbers can go as follows. Proceeding on a column-by-column basis, the first digit on the right in Eq. (13-17) is self-evident. However, for the digits in the second column we need to subtract 1 from 0, but since this is not possible as it stands, it is necessary to borrow. So we ask the question: What binary number added to 1 yields the zero of the minuend? The answer is 1 with a carry to the next column. With this carry the 0 digit of the subtrahend (third column from the right) becomes a 1. Again the same question is asked, and once again for the case at hand the response is the same. This moves the action to the fourth column, where we now have the carry from the third column added to the 1 of the fourth, thus yielding binary 10, which in turn yields zeros in the difference terms that remain.

A few more examples are warranted at this point in order that we might study the results to deduce a general rule for finding the 2s complement of a binary number without having to resort to Eq. (13-14). Thus for decimal 7 we get

$$(7)_{10} = 0111_2 = N_2$$

$$\therefore \quad \overline{N}_2 = (2^4)_{10} - N_2 = 10000 - 0111 = 1001 \tag{13-18}$$

For decimal 12 the figures are

$$(12)_{10} = 1100_2 = N_2$$

$$\therefore \quad \overline{N}_2 = 2^4 - 1100 = 10000 - 1100 = 0100 \tag{13-19}$$

A comparison of the complemented version of the binary number with the original binary expression reveals that in each case, starting from the right, all digits are the same up to and including the first encountered logic 1. Then, beyond that first 1, all digits are complemented. The procedure contained in this observation can also be described in an alternative fashion which a digital computer finds very easy to implement. It is a two-step process: (1) invert all digits of the subtrahend, and (2) add 1 to the least significant bit in (1). *These are very simple operations for a digital system to execute.* Accordingly, when an adder is modified to include this capability, it can also serve as a subtractor. The foregoing procedure is now illustrated by several examples.

EXAMPLE 13-1   The decimal number 9 is to be subtracted from decimal 24 using a binary adder. Outline the procedure.

Solution: Expressing each number in binary form, we get

$$M = (24)_{10} = 11000_2$$

$$N = (9)_{10} = 01001_2$$

Note here that the subtrahend is given as many digits as the minuend. Next, write the 2s complement of N by keeping the rightmost 1 and complementing all remaining digits.

$$\therefore \quad \overline{N} = 10111$$

Finally, perform a binary addition of M and $\overline{N}$ remembering to ignore any overflow carry beyond five digits. Thus

$$\begin{array}{r} M = 11000 \\ \text{plus} \quad \overline{N} = \underline{10111} \\ 1|01111 \end{array}$$

Hence the answer in binary form is 1111. Clearly this is 15, which is the number expected.

EXAMPLE 13-2   Outline the procedure by which 24 is subtracted from 9 using binary addition.

Solution: Let

$$M = (9)_{10} = (01001)_2$$

$$N = (24)_{10} = (11000)_2$$

Therefore,

$$\overline{N} = \text{2s complement of 24} = 01000$$

Summing yields

$$\begin{array}{r} M = 01001 \\ \text{plus} \quad N = \underline{01000} \\ 10001 \end{array}$$

Because the subtrahend exceeds the minuend, this sum necessarily represents the result in complemented form. The actual number, as it would appear in response to a print command, is the negative of the 2s complement of this sum. Or

$$-(01111)_2 = -15_{10}$$

which is the expected result.

An alternative procedure can be used in these arithmetical operations involving subtraction, which puts emphasis on the most significant bit as a representation of the sign of the number. To illustrate the point with this example, the number 24 can be more precisely written in binary as

$$0_{\wedge}11000 = 24_{10}$$

where the leftmost digit indicates that the number is positive. Correspondingly, the 2s complement version becomes

$$1_{\wedge}01000 = -24_{10}$$

Observe that six bits are needed to represent -24 in contrast to the five that are required for $+24$. The leftmost 1 denotes a negative number.

The binary addition of 9 and the 2s complement of 24 yields the following result:

$$-24_{10} = 1_{\wedge}01000$$

$$\text{plus} \quad +$$

$$9_{10} = 0_{\wedge}01001$$

$$\overline{1_{\wedge}10001}$$

The leftmost 1 is now an indication that the result is negative and is expressed in 2s complement form. The equivalent decimal number is obtained by writing the complement of the bits to the right of $_\wedge$ and replacing the 1 to the left of this symbol with a negative sign. Thus

$$\overline{1_\wedge 10001} = -(\ 01111\ ) = -15_{10}$$

EXAMPLE 13-3 Find the result of adding two negative numbers using the method of 2s complements. Assume that $M = -24$ and $N = -9$.

Solution: Although only five digits are needed to obtain the binary expression of 24, the sum will require six digits. Therefore, each binary number must have a six-digit binary representation. (The use of more digits than are actually needed does not alter the result.) We get

$$M = (24)_{10} = 011000$$
$$\therefore\ \overline{M} = 101000$$

Also,

$$N = (9)_{10} = 001001$$
$$\therefore\ \overline{N} = 110111$$

Then

$$\overline{M} =\ \ 101000$$
$$\text{plus}$$
$$\overline{N} =\ \ \underline{110111}$$
$$\text{sum} =\ 1011111$$

Neglecting the leftmost (seventh digit), 2s complementing the sum and attaching a negative sign yields

$$-(100001) = -33_{10}$$

We provide the alternative solution for this case too. The use of the leftmost digit to convey information about the sign of the number calls for seven bits to represent the result, -33_{10}, despite the need of only six for $+33_{10}$. Accordingly, each number is represented with seven bits. The leftmost bit is zero or 1 depending on whether the number is positive or negative. Thus

$$9_{10} = 0_\wedge 001001 \quad \text{and} \quad -9_{10}\ = 1_\wedge 110111$$
$$24_{10} = 0_\wedge 011000 \quad \text{and} \quad -24_{10} = 1_\wedge 101000$$
$$\text{sum}\ =\ 1\!|\overline{1_\wedge 011111}$$

The 1 to the left of $_\wedge$ means that the result is negative and is expressed in the 2s complement form. Hence the decimal equivalent is

$$\overline{1_\wedge 011111} = -(100001) = -33_{10}$$

13-4 DECODERS AND DEMULTIPLEXERS

The decoder is basically a combinational logic circuit that provides code conversion generally. One obvious application of the decoder is to provide the inverse operation of the encoder, which appears in Fig. 13-2. In such a case the input to the decoder is in the form of binary input lines and the output signals represent the decimal integers from 0 to 7. Since each decimal integer is represented by a specific coded combination of the input signals, it follows that one feature of the decoder is that only one output line can be active at any given time. For example, if the input signals are at the logic values 011 (i.e., $\overline{A}BC$ where A, B, C are used to denote the binary input signals), then only the line that represents decimal integer 3 must be activated. The truth table in this application of the decoder is the same as that used for the encoder of Sec. 13-1 except that the columns for the inputs and outputs are exchanged. The truth table for the decoder therefore looks as follows:

Inputs			Outputs								Logical minterms
A	B	C	D_0	D_1	D_2	D_3	D_4	D_5	D_6	D_7	
0	0	0	1	0	0	0	0	0	0	0	$D_0 = \overline{A}\,\overline{B}\,\overline{C} = m_0$
0	0	1	0	1	0	0	0	0	0	0	$D_1 = \overline{A}\,\overline{B}C = m_1$
0	1	0	0	0	1	0	0	0	0	0	$D_2 = \overline{A}B\overline{C} = m_2$
0	1	1	0	0	0	1	0	0	0	0	$D_3 = \overline{A}BC = m_3$
1	0	0	0	0	0	0	1	0	0	0	$D_4 = A\overline{B}\,\overline{C} = m_4$
1	0	1	0	0	0	0	0	1	0	0	$D_5 = A\overline{B}C = m_5$
1	1	0	0	0	0	0	0	0	1	0	$D_6 = AB\overline{C} = m_6$
1	1	1	0	0	0	0	0	0	0	1	$D_7 = ABC = m_7$

Included in this table as well are the minterm designations corresponding to logic 1 on each row. The notation D or lowercase m with an appropriate subscript is used to denote the specific minterm for a given coded input combination.

The logic gate diagram of this decoder can be determined from the logical expression that is represented by each minterm. Thus the decoded number 3 is represented by minterm m_3 in the form $\overline{A}BC$. Hence to generate this output signal, it is necessary merely to use an AND gate at which $\overline{A}$, B, and C appear. By extending this reasoning to each row of the truth table, it follows that the decoder of the encoder of Fig. 13-2 must contain a total of eight AND gates, i.e., one for each minterm. The complete circuit is depicted in Fig. 13-8. In general, a decoder with n binary input lines will have 2^n output lines. In the case at hand there are, of course, eight output lines since $n = 3$. It is useful to note too that the number of minterms is also equal to 2^n.

The foregoing truth table shows that all the minterms associated with a given set of binary input signals are available at the output terminals of the decoder. Because any Boolean function can be expressed as a sum of minterms, it is entirely possible to use such decoders as building blocks in designing logic circuits for other applications. This is particularly so since decoders of this type are generally available as MSI packaged units. Appearing in Fig. 13-9 is an illustration of the decoder arranged to behave as a full adder. Here the packaged 3×8 decoder is represented by the rectangle with three input lines and eight

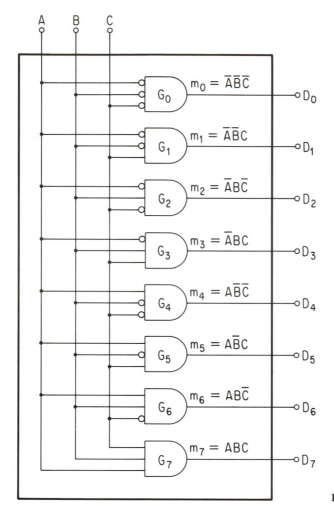

Fig. 13-8 A 3-line-to-8-line decoder.

output lines. The specific minterms appearing at each output line is described in the truth table. Thus, at terminal 1 of the 3 × 8 decoder the corresponding minterm is m_1, as listed in the table. The logical expressions for the *sum* and *carry* outputs of the full-adder are displayed in Eqs. (13-5) and (13-6) and each contains four minterms. These minterms may be replaced by their corresponding *m* notations as listed in the foregoing truth table. Hence alternative expressions for the sum and carry bits of the full-adder can then be written as

$$S_o = m_1 + m_2 + m_4 + m_7 \qquad (13\text{-}20)$$

and

$$C_0 = m_3 + m_5 + m_6 + m_7 \qquad (13\text{-}21)$$

The advantage of this representation is that the subscripts of the minterms provides the instructions about how the output terminals of the decoder are to

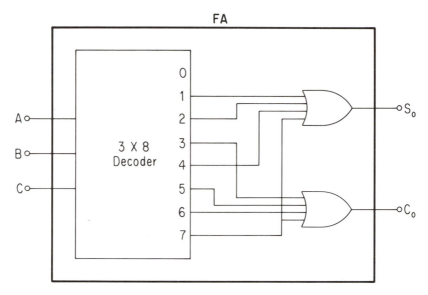

Fig. 13-9 Application of a 3 × 8 decoder to simulate a full-adder.

be connected to the two SSI OR gates where the OR'ing of the minterms, as required by Eqs. (13-20) and (13-21), takes place.

Demultiplexers. When a decoder chip is equipped with an additional input line (called the *enable* input), the unit can function as a *demultiplexer.* Internally, the enable signal is applied as an input to each of the AND gates of the decoder. As a matter of fact, packaged IC decoders are fabricated to include this modification. Therefore, when so equipped, the circuit of Fig. 13-8 would not be able to function as a decoder unless the enable input were kept at logic 1 during operation. Recall that *all* inputs to a logic AND gate must be at logic 1 for a switch to occur. It is for this reason that this terminal is called an enable input. However, the function of the enable input terminal is different when the enable-equipped decoder is used as a demultiplexer.

A demultiplexer is a combinational logic circuit which takes data applied to one input line and distributes it to one of many output lines in accordance with the instruction contained in a binary address code. The situation is conveniently illustrated by examining the 3 × 8 decoder of Fig. 13-9 when it is equipped with an enable input line E as depicted in Fig. 13-10. The input lines in Fig. 13-10 are arranged so that the enable input terminal is placed in a position to emphasize its role as the data input line. The remaining three input lines that are at terminals A, B, and C now serve as address codes for selection of a particular output line to which the data input line can be joined. The output lines thus serve as destination points or receivers. To illustrate the operation, assume that it is desired to send the data on the E terminal to destination line D_3. This can be achieved by applying the address selection signal $\overline{A}BC = 011$. This circuit is also called a *data distributor* for apparent reasons.

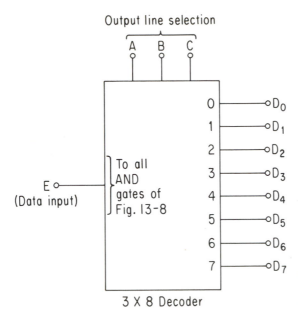

Output line selection

A B C

0 ——o D_0
1 ——o D_1
2 ——o D_2
3 ——o D_3
4 ——o D_4
5 ——o D_5
6 ——o D_6
7 ——o D_7

E o
(Data input)

} To all
  AND
  gates of
  Fig. 13-8

3 X 8 Decoder

Fig. 13-10 Modification of the 3 × 8 decoder of Fig. 13-8 to operate as a demultiplexer.

13-5 MULTIPLEXERS (DATA SELECTORS)

From the viewpoint of the orientation of input and output lines the *multiplexer* (abbreviated MUX) is essentially a mirror image of the demultiplexer. In the case where there are three selection variables (A, B, C) the number of *data input* lines is $2^n = 2^3 = 8$. Moreover, there is a single output line that is joined to a unique input line in a fashion that is dictated by the selection logical variables A, B, and C. A multiplexer can be constructed from the basic decoder of Fig. 13-8 by making the following modifications. (1) Connect the first data input line I_0 to the first AND gate G_0, then I_1 to G_1, etc., until each input line is connected to a corresponding AND gate in succession; and (2) add an OR gate and apply the output of each AND gate as an input to the OR gate. The output line is simply the output of the OR gate. A block diagram of the arrangement is displayed in Fig. 13-11.

The state of the selection variables serve jointly to determine which of the AND gates is to be put in an enabled condition. Thus if the control variables are at a state that is described by $\overline{A}BC = 011$, gate G_3 is enabled. In effect this selection connects input I_3 to the output line so that $Z = I_3$. By changing the selection code to the appropriate minterm designation, the output line can be switched to any input line. This process is called *multiplexing* because it permits inputs from many (multi) sources to be channeled into one route (or common bus).

In addition to the important task of data selection the multiplexer can be used to generate as well arbitrary logical expressions of the selection variables. This capability is to be entirely expected in view of the presence of minterms

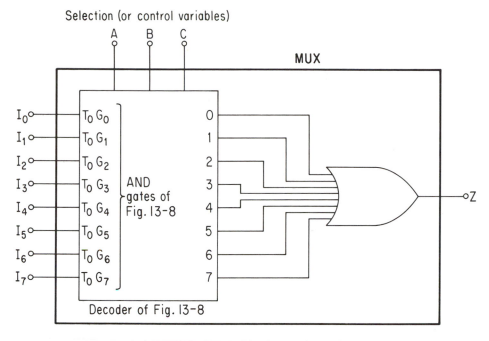

Fig. 13-11 Illustrating the modifications to be made to decoder circuit of Fig. 13-8 to obtain an 8-to-1 multiplexer (MUX).

at the output terminals of the AND gates and the tandem placement of the OR gate to perform the summing action needed for a sum-of-products implementation of logical functions. The procedure involved is best described by an example.

EXAMPLE 13-4 A process involves three logical variables. The description of the desired logical function is given by

$$Z = AC + \overline{B}\overline{C}$$

Determine the logic levels of the input lines in order that the ouput of a MUX be consistent with the given logical function.

Solution: It is necessary to put Z in the canonical form of the sum of products. More simply stated, we need a sum of minterms. Accordingly, variable B must be introduced into the first term of Z and variable A into the second term. Thus we rewrite Z in terms of an equivalent expression:

$$Z = AC(B + \overline{B}) + \overline{B}\overline{C}(A + \overline{A})$$
$$= \overline{A}\overline{B}\overline{C} + A\overline{B}\overline{C} + A\overline{B}C + ABC$$

A comparison of these minterms with the listing shown in Fig. 13-8 permits the last equation to be expressed as

$$Z = m_0 + m_4 + m_5 + m_7$$

The subscript numbers correspond as well to the data input lines. Accordingly, it is necessary for input lines I_0, I_4, I_5, and I_7 to be at logic 1 and the remaining lines at logic 0. Thus

$$I_7I_6I_5I_4I_3I_2I_1I_0 = 10110001$$

13-6 READ-ONLY MEMORY UNITS (ROMs)

The basic composition of a read-only memory consists of a decoder section, such as the one in Fig. 13-8 connected in a permanent and prespecified fashion to an encoder section such as the one in Fig. 13-2. The interconnections between the decoder and encoder can be made in several ways. One is to make specific permanent connections at the time of fabrication of the IC package. It is the establishment of the permanent connections that gives this digital device its name because a particular code of input signals generates one, and only one, repeatable, output, which is called a *word*. In some ROM units the interconnections can be *programmed* by the user. The programming consists of preserving or breaking fusible links that join the decoder to the encoder section. However, once these connections are made, the pattern of outputs corresponding to the inputs is fixed. Such devices are more commonly called PROMs. A third type of ROM is one that incorporates an *erasable programming* feature which is achieved by using gates in the interconnections that are sensitive to short-wave radiation such as ultraviolet light. The radiation can be used to make or break these connections as often as desired. These devices are identified by the acronym EPROM.

In the interest of continuity and consistency, let us design our elementary ROM based on the decoder and encoder gate circuits previously treated. Because it is the purpose of the ROM to provide a specific output for a specific input, the suggestion arises that the ROM device essentially simulates the truth table of a given application. Recalling that the output of each AND gate of the decoder represents one of the 2^n minterms associated with the n input logical variables, such a simulation reduces to a routine procedure. By way of illustration assume that the truth table for a particular application reads as follows:

Inputs			Outputs: 4-bit data words			
A	B	C	D_3	D_2	D_1	D_0
0	0	0	0	0	1	1
0	0	1	0	1	0	1
0	1	0	1	0	1	0
0	1	1	1	1	0	0
1	0	0	0	0	1	0
1	0	1	0	1	1	0
1	1	0	0	0	0	1
1	1	1	1	1	1	0

For a given input code the ROM is to provide a 4-bit output word in accordance with the specifications of this table. For example, whenever the input logic appears as, say, 011, the corresponding output word that is "read" will be 1100 and repeatedly so. The choice of a 4-bit output word means that an encoder with four gates is required. In order to complete the ROM design for this application, it is now necessary to specify the interconnections between the eight (2^3) AND gates of the decoder and the four OR gates of the encoder. Each OR gate of the encoder is responsible for generating a bit of the output word. The truth table indicates that the least significant bit of the output word, D_0, is generated from the sum of three minterms, namely, those corresponding to the 1s in the column for D_0. The pertinent logical expression is therefore

$$D_0 = m_0 + m_1 + m_6 \tag{13-22}$$

where these minterms correspond to those identified at the output terminals 0, 1, and 6 of Fig. 13-8. Similarly, the logical expressions for the remaining bits are

$$D_1 = m_0 + m_2 + m_4 + m_5 + m_7 \tag{13-23}$$

$$D_2 = m_1 + m_3 + m_5 + m_7 \tag{13-24}$$

and

$$D_3 = m_2 + m_3 + m_7 \tag{13-25}$$

The completed ROM design is depicted in Fig. 13-12. Here the block identified

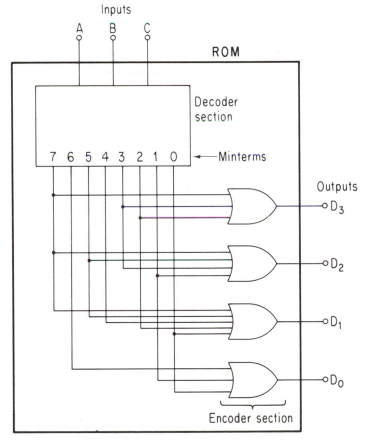

Fig. 13-12 Illustrating the three basic elements in the construction of a ROM: decoder section, encoder section, and interconnections.

as the decoder section has the details that are displayed in Fig. 13-8. Only the connections required by the logical functions of Eqs. (13-22) to (13-25) are shown. If these connections were to be made by the use of a PROM, then current pulses would be sent through the fusible links whose minterms do not appear in these logical functions. The current pulse is strong enough to break the links.

The three input variables A, B, C essentially serve as a coded address to call out a specific 4-bit output word. Thus, if the logic signals appearing at the input lines are $\overline{A}BC = 011$, the following sequence of events occurs. The logic 0 input at terminal A passes through an inverter before it is applied to gate G_3 of the decoder (see Fig. 13-8). Here it combines with the logic 1 signals at terminals B and C to produce a logic 1 level on line 3 at the output of the encoder. In turn, this signal appears simultaneously at gates D_3 and D_2 of the encoder but not at gates D_1 and D_0. Hence the output word is read as 1100, which is exactly consistent with the requirement of the truth table.

The capacity of a ROM package is identified by the notation $2^n \times m$, where n is the number of input variables and m denotes the number of bits in the output words. In short, 2^n gives the number of minterms provided by the decoder and m tells how many output gates are used. Thus, in the case of the foregoing illustration, we have a very simple 8×4 ROM, which involves a mere total of 32 bits and is consistent with the number of entries in the truth table. However, in practical, commercial-type packaged IC ROMs, the total number of bits is far greater. This is true even with relatively small capacity commercial ROMs. For example, a ROM that is designed to accommodate an 8-bit address (input) and to furnish an 8-bit output word has a capacity of $2^n \times m = 2^8 \times 8 = 256 \times 8 = 2048$ bits. This ROM is commonly called a 2K ROM after round-off to units of thousands and K denoting that unit. In light of our experience so far in this chapter we cannot help but be impressed by these numbers. Especially notable is the staggering number of minterms that are available to the designer. Even in this low-capacity ROM there are 256 minterms for the 8-digit input address. For a 9-digit address this figure rises to 512 and becomes 4096 when a 12-digit address is used. Clearly, there is represented here an enormous flexibility in the generation of a host of logical functions in connection with the design of combinational circuits. The tremendous capacity of available minterms permits many complex combinational tasks to be implemented virtually independently of the concerns for minimization. This is a fortunate development because the execution of minimization procedures for six or more variables is a laborious and time-consuming task. When this distinctive feature of the ROM is coupled with the high degree of reliability associated with the integrated-circuit fabrication of these units, it is easy to understand why the ROM is so popular among designers of combinational circuits.

Synchronous sequential logic circuits

The sequential logic circuits differ from the straight combinational type in two important details. (1) They possess memory in the sense that the output state of the circuits can be stored as binary bits and remain in their state until instructed to do otherwise. (2) They are equipped with feedback connections that serve to apply these stored outputs to logic gates at appropriate places in the circuitry. As a result, the sequential logic circuits are driven not only by the externally applied inputs but by the feedback signals as well. In combinational circuits the output at a given instant of time is solely dependent upon the inputs at the very same time instant. In sequential circuits the current state of the output is dependent not only upon the current input values but also upon the *sequence* of logic values that lead to the current input values.

Sequential circuits can be one of two types: synchronous or asynchronous. The output states of synchronous sequential circuits are permitted to change at discrete clocked instants of time. In asynchronous circuits the output states depend on the order in which the input signals change. Attention here is directed exclusively to synchronous sequential circuits, which are by far the more dominant.

A word is in order at this point concerning the read-only memory (ROM). It is important to understand that the ROM does not possess memory in the sense just described. The ROM exhibits a kind of "memory" because it can read out a specific output word (minterm) corresponding to a unique input code. However, the current output of the ROM is solely dependent upon the current input. There is no component of input signals that comes from memory feedback. The ROM is a true combinational logic circuit.

The memory elements that are used in synchronous sequential circuits are called *flip-flops*. They can be of several different types accommodating one or more input variables and providing two, complementary, output states. The more commonly encountered types of flip-flops are described below. Although flip-flops are the simplest circuits that possess memory, they can readily be arranged in groups to perform as counters, shift registers, and as general memory units.

13-7 FLIP-FLOPS

A salient feature of the flip-flop is that the output can exist in one of two stable states, logic 1 and logic 0, simultaneously. This is ensured by the appropriate crossed feedback connections associated with the most elementary form of the flip-flop known as a *latch*.

Latch. The circuit arrangement of a *latch* composed of NAND gates is shown in Fig. 13-13(a). Two stable states exist for this circuit. For example, if the output of gate $N1$ is assumed to be at logic 1, it follows that the input to gate $N2$ is 1, which in passing through the NAND gate $N2$ yields an output $\bar{Q} = 0$. But this output serves as the input to gate $N1$, which in turn calls for the output of gate $N1$ to be 1. Since this logic level corresponds to the assumed initial level, we have a stable condition. Similarly, if the output of $N1$ is initially assumed to be $Q = 0$, then the output of $N2$ is 1, which as the input to $N1$ generates an output at $N1$ of 0. So again a stable state prevails. Whatever state

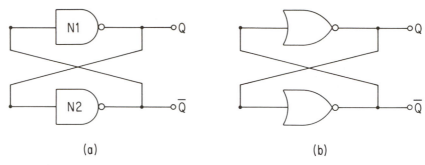

Fig. 13-13 Basic latch circuit: (a) NAND latch circuit; (b) NOR latch gate.

is assumed, however, the operation is such that the information is locked or latched until further action is taken to produce a change.

The behavior of the latch circuit depicted in Fig. 13-13(b) is described in the same way as for the circuit in Fig. 13-13(a). The difference lies in the use of NOR gates rather than NAND gates. The latching characteristic is identical.

Basic Flip-Flop Circuits. The basic flip-flop logic circuit is obtained by modifying the latch circuit so that it can accommodate externally applied input signals for the purpose of providing control of the output states. The arrangement for the basic NOR flip-flop is depicted in Fig. 13-14(a). The corresponding basic NAND flip-flop appears in Fig. 13-14(b). These circuits are sometimes called *direct-coupled* SR flip-flops or SR latches.

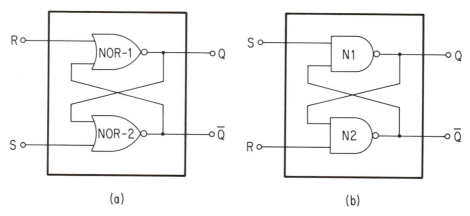

Fig. 13-14 The basic flip-flop circuit: (a) with NOR gates; (b) with NAND gates. Note the orientation of the output terminals relative to the S, R inputs.

The operation of these gate circuits may be analyzed by examining the state at the output terminals for the four combinations of the two input variables, which are called S and R. The choice of these letters becomes evident presently.

For the sake of illustration we focus attention just on the basic NOR flip-flop. Similar conditions prevail for the NAND version.

1. $R = 0$, $S = 0$: Under these conditions the circuit of Fig. 13-14(a) behaves like the latch of Fig. 13-13(b). Thus, if initially $Q = 0$, the output of NOR-2 is 1, which makes the input of NOR-1 equal to 1. In turn, this commands the output of NOR-1 to be 0, which agrees with the original assumption. For $Q = 1$, the output of NOR-2 is 0, so that the input to NOR-1 is 0, which makes $Q = 1$; i.e., the previous output state remains unchanged.

2. $S = 1$, $R = 0$: Assume initially that $Q = 0$; then $\overline{Q} = 1$. The inputs at NOR-2 are $S = 1$, $Q = 0$, which for a NOR gate yield an output at NOR-2 which is equal to $0 = \overline{Q}$ or $Q = 1$. Remember that the crossed feedback connections ensure that the output states are complementary.

   Now assume that $Q = 1$ initially, so that $\overline{Q} = 0$. The inputs at NOR-2 are now $S = 1$, $Q = 1$, which for a NOR gate yield an output of $0 = \overline{Q}$. Hence $Q = 1$.

   A review of these results reveals that whenever $S = 1$ and $R = 0$ the output of the flip-flop at terminal Q is at logic 1 *independently* of the initial value of Q. Therefore, we can say that for this combination of inputs the flip-flop is always *set* at logic 1 at the Q terminal. This explains the choice of S to identify this logical variable.

   The memory feature of the flip-flop can be understood by noting that once the output Q terminal is set at logic 1 in response to inputs $S = 1$ and $R = 0$, an immediate next occurrence of these same logical values produces no change in the output states of the flip-flop.

3. $S = 0$, $R = 1$: Again we assume that $Q = 0$ initially, which makes $\overline{Q} = 1$. The inputs at NOR-1 are then $R = 1$, $\overline{Q} = 1$, which cause the output of NOR-1 to be $0 = Q$. In other words, the logic circuit acts to keep the Q-output *reset* at 0.

   If we assume initially that $Q = 1$, then $Q = 0$. The inputs at NOR-1 then become $R = 1$, $\overline{Q} = 0$, which cause the output of NOR-1 to be $0 = Q$. In this case, because the output at the Q terminal was at logic 1, the action of $S = 0$, $R = 1$ is to *reset* the value to zero.

   In summary we can say that this combination of input signals always serves to *reset* the output at the Q terminal to the 0 logic level.

4. $S = 1$, $R = 1$: An application of the foregoing analysis discloses that for this input condition both gates are attempting to generate the same logic level, which makes the output state indeterminate. What makes the output indeterminate, is that which takes place when the inputs are removed. If S is removed before R, the flip-flop resets. If R is removed before S, the flip-flop sets. Consequently, in the basic SR flip-flop this condition of inputs is *not allowed;* however, as is explained later, modifications to this circuit can be made which remove the ambiguity. The foregoing performance of the basic SR flip-flop can be neatly summarized in the following truth table:

S_i	R_i	Q_{i+1}
0	0	Q_i
1	0	1
0	1	0
1	1	—

Here the subscript i is used to denote the ith timing instant at which the logical variables S_i and R_i are applied as inputs. Also, Q_i is the current state of the flip-flop at the ith timing interval and Q_{i+1} denotes the value of the Q output terminal at the next clocked time interval. In short, it is the next-state value of the Q output terminal.

Clocked SR Flip-Flop. The basic flip-flop circuits diagrammed in Fig. 13-14 are essentially asynchronous sequential circuits. These circuits can be modified to operate as synchronous circuits by introducing two AND gates and a clock pulse train in the manner illustrated in Fig. 13-15(a). The symbolic logic representation is shown in Fig. 13-15(b). The clock input serves as an *enabling*

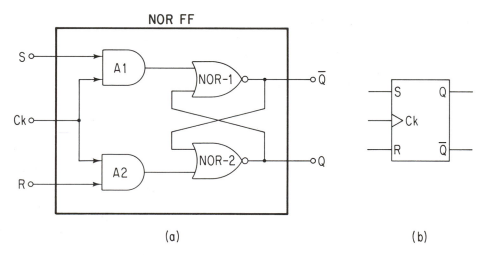

(a)   (b)

Fig. 13-15 Clocked (or gated) *SR* flip-flop: (a) logic circuit diagram; (b) logic symbol.

signal. Whenever this signal is at logic 0, the AND gates are disabled and the two OR gates then behave as simple latches. The use of the clock train pulses accordingly serves the important function of synchronizing the operations that take place in digital sequential circuits. They avoid problems created by unequal propagation delay times associated with gates. When the clock signal is present, the truth table of the clocked *SR* flip-flop becomes as shown below.

Ck	S_i	R_i	Q_{i+1}
0	X	X	
1	0	0	Q_i
1	0	1	0
1	1	0	1
1	1	1	—

Note that when Ck = 0, operation is not possible as a controlled flip-flop no matter what the values of S_i and R_i are. Only when Ck is put at logic 1 can the circuit of Fig. 13-15 behave as a flip-flop.

An explanation of the operation of this circuit proceeds in a fashion not unlike that for the basic SR flip-flop.

1. $Ck = 1$, $S_i = 0$, $R_i = 0$: In this case the outputs of the AND gates A_1 and A_2 are zero. Hence the states of the output terminals do not change, so that $Q_{i+1} = Q_i$.
2. $Ck = 1$, $S_i = 1$, $R_i = 0$: Initially let $Q = 0$ ($\overline{Q} = 1$). The input signals to NOR-1 are then $S_i = 1$, $Q = 0$, which yield an output at NOR-1 of $0 = \overline{Q}$. Hence the output at NOR-2 is the complemented quantity or $Q = 1$.

    When $Q = 1$ ($\overline{Q} = 0$) initially, the inputs to NOR-1 are $S_i = 1$, $Q = 1$ since AND gate A_1 is active. The output of NOR-1 is then $0 = \overline{Q}$. Thus again $Q = 1$.

    Therefore, independent of the value of Q, this combination of inputs always *sets* the Q output to logic 1.
3. $Ck = 1$, $S_i = 0$, $R_i = 1$: Because it is R_i that is driven high in this case, the emphasis shifts to AND gate A_2.

    If initially $Q = 0$ ($\overline{Q} = 1$), the inputs to NOR-2 are $R_i = 1$ (A2 is active) and $\overline{Q} = 1$, so that the output of NOR-2 is $0 = \overline{Q}$.

    When $Q = 1$ ($\overline{Q} = 0$), the inputs to NOR-2 are $R_i = 1$, $\overline{Q} = 0$ so the output of NOR-2 goes to $0 = Q$. Thus this combination of signals *resets* the Q output to logic zero.
4. $Ck = 1$, $S_i = 1$, $R_i = 1$: The situation here is the same as for the basic asynchronous SR flip-flop.

The logic diagram of a clocked (or gated) *SR* flip-flop that uses all NAND gates is displayed in Fig. 13-16. The operation of this circuit can be explained by a procedure similar to that which applies to Fig. 13-15. See Prob. 13-24.

Trailing-Edge Triggering and Timing Diagrams. Synchronous clocking helps to remove some of the difficulties associated with the variations encountered in propagation delays as input and feedback signals work their way through the various gates encountered in combinational and sequential logic circuits. However, this strategy does not address another problem that can cause difficulties. It is the presence of spurious signals (or noise or spikes) that can be found on the

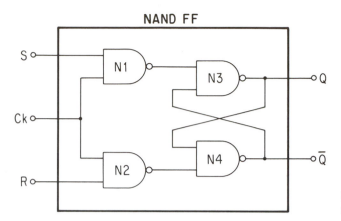

Fig. 13-16 Clocked *SR* flip-flop implemented with all NAND gates.

input lines where the useful data appears. These spikes, which are called *glitches*, can cause incorrect outputs to occur if they appear during the time the clock pulse is high. One effective solution to this problem is to apply the clock pulse to a trigger circuit the output of which generates a short pulse that serves to enable the input gate. The trigger responds to changes in the logic level of the clock pulse train. It can be arranged to provide the enabling short pulse either at the leading edge when it goes from 0 to 1 or at the trailing edge when it returns to zero from 1, i.e., a negative-going transition. In the latter case it is called *trailing-edge triggering*. The duration of the enabling pulse persists for a period which is just a fraction of the cycle period of the clock following a drop from logic 1.

An illustration of trailing-edge triggering and its discriminating capability against glitches is displayed in Fig. 13-17. The logic variations in the variables *S* and *R* are assumed to have the waveshapes indicated with a noise spike shown for the set input signal *S*. At the end of pulse 2 of the clock waveform the logic 1 signal at the *S* input is enabled and becomes effective in setting the logic level

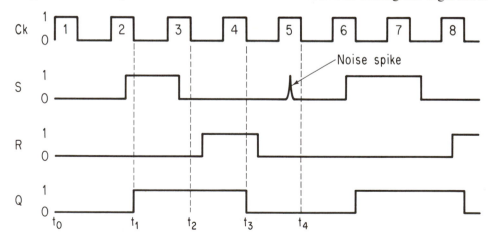

Fig. 13-17 Timing diagram for a clocked *SR* flip-flop. Trailing-edge triggering eliminates the effect of the noise spike shown.

of Q at 1. This is indicated to occur at time t_1. As a practical matter there is a slight delay incurred in getting Q to rise to this level, but for convenience this is omitted. At the end of pulse 3 both S and R are at logic 0 and so no change in the output states of the SR flip-flop takes place. The state of Q continues at logic 1 thus exhibiting memory. At the end of pulse 4 the inputs are $S = 0$ and $R = 1$ and so Q is reset to 0. Observe next the presence of a noise spike during the time that pulse 5 is at logic 1. Without edge triggering this spurious signal would serve to set Q to logic 1, but in this instance trailing-edge triggering causes the glitch to be ignored. Only in the rare situations where the glitch occurs coincident with the trailing edge would a false registration occur.

The discussion of the timing diagram was introduced at this point because it also provides an important background for describing the operation of another flip-flop type, namely, the D *flip-flop*. The D can be taken to denote either *delay* or *data* for the D flip-flop implies both meanings.

The D Flip-Flop. The D flip-flop is an SR flip-flop modified so that an inverter appears between the S and R terminals. The circuit arrangement is depicted in Fig. 13-18. Operation of the D flip-flop is obviously dependent upon the signals that appear at the SR terminals of the SR flip-flop. When Ck is at logic 1 and the data input signal, D_i, is at logic 0, it follows that $S_i = 0$ and $R_i = 1$. Our familiarity with the truth table of the SR flip-flop then tells us that $Q_{i+1} = 0$, which is the value of the D_i as well. For the situation where Ck = 1 and $D_i = 1$, the corresponding values of S_i and R_i are, respectively, 1 and 0.

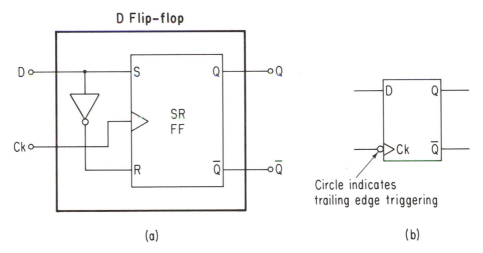

Fig. 13-18 D flip-flop: (a) logic circuit configuration; (b) logic symbol.

Accordingly, the output Q_{i+1} is set at 1, which again is the value of the data input D_i. In short,

$$Q_{i+1} = D_i$$

for the D flip-flop. Thus the next-state output, Q_{i+1}, is equal to the current state input, D_i, except for a delay of one clock period. A graphic demonstration of this feature of the data flip-flop is displayed in Fig. 13-19. The diagram is drawn for trailing-edge triggering. Observe that the Q logic wave pattern is an exact duplicate of the data input D waveform except for a *delay* of one clock period T.

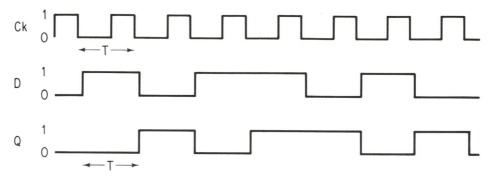

Fig. 13-19 Timing diagram that illustrates the *data* tracking and one-clock-period *delay* features of the D flip-flop.

The JK Flip-Flop. The *JK* flip-flop is yet another modified version of the clocked *SR* flip-flop. The circuit arrangement is illustrated in Fig. 13-20. Note especially the combination of the complemented output $\overline{Q}$ with the input J, which now provide the input to the S terminal through the AND gate A1. Similarly, the Q output is combined with the input K through AND gate A2 and furnish a logical signal at the R terminal. The purpose of the additional gates, A1 and A2, is to provide a convenient means by which the signals that are made to appear at the S and R terminals are dependent not only on the input signals *but on the state of the flip-flop as well*. The benefit of the arrangement lies in the

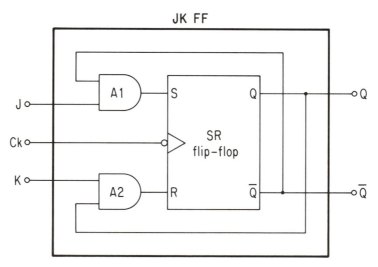

Fig. 13-20 *JK* flip-flop obtained by the indicated modifications of the *SR* flip-flop.

removal of the restriction of the SR flip-flop which prohibits $S = 1$ and $R = 1$ to exist simultaneously. In the new circuit configuration J replaces S as the input terminal and K replaces R. In fact, J and K serve the same roles and provide the same output results as for the SR flip-flop except for the condition just noted. As is explained presently, the indeterminacy state of the SR flip-flop is now replaced by a toggle action which switches the outputs to their complemented states.

The operation of the JK flip-flop can be described by examining the state of the signals applied to the S and R terminals of the circuit of Fig. 13-20 as they are made available from the two steering gates A_1 and A_2. All switching is assumed to occur consistent with trailing-edge triggering. Keep in mind that the AND gates produce logic 1 output only when both inputs are high, otherwise the gate is disabled and no transmission takes place through it. The step-by-step analysis proceeds as follows:

1. $Ck = 1$, $J = 0$, $K = 0$: Assume that initially $Q_i = 0$ (and $\overline{Q}_i = 1$). With $J = 0$ and $K = 0$ both AND gates are off; hence $S_i = 0$ and $R_i = 0$, which for the SR flip-flop means that $Q_{i+1} = 0$.

   When $Q_i = 1$ (and $\overline{Q}_i = 0$), both AND gates continue in the off position and so the output states remain unchanged, i.e., now $Q_{i+1} = 1$.

   Therefore, as long as $J = 0$, $K = 0$ we have $Q_{i+1} = Q_i$.

2. $Ck = 1$, $J = 1$, $K = 0$: Proceeding on the assumption that Q_i is initially 0 and $\overline{Q}_i = 1$, we now find gate $A1$ conducts, thus making $S_i = 1$. Gate $A2$ is nonconducting, hence $R_i = 0$. From our knowledge of the SR flip-flop this yields $Q_{i+1} = 1$.

   Similarly, for $Q_i = 1$ (and $\overline{Q}_i = 0$) gate $A1$ is off and so is gate $A2$. Thus $S_i = 0$ and $R_i = 0$. The output therefore remains at 1.

   The output thus is *set* to logic 1 whenever $J = 1$ and $K = 0$ independently of the initial value of Q.

3. $Ck = 1$, $J = 0$, $K = 1$: For $Q_i = 0$ (or $\overline{Q}_i = 1$) gate $A1$ is off, so that $S_i = 0$, and gate $A2$ is off, so that $R_i = 0$. Hence $Q_{i+1} = 0$.

   For $Q_i = 1$ (or $\overline{Q}_i = 0$) gate $A1$ is off, giving $S_i = 0$, but gate $A2$ is on, yielding $R_i = 1$. With these combinations of SR inputs the output is *reset* to zero. Thus $Q_{i+1} = 0$.

   Therefore, the conclusion is that for $J = 0$ and $K = 1$ the Q output is always reset to 0.

4. $Ck = 1$, $J = 1$, $K = 1$: For $Q_i = 0$ (or $\overline{Q}_i = 1$) we find that $A1$ conducts, to yield $S_i = 1$, and $A2$ is off, to give $R_i = 0$. Hence this is the signal combination that *sets* the output to 1. Thus the output is made to complement rather than to be indeterminant as with the SR flip-flop. This action is also called *toggling*.

   For $Q_i = 1$ (or $\overline{Q}_i = 0$) it is gate $A1$ that is off and $A2$ that conducts. Hence the inputs are $S_i = 0$ and $R_i = 1$ and this combination yields $Q_{i+1} = 0$. Observe that the result is again the complement of the initial state of Q_i.

In summary it can be stated that the *JK* flip-flop provides a toggling (complementing) result whenever $J = 1$ and $K = 1$ simultaneously at the negative-going transition of the clock pulses.

The foregoing description is summarized succinctly in the following truth table:

J	K	Q_{i+1}
0	0	Q_i
1	0	1
0	1	0
1	1	$\overline{Q}_i$

The *JK* flip-flop is a widely used memory element that finds applications in such devices as counters, registers, arithmetic logic units, and other digital systems.

The T Flip-Flop. The *T* here stands for toggle and, as might be expected, this type flip-flop is obtained by joining the *J* and *K* terminals of the *JK* flip-flop in the manner illustrated in Fig. 13-21(a). A single input *T* is applied to these connected terminals. The logic symbol for the *T* flip-flop is shown in Fig. 13-21(b).

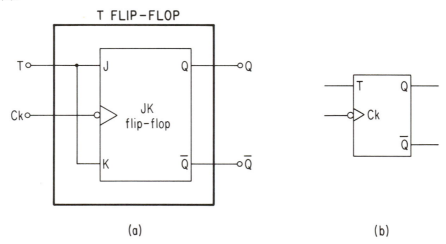

(a)   (b)

Fig. 13-21 (a) Illustrating how the *T* flip-flop is derived from the *JK* flip-flop by joining the *J* and *K* input terminals; (b) the logic symbol.

By applying the same input to the *J* and *K* terminals only rows 1 and 4 of the foregoing truth table of the *JK* flip-flop are relevant. Thus the defining relationship is

$$Q_{i+1} = \begin{cases} Q_i & \text{for } T = 0 \\ \overline{Q}_i & \text{for } T = 1 \end{cases}$$

and the behavior of such a circuit is represented in the timing diagrams of Fig. 13-22. Observe that at the end of pulse 1 and $T = 0$ the initial state of Q is preserved. At the end of the second pulse T has a value of 1 and so the outputs toggle, i.e., Q goes to 1 and $\overline{Q}$ complements to 0. The enabling pulse generated at the end of pulse 3 produces yet another toggle because T still is at logic 1. It is characteristic of the T flip-flop to produce a toggle action at each enabling pulse as long as T is at logic 1. If T is left at logic 1 continuously, the net effect is to produce a wave train that resembles the clock pulse train but differs from it in frequency. The frequency of the periodic signals appearing at the output terminals of the T flip-flop is at one-half that of the clock wave. This is apparent by noting that the period of the former is double that of the latter. In essence the T flip-flop can be treated as a frequency divider or more generally as a device which takes the input frequency at the clock terminal and divides it by 2.

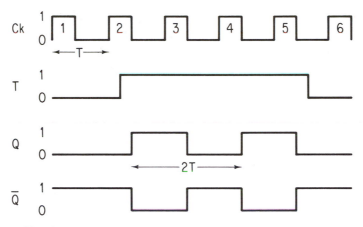

Fig. 13-22 Illustrating the toggling action of a T flip-flop in response to $T = 1$.

13-8 COUNTERS

Digital counters are important elements of many digital systems. Besides performing the obvious counting function, these devices are capable of exploiting this capability to permit the digital measurement of important quantities such as time, speed, frequency, and distance. They can be used as well to generate the timing sequences that provide control of the sequential operations that are so crucial to the proper operation of digital computers. Counters also find wide applications in digital instrumentation.

A binary counter is a group of flip-flops that is arranged to provide a predetermined, sequenced output in response to events appearing at the clock input terminal of the flip-flops. The events are essentially *count pulses* which may be obtained in a regular fashion from the normal clock waveform or may be random events that are introduced as inputs at the clock terminal. Since a single flip-flop has two output states, it follows that an array of n flip-flops has

2^n states and can count in binary from 0 to $2^n - 1$. Thus a counter constructed with four flip-flops has 16 states and can count in binary from 0 to 15. Once the last number is reached, the counter returns to its original state and repeats the cycle in response to applied count pulses. The number of states through which the counter cycles is called the *modulo* of the counter. Thus, in the example just cited, the modulo of a 4-bit counter is 16. In general, the maximum value modulo of an *n*-bit counter is 2^n, although the counter can be arranged through the design of appropriate combinational circuits to have a modulo less than 2^n. Described below are the operational characteristics of the more popular counters which can be designed using flip-flops of the *JK*, *T*, and *D* variety. Counters that are fabricated using medium-scale integration are available either as ripple counters or synchronous counters.

Ripple Counter. The circuit arrangement of a 3-bit ripple counter that uses *T* flip-flops to do the binary counting is shown in Fig. 13-23. Observe that the *T* input is held fixed at logic 1 for each flip-flop. Consequently, each time a count pulse is made to move from logic 1 to logic 0, the flip-flop toggles. The feature that distinguishes this counter from others is the method used to apply a count pulse to each succeeding flip-flop after the first one. In the ripple counter these inputs are obtained from the *Q*-output terminal of the preceding flip-flop. As a result of this arrangement, when the counter is at its full count and a new event occurs, the recycling takes place in a *rippling* fashion as the 1s change to 0s in succession starting at flip-flop FF_0. If each flip-flop has a propagation delay of t_{pd}, the total time required for resetting is $3t_{pd}$ in this case. Clearly, this time must be less than the period of the count pulse or else the counter will show an incorrect output at the next Ck. This is a shortcoming of the ripple counter.

It is helpful to refer to the timing diagram of the 3-bit ripple counter depicted in Fig. 13-24 in order to describe its operation. Initially the output of each flip-flop is assumed reset to 0 by action of the reset input. The count pulses are assumed to originate from the periodic clock wave train. During the first period of this waveform, up to but not including the drop from logic 1 to 0, all flip-flop states stay at 0, which corresponds to count 0. Then as the first pulse moves into the second clock period, the output of flip-flop FF_0 toggles since $T = 1$. Hence the signal at Q_0 moves in the positive direction from 0 to 1. This positive-going transition, of course, produces no enabling signal to flip-flop FF_1. Hence during this second clock period the output of FF_1 continues at 0. As the second period ends, FF_0 again toggles but this time from 1 to 0. Now the negative-going transition of Q_0 serves to toggle FF_1, thus changing Q_1 from 0 to 1 at the start of the third clock period. When the third clock pulse drops from 1 to 0, FF_0 again toggles. Since the action at Q_0 is from 0 to 1, FF_1 is not affected and now Q_1 continues at logic 1 for another clock period. As pulse 4 goes from 1 to 0, FF_0 toggles from 1 to 0 and thus FF_1 toggles from 1 to 0. The negative-going toggle at the output of FF_1 triggers flip-flop FF_2 to move from 0 to 1. Observe that during the fifth clock period, Q_0 is at 0, Q_1 is at 0 and Q_2 stands at 1, i.e., $Q_2Q_1Q_0 = 100$, which corresponds to count 4. This process continues until the eighth clock period during which time each flip-flop output is at logic 1.

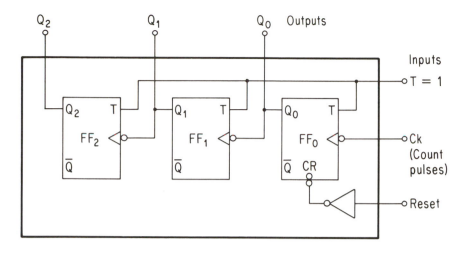

Q_2	Q_1	Q_0	Count pulses
0	0	0	0
0	0	1	1
0	1	0	2
0	1	1	3
1	0	0	4
1	0	1	5
1	1	0	6
1	1	1	7

Fig. 13-23 Three-bit ripple counter with associated state table.

It is instructive to note that the triggering schedule is such that FF_0, which is the lowest significant bit, toggles with every clock pulse. This generates the Q_0 column of the table shown in Fig. 13-23. Moreover, FF_1, which furnishes the second significant bit, complements with every pair of count pulses, i.e., it holds its logic value for two clock periods and so generates the middle (Q_1) column of binary values appearing in the table. Finally, FF_2, which provides information about the most significant bit, complements with the passage of every four count pulses.

Consider next the situation just before the end of pulse 8 in the timing diagram. Each flip-flop is at logic 1, which is the full count. Now let us examine the behavior of this counter as the eighth pulse returns to zero. After a propagation delay of t_{pd}, the output of FF_0 returns to zero. In turn, this triggers the output of FF_1 to return to zero after a delay also of t_{pd}. Finally, the output of FF_2

to AND gate A1 are $T = 1$ and $Q_0 = 1$. Hence A1 turns on and places a logic 1 signal at the T terminal of FF_1, thus preparing FF_1 to toggle at the very next negative-going clock signal. This clearly occurs as pulse 2 drops to 0. Hence Q_1 now goes to logic 1. Then, as the timing diagram indicates, during the fourth period Q_0 is once again at logic 1. Hence A1 is on again and so is A2 because Q_1 is at logic 1 as well. Accordingly, a $T = 1$ signal is present on FF_2 so that as pulse 4 returns to zero Q_2 complements to 1 and Q_1 toggles to the 0 state. Finally, note that during the eighth clock period (i.e., at count 7) each flip-flop is at logic 1, which puts both AND gates on. Therefore, the condition now exists where each flip-flop has $T = 1$ and in addition each flip-flop now receives the negative-going triggering pulse simultaneously. It is this latter action that makes this counter synchronous because the resetting to the initial state occurs at the same time. Only slight differences in the propagation delay of the individual flip-flops can cause a usually negligible deviation from synchronism. However, the delay associated with the cascaded AND gates is also a factor.

Any counter can be made to count *down* rather than up by using the complemented output terminal $\overline{Q}$ of each flip-flop in place of the normal terminal Q. In fact, by providing a set of connections that duplicates those used in Fig. 13-25 for up-counting, the synchronous counter can be made to function either as an up or a down counter. In such a case it is called an *up-down synchronous counter*. The circuit arrangement is displayed in Fig. 13-26. The OR gates serve to select either the up or down mode of operation.

Ring Counter. The structure of a 4-bit ring counter is illustrated in Fig. 13-27. The ring counter uses D flip-flops. It gets its name from the fact that the connection between the Q output terminal of a flip-flop to the D input terminal of a successive flip-flop is executed in a way that forms a ring. A glance at Fig. 13-27 makes this obvious. Although in structure this device is a counter, its presence in digital systems is for purposes other than counting. Chiefly, it is used for generating timing sequences for the purpose of controlling the operations of the digital system of which it is part. We remark on this matter further after developing an understanding of the operation of the ring counter and the waveshapes generated at the flip-flop output terminals.

The timing diagram of Fig. 13-28 is a good place to begin. Keep in mind that the governing relationship of the D flip-flop is $Q_{i+1} = D_i$, which states that the output state of a flip-flop at the next count pulse is equal to the current value of the state at the D input terminal. The desired initial state of the flip-flops in Fig. 13-28 is described by $Q_3Q_2Q_1Q_0 = 0001$, i.e., a preset command is applied to the FF_0 flip-flop to set the output at logic 1 while simultaneously resetting the remaining flip-flops. The input D_0 of FF_0 is 0 since it is obtained from the output of FF_3. Following these preset-reset commands, a clock pulse train is applied to each flip-flop at the Ck terminal. During the first period of this pulse train (see Fig. 13-28), the input at the D terminal of FF_1 is at logic 1. Hence on the negative-edge transition of pulse 1 the output of FF_1, namely Q_1, goes high. At the same time the output of FF_0, Q_0, goes low because $D_0 = 0$ during the first clock period. Thus the output states during the second clock period is

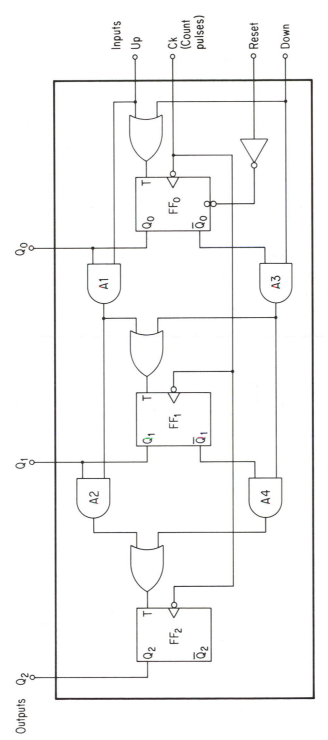

Fig. 13-26 Up-down synchronous counter.

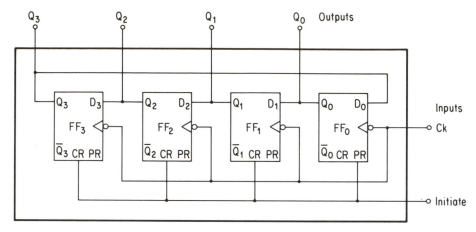

Fig. 13-27 Four-bit ring counter using *D* flip-flops.

described by $Q_3Q_2Q_1Q_0 = 0010$. Observe that, as with the initial condition, only one digit is at logic 1 and with the expiration of one clock period it has *shifted* to the next adjacent position to the left.

With Q_1 high the input signal at D_2 of FF$_2$ is high, therefore as pulse 2 returns to 0 the output Q_2 of FF$_2$ goes to logic 1 and simultaneously Q_1 is restored to zero because Q_0 was at logic zero in the second period just before switching. Now the output states of the counter are described by $Q_3Q_2Q_1Q_0 = 0100$. Again,

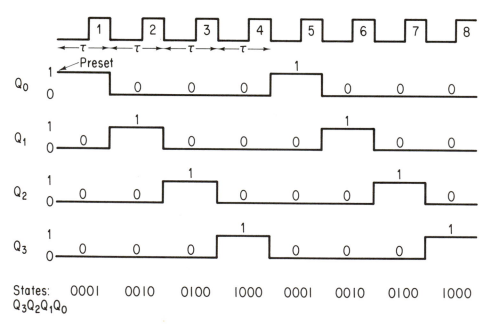

Fig. 13-28 Timing sequences appearing at flip-flop output terminals of the ring counter. Only one flip-flop output is at logic 1 at any given time and it shifts with each clock pulse.

only one digit is high and another left shift has taken place. Clearly, then, in the next clock period the output state becomes $Q_3Q_2Q_1Q_0 = 1000$. The corresponding pulse waves for each flip-flop output are properly displayed in Fig. 13-28. After four triggering operations (this is a modulo 4 counter) the initial state is reproduced.

The sequenced waveforms appearing at the output terminals of the ring counter have important applications in digital systems where it is necessary to have control over the *sequencing* of a procedure. The desired sequence can be ensured by applying the sequenced outputs of a ring counter in proper order as enabling signals on appropriate AND gates.

Switchtail Counter (Johnson Counter). An interesting modification of the ring counter results when the connection from the last flip-flop to the first is made from the complemented terminal rather than from the normal terminal. Figure 13-29 depicts the situations. This switching of the end connection gives

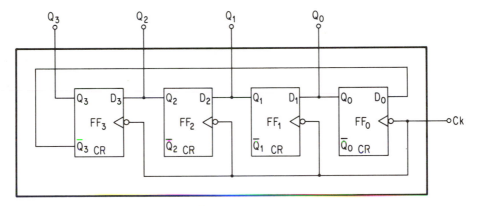

Fig. 13-29 Switchtail counter.

this counter its name. The reversal of the connections at the output of the last flip-flop is the sole distinction between this counter and the ring counter. The ensuing time variations at the normal output terminals of the flip-flops, however, are changed. For example, it is no longer true that only one flip-flop output is high at any given time. The characteristics of the switchtail counter can best be understood by again resorting to the timing diagram. This is displayed in Fig. 13-30.

The initial (reset) state for the switchtail timer is $Q_3Q_2Q_1Q_0 = 0000$. However, note that $\overline{Q}_3$ is correspondingly at logic 1 and so upon application of the first count pulse at FF_0 the D input is a logic 1. Consequently, the first negative-edge transition puts Q_0 at logic 1 so the states change from 0000 to 0001 during the second clock period. With Q_0 at 1, D_1 of FF_1 is also high so that when clock pulse 2 returns to zero Q_1 goes high. It is instructive to note that Q_0 continues high during the third clock period because D_0 remains at 1. Keep in mind that FF_3 has not yet been affected by any of the actions so far. This happens, however, at the negative-edge transition of the fourth clock pulse. Since Q_2 is high during

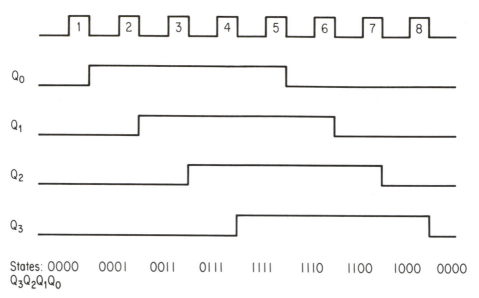

States: 0000　0001　0011　0111　1111　1110　1100　1000　0000
$Q_3Q_2Q_1Q_0$

Fig. 13-30 Timing diagram for the switchtail counter.

the preceding third period, the return of the fourth clock pulse to 0 puts Q_3 high and therefore $\overline{Q}_3$ low. Then at the subsequent switch signal (clock pulse 5) Q_0 goes low. The sequence of states for each successive count is summarized in the following table

Count	Q_3	Q_2	Q_1	Q_0	Decode logic
0	0	0	0	0	$\overline{Q}_3\overline{Q}_0$
1	0	0	0	1	$\overline{Q}_1Q_0$
2	0	0	1	1	$\overline{Q}_2Q_1$
3	0	1	1	1	$\overline{Q}_3Q_2$
4	1	1	1	1	Q_3Q_0
5	1	1	1	0	$Q_1\overline{Q}_0$
6	1	1	0	0	$Q_2\overline{Q}_1$
7	1	0	0	0	$Q_3\overline{Q}_2$

Several points are worth noting. First recall that the ring counter has n states for n flip-flops. But this table makes it quite clear that the switchtail counter has double the number of states, i.e., for $n = 4$ there are eight states as against just four for the ring counter. Consequently, it is possible to get twice as many timing sequences for a given number of flip-flops merely by switching one connection. Hence the flip-flops are used more effectively in this mode. Of course, it is necessary to provide a decoding logic that ensures that only one gate is enabled for a specific state. The required logic is listed in the rightmost column of the table. Notable too is the extended time that each flip-flop output stays high. A glance at the timing diagram discloses that for a switchtail counter with four flip-flops each output stays high for four clock pulses over the mod 8 counting. Such timing sequences can be used in control applications where an enabling signal is needed to provide sufficient *wordtime* to permit appropriate processing of binary data.

EXAMPLE 13-5 Develop the basic building block to be used in constructing a *decade* counter.

Solution: A decade counter requires a modulo 10 counter since the digits span from 0 to 9. These numbers represent the 10 states through which the counter must cycle. We must choose an n-bit counter where n is large enough to accommodate the 10 needed states. Therefore, a 4-bit counter is the choice because it provides as many as 16 states. The next step in the design process is to introduce a combinational circuit which serves to recycle the counter after 10 counts. This is easily achieved by monitoring the output bits of the 4-bit counter with an AND gate and on reaching the state $Q_3Q_2Q_1Q_0 = 1010$ (which is decimal 10) applying the output of the AND gate to the clear (CR) terminal of the counter to reset it. The AND gate output can also be used as a carry input to the adjacent decade. Figure 13-31 depicts the basic building block to be used for each required decade of the counter. It is useful to note too that the AND gate output serves as one bit of the BCD code of the decimal value of the count.

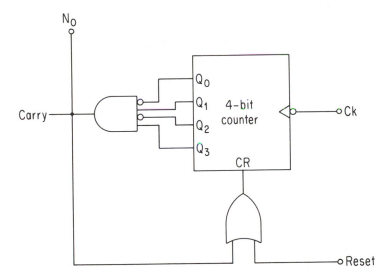

Fig. 13-31 Basic building block for each decade of a decade counter.

EXAMPLE 13-6 Design a digital device that provides a zero output until a count of 60 is reached, at which time the logic circuit is made to register a momentary logic high for enabling purposes and resets the counter as well.

Solution: A count of 60 requires at least 60 states. Since $2^6 = 64$ it follows that a 6-bit counter is necessary. Moreover, decimal 60 is represented in binary as 111100. The corresponding 6-bit output from the counter for this number is represented literally as $Q_5Q_4Q_3Q_2Q_1Q_0$. The application of these signals to an AND gate in the manner indicated in Fig. 13-32 meets the requirements of the problem. Observe that if the input is a precise

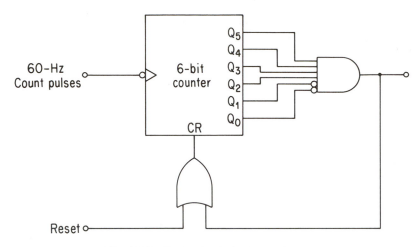

Fig. 13-32 Logic circuit for Example 13-6.

60 Hz count pulse the output of the AND gate provides a correspondingly precise time elapse of one second. Consequently, we have here essentially the basis of a digital clock.

EXAMPLE 13-7   Construct a logic device that furnishes word-time control for 16-bit binary data words.

Solution: A modulo 16 counter is needed, which in turn requires a 4-bit counter. This counter can then be used with an *SR* flip-flop together with an AND gate to provide the specified control. The circuit configuration is illustrated in Fig. 13-33.

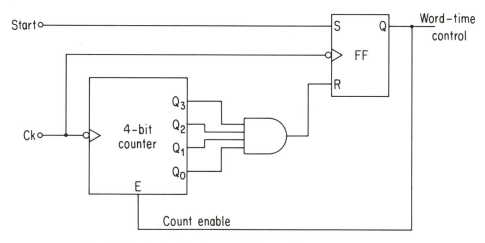

Fig. 13-33   Logic circuit to provide word-time control for 16-bit binary words.

With the counter initially cleared, a start signal puts the *SR* flip-flop output at logic 1 and the count enable then permits counting to begin. The flip-flop output Q remains high until the counter reaches the binary state $Q_3Q_2Q_1Q_0 = 1111$. Sixteen clock pulses must elapse to reach this state, at which time the AND gate output rises to logic 1 and resets the flip-flop.

13-9 REGISTERS

In the digital processing of data it is often necessary to place the data during intermediate points in the processing into temporary storage for the purpose of performing certain required manipulations after which the modified data can be directed to another but similar location. The digital devices where such temporary storage takes place are called *registers*. Because memory is an indispensable characteristic, a register is a sequential circuit that is composed of flip-flops— one flip-flop for each bit of the binary word it is designed to handle. The simplest register, therefore, can consist solely of flip-flops. As a practical matter, however, registers also contain appropriate gates to provide control. In terms of structural components the register is not unlike the counter. The distinction lies in the design of the combinational circuitry of the counter to ensure a predetermined sequence of states. Many registers use the D-type flip-flop, although the JK flip-flop is commonly used as well. Both types are readily available as commercial MSI units. Operationally, registers exhibit two notable characteristics: they are edge-triggered devices and all switching of the flip-flops is synchronized by applying the clock pulse to each flip-flop simultaneously. Activation of the register itself is achieved by means of an appropriate control signal.

Registers are classified into two basic types, although combined versions are also possible. One type is called the *parallel register* because all the binary data that appear at the input terminals of the flip-flops are transferred to the output terminals in a single clock pulse. This makes the operation of the register very fast; it is the reason for its preference in digital computers. The second type is called the *serial* or *shift register* because it processes each bit of the word in succession. Obviously, such a register is much slower than the parallel type for it requires a transfer time from one register to another which is n times greater than the parallel register, where n denotes the number of bits in the word. However, the shift register does offer the compensating advantage of requiring less equipment when it is selected for use in digital systems. We now turn attention to a description of the circuit configuration and operational behavior of each type.

Parallel Register.   The essential features of a simplified 4-bit parallel register which is fabricated with D flip-flops is depicted in Fig. 13-34. The operation is quite straightforward. The four input bits to the register are identified as $B_3B_2B_1B_0$ and can properly be assumed to originate from another parallel register. The intention now is to *write* this word into the register shown. It is instructive to observe the presence of the AND gates which are used to apply each bit of the input word to the corresponding D input terminals of the register flip-flops. Moreover, the clock pulses are applied directly to each flip-flop. However, no transfer of the input bits to the flip-flops can take place until the AND gates are switched to a conductive state, which is controlled by the signal that appears at the *load* terminal. When the load terminal is activated high, each AND gate places its respective bit input at the flip-flop D terminal. On the very next negative-edge transition of the clock pulse these binary bits are transferred to

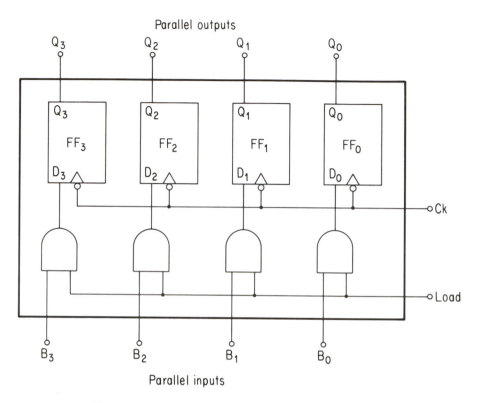

Fig. 13-34 Simplified 4-bit parallel register with D flip-flops.

the output terminals of the flip-flops in synchronism. This transference is also described as *writing into the register*. If the binary data that appear at the Q output terminals of the register are next transferred to still another register, we can describe the action as *reading* the register.

In a more elaborate version of the parallel register than that of Fig. 13-34 additional combinational circuits are included to preserve the binary information appearing at the output terminals of the flip-flops under circumstances where the load signal is deactivated.

Shift Register. The structural features of the shift register are illustrated in Fig. 13-35. The essential difference between the arrangement of the flip-flops in this configuration compared to that of the parallel register is that the data applied to the flip-flops in the shift register are sequenced with a separation of one clock pulse (CP). This is a result of the cascade connection of the flip-flops for receiving data. The information that is to be written into this shift register appears at the SI (serial input) terminal in the form of logic 1s and 0s that are fed into the terminal one at a time with a separation of one clock period. Thus if $B_3B_2B_1B_0 = 1101$ is to be transferred to the register, then in successive clock pulses (following an activation or shift signal at the S terminal) there is applied to the SI terminal a high, another high, a low, and a high sequence of pulses.

Components of Digital Systems   Chap. 13

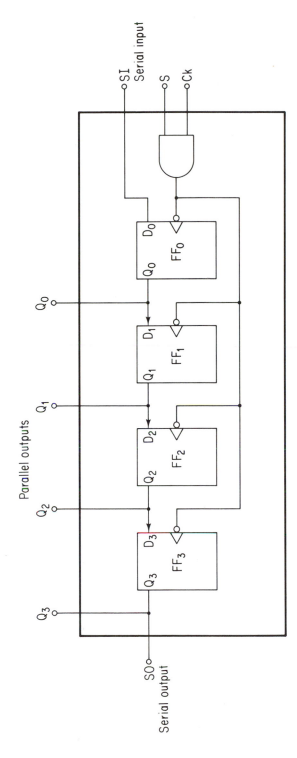

Fig. 13-35 Four-bit shift register using D flip-flops. Here S is a shift command that enables the clock pulses applied to the flip-flops.

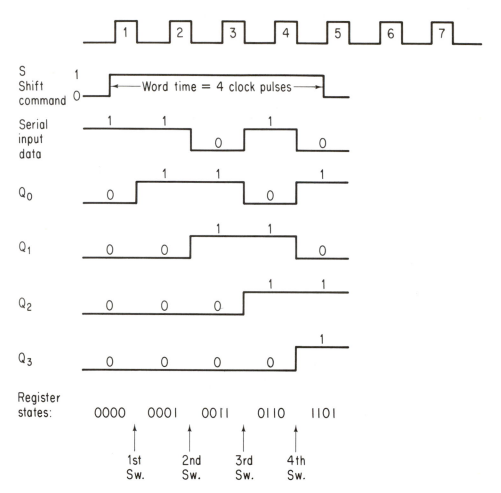

Fig. 13-36 Illustrating the shifting produced by a serial input of 1101 in response to the shift command S in the 4-bit shift register of Fig. 13-35.

Observe that the most significant bit is applied first; the least significant bit appears last at the serial input terminal. The timing diagram displayed in Fig. 13-36 illustrates the manner in which the input data are propagated through the flip-flops of the shift register. The clock pulses are applied to each flip-flop simultaneously to ensure synchronous operation. However, because of the AND gate in the control section of the register these pulses are active only when a suitable shift command is applied. The shift command is a timing sequence that must span over a number of successive clock periods equal to the word size of the register. In this case the figure is four. Hence with S assumed to go high during the first clock period shown in Fig. 13-36, Q_0 goes high in response to the appearance of logic 1 (i.e., B_3) at D_0 of flip-flop FF_0 on the very next negative-edge transition of the clock pulse. This occurs as pulse 1 returns to zero and the clock train enters its second clock period. During this second clock period the output states of the flip-flops are $Q_3Q_2Q_1Q_0 = 0001$ and $Q_0 = 1$ now appears at D_1 of flip-flop FF_1. Following the negative-edge transition of pulse 2, Q_0 continues at 1 and Q_1 rises to 1, so that during the third clock period the flip-flop states are given by $Q_3Q_2Q_1Q_0 = 0011$. Observe that after two switching clock pulses the most significant bit and the next MSB of the serial input word have been made to occupy positions at the output terminals of flip-flops FF_1 and FF_0, respectively. In preparation for the occurrence of the third consecutive switching clock pulse, note that $B_1 = 0$ is at D_0, $Q_0 = 1$ is at D_1 and Q_1 is at D_2. Hence with the negative-edge transition of pulse 3, the flip-flop states become $Q_3Q_2Q_1Q_0 = 0110$. Again note a further bit shift to the left of B_3 and B_2 of the input word and also the new arrival of $B_1 = 0$ as the rightmost digit. At the end of the fourth switching clock pulse the output state of the register reads

$$Q_3Q_2Q_1Q_0 = 1101$$

which is exactly the input word applied to the serial input terminal. The return of the shift command to zero prevents any further shifting from occurring in the flip-flops. Consequently, the register has read the serial input and the information remains recorded in the memory of the flip-flops.

The shifting action associated with the circuit configuration of Fig. 13-36 is the reason these registers are called *shift registers*. Specifically, this register was connected to provide shifting to the left. But a right shift is equally easy to arrange by simply allowing the data input to enter on the left rather than the right side and reversing the input/output connection between flip-flops.

The results described in connection with the timing diagram can be conveniently summarized in the following table.

Flip-flop state	Q_3	Q_2	Q_1	Q_0	Serial input data
Initial	0	0	0	0	1
First switch	0	0	0	1	1
Second switch	0	0	1	1	0
Third switch	0	1	1	0	1
Fourth switch	1	1	0	1	

After the first switching clock pulse, there is an entry of the serial data starting with the MSB. Each successive clock pulse shifts the data left by one bit position and permits entry of the next scheduled bit of the word. At the end of the *n*th switching clock pulse (which for the example at hand is 4) the last line of the table then represents in parallel form the data applied in serial fashion at the SI terminal.

It is instructive to note in this shift register that the four switching clock pulses serve to transfer the bits of the serial data word into their proper positions in the shift register. During this time there is no serial output taking place at the SO terminal of the shift register. If it is necessary to read out the information recorded at the flip-flops, a second shift command is needed to accomplish the task.

EXAMPLE 13-8   Using two shift registers, develop a logic circuit that permits the contents of the first register (RA) to be transferred to a second register (RB) without losing the contents of RA. Assume that the initial state of RA is 1101 and that of RB is 0000.

Solution: The required circuit configuration is depicted in Fig. 13-37. Observe that a feedback connection is introduced for register A, which serves to keep the initial stored word intact by the end of the shift command. The following table summarizes the results of the shifting operation.

	RB			Serial input to RB from RA	RA			
	Q_3	Q_2	Q_1	Q_0		Q_3	Q_2	Q_1	Q_0
Initial state → 0	0	0	0	1	1	1	0	1
0	0	0	1	1	1	0	1	1
0	0	1	1	0	0	1	1	1
0	1	1	0	1	1	1	1	0
Final state → 1	1	0	1		1	1	0	1

Initially, Q_3 of RA appears as the input to the first flip-flop of RB. Hence the first switching clock pulse following the shift command moves the initial Q_3 of RA into Q_0 of RB. At the same time the feedback connection from Q_3 of RA to D_0 of RA puts the initial value Q_3 at Q_0. Successive switching then produces the results displayed in the last line of the table. Accordingly, the contents of RA are transferred to RB while it is kept intact in RA following termination of the shift command.

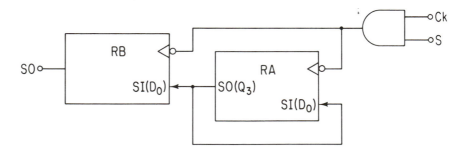

Fig. 13-37   Circuit arrangement for transferring data from register RA into RB.

13-10 RANDOM ACCESS MEMORY (RAM)

Unlike the ROM a random access memory can be written into freely. Of course, information can also be read out of the RAM; however, what is read depends entirely on the bit data that reside in the memory at the time of extraction. (In the ROM the data read out at a particular address are predetermined and unalterable, once fabricated.) Because of this double feature of the RAM, it is alternatively described as a *read-and-write memory*.

The RAM is an extremely important part of many digital systems and this is especially true of digital computers. The RAM is basically a huge collection of memory cells (flip-flops) which can be arranged to manage many words of different bit lengths. Moreover, each cell of the RAM is designed to accept three logical input signals and to provide one binary output signal. Two of the inputs furnish *control* of each memory flip-flop and the third is the binary input signal to the cell. One signal controls the read/write capability, while the other selects which of the many cells is to be so accessed.

Registers, which use arrays of flip-flops, constitute an important subdivision of the random-access-memory unit. In fact, if a digital system is characterized by 8-bit words, each word can be treated as a register. These are more properly called *storage* or *memory registers,* to distinguish them from two other registers which serve special functions in relation to the memory unit. Each memory register is assigned a specific address to facilitate its location. In the case of the 4K memory unit, for example, there are 4096 memory registers. To address each of these registers, it is necessary to have another register containing 12 bits because 2^{12} = 4096. For obvious reasons this register is called the *memory address register* (MAR). The second of the special registers that is essential in communicating with the memory unit is called the *memory buffer register.* Whenever a specific memory register is to be written into or read, the MAR is used to locate it. A subsequent read command then transfers the contents of the memory register into the buffer register to await further instructions. If, instead, a write command is issued, the contents of the buffer register are transferred to the addressed memory register. Clearly, the buffer register need have only as many bits as the memory register.

The essential operational features of the RAM memory cell are represented by the sequential circuit of Fig. 13-38(a). The actual electronic circuitry that provides the logical operations indicated here consists of two transistors equipped to accommodate multiple inputs. The symbol for this basic 1-bit cell is shown in Fig. 13-38(b). Note the three input signals, previously identified, and the single output terminal. Clearly, the output state of the cell is 0 or 1 depending upon the state of the Q output terminal of the SR flip-flop. When a write command is issued (i.e., $\overline{W} = 0$) following the selection of a particular cell ($S = 1$), observe that AND gate $A3$ goes off which inhibits reading from occurring. Furthermore, if the state of the input is high ($I = 1$), $A2$ conducts, which in turn provides logic 1 to the S terminal of the flip-flop. Accordingly, the flip-flop output goes

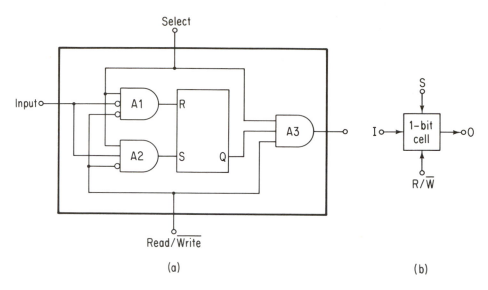

Fig. 13-38 One-bit memory cell of RAM: (a) illustrating the equivalent operational logic of the cell; (b) symbol.

high. For a low input ($I = 0$) it is gate $A1$ that becomes conductive, which provides the flip-flop with a reset signal, thus putting the output at 0 in conformity with the input state. Reading of the input data bit is thereby achieved. Note the absence of clock pulses, which are unnecessary while performing reading and writing operations.

The 1-bit memory cell is the basic building block of the random-access memory. The memory unit of the digital system is composed entirely of these cells along with the appropriate control and addressing circuitry. Hence a RAM unit that is equipped with 4096 bytes has a total of 32,768 memory register cells. With numbers of this magnitude one can easily appreciate the need to duplicate the logical operations of Fig. 13-38(a) by means of IC transistors designed to require very little space on the memory chips.

The basic fabrication structure of a RAM unit is illustrated in Fig. 13-39. This is an 8-word by 4-bit memory. A 3×8 decoder is needed to address the eight words (rows). For simplicity only the first and last words are shown. Each memory cell must have three input connections as indicated. The input bits appear at the I terminal of the cell and the same bit is applied to all cells along a column. Furthermore, an output line connection is established from each cell output along a given column to an OR gate. This procedure is duplicated for each column. When an input word, say $B_3B_2B_1B_0 = 1101$, is to be written into the memory register represented by word 3 (i.e., row 4), it is necessary to use our input address code of 011 and a logic 0 at the $R/\overline{W}$ terminal. Although two of the three input terminals of each cell are activated by this procedure, the

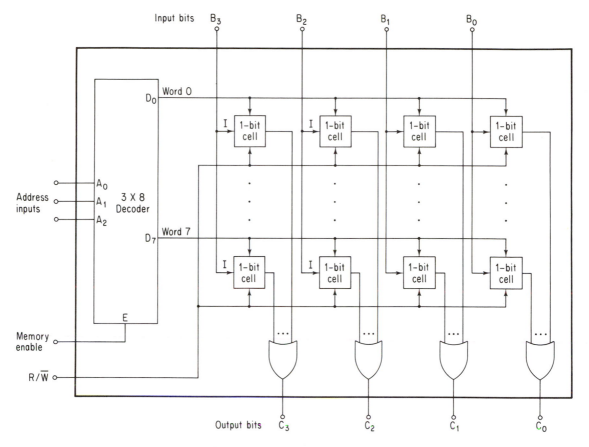

Fig. 13-39 Basic structure of a 4-bit 8-word random-access memory. Only the details of the first and eighth rows (words) are shown.

required third input signal appears only for those cells on the fourth row which have the correct address.

The 8-word by 4-bit RAM chip displayed in Fig. 13-39 can be expanded to an 8-word by 8-bit memory by paralleling the address lines as shown in Fig. 13-40. Note the inclusion of an additional control identified as the *chip select* line. Alternatively, the two 8×4 chips could be used to provide a 16×4 memory. This is achieved by paralleling all the input, output, and address lines and then inserting an inverter between the CS input terminal and the CS pin on the first 8×4 chip. In such a configuration placing a logic 0 signal at CS keeps the first chip active. The second chip is inhibited. When the CS terminal is allowed to go high, the second chip becomes active while the first is inhibited. For all practical purposes the presence of the inverter in the chip select line is made to serve as an additional address bit, thus yielding $2^4 = 16$ words.

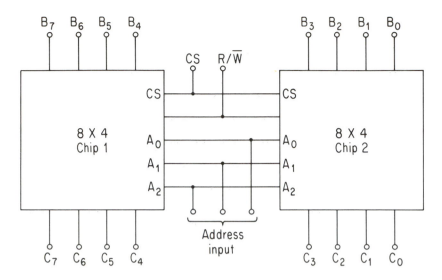

Fig. 13-40 Illustrating the parallel connection of the address lines of two 8 × 4 RAM chips to yield an 8 × 8 memory.

Summary review questions

1. Describe the basic principle that underlies the design of an encoder. How many input lines are there for an n-bit code?

2. What type of integrated circuit (SSI, MSI, or LSI) is used in the construction of encoders?

3. Describe the half-adder. Why is it so named?

4. Draw the half-adder gate circuit that is based on the sum of minterms.

5. Draw the half-adder gate circuit that is based on the product of maxterms.

6. Draw the half-adder gate circuit that is based on the exclusive-OR operation.

7. How is the full-adder distinguished from the half-adder?

8. What are the input variables and output quantities of the full adder? Show the truth tables showing how the logic levels of the outputs are determined.

9. What is the highest numerical quantity that a full-adder must be capable of accommodating in any digit position? Demonstrate.

10. Develop the Karnaugh map for the output sum digit of the full-adder. Repeat for the carry digit.

11. Based on the results of Question 10, draw the logic diagram of a full-adder after performing minimization of the logical expressions.

12. Show how a full-adder may be implemented with two half-adders.

13. How many full-adders and half-adders are needed to sum two n-bit numbers?

14. Draw the block diagram of a 4-bit parallel adder and indicate the interconnections between the adders used for each bit. Why is this device called a parallel adder?

15. Describe what is meant by carry ripple in a parallel adder? Why is it advantageous to reduce this effect?

16. Describe the strategy that is used to subtract two binary numbers. Why is this scheme preferred?

17. Define the r's complement in a number system with a radix r. What is the 10s complement of the decimal number 576.3?

18. Outline in the decimal system how subtraction can be achieved through a process of addition by using the tens-complement of the subtrahend. Assume that the minuend exceeds the subtrahend.

19. Repeat Question 18 for the case where the subtrahend is larger than the minuend.

20. Outline in the binary system how subtraction can be achieved through a process of addition by using the 2s complement of the subtrahend.

21. Describe by illustration how two binary numbers are subtracted using the normal rules associated with borrowing.

22. State the general rule that is used by digital systems to subtract two binary numbers using adders.

23. What is a decoder? How is it different from an encoder?

24. A decoder has six binary input lines. How many output lines does it have? How many AND gates are needed? How many minterms are there?

25. Draw a block diagram to illustrate how a 3×8 decoder can be used to build a full-adder. What feature of the decoder is responsible for such an implementation of a full-adder? Why is the 3×8 decoder specifically the one to use?

26. What is a demultiplexer?

27. What modification needs to be made to a decoder to allow it to operate as a demultiplexer?

28. What is the purpose of the enable input of a demultiplexer?

29. What is a multiplexer? Distinguish among the multiplexer, the demultiplexer, and the decoder regarding operation, application, and construction.

30. Besides the important function of data selection, identify and describe another useful way in which the multiplexer may be used.

31. What is a ROM? Name the basic digital elements that can be used to build a ROM.

32. Distinguish among the ROM, the PROM, and the EPROM.

33. Discuss how a ROM can be designed to provide a specific relationship between the input lines and the output lines.

34. A ROM is said to be an implementation of a truth table. Explain what this means.

35. What is meant by the phrase a *2K ROM?*

36. How many bits are there in a ROM with a 10-bit address and eight output lines?

37. What is a sequential logic circuit? How does it differ from the combinational type?

38. Explain the difference between synchronous and asynchronous sequential circuits.

39. What is the name given to the basic memory element that is found in sequential logic circuits? Why is it so named?

40. Describe the latching characteristic of a latch gate. Is the latching operation stable? Explain.

41. Draw the basic NAND flip-flop. Draw the basic NOR flip-flop. How do these differ from the basic latch?

42. Describe completely the operation of the NOR SR flip-flop for all four combinations of the two input logical variables S and R. Draw the appropriate truth table.

43. Repeat Question 42 for the NAND flip-flop.

44. What shortcoming does an SR flip-flop possess?

45. Why is it useful to operate flip-flops in a synchronous manner? Describe the operation of the clocked SR flip-flop.

46. What is a glitch? How is the effect of the presence of glitches in digital data minimized? Illustrate.

47. Explain what is meant by trailing-edge triggering.

48. What is the D flip-flop? How is it different from the SR flip-flop? How is the uncomplemented output related to the input?

49. What is the JK flip-flop? How does it differ from the SR flip-flop in construction and operation? Be sure to address the specific benefit derived by the inclusion of the AND gates in the JK flip-flop.

50. Describe the toggling action that occurs in the JK flip-flop when both J and K inputs are at logical 1. Draw the truth table of the JK flip-flop.

51. Draw a block schematic diagram of the T flip-flop, taking care to emphasize how it differs from the JK flip-flop.

52. What is the distinctive feature of the T flip-flop? What ensures this characteristic?

53. Describe the frequency division capability of the T flip-flop.

54. Name some applications of counters in digital systems.

55. Describe the general construction details of a digital counter and how it can be made to do counting.

56. How is the modulo of a counter defined? How is the number of states of a counter determined? Illustrate.

57. Describe how the ripple counter works. Why is it so named? What disadvantage can you name for this counter?

58. Explain the rippling action that occurs in the ripple counter.

59. How does the synchronous counter differ in construction and circuit arrangement from the ripple counter? What advantage does this offer? At what cost?

60. Describe the synchronous switching feature of the synchronous counter as it gets ready to recycle following completion of its modulo.

61. How is a counter made to count down rather than up?

62. Explain what is meant by the phrase a *modulo 8 counter*.

63. What is a ring counter? What type of flip-flop is used in such counters? What distinguishing feature does the ring counter display? Of what use is it?

64. How does the switchtail counter differ in construction and connections from the ring counter? What important consequences flow from this?

65. Describe the operation of the switchtail counter.

66. How many states does the switchtail counter have in comparison to the ring counter?

67. Compare the timing diagrams of the switchtail and ring counters, stressing particularly the duration times of the logic 1 signal levels appearing at the flip-flop output terminals. Comment on the application aspects of these signals in digital systems.

68. What is a register? How does it differ from a counter? What purpose does it serve in digital systems?

69. Describe how a parallel register is constructed and operates.

70. Describe the construction and operational features of the shift register.

71. Compare the parallel and shift registers, pointing out their advantages and limitations.

72. Explain the role of the word-time shift command in shift registers. Where can such signals be obtained?

73. Describe how the RAM differs from the ROM.

74. Draw the basic cell structure of the RAM. Identify the number of input signals to each cell of the RAM and the functions of each.

75. How are registers related to the RAM?

76. What is a memory address register? What function does it serve? What determines the word length of this register?

77. Show the block diagram of a simple RAM unit and describe how reading into memory and writing out of memory is achieved.

Problems _____

GROUP I

13-1. Design an encoder which converts all the decimal digits to binary form. Write the binary logical expression for each output line.

13-2. Design an encoder which converts all the hexadecimal digits to binary form. Write the logical expression for each output line.

13-3. Design a 4-line-to-2-line encoder and write the logical expression for each output line.

13-4. Show how the half-adder logic circuit configuration of Fig. 13-3(b) can be implemented with the universal NAND gate.

13-5. Obtain the sum in binary form of the following pairs of binary numbers:
(a) 101100 and 011011.
(b) 11100011 and 00111011.

13-6. Obtain the difference of the binary numbers listed in Prob. 13-5 using the conventional rules of binary subtraction.

13-7. In the circuit configuration of Fig. P13-7 (p. 580), determine the operation that is being performed when (a) $M = 0$ and (b) $M = 1$.

13-8. A half-subtractor is a combinational circuit that subtracts two bits and produces the difference. Develop the truth table for the half-subtractor and write the logical expressions for the difference and borrow logical expressions. Compare these results with those found for the half-adder and comment.

13-9. Find the 10s complement of the following numbers: (a) 0.7, (b) 24.35, and (c) 175.692.

13-10. Subtract the decimal number 27 from decimal 44 using a process of addition.

13-11. Determine the 2s complement representation in the binary system of the following decimal numbers: 47; 24.

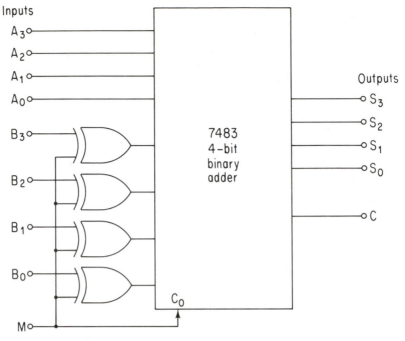

Inputs

A_3

A_2

A_1

A_0

B_3

B_2

B_1

B_0

M

7483
4-bit
binary
adder

C_0

Outputs

S_3

S_2

S_1

S_0

C

Fig. P13-7

13-12. Two numbers, A and B, are to be processed in a binary full adder.
(a) Find A plus B when $A = 44$ and $B = 27$. Express the result in a sign-magnitude binary form.
(b) Repeat part (a) for $A = 44$ and $B = -27$.
(c) Repeat part (a) for $A = -44$ and $B = 27$.
(d) Repeat part (a) for $A = -44$ and $B = -27$.

13-13. Determine the 2s complement of the following binary numbers without resorting to the associated formal procedure: (a) 1010011, (b) 1000101, (c) 011100, and (d) 10000.

13-14. Design the logic circuitry for a 2×4 decoder. Draw the truth table for inputs A, B and outputs D_0, D_1, D_2, D_3. Include as well the logical expressions for each minterm.

13-15. A LED (light-emitting diode) display uses seven segments to generate the decimal numerals as indicated in Fig. P13-15.

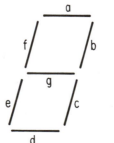

a

f

b

g

e

c

Fig. P13-15

d

(a) Identify those segments which must be activated (lighted) in order to display numerals 2 and 3.

(b) Repeat part (a) for numeral 4.

The display segments are activated from the output of a 4 × 7 decoder, where the input lines represent the BCD word $(B_3B_2B_1B_0)$ and the output lines denote each of the seven segments of the display (a, b, c, d, e, f, g).

(c) Identify all the decimal numerals that involve segment g.

(d) Determine the logical expression for g as a sum of minterms obtained from the truth table.

(e) Design the gate circuit for segment g.

13-16. A combinational circuit is defined by the following logical expressions:

$$Y = \overline{A}B$$

$$Z = AB + A\overline{C}$$

Design the circuit using a 3 × 8 decoder and any other gates that may be needed.

13-17. A process is characterized by three input logical variables A, B, and C and the following logical expression for the output variables X, Y, Z:

$$X = A\overline{C} + B\overline{C}$$

$$Y = \overline{B}$$

$$Z = A + B\overline{C}$$

Design the required combinational circuit using an appropriate decoder and gates.

13-18. The arrangement shown in Fig. P13-18 can operate as a 4 × 16 decoder. Terminal E denotes the enable input. Describe how the 4-line-to-16-line decoding result is achieved. What is the role of logic variable M?

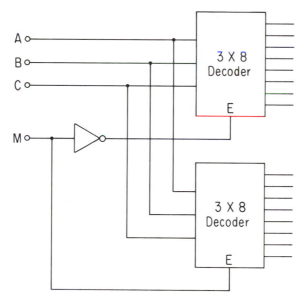

Fig. P13-18

13-19. Develop the logic gate circuitry for a 1 × 4 demultiplexer. How many address control lines are there? Where is such a circuit useful?

13-20. Show how the multiplexer of Fig. 13-11 can be used to generate the logical expression
$$Z = A\overline{B} + B\overline{C}$$

13-21. Repeat Prob. 13-21 for the case where $Z = \overline{B}$.

13-22. A process is described by the following truth table, where A and B are the input variables and the outputs are 4-bit words.

Address		Data word			
A	B	D_3	D_2	D_1	D_0
0	0	1	0	1	1
0	1	0	1	0	1
1	0	0	0	1	1
1	1	1	1	0	0

Design a ROM that generates this truth table. Be sure to specify the size of the decoder section and the number of gates in the encoder action as well as all the interconnections.

13.23. A ROM is to be used to generate the following logical functions: $Z_0 = A + B + C$ and $Z_1 = ABC$, where A, B, and C represent the three input logical variables. Design the ROM and show all interconnections between the decoder and the encoder.

13-24. Explain the operation of the clocked SR flip-flop of Fig. 13-16, which uses all NAND gates. The explanation should include each combination of input states.

13-25. A clocked SR flip-flop is supplied with the S and R logical signals depicted in Fig. P13-25 over the first eight clock pulses. The flip-flop is designed for trailing-edge triggering. Sketch the corresponding variation of the logical signal that appears at terminal Q of the flip-flop.

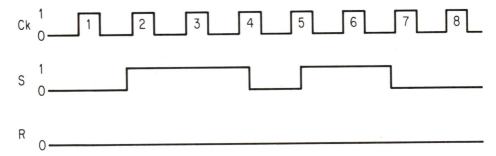

Fig. P13-25

13-26. Repeat Prob. 13-25 for the situation depicted in Fig. P13-26.

13-27. Shown in Fig. P13-27 is the circuit configuration of a master-slave flip-flop. Both the master and the slave are clocked SR flip-flops.

(a) Assuming that Q_M and Q are initially at 0, describe and plot the output at terminals Q_M and Q as the clock signal passes through one period of logic 0-to logic 1-to logic 0 while $S = 1$, $R = 0$.

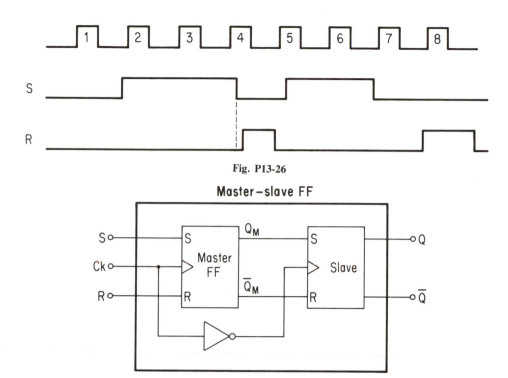

Fig. P13-26

Master–slave FF

Fig. P13-27

(b) What conclusion can you draw from the results of part (a) concerning the time at which Q switches states?

13-28. A D flip-flop can be made to operate as a JK flip-flop by adding two AND gates, an OR gate, and inversion in the manner illustrated in Fig. P13-28. Demonstrate the truth of this statement.

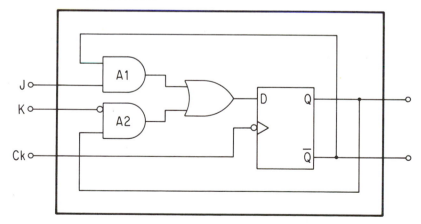

Fig. P13-28

13-29. The gate circuitry displayed in Fig. P13-29 employs the universal NAND gate. Which type of flip-flop is this?

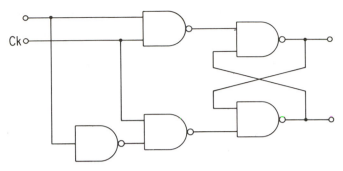

Fig. P13-29

13-30. A *JK* flip-flop is wired as shown in Fig. P13-30(a). If the input signal has the variation shown in Fig. P13-30(b), sketch the corresponding waveshape that appears at the *Q* terminal. Assume that *Q* is at logic zero initially and that trailing-edge triggering is effective.

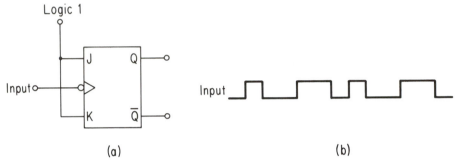

Fig. P13-30

13-31. Refer to the *JK* flip-flop of Fig. P13-30. Assume that its output at the *Q* terminal is made to serve as the input to a second *JK* flip-flop wired exactly as the first one. Sketch the *Q* output of the tandem arrangement for the input signal of Fig. P13-30(b).

13-32. **(a)** Sketch the variation of the logical signal that appears at the *Q* output terminal of the *JK* flip-flop for the arrangement shown in Fig. P13-32. Assume that the input signal is that of Fig. P13-30(b) and that initially *Q* = 0 and the switching occurs at the trailing edge.

(b) How is the signal variation found in part (a) related to the input signal of Fig. P13-30(b)?

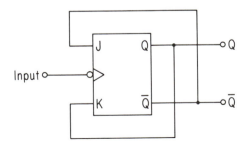

Fig. P13-32

13-33. Draw the block logic diagram of a binary counter that can count up to 100.

13-34. Design a logic circuit that endows a digital system with word-time control for 8-bit data words.

13-35. In the circuit configuration of Fig. P13-35, assume that the clock is a wave train of equally spaced pulses. With the aid of an appropriate timing diagram, describe the operation of this digital device when the CTE signal is held at logic 1.

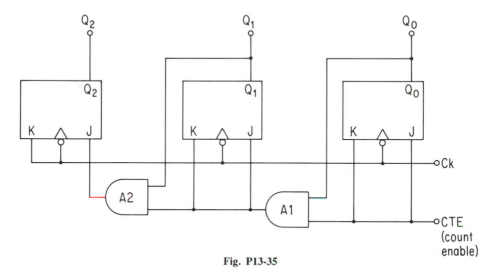

Fig. P13-35

13-36. The binary content of a 4-bit shift register is initially 1010. A serial input of 110101 is applied to the register by shifting left six times. Itemize the content of the register at each shift.

13-37. Repeat Prob. 13-36 for the case where the specific serial input is shifted to the right six times.

13-38. In the 4-bit register of Fig. 13-35, assume that $Q_3Q_2Q_1Q_0 = 1100$. Draw the waveform of each FF output if the input sequence of 01011 is applied to D_0. Assume that the shift command is at logic 1 for the required length of time.

13-39. A 16K RAM is equipped with memory registers which have 8-bit lengths. How many bits must the associated memory address register (MAR) have?

13-40. Draw the complete wiring diagram of a 4-bit four-word random-access memory. Use the basic 1-bit cell and show all interconnections. Describe how a word is written into memory and how it is read out of memory.

13-41. It is required that a random-access memory be constructed that has 256 words with 12 bits per word. Proceeding on the assumption that 16×12 RAM chips are available, find:
(a) The number of 16×12 RAM chips that are needed to build the memory.
(b) The number of flip-flops needed in the memory address register.

GROUP II

13-42. Refer to the circuit of Prob. 13-7 and discuss how this digital device can be used to provide a comparison of the two binary quantities A and B.

13-43. Develop the truth table for the full-subtractor and write the simplified logical expressions for the difference and borrow variables. Compare these results with those of the full-adder and suggest a simple modification of the full-adder that serves to convert it into a full subtractor.

13-44. Discuss some of the problems that arise in the design of a BCD-to-decimal decoder.

13-45. Repeat Prob. 13-23 for the case where the logical function to be generated is $Z = A \oplus (B \oplus C)$.

13-46. Draw the block arrangement of a 4-bit ripple counter and draw the timing diagram that displays the Q-output of each of the flip-flops. Describe the rippling effect that occurs in this counter after the negative-going transition of the sixteenth clock pulse.

13-47. Draw the interconnection wiring diagram which shows the details of how to place two 8×4 RAM chips in parallel so that the result is a RAM with sixteen 4-bit words.

13-48. It is desirable to arrange four 8×4 RAM chips to construct a 32-word, 4-bit/word random-access memory. Through the use of AND gates for chip selection purposes, devise a strategy to achieve this goal.

chapter fourteen

Microprocessor Computer Systems

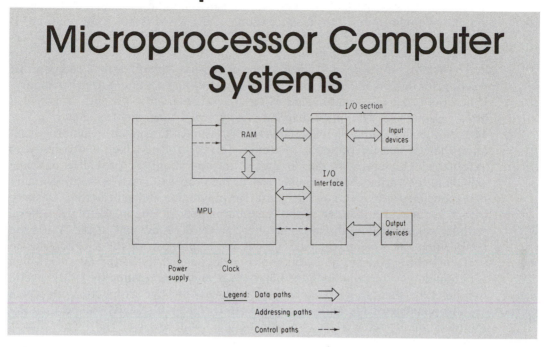

Although many of the components that are described in the preceding chapter are to be found doing important tasks in digital systems generally, it is in digital computers that they find their greatest applications. In this chapter we give attention first to the description of an elementary stored-program computer in order to illustrate how digital components can be coordinated and managed to achieve a prescribed objective. This exposition helps to bring out the inter-relationship that exists between various important elements of the computer and especially the critical role of the controller in providing the properly sequenced enabling signals that makes the system function in an ordered manner. Once this background is established and the basic notions of assembly language programming are understood, attention is next directed to a brief study of a commercial-type microcomputer system that is built around the Intel 8080 microprocessor unit (MPU). The increased complexity and proliferation of digital components in the 8080 MPU makes available a far larger instruction set than that of the elementary computer. Accordingly, it is capable of performing much more difficult tasks.

14-1 AN ELEMENTARY COMPUTER: BASIC ARCHITECTURE

It is our purpose here to describe an elementary computer which is capable of adding or subtracting numbers in an automatic fashion through a stored program.

587

The desire to have a stored program and many operands (i.e., the binary data to be manipulated in a prescribed fashion) makes it necessary to include a random-access memory as part of the system. Actually, for the level of operations to which we confine ourselves in this treatment, the RAM is not truly required. It is included in order better to appreciate its role in commercial versions of computers which are much more powerful.

The basic structure of a digital computer consists of two principal components: a processor that involves general-purpose registers and a random-access memory. When the processor is fabricated as an integrated-circuit on a chip, it is called a *microprocessor*. The block diagram of a typical arrangement is shown in Fig. 14-1. In this illustration the microprocessor is assumed to contain several registers, an arithmetic unit (AU), and a controller. The controller, which is driven by an external clock, serves as the director of all operations by providing enabling signals to the various MPU components at the proper times. The arithmetic unit is responsible for performing the operations of addition and subtraction. Observe that it includes a full-adder of the required number of bits as well as a general purpose register called the accumulator. The third component of the AU is the complement/increment register (C/I). It is the purpose of the C/I register to perform subtraction by the addition of the 2s complement of the number which is obtained by complementing the binary form and incrementing by 1. It is useful to note at this point that commercial microprocessors have arithmetic units which are also capable of the logic operations AND, OR, and X-OR, and so they are more commonly referred to as arithmetic-logic units (ALU). The PC register is a program counter that is assigned the task of providing the memory address register (MAR) with the address of the memory register of the RAM for each successive step of the program starting with the first. Finally, there is the *instruction register* (IR), which provides a double function. It supplies the controller with information about the kind of operation that is to be performed, for example, addition or subtraction. It also provides the address of the operand; the operand is copied from RAM to the C/I register of the AU. A line that is drawn in Fig. 14-1 with arrowheads pointing to opposite directions denotes a bidirectional flow of information along the indicated path.

Since our aim in treating this simple computer is to provide an opportunity to show how the data and instructions that are stored in the random-access memory are coordinated by the controller to achieve a desired result, for convenience, we assume a very simple memory that contains 16 words with a 6-bit data content. Consequently, the memory address register needs to be equipped with 4 bits to address each memory location. In Chapter 13 these memory locations were also called memory (or storage) registers. A glance at Fig. 14-1 discloses that the MAR furnishes the binary signals to the address decoder of the RAM, which in turn selects the appropriate location. To deliver the information that resides at this address to the proper special-purpose register for further processing, the contents must be transferred to the interconnecting bus. This is achieved under the direction of the controller by a signal to the enabling terminal E of the RAM. When $E = 1$, the chosen address content appears on the bus

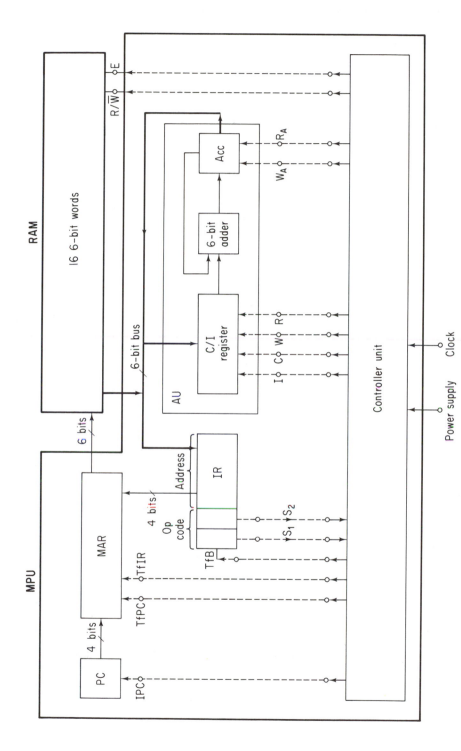

Fig. 14-1 Elementary computer composed of a microprocessor unit and a random-access memory. (After H. Taub)

and is received by one or the other of the processing registers (IR or C/I). When $E = 0$, the RAM is disconnected from the distributing bus.

The number of bits to be accommodated by the IR register always exceeds those required for addressing by the number required for the operation codes. In our simple computer four operation codes suffice. We need a binary code for each of the following operations: adding, subtracting, transferring, and halting. The transfer code is included to have the capability of placing the result of the specified operations at a particular memory location. Of course, four operation codes require a binary code composed of 2 bits. Let us agree upon the following code assignments:

Code	Instruction
00	Halt
01	Add to accumulator
10	Subtract from accumulator
11	Transfer accumulator to

It follows then that the instruction register must be designed to manage 6-bit words—2 bits for the operation code and 4 bits for the address code. Moreover, the system bus must be equipped with six lines in order to transmit the 6-bit binary data and each memory register (MR) of the RAM must contain 6 bits as well.

The restriction to 6-bit data words means that the total range over which this computer can operate is from -32 to 31. Of course, this restriction needs to be considered when writing a program for adding and subtracting numbers. A result that exceeds the lower or upper limit would have to be avoided. However, it is important to keep in mind in dealing with this fictitious computer that the objective is to convey an understanding of how the microcomputer operates. Consequently, restrictions, such as this one on the range of allowable numbers, are essentially irrelevant because in the far more powerful practical versions of this computer the range is greatly increased by increasing the number of bits per word. But, in contrast, the principles of operation remain intact.

Programming. By way of illustration, let us assume that it is required that the difference of 5 and 3 be found using a stored program that performs the operations automatically following the entry of the data and instructions into memory. When organizing the memory it is customary to divide the memory locations into three sections. The first section is the place where a full list of program instructions is written in successive order consistent with the task to be achieved. The use of a successive order simplifies the job of the MPU since it merely requires a command to increment the program counter in order to move to the next instruction. We start our program at memory address 0000. The second section of the RAM is reserved for the data, which are the operands to be used in obtaining the desired results. The last section consists of a group of memory registers where the results can be stored.

To program our simple computer to solve the equation

$$S = 5 - 3 \qquad (14\text{-}1)$$

only three programming steps are needed provided that both the accumulator and the program counter are initially cleared. The first instruction, which is identified as step 0 in Table 14-1, is to add the contents of the memory location at the first address to the accumulator. Observe that the contents of the memory register (MR) is 011010. The first two bits, 01, identify the operation code for addition. When the controller receives these bits, it is alerted to prepare the arithmetic unit to *add* the contents at memory address 1010 (the remaining four digits) to the contents already in the accumulator (which is zero for the case at hand because the accumulator is assumed to be cleared initially). The address part of the contents of a memory register of an instruction statement leads to the data section, where the operand is found. Thus 1010 is the first address of the data section (where the number $5 = 000101_2$ is found) whose contents are to be added to the accumulator. The result stays in the accumulator.

Upon completion of the first instruction, a signal is issued to increment the program counter PC, which then sets in motion a series of events that carries out the second instruction. Note that the operation code is now 10, which means that the contents of the register MR at address 1011 is to be subtracted from the contents already present in the accumulator. The content of the memory register at address 1011 is the number 3. Finally, when this statement is executed, the PC register is incremented once again and the third instruction is loaded into

TABLE 14-1 STORED-PROGRAM SOLUTION OF EQ. (14-1)

Step	Address (in binary)	Op code	Address code	Instruction	Mnemonic code
		Contents of RAM			
Instruction section					
0	0000	01	1010	Add contents of MR to Acc	ADD MR
1	0001	10	1011	Subtract contents of MR from Acc	SUB MR
2	0010	11	1111	Transfer contents of Acc to MR 15	TR MR
3	0011	00	XXXX	Halt	HLT
.	.	. .		.	.
.	.	. .		.	.
.	.	. .		.	.
Data section					
10	1010	00	0101	5	
11	1011	00	0011	3	
.	.	. .		.	.
.	.	. .		.	.
.	.	. .		.	.
Output section					
15	1111	00	0010	2	

the instruction register. Operation code 11 means that a transfer of the accumulator contents occurs. Line 3 of the table identifies the address, at which the result ($S = 2$) is to be transferred, as binary 1111. This was chosen as the last memory register of the RAM.

Program Execution via Controller Unit. The point has already been made that it is the function of the controller to direct and coordinate all the moves that are necessary in fetching data and executing the specified operations. It is instructive now to examine in detail the manner in which the controller achieves this task, particularly in view of the background of the preceding chapter. The controller for this elementary computer can be built of AND gates, OR gates, D flip-flops, an inverter, and a 2×4 decoder. The logic diagram is depicted in Fig. 14-2(a). Besides the *start* button and the clock signal for the flip-flops, the only other inputs to this controller are the signals applied to the instruction decoder. These signals, $S_1 S_2$, are obtained as the op-code bits of the instruction words (see Fig. 14-1). Correspondingly, the output signals of the controller are a sequence of enabling signals that provides for the movement of information between the RAM and the registers.

We begin on the assumption that the stored program of Table 14-1 already resides in memory and that both the PC and ACC registers are cleared. Recall too that all flip-flops are connected to the clock's wave train. At the first negative-edge transition of the clock following activation of the start button in Fig. 14-2(a), the output of D flip-flop 1 (FF$_1$) goes high (i.e., TfPC = 1). This acronym denotes a *transfer from program counter* register. Thus the memory address register (MAR) is loaded with address 0000. Hence lines are opened to read or write at this memory register. At the next clock transition the second flip-flop (FF$_2$) goes high. As indicated in Fig. 14-2(a), logic 1 control signals are transferred simultaneously to terminals E, $R/\overline{W}$, TfB, and IPC. It is common practice in the design of controllers to provide more than one enabling signal whenever conditions warrant. With E and $R/\overline{W}$ of the RAM at logic 1 the binary data (011010), which is in the memory register at address 0000, is placed on the 6-bit bus. Moreover, with the enabling signal at terminal TfB of the instruction register (IR) also high, these binary data go into the IR register. The acronym TfB in this case denotes *transfer from bus*. Once this information appears in the IR, the so-called *fetch* part of the instruction cycle is completed. Note that it takes two clock cycles to complete the fetching mode.

The *execute mode* of the instruction cycle begins with the application of the op-code signals, $S_1 S_2$, to the instruction decoder. Here $S_1 S_2 = 01$, so that the line that leads to AND gate $A1$ in Fig. 14-2(a) goes high and this switches $A1$ on, which in turn puts the D terminal of FF$_3$ high. At the next clock transition the enabling terminal TfIR (transfer from instruction register) of the MAR goes high, which permits the MAR to read address 1010 from the address portion of the instruction register. In turn, the appropriate paths in the RAM are opened so that the data that appear in the memory register at this address can be transferred to the arithmetic unit. This happens at the very next clock transition when enabling signals are applied to terminals W, E, and $R/\overline{W}$. With $W = 1$ the data (000101) representing decimal 5 are written into the C/I register. However, because the operand is a positive number, the complement and increment operations are bypassed. Hence the data are transferred to the adder for addition, which occurs on the next clock transition. At this same time the output of flip-flop FF_5 goes high. Consequently, there appears an enabling signal, W_A, which writes the result into the accumulator. This sequence of control operations is summarized in Fig. 14-2(b).

The active level of the W_A signal is also responsible for providing the transition to the second instruction in the stored program. Recall that the PC was already incremented during the fetch part of the first instruction. A glance at Fig. 14-2(a) discloses that W_A also appears at the OR gate. Therefore, after W_A goes high, there is an exact repeat of the fetch portion of the first instruction cycle. However, the op code is now 10, which alerts the controller to prepare for subtraction. In response, the instruction decoder activates AND gate $A2$, which is associated with a logic circuit that now contains two additional flip-flops (FF_8 and FF_9). These provide the enabling signals to the C/I register to perform complementing and incrementing in the generation of the 2s complement of the number to be subtracted from the operand that currently occupies the accumulator. The timing signals generated by the controller during the subtraction instruction cycle proceed as with the instruction cycle for addition. However, the execution time for subtraction is five clock cycles rather than three, thus requiring a total for subtraction of seven clock cycles.

After the usual fetch action, the operation code of the third instruction commands a transfer of the contents of the accumulator to the memory register address 1111. Hence when the instruction decoder receives the 11 op-code signal, it causes AND gate $A3$ to conduct. The subsequent active level of FF_{12} and the attached inverter serve to put the accumulator contents on the bus and to place this value into the specified output memory location. The final instruction bears op code 00 that turns on AND gate $A4$, which stops the computer.

A comment is in order at this point concerning the use of mnemonic codes to simplify the programming of digital computers. The rightmost column of Table 14-1 shows arbitrarily selected mnemonic codes that are used to call into action the specific operations required of the elementary computer. As an illustration of the usefulness of the mnemonic code, consider the first entry in Table 14-1, which is ADD MR. Here MR refers to a memory register of the RAM and ADD is used to denote the addition of the contents of the specified MR to the accumulator. Hence, when a programmer uses the mnemonic statement, ADD 1010, there is called into activity appropriate controller logic circuitry that proceeds to take the contents of the memory register at address 1010 and adds it to the contents already residing in the accumulator. Programming that is accomplished with such mnemonic codes is called *assembly language programming*. The system of logic circuits that converts the mnemonic codes to machine language (i.e., binary digits) is called an *assembler*.

The controller of microcomputers is equipped with a read-only memory

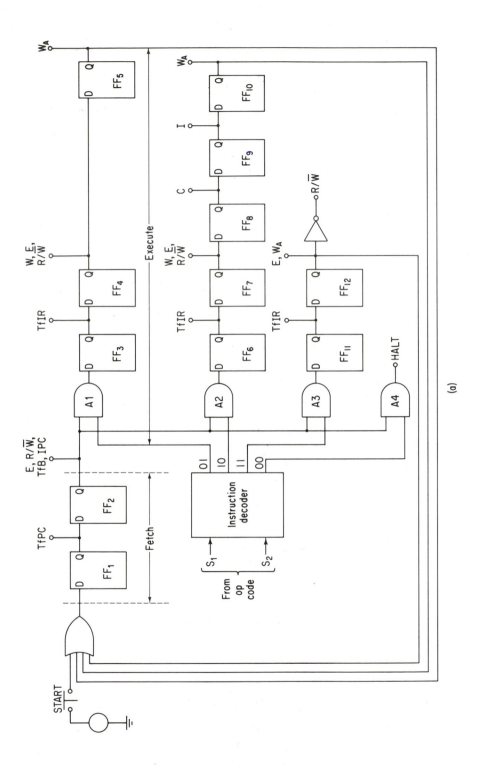

(a)

	Clock cycle	Enabled control lines	Operation in symbolic terms	Comment
Fetch	1	TfPC	PC → MAR	Select memory register (MR) for op code and operand address
Fetch	2	E, $R/\overline{W}$, TfB, IPC	MR → BUS → IR	Transfer contents of MR into instruction register (IR)
Execute	3	TfIR	IR(addr.) → MAR	Preparing to place operand on bus
Execute	4	W, E, $R/\overline{W}$	MAR → AU	Transfer operand from bus to arithmetic unit (AU)
Execute	5	W_A	Adder of AU → Acc	After addition, place result into accumulator

(b)

Fig. 14-2 The controller unit for the elementary computer of Fig. 14-1: (a) logic diagram; (b) control sequence for addition instruction cycle. (After H. Taub)

which provides the properly sequenced signals that paces the computer through the required steps to meet the needs of a program. When a computer is so driven, it is called a microprogrammed computer.

Microcomputer System. The combination of a microprocessor unit with a random-access memory furnishes the essential elements of a computing system. However, to make it a *practical* computing system, it is necessary to make provisions to allow people to communicate with the machine. This is achieved by adding a third section, called the input/output section. It consists of an input/output interface that is connected to appropriate input and output devices such as cathode-ray-tube terminals, teletypewriters, line printers, and the like. A block diagram of the arrangement is illustrated in Fig. 14-3.

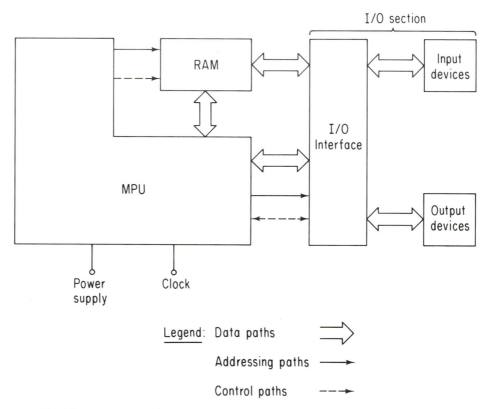

Fig. 14-3 Microcomputer equipped with input-output interface and associated peripheral devices.

14-2 MICROPROCESSORS

The essential elements of an elementary microprocessor are described in the preceding section. However, the practical forms of the microprocessor are far more powerful and sophisticated, and well they should be in view of the fact that they involve on a single IC chip many thousands of electronic components. In turn, these components are organized into hundreds of AND and OR gates and flip-flops in order to produce a very sophisticated controller unit, many special-purpose and general-purpose registers, multiplexers, decoders, and other necessary digital circuits. The 8080 microprocessor was the first commercially successful 8-bit unit. It was developed and manufactured by the Intel Corporation.† Two other widely received microprocessors evolved from the Intel 8080. These are the Motorola 6800 and the Zilog Z80 microprocessor. The Z80 unit was designed to execute 178 instructions, including the entire complement of 78 instructions of the 8080 units. Fortunately, in the interest of compatibility, the Z80 uses the same machine language for the instructions it shares with the Intel 8080. However, the Z80 does use different mnemonic codes to identify these instructions. Consequently, the assembler must be different. We choose in the remainder of this chapter to focus attention only on one of these microprocessors (the Intel 8080) because our objective is chiefly to convey a sense of how these units are organized, the degree of flexibility that is inherent in their instruction set and the procedure of assembly language programming based on the available instructions.

As already mentioned, the basic word length of the three cited microprocessors is one byte or 8 bits long. At this writing microprocessors that have basic word lengths of two bytes (16 bits) and four bytes (32 bits) are now available and in widespread use. The first personal computer to employ a microprocessor with a 32-bit word length was the Macintosh, which is manufactured by Apple, Inc. This computer is built around the Motorola 68000, a truly powerful microprocessor.

The basic structure of the Intel 8080 microprocessor is represented in block form in Fig. 14-4. It is equipped with an 8-bit distribution bus which provides the linkage between the various system subassemblies. There is a total of seven general-purpose registers including the accumulator (A). The accumulator is by far the most versatile of the group. When it is used in conjunction with the arithmetic logic unit (ALU), it becomes the seat of many general-purpose operations. These include addition and subtraction as well as the logical operations of AND, OR, X-OR, and NOT. Moreover, the use of shift registers permits bits to be shifted to the right and to the left, which serves to facilitate the execution of multiplication and division. The remaining six general-purpose registers (B, C, D, E, H, L) can be arranged to function as individual 8-bit registers or as paired registers capable of handling 16-bit words. The pairing order is B with C, D with E, and H with L. The last pair (H/L) is commonly used to provide the ALU with the address of the memory registers of the RAM. The program counter, as a 16-bit register, offers communication with as many as $2^{16} = 65,536$ addresses of the RAM. A *status register* is made part of the microprocessor in order to invest it with conditional control capability. The several flip-flop outputs of this register serve as *flags* that perform on an individual basis in response to operations

---

† Intel also developed the 8086 and 8088 microprocessors that succeeded the 8080. More recent additions of the 8086 family include the third generation units identified by the numbers 80186 and 80286.

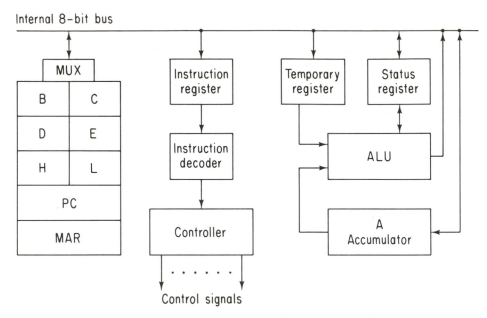

Fig. 14-4 Block diagram representation showing the principal elements and registers of the Intel 8080 microprocessor.

that take place in the ALU. They monitor the ALU output and the signal so generated can be used to provide alterations to the program. The primary flags are Z for zero, S for sign, CY for carry, and P for parity. Whenever the output of the ALU is zero, the Z flag goes to logic 1; otherwise, it is zero. Similarly, if the sign of the ALU output is negative, S is set to logic 1; otherwise, it is at 0. Moreover, if in the process of manipulating data, a carry or borrow occurs, the CY flip-flop goes to 1; otherwise, it is 0. The parity flag conveys information about whether the ALU output has odd or even parity, i.e., the number of 1s in a word is even or odd.

It is instructive and important to appreciate the fact that the great versatility of the microprocessor makes it an important element in the design of all kinds of digital systems, the computer being one outstanding example. In essence the microprocessor makes available a virtual arsenal of data processing tools that can be exploited to accomplish many different tasks. Among these tools are to be found such generally applicable procedures as adding, subtracting, shifting, up and down counting, delaying, combining words logically or numerically, ANDing, ORing, complementing, and negation. All sorts of combinations of these capabilities can be arranged to meet the objectives of a specific task. Although it is possible to meet the objectives of any specific task by designing a custom-made set of appropriate combinational and sequential circuits solely dedicated for the purpose, the use of a microcomputer, a RAM, and input-output devices offers a more attractive approach. It often means lower cost, less space, and a design technique that readily lends itself to modification with the least of inconvenience. In short, by making available hardware that provides generally applicable computing procedures, digital design is greatly simplified by developing the programming (software) to organize, coordinate, and manage those procedures to achieve desired results. This design strategy replaces the specifically dedicated hard-wired logical circuits (i.e., hardware) by a general-purpose device (i.e., the microprocessor) and suitable software.

The 8080 microprocessor instruction set is divided into three categories: the transfer instructions, the instructions for operation of the data, and conditional control instructions. The salient features of each of these groups is treated in the following sections.

14-3 TRANSFER INSTRUCTIONS

It is the purpose of the transfer instructions to move data between the various registers of the microprocessor, the associated random-access memory and the input-output devices without in any way altering the binary data that are transmitted. A listing of these instructions is displayed in Table 14-2.

Some comments are in order concerning the interpretation of the mnemonics that are used in this instruction set. But first let it be said that similar interpretations apply to the instruction set of other microprocessors so that familiarity with one microprocessor type is helpful in the use of the others. Whenever a mnemonic includes two registers separated by a comma, the register on the right of the comma is the source and the one on the left is the destination register. Thus the mnemonic MOV r1, r2 means transfer the contents of register r2 into register r1. The register r2 is the source register and may be any of the general-purpose registers A, B, C, D, E, H, or L, and the destination register r1 may also be any one of these same registers. More specifically, the instruction MOV A, B means that the contents of register B are to be transferred to the accumulator (register A). A symbolic representation of this statement is given in column 3 of the table. The column is helpful in clarifying the meaning of any instruction.

How does transfer take place when either the source or the destination register is to be a memory register of the RAM? Because a 64K RAM contains in excess of 65,000 addresses, it is necessary first to make provisions for assigning a 16-bit address to each memory register so that it may be accessed. The 8080 microprocessor accomplishes this result by dedicating the register pair H,L for this purpose. The 16-bit word that is placed in the H,L registers of the micro-processor is the address of the corresponding memory register of the RAM. The letter M is used to represent the *contents* of the memory location whose address appears as a 16-bit word in the H,L registers. A clearer notation would be to use M(H,L), where the quantity in parentheses serves to remind us that the memory register of interest has the address specified in the H,L registers. Thus an instruction such as MOV A,M means that the contents of the memory register at address given by H,L are to be transferred to the accumulator. The symbolic representation of this instruction is $M(H,L) \rightarrow A$. Of course, it should be understood that transfer instructions leave the source register unaltered.

TABLE 14-2 LIST OF TRANSFER INSTRUCTIONS OF THE 8080 MICROPROCESSOR

Item	Mnemonic	Description	Symbolic representation	Clock cycles
1	MOV r1, r2	Move register to register	$r2 \rightarrow r1$	5
2	MOV M, r	Move register to memory	$r \rightarrow M(H,L)$	7
3	MOV r, M	Move memory to register	$M(H,L) \rightarrow r$	7
4	MVI r $\langle B_2 \rangle$	Move immediate to register	$\langle B_2 \rangle \rightarrow r$	7
5	MVI M $\langle B_2 \rangle$	Move immediate to memory	$\langle B_2 \rangle \rightarrow M(H,L)$	10
6	LDA $\langle B_2 \rangle$ $\langle B_3 \rangle$	Load A direct	$M(B_3B_2) \rightarrow A$	13
7	STA $\langle B_2 \rangle$ $\langle B_3 \rangle$	Store A direct	$A \rightarrow M(B_3B_2)$	13
8	LDAX B	Load A indirect	$M(B) \rightarrow A$	7
9	LDAX D	Load A indirect	$M(D) \rightarrow A$	7
10	STAX B	Store A indirect	$A \rightarrow M(B)$	7
11	STAX D	Store A indirect	$A \rightarrow M(D)$	7
12	LXI B $\langle B_2 \rangle$ $\langle B_3 \rangle$	Load immediate register pair B and C	$\langle B_2 \rangle \rightarrow C; \langle B_3 \rangle \rightarrow B$	10
13	LXI D $\langle B_2 \rangle$ $\langle B_3 \rangle$	Load immediate register pair D and E	$\langle B_2 \rangle \rightarrow E; \langle B_3 \rangle \rightarrow D$	10
14	LXI H $\langle B_2 \rangle$ $\langle B_3 \rangle$	Load immediate register pair H and L	$\langle B_2 \rangle \rightarrow L; \langle B_3 \rangle \rightarrow H$	10
15	LHLD $\langle B_2 \rangle$ $\langle B_3 \rangle$	Load H and L direct	$M(B_3B_2) \rightarrow L$ $M(B_3B_2 + 1) \rightarrow H$	16
16	SHLD $\langle B_2 \rangle$ $\langle B_3 \rangle$	Store H and L direct	$L \rightarrow M(B_3B_2)$ $H \rightarrow M(B_3B_2 + 1)$	16
17	XCHG	Exchange D and E, H and L registers	$H \leftrightarrow D; L \leftrightarrow E$	4

The first three entries of Table 14-2 are representative of one-byte instructions. It is helpful to keep in mind that the mnemonic is simply a memory aid for human beings, which in the microprocessor has a specific 8-bit binary pattern. The mnemonic is a simple way of calling this 8-bit word into action for it tells the assembler which machine language instruction to generate. However, the 8080 μP† also has provisions for specifying two-byte and three-byte words. Entry 4 of the table is an example of a two-byte word. As usual, the first byte is the

† Shorthand for microprocessor.

operational part of the instruction. However, because this instruction allows for moving *data next in order* into the specified register, the second byte (denoted by $\langle B_2 \rangle$) is the binary form of these data. Accordingly, the instruction MVI C, 0101 1011 places the number 91 directly into register C. The symbolic representation of this procedure is $\langle B_2 \rangle \rightarrow$ C. When it is desired to place an 8-bit word immediately into a memory register of the RAM, the instruction on line 5 of Table 14-2 is used. Thus MVI M, $\langle B_2 \rangle$ means that the information contained in byte B_2 is to be transferred into the memory register of the RAM whose address is specified in the H,L registers of the μP.

An example of a three-byte instruction is illustrated by entry 6 of the table. Once again the first byte is the operational part of the instruction, which is a command to load the accumulator directly (LDA). The second and third bytes (B_2, B_3) identify the 16-bit address of a memory register of the RAM the contents of which are to be placed into the accumulator. The symbolic representation is $M(B_3 B_2) \rightarrow$ A. The use of M( ) means that a memory register of the RAM is involved and the quantity in the parentheses designates its address. Therefore, this instruction can be used to place the contents of any memory register of the RAM directly into the accumulator. Entry 7 of the table is a similar instruction with the difference that the flow of information is reversed.

Instructions 8 and 9 of Table 14-2 furnish the means by which the contents of register pairs B and C as well as D and E represent the addresses of memory registers of the RAM, the contents of which are to be loaded into the accumulator. This is an indirect loading procedure because the B and D register pairs of the μP furnish the addresses of the memory and not the data itself that is to be entered into the accumulator. The symbolic representation helps to remove any ambiguity for $M(B) \rightarrow$ A means that the contents of the main memory with address in register pair B and C are to be transferred to the accumulator. Entries 10 and 11 are similar except for a reversal of the flow of information.

Instructions 12 to 14 are further examples of three-byte instructions. These offer the designer the option of loading the general-purpose register pairs B,C and D,E and H,L directly with the data represented by the two-byte word $B_3 B_2$, where B_3 denotes the higher-order byte. It is useful to note too that the general-purpose register pairs of the μP are involved whenever the uppercase letter X is used in the instruction mnemonic.

Instruction 15 allows the H,L register pair to be loaded directly with the contents appearing at two consecutive 8-bit memory registers of the RAM. The address of the first memory register is specified by the second and third bytes, $B_3 B_2$, and the contents at this address are transferred to the L register. Then the contents of the memory register at address $(B_3 B_2 + 1)$ are moved to register H. Symbolically, we can write simply that $M(B_3 B_2) \rightarrow$ L and $M(B_3 B_2 + 1) \rightarrow$ H. Entry 16 allows the programmer to place the contents of the L and H registers at any two consecutive memory registers of the RAM. It is the reverse of entry 15.

Finally, entry 17 is a one-byte instruction mnemonic that permits the information contained in register pairs D,E and H,L to be interchanged.

14-4 OPERATION INSTRUCTIONS

These instructions are designed to perform such operations as addition, subtraction, shifting, incrementing, decrementing, logic, complementing, and comparisons on the data that are stored in the general-purpose registers of the microprocessor as well as the words stored in the memory registers of the RAM. An abridged listing of these instructions is displayed in Table 14-3. An explanation once again is in order regarding the interpretation of these instruction mnemonics.

First it should be understood that in the interest of simplifying the assembly language programming one of the operands in an arithmetical or logical operation is to be found in the accumulator. Moreover, it is also understood that the result is to reside in the accumulator to await further processing. The first group of operation instruction in the table has reference to addition. Observe that the mnemonic is written with the implication that one of the operands is already present in the accumulator. Thus the instruction ADD r states that the contents of register r are to be added to the quantity that is already in the accumulator. The symbolic representation of this instruction is A plus r $\rightarrow$ A, which is to be read as the contents of register A plus those of register r *replaces* ($\rightarrow$) the current contents of A. The arrow is used to denote the action of replacement. Note too that we have avoided using the $+$ sign to denote addition in order not to confuse it with the logical operation AND. Entry 2 indicates that data may be added directly to the contents of the accumulator by using a two-byte instruction. Furthermore, entry 5 discloses that the contents of a memory register of the RAM can also be added to the accumulator provided that the address of the given memory register is present in the H,L register pair of the microprocessor. Instructions 3, 4, and 6 make provisions for performing addition with carry. The instructions for subtraction listed in entries 10 to 15 are similar to those for addition. However, now a difference is generated along with a borrow to replace the carry when necessary.

The shift instructions are listed as items 16 to 19. The instruction mnemonic RLC causes the contents of the accumulator to shift one bit to the left so that A_0 goes to position A_1, A_1 to A_2, etc., until A_7 is reached. This most significant bit is then placed at position A_0 as well as in the carry bit CY of the status register. No other flags are affected by this instruction. In contrast, the most significant bit of the RAL instruction is treated differently. Although the bits are also shifted to the left, in this case the last bit, A_7, is shifted to the carry position only and the carry bit is then put into the lowest significant bit position of the accumulator A_0.

Entries 20 to 29 describe the instruction mnemonics that are used to increment or decrement the contents of a register, a register pair, or a memory register of the RAM. Again note that to accomplish this result for the contents of a memory register, it is necessary to specify its address in the H,L register pair of the microprocessor. Observe too that the presence of X in the mnemonic means the involvement of the register pairs (except for XOR-type operations).

The logic operations of AND, OR, and Exclusive-OR appear in items 30 to 38. To perform an immediate AND, OR, or Exclusive-OR with the contents of the accumulator, it is necessary to employ a two-byte instruction word. The AND, OR, and SHIFT instructions, among other things, can prove to be very useful in manipulating numbers in the accumulator.

There are two instructions to do complementing. The instruction on line 39 of Table 14-3 complements the contents of the accumulator, while that on line 40 is used to complement the carry bit of the status register.

14-5 CONTROL INSTRUCTIONS

Programming flexibility and versatility can be increased considerably by monitoring key variables. Then, upon reaching some specified condition, the microprocessor can be made to issue a command that changes the program sequence. Table 14-4 displays a partial list of such commands. The first group provides for a comparison of the contents of a register, a memory location, or a directly addressable byte with the contents of the accumulator. Although the CMP instruction performs subtraction, the result of this operation is not registered anywhere except in the status register. Hence the Z flag of the status register is set to 1 for a result in the ALU that is zero. If the result of subtraction is not zero, the Z flag is cleared to 0. Depending on the nature of the program's objective, the sequence of the subsequent steps of the program can be accordingly altered.

A common method of changing the program sequence in response to a conditional control signal is to invoke the jump instructions which are listed as items 4 to 9 in Table 14-4. We give attention here only to three of the five flags of the status register, namely, the zero (Z) flag, the carry flag (CY), and the sign (S) flag. Since each flag cited can assume one of two binary states, there is a total of six conditional jump instructions. As a glance at the table quickly discloses, each jump instruction consists of three bytes. The first byte is for the instruction and the remaining two bytes provide the address to be entered into the program counter. A detour to a new address effectively changes the progam sequence. Thus, by instruction 4, if the Z flag of the status register is set, i.e., $Z = 1$, the program will be made to continue with the program counter pointing at the new address, B_3B_2, which is the one supplied with the instruction. On the other hand, if the current program instruction is JNZ, i.e., jump on no zero, the program counter responds by moving to a new address if the zero flag is zero. A similar interpretation is applicable for the sign and carry flags. The explanations are self-evident from the symbolic representations of the instruction mnemonics.

The last entry of Table 14-4 is the unconditional jump, JMP. Once this instruction is reached in a program and is executed, the very next instruction to be executed is the one found at the address specified by $\langle B_3B_2 \rangle$.

The repertory of the 8080 instructions also includes Call and Return commands, which are activated conditionally in the same manner as for the jump instructions. Often, a Call instruction is invoked at a particular point in the main program in order to enter a subroutine for the purpose of procuring results that are needed for the main program to continue. The Return instructions serve to permit the

TABLE 14-3 ABRIDGED LIST OF OPERATION INSTRUCTIONS OF THE 8080 μP

Item	Mnemonic	Description	Symbolic representation	Clock cycles
Addition				
1	ADD r	Add register to A	A plus r $\rightarrow$ A	4
2	ADI $\langle B_2 \rangle$	Add immediate to A	A plus $\langle B_2 \rangle \rightarrow$ A	7
3	ADC r	Add register to A with carry	A plus r plus CY $\rightarrow$ A	4
4	ACI $\langle B_2 \rangle$	Add immediate to A with carry	A plus $\langle B_2 \rangle$ plus CY $\rightarrow$ A	7
5	ADD M	Add memory to A	A plus M(H,L) $\rightarrow$ A	7
6	ADC M	Add memory to A with carry	A plus M(H,L) plus CY $\rightarrow$ A	7
7	DAD B	Add B and C to H and L	B,C plus H,L $\rightarrow$ H,L	10
8	DAD D	Add D and E to H and L	D,E plus H,L $\rightarrow$ H,L	10
9	DAD H	Add H and L to H and L	H,L plus H,L $\rightarrow$ H,L	10
Subtraction				
10	SUB r	Subtract register from A	A $-$ r $\rightarrow$ A	4
11	SUI $\langle B_2 \rangle$	Subtract immediate from A	A $-$ $\langle B_2 \rangle \rightarrow$ A	7
12	SBB r	Subtract register from A with borrow	A $-$ r $-$ CY $\rightarrow$ A	4
13	SBI $\langle B_2 \rangle$	Subtract immediate from A with borrow	A $-$ $\langle B_2 \rangle$ $-$ CY $\rightarrow$ A	7
14	SUB M	Subtract memory from A	A $-$ M(H,L) $\rightarrow$ A	7
15	SBB M	Subtract memory from A with borrow	A $-$ M(H,L) $-$ CY $\rightarrow$ A	7
Shifting				
16	RLC	Rotate A left	$A_n \rightarrow A_{n+1}$; $A_7 \rightarrow A_0$; $A_7 \rightarrow CY$	4
17	RRC	Rotate A right	$A_n \rightarrow A_{n-1}$; $A_0 \rightarrow A_7$; $A_0 \rightarrow CY$	4
18	RAL	Rotate A left through carry	$A_n \rightarrow A_{n+1}$; $A_7 \rightarrow CY$; $CY \rightarrow A_0$	4
19	RAR	Rotate A right through carry	$A_n \rightarrow A_{n-1}$; $A_0 \rightarrow CY$; $CY \rightarrow A_7$	4

Incrementing/decrementing

20	INR r	Increment register	$r + 1 \to r$	5
21	INR M	Increment memory	$M(H,L) + 1 \to M(H,L)$	10
22	DCR r	Decrement register	$r - 1 \to r$	5
23	DCR M	Decrement memory	$M(H,L) - 1 \to M(H,L)$	10
24	INX B	Increment B and C registers	$BC + 1 \to BC$	5
25	INX D	Increment D and E registers	$DE + 1 \to DE$	5
26	INX H	Increment H and L registers	$HL + 1 \to HL$	5
27	DCX B	Decrement B and C registers	$BC - 1 \to BC$	5
28	DCX D	Decrement D and E registers	$DE - 1 \to DE$	5
29	DCX H	Decrement H and L registers	$HL - 1 \to HL$	5

Logic

30	ANA r	AND register with A	$A \cdot r \to A$	4
31	ANI $\langle B_2 \rangle$	AND immediate with A	$A \cdot \langle B_2 \rangle \to A$	7
32	ANA M	AND memory with A	$A \cdot M(H,L) \to A$	7
33	ORA r	OR register with A	$A + r \to A$	4
34	ORI $\langle B_2 \rangle$	OR immediate with A	$A + \langle B_2 \rangle \to A$	7
35	XRA r	Exclusive-OR register with A	$A \oplus r \to A$	4
36	XRI $\langle B_2 \rangle$	Exclusive-OR immediate with A	$A \oplus \langle B_2 \rangle \to A$	7
37	ORA M	OR memory with A	$A + M(H,L) \to A$	7
38	XRA M	Exclusive-OR memory with A	$A \oplus M(H,L) \to A$	7

Complementing/setting

39	CMA	Complement A	$\overline{A} \to A$	4
40	CMC	Complement carry	$\overline{CY} \to CY$	4
41	STC	Set carry	$1 \to CY$	4

TABLE 14-4 PARTIAL LIST OF CONTROL INSTRUCTIONS OF THE 8080 μP

Item	Mnemonic	Description	Symbolic representation	Clock cycles
Comparisons				
1	CMP r	Compare register with A	$A - r$	4
2	CMP M	Compare memory with A	$A - M(H,L)$	7
3	CPI $\langle B_2 \rangle$	Compare immediate with A	$A - \langle B_2 \rangle$	7
Jumps				
4	JZ $\langle B_2 \rangle$ $\langle B_3 \rangle$	Jump on zero	If $Z = 1$, $\langle B_3 B_2 \rangle \rightarrow PC$	10
5	JNZ $\langle B_2 \rangle$ $\langle B_3 \rangle$	Jump on no zero	If $Z = 0$, $\langle B_3 B_2 \rangle \rightarrow PC$	10
6	JC $\langle B_2 \rangle$ $\langle B_3 \rangle$	Jump on carry	If $CY = 1$, $\langle B_3 B_2 \rangle \rightarrow PC$	10
7	JNC $\langle B_2 \rangle$ $\langle B_3 \rangle$	Jump on no carry	If $CY = 0$, $\langle B_3 B_2 \rangle \rightarrow PC$	10
8	JP $\langle B_2 \rangle$ $\langle B_3 \rangle$	Jump on positive	If $S = 0$, $\langle B_3 B_2 \rangle \rightarrow PC$	10
9	JM $\langle B_2 \rangle$ $\langle B_3 \rangle$	Jump on minus	If $S = 1$, $\langle B_3 B_2 \rangle \rightarrow PC$	10
10	JMP $\langle B_2 \rangle$ $\langle B_3 \rangle$	Jump unconditional	$\langle B_3 B_2 \rangle \rightarrow PC$	10

program to revert to the place of interruption. The details of these instructions are not treated here because they are more properly dealt with in a follow-up course that is specifically devoted to these topics.

14-6 ASSEMBLY LANGUAGE PROGRAMMING: EXAMPLES

Our interest in this section is to provide a few examples that serve to convey a sense of how the 8080 instruction set can be used to manipulate, manage, and process information.

EXAMPLE 14-1   Write a program in assembly language that adds the decimal number 25 to the contents of the memory register whose address is 163. The new result is to reside at the same address.

Solution: The operating strategy is to transfer the contents of the memory register at address 163 to the accumulator and then to issue an *add immediate* instruction in order to get the new contents. The program follows.

Step	Mnemonic	Explanation	Clock cycles
1	LXI H ⟨1010 0011⟩ ⟨0000 0000⟩	The address $163_{10} = 1010\ 0011_2$ is placed in register pair, H,L	10
2	MOV A, M	The contents at memory location 163 are transferred to the A	7
3	ADI ⟨0001 1001⟩	This instruction causes $25_{10} = 0001\ 1001_2$ to be added to the contents of A	7
4	MOV M, A	The contents of the Acc are transferred to the memory register whose address appears in register pair H,L	7

Observe that before it is possible to transfer the contents of a memory register of the RAM to the accumulator the address of that register is placed in the H,L register pair. This is the purpose of the first instruction. Note too that the total number of clock cycles that is required to perform the steps of this instruction is 31. Accordingly, if the basic clock frequency is $f = 2$ MHz $= 2 \times 10^6$ Hz, the time for each clock cycle is

$$T = \frac{1}{f} = \frac{1}{2 \times 10^6} = 0.5 \times 10^{-6} = 0.5\ \mu s$$

Hence the total time that elapses in executing the steps of this program is

$$t_p = 31T = 15.5\ \mu s$$

EXAMPLE 14-2 Identify two instructions that serve to clear the accumulator.

Solution: One method is to invoke the instruction mnemonic for subtraction. Thus SUB A subtracts the contents of the accumulator from itself thus yielding zero.

Another procedure is to issue the two-byte command ANI ⟨0000 0000⟩. Clearly, by ANDing the contents of the accumulator immediately with the second byte of the instruction mnemonic all the bits of which are zero, the contents of the accumulator are cleared.

EXAMPLE 14-3 Repeat Example 14-1 with the modification that the sum is to be complemented and returned to address 164.

Solution: The first three steps are identical to those of Example 14-1. Two intervening steps are then required. One is for complementing the contents of the accumulator; the other is for incrementing the contents of the H,L register pair in order to establish the new address. The complete program follows.

Step	Mnemonic	Explanation	Clock cycles
1 2 3	as in Example 14-1		
4	CMA	This instruction complements the contents of the A	4
5	INX H	The contents of register pair H,L are increased by 1; thus the new address is 164	5
6	MOV M,A	The contents of the A are transferred to M(HL)	7

The total time needed to execute the six steps of this program for a 2-MHz clock is $40/2 = 20\ \mu s$.

EXAMPLE 14-4 Write a program that performs the following data processing: The contents at memory location 163 to be shifted to the left by one bit, the rightmost bit to be left cleared, and the result returned to memory location 164.

Solution: First it is necessary to transfer the contents of memory location 163 to the accumulator. This is done by issuing an immediate load command to register pair H,L. The contents are then moved to the accumulator, following which a rotate left command (RLC) is performed. To ensure that the rightmost bit is at 0, the accumulator is next ANDed with $\langle 1111\ 1110\rangle$ before being returned to the new memory location. The list of program steps follows together with the symbolic explanations of the instructions.

Step	Mnemonic	Symbolic representation
1	LXI H $\langle 1010\ \ 0011\rangle$ $\langle 0000\ \ 0000\rangle$	$\overbrace{\langle 0000\ 0000}^{B_3}\ \ \overbrace{1010\ 0011\rangle}^{B_2} \to$ H,L
2	MOV A,M	$M(H,L) \to A$
3	RLC	$A_n \to A_{n+1};\ A_7 \to A_0;\ A_7 \to CY$
4	ANI $\langle 1111\ \ 1110\rangle$	$A \cdot \langle 1111\ 1110\rangle \to A$
5	INX H	$H,L + 1 \to$ H,L
6	MOV M,A	$A \to M(HL)$

What arithmetic operation is performed by this routine?

EXAMPLE 14-5 Write a program that serves to clear the contents of the memory registers at locations 163 and 164.

Solution: First load the address 163 into the H,L registers. Next, the accumulator is cleared. The cleared accumulator is then transferred to address 163. Finally, following an increment command on the H,L register pair a second transfer command of the cleared accumulator is issued. The program follows.

Step	Mnemonic	Symbolic representation
1	LXI H $\langle 1010\ 0011\rangle$ $\langle 0000\ 0000\rangle$	$\langle B_3 B_2\rangle \to$ H,L
2	SUB A	$A\text{-}A \to A$
3	MOV M,A	$0 \to M(H,L)$
4	INX H	$HL + 1 \to HL$
5	MOV M,A	$0 \to M(H,L)$

EXAMPLE 14-6 The purpose of this example is to illustrate the use of the *compare* and *jump* mnemonics of the 8080 instruction set.

Assume that two unsigned numbers, N_1 and N_2, are present in memory locations with addresses 163 and 164, respectively. Write a program that takes the smaller of these numbers and puts it at address 165.

Solution: A common destination address that is associated with the jump instruction mnemonic is the "symbolic address." In the 8080 microprocessor (as well as the Z80) this symbolic address is called LABEL and it simply refers to an instruction to be found in a memory register of the RAM. The program that executes the assigned task reads as follows. A column with clarifying comments is also included.

Step	Label	Mnemonic	Symbolic representation	Comment
1		LXI H $\langle$1010 0011$\rangle$ $\langle$0000 0000$\rangle$	$\langle B_3B_2 \rangle \rightarrow$ H,L	Address 163 is put into the H,L register pair
2		MOV A,M	M(163) $\rightarrow$ A	N_1 is put into the A
3		INX H	H,L + 1 $\rightarrow$ H,L	Address 164 is in H,L register
4		CMP M	A-M(H,L)	Forms the difference: $N_1 - N_2$
5		JC END	If CY = 1, jump to END	If carry flag sets (CY = 1), it means that N_1 is the smaller number
			If CY = 0, continue with next program step	Flag CY = 0 means N_1 is the larger number; hence N_2 must be moved to the A
6		MOV A,M	M(164) $\rightarrow$ A	N_2 now occupies the A
7	END	INX H	HL + 2 $\rightarrow$ HL + 1	Address 165 is in H,L register
8		MOV M,A	A $\rightarrow$ M(165)	The smaller number occupies memory location 165

Step 5 in the program, which is *jump* on *carry* (JC), is the crucial one. Keep in mind that N_1 is in the accumulator and N_2 is at memory location 164 at this juncture. When the difference is formed by the *compare* instruction (CMP), the carry flag (CY) is set if $N_1 < N_2$ because the process of subtraction makes it necessary for a borrow to occur. Accordingly, CY = 1 establishes that N_1 is the smaller number and so the program next proceeds to place N_1, which is already in the accumulator, to memory location 165 by first incrementing the H,L register pairs and then issuing a move command. The status flag, CY = 1, thus jumps the program from step 5 to steps 7 and 8, which in turn act to put N_1 into memory location 165. On the other hand, if the carry status flag, following comparison, is zero, then it follows that N_2 is the smaller of the two numbers. Consequently, it is the purpose of step 6 to transfer the contents of memory location 164, which now holds N_2, into the accumulator in preparation for transferral to address 165. Steps 7 and 8 then take care of the transfer.

We conclude this chapter with a final remark. It should be apparent from the foregoing examples that the manipulation and processing of data in assembly language can become a laborious task because of the attention to detail that is necessary at each step. However, it is possible to encode a program for specific applications by resorting to high-level languages such as Basic, Fortran, or Pascal.

Here easily understood commands are used to invoke quite complex machine language operations. But the execution of the many operations that lead from these simple instructions of the high-level languages to the corresponding machine language formulations must be encoded in a dedicated program for this purpose. Such a program is called a *compiler*. Assembly language programming is by far more efficient in the use of memory space and faster as well. Of course, as the capacity of memory units increases and their cost diminishes, the importance of this distinction loses urgency.

Summary review questions

1. List the chief components of a digital computer.
2. What is a microprocessor? What are its principal components? How does it differ from a digital computer?
3. Describe in some detail the important role of the controller unit of the microprocessor.
4. The program counter (PC) is a constituent part of the microprocessor. What is its function?
5. Another important component of the microprocessor is the instruction register (IR). What important functions does this digital component serve in the operation of the microprocessor?
6. Describe the bit contents of the instruction register, specifically its double function and the length of its word in bits.
7. How is the number of bits that is required for each memory register of the RAM determined?
8. What is the composition and function of the arithmetic unit that is used in the microprocessor of the elementary computer of Fig. 14-1? In what respects does this unit differ from the corresponding unit of commercial-type microprocessors?
9. Describe the manner in which the RAM is customarily organized. Why is it advantageous to place all the program instructions, properly ordered, into one specific part of the random-access memory?
10. Describe what is meant by the stored-program solution of a problem.
11. Explain the meaning of an operand.
12. Describe in proper sequence the role of the following devices in the addition process of two binary numbers: the program counter, the memory address register, the RAM, the instruction register, the controller unit, and the accumulator.
13. What additional digital devices need to be included in answer to Question 12 when the process involves subtraction?
14. Distinguish between the fetch and execute cycles of an operation.
15. List the kinds of digital devices that make up the controller unit of the elementary computer of Fig. 14-1.
16. Describe the mechanism that is used in the elementary computer of Fig. 14-1 to transfer the information that resides at a particular memory register (MR) to the instruction register (IR).

17. Explain how the controller unit of the elementary computer of Fig. 14-1 permits transition to take place from one instruction to the next instruction in the program.

18. Describe how the 2s-complement of a number is produced in the elementary computer of Fig. 14-1 in order to do subtraction. What effect does this have on the number of clock cycles required to perform subtraction compared to addition?

19. Describe how the transfer command is executed in the elementary computer of Fig. 14-1. How is the memory location of the result of an operation determined?

20. What is the role of the use of mnemonic codes in the programming process? How are the mnemonic codes selected?

21. Describe and illustrate what is meant by assembly language programming.

22. What is the purpose of an assembler?

23. What is the meaning of the phrase *microprogrammed computer*?

24. Name three commercially available microprocessor units which use 8-bit words.

25. How many general-purpose registers are included in the Intel 8080 microprocessor? Identify them and state their functions.

26. Describe the purpose of the status register of a microprocessor.

27. What is a flag? Name four of the five flags of the 8080 μP.

28. Comment about the microprocessor as a general-purpose processing tool. Illustrate.

29. Describe what is meant by the phrase *wiring by software*.

30. Describe in general but inclusive terms the objective of the transfer instructions of a commercial-type microprocessor. Give one or two illustrations.

31. In the Intel 8080 microprocessor distinguish between a regular transfer instruction and one that is identified by the word immediate.

32. When the information that resides in the accumulator of the 8080 μP is to be stored in a memory register of the RAM, where is the address for this register specified?

33. Write the mnemonic which is used in the 8080 to denote the movement of the contents of register X into register Y.

34. Describe the meaning of the transfer instruction MOV A,M for the 8080 microprocessor.

35. Some instructions in the 8080 instruction set have two or three bytes. Explain the meaning of each byte for the two-byte case. Repeat for the three-byte case.

36. Write the mnemonic instruction which is used in the 8080 μP to move a specific 8-bit word into a specific memory location of the RAM.

37. Name the mnemonic of the 8080 instruction set that allows the transfer of the contents of any memory register of the RAM into the accumulator.

38. What useful information is conveyed whenever a mnemonic instruction (other than logic) of the 8080 μP involves the letter X?

39. State the purpose of the operation instructions of a commercial-type microprocessor.

40. Explain the meaning of the operation instruction of the 8080 μP which is expressed by the mnemonic ADD M.

41. Distinguish between the two operation mnemonics ADD r and ADI $\langle B_2 \rangle$.

42. Describe the operation that is represented by the mnemonic RRC. By RAR.

43. State the mnemonic that is used in the 8080 μP to increment the contents of a specific memory register. Explain it.

44. State the mnemonic that is used to AND the contents of a memory register with the contents of the accumulator.

45. State the two mnemonics of the 8080 μP that are used to complement binary data.

46. Explain the meaning of the mnemonics CPI $\langle B_2 \rangle$ and CMP M.

47. Explain the relationship between the CMP instruction and the Z flag of the status register.

48. Describe the provisions that are made in the 8080 μP for altering the progress of a program.

49. How many bytes are there for each jump instruction of the 8080 μP? Explain the purpose of each byte.

50. Differentiate between conditional and unconditional jumps.

51. What is the purpose of a compiler?

Problems

GROUP I

(N.B.: Because of the descriptive nature of this chapter, additional problem assignments can properly be made from the Summary Review Questions.)

14-1. Write a program in the manner outlined in Table 14-1 which causes the accumulator of the simple computer of Fig. 14-1 to be cleared.

14-2. Write a stored program for the computer of Fig. 14-1 which performs a step-by-step solution of the following operation: $S = 2 + 7 - 4 + 1$.

14-3. Write a stored program for the computer of Fig. 14-1 which performs a solution to $S = A(B - C)$, where specifically $A = 2$, $B = 5$, and $C = 3$.

14-4. Examine Figs. 14-1 and 14-2 and describe how a program can be written that serves to clear the C/I register.

The problems listed below are to be answered for a microcomputer that uses the instruction set of the 8080 microprocessor as summarized in Tables 14-2, 14-3, and 14-4. Assume a 64K RAM.

14-5. Write an assembly language program that serves to complement the contents of memory register 206 and then places the result in memory register 207.

14-6. A program is to be written which places the most significant 4 bits of the contents of memory register at address 25 into address location 26. The least significant bits are to be cleared.

14-7. Write a program that adds the contents of register pair B to the contents of register pair D and then saves the result in register pair B.

14-8. Using assembly language write a program that takes the contents of memory location 35 and replaces it with the negative version of the same number.

14-9. A 16-bit number has its more significant byte located at memory register 81 and its less significant byte at memory location 82. Write a program to place the 2s complement of this number into memory locations 83 and 84 with location 84 holding the less significant byte.

14-10. A series of numbers appears in consecutively numbered memory locations starting with 102. Information about the length of the series appears in memory location

101. Write a program that generates the sum of this series of numbers and then places the sum in memory location 100. For simplicity assume that the sum does not exceed 8 bits.

14-11. There are four numbers residing in four consecutive memory registers beginning with location 12. Write a program for the sum of these numbers and place the result in memory register 16.

14-12. Ten numbers are stored in memory registers 11 through 20. Write a program that examines these numbers and records the number of positive and negative numbers in register pair D,E, respectively.

GROUP II

14-13. Assume that the accumulator of the computer of Fig. 14-1 is equipped with an output terminal that signals when the accumulator content is zero. If it is desirable to halt the computer whenever this condition occurs, show how this output signal can be used by the controller of Fig. 14-2 to effect this result.

14-14. Two two-byte numbers are expressed as $N_1 = N_1(m)N_1(l)$ and $N_2 = N_2(m)N_2(l)$, where $N(m)$ denotes the *more* significant byte and $N(l)$ represents the *less* significant byte. Write a program showing how these two-byte numbers can be added.

14-15. Repeat Example 14-6 for the case where the *larger* of the two numbers is to be placed into memory register 165.

14-16. Repeat Prob. 14-10 for the case where the generated sum exceeds the capacity of 8 bits, thereby making it necessary to use two bytes to accommodate the sum.

chapter fifteen

Magnetic Theory and Circuits

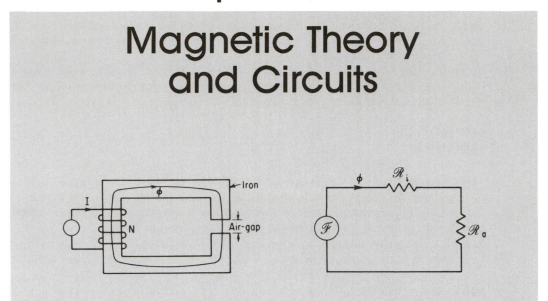

An understanding of electromagnetism is essential to the study of electrical engineering because it is the key to the operation of a great part of the electrical apparatus found in industry as well as the home. All electric motors and generators, ranging in size from the fractional horsepower units found in home appliances to the 25,000-hp giants used in some industries, depend upon the electromagnetic field as the coupling device permitting interchange of energy between an electrical system and a mechanical system and vice versa. Similarly, static transformers provide the means for converting energy from one electrical system to another through the medium of a magnetic field. Transformers are to be found in such varied applications as radio and television receivers and electrical power distribution circuits. Other important devices—for example, circuit breakers, automatic switches, and relays—require the presence of a confined magnetic field for their proper operation. It is the purpose of this chapter to provide the reader with background so that he can identify a magnetic field and its salient characteristics and more readily understand the function of the magnetic field in electrical equipment.

As has been previously pointed out, the science of electrical engineering is founded on a few fundamental laws derived from basic experiments. In the area of electromagnetics, it is Ampère's law that concerns us, and, in fact, serves as the starting point of our treatment.† On the basis of the results obtained by

† It should not be here inferred that this is the only starting point in developing a quantitative

Ampère in 1820, in his experiments on the forces existing between two current-carrying conductors, such quantities as magnetic flux density, magnetic field intensity, permeability, and magnetic flux are readily defined. Once this base is established, attention is then directed to a discussion of the magnetic properties of certain useful engineering materials as well as to the idea of a "magnetic circuit" to help simplify the computations involved in analyzing magnetic devices.

15-1 AMPÈRE'S LAW—DEFINITION OF MAGNETIC QUANTITIES

Appearing in Fig. 15-1 is a simplified modification of Ampère's experiment. The configuration consists of a very long conductor 1 carrying a constant-magnitude current I_1 and an elemental conductor of length l carrying a constant-magnitude current I_2 in a direction opposite to I_1. When taken together the elemental conductor and the current I_2 constitute a *current element* $I_2 l$. The elemental conductor 2 is actually part of a closed circuit in which I_2 flows, but for simplicity and convenience the details of the circuit are omitted except for the length l. Moreover, it is assumed that conductors 1 and 2 lie in the same horizontal plane and are parallel to each other. In accordance with Ampère's law it is found that with this configuration there exists a force on the elemental conductor directed to the right. Furthermore, the magnitude of the force is found to be directly proportional to I_1, I_2, l, and the medium surrounding the conductors as well as inversely proportional to the distance between them. In terms of MKS units the magnitude of this force can be shown to be given by

$$F = \frac{\mu I_1}{2\pi r} I_2 l \quad \text{N} \tag{15-1}$$

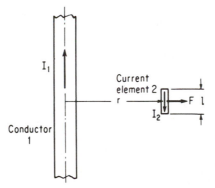

Fig. 15-1 Illustrating the force existing between a current element $I_2 l$ and a very long conductor carrying current, I_1 as described by Ampère's law.

theory of the magnetic circuit. Faraday's law of induction is equally valid as a starting point, and is preferred when the goal is the development of an electromagnetic wave theory rather than a theory leading to the treatment of electromechanical energy conversion.

where I_1 and I_2 are expressed in amperes, l and r in meters, and μ is a property of the medium. A further interesting revelation about this experiment is that if the elemental conductor 2 is used as an exploring device to find those points in space where the force is of constant magnitude and outwardly directed, the locus is found to be a circle of radius r and centered along the axis of conductor 1. In other words it is possible to identify a field having constant *lines of force*. In this connection it is useful at this point to rewrite Eq. (15-1) as follows:

$$F = I_2 l B \tag{15-2}$$

where

$$B \equiv \frac{\mu I_1}{2\pi r} \tag{15-3}$$

From Eq. (15-2) it is obvious that the defined quantity B carries the units of force per current-length. In rationalized MKS units this is expressed as

$$B \propto \frac{\text{newtons}}{\text{ampere} \times \text{meter}} \tag{15-4}$$

where $\propto$ denotes "is proportional to." A closer examination of this relationship reveals an interesting and very useful aspect of the significance of B. Recalling that

$$\text{newton} = \frac{\text{energy}}{\text{length}} = \frac{\text{watt-seconds}}{\text{meter}} = \frac{\text{volts} \times \text{amperes} \times \text{seconds}}{\text{meter}} \tag{15-5}$$

and inserting into expression (15-4) allows B to be expressed alternatively as

$$B \propto \frac{\text{volts} \times \text{seconds} \times \text{amperes}}{\text{ampere} \times \text{meter}^2} = \frac{\text{volts} \times \text{seconds}}{\text{meter}^2} \tag{15-6}$$

From Faraday's law ($d\phi = e\, dt$) we note further that the unit of (volts × sec) is equivalent to that of flux. Therefore, we can properly conclude that B is in fact a *flux density* since it has the units of flux per square meter. When the flux is expressed in webers, B then has the units of webers/meter^2 or *teslas*. Equation (15-3) indicates the factors that determine the magnitude of B but it serves also as a means of identifying the manner in which the current I_1 influences the force field about the current element. It is significant to note that as long as I_2 is not zero, the force field and the magnetic field have the same characteristics— both have a circular locus and both are vector quantities possessing magnitude and direction. However, because of the way B is defined, the magnetic field exists as long as I_1 is not zero irrespective of the value of I_2.

In our study of electrical machinery, which comes later, conductor 1 will be referred to as the *field winding* because it sets up the working magnetic field, whereas the total circuit of which the elemental conductor is a part is called the *armature winding*.

The direction of the magnetic field is readily determined by the *right-hand rule* which states that if the field winding conductor (1 in this case) is grasped in the right hand with the thumb pointing in the direction of current flow, the lines of flux (or flux density) will be in the direction in which the fingers wrap around the conductor.

Permeability. A glance at Eq. (15-1) shows that all the factors are known in this equation with the exception of the proportionality factor μ, which is a characteristic of the surrounding medium. Upon repeating the experiment of Fig. 15-1 in iron rather than air, it is found that the force is many times greater for the same values of I_1, I_2, l, and r. Therefore, it follows that μ may be defined from Eq. (15-1) for various media because it is the only unknown quantity. Moreover, because of the way magnetic flux density was defined, Eq. (15-3) indicates that the effect of the surrounding medium may be described in terms of the degree to which it increases or decreases the magnetic flux density for a specified current I_1. Thus, when iron rather than free space is the medium, it can be said that the iron provides a greater penetration of the magnetic field in a given region, i.e., there is a greater flux density. This property of the surrounding medium in which the conductors are embedded is called *permeability*.

When the conductors of Fig. 15-1 are assumed placed in a vacuum (free space) and the force is measured for specified values of I_1, I_2, l, and r, the solution for the permeability of free space obtained from Eq. (15-1) and expressed in SI units comes out to be

$$\mu_0 = 4\pi \times 10^{-7} \qquad\qquad (15\text{-}7)$$

The unit of permeability also follows from Eq. (15-1). Thus

$$\mu_0 \propto \frac{\text{newtons}}{\text{ampere}^2}$$

But

$$\text{newton-meter} = \text{joule} = \text{volts} \times \text{amperes} \times \text{second}$$

or

$$\mu_0 \propto \frac{\text{volts} \times \text{amperes} \times \text{seconds}}{\text{ampere}^2 \times \text{meter}} = \frac{\text{volts} \times \text{seconds}}{\text{ampere} \times \text{meter}}$$

However, volts $\times$ seconds/ampere is the unit of inductance expressed in henrys. Accordingly, permeability is expressed in units of henrys/meter.

In those cases where the surrounding medium is other than free space, the absolute permeability is again readily found from Eq. (15-1). A comparison with the result obtained for free space then leads to a quantity called *relative permeability*, μ_r. Expressed mathematically, we have

$$\mu_r = \frac{\mu}{\mu_0} \qquad\qquad (15\text{-}8)$$

Equation (15-8) clearly indicates that relative permeability is simply a numeric which expresses the degree to which the magnetic flux density is increased or decreased over that of free space. For some materials, such as Deltamax, the value of μ_r can exceed one hundred thousand. Most ferromagnetic materials, however, have values of μ_r in the hundreds or thousands.

Magnetic Flux ϕ. It is reasonable to expect that since B denotes magnetic flux density, multiplication by the effective area that B penetrates should yield the total *magnetic flux*. To illustrate this point refer to Fig. 15-2, which shows a coil of area *ab* lying in the same horizontal plane containing conductor 1. We already know that when a current I_1 flows through this conductor a magnetic field is created in space and specifically it is described by Eq. (15-3). To find the total flux penetrating the coil it is necessary merely to perform an integration of *B* over the surface area involved. Of course, if *B* were a constant over the area of concern, the flux would be simply the product of *B* and the area *ab*. Next consider that the plane of the coil is tilted with respect to the plane of conductor 1 by 60°, as depicted in Fig. 15-3. Clearly now the total flux penetrating the coil is reduced by a factor of one-half. If the coil is oriented to a position of 90° with respect to the horizontal plane, no flux threads the coil.

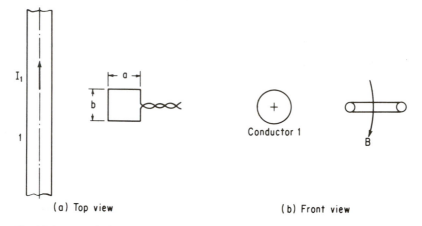

Fig. 15-2   Associating an area with a magnetic field *B* to identify a magnetic flux.

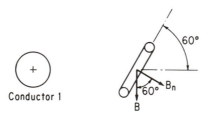

Fig. 15-3   Same as Fig. 15-2(b) except that the coil is tilted 60° relative to the horizontal plane.

On the basis of these observations, then, the magnetic flux through any surface is more rigorously defined as the surface integral of the normal component of the vector magnetic field *B*. Expressed mathematically, we have

$$\phi = \int_s B_n \, dA \qquad (15\text{-}9)$$

where *s* stands for surface integral, *A* represents the area of the coil, and B_n is the normal component of *B* to the coil area. From expression (15-6) we know that magnetic flux must have the dimensions of *volt-seconds*. However, this is more commonly called *webers*. The volt-seconds unit of flux is better understood in terms of Faraday's law of induction.

Magnetic Field Intensity H.   Often in magnetic circuit computations it is helpful to work with a quantity representing the magnetic field which is independent of the medium in which the magnetic flux exists. This is especially true in situations such as are found in electrical machinery where a common flux penetrates several different materials, including air. A glance at Eq. (15-3) discloses that division of B by μ identifies such a quantity. Accordingly, magnetic field intensity is defined as

$$H \equiv \frac{B}{\mu} \qquad (15\text{-}10)$$

and has the units of

$$\frac{\dfrac{\text{newtons}}{\text{ampere} \times \text{meter}}}{\dfrac{\text{newtons}}{\text{ampere}^2}} = \frac{\text{amperes}}{\text{meter}}$$

Thus H is dependent upon the current that produces it and also on the geometry of the configuration but not the medium. For the system of Fig. 15-1 the value of the magnetic field intensity immediately follows from Eq. (15-3) and is given by

$$H = \frac{B}{\mu} = \frac{I_1}{2\pi r} \qquad (15\text{-}11)$$

Because H is independent of the medium, it is frequently looked upon as the intensity that is responsible for driving the flux density through the medium. Actually, though, H is a derived result.

More generally the units for H are ampere-turns/meter rather than amperes/meter. This is apparent whenever the field winding is made up of more than just a single conductor.

Ampère's Circuital Law.   Now that the magnetic field intensity has been defined and shown to have dimensions of ampere-turns/meter, we shall develop a very useful relationship. Recall that H is a vector having the same direction as the magnetic field B. For the configuration of Fig. 15-1 H has the same circular locus as B. A line integration of H along any given closed circular path proves interesting. Of course the line integral is considered because H involves a per unit length dimension. Thus

$$\oint H \, dl = \int_0^{2\pi r} \frac{I_1}{2\pi r} \, dl = I_1 \qquad \text{amperes} \qquad (15\text{-}12)$$

(Again keep in mind that the units here would be ampere-turns if more than one conductor were involved in Fig. 15-1.) Equation (15-12) states that the closed line integral of the magnetic field intensity is equal to the enclosed current (or ampere-turns) that produces the magnetic field lines. This relationship is called *Ampère's circuital law* and is more generally written as

$$\oint H \, dl = \mathscr{F}$$

(15-13)

where $\mathscr{F}$ denotes the ampere-turns enclosed by the assumed closed flux line path. The quantity $\mathscr{F}$ is also known as the *magnetomotive force* and frequently abbreviated as mmf. This relationship is useful in the study of electromagnetic devices and is referred to in subsequent chapters.

   Derived Relationships.   In the preceding pages the fundamental magnetic quantities—flux density, flux, field intensity and permeability—are defined starting with Ampère's basic experiment involving two current-carrying conductors. By the appropriate manipulation of these quantities additional useful results can be obtained.

   Equation (15-10) is a vector equation describing the magnetic field intensity for a given geometry and current. If the total path length of a flux line is assumed to be L, then the total magnetomotive force associated with the specified flux line is

$$\mathscr{F} = HL = \frac{B}{\mu} L$$

(15-14)

Now in those situations where B is a constant and penetrates a fixed, known area A, the corresponding magnetic flux may be written from Eq. (15-9) as

$$\phi = BA$$

(15-15)

Inserting Eq. (15-15) into Eq. (15-14) yields

$$\mathscr{F} = HL = \phi\left(\frac{L}{\mu A}\right)$$

(15-16)

The quantity in parentheses in this last expression is interesting because it bears a very strong resemblance to the definition of resistance in an electric circuit. Refer to Eq. (2-16). Recall that the resistance in an electric circuit represents an impediment to the flow of current under the influence of a driving voltage. An examination of Eq. (15-16) provides a similar interpretation for the magnetic circuit. We are already aware that $\mathscr{F}$ is the driving magnetomotive force which creates the flux ϕ penetrating the specified cross-sectional area A. However, this flux is limited in value by what is called the *reluctance* of the magnetic circuit, which is defined as

$$\mathscr{R} = \frac{L}{\mu A}$$

(15-17)

No specific name is given to the dimension of reluctance except to refer to it as so many units of reluctance.

Equation (15-17) reveals that the impediment to the flow of flux which a magnetic circuit presents is directly proportional to the length and inversely proportional to the permeability and the cross-sectional area—results which are entirely consistent with physical reasoning.

Inserting Eq. (15-17) into Eq. (15-16) yields

$$\boxed{\mathscr{F} = \phi \mathscr{R}}$$

(15-18)

which is often referred to as the Ohm's law of the magnetic circuit. It is important to keep in mind, however, that these manipulations in the forms shown are permissible as long as B and A are fixed quantities.

Ampère's Law for Various Orientations of the Current Element. In Fig. 15-1 the assumption was made that the current element was located parallel to conductor 1 and lying in the same plane. Because this orientation was sufficient for the purpose at hand—to define the fundamental magnetic quantities—it was pursued as a matter of convenience. However, in the interest of furnishing a more complete picture of the experiment, we shall now consider the effect on the force of placing the current element, $I_2 l$, in two additional different orientations. Consider first that the current element is no longer placed parallel to conductor 1 but continues to be located in the same horizontal plane. Refer to Fig. 15-4. The dots in this figure indicate that the magnetic field is directed outward on the left side of conductor 1 and inward (with respect to the plane of the paper) on the right side of the conductor as indicated by the right-hand rule. The results of this experiment show that the magnitude of the force is the same as that found by using the configuration of Fig. 15-1.† This conclusion is not surprising because the value of the magnetic flux density as well as I_2 and l remain unchanged so that Eq. (15-2) is still valid in describing the force. The only change is the direction of the force. However, as Fig. 15-4 indicates, the force continues to be normal to the current element. It is worthwhile to keep this point in mind. Presently, a general rule for establishing the force direction for all configurations will be described.

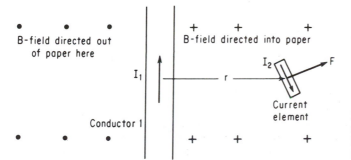

Fig. 15-4 Showing the direction of the force when the current element is no longer located parallel to conductor 1 but remains in the same plane.

† Because of the infinitesimal nature of the current element, the value of B is assumed to be constant about $I_2 l$ irrespective of its orientation.

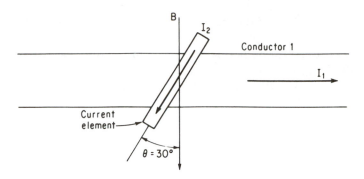

Fig. 15-5 Side-view projection of Fig. 15-1 but with the current element tilted relative to the horizontal plane. Force is directed out of paper.

Next let us consider that orientation of the current element which places it parallel to conductor 1 but inclined at an angle $\theta = 30°$ with respect to the vertical. A side-view projection of the configuration is depicted in Fig. 15-5. Note that the magnetic field is directed downward along the vertical for this view. Actually, of course, the locus of B is circular, but in Fig. 15-5 we are looking at just that small portion of the B-field about the plane containing conductor 1. With this configuration the force on the current element is found to have the same direction but one half the magnitude of that obtained with the orientation of Fig. 15-1. It follows, then, that the angle between the current vector $I_2 l$ and the flux density B affects the magnitude of the force. As a matter of fact, further experimentation reveals that the general expression for the force is

$$F = I_2 l B \sin \theta \quad \text{N} \tag{15-19}$$

Equation (15-19) conveys information solely about the magnitude of the force and not its direction. It is possible, however, by employing the notation of vector analysis, to rewrite Eq. (15-19) so that information about magnitude as well as direction is present. This result is readily accomplished by the use of the *cross-product* notation between two vectors, yielding a third vector having magnitude and direction. Thus Eq. (15-19) is more completely expressed as

$$\overline{F} = I_2 \overline{l} \times \overline{B} \quad \text{N} \tag{15-20}$$

The cross symbol must always be understood to involve the *sine* of the angle between the two vectors $\overline{B}$ and $\overline{l}$ (or the direction of I_2, which is determined by the orientation of $\overline{l}$). Moreover, whenever the cross product is involved, the direction of the resultant vector is always normal to the plane containing the vectors $\overline{B}$ and $\overline{l}$ in the sense determined by the direction of advance of a right-hand screw as $\overline{l}$ is turned into $\overline{B}$ through the smaller of the two angles made by the vectors. Accordingly, in the configuration of Fig. 15-5 the direction of the force is found by turning $I_2 \overline{l}$ into $\overline{B}$ and then noting that this would cause a right-hand screw to advance out of the plane of the paper. Hence the force is directed outward, and this corresponds with the experimentally established result.

Magnetic Theory and Circuits   Chap. 15

It is also possible to determine the direction of the force by means of another right-hand rule, which requires that the forefinger be put in the direction of the current and the middle finger in the direction of $\overline{B}$ and the two assumed lying in the same plane. The thumb of the right hand then points in the direction of the force when placed perpendicular to the other two fingers.

15-2 THEORY OF MAGNETISM

In order to understand the magnetic behavior of materials, it is necessary to take a microscopic view of matter. A suitable starting point is the composition of the atom, which Bohr described as consisting of a heavy nucleus and a number of electrons moving around the nucleus in specific orbits. Closer investigation reveals that the atom of any substance experiences a torque when placed in a magnetic field; this is called a *magnetic moment*. The resultant magnetic moment of an atom depends upon three factors—the positive charge of the nucleus spinning on its axis, the negative charge of the electron spinning on its axis, and the effect of the electrons moving in their orbits. The magnetic moment of the spin and orbital motions of the electron far exceeds that of the spinning proton. However, this magnetic moment can be affected by the presence of an adjacent atom. Accordingly, if two hydrogen atoms are combined to form a hydrogen *molecule*, it is found that the electron spins, the proton spins, and the orbital motions of the electrons of each atom oppose each other so that a resultant magnetic moment of zero should be expected. Although this is almost the case, experiment reveals that the relative permeability of hydrogen is not equal to one but rather is very slightly less than unity. In other words, the molecular reaction is such that when hydrogen is the medium there is a slight decrease in the magnetic field compared to free space. This behavior occurs because there is a precessional motion of all rotating charges about the field direction, and the effect of this precession is to set up a field opposed to the applied field regardless of the direction of spin or orbital motion. Materials in which this behavior manifests itself are called *diamagnetic* for obvious reasons. Besides hydrogen, other materials possessing this characteristic are silver and copper.

Continuing further with the hydrogen molecule, let us assume next that it is made to lose an electron, thus yielding the hydrogen ion. Clearly, complete neutralization of the spin and orbital electron motions no longer takes place. In fact when a magnetic field is applied, the ion is so oriented that its net magnetic moment aligns itself with the field, thereby causing a slight increase in flux density. This behavior is described as *paramagnetism* and is characteristic of such materials as aluminum and platinum. Paramagnetic materials have a relative permeability slightly in excess of unity.

So far we have considered those elements whose magnetic properties differ only very slightly from those of free space. As a matter of fact, the vast majority of materials fall within this category. However, there is one class of materials—principally iron and its alloys with nickel, cobalt, and aluminum—for which the relative permeability is very many times greater than that of free space. These materials are called *ferromagnetic*† and are of great importance in electrical engineering. We may ask at this point why iron (and its alloys) is so very much more magnetic than other elements. Essentially, the answer is provided by the *domain* theory of magnetism. Like all metals, iron is crystalline in structure with the atoms arranged in a space lattice. However, domains are subcrystalline particles of varying sizes and shapes containing about 10^{15} atoms in a volume of approximately 10^{-9} cm³. *The distinguishing feature of the domain is that the magnetic moments of its constituent atoms are all aligned in the same direction.* Thus in a ferromagnetic material not only must there exist a magnetic moment due to a nonneutralized spin of an electron in an inner orbit, but also the resultant spin of all neighboring atoms in the domain must be parallel.

It would seem by the explanation so far that if iron is composed of completely magnetized domains then the iron should be in a state of complete magnetization throughout the body of material even without the application of a magnetizing force. Actually, this is not the case, because the domains act independently of each other, and for a specimen of unmagnetized iron these domains are aligned haphazardly in all directions so that the net magnetic moment is zero over the specimen. Figure 15-6 illustrates the situation diagrammatically in a simplified fashion. Because of the crystal lattice structure of iron the "easy" direction of domain alignment can take place in any one of six directions—left, right, up, down, out or in—depending upon the direction of the applied magnetizing force. Figure 15-6(a) shows the unmagnetized configuration. Figure 15-6(b) depicts the result of applying a force from left to right of such magnitude as to effect alignment of all the domains. When this state is reached the iron is said to be *saturated*— there is no further increase in flux density over that of free space for further increases in magnetizing force.

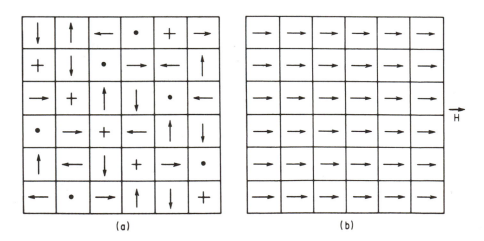

(a)   (b)

Fig. 15-6 Representation of a ferromagnetic crystal: (a) unmagnetized and (b) fully magnetized by the field *H*.

† Derived from the Latin word for iron—*ferrum*.

Large increases in the temperature of a magnetized piece of iron bring about a decrease in its magnetizing capability. The temperature increase enforces the agitation existing between atoms until at a temperature of 750°C the agitation is so severe that it destroys the parallelism existing between the magnetic moments of the neighboring atoms of the domain and thereby causes it to lose its magnetic property. The temperature at which this occurs is called the *curie point*.

15-3 MAGNETIZATION CURVES OF FERROMAGNETIC MATERIALS

If the experiment of Fig. 15-1 is repeated with iron or steel as the medium for increasing values of the field winding current I_1 and the corresponding values of μ computed, it is found that the relative permeability varies considerably with the magnetizing force that establishes the operating flux density. A typical variation of μ_r for cast steel appears in Fig. 15-7. Here μ_r is plotted versus flux density rather than the magnetizing force because it is the onset of the realignment of more and more domains that brings about the change in permeability. Unfortunately, the state of development of the theory of magnetism is not so far advanced that it allows the prediction of the magnetic properties of a material on a purely theoretical basis even though the exact composition of the material is known. For example, with the present theory it is not possible to say exactly what the flux density will be in a given specimen of iron for a specified value of the magnetizing force. Rather, it is customary to obtain this information by consulting technical and descriptive bulletins where the measured magnetic properties of a representative sample of the specimen are published. These bulletins are made available to users by the manufacturers of magnetic steels and they include information on such varied shapes and forms as sheets, wires, bars, and even castings weighing up to hundreds of tons.

Usually, the published magnetic characteristics of the various iron and steel samples are presented as plots of flux density B as a function of the magnetic

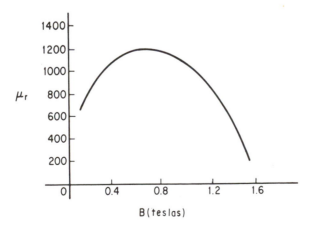

Fig. 15-7 Graph of relative permeability versus flux density.

field intensity H. In the interest of presenting a complete graphical picture of the functional relationship existing between these two quantities as well as to define additional terms used in this connection, refer to Fig. 15-8. Assume that the steel specimen is initially unmagnetized and is in the form of a toroidal ring with a coil of N turns wrapped around it. Assume too that the coil can be energized from a variable voltage source capable of furnishing current flow in either direction in the coil. As the current I is increased from zero in the positive direction (current flowing into the top terminal of the coil), an increasing magnetic flux ϕ can be measured as taking place within the body of the toroid in the clockwise direction. For any fixed value of I there is a specific value of flux. Then by Eq. (15-15) the corresponding flux density is determined since the toroidal cross-sectional area is known. Moreover, the magnetomotive force NI can be replaced by HL_m in accordance with Ampère's circuital law [Eq. (15-13)] where L_m is the mean length of path of the toroid (as shown by the broken line circle in Fig. 15-8). The two fundamental quantities involved in this arrangement then are B in teslas and H expressed in ampere-turns per meter. H is the quantity we want to deal with rather than magnetomotive force because for the same flux density, doubling the mean magnetic length will not change H but will require doubling the magnetomotive force. The conclusion to be drawn is that a plot of B versus H is a universal plot for the given material because it can be extended to any geometry of cross-sectional area and length. In contrast, a plot of ϕ versus NI is limited to a single geometrical configuration. Therefore, a plot of the magnetic characteristics of a material always involves plotting B versus H. For the virgin sample of Fig. 15-8 the graph of B versus H follows the curve Oa of Fig. 15-9 for field intensities up to H_a. Take note of the nonlinear relationship existing between these two quantities.

Another interesting characteristic of ferromagnetic materials is revealed when the field intensity, having been increased to some value, say H_a, is subsequently decreased. It is found that the material opposes demagnetization and, accordingly, does not retrace along the magnetizing curve Oa but rather along

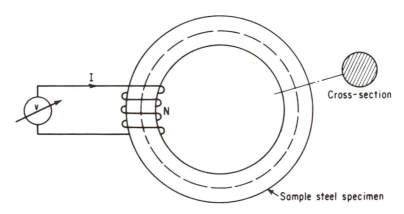

Fig. 15-8 Obtaining the magnetization curve of a sample steel specimen.

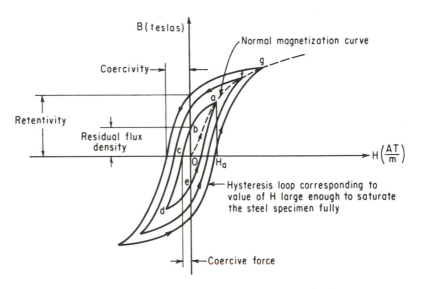

Fig. 15-9 Typical hysteresis loops and normal magnetization curve.

a curve located above *Oa*. See curve *ab* in Fig. 15-9. Furthermore, it is seen that when the field intensity is returned to zero, the flux density is no longer zero as was the case with the virgin sample. This happens because some of the domains remain oriented in the direction of the originally applied field. The value of *B* that remains after the field intensity *H* is removed is called *residual flux density*. Moreover, its value varies with the extent to which the material is magnetized. The maximum possible value of the residual flux density is called *retentivity* and results whenever values of *H* are used that cause complete saturation.

Frequently, in engineering applications of ferromagnetic materials, the steel is subjected to cyclically varying values of *H* having the same positive and negative limits. As *H* varies through many identical cycles, the graph of *B* versus *H* gradually approaches a fixed closed curve as depicted in Fig. 15-9. The loop is always traversed in the direction indicated by the arrows. Since time is the implicit variable for these loops, note that *B* is always lagging behind *H*. Thus, when *H* is zero, *B* is finite and positive, as at point *b*, and when *B* is zero, as at *c*, *H* is finite and negative, and so forth. This tendency of the flux density to *lag behind* the field intensity when the ferromagnetic material is in a symmetrically cyclically magnetized condition is called *hysteresis*† and the closed curve *abcdea* is called a *hysteresis loop*. Moreover, when the material is in this cyclic condition, the amount of magnetic field intensity required to reduce the residual flux density to zero is called the *coercive force*. Usually, the larger the residual flux density, the larger must be the coercive force. The maximum value of the coercive force is called the *coercivity*.

A glance at the hysteresis loops of Fig. 15-9 makes it quite evident that the flux density corresponding to a particular field intensity is not single-valued.

† Derives from the Greek *hysterien* meaning to be behind or to lag.

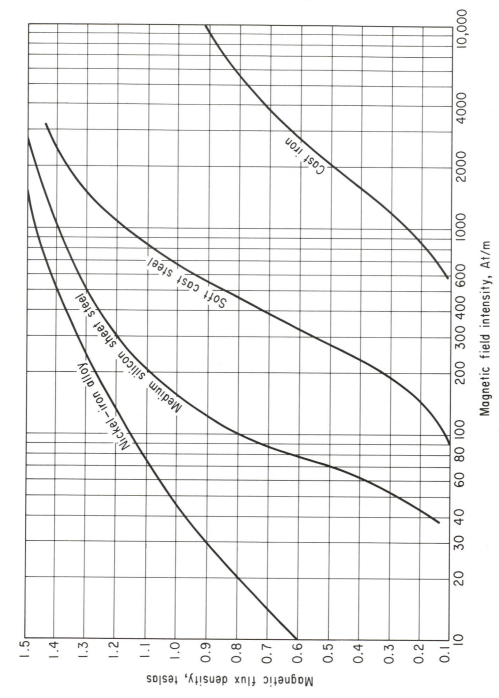

Fig. 15-10 Magnetization curves of typical ferromagnetic materials.

Its value lies between certain limits depending upon the previous history of the ferromagnetic material. However, since in many situations involving magnetic devices this previous history is unknown, a compromise procedure is used in making magnetic calculations by working with a single-valued curve called the *normal magnetization curve*. This curve is found by drawing a curve through the tips of a group of hysteresis loops generated while in a cyclic condition. Such a curve is *Oafg* in Fig. 15-9. Typical normal magnetization curves of commonly used ferromagnetic materials appear in Fig. 15-10.

A final observation is in order at this point. By Eq. (15-10) the permeability of a material may be expressed as a ratio of B to H. Coupling this with the nonlinear variation existing between B and H (see Fig. 15-9), the variation of permeability with flux density as already cited in connection with Fig. 15-7 is verified. As a matter of fact, for a material in a cyclic condition, the permeability is nothing more than the ratio of B to H for the various points along the hysteresis loop.

15-4 THE MAGNETIC CIRCUIT: CONCEPT AND ANALOGIES

In general, problems involving magnetic devices are basically field problems because they are concerned with quantities such as ϕ and B which occupy three-dimensional space. Fortunately, however, in most instances the bulk of the space of interest to the engineer is occupied by ferromagnetic materials except for small air gaps which are present either by intention or by necessity. For example, in electromechanical energy-conversion devices the magnetic flux must permeate a stationary as well as a rotating mass of ferromagnetic material, thus making an air gap indispensable. On the other hand, in other devices an air gap may be intentionally inserted in order to mask the nonlinear relationship existing between B and H. But in spite of the presence of air gaps it happens that the space occupied by the magnetic field and the space occupied by the ferromagnetic material are practically the same. Usually, this is because air gaps are made as small as mechanical clearance between rotating and stationary members will allow and also because the iron by virtue of its high permeability confines the flux to itself as copper wire confines electric current or a pipe restricts water. On this basis the three-dimensional field problem becomes a one-dimensional circuit problem and in accordance with Eq. (15-18) leads to the idea of a *magnetic circuit*. Thus we can look upon the magnetic circuit as consisting predominantly of iron paths of specified geometry which serves to confine the flux; air gaps may be included. Figure 15-11 shows a typical magnetic circuit consisting chiefly of iron. Note that the magnetomotive force of the coil produces a flux which is confined to the iron and to that part of the air having effectively the same cross-sectional area as the iron. Furthermore, a little thought reveals that this magnetic circuit may be replaced by a single-line equivalent circuit as depicted in Fig. 15-12. As suggested by Eqs. (15-17) and (15-18) the equivalent circuit consists of

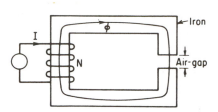

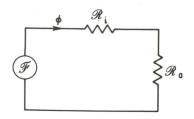

Fig. 15-11 Typical magnetic circuit involving iron and air.

Fig. 15-12 Single-line equivalent circuit of Fig. 15-11.

the magnetomotive force driving flux through two series-connected reluctances—$\mathcal{R}_i$, the reluctance of the iron, and $\mathcal{R}_a$, the reluctance of the air.

This analogy of the magnetic circuit with the electric circuit carries through in many other respects. For the sake of completeness these details are presented below for the case of a toroidal copper ring and a toroidal iron ring having the same mean radius r and cross-sectional area A.

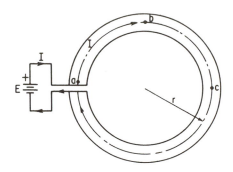

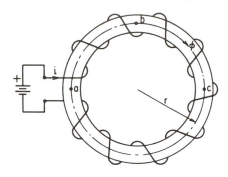

ELECTRIC CASE

The toroidal copper ring is assumed open by an infinitesimal amount with the ends connected to a battery; a current of I amperes flows through the ring.

MAGNETIC CASE

The toroidal iron ring is assumed wound with N turns of wire so that with a current i flowing through it the magnetomotive force creates the flux ϕ.

Driving Force

applied battery voltage = E   applied ampere-turns = $\mathcal{F}$

Response

$$\text{current} = \frac{\text{driving force}}{\text{electric resistance}}$$ $$\text{flux} = \frac{\text{driving force}}{\text{magnetic reluctance}}$$

or

$$I = \frac{E}{R}$$ $$\phi = \frac{\mathcal{F}}{\mathcal{R}} \qquad (15\text{-}21)$$

Impedance is a general term used to indicate the impediment to a driving force in establishing a response.

$$\text{resistance} = R = \rho\frac{L}{A}$$

$$\text{reluctance} = \mathcal{R} = \frac{L}{\mu A} \quad (15\text{-}22)$$

where $L = 2\pi r$ = mean length of turn of the toroid and A is the toroidal cross-sectional area.

where $L = 2\pi r$ = mean length of turn of the toroid and A is the toroidal cross-sectional area.

Equivalent Circuit

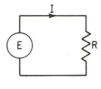

$$E = IR$$

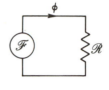

$$\mathcal{F} = \phi\mathcal{R}$$

Electric Field Intensity

With the application of the voltage E to the homogeneous copper toroid, there is produced within the material an electric potential gradient given by

$$\mathcal{E} \equiv \frac{E}{L} = \frac{E}{2\pi r} \quad \text{V/m}$$

This electric field must occur in a closed path if it is to be maintained. It then follows that the closed line integral of $\mathcal{E}$ is equal to the battery voltage E. Thus

$$\oint \mathcal{E}\, dl = E$$

Magnetic Field Intensity

When a magnetomotive force is applied to the homogeneous iron toroid, there is produced within the material a magnetic potential gradient given by

$$H \equiv \frac{\mathcal{F}}{L} = \frac{\mathcal{F}}{2\pi r} \quad \text{At/m} \quad (15\text{-}23)$$

As already pointed out in connection with Ampère's circuital law, the closed line integral of H equals the enclosed magnetomotive force. Thus

$$\oint H\, dl = \mathcal{F} \quad (15\text{-}24)$$

Voltage Drop

If it is desired to find the voltage drop occurring between two points—a and b—of the copper toroid, we may write:

$$V_{ab} = \int_a^b \mathcal{E}\, dl = \frac{E}{L}\int_a^b dl = \frac{IR}{L}l_{ab}$$

$$= \frac{I}{L}\rho\frac{L}{A}l_{ab} = I\rho\frac{l_{ab}}{A} = IR_{ab}$$

mmf Drop

The letters mmf are used to represent magnetomotive force. The portion of the total applied mmf appearing between points a and b is found similarly:

$$\mathcal{F}_{ab} = \int_a^b H\, dl = \frac{\mathcal{F}}{L}l_{ab} = \frac{\phi\mathcal{R}}{L}l_{ab}$$

$$= \frac{\phi}{L}\frac{L}{\mu A}l_{ab} = \phi\frac{l_{ab}}{\mu A} = \phi\mathcal{R}_{ab}$$

i.e.,

$$V_{ab} = IR_{ab} \qquad\qquad \mathscr{F}_{ab} = \phi\mathscr{R}_{ab} \qquad (15\text{-}25)$$

where R_{ab} is the resistance of the copper toroid between points a and b.

where $\mathscr{R}_{ab}$ is the reluctance of the iron toroid between points a and b.

Current Density

By definition, current density is the amount of amperes per unit area. Thus

$$J \equiv \frac{I}{A} = \frac{E}{AR} = \frac{\mathscr{E}L}{A\rho(L/A)} = \frac{\mathscr{E}}{\rho}$$

or

$$\mathscr{E} = \rho J$$

This last expression is often referred to as the *microscopic* form of Ohm's law.

Flux Density

Flux density is expressed as webers per unit area. Thus

$$B = \frac{\phi}{A} = \frac{\mathscr{F}}{A\mathscr{R}} = \frac{HL}{A(L/\mu A)} = \mu H \qquad (15\text{-}26)$$

or

$$H = \frac{B}{\mu} \qquad (15\text{-}27)$$

It should not be inferred from the foregoing that electric and magnetic circuits are analogous in *all* respects. For example, there are no magnetic insulators analogous to those known to exist for electric circuits. Also, when a direct current is established and maintained in an electric circuit, energy must be continuously supplied. An analogous situation does not prevail in the magnetic case, where a flux is established and maintained constant.

15-5 UNITS FOR MAGNETIC CIRCUIT CALCULATIONS

Magnetic circuit calculations can be carried out by use of any one of several different systems of units. These various systems arose initially because it was thought that the phenomena of electricity and magnetism were unrelated—thereby leading to the development of a separate system of units for each—and secondly, because of the desire to deal with practical values of the units once the relationship was discovered. Up to now attention has been given exclusively to the mks (meter-kilogram-second) system of units as developed by Giorgi about the turn of the twentieth century. This policy is prompted by the acceptance in 1960 of the mks system of units as the standard for scientific work and now referred to as SI units (System International Unite's). However, a good part of the past literature is written in terms of the units of the CGS (centimeter-gram-second) system. Furthermore, many of the present-day computations are carried on in

TABLE 15-1 MAGNETIC UNITS

Quantity	Symbol	CGS Unit	CGS Relation	SI units Unit	SI Symbol	SI Relation	Mixed English Unit	Mixed English Relation	Conversion factors
mmf	$\mathscr{F}$	gilberts	$\mathscr{F} = 0.4\pi NI$	ampere-hour	At	$\mathscr{F} = NI$	At	$\mathscr{F} = NI$	pragilbert = 10 gilberts At = 0.4π gilberts = 1.257 gilberts
Permeability: Free space μ_0 Abs. norm. perm. μ	μ_0 μ		$\mu_0 = 1$ $\mu = \mu_0\mu_r = \mu_r$			$\mu_0 = 4\pi10^{-7}$ $\mu = 4\pi10^{-7}\mu_r$		$\mu_0 = 3.19$ $\mu = 3.19\mu_r$	
Length	l	cm		meter	m		inch		1 centimeter = 0.01 meter 1 meter = 39.4 inches
Area	A	cm^2		meter^2	m^2		inch^2		1 meter^2 = 1550 inches^2
Reluctance	$\mathscr{R}$		$\mathscr{R} = \dfrac{l}{\mu_r A}$			$\mathscr{R} = \dfrac{l}{4\pi10^{-7}\mu_r A}$		$\mathscr{R} = \dfrac{l}{3.19\mu_r A}$	
Flux	Φ	maxwells = lines	$\Phi = \dfrac{0.4\pi NI}{\dfrac{l}{\mu_r A}}$	weber	Wb	$\Phi = \dfrac{NI}{\dfrac{l}{\mu_r 4\pi10^{-7}A}}$	maxwells = lines	$\Phi = \dfrac{NI}{\dfrac{l}{3.19\mu_r A}}$	1 weber = 10^8 lines
Magnetic field intensity	H	gilberts/cm = oersteds	$H = \dfrac{0.4\pi NI}{l}$	$\dfrac{At}{meter}$	$\dfrac{At}{m}$	$H = \dfrac{NI}{l}$	$\dfrac{At}{inch}$	$H = \dfrac{NI}{l}$	1 oersted = 79.6 At/m 1 praoersted = 1000 oersteds $1\dfrac{At}{in} = \dfrac{0.4\pi}{2.54} = 0.495$ oersted $1\dfrac{At}{in} = \dfrac{2.02}{1000}$ praoersted
Flux density	B	lines/cm^2 = gauss	$B = \mu_r H$	$\dfrac{webers}{meter^2}$ = tesla	T	$B = \mu_r\mu_0 H$	lines/inch^2	$B = 3.19\mu_r H$	1 gauss = 6.45 lines/in^2 1 tesla = 64,500 lines/in^2 1 tesla = 10,000 gauss

terms of the *mixed* system employing such units as ampere-turns/inch, maxwells/inch², and ampere-turns because of the convenience they offer in dealing with dimensions that are expressed in inches. For these reasons the units of all three systems are shown in Table 15-1.

The weber, which is the unit of flux in the SI system, is equal to 10^8 *maxwells* (or lines) where the maxwell is the unit of flux in the CGS system. The *gilbert* is the CGS unit for mmf and is equal to 0.4π times the number of ampere-turns. The CGS unit for magnetic field intensity H is the *oersted* (or gilbert/cm) and the CGS unit for flux density B is the *gauss* (or lines/cm²). The relationships existing for the same quantity between the various systems of units are given in the last column of the table.

15-6 MAGNETIC CIRCUIT COMPUTATIONS

Basically magnetic circuit calculations involving ferromagnetic materials fall into two categories. In the first the value of the flux is known and it is required to find the magnetomotive force to produce it. This is the situation typical of the design of a-c and d-c electromechanical energy converters. On the basis of the desired voltage rating of an electric generator or the torque rating of an electric motor information about the required magnetic flux is readily obtained. Then with this knowledge and the configuration of the magnetic circuit the total mmf needed to establish the flux is determined straightforwardly. In the second case it is the flux for which we must solve, knowing the geometry of the magnetic circuit and the applied mmf. An engineering application in which this situation prevails is the magnetic amplifier, where it is often necessary to find the resultant magnetic flux caused by one or more control windings. Because the reluctance (or permeability) of the ferromagnetic material is not constant, the solution of this problem is considerably more involved than that of the first category as illustrated by the examples below.

EXAMPLE 15-1   A toroid is composed of three ferromagnetic materials and is equipped with a coil having 100 turns as depicted in Fig. 15-13. Material *a* is a nickel-iron alloy having a mean arc length L_a of 0.3 m. Material *b* is medium silicon steel and has a mean arc length L_b of 0.2 m. Material *c* is of cast steel having a mean arc length equal to 0.1 m. Each material has a cross-sectional area of 0.001 m².

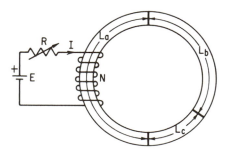

Fig. 15-13 Toroid composed of three different materials.

(a) Find the magnetomotive force needed to establish a magnetic flux of $\phi = 6 \times 10^{-4}$ Wb = 60,000 lines.

(b) What current must be made to flow through the coil?

(c) Compute the relative permeability and reluctance of each ferromagnetic material.

Solution: (a) To obtain the total mmf of the coil all we need to do is to apply Ampère's circuital law. Thus

$$\mathcal{F} = \mathcal{F}_a + \mathcal{F}_b + \mathcal{F}_c = H_a L_a + H_b L_b + H_c L_c$$

The unknown quantities here are H_a, H_b, and H_c. These can readily be found from a knowledge of the flux density, which here is the same for each section because the flux is common and the cross-sectional areas are the same. Hence

$$B_a = B_b = B_c = \frac{\phi}{A} = \frac{0.0006}{0.001} = 0.6 \text{ T}$$

Now H_a is found by entering the *B-H* curve of the nickel-iron alloy of Fig. 15-10 corresponding to $B_a = 0.6$. This yields

$$H_a = 10 \text{ At/m}$$

Similarly,

$$H_b = 77 \text{ At/m}$$
$$H_c = 270 \text{ At/m}$$

Accordingly, the total required mmf is

$$\mathcal{F} = H_a L_a + H_b L_b + H_c L_c$$
$$= (10)(0.3) + 77(0.2) + 270(0.1)$$
$$= 3 + 15.4 + 27 = 45.4 \text{ At}$$

Note that although the path length of cast steel is the smallest, it nonetheless requires the greatest portion of the mmf to force the specified flux through. This happens because of its much lower permeability as shown in part (c).

(b) In the SI system the mmf is equal to the number of ampere-turns. Hence

$$I = \frac{\mathcal{F}}{N} = \frac{45.4}{100} = 0.454 \text{ A}$$

(c) From Eq. (15-10)

$$\mu_a = \frac{H_a}{B_a} = \frac{0.6}{10} = 0.06 \text{ H/m}$$

Also,

$$\mu_{ra} \mu_0 = \mu_a$$

$$\therefore \quad \mu_{ra} = \frac{\mu_a}{\mu_0} = \frac{0.06}{4\pi \times 10^{-7}} = 47,746$$

Furthermore, from Eq. (15-16) the reluctance is found to be

$$\mathcal{R}_a = \frac{\mathcal{F}_a}{\phi} = \frac{3}{6 \times 10^{-4}} = 5000 \text{ rationalized mks units of reluctance.}$$

Proceeding in a similar fashion for materials *b* and *c* leads to the following results:

$$\mu_{rb} = 6207 \qquad \mathcal{R}_b = 25,667$$
$$\mu_{rc} = 1768 \qquad \mathcal{R}_c = 45,000$$

Next we consider the more difficult problem: that of finding the flux in a given magnetic circuit corresponding to a specified mmf. The solution cannot be arrived at directly because, as a result of the nonlinear relationship between B and H, there are too many unknowns. The easiest way of finding the solution is to employ a cut-and-try procedure guided by a knowledge of the permeability characteristics of the materials such as appears in their magnetization curves. The following example illustrates the technique involved.

EXAMPLE 15-2   For the toroid of Example 15-1, shown in Fig. 15-13, find the magnetic flux produced by an applied magnetomotive force of $\mathscr{F} = 35$ At.

Solution: The solution cannot be determined directly because to do so we must know the reluctance of each part of the magnetic circuit, which can be known only if the flux density is known—which means that ϕ must be known right at the start. This is clearly impossible.

To obtain the solution by the cut-and-try procedure, we begin by first assuming that all of the applied mmf appears across the material having the highest reluctance. This yields an approximate value of ϕ which can subsequently be refined. A glance at Fig. 15-10 shows that the poorest magnetic "conductor" is cast steel. Hence by assuming the entire mmf to appear across material c we can find H, from which B follows, which in turn yields ϕ. Thus

$$H_c = \frac{\mathscr{F}_c}{L_c} = \frac{\mathscr{F}}{L_c} = \frac{35}{0.1} = 350 \text{ At/m}$$

From Fig. 15-10

$$B_c = 0.65 \text{ T}$$

$$\therefore \quad \phi_1 = B_c A_c = (0.65)(0.001) = 0.00065 \text{ Wb}$$

This value represents the first approximation for the flux as indicated by the subscript.

Also, since the cross-sectional area is the same for each material, it follows that

$$B_a = B_b = B_c = 0.65 \text{ T}$$

Reference to the nickel-iron magnetization curve reveals that the value of H_a corresponding to B_a is negligibly small compared to H_c. Hence for all practical purposes its effect can be neglected. However, note that for medium silicon steel the value of H_b is almost 90 At/meter. This, coupled with the fact that $L_b = 2L_c$, indicates that material b takes about half as much mmf as material c in maintaining the flow of flux. In other words, at this point in our analysis we can make a refinement on our original assumption of assigning the entire mmf to material c. Now we see that about 50% of that assigned to c should be assigned to b. Thus

$$\mathscr{F}_c + \mathscr{F}_b = \mathscr{F} \qquad \text{(assumed)}$$

but

$$\mathscr{F}_b = 0.5\,\mathscr{F}_c \qquad \text{(assumed)}$$

Hence

$$1.5\mathscr{F}_c = \mathscr{F} = 35$$

$$\therefore \quad \mathscr{F}_c = 23.3 \text{ At}$$

Accordingly, a second approximation for the solution can be obtained. Therefore,

$$H_c = \frac{23.3}{0.1} = 233 \text{ At/m}$$

which in turn yields
$$B_c = 0.4 \text{ T}$$
so that the value of the flux now becomes
$$\phi_2 = B_c A_c = 0.0004 \text{ Wb}$$

To determine whether or not this is the correct answer we must at this point compute the mmf drops for each material and add to see whether they yield a value equal to the applied mmf. If not, the foregoing procedure must be repeated until Ampère's circuital law is satisfied. Making this check for the second approximation we have

$$H_b = 62 \text{ At/m} \quad \text{corresponding to } B_a = 0.4$$

and

$$H_a = 5.7 \text{ At/m} \quad \text{for } B_a = 0.4$$

Accordingly,

$$HL = H_a L_a + H_b L_b + H_c L_c$$
$$= (5.7)(0.3) + (62)(0.2) + 233(0.1)$$
$$= 1.7 + 12.4 + 23.3 = 37.4 \text{ At}$$

Obviously this is too high by about 7%. Hence, as a third try, reduce the biggest contributor to the mmf by a factor of 5%. That is, assume that now

$$\mathcal{F}_c = 22 \text{ At}$$

Then

$$H_c = 220 \text{ At/m} \quad \text{and} \quad B_c = 0.375$$
$$\therefore \quad \phi_3 = 0.000375 \text{ Wb}$$

Corresponding to this flux we find

$$H_b = 59 \quad \text{and} \quad H_a = 5$$

Hence

$$\text{mmf} = (5)(0.3) + (59)(0.2) + 22 = 35.3 \text{ At}$$

Since this summation of mmf's agrees with the applied mmf of 35 At, the correct solution for the flux is

$$\phi = 0.000375 \text{ Wb} = 37,500 \text{ lines}$$

When making magnetic circuit computations of the kind just illustrated, it is common practice to accept as valid any solution that comes within ±5% of the exact solution. The reason is that we are dealing with normal magnetization curves which neglect hysteresis and which are after all only *typical* of the material actually being used in the specified circuit. Deviations can and often do exist.

As a final example to illustrate magnetic circuit computations, we shall solve a problem involving parallel magnetic paths as well as the presence of an air gap. Moreover, we will consider only the first type where the mmf needed to establish a specified flux is to be found. This is justified not only because the solution is straightforward but also because it is by far more representative of the kind of magnetic circuit problem the engineer is likely to be concerned with.

EXAMPLE 15-3 A magnetic circuit having the configuration and dimensions shown in Fig. 15-14 is made of cast steel having a thickness of 0.05 m and an air gap of 0.002 m

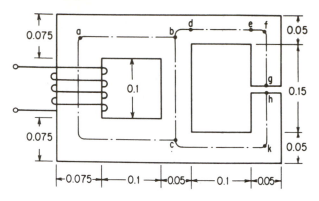

length appearing between points g and h. The problem is to find the mmf to be produced by the coil in order to establish an air gap flux of 4×10^{-4} Wb (or 40,000 lines).

Solution: The method of solution can be readily ascertained by referring to the equivalent circuit of this magnetic circuit as shown in Fig. 15-15. Knowledge of ϕ_g enables us to find the mmf drop appearing across b and c. From this information the flux in leg bc can be determined and, upon adding it to ϕ_g, we find the flux in leg cab. In turn, the mmf needed to maintain the total flux in leg cab can be computed, and when we add it to the mmf drop across bc we obtain the resultant mmf.

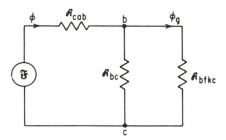

Fig. 15-15 Equivalent circuit of Fig. 15-14.

The computations involved for the various parts of the magnetic circuit are as follows.

Part gh. This is the air gap for which the flux is specified as 4×10^{-4} Wb. The cross-sectional area of the gap is $(0.05)(0.05) = 0.0025$ m². Normally, however, this area is slightly higher because of the tendency of the flux to bulge outward along the edges of the air gap—which is often referred to as *fringing.* For convenience this effect is neglected. Thus, the air gap flux density is found to be

$$B_g = \frac{\phi_g}{A_g} = \frac{4 \times 10^{-4}}{0.0025} = 0.16 \text{ T}$$

Since the permeability of air is practically the same as that of free space, we have

$$H_g = \frac{B_g}{\mu_0} = \frac{0.16}{4\pi \times 10^{-7}} = 127,300 \text{ At/m}$$

and

$$\mathscr{F}_g = H_g L_g = (127,300)(0.002) = 255 \text{ At}$$

Part bg and hc. The length of ferromagnetic material involved here is

$$L_{bg} + L_{hc} = 2\left(L_{bd} + L_{de} + L_{ef} + \frac{L_{fk}}{2}\right) - L_g$$

$$= 2(0.025 + 0.1 + 0.025 + 0.1) - 0.002$$

$$= 0.498 \text{ m}$$

Moreover, corresponding to a flux density in the cast steel of 0.16 T, the field intensity H is found to be 125 At/m. Hence

$$\mathscr{F}_{bg+hc} = 125(0.498) = 62.2 \text{ At}$$

Part bc. Because path *bfkc* is in parallel with path *bc*, the total mmf across path *bfkc* also appears across path *bc*. Hence

$$\mathscr{F}_{bc} = 255 + 62.2 = 317.2 \text{ At}$$

Also,

$$L_{bc} = 0.1 + 0.075 = 0.175 \text{ m}$$

$$\therefore \quad H_{bc} = \frac{317.5}{0.175} = 1810 \text{ At/m}$$

and from Fig. 15-10 for cast steel the corresponding flux density is found to be

$$B_{bc} = 1.38 \text{ T}$$

Hence

$$\phi_{bc} = 1.38(0.0025) = 0.00345 \text{ Wb}$$

Part cab. Accordingly, the total flux existing in leg *cab* is

$$\phi_{cab} = \phi_{bc} + \phi_g = 0.00345 + 0.0004 = 0.00385 \text{ Wb}$$

Knowledge of this flux then leads to determination of the mmf needed in leg *cab* to sustain it. Thus

$$B_{cab} = \frac{0.00385}{0.00375} = 1.026 \text{ T}$$

From which

$$H_{cab} = 690 \text{ At/m}$$

Hence

$$\mathscr{F}_{cab} = H_{cab}L_{cab} = (690)(0.5) = 345 \text{ At}$$

Therefore the total mmf required to produce the desired air gap flux is

$$\mathscr{F} = \mathscr{F}_{cab} + \mathscr{F}_{bc} = 345 + 317.2 = 662.2 \text{ At}$$

15-7 HYSTERESIS AND EDDY-CURRENT LOSSES IN FERROMAGNETIC MATERIALS

The process of magnetization and demagnetization of a ferromagnetic material in a symmetrically cyclic condition involves a storage and release of energy which is not completely reversible. As the material is magnetized during each half-cycle, it is found that the amount of energy stored in the magnetic field exceeds that which is released upon demagnetization. The background for understanding this behavior was provided in Sec. 15-3. There the hysteresis loop was identified as the variation of flux density as a function of the magnetic field intensity for a ferromagnetic material in a cyclic condition. The salient feature of the hysteresis loop is the delayed reorientation of the domains in response to a cyclically varying magnetizing force. A single hysteresis loop is depicted in Fig. 15-16. The direction of the arrows on this curve indicates the manner in which B changes as H varies from zero to a positive maximum through zero to a negative maximum and back to zero again, thus completing the loop. To appreciate the meaning of the various shaded areas shown in Fig. 15-16, let us look at the units associated with the product of B and H. Thus

$$\text{units of } (HB) = \frac{\text{amperes}}{\text{meter}} \times \frac{\text{newtons}}{\text{ampere-meter}} = \frac{\text{newtons}}{\text{meter}^2} = \frac{N}{m^2}$$

But

$$\text{newton-meter} = \text{joule}$$

Hence

$$\text{units of } (HB) = \frac{\text{joule}}{\text{meter}^3} = \frac{J}{m^3}$$

which is clearly recognized as an energy density. Therefore, in dealing with areas involving B and H in connection with a hysteresis loop we are really dealing with energy densities expressed on a per cycle basis because the hysteresis loop is repeatable for each cyclic variation of H.

The energy stored in the magnetic field during that portion of the cyclic variation of H when it increases from zero to its positive maximum value (assuming the material is already in a cyclic state) is given by

$$w_1 = \int_{B_a}^{B_b} H \, dB \qquad J/m^3 \qquad (15\text{-}28)$$

when mks units are used; i.e., H must be expressed in At/m and B in teslas. Note that the axes of Fig. 15-16 are so labeled. Moreover, during that portion of its cyclic variation when H decreases from its positive maximum value to zero (as it follows along curve bc of the hysteresis loop) energy is being released by the magnetic field and returned to the source, and this quantity can be represented as

$$w_2 = \int_{B_b}^{B_c} H \, dB \qquad J/m^3 \qquad (15\text{-}29)$$

In this equation, since $B_b > B_c$, the quantity w_2 will be negative, indicating that the energy is being released rather than stored by the magnetic field.

A graphical interpretation of Eq. (15-28) leads to the result that the energy absorbed by the field, when H is increasing in the positive direction, can be represented by the area $abdca$. Similarly, the energy released by the field as H varies from H_{max} to zero can be represented by area $bdcb$. The difference between these two energy densities represents the amount of energy which is not returned to the source but rather is dissipated as heat as the domains are realigned in response to the changing magnetic field intensity. This dissipation of energy is

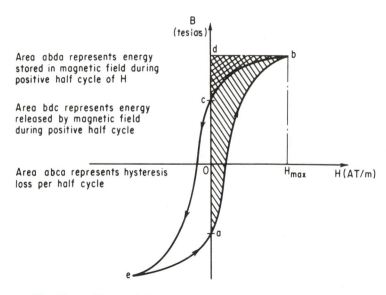

Area *abda* represents energy
stored in magnetic field during
positive half cycle of H

Area *bdc* represents energy
released by magnetic field
during positive half cycle

Area *abca* represents hysteresis
loss per half cycle

Fig. 15-16   Hysteresis loop and energy relationships per half-cycle.

called *hysteresis loss*. Keep in mind that Fig. 15-16 depicts this energy density loss for a one-half cycle variation of H. Hence area *abca* represents the hysteresis loss per half-cycle. It certainly follows from symmetry that upon completion of the negative half-cycle variation of H an equal energy loss occurs. Therefore, as H varies over the complete cycle, the total energy loss per cubic meter is represented by the area of the hysteresis loop. More specifically, this energy loss per cycle can be expressed mathematically as

$$w_h = \text{(area of hysteresis loop)} \quad \frac{\text{J}}{\text{m}^3 \times \text{cycle}} \qquad (15\text{-}30)$$

where rationalized mks units are used for H and B.

It is frequently desirable to express the hysteresis loss of ferromagnetic materials in watts—the unit of power. A little thought about the units of w_h in Eq. (15-30) shows how this can be directly accomplished. Thus

$$w_h = \frac{\text{energy}}{\text{vol} \times \text{cycles}} = \frac{\text{power} \times \text{seconds}}{\text{vol} \times \text{cycles}} = \frac{\text{power}}{\text{vol} \times \text{cycles/second}} \qquad (15\text{-}31)$$

Now let P_h = power loss in watts

v = volume of ferromagnetic material

f = cycles/seconds = frequency of variation of H

Then Eq. (15-31) becomes

$$w_h = \frac{P_h}{vf} \qquad (15\text{-}32)$$

or

$$P_h = w_h vf \qquad (15\text{-}33)$$

where w_h—the energy density loss—is determined from Eq. (15-30).

To obviate the need of finding the area of the hysteresis loop in order to compute the hysteresis loss in watts from Eq. (15-33), Steinmetz obtained an empirical formula for w_h based on a large number of measurements for various ferromagnetic materials. He expressed the hysteresis power loss as

$$P_h = vfK_h B_m^n \qquad (15\text{-}34)$$

where B_m is the maximum value of the flux density and n lies in the range $1.5 \leqslant n \leqslant 2.5$ depending upon the material used. The parameter K_h also depends upon the material. Some typical values are: cast steel 0.025, silicon sheet steel 0.001, and permalloy 0.0001.

In addition to the hysteresis power loss, another important loss occurs in ferromagnetic materials that are subjected to time-varying magnetic fluxes—the *eddy-current* loss. This term is used to describe the power loss associated with the circulating currents that are found to exist in closed paths within the body of a ferromagnetic material, causing an undesirable heat loss. These circulating currents are created by the differences in potential existing throughout the body of the material owing to the action of the changing flux. If the magnetic circuit is composed of solid iron, the ensuing power loss is appreciable because the circulating currents encounter relatively little resistance. To increase significantly the resistance encountered by these eddy currents, the magnetic circuit is invariably composed of very thin *laminations* (usually 14 to 25 mils thick) whenever the electromagnetic device is such that a varying flux permeates it in normal operation. This is the case with transformers and all a-c electric motors and generators. An empirical equation for the eddy-current loss is

$$P_e = K_e f^2 B_m^2 \tau^2 v \quad \text{W} \qquad (15\text{-}35)$$

where K_e = a constant dependent upon the material
$\qquad f$ = frequency of variation of flux in Hz
$\quad B_m$ = maximum flux density
$\qquad \tau$ = lamination thickness
$\qquad v$ = total volume of the material

A comparison of this equation with Eq. (15-34) reveals that eddy-current losses vary as the square of the frequency, whereas the hysteresis loss varies directly with the frequency.

Taken together the hysteresis and eddy-current losses constitute what is frequently called the *core losses* of electromagnetic devices that involve time-varying fluxes for their operation. More than just passing attention is devoted to these losses here because, as will be seen, core losses have an important bearing on temperature rise, efficiency and rating of electromagnetic devices.

15-8 RELAYS—AN APPLICATION OF MAGNETIC FORCE

A *relay* is an electromagnetic device which can often be activated by relatively little energy, causing a movable ferromagnetic armature to open or close one or several pairs of electrical contact points located in another control circuit or in a main circuit handling large amounts of energy. A-c and d-c motor starters are equipped with relays designed to ensure proper operation of motors during starting and running conditions. These devices are found in many applications in all fields of engineering, especially in situations where control of a process or machine is involved. Our objective in this section is to describe the principles that underlie the operation of these electromagnetic devices. Besides the knowledge it offers, this treatment gives the motivation for studying the theory of magnetic fields and circuits.

The derivation of a magnetic force equation is our primary interest because it shows how it is possible to do mechanical work—moving a relay armature—by abstracting energy from that stored in the magnetic field. Moreover, since the emphasis is on the principles involved, the simplifying assumptions of no saturation or no losses are imposed; i.e., linear analysis is used throughout. Accordingly, the magnetization curve of the ferromagnetic material is assumed to be a straight line as depicted in Fig. 15-17(a). (Only the curve corresponding to positive values of H is shown.) Now, by appropriately modifying the axes of the curve plotted in Fig. 15-17(a), a significant and useful thing happens. First recall that the area between the B-axis and the magnetization curve represents the energy absorbed from the source and stored in the magnetic field on a per unit volume basis. Equation (15-28) states this result mathematically. Repeating we have

$$w_f = \int_{B_a}^{B_b} H \, dB \qquad \text{J/m}^3 \qquad (15\text{-}36)$$

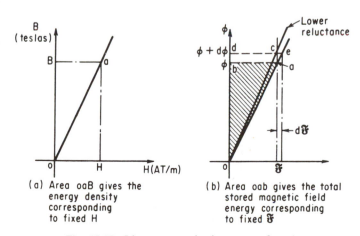

(a) Area oaB gives the energy density corresponding to fixed H

(b) Area oab gives the total stored magnetic field energy corresponding to fixed $\mathfrak{F}$

Fig. 15-17 Linear magnetization curve of a relay.

In the simplified situation of Fig. 15-17(a), Eq. (15-36) becomes merely the area of triangle OaB. Thus

$$w_f = \tfrac{1}{2}BH \quad \text{J/m}^3 \tag{15-37}$$

where H is assumed fixed at the value shown.

Moreover, since w_f is an energy density, the total energy stored in the magnetic field is found by multiplying Eq. (15-28) by the volume. Thus

$$W_f = w_f v = w_f AL \quad \text{J} \tag{15-38}$$

where L is the length and A the cross sectional area of the magnetic circuit. Inserting Eq. (15-37) into Eq. (15-38) yields

$$\boxed{W_f = \tfrac{1}{2}(BA)(HL) = \tfrac{1}{2}\phi\mathscr{F}} \quad \text{J} \tag{15-39}$$

Note that in this equation A is combined with B to identify the flux ϕ and H is combined with L to identify the magnetomotive force $\mathscr{F}$. A graphical representation of Eq. (15-39) appears in Fig. 15-17(b). It should be apparent that Fig. 15-17(b) derives from Fig. 15-17(a) by multiplying the ordinate axis by A and the abscissa axis by L, thus plotting ϕ versus $\mathscr{F}$. Then for a fixed H (or $\mathscr{F}$) area OaB of Fig. 15-17(a) gives the energy density whereas the corresponding area Oab of Fig. 15-17(b) gives the total energy stored in the magnetic field.

To understand how mechanical work can be done by the abstraction of energy stored in the magnetic field, consider the circuitry appearing in Fig. 15-18, which depicts the basic composition of an electromagnetic relay. It consists of an exciting coil placed on a fixed ferromagnetic core equipped with a movable element called the *relay armature*. The relay is energized from a constant voltage source through an adjustable resistor R. To begin with, consider that R is fixed at that value which makes the coil mmf equal to $\mathscr{F}$ and producing the flux ϕ as shown in Fig. 15-17(b). Then adjust R to increase the mmf by $d\mathscr{F}$ and thereby the flux by $d\phi$, assuming the relay armature is held fast to keep the reluctance invariant. Figure 15-17(b) shows that an additional amount of energy is absorbed from the source and stored in the magnetic field virtually equal to the area of rectangle $abdc$. Next readjust R to make $d\mathscr{F}$ zero, and then, keeping the mmf fixed at $\mathscr{F}$, release the armature, thus allowing it to move in the direction to decrease the air gap. Hold the armature fast again when the decreased reluctance causes the flux to increase by the amount $d\phi$ obtained previously.

It is important to note here that, neglecting second-order effects, the new (lower-reluctance) position of the armature is also responsible for permitting the source to supply an additional amount of energy virtually equal to area $abdc$ as

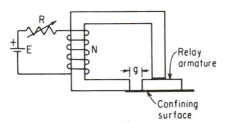

Fig. 15-18 Basic composition of an electromagnetic relay.

before but with one significant difference. Whereas when the armature is held fixed, the additional energy supplied by the source is converted entirely to stored magnetic energy, on the other hand, when the armature is allowed to move towards a lower-reluctance position, only half of the same amount of additionally supplied energy is stored in the magnetic field. The other half is consumed in doing the mechanical work involved in moving the relay armature from the higher- to the lower-reluctance position. That the amount is one-half, readily follows from a glance at Fig. 15-17(b) if we note that the area of triangle *Oac* is one-half the area of rectangle *abdc*. The slope of the magnetization curve is the permeability and so can be used as a measure of the reluctance. Hence the higher the slope of the magnetization curve, the higher μ and the lower the reluctance. For the magnetization curve in Fig. 15-17(b) to go from position *Oa* to *Oc*, it is necessary for the relay armature to be moved to a position corresponding to a smaller air gap. The mechanical work involved in accomplishing this is represented by area *Oac* in Fig. 15-17(b).

The conclusions described in the foregoing can now be expressed mathematically. It is important to keep in mind, however, that the interchange of energy between the magnetic field and the mechanical system (the relay armature) necessarily involves a *change in reluctance*. In other words the change in energy to do mechanical work, dW_m, is equal to the change in magnetic field energy associated with a change in reluctance $d\mathscr{R}$. Thus by Eqs. (15-18) and (15-39)

$$dW_m = -\tfrac{1}{2}\phi^2 d\mathscr{R} \tag{15-40}$$

where the negative sign emphasizes that mechanical work is done through a *decrease* in reluctance.

An expression for the magnetic force developed on the relay armature is readily obtained by recalling that

$$F\,dx = dW_m \tag{15-41}$$

where F is the force in newtons. Inserting this expression into Eq. (15-40) then yields

$$\boxed{F = -\tfrac{1}{2}\phi^2 \frac{d\mathscr{R}}{dx}} \quad \text{N} \tag{15-42}$$

Hence the magnitude of the instantaneous magnetic force is dependent upon the value of the flux as well as the rate of change of reluctance. Moreover, the direction of this force is always such as to bring about a decrease in reluctance as indicated by the minus sign.

EXAMPLE 15-4 In the relay circuit of Fig. 15-18 assume the cross-sectional area of the fixed core and the relay armature to be A, and the air gap flux to be ϕ. Neglecting the reluctance of the iron, find the expression for the magnetic force existing on the relay armature.

Solution: We must find the rate of change of reluctance with distance along the sliding surface. Thus

$$\mathscr{R} = \frac{g}{\mu_0 A} \tag{15-43}$$

where g = air gap length.

$$\frac{d\mathcal{R}}{dx} = \frac{1}{\mu_0 A}\frac{dg}{dx} \tag{15-44}$$

But

$$dg = -dx \tag{15-45}$$

Hence

$$\frac{d\mathcal{R}}{dx} = -\frac{1}{\mu_0 A} \tag{15-46}$$

Inserting this last expression into Eq. (15-42) yields the desired result.

$$F = \frac{1}{2}\frac{\phi^2}{\mu_0 A} \quad N \tag{15-47}$$

EXAMPLE 15-5 The cross-sectional view of a cylindrical plunger magnet appears in Fig. 15-19. The plunger (or armature) is free to move inside a nonferromagnetic guide around which the coil is surrounded by a cylindrically shaped steel shell. The plunger is separated from the shell by an air gap of length g.

(a) Derive the expression for the magnetic force exerted on the plunger when it is in the position shown in Fig. 15-19. The length of the plunger is at least equal to that of the shell. Also it has a radius of a meters. Neglect the reluctance of the steel.

(b) Find the magnitude of the force when the mmf is 1414 At and the plunger magnet dimensions are

$$x = 0.025 \text{ m} \qquad h = 0.05 \text{ m}$$
$$a = 0.025 \text{ m} \qquad g = 0.00125 \text{ m}$$

Solution: Before proceeding with the calculations let us review the principle that gives rise to the force. For the direction of coil current assumed a typical flux path is that shown in Fig. 15-19. Note that the flux path must cross the air gap twice. It is particularly important to note too that the reluctance "seen" by the flux as it crosses the bottom gap is less than that seen by the same flux crossing the gap at the upper part of the plunger because the cross-sectional area of the magnetic circuit is smaller at the upper part than it is at the lower part of the plunger. As a matter of fact, at the lower part of the plunger the cross-sectional area seen by the flux is fixed for all positions of the plunger lying in

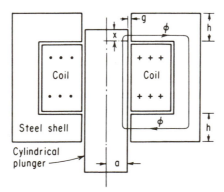

Fig. 15-19 Plunger magnet for Example 15-5.

the range $0 \leqslant x \leqslant h$, and specifically is equal to $2\pi ah$ (assuming g is small compared to a). Accordingly, a force is created on the plunger so directed as to decrease the reluctance. This means that the magnetic force acts to move the plunger upward.

(a) The solution of this part is obtained from Eq. (15-42), but first we need to identify the correct expression for the reluctance as seen by the flux. The reluctance of the steel is negligible. Hence the reluctance associated with the typical flux shown in Fig. 15-19 is merely the sum of the reluctances associated with each air gap. Thus

$$\mathcal{R} = \frac{g}{\mu_0 2\pi ah} + \frac{g}{\mu_0 2\pi ax} = \frac{g}{2\pi a\mu_0}\left(\frac{1}{h} + \frac{1}{x}\right) \tag{15-48}$$

Then

$$\frac{d\mathcal{R}}{dx} = \frac{g}{2\pi a\mu_0}\left(-\frac{1}{x^2}\right) \tag{15-49}$$

Also

$$\phi = \frac{\mathcal{F}}{\mathcal{R}} = \frac{\mathcal{F}}{g/\mu_0 2\pi a}\frac{xh}{h + x} \tag{15-50}$$

Inserting Eqs. (15-49) and (15-50) into Eq. (15-42) yields

$$F = \pi a\mu_0 \frac{\mathcal{F}^2 h^2}{g}\left(\frac{1}{x + h}\right)^2 = 3.94 \times 10^{-6}\frac{ah^2}{g}\mathcal{F}^2\left(\frac{1}{x + h}\right)^2 \text{ N} \tag{15-51}$$

(b) Inserting the specified dimensions into Eq. (15-51) gives the magnetic force as

$$F = 70 \text{ N} = 15.74 \text{ lb}$$

Summary review questions

1. State Ampère's Law and comment about the motion of lines of force.
2. Describe how the magnetic quantity—flux density—is defined from Ampère's Law. Demonstrate why this quantity is aptly named.
3. Define permeability and show how this quantity can be experimentally determined for a particular medium. What is relative permeability?
4. Explain the meaning of magnetic flux and show how it is related to magnetic flux density.
5. What is magnetic field intensity? How is it different from magnetic flux density?
6. Describe Ampère's circuital law and illustrate its usefulness in magnetic circuit computations.
7. What is magnetomotive force? How does it differ from electromotive force? How is it similar?
8. How does the notion of reluctance arise in dealing with magnetic circuits? Why is this property useful? Name the physical parameters that influence this quantity.
9. Ampère's law deals with the force that exists between two current-carrying conductors. What influence, if any, does the orientation of one conductor relative to the other have on this force? Illustrate.
10. Write Ampère's law expressed in terms of the notation of vector analysis.
11. Describe how the direction of the force between two current-carrying conductors is determined.

12. Explain the following terms: diamagnetic, paramagnetic, ferromagnetic, magnetic moment.

13. Describe briefly the domain theory of magnetism.

14. What is saturation as it applies to ferromagnetism?

15. Magnetic circuits are basically nonlinear. Explain what this statement means and why it is so.

16. Define the following: hysteresis, retentivity, coercivity, residual flux density, coercive force, normal magnetization curve.

17. Describe the premise on the basis of which it is possible to represent the three-dimensional field problems of magnetism by a magnetic circuit.

18. Describe the analogies that can be made between electric and magnetic circuits regarding the following items: driving force, field intensity, impedance drops, equivalent circuits.

19. What is the relationship between the tesla and the number of lines/in.^{2}? Between the gauss and the number of lines/in.^{2}? Between the tesla and the gauss?

20. Describe in some detail the two types of magnetic circuit problems that confront the designer of such circuits.

21. Demonstrate through dimensional analysis why the hysteresis loop represents an energy loss per cycle. What can be done to diminish this loss?

22. Cite an empirical formula for the evaluation of hysteresis loss expressed in watts. Repeat for eddy-current loss.

23. Describe how mechanical work can be done through the extraction of energy that is stored in a magnetic field.

24. Cite the formula for the force produced in a relay and explain the meaning of each term.

Problems _____

GROUP I

15-1. A long straight wire located in air carries a current of 4 A. Assume the relative permeability of air is unity.
   (a) Find the value of the magnetic field intensity at a distance 0.5 m from the center of the wire.
   (b) A second long straight wire carrying a current of 2 A is placed parallel to the first one at a distance of 0.5 m with the current flowing in the same direction. Find the direction and magnitude of the force per meter existing between the wires.
   (c) Repeat part (b) for the case where the wires are imbedded in iron having a relative permeability of 10,000 and a spacing of 0.05 m.

15-2. The wires shown in Fig. P15-2 are long, straight, and parallel and are completely embedded in iron having a relative permeability of 1000. Each wire carries a current of 10 A.

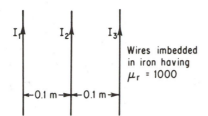

Fig. P15-2

(a) Compute the magnitude and direction of the resultant force per meter on the wire in which I_2 flows.

(b) Compute the magnitude and direction of the force per meter on the wire in which I_1 flows.

(c) Repeat part (a) for the third wire.

15-3. Repeat Prob. 15-2 for the case where the current I_2 flows opposite to I_1 and I_3.

15-4. A uniform magnetic field of 0.7 T in the iron is applied to the configuration of Fig. P15-2.

(a) Compute the direction and magnitude of the resultant force per meter when the magnetic field is directed perpendicularly into the plane of the paper.

(b) What is the value of this resultant force per meter when the magnetic field is applied in the plane of the wires and directed from right to left?

(c) Compute the resultant force per meter when the magnetic field is applied at an angle of 45° relative to the plane of the paper and directed into it from right to left.

(d) What is the resultant force per meter when the magnetic field is applied at an angle of 60° relative to the plane of the paper and directed into it from top to bottom?

15-5. In the configuration of Fig. P15-5 the current I_1 has a value of 40 A. Find the value of I_2 that causes the magnetic field intensity at point P to disappear.

15-6. A circular loop of wire of radius r meters and consisting of a single turn carries the current I as shown in Fig. P15-6. Derive the expression for the magnetic field intensity at the center.

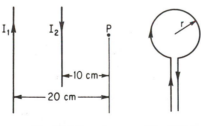

Fig. P15-5   **Fig. P15-6**

15-7. A magnetic circuit composed of silicon sheet steel has the square construction shown in Fig. P15-7.

(a) Find the mmf required to produce a core flux of 25×10^{-4} Wb.

(b) If the coil has 80 turns, how much current must be made to flow through the coil?

15-8. The magnetic circuit of Prob. 15-7 has an air gap of 0.1 cm cut in the right leg.

For a coil having 100 turns find the current that must be allowed to flow in order that the core flux be 0.0025 Wb.

15-9. In the magnetic circuit of Fig. P15-9 determine the coil mmf needed to produce a flux of 0.0014 Wb in the right leg. The thickness of the magnetic circuit is 0.04 m and is uniform throughout. Medium silicon steel is used.

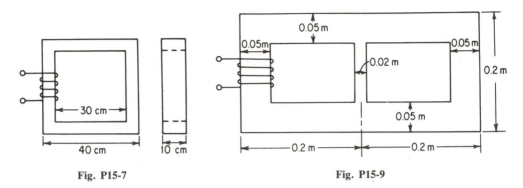

Fig. P15-7   Fig. P15-9

15-10. Repeat Prob. 15-9 for the case where the coil is placed on the center leg.

15-11. In the magnetic circuit of Prob. 15-7 find the core flux produced by a coil current of 200 At.

15-12. The core shown in Fig. P15-12 has a uniform cross-sectional area of 2 in.^{2} and a mean length of 12 in. Also, coil A has 200 turns and carries 0.5 A, coil B has 400 turns and carries 0.75 A, and coil C carries 1.00 A. How many turns must coil C have in order that the core flux be 120,000 lines? The coil currents have the directions indicated in the figure. The core is made of silicon sheet steel.

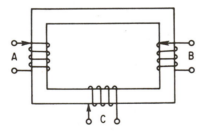

Fig. P15-12

15-13. A sample of iron having a volume of 16.4 cm^3 is subjected to a magnetizing force sinusoidally varying at a frequency of 400 Hz. The area of the hysteresis loop is found to be 64.5 cm^2 with flux density plotted in kilolines per square inch and magnetizing force in ampere-turns per inch. The scale factors used are: 1 in. = 5 kilolines/in.^{2} and 1 in. = 12 At/in. Find the hysteresis loss in watts.

15-14. A ring of ferromagnetic material has a rectangular cross section. The inner diameter is 7.4 in., the outer diameter is 9 in., and the thickness is 0.8 in. There is a coil of 600 turns wound on the ring. When the coil carries a current of 2.5 A, the flux produced in the ring is 1.2×10^{-3} Wb. Find the following quantities expressed in mks units: (a) magnetomotive force; (b) magnetic field intensity; (c) flux density; (d) reluctance; (e) permeability; (f) relative permeability.

15-15. In plotting a hysteresis loop the following scales are used: 1 cm = 10 At/in., and

1 cm = 20 kilolines/in.^{2}. The area of the loop for a certain material is found to be 6.2 cm^2. Calculate the hysteresis loss in joules per cycle for the specimen tested if the volume is 400 cm^3.

15-16. The flux in a magnetic core is alternating sinusoidally at a frequency of 500 Hz. The maximum flux density is 50 kilolines/in.^{2}. The eddy-current loss then amounts to 14 W. Find the eddy-current loss in this core when the frequency is 750 Hz and the flux density 40 kilolines per square inch.

15-17. A flux ϕ penetrates the complete volume of the iron bars as shown in Fig. P15-17. When the bars are assumed separated by g meters, find the expression for the force existing between the two parallel plane faces. Neglect the reluctance of the iron.

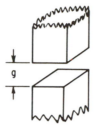

Fig. P15-17

GROUP II

15-18. In the magnetic circuit of Prob. 15-8 find the core flux produced by a coil mmf of 600 At.

15-19. In the magnetic circuit shown in Fig. P15-19 the coil F_1 is supplied with 350 At in the direction indicated. Find the direction and magnitude of the mmf required in coil F_2 in order that the air-gap flux be 180,000 lines. The core has an effective cross-sectional area of 9 in.^{2} and is made of silicon sheet steel. The length of the air gap is 0.05 in.

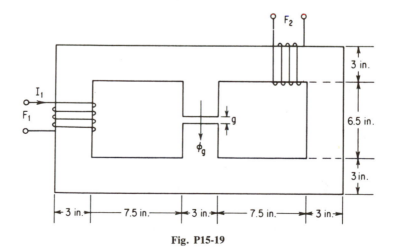

Fig. P15-19

15-20. In the magnetic circuit of Fig. P15-19 the coil F_1 is supplied with 200 At in the direction shown. Find the direction and magnitude of the mmf required of coil F_2 in order that the air-gap flux be 90,000 lines.

15-21. The total core loss (hysteresis plus eddy current) for a specimen of magnetic sheet steel is found to be 1800 W at 60 Hz. If the flux density is kept constant and the frequency of the supply increased 50%, the total core loss is found to be 3000 W. Compute the separate hysteresis and eddy-current losses at both frequencies.

15-22. A magnetic flux ϕ penetrates a movable magnetic core which is vertically misaligned relative to the north and south poles of an electromagnet as depicted in Fig. P15-22. The depth dimension for the core and the electromagnet is b. Determine the expression for the force that acts to bring the core into vertical alignment. Express the result in terms of the air-gap flux density and the physical dimensions. Neglect the reluctance of the iron.

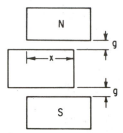

Fig. P15-22

15-23. The magnetization curve of a relay is shown in Fig. P15-23.

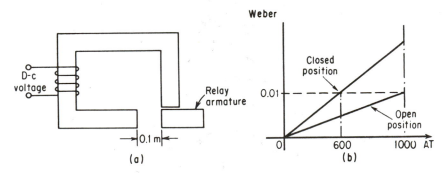

Fig. P15-23

   (a) Find the energy stored in the field with the relay in the open position.
   (b) Assume the relay armature moves rapidly under conditions of constant flux. Compute the work done in going from the open to the closed position.
   (c) Calculate the force in newtons exerted on the armature in part (b).
   (d) Assume the relay armature moves slowly at constant mmf. Compute the work done in going from the open to the closed position.
   (e) Does the energy of part (d) come from the original stored field energy? Explain.

chapter sixteen

Transformers

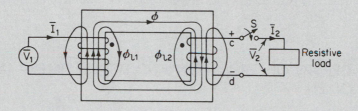

An understanding of the transformer is essential to the study of electromechanical energy conversion. Although electromechanical energy conversion involves interchange of energy between an electrical system and a mechanical system whereas the transformer involves the interchange of energy between two or more electrical systems, nonetheless the coupling device in both cases is the magnetic field and its behavior in each case is fundamentally the same. As a result we find that many of the pertinent equations and conclusions of transformer theory have equal validity in the analysis of a-c machinery and some aspects of d-c machinery. For example, the equivalent circuit of single-phase and three-phase induction motors is found to be identical in form to that of the transformer. Other examples can be cited, as a glance through the next chapter readily reveals.

In addition to serving as a worthwhile prelude to the study of electromechanical energy conversion, an understanding of transformer theory is important in its own right because of the many useful functions the transformer performs in prominent areas of electrical engineering. In communication systems ranging in frequency from audio to radio to video, transformers are found fulfilling widely varied purposes. For example, *filament transformers* are used to supply heater power to the filaments of vacuum tubes, including cathode ray tubes that are used in television receivers and computer display terminals. *Input transformers* (to connect a microphone output to the first stage of an electronic amplifier), *interstage transformers*, and *output transformers* are to be found in radio and television circuits. Transformers are also used in communication circuits as impedance transformation devices which allow maximum transfer of power from the input circuit to the coupled circuit. Telephone lines and control circuits are two more areas in which the transformer is used extensively. In electric power distribution systems the transformer makes it possible to convert electric power from a generated voltage of about 15 to 20 kV (as determined by generator design limitations) to values of 380 to 750 kV, thus permitting transmission over long distances to appropriate distribution points (e.g., urban areas) at tremendous savings in the cost of copper as well as in power losses in the transmission lines. Then at the distribution points the transformer is the means by which these dangerously high voltages are reduced to a safe level (208/120 V) for use in homes, offices, shops, and so on. In short, the transformer is a useful device to be found in many phases of electrical engineering and therefore merits the attention given to it in this chapter.

16-1 THEORY OF OPERATION AND DEVELOPMENT OF PHASOR DIAGRAMS

In the interest of clarity and motivation the theory of operation of the transformer is developed in five steps, starting with the ideal iron-core reactor and ending with the transformer phasor diagram at full-load. In the process the important transformer equations are developed along with a physical explanation of how the transformer operates. Emphasis on complete understanding of the physical behavior of a device is stressed, because once this is grasped, the theory as well as the associated mathematical descriptions readily follow.

In its simplest form a transformer consists of two coils which are mutually coupled. When the coupling is provided through a ferromagnetic ring (circular or otherwise), the transformer is called an *iron-core* transformer. When there is no ferromagnetic material but only air, the device is described as an *air-core transformer*. In this chapter attention is confined exclusively to the iron-core type. Of the two coils the one that receives electric power is called the *primary winding* whereas the other, which can deliver electric power to a suitable load circuit, is called the *secondary winding*.

Ideal Iron-Core Reactor. As the first step in the development of the theory of transformers refer to Fig. 16-1 which depicts an ideal iron-core reactor energized

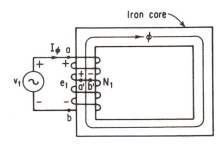

Fig. 16-1 Ideal iron-core reactor—zero winding resistance and no core losses. Polarities are instantaneous.

from a sinusoidal voltage source v_1. The reactor is ideal because the coil resistance and the iron losses are assumed zero and the magnetization curve is assumed to be linear.

Our interest at this point is to develop the phasor diagram that applies to the circuitry of Fig. 16-1, because thereby much is revealed about the theory. As v_1 increases in its sinusoidal variation, it causes a sinusoidal magnetizing current i_ϕ† to flow through the N_1 turns of the coil, which in turn produces a sinusoidally varying flux in time phase with the current. In other words, when the current is zero the flux is zero and when the current is at its positive maximum value, so too is the flux. Applying Kirchhoff's voltage law to the circuit, we have

$$v_1 - e_1 = 0 \qquad (16\text{-}1)$$

where e_1 is the *voltage drop* associated with the flow of current through the coil and can be written in terms of i_ϕ as

$$e_{ab} = e_1 = L\frac{di_\phi}{dt} \qquad (16\text{-}2)$$

where L is the inductance of the coil and $i_\phi = i_{ab}$. Inserting Eq. (16-2) into Eq. (16-1) and rearranging yields

$$v_1 = e_1 = L\frac{di_\phi}{dt} = e_{ab} \qquad (16\text{-}3)$$

We also know that the magnetizing current is sinusoidal because the applied voltage is sinusoidal and the magnetization curve is assumed linear. Consequently, we can write for the magnetizing current the expression

$$i_\phi = \sqrt{2}\,I_\phi \sin \omega t \qquad (16\text{-}4)$$

where I_ϕ is the rms value of the current.

Introducing Eq. (16-4) into Eq. (16-3) and performing the differentiation called for yields

$$v_1 = e_1 = \sqrt{2}(\omega L)I_\phi \cos \omega t = \sqrt{2}\,(\omega L)I_\phi \sin\left(\omega t + \frac{\pi}{2}\right) \qquad (16\text{-}5)$$

The maximum value of this quantity occurs when the $\cos \omega t$ has the value of unity. Then

$$E_{1\max} = \sqrt{2}\,(\omega L)I_\phi = \sqrt{2}\,E_1 \qquad (16\text{-}6)$$

where

$$E_1 \equiv \omega L I_\phi = \text{rms value of the inductive reactance drop} \qquad (16\text{-}7)$$

A comparison of Eqs. (16-4) and (16-5) shows that the magnetizing current lags the voltage appearing across the coil by 90 electrical degrees. The phasor diagram of Fig. 16-2(a) depicts this condition. Since all the quantities involved here are sinusoidal, the phasors shown represent the rms value of the quantities.

Another approach that can be used in arriving at the phasor diagram for the ideal iron-core reactor is worth considering because of its use later. The

† Lowercase letters are used to represent instantaneous values.

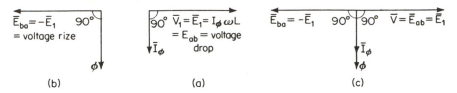

Fig. 16-2 Phasor diagrams of the ideal iron-core reactor: (a) circuit viewpoint; (b) field viewpoint; (c) combined diagram.

approach just described can be referred to as a *circuit viewpoint* because it deals solely with the electric circuit in which I_ϕ flows as well as the circuit parameter L. No direct use is made of the magnetic flux ϕ, which is a space (or field) quantity. In contrast the second approach starts with the magnetic field as it appears in Faraday's law of induction. This is one of the fundamental laws upon which the science of electrical engineering is based. Faraday's law states that the emf induced in a coil is proportional to the number of turns linking the flux as well as the time rate of change of the linking flux. In applying Faraday's law to the circuit of Fig. 16-1 the expression may be written either as a voltage drop (*a* to *b*) or as a voltage rise (*b* to *a*). By the former description we can write

$$e_{ab} = e_1 = +N\frac{d\phi}{dt} \tag{16-8}$$

Expressed as a voltage rise, the induced emf equation becomes

$$e_{ba} = -e_1 = -N\frac{d\phi}{dt} \tag{16-9}$$

In either case, the sign is attributable to Lenz's law. The minus sign states that the emf induced by a changing flux is always in the direction in which current would have to flow to oppose the changing flux. On the other hand, the plus sign in Eq. (16-8) denotes that the polarity of *a* is positive whenever the time rate of change of flux is positive as determined by the direction of current flow.

Equation (16-9) then represents the starting point for the second approach in analyzing the circuit of Fig. 16-1. Corresponding to the sinusoidal variation of the magnetizing current as expressed by Eq. (16-4), there is also a sinusoidal variation of flux which may be expressed as

$$\phi = \Phi_m \sin \omega t \tag{16-10}$$

where Φ_m denotes the maximum value of the magnetic flux. Putting this expression into Eq. (16-9) and performing the required differentiation leads to

$$e_{ba} = -e_1 = -N_1\frac{d\phi}{dt}$$

$$= -N_1\Phi_m\omega \cos \omega t = E_{1\text{max}} \sin\left(\omega t - \frac{\pi}{2}\right) \tag{16-11}$$

where

$$E_{1\text{max}} = N_1\Phi_m\omega \equiv \sqrt{2}\, E_1 \tag{16-12}$$

and

$$E_1 = \text{rms value of the induced emf}$$

A comparison of Eqs. (16-10) and (16-11) makes it clear that the reaction emf, $-e_1$, *lags* the changing flux that produces it by 90°. This situation is depicted in Fig. 16-2(b). Basically, the emf, $-e_1$, is a voltage rise (or generated emf) which has such a direction that, if it were free to act, would cause a current to flow opposing the action of I_ϕ. To verify this solely in terms of the action-reaction law consider any two adjacent points such as a' and b' on coil N_1 in Fig. 16-1. If the flux is assumed increasing in the direction shown, then the reaction in the coil must be such that current would have to flow from point b' to a' in order to oppose the increasing flux. Recall that by the *right-hand rule* any current assumed to be flowing from b' to a' produces flux lines opposing the increasing flux. In order for this condition to prevail, it is therefore necessary for point a' to be positive with respect to b' for the instant being considered. This line of reasoning when extended to the full coil, leads to the polarity markings shown in Fig. 16-1. The interesting thing to note here is that on the basis of the flux viewpoint the emf $-e_1$ in Fig. 16-1 is treated as a reaction (or generated) voltage rise in progressing from b' to a'. On the other hand, from the circuit viewpoint the voltage e_1, and its polarity when considered in the direction of flow of the current I_ϕ, appears as a voltage drop. The voltage e_1 remains the same; it is merely the point of view that changes.

Because I_ϕ and the flux ϕ are in time phase, it is customary to combine the two viewpoints into a single phasor diagram as depicted in Fig. 16-2(c). Since $E_{ab} = E_1$ represents the voltage drop, it follows that the same induced emf viewed on the basis of a reaction to the changing flux, being a voltage rise, is equal and opposite to E_1. That is, $E_{ba} = -E_1$. Another way of describing this is to say that the terminal voltage must always contain a component equal and opposite to the voltage rise.

It should be apparent up to this point that the reaction voltage viewed from b' to a' in Fig. 16-1 does not succeed in actually establishing a reverse current to oppose the increasing flux. This never happens in the primary winding of a two-winding transformer, but it does happen in the secondary winding as is presently described.

In the work that follows in this chapter preference is given to treating the induced emf as a voltage drop. That is, in applying Faraday's law, the version described by Eq. (16-8) will be used. Due regard is given to Lenz's law in this equation by noting that, when the time rate of change of flux in the configuration of Fig. 16-1 is positive, the polarity of point a is also positive.

Practical Iron-Core Reactor. As the second step in the development of transformer theory, let us consider how the phasor diagram of Fig. 16-2(a) must be modified in order to account for the fact that a practical reactor has a winding-resistance loss as well as hysteresis and eddy-current losses. A linear magnetization curve will continue to be assumed because the effect of the nonlinearity is to cause higher harmonics of fundamental frequency to exist in the magnetizing current—which, of course, cannot be included in the phasor diagram. Only quantities of the same frequency can be shown.

To develop the phasor diagram in this case we start with the flux phasor ϕ and then draw the induced emf E_1 90° leading. This much is shown in Fig. 16-3(a). Next we must consider the location of the current. Keep in mind that the iron is in a cyclic condition, i.e., the flux is sinusoidally varying with time. Moreover, this cyclic variation takes place along a hysteresis loop of finite area because the core losses are no longer assumed zero. Now, as pointed out in Chapter 15, it is characteristic of ferromagnetic materials in a cyclic condition to have the flux *lag* behind the magnetizing current by an amount called the *hysteretic angle* γ. See Fig. 16-3(b). Since the positive direction of rotation is counterclockwise in a phasor diagram, $\bar{I}_m$ is shown leading ϕ to bear out the physical fact. Note, too, that the magnetizing current $\bar{I}_m$ is now considered made up of two fictitious components $\bar{I}_\phi$ and $\bar{I}_c$. The quantity $\bar{I}_\phi$ is the purely reactive component and is the current that would flow if the core losses were zero. The quantity $\bar{I}_c$ is that fictitious quantity which, when multiplied by the voltage E_1 produced by the changing flux, represents exactly the power needed to supply the core losses caused by the same changing flux. Thus we can write

$$I_c E_1 \equiv P_c \qquad \text{W} \tag{16-13}$$

or

$$I_c \equiv \frac{P_c}{E_1} \qquad \text{A} \tag{16-14}$$

where P_c denotes the sum of the hysteresis and eddy-current losses. This technique for handling the core loss of an electromagnetic device is a useful one and is frequently applied in the study of a-c machinery.

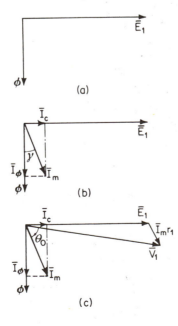

Fig. 16-3 Development of the phasor diagram of a practical iron-core reactor: (a) building on the flux phasor; (b) accounting for core losses; (c) accounting for core losses and winding resistance.

To complete the phasor diagram for the practical iron-core reactor, it is only necessary to account for the winding-resistance drop. This is readily accomplished by applying Kirchhoff's voltage law to the circuit. Hence

$$\overline{V}_1 = \overline{I}_m r_1 + \overline{E}_1 \quad \text{V} \tag{16-15}$$

where r_1 is the winding resistance in ohms. Since r_1 is a constant, it follows that the $\overline{I}_m r_1$ drop must be in phase with $\overline{I}_m$ or located parallel to $\overline{I}_m$ as depicted in Fig. 16-3(c). Note that the component of $\overline{I}_m$ that is in phase with the applied voltage $\overline{V}_1$ exceeds $\overline{I}_c$, as well it should because this component must supply not only the core losses but the winding copper losses too. Thus the input power may be expressed as

$$P_{in} = V_1 I_m \cos \theta_0 = I_m^2 r_1 + P_c \quad \text{W} \tag{16-16}$$

where θ_0 denotes the power-factor angle.

The Two-Winding Transformer. By placing a second winding on the core of the reactor of Fig. 16-1, we obtain the simplest form of a transformer. This is depicted in Fig. 16-4. The rms value of the induced emf E_1 appearing in the primary winding *ab* readily follows from Eq. (16-12) as

$$E_1 = \frac{N_1 \Phi_m \omega}{\sqrt{2}} = \frac{2\pi f}{\sqrt{2}} \Phi_m N_1 = 4.44 f \Phi_m N_1 \tag{16-17}$$

Keep in mind that this result derives directly from Faraday's law and is very important in machinery analysis. One significant result deducible from this equation is that for transformers having relatively small winding resistance (i.e., $E_1 \approx V_1$), the value of the maximum flux is determined by the applied voltage.

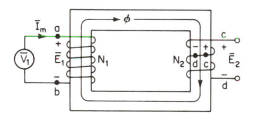

Fig. 16-4 The two-winding transformer.

The induced emf appearing across the secondary winding terminals is produced by the same flux that causes E_1. Hence the only difference in the rms values is brought about by the difference in the number of turns. Thus we can write

$$E_2 = 4.44 f \Phi_m N_2 \tag{16-18}$$

Dividing Eq. (16-17) by (16-18) leads to

$$\frac{E_1}{E_2} = \frac{N_1}{N_2} \equiv a = \text{ratio of transformation} \tag{16-19}$$

Another fact worth noting is that E_1 and E_2 are in time phase because they are induced by the same changing flux.

   Transformer Phasor Diagram at No-Load.   When a transformer is at no-load, no secondary current flows. Figure 16-5 depicts the situation. A study of Fig. 16-5 should make it apparent that the development of the phasor diagram follows from Fig. 16-3(c) by making two modifications. One is to identify the secondary induced emf in the diagram. But, as already observed in the preceding step, the emf induced in the secondary winding must be in phase with the corresponding voltage E_1 of the primary winding. For convenience call this secondary induced emf E_2. Then E_2 and E_1 must both lead the flux phasor by 90° as shown in Fig. 16-6. For convenience the turns ratio a is assumed to be unity. The second modification is concerned with accounting for the fact that not all of the flux produced by the primary winding links the secondary winding. The difference between the total flux linking the primary winding and the mutual flux linking both windings is called the *primary leakage flux* and is denoted by ϕ_{l1}. Note that some of the paths drawn to represent ϕ_{l1} encircle only several turns and not all N_1 turns. As long as such closed paths encircle an mmf different from zero, such a flux path does in fact exist. This is borne out by Ampère's circuital law.

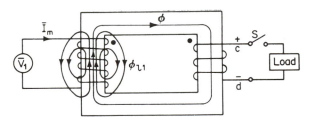

Fig. 16-5   Two-winding transformer at no-load; S open.

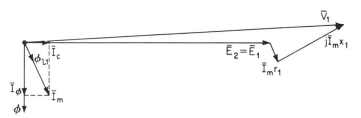

Fig. 16-6   Phasor diagram of transformer at no-load.

   An important distinction exists between the mutual flux ϕ and the primary leakage flux ϕ_{l1}. It is this. The mutual flux exists wholly in iron and so involves a hysteresis loop of finite area. The leakage flux, on the other hand, always involves appreciable air paths, and, although some iron is included in the closed path, the reluctance experienced by the leakage flux is practically that of the air. Consequently, the cyclic variation of the leakage flux involves no hysteresis (or lagging) effect. Hence in the phasor diagram the primary leakage flux must be placed in phase with the primary winding current as shown in Fig. 16-6. Now this leakage flux induces a voltage E_{l1} which must lead ϕ_{l1} by 90°. This emf due to leakage flux may be replaced by an equivalent *primary leakage reactance drop*. Thus

$$\boxed{I_m x_1 \equiv E_{l1}} \qquad (16\text{-}20)$$

This quantity must lead $\bar{I}_m$ by 90°. The phasor addition of this drop and the primary winding resistance drop as well as the voltage drop $\bar{E}_1$ associated with the mutual flux yields the primary applied voltage $\bar{V}_1$ as depicted in Fig. 16-6.

The quantity x_1 of Eq. (16-20) is called the *primary leakage reactance*; it is a fictitious quantity introduced as a convenience in representing the effects of the primary leakage flux.

Transformer under Load. In this last step in the development of the theory of operation of the transformer, we direct attention first to the phasor diagram representation of Kirchhoff's voltage law as it applies to the secondary circuit. For simplicity consider that the load appearing in the secondary circuit of Fig. 16-7 is purely resistive and further that the flux has the direction indicated with the flux increasing. By Lenz's law there is an emf induced in the secondary winding which instantaneously makes terminal c positive with respect to terminal d. When switch S is closed, a current $\bar{I}_2$ flows instantaneously from c through the load to d. For convenience the load resistor is assumed adjusted to cause rated secondary current to flow. $\bar{E}_2$ establishes the secondary load voltage besides accounting for other voltage drops existing in the secondary circuit. When equilibrium is established after the load switch is closed, the appropriate phasor diagram must show $\bar{V}_2$ and $\bar{I}_2$ in phase as depicted in Fig. 16-8(a). Since it is physically impossible for the secondary winding to occupy the same space as the primary winding, a flux is produced by the secondary current which does not link with the primary winding. This is called the *secondary leakage flux*. The voltage E_{l2} induced by this secondary leakage flux leads ϕ_{l2} by 90°. Again as a matter of convenience this secondary leakage flux voltage is replaced by a secondary leakage reactance drop. Thus

$$\boxed{E_{l2} \equiv I_2 x_2} \qquad (16\text{-}21)$$

where x_2 is the fictitious secondary leakage reactance which represents the effects of the secondary leakage flux. Also, because leakage flux has its reluctance predominantly in air, Eq. (16-21) is a linear equation. Thus doubling the current doubles the leakage flux, which doubles $\bar{E}_{l2}$, and by Eq. (16-21) this is represented

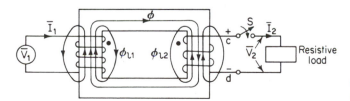

Fig. 16-7 Two-winding transformer under load; S closed.

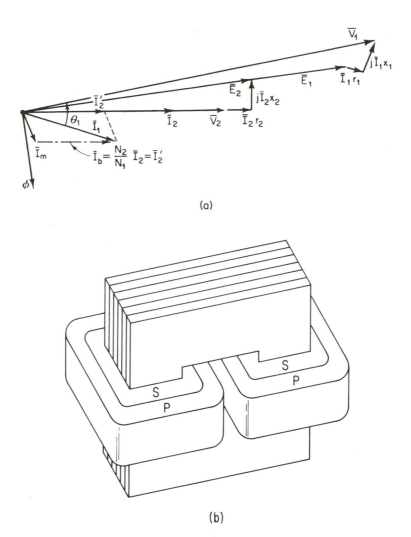

(a)

(b)

Fig. 16-8 (a) Complete phasor diagram of transformer under load for the case where $a = N_1/N_2 > 1$; (b) illustrating the construction features of the two-winding transformers where P and S denote one-half the primary and secondary turns respectively.

by a doubled reactance drop. The phasor sum of the terminal voltage, the secondary winding resistance drop, and the secondary leakage reactance drop adds up to the source voltage $\bar{E}_2$ as demonstrated in Fig. 16-8(a).

When a phasor diagram shows $\bar{V}_2$ and $\bar{I}_2$ in phase, the implication is that power is being supplied to the load equal to $V_2 I_2$. Clearly, this power must come from the primary circuit. But by what mechanism does this come about? To understand the answer to this question, consider the situation depicted in Fig. 16-7. With the flux increasing in the direction shown the secondary emf has the polarity indicated, so that when the switch is closed the action of the secondary current is to cause a decrease in the mutual flux. This tendency for the flux to decrease owing to the action of $N_2 I_2$ causes the voltage drop E_1 [or $e_1 = N_1(d\phi/dt)$] to decrease. However, since the applied voltage $\overline{V}_1$ is constant, the slight decrease in $\overline{E}_1$ creates an appreciable increase in primary current. In fact the primary current is allowed to increase to that value which, when flowing through the primary turns N_1, represents sufficient mmf to neutralize the demagnetizing action of the secondary mmf. Expressed mathematically we have

$$\overline{I}_b N_1 - N_2 \overline{I}_2 = 0 \tag{16-22}$$

where $\overline{I}_b$ denotes the increase in primary current needed to cancel out the effect of the secondary mmf to reduce the flux. $\overline{I}_b$ is often called the *balance* current. This is above and beyond the value $\overline{I}_m$ needed to establish the operating flux. Equation (16-22) reveals that $N_2 \overline{I}_b$ must be oppositely directed to $N_2 \overline{I}_2$. In short, then, it is through the medium of the flux and its small attendant changes that power is transferred from the primary to the secondary circuit of a transformer. Of course, the total primary current is the phasor sum of $\overline{I}_b$ and $\overline{I}_m$. In this case the primary leakage flux is in phase with $\overline{I}_1$ rather than $\overline{I}_m$, which is the case at no-load.

For transformers in which the winding resistance drop and the leakage reactance drop are negligibly small, the magnetizing mmf is the same at no-load as under load. In other words, the phasor sum of the total primary mmf $N_1 \overline{I}_1$ and the secondary mmf $N_2 \overline{I}_2$ must yield $N_1 \overline{I}_m$. Thus

$$\boxed{N_1 \overline{I}_1 - N_2 \overline{I}_2 = N_1 \overline{I}_m} \tag{16-23}$$

Dividing through by N_1 leads to

$$\overline{I}_1 - \frac{N_2}{N_1} \overline{I}_2 = \overline{I}_m \tag{16-24}$$

Inserting Eq. (16-22) then yields

$$\overline{I}_1 - \overline{I}_b = \overline{I}_m = \overline{I}_1 - \overline{I}_2' \tag{16-25}$$

where $\overline{I}_2'$ is equivalent to $\overline{I}_b$. These equations are readily verified in the phasor diagram of Fig. 16-8(a).

Finally, a comment regarding the construction features of practical transformers. Contrary to the schematic representation of Fig. 16-7. it should not be implied that the total number of primary winding turns are placed on one leg of the ferromagnetic core and all the turns of the secondary winding are placed on the other leg. Such an arrangement leads to an excessive amount of leakage flux, so that for a given source voltage the mutual flux is correspondingly smaller. The usual procedure in transformer construction is to put half the primary and secondary turns atop one another on each leg. Figure 16-8(b) illustrates the construction details. This keeps the leakage flux within a few per cent of the mutual flux. Of course still greater reduction in leakage flux can be achieved by further subdividing and sandwiching the primary and secondary turns, but this is clearly obtained at an appreciable increase in the cost of assembly.

16-2 THE EQUIVALENT CIRCUIT

Derivation. It is customary in analyzing devices in electrical engineering to represent the device by means of an appropriate equivalent circuit. In this way further analysis and design as well as computational accuracy is facilitated by the direct application of the techniques of electric circuit theory. This procedure is followed whenever new devices are investigated. Accordingly, equivalent circuits are derived not only for the transformer but for all a-c and d-c motors as well as important electronic devices such as the transistor. Generally, the equivalent circuit is merely a circuit interpretation of the equation(s) that describe the behavior of the device. For the case at hand there are two equations—Kirchhoff's voltage equations for the primary winding and for the secondary winding. Thus, for the primary winding we have

$$\overline{V}_1 = \overline{I}_1 r_1 + j\overline{I}_1 x_1 + \overline{E}_1 \tag{16-26}$$

and for the secondary winding

$$\overline{E}_2 = \overline{V}_2 + \overline{I}_2 r_2 + j\overline{I}_2 x_2 \tag{16-27}$$

Both equations are represented in the phasor diagram of Fig. 16-8, and the corresponding circuit interpretation appears in Fig. 16-9(a). The dot notation is used to indicate which sides of the coils instantaneously carry the same polarity. The schematic diagram of Fig. 16-9(a) can be replaced by the equivalent circuit of Fig. 16-9(b) by recognizing that $\overline{I}_1$ is composed of $\overline{I}_b$ and $\overline{I}_m$. Moreover, since by definition $\overline{I}_m$ consists of two components, $\overline{I}_c$ and $\overline{I}_\phi$, it can be considered as splitting into two parallel branches. One branch is purely resistive and carries the current $\overline{I}_c$, which was defined in-phase with $\overline{E}_1$. The other branch must be purely reactive because it carries $\overline{I}_\phi$, which was defined 90 degrees lagging $\overline{E}_1$ as depicted in Fig. 16-3(b). The ideal transformer is included in Fig. 16-9(b) to account for the transformation of voltage and current that occurs between primary and secondary windings. Thus, with the ratio of transformation $a = 2$, the ideal transformer conveys the information that $\overline{E}_2 = \frac{1}{2}\overline{E}_1$ and that $\overline{I}_2 = 2\overline{I}_b$. The fact that the actual transformer has primary winding resistance and leakage flux and must carry a magnetizing current is taken care of by the circuitry preceding the symbol for the ideal transformer in Fig. 16-9(b).

The resistive element r_c must have such a value that when the voltage E_1 appears across it, it permits the current I_c to flow. Thus

$$r_c \equiv \frac{E_1}{I_c} = \frac{P_c}{I_c^2} \tag{16-28}$$

The second form of this equation is obtained by replacing E_1 by the expression of Eq. (16-13). Hence, r_c may also be looked upon as that resistance which accounts for the transformer core loss. Similarly, the magnetizing reactance of the transformer may be defined as that reactance which, when E_1 appears across

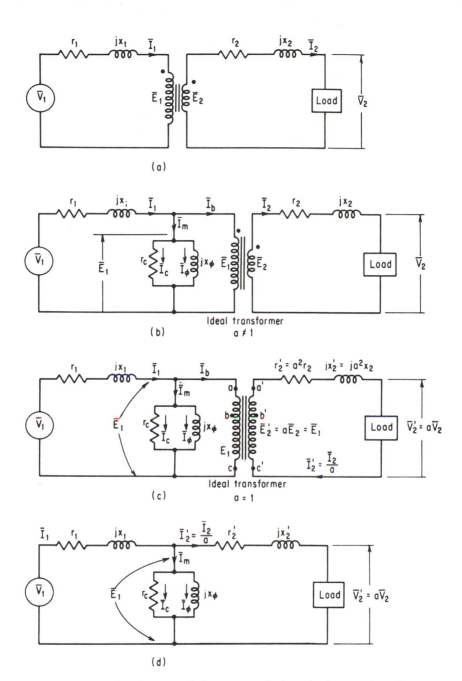

Fig. 16-9 Development of the exact equivalent circuit: (a) schematic diagram; (b) two-winding equivalent circuit; (c) replacing the actual secondary winding with an equivalent winding having N_1 turns; (d) final complete form.

it, causes the current I_ϕ to flow. Thus

$$x_\phi \equiv \frac{E_1}{I_\phi} \qquad (16\text{-}29)$$

A single-line equivalent circuit of the transformer is a desirable goal. However, as long as $E_2 \neq E_1$ this is impossible to accomplish because the potential distribution along the primary and secondary windings can never be identical. Only when $a = 1$ is this condition possible. Therefore, in the interest of achieving the desired simplicity in the equivalent circuit, let us investigate replacing the actual secondary winding by an equivalent winding having the same number of turns as the primary winding. A little thought should make it reasonable to expect that if an equivalent secondary winding having N_1 turns is made to handle the same kVA,† the same copper loss and leakage flux and the same output power as the actual secondary winding, then insofar as the applied voltage source is concerned it can detect no difference between the actual secondary winding of N_2 turns and the equivalent secondary winding of N_1 turns. That is, one winding can replace the other without causing any changes in the primary quantities.

Because the equivalent secondary winding is assumed to have N_1 turns, the corresponding induced emf must be different by the factor a. We have

$$E_2' = aE_2 = E_1 \qquad (16\text{-}30)$$

where the prime notation is used to refer to the equivalent winding having N_1 turns. For the kVA of the equivalent secondary winding to be the same as that of the actual secondary winding, the following equation must be satisfied:

$$E_2'I_2' = E_2I_2 \qquad (16\text{-}31)$$

Inserting Eq. (16-30) into Eq. (16-31) shows that the current rating of the equivalent secondary winding must be

$$I_2' = \frac{I_2}{a} \qquad (16\text{-}32)$$

Since in general $a \neq 1$ it follows that the winding resistance of the equivalent secondary must be adjusted so that the copper loss is the same in both. Accordingly,

$$(I_2')^2 r_2' = I_2^2 r_2 \qquad (16\text{-}33)$$

This equation, together with Eq. (16-32), establishes the value of the winding resistance of the equivalent secondary. Thus

$$r_2' = \left(\frac{I_2}{I_2'}\right)^2 r_2 = a^2 r_2 \qquad (16\text{-}34)$$

† kVA = kilovolt-amperes.

Often the quantity r_2' is described as the secondary resistance referred to the primary. The phrase "referred to the primary" is used because we are dealing with an equivalent secondary winding having the same number of turns as the primary winding.

By proceeding in a similar fashion the secondary leakage reactance referred to the primary can be expressed as

$$\boxed{x_2' = a^2 x_2} \tag{16-35}$$

Moreover, since the load kVA must be the same for the two windings, we have

$$V_2' I_2' = V_2 I_2 \tag{16-36}$$

from which it follows that the load terminal voltage in the equivalent secondary winding is

$$V_2' = \frac{I_2}{I_2'} V_2 = a V_2 \tag{16-37}$$

Appearing in Fig. 16-9(c) are all of the foregoing results in addition to an ideal transformer having unity turns ratio. Keep in mind that Fig. 16-9(c) is drawn with the equivalent secondary winding having N_1 turns whereas Fig. 16-9(b) is drawn with the actual secondary winding. As far as the primary is concerned it cannot distinguish between the two. For the ideal transformer of Fig. 16-9(c) note that joining point a to a', b to b', and c to c' causes no disturbance because instantaneously these points are at the same potential. Accordingly, this ideal transformer may be entirely dispensed with, in which case Fig. 16-9(c) becomes Fig. 16-9(d), which is called the *complete equivalent circuit* of the transformer. This is the single-line diagram we set out to derive.

Approximate Equivalent Circuit. In most constant-voltage, fixed-frequency transformers such as are used in power and distribution applications, the magnitude of the magnetizing current I_m is only a few per cent (2 to 5) of the rated winding current. This happens because high-permeability steel is used for the magnetic core and so keeps I_ϕ very small. Furthermore, steel with a few percent of silicon added is often used and this helps to reduce the core losses, thereby keeping I_c small. Consequently, little error is introduced and considerable simplification is achieved by assuming I_m so small as to be negligible. This makes the equivalent circuit of Fig. 16-9(d) take on the configuration shown in Fig. 16-10. Note that

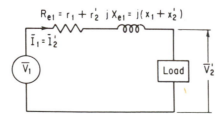

Fig. 16-10 Approximate equivalent circuit referred to the primary.

the primary current $\bar{I}_1$ is now equal to the referred secondary current $\bar{I}_2'$. Also, we can now speak of an equivalent resistance referred to the primary

$$R_{e1} \equiv r_1 + r_2' = r_1 + a^2 r_2 \qquad (16\text{-}38)$$

and an equivalent leakage reactance referred to the primary

$$X_{e1} \equiv x_1 + x_2' = x_1 + a^2 x_2 \qquad (16\text{-}39)$$

The subscript 1 denotes the primary.

The corresponding phasor diagram drawn for the approximate equivalent circuit appears in Fig. 16-11. Note the considerable simplification over the phasor diagram of the exact equivalent circuit which appears in Fig. 16-8.

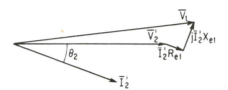

Fig. 16-11 Phasor diagram of the approximate equivalent circuit for a lagging power factor angle θ_2.

The analysis of the preceding pages concerns itself with replacing the actual secondary winding by an equivalent winding having the same number of turns as the primary winding. A little thought should make it apparent that a similar procedure may be employed to draw a single-line equivalent circuit with all quantities referred to the secondary winding. In other words, the actual primary winding can be replaced by an equivalent primary winding having N_2 turns. The resulting equivalent circuit then assumes the form depicted in Fig. 16-12. It is interesting to note that to refer the primary winding resistance r_1 to the secondary, it is necessary to divide by a^2. Thus

$$r_1' = \frac{r_1}{a^2} \qquad (16\text{-}40)$$

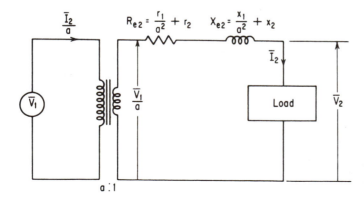

Fig. 16-12 Approximate equivalent circuit referred to the secondary.

Furthermore, the equivalent winding resistance referred to the secondary is

$$R_{e2} = r_1' + r_2 = \frac{r_1}{a^2} + r_2 \qquad (16\text{-}41)$$

and the equivalent leakage reactance referred to the secondary is

$$X_{e2} = x_1' + x_2 = \frac{x_1}{a^2} + x_2 \qquad (16\text{-}42)$$

Note too that the ideal transformer is included in Fig. 16-12, merely to emphasize the transformation that takes place from the primary side to the secondary side. Normally, the primary circuit is omitted entirely leaving only the single-line secondary circuit.

Meaning of Transformer Nameplate Rating. A typical transformer carries a nameplate bearing the following information: 10 kVA 2200/110 V. What are the meanings of these numbers in terms of the terminology and analysis developed so far? To understand the answer refer to Fig. 16-13, which is really the approximate equivalent circuit drawn to include the ideal transformer in order to bring into evidence the transformation action existing between the primary and secondary circuits. The figure 110 refers to the rated secondary voltage. It is the voltage appearing across the load when rated current flows. We call this V_2. The 2200 figure refers to the primary winding voltage rating, which is obtained by taking the rated secondary voltage and multiplying by the turns ratio a between primary and secondary. In our terminology this is V_2'. By specifying the ratio $V_2'/V_2 = 2200/110$ on the nameplate, information about the turns ratio is given. Finally, the kVA figure always refers to the *output* kVA appearing at the secondary load terminals. Of course the input kVA is slightly different because of the effects of R_{e1} and X_{e1}.

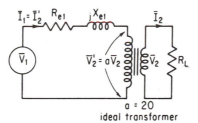

Fig. 16-13 Approximate equivalent circuit drawn with the ideal transformer.

Transfer Function. The transfer function of the transformer may be specified with respect either to voltage or to current. The voltage transfer function is defined as the ratio of the output voltage $\overline{V}_2$ to the input voltage $\overline{V}_1$ with the load connected. Thus

$$T_V = \frac{\overline{V}_2}{\overline{V}_1} = \frac{\overline{V}_2}{a\overline{V}_2 + \overline{I}_2 R_{e1} + j\overline{I}_2 X_{e1}} \approx \frac{1}{a} \qquad (16\text{-}43)$$

Equation (16-43) indicates that, when the winding resistances and leakage fluxes are negligibly small, the voltage transfer function T_V is approximately the reciprocal of the ratio of transformation. When these quantities cannot be neglected, T_v will be a complex number which varies with the load.

The current transfer function T_I is defined as the output current divided by the input current. For the approximate equivalent circuit of Fig. 16-13 we have

$$T_I = \frac{I_2}{I_2'} = a \qquad (16\text{-}44)$$

Here the current transfer function is exactly equal to a because the magnetizing current is assumed negligible. When $\bar{I}_m$ is not negligible, T_I will be somewhat smaller than a and slightly complex.

Input Impedance. The input impedance of a transformer is the impedance measured at the input terminals of the primary winding. In the case of the approximate equivalent circuit shown in Fig. 16-13 we have

$$\bar{Z}_i = \frac{\bar{V}_1}{\bar{I}_1} = R_{e1} + a^2 R_L + j X_{e1} \qquad (16\text{-}45)$$

A glance at this result indicates that the input impedance is dependent upon the load impedance. In the case of power applications the input impedance is of no particular concern because the internal impedance of the source is negligibly small. In communication applications, however, where V_1 is obtained from a vacuum tube or transistor, input impedance is a useful quantity because it describes the effect that the transformer has on the voltage source.

Frequency Response. One of the things we are interested in, when referring to the frequency response of a transformer, is the manner in which the voltage transfer function varies for a fixed load as the frequency of the source voltage V_1 is varied. The amplitude of V_1 is assumed fixed. For power transformers frequency response is of no serious interest because these units are operated at a single fixed frequency—usually 60 Hz and sometimes 50 Hz. However, in communication circuits the frequency of the source voltage is likely to vary over a wide range. For example, in the case of an output transformer which couples the power stage of an audio amplifier to a loudspeaker the frequency may vary over the entire audio range. What, then, is the effect of this varying frequency on the voltage transfer function? The answer follows from an inspection of Fig. 16-9(d). Keep in mind that inductive reactance is directly proportional to the frequency, i.e.,

$$x = \omega L = 2\pi f L \qquad (16\text{-}46)$$

Hence at low frequencies the reactance x_ϕ, which normally is so high that it may be removed from the circuit, can now no longer be neglected. As a matter of fact, this low magnetizing reactance can effectively shunt the fixed load impedance thereby causing a severe dropoff of output voltage as depicted in Fig. 16-15. Figure 16-14(a) shows the configuration of the equivalent circuit as it applies at very low frequencies. Note that x_1 and x_2 are so small at these frequencies that they may be omitted entirely from the diagram.

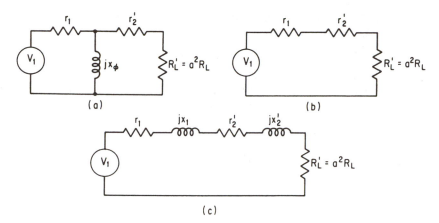

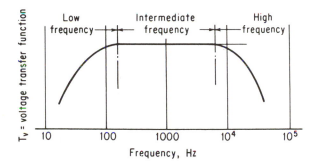

Fig. 16-14 Equivalent circuit for various portions of the frequency spectrum: (a) low frequencies; (b) intermediate frequencies; (c) high frequencies.

In the intermediate frequency range the design of these transformers for communication systems is such that x_1 and x_2 are still quite small while x_ϕ is sufficiently large that it may be omitted. The appropriate circuitry appears in Fig. 16-14(b). For a resistive load the equivalent circuit consists solely of resistive elements so that the transfer function remains constant over the band of intermediate frequencies. Refer to Fig. 16-15.

Fig. 16-15 Voltage transfer function V_2/V_1 versus frequency.

At very high frequencies x_1 and x_2' are no longer negligible and therefore must be included in the circuit. However, x_ϕ may continue to be omitted. The equivalent circuit now takes the form depicted in Fig. 16-14(c). As the frequency is allowed to get higher and higher, more and more of the source voltage V_1 appears across x_1 and x_2' and less and less appears across the fixed load resistance. Therefore again a dropoff occurs in the value of the voltage transfer function as depicted in Fig. 16-15. In communication transformers a suitable frequency response such as that shown in Fig. 16-15 is very often the most important characteristic of the device.

EXAMPLE 16-1 A distribution transformer with the nameplate rating of 100 kVA, 1100/220 V, 60 Hz has a high-voltage winding resistance of 0.1 Ω and a leakage reactance of 0.3 Ω. The low-voltage winding resistance is 0.004 Ω and the leakage reactance is 0.012 Ω. The source is applied to the high-voltage side.

(a) Find the equivalent winding resistance and reactance referred to the high-voltage side and the low-voltage side.

(b) Compute the equivalent resistance and equivalent reactance drops in volts and in per cent of the rated winding voltages expressed in terms of the primary quantities.

(c) Repeat (b) in terms of quantities referred to the low-voltage side.

(d) Calculate the *equivalent leakage impedances* of the transformer referred to the primary and secondary sides.

Solution: From the nameplate data the turns ratio is $a = 1100/220 = 5$.

(a) Hence the equivalent winding resistance referred to the primary by Eq. (16-38) is

$$R_{e1} = r_1 + a^2 r_2 = 0.1 + (5)^2(0.004) = 0.2 \ \Omega$$

Also,

$$X_{e1} = x_1 + a^2 x_2 = 0.3 + 25(0.012) = 0.6 \ \Omega$$

The corresponding quantities referred to the secondary are

$$R_{e2} = \frac{0.1}{25} + 0.004 = 0.008 \ \Omega$$

$$X_{e2} = \frac{0.3}{25} + 0.012 = 0.024 \ \Omega$$

These computations reveal that $r_1 = r_2'$ and $x_1 = x_2'$. This is a very common occurrence in well-designed transformers. A well-designed transformer is one that uses a minimum amount of iron and copper for a specified output. Note, too, that the resistance of the secondary winding is considerably smaller than that of the primary winding. This is consistent with the larger current rating of the former.

(b) The primary winding current rating in this case is

$$I_2' = \frac{100 \ \text{kVA}}{1.1 \ \text{kV}} = 91 \ \text{A}$$

Hence

$$I_2' R_{e1} = 91(0.2) = 18.2 \ \text{V}$$

Expressed as a percentage of the rated voltage of the high-tension winding we have

$$\frac{I_2' R_{e1}}{V_2'}(100) = \frac{18.2}{1100}(100) = 1.65\%$$

Also,

$$I_2' X_{e1} = 91(0.6) = 54.6 \ \text{V}$$

and

$$\frac{I_2' X_{e1}}{V_2'}(100) = \frac{54.6}{1100} = 4.96\%$$

(c) The secondary winding current rating is

$$I_2 = a I_2' = 5(91) = 455 \ \text{A}$$

Hence

$$I_2 R_{e2} = 455(0.008) = 3.64 \ \text{V}$$

and

$$\frac{I_2 R_{e2}}{V_2}(100) = \frac{3.64}{2200}(100) = 1.65\%$$

Note that the percentage value of the winding resistance drop on the secondary side in this case is identical to the value on the primary side. Similarly for the leakage reactance,

$$I_2 X_{e2} = 455(0.024) = 10.9 \text{ V}$$

$$\frac{I_2 X_{e2}}{V_2}(100) = \frac{10.9}{220}(100) = 4.96\%$$

(d) The equivalent leakage impedance referred to the primary is

$$\overline{Z}_{e1} \equiv R_{e1} + jX_{e1} = 0.2 + j0.6 = 0.634\underline{/71.6°}$$

Referred to the secondary we have

$$\overline{Z}_{e2} = \frac{\overline{Z}_{e1}}{a_2} = 0.0253\underline{/71.6°}$$

16-3 PARAMETERS FROM NO-LOAD TESTS

The exact equivalent circuit of the transformer has a total of six parameters as Fig. 16-9(d) shows. A knowledge of these parameters allows us to compute the performance of the transformer under all operating conditions. When the complete design data of a transformer are available, the parameters may be calculated from the dimensions and properties of the materials involved. Thus, the primary winding resistance r_1 may be found from the resistivity of copper, the total winding length, and the cross-sectional area. In a similar fashion a parameter such as the magnetizing reactance x_ϕ may be determined from the number of primary turns, the reluctance of the magnetic path, and the frequency of operation. The calculation of the leakage reactance is a bit more complicated because it involves accounting for partial flux linkages. However, formulas are available for making a reliable determination of these quantities.

   A more direct and far easier way of determining the transformer parameters is from tests that involve very little power consumption, called no-load tests. The power consumption is merely that which is required to supply the appropriate losses involved. The primary winding (or secondary winding) resistance is readily determined by applying a small d-c voltage to the winding in the manner shown in Fig. 16-16. The voltage appearing across the winding must be large enough

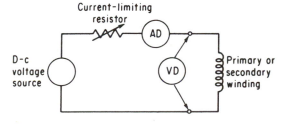

Fig. 16-16 Circuitry for finding winding resistance.

to cause approximately rated current to flow. Then the ratio of the voltage drop across the winding as recorded by the voltmeter, VD, divided by the current flowing through it as recorded by the ammeter, AD, yields the value of the winding resistance.

Open-Circuit Test. Information about the core-loss resistor r_c and the magnetizing reactance x_ϕ is obtained from an *open-circuit test*, the circuitry of which appears in Fig. 16-17. Note that the secondary winding is open. Therefore, as Fig. 16-9(d) indicates, the input impedance consists of the primary leakage impedance and the magnetizing impedance. In this case the ammeter and voltmeter are a-c instruments and the second letter A is used to emphasize this. Thus AA denotes a-c ammeter. The wattmeter shown in Fig. 16-17 is used to measure the power drawn by the transformer. The open circuit test is performed by applying voltage either to the high-voltage side or the low-voltage side, depending upon which is more convenient. Thus, if the transformer of Example 16-1 is to be tested, the voltage would be applied to the low-tension side because a source of 220 V is more readily available than 1100 V. The core loss is the same whether 220 V is applied to the winding having the smaller number of turns or 1100 V is applied to the winding having the larger number of turns. The maximum value of the flux, upon which the core loss depends, is the same in either case as indicated by Eq. (16-17). Now for convenience assume that the instruments used in the open-circuit test yield the following readings:

$$\text{wattmeter reading} = P$$
$$\text{ammeter reading} = I_m$$
$$\text{voltmeter reading} = V_L$$

The voltmeter reading carries the subscript L to emphasize that the test is performed with the source and instruments on the low-voltage side.

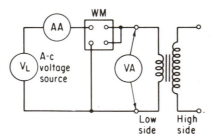

Low   High   **Fig. 16-17** Wiring diagram for trans-
side   side   former open-circuit test.

Neglecting instrument losses, the wattmeter reading may be taken entirely equal to the core loss, i.e. $P_c = P$. This is because the attendant copper loss is negligibly small. Moreover, since I_m is very small, the primary leakage impedance drop may be neglected, so that for all practical purposes the induced emf is equal to the applied voltage, i.e. $E_1 = V_L$. In accordance with these simplifications the open-circuit phasor diagram is represented by Fig. 16-3(b). The no-load power factor angle θ_0 is computed from

$$\theta_0 = \cos^{-1} \frac{P}{V_L I_m} \tag{16-47}$$

It then follows that

$$I_c = I_m \cos \theta_0 \tag{16-48}$$

and

$$I_\phi = I_m \sin \theta_0 \tag{16-49}$$

It is important to keep in mind that these currents are referred to the low-tension side. The corresponding values on the high-voltage side would be less by the factor $1/a$. The low-side core-loss resistor becomes

$$r_{cL} = \frac{P}{I_c^2} = \frac{P}{(I_m \cos \theta_0)^2} \tag{16-50}$$

The corresponding high-side core-loss resistor is

$$r_{cH} = a^2 r_{cL} \tag{16-51}$$

The magnetizing reactance referred to the low side follows from Eq. (16-29). Thus

$$x_{\phi L} = \frac{E_1}{I_\phi} = \frac{V_L}{I_m \sin \theta_0} \tag{16-52}$$

On the high side this becomes

$$x_{\phi H} = a^2 x_{\phi L} \tag{16-53}$$

Although the manner of finding the magnetizing impedance is treated in the foregoing, the motivation is not so much necessity as the desire for completeness. Invariably, the use of the approximate equivalent circuit is sufficient for all but a few special applications of transformers. Accordingly, the one useful bit of information obtained from the open-circuit test is the value of the core loss, which helps to determine the efficiency of the transformer.

Short-Circuit Test. Of the six parameters of the exact equivalent circuit two remain to be determined: the primary and secondary leakage reactances x_1 and x_2. This information is obtained from a short-circuit test which involves placing a small a-c voltage on one winding and a short circuit on the other as depicted in Fig. 16-18(a). Appearing in Fig. 16-18(b) is the approximate equivalent circuit with the secondary winding short-circuited. Note that the input impedance is merely the equivalent leakage impedance.

When performing the short-circuit test, a reduced a-c voltage is applied to the high side usually because it is more convenient to do so. From the computations of Example 16-1 recall that the leakage impedance drop is only about 5% of the winding voltage rating. Hence 5% of 1100 V, which is 55 V, is easier and more accurate to deal with than 5% of 220 V, which is 11 V. Also, the wattmeter reading in the circuitry of Fig. 16-18(a) can be taken entirely equal to the winding-copper losses. This follows from the fact that the greatly reduced voltage used

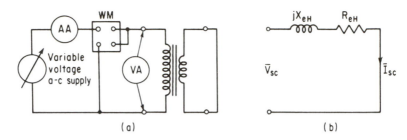

Fig. 16-18 Transformer short-circuit test: (a) wiring diagram; (b) equivalent circuit.

in the short-circuit test makes the core loss negligibly small. Calling the wattmeter reading P_{sc} and the ammeter reading I_{sc}, we can determine the equivalent winding resistance as

$$R_{eH} = \frac{P_{sc}}{I_{sc}^2} \qquad (16\text{-}54)$$

The subscript H is used in place of L since it is known from the wiring diagram that the primary side is the high-voltage side. The equivalent resistance referred to the low side is then

$$R_{eL} = \frac{R_{eH}}{a^2} \qquad (16\text{-}55)$$

where

$$a = \frac{N_H}{N_L} = \frac{\text{high-side turns}}{\text{low-side turns}} \qquad (16\text{-}56)$$

Before the equivalent leakage reactance can be determined, it is first necessary to find the equivalent leakage impedance, which a glance at Fig. 16-18(b) reveals to be

$$\overline{Z}_{eH} = \frac{\overline{V}_{sc}}{\overline{I}_{sc}} \qquad (16\text{-}57)$$

During the test the applied voltage, V_{sc}, is adjusted so that I_{sc} is at least equal to the rated winding current. The equivalent leakage reactance follows from Eqs. (16-54) and (16-57). Thus

$$X_{eH} = \sqrt{Z_{eH}^2 - R_{eH}^2} \qquad (16\text{-}58)$$

It is important to note that this computation provides the *sum* of the primary and secondary leakage reactances. It gives no information about the individual breakdown between x_1 and x_2. Whenever the approximate equivalent is used in analysis such a breakdown is unnecessary. On the few occasions when the individual breakdown is needed, it is customary to assume that

$$x_1 = x_2' \qquad (16\text{-}59)$$

This statement is based on the assumption that the transformer is well designed.

One final note. It should be apparent from the theory of transformer action that the establishment of rated current in the high-side winding requires a corresponding flow of rated current in the secondary circuit. It is for this reason that the *total* winding copper loss is measured in the short-circuit test.

EXAMPLE 16-2 The following data were obtained on a 50-kVA, 2400/120-V transformer.
Open-circuit test, instruments on low side:

$$\text{wattmeter reading} = 396 \text{ W}$$
$$\text{ammeter reading} = 9.65 \text{ A}$$
$$\text{voltmeter reading} = 120 \text{ V}$$

Short-circuit test, instruments on high side:

$$\text{wattmeter reading} = 810 \text{ W}$$
$$\text{ammeter reading} = 20.8 \text{ A}$$
$$\text{voltmeter reading} = 92 \text{ V}$$

Compute the six parameters of the equivalent circuit referred to the high and low sides.

Solution: From the open-circuit test

$$\theta_0 = \cos^{-1} \frac{P}{V_L I_m} = \cos^{-1} \frac{396}{120(9.65)} = 70°$$

$$\therefore \ I_{cL} = I_m \cos \theta_0 = 9.65(0.342) = 3.3 \text{ A}$$

$$I_{\phi L} = I_m \sin \theta_0 = 9.65 \sin 70° = 9.07 \text{ A}$$

Hence

$$r_{cL} = \frac{396}{3.3^2} = 36.3 \ \Omega$$

This can also be found from

$$r_{cL} = \frac{E_1}{I_c} = \frac{V_L}{I_c} = \frac{120}{3.3} = 36.3 \ \Omega$$

On the high side this quantity becomes

$$r_{cH} = a^2 r_{cL} = 400(36.3) = 14,520 \ \Omega$$

The magnetizing reactance referred to the low side is

$$x_{\phi L} = \frac{E_1}{I_\phi} = \frac{V_L}{I_\phi} = \frac{120}{9.07} = 13.2 \ \Omega$$

The corresponding high-side value is

$$x_{\phi H} = a^2 x_{\phi L} = 400(13.2) = 5280 \ \Omega$$

From the data of the short-circuit test we get directly

$$Z_{eH} = \frac{92}{20.8} = 4.42 \ \Omega \qquad \therefore \ Z_{eL} = \frac{4.42}{400} = 0.011 \ \Omega$$

Also,

$$R_{eH} = \frac{810}{(20.8)^2} = 1.87 \qquad \therefore \ R_{eL} = \frac{1.87}{400} = 0.0047$$

and

$$X_{eH} = \sqrt{Z_{eH}^2 - R_{eH}^2} = \sqrt{4.42^2 - 1.87^2} = 4 \ \Omega$$

$$\therefore \ X_{eL} = \frac{4}{400} = 0.01 \ \Omega$$

16-4 EFFICIENCY AND VOLTAGE REGULATION

The manner of describing the performance of a transformer depends upon the application for which it is designed. It has already been pointed out that in communication circuits the frequency response of the transformer is very often of prime importance. Another important characteristic of such transformers is to provide matching of a source to a load for maximum transfer of power. The efficiency and regulation is usually of secondary significance. This is not so, however, with power and distribution transformers, which are designed to operate under conditions of constant rms voltage and frequency. Accordingly, the material that follows has relevance chiefly to these transformers.

As is the case with other devices, the efficiency of a transformer is defined as the ratio of the useful output power to the input power. Thus

$$\eta = \frac{\text{output watts}}{\text{input watts}} \tag{16-60}$$

Usually the efficiency is found through measurement of the losses as obtained from the no-load tests. Therefore by recognizing that

$$\text{output watts} = \text{input watts} - \sum \text{losses} \tag{16-61}$$

and inserting into Eq. (16-60), a more useful form of the expression for efficiency results. Thus

$$\boxed{\eta = 1 - \frac{\sum \text{losses}}{\text{input watts}}} \tag{16-62}$$

where

$$\sum \text{losses} = \text{core loss} + \text{copper loss} = P_c + I_2^2 R_{e2} \tag{16-63}$$

Equation (16-62) also leads to a more accurate value of the efficiency.

Power and distribution transformers very often supply electrical power to loads that are designed to operate at essentially constant voltage regardless of whether little or full load current is being drawn. For example, as more and more light bulbs are placed across a supply line, which very likely originates from a distribution transformer, it is important that the increased current being drawn from the supply transformer does not cause a significant drop (more than 10%) in the load voltage. If this should happen the illumination output from the bulbs is sharply reduced. Similar undesirable effects occur in television sets that are connected across the same supply lines. The reduced voltage means reduced picture size. An appreciable drop in line voltage with increasing load demands can also cause harmful effects in connected electrical motors such as those found in refrigerators, washing machines, and so on. Continued operation at low voltage can cause these units to overheat and eventually burn out. The way to prevent the drop in supply voltage with increasing load in distribution circuits is to use a distribution transformer designed to have small leakage impedance. The figure of merit used to identify this characteristic is the voltage regulation.

The *voltage regulation* of a transformer is defined as the change in magnitude of the secondary voltage as the current changes from full-load to no-load with the primary voltage held fixed. Hence in equation form,

$$\text{voltage regulation} = \frac{|\overline{V}_1| - |\overline{V}_2'|}{|\overline{V}_2'|} = \frac{\left|\dfrac{\overline{V}_1}{a}\right| - |\overline{V}_2|}{|\overline{V}_2|} \tag{16-64}$$

where the symbols denote the quantities previously defined; absolute signs are used in order to emphasize that it is the *change in magnitude* that is important. Equation (16-64) expresses the voltage regulation on a per unit basis. To convert to percentage, it is necessary to multiply by 100. The smaller the value of the voltage regulation, the better suited is the transformer for supplying power to constant-voltage loads (constant voltage is characteristic of most commercial and industrial loads).

In determining the voltage regulation it is customary to assume that $\overline{V}_1$ is adjusted to that value which allows rated voltage to appear across the load when rated current flows through it. Under these conditions the necessary magnitude of $\overline{V}_1$ is found from Kirchhoff's voltage law as it applies to the approximate equivalent circuit. Refer to Figs. 16-10 and 16-11. For the assumed lagging power factor angle θ_2 of the secondary circuit the expression for $\overline{V}_1$ is

$$\overline{V}_1 = V_2' \underline{/0^\circ} + I_2' \underline{/-\theta_2} Z_{e1} \underline{/\theta_z} \tag{16-65}$$

where

$$\theta_z = \tan^{-1} \frac{X_{e1}}{R_{e1}} \tag{16-66}$$

Equation (16-65) is written using the rated secondary terminal voltage referred to the primary as the reference.

EXAMPLE 16-3   For the transformer of Example 16-2:
  (a) Find the efficiency when rated kVA is delivered to a load having a power factor of 0.8 lagging.
  (b) compute the voltage regulation.

Solution: (a) The losses at rated load current and rated voltage are

$$\sum \text{losses} = 396 + 810 = 1206 \text{ W} = 1.2 \text{ kW}$$

Also,

$$\text{output kW} = 50(0.8) = 40 \text{ kW}$$
$$\text{input kW} = 40 + 1.2 = 41.2 \text{ kW}$$

Hence

$$\eta = 1 - \frac{\sum \text{losses}}{\text{input kW}} = 1 - \frac{1.2}{41.2} = 1 - 0.029 = 0.971$$

or

$$\eta = 97.1\%$$

(b) From Eq. (16-65) we have

$$\overline{V}_1 = \overline{V}_H = 2400 + 20.8 \underline{/-37°} \; 4.42 \underline{/65°}$$

where

$$\theta_z = \tan^{-1} \frac{X_{eH}}{R_{eH}} = \tan^{-1} \frac{4}{1.87} = 65°$$

Therefore,

$$\overline{V}_1 = 2400 + 92 \underline{/28°} = 2400 + 81.1 + j43.1$$

$$V_1 \approx 2481.1$$

Note that the *j*-component is negligible in comparison to the in-phase component. This calculation indicates that in order for rated voltage to appear across the load when rated current is drawn, the primary voltage V_1 must be slightly different from the primary winding voltage rating of 2400 V. The percent voltage regulation is accordingly found to be

$$\text{percent voltage regulation} = \frac{2481.1 - 2400}{2400} \times 100 = 3.38\%$$

Thus there is approximately a 3.4% change in voltage across the load as increased current is drawn from the transformer.

16-5 MUTUAL INDUCTANCE

Analysis of the two-winding transformer by the methods of electric circuit theory involves the use of self- and mutual-inductance. So far, Kirchhoff's voltage equation applied to the primary winding has led to Eq. (16-26). In the material that follows this result will be shown to be identical with that obtained by the classical approach.

For simplicity assume the ferromagnetic material has a linear magnetization curve with no core losses. Refer to Fig. 16-19 where for the indicated current directions Kirchhoff's voltage equation becomes

$$v_1 = i_1 r_1 + L_1 \frac{di_1}{dt} - M \frac{di_2}{dt} \tag{16-67}$$

where lowercase letters for current and voltage denote instantaneous values. Here the quantity $L_1 \, di_1/dt$ denotes the voltage drop associated with the total flux produced by i_1 and linking the primary turns N_1. Similarly, the quantity $-M \, di_2/dt$ refers to the *voltage rise* associated with the flux produced by i_2 and linking the primary turns. The existence of a voltage rise in the primary circuit for the assumed direction of i_2 is easily verified. Note that the secondary mmf

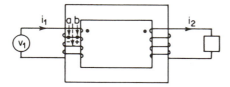

Fig. 16-19 Elementary transformer for analysis by the classical approach.

$N_2 i_2$, if allowed to act freely, produces a flux which threads counterclockwise in the configuration of Fig. 16-19. Then the reaction in the primary coil between two points such as a and b by Lenz's law requires that point b should be at a higher potential than a. Accordingly, this induced emf becomes a voltage rise with respect to the specified direction of i_1.

Assuming that the applied primary voltage is a sinusoid, Eq. (16-67) may be written in rms quantities as

$$\overline{V}_1 = \overline{I}_1 r_1 + j\omega L_1 \overline{I}_1 - j\omega M \overline{I}_2 \tag{16-68}$$

Inserting the relationship

$$\overline{I}_2 = a\overline{I}_2' \tag{16-69}$$

wherein a is the ratio of transformation, we get

$$\overline{V}_1 = \overline{I}_1 r_1 + j\omega L_1 \overline{I}_1 - j\omega a M \overline{I}_2' \tag{16-70}$$

But by Eq. (16-24)

$$\overline{I}_2' = \overline{I}_1 - \overline{I}_m = \overline{I}_1 - \overline{I}_\phi$$

so that

$$\overline{V}_1 = \overline{I}_1 r_1 + j\omega(L_1 - aM)\overline{I}_1 + j\omega a M \overline{I}_\phi \tag{16-71}$$

Equation (16-71) is now in a form where correspondence with Eq. (16-26) is possible after some algebraic manipulation.

The expression for the self-inductance of the primary winding may be written as

$$L_1 = N_1 \frac{d\phi_1}{di_1} = N_1 \frac{\Phi_1}{i_1} \tag{16-72}$$

The equivalence of the differential form with the total quantities in this last expression is allowed by the linearity condition. Recalling that the total flux linking the primary coil consists of the mutual flux Φ_m plus the primary leakage flux Φ_{l1}, Eq. (16-72) can be rewritten as

$$L_1 = N_1 \frac{\Phi_m + \Phi_{l1}}{i_1} \tag{16-73}$$

Also,

$$\Phi_m = \frac{N_1 i_1}{\mathscr{R}_m} \quad \text{and} \quad \Phi_{l1} = \frac{N_1 i_1}{\mathscr{R}_{l1}} \tag{16-74}$$

where $\mathscr{R}_m$ is the reluctance experienced by the mutual flux and consists almost exclusively of iron and $\mathscr{R}_{l1}$ is the net reluctance experienced by the leakage flux of the primary in air. Accordingly, Eq. (16-73) becomes

$$L_1 = \frac{N_1^2}{\mathscr{R}_m} + \frac{N_1^2}{\mathscr{R}_{l1}} = L_m + L_{l1} \tag{16-75}$$

where L_m denotes an inductance associated with the mutual flux and L_{l1} is the inductance associated with the primary leakage flux. It is important to note here that L_m is not the mutual inductance.

The classical definition of mutual inductance is given in this case by

$$M_{21} = N_2 \frac{d\phi_{m1}}{di_1} = N_2 \frac{\Phi_{m1}}{i_1} \tag{16-76}$$

Thus the mutual inductance is related to the number of turns of winding 2 and the change of flux with respect to the current in winding 1, which produces the mutual flux Φ_{m1}. Upon inserting $N_1 i_1 = \Phi_{m1} \mathcal{R}_m$ into Eq. (16-76), there results

$$M_{21} = \frac{N_2 N_1}{\mathcal{R}_m} \tag{16-77}$$

The expression for mutual inductance may also be found by considering a change in mutual flux corresponding to a change in secondary current i_2 linking N_1 turns. Thus

$$M_{12} = N_1 \frac{d\phi_{m2}}{di_2} = N_1 \frac{\Phi_{m2}}{i_2} \tag{16-78}$$

But

$$\Phi_{m2} \mathcal{R}_m = N_2 i_2 \tag{16-79}$$

Inserting this into Eq. (16-78) yields

$$M_{12} = \frac{N_1 N_2}{\mathcal{R}_m} = M \tag{16-80}$$

A comparison of Eq. (16-80) with Eq. (16-77) shows them to be the same, and from here on either expression will be denoted by M.

Returning to Eq. (16-71) we are now in a position to evaluate the significance of the term in parentheses. Thus, by Eqs. (16-75) and (16-77)

$$L_1 - aM = \frac{N_1^2}{\mathcal{R}_m} + \frac{N_1^2}{\mathcal{R}_{l1}} - \frac{N_1}{N_2} \frac{N_2 N_1}{\mathcal{R}_m} = \frac{N_1^2}{\mathcal{R}_{l1}} = L_{l1} \tag{16-81}$$

Clearly this quantity is the primary leakage inductance, and it represents the effect of the primary leakage flux. Of course, the total quantity

$$\omega(L_1 - aM)I_1 = \omega L_{l1} I_1 = x_1 I_1 \tag{16-82}$$

is nothing more than the primary leakage reactance drop and is the counterpart of the second term of Eq. (16-26).

Let us now examine the last term of Eq. (16-71). Thus

$$\omega aMI_\phi = \omega \frac{N_1}{N_2} \frac{N_2 N_1}{\mathcal{R}_m} I_\phi = \omega N_1 \frac{N_1 I_\phi}{\mathcal{R}_m} \tag{16-83}$$

The fraction quantity on the right side of Eq. (16-83) is the expression for mutual flux. Specifically, however, since it involves the reluctance and the rms value I_ϕ of the magnetizing current, this expression represents the magnitude of the maximum value of the mutual flux divided by $\sqrt{2}$. That is,

$$\frac{N_1 I_\phi}{\mathcal{R}_m} = \frac{\Phi_m}{\sqrt{2}} \tag{16-84}$$

where Φ_m denotes the maximum value of the mutual flux. Accordingly, Eq. (16-

83) becomes

$$\omega a M I_\phi = \omega N_1 \frac{\Phi_m}{\sqrt{2}} = \frac{2\pi}{\sqrt{2}} f N_1 \Phi_m = E_1 \qquad (16\text{-}85)$$

It therefore follows from Eq. (16-85) that the mutual inductance parameter so commonly encountered in linear circuit analysis is implied in the expression for the induced emf. Clearly, both the induced emf and the mutual inductance are related by the same mutual flux.

Summary review questions

1. What is a transformer? List some of the important functions it fulfills.
2. Distinguish between the primary and secondary windings of a transformer. Distinguish between the air-core and the iron-core transformers. Which draws the larger magnetizing current?
3. In a transformer with negligible core loss, what is the time-phase relationship between the core flux and the magnetizing current that produces it?
4. Name the single most important factor that determines the maximum value of the flux in a given transformer.
5. What is the time-phase relationship between the time-varying flux and the counter (or reaction) emf which it produces in the windings of a transformer?
6. Describe the effect which the presence of core losses has on the time-phase relationship of the magnetizing current and its associated emf viewed as a voltage drop.
7. What is the time-phase relationship between the time-varying flux of the transformer and the *generated* emfs in the primary and secondary windings? Explain.
8. What is meant by the ratio of transformation of the transformer? Can this ratio be expressed as a ratio of voltages? Which voltages? Why?
9. Describe the mechanism which comes into play that permits the electrically isolated secondary winding of the transformer to supply energy to a load.
10. What is leakage flux as it applies to the iron-core transformer? How is it taken into account in the analysis of the transformer?
11. What is leakage reactance? Does this reactance have a separate identity associated with its own specific coil? Explain.
12. Draw the complete phasor diagram of the transformer under no-load conditions.
13. Draw the complete phasor diagram of the transformer under load conditions and explain the presence and position of each phasor.
14. What gives rise to the balance current in the primary winding of a transformer? Be specific.
15. State the advantages of the equivalent circuit representation of a physical device.
16. In deriving the equivalent circuit of the transformer explain the importance of replacing the actual secondary winding by an equivalent winding having the same number of turns as the primary winding. List the specific conditions that must be satisfied by this secondary winding for equivalency to the actual winding.

17. Draw the approximate equivalent circuit of the transformer referred to the primary side and indicate how it differs from the exact version.

18. Describe what is meant by the phrase *secondary winding resistance referred to the primary*.

19. Describe the meaning of the equivalent leakage reactance of a transformer referred to the secondary winding.

20. How is the voltage transfer function of the transformer defined? What makes this ratio different from the ratio of transformation?

21. Using a semilogarithmic scale, sketch the frequency response of the transformer as it relates to the voltage transfer function. Explain why the curve has the shape drawn.

22. Name the parameters of the exact equivalent circuit of the transformer that can be determined from an open-circuit test. Identify the data that are measured and how these data are used to obtain these parameters.

23. Repeat Question 22 as it relates to the short-circuit test.

24. Why is the core loss considered to be negligible in the short-circuit test of a transformer?

25. Why is the copper loss considered to be negligible in the open-circuit test of an iron-core transformer?

26. Explain how the voltage regulation of a transformer is defined. Is this quantity large or small for the iron-core transformer? Explain.

27. Is the voltage regulation of an air-core transformer larger than that of the iron-core transformer? Assume the same construction form.

28. Why is it important to maintain high efficiency of operation and low values of voltage regulation for power transformers?

29. Sketch a two-winding transformer and indicate voltage polarities and current directions. Write the Kirchhoff voltage equation that applies to your circuit using mutual and self-inductances.

30. Analysis of the power transformer from the field viewpoint makes no mention of the mutual inductance parameter. Where is the effect of mutual inductance taken into account in such an analysis?

Problems

GROUP I

16-1. A reactor coil with an iron core has 400 turns. It is connected across the 115-V 60-Hz power line.
  (a) Neglecting the resistance voltage drop, calculate the maximum value of the operating flux.
  (b) If the flux density is not to exceed 75 kilolines/in.^{2}, what must be the cross-sectional area of the core?

16-2. A fixed sinusoidal voltage is applied to the circuit shown in Fig. P16-2. If the voltage remains connected and the shaded portion of the iron removed, state what happens to the maximum value of the flux and the maximum value of the magnetizing current. Justify your answer. Assume negligible leakage impedance and winding resistance.

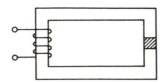

Fig. P16-2

16-3. Repeat Prob. 16-2 assuming the voltage applied to the core is a fixed d-c quantity. Assume the d-c current is limited by an external resistor.

16-4. A transformer coil rated at 200 V and having 100 turns is equipped with a 0.5 tap. If 50 V are applied to half the number of turns, find the change in the maximum value of the flux compared to the rated value.

16-5. In an open-circuit test of a 25-kVA 2400/240-V transformer made on the low side the corrected readings of amps, volts, and watts are respectively 1.6, 240, and 114. In the short-circuit test the low side is short-circuited and the current, voltage, and power to the high side are measured to be respectively 10.4 A, 55 V, and 360 W.

  (a) Find the core loss.
  (b) What is the full-load copper loss?
  (c) Find the efficiency for a full-load of 0.8 power factor leading.
  (d) Compute the percent voltage regulation for part (c).

16-6. A 50-kVA 2300/230-V 60-Hz transformer has a high-voltage winding resistance of 0.65 Ω and a low-voltage winding resistance of 0.0065 Ω. Laboratory tests showed the following results:

$$\text{Open-circuit test: } V = 230 \text{ V} \qquad I = 5.7 \text{ A} \qquad P = 190 \text{ W}$$
$$\text{Short-circuit test: } V = 41.5 \text{ V} \qquad I = 21.7 \text{ A} \qquad P = \text{no WM used}$$

  (a) Compute the value of primary voltage needed to give rated secondary voltage when the transformer is connected *step-up* and is delivering 50 kVA at a power factor of 0.8 lagging.
  (b) Compute the efficiency under the conditions of part (a).

16-7. The following test data were taken on a 15-kVA 2200/440-V 60-Hz transformer:

$$\text{Short-circuit test: } P = 620 \text{ W} \qquad I = 40 \text{ A} \qquad V = 25 \text{ V}$$
$$\text{Open-circuit test: } P = 320 \text{ W} \qquad I = 1 \text{ A} \qquad V = 440 \text{ V}$$

  (a) Calculate the voltage regulation of this transformer when it supplies full load at 0.8 pf lagging. Neglect the magnetizing current.
  (b) Find the efficiency at the load condition of part (a).

16-8. The following test data were taken on a 110-kVA 4400/440-V 60-Hz transformer:

$$\text{Short-circuit test: } P = 2000 \text{ W} \qquad I = 200 \text{ A} \qquad V = 18 \text{ V}$$
$$\text{Open-circuit test: } P = 1200 \text{ W} \qquad I = 2 \text{ A} \qquad V = 4400 \text{ V}$$

Calculate the voltage regulation of this transformer when it supplies rated current at 0.8 pf lagging. Neglect the magnetizing current.

16-9. A 50-kVA 2300/230-V 60-Hz distribution transformer takes 360 W at a power factor of 0.4 with 2300 V applied to the high-voltage winding and the low-voltage winding open-circuited. If this transformer has 230 V impressed on the low side with no load on the high side, what will be the current in the low-voltage winding? Neglect saturation.

16-10. A 200/100-V 60-Hz transformer has an impedance of 0.3 + j0.8 Ω in the 200-V winding and an impedance of $0.1 + j0.25\ \Omega$ in the 100-V winding. What are the currents on the high and low side, if a short circuit occurs on the 100-V side with 200 V applied to the high side?

16-11. A 10-kVA 500/100-V transformer has $R_{eH} = 0.3\ \Omega$ and $X_{eH} = 5.2\ \Omega$, and it is used to supply power to a load having a lagging power factor. When supplying power to this load an ammeter, wattmeter, and voltmeter placed in the high-side circuit read as follows:

$$I_1 = 20\ \text{A} \qquad V_1 = 500\ \text{V} \qquad WM = 8\ \text{kW}$$

For this condition calculate what a voltmeter would read if placed across the secondary load terminals. Assume the magnetizing current to be negligibly small.

16-12. The following data are taken on a 30-kVA 2400/240-V 60-Hz transformer:

$$\text{Short-circuit test: } V = 70\ \text{V} \qquad I = 18.8\ \text{A} \qquad P = 1050\ \text{W}$$
$$\text{Open-circuit test: } V = 240\ \text{V} \qquad I = 3.0\ \text{A} \qquad P = 230\ \text{W}$$

(a) Determine the primary voltage when 12.5 A at 240 V are taken from the low-voltage side supplying a load of 0.8 pf lagging.

(b) Compute the efficiency in part (a).

16-13. The two windings of a 2400/240-V 48-kVA 60-Hz transformer have resistances of 0.6 and 0.025 Ω for the high- and low-voltage windings, respectively. This transformer requires that 238 V be impressed on the high-voltage coil in order that rated current be circulated in the short-circuited low-voltage winding.

(a) Calculate the equivalent leakage reactance referred to the high side.

(b) How much power is needed to circulate rated current on short circuit?

(c) Compute the efficiency at full-load when the power factor is 0.8 lagging. Assume that the core loss equals the copper loss.

GROUP II

16-14. A 60-Hz three-winding transformer is rated at 2300 V on the high side with a total of 300 turns. Of the two secondary windings, each designed to handle 200 kVA, one is rated at 575 V and the other at 230 V. Determine the primary current when rated current in the 230-V winding is at unity power factor and the rated current in the 575-V winding is at 0.5 pf lagging. Neglect all leakage impedance drops and magnetizing current.

16-15. An "ideal" transformer has secondary winding tapped at point b. The number of primary turns is 100 and the number of turns between a and b is 300 and between b and c is 200. The transformer supplies a resistive load connected between a and c and drawing 7.5 kW. Moreover, a load impedance of $10\underline{/45°}$ ohms is connected between a and b. The primary voltage is 1000 V. Find the primary current.

16-16. Draw a neat phasor diagram of a transformer operating at rated conditions. Assume $N_1/N_2 = 2$ and

$$I_1 r_1 = 0.1E_1 \qquad I_2 r_2 = 0.1V_2 \qquad I_m = 0.3I_2'$$
$$I_1 x_1 = 0.2E_1 \qquad I_2 x_2 = 0.2V_2 \qquad I_c = 0.1I_2'$$

Consider the load power factor to be 0.6 leading. Use V_2 as the reference phasor and show all currents and voltages drawn to scale.

16-17. A 30-kVA 240/120-V 60-Hz transformer has the following data:

$$r_1 = 0.14 \ \Omega \qquad r_2 = 0.035 \ \Omega$$

$$x_1 = 0.22 \ \Omega \qquad x_2 = 0.055 \ \Omega$$

It is desired to have the primary induced emf equal in magnitude to the primary terminal voltage when the transformer carries the full-load current. Neglect the magnetizing current. How must the transformer be loaded to achieve this result?

16-18. A coil located on an iron core has a constant 60-Hz voltage applied to it. The total inductance is found to be 10 H. A second identical coil is placed in parallel with the first one and is so wound that it produces aiding flux. What is the total inductance of the parallel combination? Neglect the winding resistance and leakage flux of each winding.

16-19. Show the derivation that allows Eq. (16-68) to be written from Eq. (16-67) for a sinusoidal forcing function and no saturation.

_____chapter seventeen_____

Electromechanical Energy Conversion

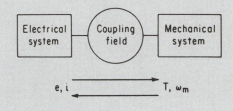

Electromechanical energy conversion involves the interchange of energy between an electrical system and a mechanical system through the medium of a coupling magnetic field. The process is essentially reversible except for a small amount which is lost as heat energy. When the conversion takes place from electrical to mechanical form the device is called a *motor*. When mechanical energy is converted to electrical energy, the device is called a *generator*. Moreover, when the electrical system is characterized by alternating current the devices are referred to as *a-c* motors and *a-c* generators, respectively. Similarly, when the electrical system is characterized by direct current the electromechanical conversion devices are called *d-c* motors and *d-c* generators.

The same fundamental principles underly the operation of both a-c and d-c machines, and they are governed by the same basic laws. Thus, in the computation of the developed torque of an electromechanical energy-conversion device, one basic torque formula (which derives directly from Ampère's law) applies whether the machine is a-c or d-c. The ultimate forms of the torque equations appear different for the two types of machines only because the details of mechanical construction differ. In other words, starting with the same basic principles for the production of electromagnetic torque, the final forms of the torque equations differ to the extent that the mechanical details differ. These comments apply equally in the matter of generating an electomotive force in the armature winding of a machine be it a-c or d-c. Once again a simple basic relationship (Faraday's law) governs the voltage induced. The final forms of the voltage equations differ merely as a reflection of the differences in the construction of the machines. It is important that this be understood at the outset, because a prime objective of the treatment of the subject matter as it unfolds in this chapter is to impress upon the reader the fact that a-c machines are not fundamentally different from d-c machines. They differ merely in construction details; the underlying principles are the same.

This chapter is arranged to give still greater emphasis to this theme. Attention first is given to a basic analysis of the production of electromagnetic torque as well as the generation of voltages starting with the classical fundamental laws. Then the construction features of the various types of electric machines are examined with the purpose of indicating how the conditions for the production of torque and voltage are fulfilled as well as to point out why differences are necessary. Once this background is established, the fundamental equations for developed torque and induced voltage are modified to make these expressions consistent with the particular construction features of the machines involved. In this way the equations are put in a more useful form for purposes of analysis and design. Attention is then directed in succeeding chapters to an analysis of each of the four major types of electromechanical energy-conversion devices to be found performing innumerable tasks throughout industry and in homes, shops, and offices everywhere.

17-1 BASIC ANALYSIS OF ELECTROMAGNETIC TORQUE

Since electromechanical energy conversion involves the interchange of energy between an electrical and a mechanical system, the primary quantities involved in the mechanical system are torque and speed, while the analogous quantities in the electrical system are voltage and current, respectively. Figure 17-1 represents the situation diagrammatically. *Motor action* results when the electrical system causes a current i to flow through conductors that are placed in a magnetic field. Then by Eq. (17-2) a force is produced on each conductor so that if the conductors are located on a structure that is free to rotate, an electromagnetic torque T results which in turn manifests itself as an angular velocity ω_m. Moreover, with the development of this motor action the revolving conductors cut through the magnetic field, thereby undergoing an electromotive force e, which is really a reaction voltage analogous to the induced emf in the primary winding of a transformer. Note that the coupling field is involved in establishing the electromagnetic torque T as well as the reaction induced emf e. In the case of *generator*

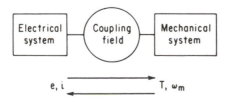

Fig. 17-1 Block representation of electro-mechanical energy conversion.

action the reverse process takes place. Here the rotating member—the *rotor*—is driven by a prime mover (steam turbine, gasoline engine, etc.), causing an induced voltage *e* to appear across the armature winding terminals. Upon the application of an electrical load to these terminals, a current *i* is made to flow, delivering electrical power to the load. The flow of this current through the armature conductors interacts with the magnetic field to produce a reaction torque opposing the applied torque originating from the prime mover.

In the interest of emphasizing further the character of this motor and generator action and its relationship to the coupling magnetic field, a mathematical description is now undertaken. Consider that we have an electromechanical energy-conversion device which consists of a stationary member—the *stator*—and the rotating member, the rotor. Refer to Fig. 17-2(a).

Assume that the stator is equipped with a coil *A-B*, which is energized to cause a constant current to flow in the direction shown by the dot-cross notation. By Flemings' right-hand rule, the flux lines that are produced by the coil ampereturns leave the stator from the upper half, cross the air gap, penetrate the rotor iron, and then once again cross the air gap to form the closed path illustrated in Fig. 17-2(a). If the rotor and stator surfaces were to be cut and laid out in a flat manner, the resulting configuration would appear as shown in Fig. 17-2(b). If the reluctance of the iron is neglected, then half the coil mmf is used for each crossing of the air gap. Accordingly, the value of the flux that appears between coil sides *A-B* in Fig. 17-2(b) is one-half the coil mmf divided by the air-gap reluctance. Since both quantities are fixed, it follows that the flux density, too, is constant from coil side *A* to *B* and *B* to *A*. This situation is depicted in Fig. 17-2(c) in a manner that emphasizes that the flux that leaves the stator (in the upper section between coil sides *A-B*) behaves like a north-pole flux, while the flux that enters the stator (in the lower section between coil sides *B-A*) behaves like a south-pole flux.

With the uniformly distributed field now established, the focus of attention is shifted to the rotor, which is assumed to be equipped with a total of *Z* conductors only two of which (*a* and *b*) are shown in Fig. 17-2(a). It is further assumed that conductor *a* is joined to conductor *b* to form a coil in the fashion illustrated in Fig. 17-2(d). Because the principle of electromechanical energy conversion is concerned with the movement of a coil relative to a magnetic field, the analysis now proceeds from the arrangement shown in Fig. 17-2(e), where only the presence of the field density *B* is given importance and not its source. It is helpful to observe that initially the flux that links rotor coil *a-b* is zero. In Fig. 17-2(e), this situation is represented by noting that coil *a-b* links one-half of the positive (north-pole) air-gap flux in addition to one-half of the negative (south-pole) air-gap flux for a resultant of zero. By Faraday's law, displacement of the coil by a differential distance *dx* to the new position *a'b'* causes a change in the flux linking the coil equal to twice the shaded area. Thus

$$d\phi = 2B \, dA \qquad (17\text{-}1)$$

The factor 2 accounts for the two areas involved; *B* is the value of the flux density at the points where conductors *a* and *b* lie, and *dA* is the differential

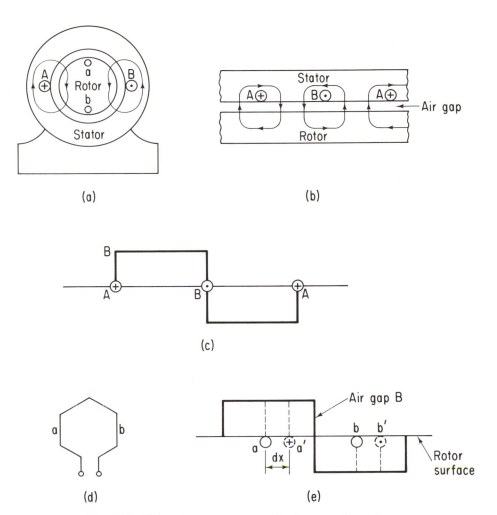

Fig. 17-2 Illustrating some construction features of an electro-mechanical energy conversion device: (a) stator and rotor; (b) a plan view; (c) a uniform air-gap flux density produced by coil A-B; (d) a full-pitch coil; (e) changing flux linkage of coil a-b with displacement dx.

area through which the B-field penetrates. A little thought reveals this area to be equal to the axial length of the rotor l times dx. Thus

$$dA = l\,dx \tag{17-2}$$

However, if it is further assumed that dx results from a linear velocity v imparted to the conductor for the differential time dt, it follows that

$$dx = v\,dt \tag{17-3}$$

Insertion of Eqs. (17-2) and (17-3) into Eq. (17-1) then leads to

$$d\phi = 2Blv\,dt \tag{17-4}$$

The corresponding emf induced in coil *ab* becomes, by Faraday's law,

$$e = \frac{d\phi}{dt} = 2Blv \tag{17-5}$$

The factor 2 appears in Eq. (17-5) because the coil involves two conductors. It then follows that if the rotor is equipped with a total of Z conductors, which are all series-connected, the total induced voltage can be expressed as

$$e = ZBlv \tag{17-6}$$

When a current i flows through conductors *a* and *b* in the configuration of Fig. 17-2(c), the direction of i and the direction of B are 90 degress apart.† Accordingly, a force exists on each conductor which is given by Eq. (15-2). Thus

$$F_c = Bli \tag{17-7}$$

where F_c denotes the force developed on an individual conductor. Since there are Z conductors on the rotor surface, it follows that the total developed force is‡

$$F = ZBli \tag{17-8}$$

Moreover, if this force is made to act through a moment arm r, which is the radius of the rotor, the resulting developed torque can be expressed as

$$T = Fr = ZBlri \tag{17-9}$$

A glance at Eqs. (17-6) and (17-9) shows that both the induced voltage and the developed electromagnetic torque are dependent upon the coupling magnetic field. The factor that determines whether the induced voltage e and the developed torque T are actions or reactions depends upon whether generator or motor action is being developed.

The two mechanical quantities, T and ω_m, and the two electrical quantities, e and i, which are coupled through the medium of the magnetic field B, are related further by the law of conservation of energy. This is readily demonstrated by dividing Eq. (17-6) by Eq. (17-9). Thus

$$\frac{e}{T} = \frac{ZBlv}{ZBlir} = \frac{\omega_m}{i} \tag{17-10}$$

where

$$\omega_m = \frac{v}{r} = \text{mechanical angular velocity of the rotor} \tag{17-11}$$

Rearranging Eq. (17-10) leads to

$$ei = T\omega_m \tag{17-12}$$

which states that the developed electrical power is equal to the developed mechanical power. This statement is valid for generator as well as motor action.

Finally, on the basis of the foregoing discussion, it should be apparent that

---

† Keep in mind that i flows through the conductors and that the conductors by the geometry of the machine will always cut the B-field at right angles.

‡ This equation is valid provided that the conductors are paired to form series-connected coils that have a span of one *pole pitch*. A pole pitch is equal to 180 electrical degrees.

a conventional electromechanical energy-conversion device must involve two components. One is the *field winding*, which by definition is that part of the machine which produces the coupling field B. The other is the *armature winding*, which by definition is that part of the machine in which the "working" emf e and the current i exist.

Developed Torque for Sinusoidal B and Ni (the Situation of A-C Machines). The expression for electromagnetic torque as it appears in Eq. (17-9) has limited application because B is assumed constant and all conductors are assumed to produce equal torques in the same direction—a situation which rarely occurs in practical machines. In fact, in most a-c machines the field winding is intentionally designed to produce an almost sinusoidally distributed flux density, and the armature winding is similarly arranged to yield an ampere-conductor distribution which is also almost sinusoidally distributed along the periphery of the rotor structure. For this reason we shall now consider the development of an electromagnetic torque equation that applies for such cases.

In differential form Eq. (17-7) can be expressed as

$$dF = Bl\,di \tag{17-13}$$

Here $l\,di$ is an assumed differential current element and its orientation is one of quadrature with the B-field. Moreover, because the field distribution is essentially sinusoidal for a-c machines, we can write for B

$$B = B_m \sin\theta \tag{17-14}$$

where B_m denotes the amplitude of the flux density wave and θ is a displacement angle measured in electrical degrees. The differential current can be expressed in terms of a spatial distribution as

$$di = J\,d\theta = J_m \sin(\theta + \psi)\,d\theta \tag{17-15}$$

where J is expressed in units of amperes per radian. Of course, J_m denotes the peak value of J. Equation (17-15) essentially states that the current distribution in the armature winding of a-c machines also assumes a sinusoidal distribution. In general, the armature winding current distribution is displaced from the field distribution by the angle ψ. See Fig. 17-3. The sine wave representing the current distribution always has the same number of poles as the field distribution because the latter induces the former. Inserting Eqs. (17-14) and (17-15) into Eq. (17-13) allows the differential force produced on the rotor of the machine to be expressed as

$$dF = Bl\,di = B_m J_m l \sin\theta \sin(\theta + \psi)\,d\theta \tag{17-16}$$

The expression for the differential torque then becomes

$$dT = r\,dF = B_m J_m lr \sin\theta \sin(\theta + \psi)\,d\theta \tag{17-17}$$

The total developed torque follows readily from this last equation by taking the summation of all torques associated with the current elements beneath one pole and then multiplying by the number of poles, p, of the machine. The result is

$$T = \pi\frac{p}{2}B_m J_m lr \cos\psi \tag{17-18}$$

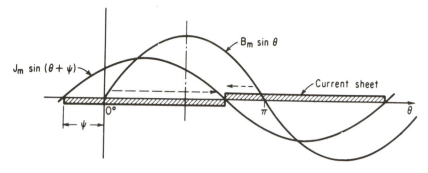

Fig. 17-3 Configuration for development of the torque equation showing the sinusoidally distributed flux density produced by the field winding and the current sheet which represents the armature winding. The dashed-line arrows denote relative magnitude and direction of the electromagnetic torque developed per pole. The magnetic field and the current sheet are stationary relative to each other.

The flux per pole, Φ, is a very useful and important quantity in machine analysis. Consequently, it is frequently desirable to express the developed electromagnetic torque in terms of this quantity. Obtaining the flux per pole is an easy matter, if the field distribution is known as we have certainly here assumed by Eq. (17-14). It merely requires finding the area of the B curve. Such a computation for the sine wave is straightforward. The final result for the developed electromagnetic torque thus becomes

$$T = \frac{\pi}{8}p^2\Phi J_m \cos\psi \quad \text{N-m} \tag{17-19}$$

A close examination of Eq. (17-19) points out clearly the three conditions that must be satisfied for the development of torque in conventional electromechanical energy-conversion devices. First, there must be a field distribution. This is represented by Φ. Second, there must be a current distribution which here is denoted by J_m. Actually, a better way to describe this is to say that there must be an *ampere-conductor distribution* that has the same number of poles as the field distribution and is stationary relative to it. After all, the current must flow in conductors and the conductors† are the elements that are distributed over the surface of the rotor, thereby becoming subject to the influence of the flux density. Finally, there must exist a favorable space displacement angle between the field distribution and the ampere-conductor distribution. One should note that it is possible to have both a field and an ampere-conductor distribution and still have zero net torque. This happens whenever the field pattern is such that equal and opposite torques beneath each pole are developed on the conductors. In other words $\psi = 90°$.

It is also worthwhile to state at this point that no really new concepts have

† The magnitude of the currents existing in successive conductors, however, varies sinusoidally.

been used to derive Eq. (17-19). This result is merely an extension of Ampère's law.

Torque for Nonsinusoidal B and Uniform Ampere-Conductor Distribution (the Situation of D-C Machines). The approach in the derivation here is the same as in the preceding case. Again we start with the expression for the differential force as given by Eq. (17-13). In the case of d-c machines, however, B is nonsinusoidal as a glance at Fig. 17-4 discloses. Moreover, an analytical expression for this field is virtually impossible to obtain because of the saturated state of the sheet steel which this flux penetrates. For our purposes at the moment we therefore introduce

$$B = B_\theta \tag{17-20}$$

where the θ subscript reminds us that because of the nonsinusoidal variation the value of the flux density for a particular θ would have to come from a field plot. The differential current on the other hand is given simply as

$$di = J \, d\theta \tag{17-21}$$

where J is a constant quantity. This particular characteristic of the d-c machine is attributable to the presence of conducting brushes placed on a commutator. More is said about this in Sec. 17-3.

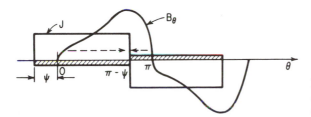

Fig. 17-4 Current-sheet and flux-density curves characteristic of d-c machines.

The differential force appearing on a differential current element located beneath a pole of the field distribution thus becomes

$$dF = B_\theta Jl \, d\theta \tag{17-22}$$

and the related torque is

$$dT = r \, dF = B_\theta Jlr \, d\theta \tag{17-23}$$

Integrating Eq. (17-23) over one pole and multiplying by the number of poles yields the total resultant electromagnetic torque. Thus

$$T = pJ \int_0^\pi B_\theta lr \, d\theta \qquad \text{N-m} \tag{17-24}$$

Note that an expression without the integral is not possible in this case as it was with the a-c machine because an analytical expression for B_θ is not known. Of course the integration called for in Eq. (17-24) could be performed graphically once the field plot for B_θ is known. However, this inconvenience can be avoided entirely by working rather in terms of the flux per pole of the machine. Moreover, information about the flux per pole is available from other considerations such as the power rating and voltage rating of the machine. Proceeding in this fashion, then, the expression for the developed electromagnetic torque for the d-c machine becomes simply

$$T = \frac{p^2}{2} J\Phi \quad \text{N-m} \qquad (17\text{-}25)$$

Here again it is worthwhile to note that the electromagnetic torque is dependent upon a field distribution represented by Φ, an ampere conductor distribution represented by J, and the angular displacement between the two distributions. The quantity ψ does not appear in Eq. (17-25) because for simplicity (and also because it is common practice) the current sheet was assumed in phase with the flux-density curve, thus assuring that all elemental parts of the current sheet experience a unidirectional torque. A glance at Fig 17-4, however, should make it plain that if ψ is changed from 0° to 90°, the resultant electromagnetic torque becomes zero.

17-2 ANALYSIS OF INDUCED VOLTAGES

The electromechanical energy-conversion process involves the interaction of two fundamental quantities which are related through the law of conservation of energy and the coupling magnetic field. These are, of course, electromagnetic torque and induced voltage. The manner in which electromagnetic torque is developed is described in the preceding section. Here we shall consider the generation of voltages in the armature winding brought about by changing flux linkages. Whether or not this generated voltage is a reaction voltage depends upon whether or not the conversion device is behaving as a motor or generator. Our objective now is to obtain a general result for the armature-winding induced emf that can be applied to either a-c or d-c type electromechanical energy-conversion devices.

Consider the configuration shown in Fig. 17-5. The flux-density curve is assumed to be sinusoidally† distributed along the air gap of the machine, and the armature winding is assumed to consist of a full-pitch coil having N turns. A full-pitch coil is one that spans π electrical radians. Just as Ampère's law serves as the starting point in the development of the torque relationships, so

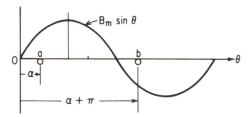

Fig. 17-5 Configuration for derivation of induced-emf equation.

† For a nonsinusoidal field distribution Fig. 17-5 represents the fundamental component.

Faraday's law serves as the starting point in establishing the induced-voltage relationships. Equation (17-5) is not the result we are seeking, because it applies for the case where the flux-density curve is uniform over the pole pitch. It is more generally useful to deal rather with a sinusoidal distribution of the flux density, as explained more fully later. Since the energy-conversion process involves relative motion between the field and the armature winding, we know that the flux linking the coil changes from a positive maximum to zero to a negative maximum and back to zero again. This cycle repeats every time the coil passes through a pair of flux-density poles. It should be clear from Fig. 17-5 that maximum flux penetrates the coil when it is in a position corresponding to $\alpha = 0°$ or $\alpha = 180°$ or multiples thereof. Minimum flux linkage occurs when $\alpha = 90°$ or $270°$ or multiples thereof. It should be apparent too that the maximum flux that links the coil is the same as the flux per pole, which for sinusoidal distributions can be shown to be given by $\Phi = \frac{4}{p}B_m lr$. Moreover, as the coil moves relative to the field, it follows that the instantaneous flux linking the coil may be expressed as

$$\phi = \Phi \cos \alpha = \frac{4}{p}B_m lr \cos \alpha \qquad (17\text{-}26)$$

The $\cos \alpha$ factor is included because of the sinusoidal distribution. Furthermore, since α increases with time in accordance with

$$\alpha = \omega t \qquad (17\text{-}27)$$

where ω is expressed in electrical radians per second, the expression for the instantaneous flux linkage becomes

$$\phi = \frac{4}{p}B_m lr \cos \omega t = \Phi \cos \omega t \qquad (17\text{-}28)$$

Then by Faraday's law the expression for the instantaneous induced voltage becomes

$$e = -N\frac{d\phi}{dt} = \omega N\Phi \sin \omega t \qquad (17\text{-}29)$$

It is important to keep in mind here that Φ represents the *maximum flux linkage*, which is the same as the flux per pole.

Equation (17-29) is applicable to d-c as well as a-c machines. In a later section the induced emf per phase in an a-c machine and the induced voltage appearing between brushes in a d-c machine are derived starting with Eq. (17-29). The fact that in the d-c machine the flux density curve is nonsinusoidal, as Fig. 17-4 shows, is relatively unimportant; what is of primary importance is the fundamental component of the nonsinusoidal distribution. It is for this reason that Eq. (17-29) is derived for sinusoidal distributions. Recall that by means of the Fourier series any nonsinusoidal periodic wave may be expressed entirely in terms of sinusoids.

17-3 CONSTRUCTION FEATURES OF ELECTRIC MACHINES

The energy-conversion process usually involves the presence of two important features in a given electromechanical device. These are the field winding, which produces the flux density, and the armature winding, in which the "working" emf is induced. In this section the salient construction features of the principal types of electric machines are described to show the location of these windings, as well as to demonstrate the general composition of such machines.

The Three-Phase Induction Motor. This is one of the most rugged and most widely used machines in industry. Its stator is composed of laminations of high-grade sheet steel. The inner surface is slotted to accommodate a three-phase winding. In Fig. 17-6(a) the three-phase winding is represented by three coils, the axes of which are 120 electrical degrees apart. Coil aa' represents all the coils assigned to phase a for one pair of poles. Similarly coil bb' represents phase b coils, and coil cc' represents phase c coils. When one end of each phase is tied together, as depicted in Fig. 17-6(b), the three-phase stator winding is said to be Y-connected. Such a winding is called a three-phase winding because the voltages induced in each of the three phases by a revolving flux density field are out of phase by 120 electrical degrees—a distinguishing characteristic of a balanced three-phase system.

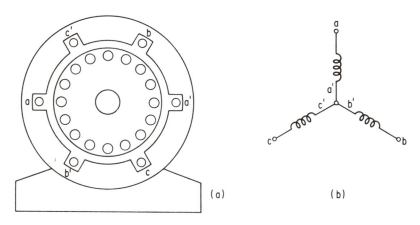

(a)   (b)

Fig. 17-6 Three-phase induction motor: (a) showing stator with three-phase winding and squirrel-cage rotor; (b) schematic representation of three-phase Y-connected stator winding.

The rotor also consists of laminations of slotted ferromagnetic material, but the rotor winding may be either the squirrel-cage type or the wound-rotor type. The latter is of a form similar to that of the stator winding. The winding terminals are brought out to three slip rings as depicted in Fig. 17-7. This allows an external three-phase resistor to be connected to the rotor winding for the purpose of providing speed control. As a matter of fact, it is the need for speed

control which in large measure accounts for the use of the wound-rotor type induction motor. Otherwise the squirrel-cage induction motor would be used. The squirrel-cage winding consists merely of a number of copper bars embedded in the rotor slots and connected at both ends by means of copper end rings as depicted in Fig. 17-8. (In some of the smaller sizes aluminum is used.) The squirrel-cage construction is not only simpler and more economical than the wound-rotor type but more rugged as well. There are no slip rings or carbon brushes to be bothered with.

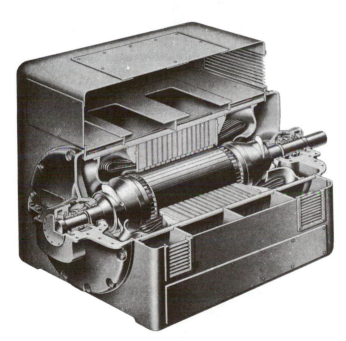

In normal operation a three-phase voltage is applied to the stator winding at points a-b-c in Fig. 17-6. As described in Sec. 18-1 magnetizing currents flow in each phase which together create a revolving magnetic field having two poles. The speed of the field is fixed by the frequency of the magnetizing currents and the number of poles for which the stator winding is designed. Figure 17-6 shows the configuration for two poles. If the pattern a-c'-b-a'-c-b' is made to span only

180 mechanical degrees and then is repeated over the remaining 180 mechanical degrees, a machine having a four-pole field distribution results. For a p-pole machine the basic winding pattern must be repeated $p/2$ times within the circumference of the inner surface of the stator.

The revolving field produced by the stator winding cuts the rotor conductors, thereby inducing voltages. Since the rotor winding is short-circuited by the end rings, the induced voltages cause currents to flow which in turn react with the field to produce electromagnetic torque—and so motor action results.

Accordingly, on the basis of the foregoing description, it should be clear that for the three-phase induction motor the field winding is located on the stator and the armature winding on the rotor. Another point worth noting is that this machine is singly excited, i.e., electrical power is applied only to the stator winding. Current flows through the rotor winding by induction. As a consequence both the magnetizing current, which sets up the magnetic field, and the power current, which allows energy to be delivered to the shaft load, flow through the stator winding. For this reason, and in the interest of keeping the magnetizing current as small as possible in order that the power component may be correspondingly larger for a given current rating, the air gap of induction motors is made as small as mechanical clearance will allow. The air-gap lengths vary from about 0.02 in. for smaller machines to 0.05 in. for machines of higher rating and speed.

Synchronous Machines. The essential construction features of the synchronous machine are depicted in Fig. 17-9. The stator consists of a stator frame, a slotted stator core, which provides a low-reluctance path for the magnetic flux, and a three-phase winding imbedded in the slots. Note that the basic two-pole pattern of Fig. 17-6(a) is repeated twice, indicating that the three-phase winding is designed for four poles. The rotor either is cylindrical and equipped with a distributed winding or else has salient poles with a coil wound on each leg as depicted in Fig. 17-9. The cylindrical construction is used exclusively for turbogenerators, which operate at high speeds. On the other hand the salient-pole

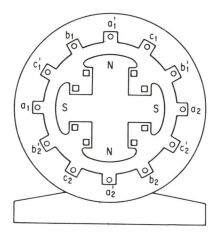

Fig. 17-9 Salient-pole synchronous machine.

construction is used exclusively for synchronous motors operating at speeds of 1800 rpm or less.

When operated as a generator the synchronous machine receives mechanical energy from a prime mover such as a steam turbine and is driven at some fixed speed. Also, the rotor winding is energized from a d-c source, thereby furnishing a field distribution along the air gap. When the rotor is at standstill and d-c flows through the rotor winding, no voltage is induced in the stator winding because the flux is not cutting the stator coils. However, when the rotor is being driven at full speed, voltage is induced in the stator winding and upon application of a suitable load electrical energy may be delivered to it.

For the synchronous machine the field winding is located on the rotor; the armature winding is located on the stator. This statement is valid even when the synchronous machine operates as a motor. In this mode a-c power is applied to the stator winding and d-c power is applied to the rotor winding for the purpose of energizing the field poles. Mechanical energy is then taken from the shaft. Note, too, that unlike the induction motor, the synchronous motor is a doubly fed machine; i.e., energy is applied to the rotor as well as the stator winding. In fact it is this characteristic which enables this machine to develop a nonzero torque at only one speed—hence the name *synchronous*. This matter is discussed further in Sec. 19-3.

Because the magnetizing current for the synchronous machine originates from a separate source (the d-c supply), the air-gap lengths are larger than those found in induction motors of comparable size and rating. However, synchronous machines are more expensive and less rugged than induction motors in the smaller horsepower ratings because the rotor must be equipped with slip rings and brushes in order to allow the direct current to be conducted to the field winding.

D-C Machines. Electromechanical energy-conversion devices that are characterized by direct current are more complicated than the a-c type. In addition to a field winding and armature winding, a third component is needed to serve the function of converting the induced a-c armature voltage into a direct voltage. Basically the device is a mechanical rectifier and is called a *commutator*.

Appearing in Fig. 17-10 are the principal features of the d-c machine. The stator consists of an unlaminated ferromagnetic material equipped with a protruding structure around which coils are wrapped. The flow of direct current through the coils establishes a magnetic field distribution along the periphery of the air gap in much the same manner as occurs in the rotor of the synchronous machine. Hence in the d-c machine the field winding is located on the stator. It follows then that the armature winding is on the rotor. The rotor is composed of a laminated core, which is slotted to accommodate the armature winding. It also contains the commutator—a series of copper segments insulated from one another and arranged in cylindrical fashion as shown in Fig. 17-11. Riding on the commutator are appropriately placed carbon brushes which serve to conduct direct current to or from the armature winding depending upon whether motor or generator action is taking place.

In Fig. 17-10 the armature winding is depicted as a coil wrapped around a

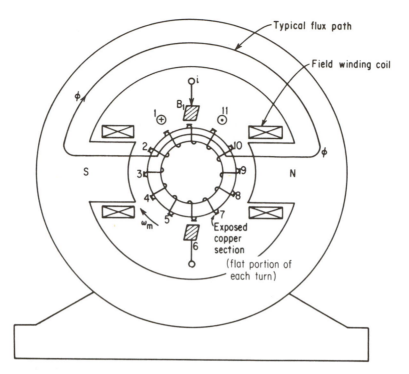

Fig. 17-10 Construction features of the d-c machine showing the Gramme-ring armature winding.

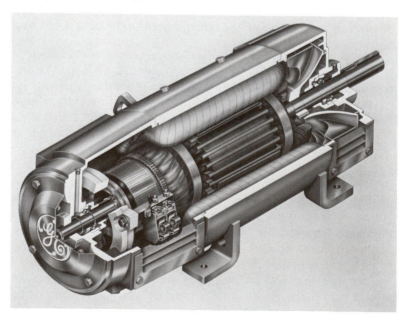

Fig. 17-11 Armature winding of a d-c machine showing the coils connected to the commutator.

toroid. This is merely a schematic convenience. In an actual winding, no conductors are wasted by placing them on the inner surface of the rotor core where no flux penetrates. In Fig. 17-9 those parts of the armature winding which lie directly below the brush width are assumed to have the insulation removed, i.e., the copper is exposed. This allows current to be conducted to and from the armature winding through the brush as the rotor revolves. In a practical winding each coil is made accessible to the brushes by connecting the coils to individual commutator segments as indicated in Fig. 17-11 and then placing the brushes on the commutator.

For motor action direct current is made to flow through the field winding as well as the armature winding. If current is assumed to flow into brush B_1 in Fig. 17-10, then note that on the left side of the rotor for the *outside* conductors current flows into the paper while the opposite occurs for the conductors located on the outside surface of the right side of the rotor. By Eq. (12-2) a force is produced on each conductor, thereby producing a torque causing clockwise rotation. Now the function of the commutator is to assure that as a conductor such as 1 in Fig. 17-10 revolves and thus goes from the left side of brush B_1 to the right side, the current flowing through it reverses, thus yielding a continuous unidirectional torque for the entire armature winding. Recall that a reversed conductor current in a flux field of reversed polarity keeps the torque unidirectional. The reversal of current comes about because the commutator always allows current to be conducted in the same directions in either side of the armature winding whether or not it is rotating.

Another point of interest in Fig. 17-10 concerns the location of the brushes. By placing the brushes on a line perpendicular to the field axis all conductors contribute in producing a unidirectional torque. If, on the other hand, the brushes were placed on the same line as the field axis, then half of the conductors would produce clockwise torque and the other half counterclockwise torque, yielding a zero net torque. In the notation of Fig. 17-4 this situation would correspond to a value of ψ equal to 90°.

17-4 PRACTICAL FORMS OF TORQUE AND VOLTAGE FORMULAS

Our objective here is to modify the basic expression for induced voltage and electromagnetic torque to a form that is more meaningful in terms of the specific data of a particular electromechanical energy-conversion device be it a-c or d-c.

A-C Machines. If N denotes the total number of turns per phase of a three-phase winding, the instantaneous value of the emf induced in any one phase can be represented by Eq. (17-29), which is repeated here for convenience. Thus

$$e = \omega N\Phi \sin \omega t \qquad (17\text{-}30)$$

Keep in mind that Φ denotes the total flux per pole and ω represents the relative cutting speed in electrical radians per second of the winding with respect to the flux-density wave. It is related to the frequency f of the a-c device by

$$\omega = 2\pi f \qquad (17\text{-}31)$$

where f is expressed in cycles per second or hertz.

The maximum value of this a-c voltage occurs when $\sin \omega t$ has a value of unity. Hence

$$E_{max} = \omega N\Phi \quad V \qquad (17\text{-}32)$$

The corresponding rms value is

$$E \equiv \frac{E_{max}}{\sqrt{2}} = \sqrt{2}\,\pi f N\Phi = 4.44 f N\Phi \quad V \qquad (17\text{-}33)$$

A comparison of this expression with Eq. (16-18) reveals that the equations have an identical form. There is a difference, however, and it lies in the meaning of Φ. In the transformer, Φ_m is the maximum flux corresponding to the peak magnetizing current consistent with the magnitude of the applied voltage. In the a-c electromechanical energy-conversion device Φ is the maximum flux per pole that links with a coil having N turns and spanning the full pole pitch.

In any practical machine the total turns per phase are not concentrated in a single coil but rather are distributed over one-third of a pole pitch (or 60 electrical degrees for each of the three phases). In addition, the individual coils that make up the total N turns are intentionally designed not to span the full pole pitch but rather only about 80 to 85% of a pole pitch. Such a coil is called a *fractional-pitch coil*. The use of a *distributed* winding employing *fractional-pitch* coils has the advantage of virtually eliminating the effects of all harmonics that may be present in the flux-density wave while only slightly reducing the fundamental component.† The reduction in the fundamental component can be represented by a winding factor denoted by K_w. Usually, K_w has values ranging from 0.85 to 0.95. Accordingly, the final practical version of the induced rms voltage equation for an a-c machine is

$$\boxed{E = 4.44 f N K_w \Phi} \quad V \qquad (17\text{-}34)$$

We turn our attention next to the practical forms of the torque equations. There are two useful forms to which the basic expression for electromagnetic torque in a-c machines as expressed by Eq. (17-19) may be converted. In one form the ampere-conductor distribution of the armature winding is emphasized. In the other, it is the conservation of energy which is stressed.

The quantity J_m in Eq. (17-19) is related to the mmf per pole by

$$J_m = \mathscr{F}_p \qquad (17\text{-}35)$$

Furthermore, for any given a-c machine the quantity $\mathscr{F}_p$ is entirely determinable from the design data as they appear in the following equation:‡

$$\mathscr{F}_p = \frac{2\sqrt{2}}{\pi} q \frac{N_2 K_{w2}}{p} I_2 = 0.9 q \frac{N_2 K_{w2}}{p} I_2 \qquad (17\text{-}36)$$

---

† Consult M. Liwschitz-Garik and C. C. Whipple, *Electric Machinery*, vol. II (Princeton, N. J.: D. Van Nostrand Co., Inc., 1946), Chap. 5.

‡ Consult V. Del Toro, *Electric Machines and Power Systems* (Englewood Cliffs, N. J.: Prentice-Hall, Inc., 1985), App. C.

where q = number of phases of the armature winding
$\quad\quad N_2$ = number of turns per phase of the armature winding
$\quad\quad K_{w2}$ = armature winding factor
$\quad\quad I_2$ = armature winding current per phase
$\quad\quad p$ = number of poles

Hence the expression for electromagnetic torque as described by Eq. (17-19) becomes

$$T = \frac{\pi}{8}p^2\Phi\mathscr{F}_p\cos\psi = \frac{\pi}{8}(0.9)p\Phi N_2 K_{w2}qI_2\cos\psi \quad\quad (17\text{-}37)$$

Moreover, qN_2 is the *total* number of turns on the armature surface. This can be expressed in terms of the total number of conductors, Z_2, by making use of the fact that it requires two conductors to make one turn. Thus

$$N_2 = \frac{Z_2}{2q} \quad\quad (17\text{-}38)$$

Inserting this into Eq. (17-37) yields

$$\boxed{T = 0.177p\Phi(Z_2 K_{w2}I_2)\cos\psi} \quad \text{N-m} \quad\quad (17\text{-}39)$$

The quantity in parentheses emphasizes the role played by the ampere-conductor distribution of the armature winding. Of course in order to compute the torque by this equation we need information about the space displacement angle ψ in addition to the design data, namely, p, Φ, Z_2, K_{w2}. Normally ψ can be determined by the same data that lead to the determination of I_2, which is described in the next chapter.

A second approach to obtaining a useful form of the basic torque equation is to replace Φ by Eq. (17-34) and J_m by Eqs. (17-35) and (17-36) in Eq. (17-19). Thus

$$T = \frac{\pi}{8}p^2\left[\frac{E_2}{\sqrt{2}\pi f N_2 K_{w2}}\right]\left[\frac{2\sqrt{2}}{\pi}q\frac{N_2 K_{w2}}{p}I_2\right]\cos\psi$$

or

$$T = \frac{p}{4\pi f}qE_2 I_2\cos\psi \quad\quad (17\text{-}40)$$

Moreover, since the mechanical angular velocity ω_m is related to the electrical angular velocity ω by

$$\omega_m = \frac{2}{p}\omega = \frac{2}{p}(2\pi f) = \frac{4\pi f}{p} \quad\quad (17\text{-}41)$$

it follows that Eq. (17-40) may be written as

$$T = \frac{1}{\omega_m}qE_2 I_2\cos\psi \quad \text{N-m} \quad\quad (17\text{-}42)$$

where E_2 and I_2 denote the induced voltage and current per phase of the armature winding.

It can be shown that the *space* displacement angle ψ in this equation is identical to the time displacement angle θ_2,[†] which is the phase angle existing between the two time phasors $\bar{E}_2$ and $\bar{I}_2$. Accordingly, the expression for the electromagneic torque may be written as

$$T = \frac{1}{\omega_m} q E_2 I_2 \cos \theta_2 \quad \text{N-m} \qquad (17\text{-}43)$$

Note the power balance, which states that the developed mechanical power is equal to the developed electrical power.

D-C Machines. The useful version of the expression for the emf induced in the armature winding of a d-c machine and appearing at the brushes also derives from Eq. (17-29). The presence of the commutator, however, makes a difference in the manner in which this equation is to be treated to obtain the desired result. We are interested in the voltage as it appears at the brushes. If we follow coil 1 in Fig. 17-10 as it leaves brush B_2 and advances to brush B_1 we notice that the emf induced in this coil is always directed out of the paper beneath the south-pole flux.[‡] In fact this is true of any coil that moves from B_2 to B_1. Having established the direction of this voltage for the coils on the left side of the armature, we then note that, upon tracing through the winding starting in the direction called for by the sign of the emf induced in coil 1, we terminate at brush B_2. When coil 1 rotates to a position on the right side of the armature winding such as position 11, the direction of the induced emf is out of the paper. If we start with this indicated direction and trace the armature winding, we again terminate at brush B_2. Therefore it is reasonable to conclude that brush B_1 is at a fixed *negative* polarity and brush B_2 is at a fixed *positive* polarity. In essence this means that, if we were to ride along as observers on coil 1 as it traverses through one revolution beginning at brush B_2, the fundamental component of the induced emf would exhibit the variation shown in Fig. 17-12. Note that *relative to brush B_1* the voltage induced in the coil as it moves from B_1 to B_2 appears *rectified* because the action of the commutator in conjunction with the brushes is to fix the armature winding in space in spite of its rotation. Accordingly,

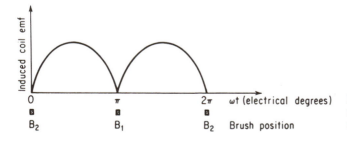

Fig. 17-12 Fundamental component of the induced emf in a coil of the armature winding of a d-c machine.

† See p. 729.

‡ For motor action which is depicted here the emf and the current in the coils are oppositely directed.

whether we look at a coil moving from B_2 to B_1 on the left side or from B_1 to B_2 on the right side, the directions of the induced emf's are such as to make B_1 assume one fixed polarity and B_2 the opposite fixed polarity.

For a single coil having N_c turns the average value of the unidirectional voltage appearing between the brushes is obtained by integrating Eq. (17-29) over π electrical radians and dividing by π. Thus

$$E_c = \frac{1}{\pi} \int_0^\pi \omega N_c \Phi \sin(\omega t) \, d(\omega t) = 4 f N_c \Phi \qquad (17\text{-}44)$$

It is customary to express the frequency in terms of the speed of the armature in *rpm* which is given by the relationship†

$$\boxed{f = \frac{pn}{120}} \qquad (17\text{-}45)$$

where p denotes the number of poles and n the speed in *rpm*. Inserting Eq. (17-45) into Eq. (17-44) and rearranging leads to

$$E_c = p\Phi(2N_c)\frac{n}{60} = p\Phi z\frac{n}{60} \qquad (17\text{-}46)$$

where z denotes the number of conductors per coil.

Equation (17-46) is the average value of the induced emf for a single coil. If many such coils are placed to cover the entire surface of the armature, the total d-c voltage appearing between the brushes can be considerably increased. If the armature winding is assumed to have a total of Z conductors and a parallel paths, then the induced emf of the armature winding appearing at the brushes becomes

$$\boxed{E_a = p\Phi\frac{Z}{a}\frac{n}{60} = \frac{pZ}{60a}\Phi n = K_E \Phi n} \qquad (17\text{-}47)$$

where K_E is a winding constant defined as

$$K_E \equiv \frac{pZ}{60a} \qquad (17\text{-}48)$$

It is worth keeping in mind that the induced-emf equation for the a-c machine and that for the d-c machine originate from the same starting point. To further emphasize this common origin let it be said that we can derive Eq. (17-47) from Eq. (17-34). It merely requires introducing the appropriate winding factor as it applies to the d-c armature winding. In fact it is the quasi-annihilating effect of the winding factor on the harmonics that allows the derivation of Eq.

---

† Rotating a coil in a two-pole field at a rate of one revolution per second results in a frequency of 1 Hz. For a four-pole field each revolution per second yields 2 Hz. For a p-pole distribution the frequency generated in hertz is

$$f = \frac{p}{2}(\text{rps}) = \frac{p}{2}\frac{\text{rpm}}{60} = \frac{pn}{120}$$

(17-47) to proceed in terms of the fundamental in spite of the nonsinusoidal field distribution characteristic of d-c machines. It is assumed throughout, however, that the fundamental component of flux is virtually the same as the total flux per pole.

A practical version of the electromagnetic torque formula readily follows from Eq. (17-25) if we replace the current-density sheet J by an equivalent expression involving the design data of the machine. In this connection refer to Fig. 17-13. Applying Ampère's circuital law to the typical flux path shown leads to

$$\mathscr{F} = \pi \text{ (rad) } J \quad \left(\frac{\text{At}}{\text{rad}}\right) \tag{17-49}$$

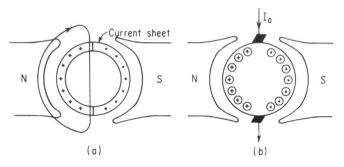

Fig. 17-13 D-c machine: (a) armature with current sheet; (b) armature with finite ampere-conductor distribution.

Hence the mmf per pole is

$$\mathscr{F}_p = \frac{\mathscr{F}}{2} = \frac{\pi J}{2} \tag{17-50}$$

This expression is applicable whenever the armature winding is represented by a current sheet. In a practical situation the armature winding is always represented by a finite ampere-conductor distribution as depicted in Fig. 17-12(b). In such cases the armature mmf per pole may be expressed in terms of the total current I_a entering or leaving a brush and the total conductors Z on the armature surface. Thus

$$\mathscr{F}_p = \left(\frac{I_a}{a}\right)\left(\frac{Z}{2}\right)\frac{1}{p} = \text{mmf/pole} = \text{At/pole} \tag{17-51}$$

Upon equating the last two expressions we obtain the equation for J. Thus

$$J = \frac{I_a Z}{\pi p a} \tag{17-52}$$

Insertion of Eq. (17-52) into Eq. (17-25) then yields the desired expression for the electromagnetic torque developed by the d-c machine. Hence

$$T = \frac{p^2}{2} J\Phi = \frac{pZ}{2\pi a}\Phi I_a \quad \text{N-m} \tag{17-53}$$

or

$$T = K_T\Phi I_a$$

(17-54)

where

$$K_T \equiv \frac{pZ}{2\pi a} = \text{torque constant}$$

(17-55)

The foregoing form of the torque formula stresses the ampere-conductor distribution taken in conjunction with the flux field. It is analogous to Eq. (17-39) for the a-c machine. Of course in Eq. (17-54) ψ does not appear because it was made zero by placing the brush axis in quadrature with the field axis. It is important to note, however, that the fundamental quantities for the production of torque are there. The equations differ only to the extent that the mechanical details of construction differ.

An alternative expression for the electromagnetic torque results upon substituting in Eq. (17-53) the expression for Φ as obtained from Eq. (17-47). Thus from Eq. (17-47)

$$\Phi = \frac{60a}{pZn}E_a$$

(17-56)

so that

$$T = \frac{pZ}{2\pi a}\left[\frac{60a}{pZn}\right]E_a I_a = \frac{60}{2\pi n}E_a I_a$$

(17-57)

But

$$\omega_m = \frac{2\pi n}{60}$$

(17-58)

Therefore,

$$T = \frac{1}{\omega_m}E_a I_a$$

(17-59)

A glance at this result again points out the power-balance relationship that underlies the operation of electromechanical energy-conversion devices. Equation (17-59) for the d-c machine is analogous to Eq. (17-43) for the a-c machine.

Summary review questions

1. Name and illustrate the two fundamental laws on which is based the study of electric machines.
2. Describe motor action from the viewpoint of energy flow and carefully identify the roles of the related electrical and mechanical quantities.
3. Repeat Question 2 for generator action.
4. List the three conditions that must exist in an electric machine in order to produce torque. Explain the role of each.

5. Is it possible to have an ampere-conductor distribution under the influence of a flux field and still fail to produce a useful net torque? Explain.

6. Why is the field distribution, B, of a-c motors very nearly sinusoidal? Why is the ampere-conductor distribution of such motors also very nearly sinusoidal?

7. Write the torque equation of the a-c motor in terms of electrical and magnetic quantities and describe the role of each term in the formula.

8. The field distribution (B) in the d-c machine is not sinusoidal and an accurate mathematical description is not easily available. How is this inconvenience avoided in dealing with the basic torque equation of such machines?

9. Describe the role of the coupling magnetic field in the electromechanical energy conversion process between an electrical and a mechanical device.

10. Show how Faraday's law is used to derive a general expression for the emf induced in a coil of N turns. Assume the field distribution is sinusoidal.

11. Why is the polyphase induction motor considered to be one of the most rugged of the integral horsepower electric motors?

12. Electric motors contain a stationary member as well as a rotating member. For each of the following machines, identify in which part of the motor the field winding and the armature winding are located: three-phase induction motor, three-phase synchronous motor, d-c motor.

13. Are both the stator and rotor of a-c polyphase motors constructed of thin laminations of ferromagnetic material? Explain.

14. Repeat Question 13 for the d-c motor.

15. Repeat Question 13 for the synchronous motor.

16. Describe how the three phase windings of the two-pole induction motor are given a space displacement of 120 electrical degrees.

17. Explain how the phase windings of the three-phase induction motor must be arranged for operation as a four-pole motor.

18. What is a squirrel-cage winding? What advantages does it offer over the wound-rotor winding? What are the disadvantages?

19. What important consequence arises from the fact that the polyphase induction motor is a singly excited machine?

20. What important consequence arises from the fact that the synchronous motor is a doubly excited machine?

21. Why is special care directed to design induction motors with air gaps that are as small as mechanical clearance allows, whereas this care is not urgent for synchronous machines?

22. Why are synchronous motors more expensive than induction motors?

23. What is a commutator and describe its function in the d-c machine.

24. By means of an appropriate sketch show how the presence of a commutator and a pair of brushes makes the armature winding of the d-c machine appear to be stationary despite its rotation.

25. Why is the placement of the brushes relative to the field poles crucial to the proper operation of the d-c machine? Illustrate.

26. Write the equation for the induced voltage per phase in the three-phase motor. Explain each quantity.

27. Write the torque equation for the polyphase induction motor that is based on the conservation of energy between the electrical part of the system (motor) and the mechanical load.

28. Are the voltages induced in the individual coils of the armature winding of the d-c machine a-c or d-c? Explain.

29. Write the basic developed torque equation of the d-c motor in terms of electrical and magnetic quantities and describe the meaning of each quantity.

30. Write the basic developed torque equation of the d-c motor that puts emphasis on the conservation of energy viewpoint.

Problems

GROUP I

17-1. The instantaneous voltage generated in a coil revolving in a magnetic field can be calculated either by the flux-linkage concept or by the "Blv" concept. With this in mind consider the following problem. A square coil 20 cm on a side has 60 turns and is so located that its axis of revolution is perpendicular to a uniform magnetic field in air of 0.06 Wb/m². The coil is driven at a constant speed of 150 rpm. Compute:

(a) The maximum flux passing through the coil.
(b) The maximum flux linkage.
(c) The time variation of the flux linkage through the coil.
(d) The maximum instantaneous voltage generated in the coil using both concepts referred to above. Indicate by a sketch the position of the coil at this instant.
(e) The average value of the voltage induced in the coil over one cycle.
(f) The voltage generated in the coil when the plane of the coil is 30° from the vertical. Compute by both methods.

17-2. A square coil of 100 turns is 10 cm on each side. It is driven at a constant speed of 300 rpm.

(a) The coil is placed so that its axis of revolution is perpendicular to a *uniform* field of 0.1 Wb/m². At time $t = 0$, the coil is in a position of maximum flux linkages. Derive an expression for the instantaneous voltage generated. Sketch the voltage waveshape for one cycle. [See Fig. P17-3(b)].
(b) The coil is now placed in a *radial* field. All other data remain the same as for part (a). Sketch the voltage waveshape for one cycle showing the numerical value of the maximum voltage. [See Fig. P17-3(a)].

17-3. In Fig. P17-3 the flux per pole is 0.02 Wb. The coil is revolving at 1800 rpm and consists of 2 turns.

(a) For the configuration of Fig. P17-3(a) derive an expression for the flux linking the coil in terms of the flux per pole and θ_0.
(b) What is the instantaneous value of the coil voltage in Fig. P17-3(a).
(c) Compute the maximum value of the voltage induced in the arrangement of Fig. P17-3(b).

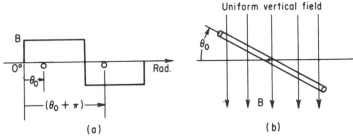

Fig. P17-3

(d) Assuming the coil of Fig. P17-3(b) is connected to a pair of commutator segments, find the d-c value of this voltage.

17-4. A 4-pole a-c induction motor is characterized by a sinusoidal rotor mmf as well as a sinusoidal flux density. Moreover, the flux per pole is 0.02 Wb. When operating at a specified load condition, the space displacement angle is found to be 53.3°. What is the amplitude of the rotor mmf wave required to produce a torque of 40 N-m?

17-5. A 60-Hz eight-pole a-c motor has an emf of 440 V induced in its armature winding, which has 180 effective turns. When this motor develops 10 kW of power, the amplitude of the sinusoidal field mmf is equal to 800 At.
(a) Find the value of the resultant air-gap flux per pole.
(b) Find the angle between the flux wave and the mmf wave.

17-6. The induced emf per phase of a 60-Hz four-pole induction motor is 120 V. The number of effective turns per phase is 100. When this motor develops a torque of 60 N-m the space displacement angle is 30°. Find the value of the total armature ampere-turns.

17-7. The three-phase armature winding of a four-pole 60-Hz a-c induction motor is equipped with 320 effective turns per phase. When the motor develops 10 kW, it has an armature current of 40 A, a space displacement angle of 37°, and a speed of rotation of 1700 rpm.
(a) Find the amplitude of the armature mmf per pole.
(b) What is the value of the developed torque?
(c) Compute the induced emf in the armature winding per phase.
(d) Find the flux per pole.

GROUP II

17-8. An electric machine has a field distribution as shown in Fig. P17-8. The coil spans a full 180° and has N turns. The machine has two poles and an axial length l and rotor radius r.

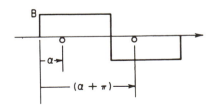

Fig. P17-8

Electromechanical Energy Conversion   Chap. 17

(a) Obtain the expression for the flux linkage per pole in terms of α expressed in electrical degrees.

(b) By using Faraday's law find the expression for the induced emf in terms of the flux per pole.

(c) Use the Blv form for the induced emf and verify the result of part (b).

17-9. A p-pole machine (Fig. P17-9) has the sinusoidal field distribution $B_p \sin \theta$, where θ is in electrical degrees. The armature consists of a uniform current sheet of value J, which has associated with it a triangular mmf having an amplitude per pole of F_A. The machine has an axial length l and a radius r.

(a) Determine the expression for the maximum flux per pole.

(b) Obtain the expression for the developed electromagnetic torque in terms of B_p, J, and δ.

(c) Express the torque found in part (b) in terms of F_A and the flux per pole Φ. Assume $\delta = \pi/2$.

(d) Is the torque found in part (b) constant at every point in the air gap? Explain.

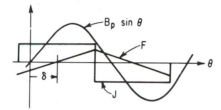

Fig. P17-9

chapter eighteen

The Three-Phase Induction Motor

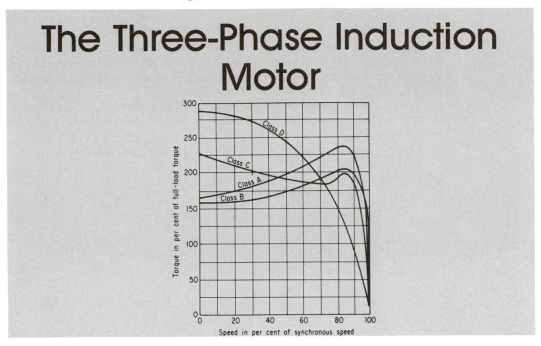

One distinguishing feature of the induction motor is that it is a *singly excited* machine. Although such machines are equipped with both a field winding and an armature winding, in normal use an energy source is connected to one winding alone, the field winding. Currents are made to flow in the armature winding by induction, which creates an ampere-conductor distribution that interacts with the field distribution to produce a net unidirectional torque. The frequency of the induced current in the conductor is affected by the speed of the rotor on which it is located; however, the relationship between the rotor speed and the frequency of the armature current is such as to yield a resulting ampere-conductor distribution that is stationary relative to the field distribution of the stator. As a result, the singly excited induction machine is capable of producing torque *at any speed below synchronous speed*.† For this reason the induction machine is placed in the class of *asynchronous machines*. In contrast, *synchronous machines* are electromechanical energy-conversion devices in which a net torque can be produced at only one‡ speed of the rotor. The distinguishing characteristic of the synchronous

---

† Synchronous speed is determined by the frequency of the source applied to the field winding and the number of poles for which the machine is designed. These quantities are related by Eq. (17-45). Thus

$$\text{synchronous speed} = \frac{120f}{p} = n_s$$

‡ Theoretically there are two rotor speeds at which a net torque different from zero can exist, but at the second speed enormous currents flow, which makes operation impractical.

machine is that it is a *doubly excited* device except when it is being used as a reluctance motor.

The salient construction features of the three-phase induction motor are described in Sec. 17-3. Because the induction machine is singly excited, it is necessary that both the magnetizing current and the power component of the current flow in the same lines. Moreover, because of the presence of an air gap in the magnetic circuit of the induction machine, an appreciable amount of magnetizing current is needed to establish the flux per pole demanded by the applied voltage. Usually, the value of the magnetizing current for three-phase induction motors lies between 25 and 40% of the rated current. Consequently, the induction motor is found to operate at a low power factor at light loads and at less than unity power factor in the vicinity of rated output.

We are concerned in this chapter with a description of the theory of operation and of the performance characteristics of the three-phase induction motor. Our discussion begins with an explanation of how a revolving magnetic field is obtained with a three-phase winding. After all it is this field that is the driving force behind induction motors.

18-1 THE REVOLVING MAGNETIC FIELD

The application of a three-phase voltage to the three-phase stator winding of the induction motor creates a rotating magnetic field, which by transformer action induces a "working" emf in the rotor winding. The rotor induced emf is called a working emf because it causes a current to flow through the armature winding conductors. This combines with the revolving flux-density wave to produce torque in accordance with Eq. (17-39). Consequently, we can view the revolving field as the key to the operation of the induction motor.

The rotating magnetic field is produced by the contributions of space-displaced phase windings carrying appropriate time-displaced currents. To understand this statement let us turn attention to Figs. 18-1 and 18-2. Appearing in Fig. 18-1 are the three-phase currents which are assumed to be flowing in phases *a*, *b*, and *c*, respectively. Note that these currents are time-displaced by 120 electrical degrees. Depicted in Fig. 18-2 is the stator structure and the three-phase winding. Note that each phase (normally distributed over 60 electrical degrees) for convenience is represented by a single coil. Thus coil *a–a'* represents

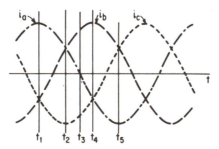

Fig. 18-1 Balanced three-phase alternating currents.

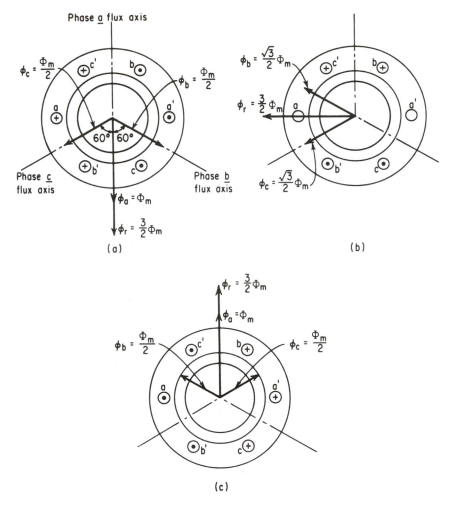

Fig. 18-2 Representing the rotating magnetic field at three different instants of time: (a) time t_1 in Fig. 18-1; (b) time t_3; (c) time t_5.

the entire phase a winding having its flux axis directed along the vertical. This means that whenever phase a carries current it produces a flux field directed along the vertical—up or down. The right-hand flux rule readily verifies this statement. Similarly, the flux axis of phase b is 120 electrical degrees displaced from phase a, and that of phase c is 120 electrical degrees displaced from phase b. The unprimed letters refer to the beginning terminal of each phase.

Let us consider the determination of the magnitude and direction of the resultant flux field corresponding to time instant t_1 in Fig. 18-1. At this instant the current in phase a is at its positive maximum value while the currents in phases b and c are at one-half their maximum negative values. In Fig. 18-2 it is arbitrarily assumed that when current in a given phase is positive, it flows into the paper with respect to the unprimed conductors. Thus since at time t_1 i_a is positive a cross is used for conductor a. See Fig. 18-2(a). Of course a dot is used for a' because it refers to the return connection. Then by the right-hand rule it follows that phase a produces a flux contribution directed downward along the vertical. Moreover, the magnitude of this contribution is the maximum value because the current is a maximum. Hence $\phi_a = \Phi_m$ where Φ_m is the maximum flux per pole of phase a. It is important to understand that phase a really produces a sinusoidal flux field with the amplitude located along the axis of phase a as depicted in Fig. 18-3. However, in Fig. 18-2(a) this sinusoidal distribution is conveniently represented by the vector ϕ_a.

In order to determine the direction and magnitude of the field contribution of phase b at time t_1, we note first that the current in phase b is negative with respect to that in phase a. Hence the conductor that stands for the beginning of phase b must be assigned a dot while b' is assigned a cross. Hence the instantaneous flux contribution of phase b is directed downward along its flux axis and the magnitude of phase b flux is one-half the maximum because the current is at one-half its maximum value. Similar reasoning leads to the result shown in Fig. 18-2(a) for phase c. A glance at the space picture corresponding to time t_1, as illustrated in Fig. 18-2(a), should make it apparent that the resultant flux per pole is directed downward and has a magnitude 3/2 times the maximum flux per pole of any one phase. Figure 18-3 depicts the same results as Fig. 18-2(a) but does so in terms of sinusoidal flux waves rather than flux vectors. Keep in mind that the resultant flux vector in Fig. 18-2 shows the direction in which flux crosses the air gap. Once across the air gap, the flux is confined to the iron in the usual fashion.

Next let us investigate how the situation of Fig. 18-2(a) changes as time passes through 90 electrical degrees from t_1 to t_3 in Fig. 18-1. Here phase a current is zero, yielding no flux contribution. The current in phase b is positive and equal to $\sqrt{3}/2$ its maximum value. Phase c has the same current magnitude

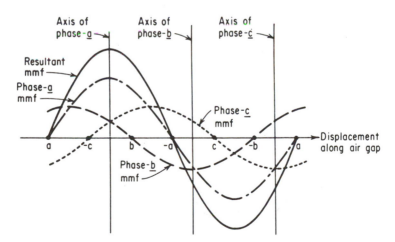

Fig. 18-3 Component and resultant field distributions corresponding to t_1 in Fig. 18-1.

but is negative. Together phases b and c combine to produce a resultant flux having the same magnitude as at time t_1. See Fig. 18-2(b). It is important to note, too, that an elapse of 90 electrical degrees in time results in a rotation of the magnetic flux field of 90 electrical degrees.

A further elapse of time equivalent to an additional 90 electrical degrees leads to the situation depicted in Fig. 18-2(c). Note that again the axis of the flux field is revolved by an additional 90 electrical degrees.

On the basis of the foregoing discussion it should be apparent that the application of three-phase currents through a balanced three-phase winding gives rise to a rotating magnetic field that exhibits two characteristics: (1) it is of constant amplitude and (2) it is of constant speed. The first characteristic has already been demonstrated. The second follows from the fact that the resultant flux traverses through 2π electrical radians in space for every 2π electrical radians of variation in time for the phase currents. For a two-pole machine, where electrical and mechanical degrees are identical, each cycle of variation of current produces one complete revolution of the flux field. Therefore, this is a fixed relationship which is dependent upon the frequency of the currents and the number of poles for which the three-phase winding is designed. In the case where the winding is designed for four poles it requires two cycles of variation of the current to produce one revolution of the flux field. It follows that for a p-pole machine the relationship is

$$f = \frac{p}{2}\,(\text{rps}) = \frac{p}{2}\left(\frac{n}{60}\right) \tag{18-1}$$

where f is in hertz and rps denotes revolutions per second. Note that Eq. (18-1) is identical to Eq. (17-45).

An inspection of the ampere-conductor distribution of the stator winding at the various time instants reveals that the individual phases cooperate in such a fashion as to produce a solenoidal effect in the stator. Thus in Fig. 18-2(a) the directions of the currents are such that they all enter on the left side and leave on the right side. The right-hand rule indicates that the flux field is then directed downward along the vertical. In Fig. 18-2(b) the situation is similar except that now the cross and dot distribution is such that the resultant flux field is oriented horizontally towards the left. Therefore, it can be concluded that the rotating magnetic field is a consequence of the revolving mmf associated with the stator winding.

In the foregoing, it is pointed out that the flow of balanced three-phase currents through a balanced three-phase winding yields a rotating field of constant amplitude and speed. If neither of these conditions is exactly satisfied, it is still possible to obtain a revolving magnetic field but it will not be of constant amplitude nor of constant linear speed. In general for a q-phase machine a rotating field of constant amplitude and constant speed results when the following two conditions are satisfied: (1) there is a *space* displacement between balanced phase windings of $2\pi/q$ electrical radians, and (2) the currents flowing through the phase windings are balanced and *time*-displaced by $2\pi/q$ electrical radians. For the three-phase machine $q = 3$ and so the now familiar 120° figure is obtained. The only exception to the rule is the two-phase machine. Because the two-phase situation is a special case of the four-phase system, a value of q equal to 4 must be used.

One final point is now in order. The speed of rotation of the field as described by Eq. (18-1) is always given relative to the phase windings carrying the time-varying currents. Accordingly, if a situation arises where the winding is itself revolving, then the speed of rotation of the field relative to inertial space is different than it is with respect to the winding.

18-2 THE INDUCTION MOTOR AS A TRANSFORMER

The three-phase induction motor may be compared with the transformer because it is a singly energized device which involves changing flux linkages with respect to the stator and rotor windings. In this connection assume that the rotor is of the wound type and Y-connected as illustrated in Fig. 18-4. With the rotor winding open-circuited no torque can be developed. Hence the application of a three-phase voltage to the three-phase stator winding gives rise to a rotating magnetic field which cuts both the stator and rotor windings at the line frequency f_1. The rms value of the induced emf per phase of the rotor winding is given by Eq. (17-34) as

$$E_2 = 4.44 f_1 N_2 K_{w2} \Phi \tag{18-2}$$

where the subscript 2 denotes rotor winding quantities. Note that the stator frequency f_1 is used here because the rotor is at standstill. Hence E_2 is a *line-frequency* emf. Of course the flux Φ is the flux per pole, which is mutual to the stator and rotor windings.

A similar expression describes the rms value of the induced emf per phase occurring in the stator winding. Thus

$$E_1 = 4.44 f_1 N_1 K_{w1} \Phi \tag{18-3}$$

From Eqs. (18-2) and (18-3) we can formulate the ratio

$$\frac{E_1}{E_2} = \frac{N_1 K_{w1}}{N_2 K_{w2}} \tag{18-4}$$

Note the similarity of this expression to the voltage transformation ratio of the transformer. The difference lies in the inclusion of the winding factors of the motor, necessitated by the use of a distributed winding for the motor as contrasted

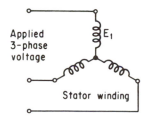

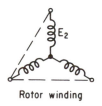

Fig. 18-4 Schematic representation of three-phase wound rotor induction motor. Dashed line indicates short-circuit links for normal operation.

to the concentrated coils used in the transformer. In essence, then, the induction motor at standstill exhibits the characteristics of a transformer wherein the stator winding is the primary and rotor winding is the secondary.

Next let us consider the behavior of the induction motor under running conditions—again with the intention of pointing out similarities to the transformer. To produce a starting torque (and subsequently a running torque) it is necessary to have a current flowing through the rotor winding. This is readily accomplished by short-circuiting the winding in the manner indicated by the dashed line in Fig. 18-4. Initially the induced emf E_2 causes a rotor current per phase I_2 to flow through the short circuit, producing an ampere-conductor distribution which acts with the flux field to produce the starting torque. The sense of this torque is always such as to cause the rotor to travel in the same direction as the rotating field. An examination of Fig. 18-5 makes this apparent. Assume that the flux field is revolving clockwise at a speed corresponding to the applied stator frequency and the poles of the stator winding. This speed is called the *synchronous speed* and is described by Eq. (18-1). Thus

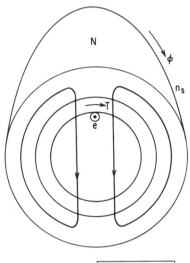

Fig. 18-5 Showing the direction of induced voltage and developed torque on a typical rotor conductor.

$$n_s = \frac{120 f_1}{p} \quad \text{rpm} \tag{18-5}$$

By the $\bar{v} \times \bar{B}$ rule the induced emf in a typical conductor lying beneath a north-pole flux is directed out of the paper as indicated in Fig. 18-5. Then for this direction of current the $\bar{I}_2 \times \bar{B}$ rule reveals the torque to be directed clockwise. Therefore the rotor moves in a direction in which it tries to catch up with the stator field.

As the rotor increases its speed, the rate at which the stator field cuts the rotor coils decreases. This reduces the resultant induced emf per phase, in turn diminishing the magnitude of the ampere-conductor distribution and yielding less torque. In fact this process continues until that rotor speed is reached which yields enough emf to produce just the current needed to develop a torque equal to the opposing torques. If there is no shaft load, the opposing torque consists chiefly of frictional losses. It is important to understand that as long as there is an opposing torque to overcome—however small or whatever its origin—*the rotor speed can never be equal to the synchronous speed*. This is characteristic of singly excited electromechanical energy-conversion devices. Since the rotor (or secondary) winding current is produced by induction, there must always be a difference in speed between the stator field and the rotor. In other words, transformer action must always be allowed to take place between the stator (or primary) winding and the rotor (or secondary) winding.

This speed difference, or *slip*, is a very important variable for the induction motor. In terms of an equation we may write

$$\text{slip} \equiv n_s - n \qquad \text{rpm} \tag{18-6}$$

when n denotes the actual rotor speed in *rpm*. The term slip is used because it describes what an observer riding with the stator field sees looking at the rotor—the rotor appears to be slipping backward. A more useful form of the slip quantity results when it is expressed on a per unit basis using synchronous speed as the reference. Thus the slip in per unit is

$$s = \frac{n_s - n}{n_s} \tag{18-7}$$

For the conventional induction motor the values of s lie between zero and unity.

It is customary in induction-motor analysis to express rotor quantities (such as induced voltage, current, and impedance) in terms of line-frequency quantities and the slip as expressed by Eq. (18-7). For example, if the rotor is assumed to be operating at some speed $n < n_s$, then the actual emf induced in the rotor winding per phase may be expressed in terms of the line-frequency quantity E_2 as sE_2. This formulation has definite advantages, as described in the next section. In a similar fashion it is possible to express the rotor-winding impedance per phase as

$$z_2 = r_2 + jsx_2 \tag{18-8}$$

where z_2 denotes the rotor phase impedance, r_2 is the rotor resistance per phase, and x_2 is the line-frequency leakage reactance per phase of the rotor winding. Of course the effective value of this reactance when the rotor operates at a speed n (or slip s) is only s times as large. Keep in mind that the frequency of the currents in the rotor is directly related to the relative speed of the stator field to the rotor winding. Accordingly we may write

$$f_2 = \frac{p(\text{slip rpm})}{120} = \frac{p(n_s - n)}{120} \tag{18-9}$$

where f_2 is the frequency of the rotor emf and current. By means of Eq. (18-7) it is possible to rewrite Eq. (18-9) as

$$f_2 = \frac{psn_s}{120} = s\frac{pn_s}{120} = sf_1 \tag{18-10}$$

which indicates that the rotor frequency f_2 is obtained by merely multiplying the stator line frequency by the appropriate per unit value of the slip. For this reason f_2 is often called the *slip frequency*.

18-3 THE EQUIVALENT CIRCUIT

It is desirable to have an equivalent circuit of the three-phase induction motor in order to direct the analysis of operation and to facilitate the computation of performance. From the remarks made in the preceding section it should not come as a surprise that the equivalent circuit assumes a form identical to that of the exact equivalent circuit of the transformer. The derivation proceeds in a similar fashion with necessary modifications introduced to account for the fact that the secondary winding (the rotor) in this instance revolves and thereby develops mechanical power.

The Magnetizing Branch of the Equivalent Circuit. All the parameters of the equivalent circuit are expressed on a per phase basis. This applies whether the stator winding is Y- or Δ-connected. In the latter case the values refer to the equivalent Y connection. Appearing in Fig. 18-6(a) is the portion of the equivalent circuit that has reference to the stator (or primary) winding. Note that it consists of a stator phase winding resistance r_1, a stator phase winding leakage reactance x_1, and a magnetizing impedance made up of the core-loss resistor r_c and the magnetizing reactance x_ϕ. There is no difference in form between this circuit and that of the transformer. The difference lies only in the magnitude of the parameters. Thus the total magnetizing current $\bar{I}_m$ is considerably larger in the case of the induction motor because the magnetic circuit necessarily includes an air gap. Whereas in the transformer this current is about 2 to 5% of the rated current, here it is approximately 25 to 40% of the rated current depending upon the size of the motor. Moreover, the primary leakage reactance for the induction motor also is larger because of the air gap as well as because the stator and rotor windings are distributed along the periphery of the air gap rather than concentrated on a core as in the transformer.

The effects of the actions that take place in the rotor (or secondary) winding must reflect themselves at the proper equivalent-voltage level at terminals *a-b* in Fig. 18-6(a). We next investigate the manner in which this comes about.

The Actual Rotor Circuit per Phase. For any specified load condition that calls for a particular value of slip s, the rotor current per phase may be expressed as

$$\bar{I}_2 = \frac{s\bar{E}_2}{r_2 + jsx_2} \qquad (18\text{-}11)$$

where $\bar{E}_2$ and x_2 are the standstill (or line-frequency) values. The circuit interpretation of Eq. (18-11) is depicted in Fig. 18-6(b). It illustrates that $\bar{I}_2$ is a slip-frequency

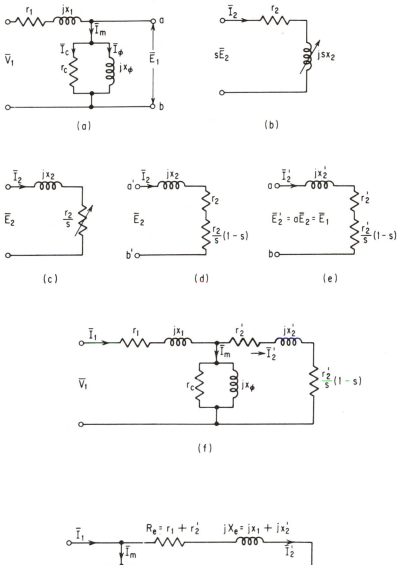

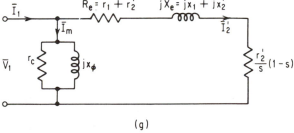

Fig. 18-6 Derivation of the equivalent circuit: (a) stator winding section; (b) actual rotor circuit; (c) equivalent rotor circuit; (d) modified equivalent rotor circuit; (e) stator-referred equivalent rotor circuit; (f) exact equivalent circuit; (g) approximate equivalent circuit.

current produced by the slip-frequency induced emf $s\overline{E}_2$ acting in a rotor circuit having an impedance per phase of $r_2 + jsx_2$. In other words, this is the current that would be "seen" by an observer riding with the rotor winding. Furthermore, the amount of real power involved in this rotor circuit is the current squared times the real part of the rotor impedance. In fact, this power represents the rotor copper loss per phase. Hence the total rotor copper loss may be expressed as

$$P_{Cu2} = qI_2^2 r_2 \qquad (18\text{-}12)$$

where q denotes the number of rotor phases.

The Equivalent Rotor Circuit.   By dividing both the numerator and denominator of Eq. (18-11) by the slip s we get

$$\overline{I}_2 = \frac{\overline{E}_2}{\dfrac{r_2}{s} + jx_2} \qquad (18\text{-}13)$$

The corresponding circuit interpretation of this expression appears in Fig. 18-6(c). Note that the magnitude and phase angle of $\overline{I}_2$ remain unaltered by this operation. However, there is a significant difference between Eqs. (18-11) and (18-13). In the latter case $\overline{I}_2$ is considered to be produced by a line-frequency voltage $\overline{E}_2$ acting in a rotor circuit having an impedance per phase of $r_2/s + jx_2$. Hence the $\overline{I}_2$ of Eq. (18-13) is a *line-frequency* current, whereas the $\overline{I}_2$ of Eq. (18-11) is a *slip-frequency* current. It is important that this distinction be understood.

Division of the numerator and denominator of Eq. (18-11) by s has enabled us to go from an actual rotor circuit characterized by constant resistance and variable leakage reactance [see Fig. 18-6(b)] to one characterized by variable resistance and constant leakage reactance [see Fig. 18-6(c)]. Moreover, the real power associated with the equivalent rotor circuit of Fig. 18-6(c) is clearly

$$P = I_2^2 \frac{r_2}{s} \qquad (18\text{-}14)$$

Hence the total power for q phases is

$$P_g = qI_2^2 \frac{r_2}{s} \qquad (18\text{-}15)$$

A comparison of this expression with Eq. (18-12) indicates that the power associated with the equivalent circuit of Fig. 18-6(c) is considerably greater. For example, in a large machine a typical value of s is 0.02. Hence P_g is greater than the actual rotor copper loss by a factor of fifty.

What is the meaning of this power discrepancy? The answer lies in the fact that by Eq. (18-13) $\overline{I}_2$ is a *line-frequency* current. This means that the point of reference has changed from the rotor (where slip-frequency quantities exist) to the stator (where line-frequency quantities exist). In accordance with the circuit representation of Fig. 18-6(c) the observer changes his point of reference from the rotor to the stator. This shift is significant because now, upon looking into the rotor, the observer "sees" not only the rotor copper loss *but the mechanical power developed as well.* The latter quantity is included because with respect to a stator-based observer the rotor speed is no longer zero as it is relative to a rotor-based observer. As a matter of fact, Eq. (18-15) gives the total power input to the rotor. It is the power transferred across the air gap from the stator to the rotor. We can rewrite Eq. (18-15) in a manner that stresses this fact:

$$P_g = qI_2^2\frac{r_2}{s} = qI_2^2\left[r_2 + \frac{r_2}{s}(1 - s)\right] \qquad (18\text{-}16)$$

In other words, the variable resistance of Fig. 18-6(c) may be replaced by the actual rotor winding resistance r_2 and a variable resistance R_m, which represents the mechanical shaft load. That is,

$$R_m \equiv \frac{r_2}{s}(1 - s) \qquad (18\text{-}17)$$

This expression is useful in analysis because it allows any mechanical load to be represented in the equivalent circuit by a resistor. Figure 18-6(d) depicts the modified version of the rotor equivalent circuit. Finally, it should be apparent, on the basis of the foregoing remarks, that the rotor equivalent circuit is equivalent only insofar as the magnitude and phase angle of the rotor current per phase are concerned.

The Stator-Referred Rotor Equivalent Circuit. The voltage appearing across terminals *a-b* in Fig. 18-6(a) is a line-frequency quantity having N_1K_{w1} effective turns. The voltage appearing across terminals *a'-b'* in Fig. 18-6(d) is also a line-frequency quantity but has N_2K_{w2} effective turns. In general $\overline{E}_1 \neq \overline{E}_2$, so that *a'-b'* in Fig. 18-6(d) cannot be joined to *a-b* in Fig. 18-6(a) to yield a single-line equivalent circuit. To accomplish this it is necessary to replace the actual rotor winding with an equivalent winding having N_1K_{w1} effective turns as was done with the transformer. In other words, all the rotor quantities must be referred to the stator in the manner depicted in Fig. 18-6(e). The prime notation is used to denote stator-referred quantities.

The Complete Equivalent Circuit. The voltage appearing across terminals *a-b* in Fig. 18-6(e) is the same as that appearing across terminals *a-b* in Fig. 18-6(a). Hence these terminals may be joined to yield the complete equivalent circuit as it appears in Fig. 18-6(f). Note that the form is identical to that of the two-winding transformer.

The Approximate Equivalent Circuit. Considerable simplification of computation with little loss of accuracy can be achieved by moving the magnetizing branch to the machine terminals as illustrated in Fig. 18-6(g). This modification is essentially based on the assumption that $V_1 \approx E_1 = E_2'$. All performance calculations will be carried out using the approximate equivalent circuit.

18-4 COMPUTATION OF PERFORMANCE

When the three-phase induction motor is running at no-load, the slip has a value very close to zero. Hence the mechanical load resistor R_m has a very large value, which in turn causes a small rotor current to flow. The corresponding electromagnetic torque, as described by Eq. (17-39), merely assumes that value which is needed to overcome the *rotational losses* consisting of friction and windage. If a mechanical load is next applied to the motor shaft, the initial reaction is for the shaft load to drop the motor speed slightly, thereby increasing the slip. The increased slip subsequently causes I_2 to increase to that value which, when inserted into Eq. (17-39), yields sufficient torque to provide a balance of power to the load. Thus equilibrium is established and operation proceeds at a particular value of s. In fact for each value of load horsepower requirement there is a unique value of slip. This can be inferred from the equivalent circuit, which shows that once s is specified then the power input, the rotor current, the developed torque, the power output, and the efficiency are all determined.

The use of a power-flow diagram in conjunction with the approximate equivalent circuit makes the computation of the performance of a three-phase induction motor a straightforward matter. Depicted in Fig. 18-7(a) is the power flow in statement form. Note that the loss quantities are placed on the left side of a flow point. Appearing in Fig. 18-7(b) is the same power-flow diagram but now expressed in terms of all the appropriate relationships needed to compute the performance. It should be clear that to calculate performance one must first compute the currents I_1 and I_2 from the equivalent circuit and then make use of the pertinent relationships depicted in Fig. 18-7(b).

EXAMPLE 18-1   A three-phase four-pole 50-hp 440-V 60-Hz Y-connected induction motor has the following parameters per phase:

$$r_1 = 0.10 \ \Omega \qquad x_1 = 0.35 \ \Omega$$
$$r_2' = 0.12 \ \Omega \qquad x_2' = 0.40 \ \Omega$$

It is known that the stator core losses amount to 1200 W and the rotational losses equal 950 W. Moreover, at no-load the motor draws a line current of 18 A at a power factor of 0.089 lagging.

When the motor operates at a slip of 2.5% find:

(a) The input line current and power factor.
(b) The developed electromagnetic torque in newton-meters.
(c) The horse-power output.
(d) The efficiency.

Solution: (a) The computations are carried out on a per phase basis. Hence the phase voltage is $440/\sqrt{3}$ or 254 V, and the equivalent circuit is depicted in Fig. 18-8. The stator-referred rotor current then follows from

$$\overline{I}_2' = \frac{\overline{V}_1}{r_1 + \dfrac{r_2'}{s} + j(x_1 + x_2')} = \frac{254}{4.9 \times j0.75}$$

$$= 51.3 \underline{/-8.7^\circ} = 50.9 - j7.8$$

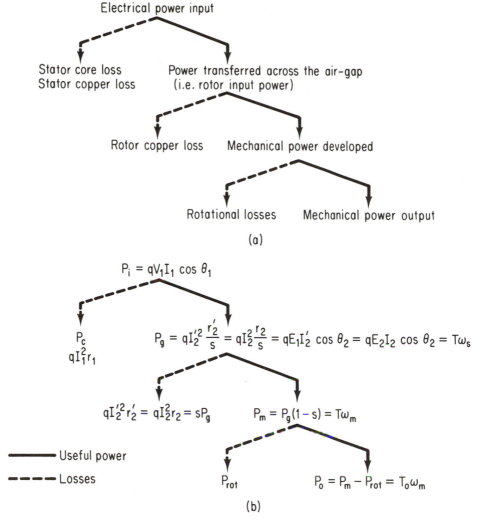

Electrical power input

Stator core loss   Power transferred across the air-gap
Stator copper loss   (i.e. rotor input power)

        Rotor copper loss   Mechanical power developed

                Rotational losses   Mechanical power output

(a)

$P_i = qV_1I_1 \cos \theta_1$

P_c
$qI_1^2 r_1$

$P_g = qI_2'^2 \dfrac{r_2'}{s} = qI_2^2 \dfrac{r_2}{s} = qE_1I_2' \cos \theta_2 = qE_2I_2 \cos \theta_2 = T\omega_s$

$qI_2'^2 r_2' = qI_2^2 r_2 = sP_g$ $P_m = P_g(1 - s) = T\omega_m$

──── Useful power

───── Losses

P_{rot} $P_o = P_m - P_{rot} = T_o\omega_m$

(b)

Fig. 18-7 Power-flow diagram: (a) statement form; (b) equation form.

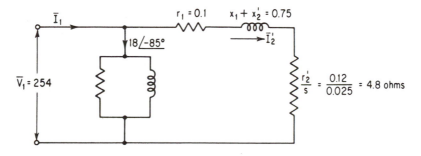

Fig. 18-8 Equivalent circuit for Example 18-1.

For all practical purposes the magnetizing current may be taken equal to the no-load current because the corresponding rotor current is negligibly small. Thus

$$\bar{I}_m = 18 \angle -85° = 1.6 - j17.95$$

Hence the input line current is

$$\bar{I}_1 = \bar{I}_m + \bar{I}_2' = (50.9 + 1.6) - j(17.95 + 7.8) \tag{18-18}$$

$$\bar{I}_1 = 52.5 - j25.75 = 58.2 \angle -26.2°$$

and

$$pf = \cos \theta_1 = \cos 26.2° = 0.895 \text{ lagging}$$

(b) The developed torque is found from

$$T = \frac{P_g}{\omega_s} \tag{18-19}$$

Also,

$$\omega_s = \frac{2\pi n_s}{60} = \frac{2\pi(1800)}{60} = 60\pi \text{ rad/s}$$

and

$$P_g = qI_2'^2 \frac{r_2'}{s} = 3(51.3)^2 4.8 = 37,896 \text{ W}$$

Therefore,

$$T = \frac{37,896}{60\pi} = 201 \text{ N-m}$$

(c) From the power-flow diagram the power output is

$$P_o = P_m - P_{rot} = P_g(1 - s) - P_{rot} = 36,948 - 950 = 35,999 \text{ W}$$

The horsepower output is therefore

$$HP_0 = \frac{35.999}{0.746} = 48.25$$

Note that this is slightly less than the rated horsepower of 50. Rated hp occurs at a slip somewhat larger than 2.5%.

(d) It is more accurate to find the efficiency from the relationship

$$\eta = 1 - \frac{\Sigma \text{ losses}}{P_i} \tag{18-20}$$

rather than from the output-to-input ratio. The tabulation of the losses is as follows:

$$P_c = \text{core loss} = 1200 \text{ W}$$

$$\text{stator copper loss} = qI_1^2 r_1 = 3(58.2)^2(0.1) = 1016 \text{ W}$$

$$\text{rotor copper loss} = qI_2'^2 r_2' = sP_g = 0.025(37,896) = 947 \text{ W}$$

$$P_{rot} = \text{rotational loss} = 950 \text{ W}$$

$$\Sigma \text{ losses} = 4113 \text{ W}$$

Also, the input power is

$$P_i = \sqrt{3}(440)(58.2)(0.895) = 39,697 \text{ W}$$

Hence

$$\eta = 1 - \frac{4113}{39,697} = 1 - 0.104 = 0.896$$

The efficiency is 89.6%.

18-5 TORQUE-SPEED CHARACTERISTIC: STARTING TORQUE AND MAXIMUM DEVELOPED TORQUE

The variation of torque with speed (or slip) is an important characteristic of the three-phase induction motor. The general shape of the curve can be identified in terms of the basic torque equation [Eq. (17-39)] and knowledge of the performance computation procedure. When the motor operates at a very small slip, as at no-load, the rotor current is practically zero so that only that amount of torque is developed which is needed to supply the rotational losses. As the slip is allowed to increase from nearly zero to about 10%, Eq. (18-11) shows that the rotor current increases almost linearly. This is because the j part of the impedance, sx_2, is small compared to r_2. Furthermore, it can be shown that the space-displacement angle ψ in Eq. (17-39) is identical to the rotor power-factor angle θ_2. That is,

$$\psi = \theta_2 = \tan^{-1}\frac{sx_2}{r_2} \qquad (18\text{-}21)$$

Therefore, for values of s from zero to 10 per cent, ψ varies over a range of about 0 to 15°. This means that the $\cos \psi$ remains practically invariant over the specified slip range, so the torque increases almost linearly in this region. Of course the quantity Φ in Eq. (17-39) is essentially fixed since the applied voltage is constant.

As the slip is allowed to increase still further, the current continues to increase but much less rapidly than at first. The reason lies in the increasing importance of the sx_2 term of the rotor impedance. In addition, the space angle ψ now begins to increase at a rapid rate which makes the $\cos \psi$ diminish more rapidly than the current increases. Since the torque equation now involves two opposing factors, it is entirely reasonable to expect that a point is reached beyond which further increases in slip culminate in decreased developed torque. In other words, the rapidly decreasing $\cos \psi$ factor predominates over the slightly increasing I_2 factor in Eq. (17-39). As ψ increases, the field pattern for producing torque becomes less and less favorable because more and more conductors that produce negative torque are included beneath a given pole flux. A glance at Fig. 17-3 makes this statement self-evident. Accordingly, the composite torque-speed curve takes on a shape similar to that shown in Fig. 18-9.

The *starting torque* is the torque developed when s is unity, i.e., the speed n is zero. Figure 18-9 indicates that for the case illustrated the starting torque is somewhat in excess of rated torque, which is fairly typical of such machines.

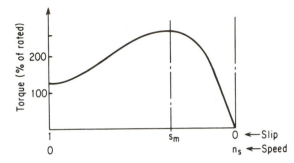

Fig. 18-9 Typical torque-speed curve of a three-phase induction motor.

The starting torque is computed in the same manner as torque is computed for any value of slip. Here it merely requires using $s = 1$. Thus the magnitude of the rotor current at standstill is

$$I_2' = \frac{V_1}{\sqrt{(r_1 + r_2')^2 + (x_1 + x_2')^2}} \tag{18-22}$$

The corresponding gap power is then

$$P_g = qI_2'^2 r_2 = \frac{qV_1^2 r_2}{(r_1 + r_2')^2 + (x_1 + x_2')^2} \tag{18-23}$$

It is interesting to note that higher starting torques result from increased rotor copper losses at standstill.

At unity slip the input impedance is very low so that large starting currents flow. Equation (18-22) makes this apparent. In the interest of limiting this excessive starting current, motors whose ratings exceed 3 hp are usually started at reduced voltage by means of line starters. This matter is discussed further in Sec. 18-8. Of course, starting with reduced voltage also means a reduction in the starting torque. In fact if 50% of the rated voltage is used upon starting, then clearly by Eq. (18-23) it follows that the starting torque is only one-quarter of its full-voltage value.

Another important torque quantity of the three-phase induction motor is the maximum developed torque. This quantity is so important that it is frequently the starting point in the design of the induction motor. The maximum (or breakdown) torque is a measure of the reserve capacity of the machine. It frequently has a value of 200 to 300% of rated torque. It permits the motor to operate through momentary peak loads. However, the maximum torque cannot be delivered continuously because the excessive currents that flow would destroy the insulation.

Since the developed torque is directly proportional to the gap power, it follows that the torque is a maximum when P_g is a maximum. Also, P_g is a maximum when there is a maximum transfer of power to the equivalent circuit resistor r_2'/s. Applying the maximum-power-transfer theorem to the appropriate equivalent circuit leads to the result that

$$\frac{r_2'}{s_m} = \sqrt{r_1^2 + (x_1 + x_2')^2} \tag{18-24}$$

That is, maximum power is transferred to the gap power resistor r_2'/s when this

The Three-Phase Induction Motor   Chap. 18

resistor is equal to the impedance looking back into the source. Accordingly, the slip s_m at which the maximum torque is developed is

$$s_m = \frac{r_2'}{\sqrt{r_1^2 + (x_1 + x_2')^2}}$$ (18-25)

Note that the slip at which the maximum torque occurs may be increased by using a larger rotor resistance. Some induction motors are in fact designed so that the maximum torque is available as a starting torque, i.e., $s_m = 1$.

With the slip s_m known, the corresponding rotor current can be found and then inserted into the torque equation to yield the final form for the breakdown torque. Thus

$$T_m = \frac{1}{\omega_s} q I_2'^2 \frac{r_2'}{s_m} = \frac{1}{\omega_s} \frac{q V_1^2}{2[r_1 + \sqrt{r_1^2 + (x_1 + x_2')^2}]}$$ (18-26)

An examination of Eq. (18-26) reveals the interesting information that the maximum torque is independent of the rotor winding resistance. Thus increasing the rotor winding resistance increases the slip at which the breakdown torque occurs but it leaves the magnitude of this torque unchanged. Figure 18-10 shows the effect of increasing the rotor resistance on a typical torque-speed curve.

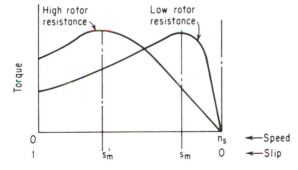

Fig. 18-10 Showing the effect of increased rotor resistance on torque-speed curve.

18-6 PARAMETERS FROM NO-LOAD TESTS

The performance computations of a three-phase induction motor presuppose knowledge of the parameters of the equivalent circuit. This information may be available either from design data or from appropriate tests. In the case where the design data are not known, a simple no-load test yields information about the magnetizing current as described in Example 18-1. It merely requires measuring the input line current and power. Note that specific knowledge about the core-loss resistor r_c and the magnetizing reactance x_ϕ is usually unnecessary. However, these parameters can readily be determined by proceeding in a fashion similar to that used in finding these same quantities for the transformer. A slight modification is necessary, though, because of the presence of the rotational losses.

To obtain information about the winding resistances and the leakage reactances a rotor-blocked test is necessary. This test is analogous to the short-circuit test of the transformer. It requires that the rotor be blocked to prevent rotation and that the rotor winding be short-circuited in the usual fashion. Furthermore, since the slip is unity, the mechanical load resistor R_m is zero and so the input impedance is quite low. Hence, in order to limit the rotor current in this test to reasonable values, a reduced voltage must be used—usually about 10 to 25% of the rated value. Also, operation at such a reduced voltage renders the core loss as well as the magnetizing current negligibly small. Accordingly, the equivalent circuit in this test takes on the configuration shown in Fig. 18-11.

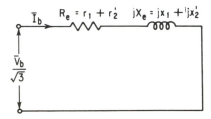

Fig. 18-11 Induction motor equivalent circuit for the rotor-blocked test.

Assume now that the following instrument readings are taken in performing a rotor-blocked test on a Y-connected three-phase induction motor:

P_b = total wattmeter reading in watts

I_b = line current of Y-connection

V_b = line voltage of Y-connection

From these measurements it follows that the equivalent winding resistance is given by

$$R_e \equiv r_1 + r_2' = \frac{P_b}{3I_b^2} \tag{18-27}$$

where R_e is the equivalent resistance per phase and the factor 3 accounts for the three phases. The equivalent phase impedance is obtained from

$$Z_e = \frac{V_b}{\sqrt{3}I_b} \tag{18-28}$$

Finally, the equivalent leakage reactance is determined from

$$X_e = \sqrt{Z_e^2 - R_e^2} = x_1 + x_2' \tag{18-29}$$

Note that as long as computations are carried on in connection with the approximate equivalent circuit, it is sufficient to deal directly with X_e without a further breakdown into x_1 and x_2'. In fact it is the chief purpose of the blocked-rotor test to make X_e available.

18-7 RATINGS AND APPLICATIONS OF THREE-PHASE INDUCTION MOTORS

Now that the theory of operation, the characteristics, and the performance of the three-phase induction motor are understood, we can study their standard ratings and typical applications. Of course, before it is possible to specify a particular motor for a given application, the characteristics of the load must be known. This includes such items as horsepower requirement, starting torque, acceleration capability, speed variation, duty cycle, and the environment in which the motor is to operate. Once this information is available it is often possible to select a general-purpose motor to do the job satisfactorily. Table 18-1 is a list of such motors which are readily available and standardized in accordance with generally accepted criteria established by the National Electrical Manufacturers Association (NEMA). The table is essentially self-explanatory. The primary distinguishing features of the various classes of squirrel-cage motors are the construction details of the rotor slots. For example, the Class A motor uses a low resistance squirrel cage with slots of medium depth. Classes B and C use much deeper bars in order to reduce the amount of full-voltage starting current and to obtain higher starting torque (for Class C). The Class D motor uses very high resistance bars for the squirrel cage. This yields very high starting torque but also poor operating efficiency. Figure 18-12 shows the differences in the torque-speed curves.

18-8 CONTROLLERS FOR THREE-PHASE INDUCTION MOTORS

After the right motor is selected for a given application, the next step is to select the appropriate controller for the motor. The primary functions of a controller are to furnish proper starting, stopping, and reversing without damage or inconvenience to the motor, other connected loads, or the power system. However, the controller fulfills other useful purposes as well, especially the following:

1. It limits the starting torque. Some connected shaft loads may be damaged if excessive torque is applied upon starting. For example, fan blades can be sheared off or gears having excessive backlash can be stripped. The controller supplies reduced voltage at the start and as the speed picks up the voltage is increased in steps to its full value.

2. It limits the starting current. Most motors above 3 hp cannot be started directly across the three-phase line because of the excessive starting current that flows. Recall that at unity slip the current is limited only by the leakage impedance, which is usually quite a small quantity, especially in the larger motor sizes. A large starting current can be annoying because it causes light to flicker and may even cause other connected motors to stall. Reduced-voltage starting readily eliminates these annoyances.

TABLE 18-1 CHARACTERISTICS AND APPLICATIONS

Type classification	HP range	Starting torque (%)	Maximum torque (%)	Starting current (%)
General-purpose, normal torque and starting current, NEMA Class A	0.5 to 200	Poles-Torque 2–150 4–150 6–135 8–125 10–120 12–115 14–110 16–105	Up to 250 but not less than 200	500–1000
General-purpose, normal torque, low starting current, NEMA Class B	0.5 to 200	Same as above or larger	About the same as Class A but may be less	About 500–550, less than average of Class A
High torque, low starting current, NEMA Class C	1 to 200	200 to 250	Usually a little less than Class A but not less than 200	About same as Class B
High torque, medium and high slip, NEMA Class D	0.5 to 150	Medium slip 350 High slip 250–315	Usually same as standstill torque	Medium slip 400–800, high slip 300–500
Low starting torque, either normal starting current, NEMA Class E, or low starting current, NEMA Class F	40 to 200	Low, not less than 50	Low, but not less than 150	Normal 500–1000, low 350–500
Wound-rotor	0.5 to 5000	Up to 300	200–250	Depends upon external rotor resistance but may be as low as 150

[a] By permission from M. Liwschitz-Garik and C. C. Whipple, *Electric Machinery,* Vol. II (Princeton, N.J.: D. Van Nostrand Co., Inc., 1946).

[b] Figures are given in percent of rated full-load values.

Slip (%)	Power factor (%)	Efficiency (%)	Typical applications
Low, 3–5	High, 87–89	High, 87–89	Constant-speed loads where excessive starting torque is not needed and where high starting current is tolerated. Fans, blowers, centrifugal pumps, most machinery tools, woodworking tools, line shafting. Lowest in cost. May require reduced voltage starter. Not to be subjected to sustained overloads, because of heating. Has high pull-out torque.
3–5	A little lower than Class A	87–89	Same as Class A—advantage over Class A is lower starting current, but power factor slightly less.
3–7	Less than Class A	82–84	Constant-speed loads requiring fairly high starting torque and lower starting current. Conveyors, compressors, crushers, agitators, reciprocating pumps. Maximum torque at standstill.
Medium 7–11, high, 12–16	Low	Low	Medium slip. Highest starting torque of all squirrel-cage motors. Used for high-inertia loads such as shears, punch presses, die stamping, bulldozers, boilers. Has very high average accelerating torque. High slip used for elevators, hoists, etc., on intermittent loads.
1 to 3½	About same as Class A or Class B	About same as Class A or Class B	Direct-connected loads of low inertia requiring low starting torque, such as fans and centrifugal pumps. Has high efficiency and low slip.
3–50	High, with rotor shorted same as Class A	High, with rotor shorted same as Class A, but low when used with rotor resistor for speed control	For high-starting-torque loads where very low starting current is required or where torque must be applied very gradually and where some speed control (50%) is needed. Fans, pumps, conveyors, hoists, cranes, compressors. Motor with speed control more expensive and may require more maintenance.

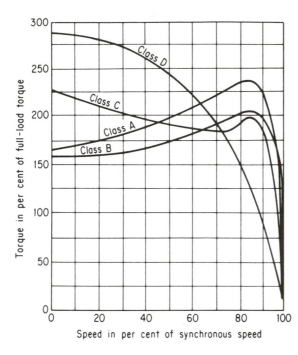

Fig. 18-12 Typical torque-slip curves for various classes of induction motors.

3. It provides overload protection. All general-purpose motors are designed to deliver full-load power continuously without overheating. However, if for some reason the motor is made to deliver, say, 150% of its rated output continuously, it will proceed to accommodate the demand and burn itself up in the process. The horsepower rating of the motor is based on the allowable temperature rise that can be tolerated by the insulation used for the field and armature windings. The losses produce the heat that raises the temperature. As long as these losses do not exceed the rated values there is no danger to the motor, but if they are allowed to become excessive, damage will result. There is nothing inherent in the motor that will keep the temperature rise within safe limits. Accordingly, it is also the function of the controller to provide this protection. Overload protection is achieved by the use of an appropriate time-delay relay which is sensitive to the heat produced by the motor line currents.

4. It furnishes undervoltage protection. Operation at reduced voltage can be harmful to the motor, especially when the load demands rated power. If the line voltage falls below some preset limit, the motor is automatically disconnected from the three-phase line source by the controller.

Controllers for electric motors are of two types—manual and magnetic. We shall consider only the magnetic type, which has many advantages over the manual type. It is easier to operate. It provides undervoltage protection. It can be remotely operated from one or several different places. Moreover, the magnetic controller is automatic and reliable whereas the manual controller requires a trained operator, especially where a sequence of operations is called for in a given application. The one disadvantage is the greater initial cost of the magnetic type.

Appearing in Fig. 18-13 is the schematic diagram of a magnetic full-voltage starter for a three-phase induction motor. The operation is simple. When the start button is pressed, the relay coil M is energized. This moves the relay armature to its closed position, thereby closing the main contactors M, which in turn apply full voltage to the motor. When the relay armature moves to its energized position, it also closes an auxiliary contactor M_a which serves as an electrical interlock, allowing the operator to release the start button without de-energizing the main relay. Of course contactors M are much larger in size than M_a. The former set must have a current rating that enables it to handle the starting motor current. The latter needs to accommodate just the exciting current of the relay coil. Figure 18-13 also shows that the motor line current flows through two overload heater elements. If the temperature rise becomes excessive, the heater element causes the overload contacts in the control circuit to open. In controller diagrams it is important to remember that all contactors are shown in their de-energized state. Thus the symbol ⊣⊢ means that the contactors M are open when the coil is not energized. Similarly, the symbol ⊣⊬ means that these contactors are closed in the de-energized state.

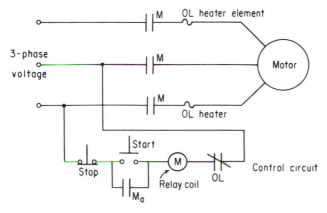

Fig. 18-13 Full-voltage magnetic starter.

Undervoltage protection is inherent in the magnetic start of Fig. 18-13. This comes about as a result of designing the coil M so that if the coil voltage drops below a specified minimum, the relay armature can no longer be held in the closed position.

A full-voltage magnetic starter equipped with the control circuitry to permit reversing is illustrated in schematic form in Fig. 18-14. A three-phase induction motor is reversed by crossing two of the three line leads going to the motor terminals. In this connection note the criss-cross of two of the R contactors. Pressing the forward (*FWD*) button energizes coil F which in turn closes the main contactors F as well as the interlock F_a. This allows the motor to reach its forward operating speed. To reverse the motor the REV button is pushed. This does two things: (1) it deenergizes coil F, thus opening the F contactors,

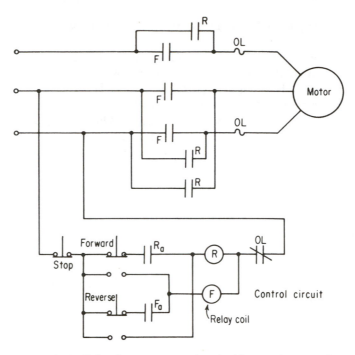

Fig. 18-14 Full-voltage starter equipped with reversing control.

and (2) it energizes the R relay coil, thus closing the R contactors which apply a reversed-phase sequence to the motor causing it to attain full speed in the reverse direction. Putting the *REV* switch in the F_a interlock circuit is a safety measure which prevents having both the R and F contactors closed at the same time.

An illustration of a reduced-voltage magnetic controller using limiting resistors in the line circuit appears in Fig. 18-15. This unit is frequently referred to as a three-step acceleration starter because the line resistors are removed in three steps. Pushing the start button energizes coil M and closes contactors M, thus applying a three-phase voltage to the motor through the full resistors. In addition to contactors M and M_a coil M is also equipped with a time-delayed contactor T_M. This contactor is so designed that it does not close until a preset time *after* the armature of coil M is closed. The delay is usually obtained through a mechanical escapement of some sort which is actuated by the relay armature. Of course the time delay is needed to permit the motor to accelerate to a speed corresponding to the reduced applied voltage. After the elapse of the preset time, the T_M contacts close, energizing coil 1A which in turn closes contactors 1A, short-circuiting the first part of the series resistor. Coil 1A is also equipped with a time-delay contactor T_{1A}, which is designed to allow the motor to accelerate to a higher speed before it closes. When contactor 1A does close, coil 2A is energized. This immediately closes contactors 2A, shorting out the second section of the line resistor. Then, after still another time delay, contactor T_{2A} closes, applying an excitation voltage to coil 3A. With the closing of contactors 3A the full line

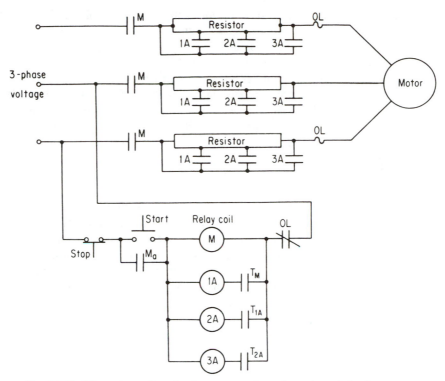

Fig. 18-15 Three-step reduced-voltage starter for a three-phase induction motor.

voltage is applied to the motor. In this manner the motor is brought up to speed in a "soft," smooth fashion without drawing excessive starting current or developing large starting torques.

Summary review questions

1. How is the synchronous speed of an a-c motor defined? What is an asynchronous motor?

2. How is a rotating magnetic field in a three-phase induction motor obtained.

3. Name the conditions that must be satisfied in order that the revolving magnetic field of a three-phase induction motor be of constant amplitude and of constant peripheral velocity.

4. Describe the ways in which the three-phase induction motor is similar to the static transformer.

5. Why is it that the rotor speed of a three-phase singly excited induction motor can never attain synchronous speed exactly?

6. Describe what is meant by the slip of an induction motor.

7. What is slip frequency?

8. How does the magnetizing current of the induction motor compare with that of the transformer? Which is larger? Explain.

9. How is the mechanical load that is attached to the three-phase induction motor treated in the equivalent circuit of the motor?

10. Show how the power that is transferred across the air gap of the three-phase induction motor is represented. Explain the terms. What portion of this is useful power?

11. Describe in detail the manner in which a three-phase induction motor responds to meet the demand for increased power to be delivered to the load.

12. Draw the complete equivalent circuit of the three-phase induction motor and explain the meaning of each parameter and electrical variable appearing in the circuit.

13. Explain the difference between the approximate version and the exact version of the equivalent circuit.

14. What constitutes the rotational losses of the induction motor? How are these supplied?

15. By means of a power-flow diagram show the flow of power in a three-phase induction motor from the electrical source to the mechanical load at the motor shaft.

16. Sketch the torque-speed curve of the induction motor and show how the basic torque equation developed in Chapter 17 can be used to explain the shape taken.

17. List the factors that determine the starting torque of the three-phase induction motor. How does this torque generally compare with the value of the rated torque?

18. What is meant by the breakdown torque of the induction motor?

19. List the factors that determine the maximum developed torque of the induction motor.

20. The maximum torque of the polyphase induction motor is often about 200% of the motor's rated torque. List reasons why such a high excess capacity is designed into the motor.

21. Describe the effect of increased rotor winding resistance on the value of the breakdown torque of the induction motor; on the value of the slip at which the maximum torque occurs.

22. Describe the information that is obtained from the rotor-blocked test of the induction motor.

23. Explain the procedure to be used to find the magnetizing reactance of a three-phase induction motor.

24. Name the four principal classes of squirrel cage induction motors. What design feature is responsible for the distinction between these types?

25. Of the four classes of squirrel-cage induction motors—A, B, C, D—identify which has the largest starting torque. Which has the greatest breakdown torque? For a constant torque load which provides the best accelerating capability? Which exhibits the poorest speed regulation? Which shows the poorest efficiency at rated horsepower?

26. Why are controllers needed for integral horsepower motors? List some of the benefits of such controllers.

GROUP I

18-1. A three-phase armature winding is shown in Fig. P18-1. Sinusoidal currents having an amplitude of 100 A flow in the three phases. Each coil consists of three turns. Carefully sketch to scale the actual mmf distribution along the air gap for a span of two poles for the indicated time instants t_1 and t_2.

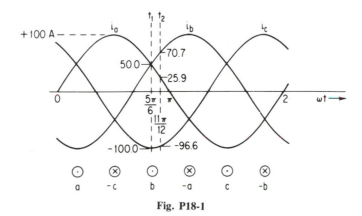

Fig. P18-1

18-2. State the general conditions under which it is possible for m alternating fields (with axes fixed in space) to yield a revolving field constant in amplitude and traveling at constant speed.

18-3. Determine the magnitude and direction of the resultant flux field in the machine configuration of Fig. 18-2 corresponding to time instants t_2 and t_4 in Fig. 18-1.

18-4. In the time variation of the phase currents depicted in Fig. 18-1 assume that the amplitude of phase b current is one-half that of phases a and c but that each is displaced by 120° from the other. Find the magnitude and direction of the resultant flux for the time instants t_1, t_3, and t_5, and compare with the results shown in Fig. 18-2.

18-5. A balanced, three-phase, 60-Hz voltage is applied to a three-phase four-pole induction motor. When the motor delivers rated output horsepower, the slip is found to be 0.05. Determine the following:
   (a) The speed of the revolving field relative to the stator structure, which accommodates the exciting winding.
   (b) The frequency of the rotor currents.
   (c) The speed of the rotor mmf relative to the rotor structure.
   (d) The speed of the rotor mmf relative to the stator structure.
   (e) The speed of the rotor mmf relative to the stator field distribution.
   (f) Are the conditions right for the development of a net unidirectional torque? Explain.

18-6. Repeat Prob. 18-5 for the case where the rotor structure is blocked, thus preventing rotation in spite of the application of a balanced three-phase voltage to the stator.

18-7. A 60-Hz polyphase induction motor runs at a speed of 873 rpm at full-load. What is the synchronous speed? Find the frequency of the rotor currents.

18-8. Answer each part briefly:
  (a) What is the effect of doubling the air gap of an induction motor on the magnitude of the magnetizing current and of the maximum value of the flux per pole? Neglect the effect of the leakage impedance.
  (b) Describe the effect of a reduced leakage reactance on the maximum developed torque, the power factor at full-load, and the starting current of a three-phase induction motor.

18-9. The shaft output of a three-phase 60-Hz induction motor is 75 kW. The friction and windage losses are 900 W, the stator core loss is 4200 W, and the stator copper loss is 2700 W. The rotor current referred to the stator (primary) is 100 A. If the slip is 3.75%, what is the percent efficiency at this output?

18-10. A 15-hp 22-V three-phase 60-Hz six-pole Y-connected induction motor has the following parameters per phase: $r_1 = 0.128$ Ω, $r_2' = 0.0935$ Ω, $x_1 + x_2' = 0.496$ Ω, $r_c = 183$ Ω, $x_\phi = 8$ Ω. The rotational losses are equal to the stator hysteresis and eddy-current losses. For a slip of 3% find: (a) the line current and power factor; (b) the horsepower output; (c) the starting torque.

18-11. A three-phase induction motor has a Y-connected rotor winding. At standstill the rotor induced emf per phase is 100 V rms. The resistance per phase is 0.3 Ω, and the leakage reactance is 1.0 Ω per phase.
  (a) With the rotor blocked what is the rms value of the rotor current? What is the power factor of the rotor circuit?
  (b) When the motor is running at a slip of 0.06, what is the rms value of the rotor current? What is the power factor of the rotor circuit?
  (c) Compute the value of the developed power in part (b).

18-12. A three-phase 12-pole 60-Hz 2200-V induction motor runs at no-load with rated voltage and frequency impressed and draws a line current of 20 A and an input power of 14 kW. The stator is Y-connected and its resistance per phase is 0.4 Ω. The rotor resistance r_2' is 0.2 Ω per phase. Also, $x_1 + x_2' = 2.0$ Ω per phase. The motor runs at a slip of 2% when it is delivering power to a load. For this condition compute (a) the developed torque and (b) the input line current and power factor.

18-13. A three-phase 440-V 60-Hz Y-connected eight-pole 100-hp induction motor has the following parameters expressed per phase:

$$r_1 = 0.06 \ \Omega \qquad x_1 = x_2' = 0.26 \ \Omega$$

$$r_2' = 0.048 \ \Omega \qquad r_c = 107.5 \ \Omega$$

$$x_\phi = 8.47 \ \Omega \qquad s = 0.03$$

The rotational losses are 1600 W. Using the approximate equivalent circuit, determine:
  (a) The input line current and power factor.
  (b) The efficiency.

18-14. A three-phase 335-hp 2000-V six-pole 60-Hz Y-connected squirrel-cage induction motor has the following parameters per phase that are applicable at normal slips:

$$r_1 = 0.2 \ \Omega \qquad x_1 = x_2' = 0.707 \ \Omega$$

$$r_2' = 0.203 \ \Omega \qquad r_c = 450 \ \Omega$$

$$x_\phi = 77 \ \Omega$$

The rotational losses are 4100 W. Using the approximate equivalent circuit, compute for a slip of 1.5%:

(a) The line power factor and current

(b) The developed torque.

(c) The efficiency.

18-15. A six-pole three-phase 40-hp 60-Hz induction motor has an input when loaded of 35 kW, 51 A, 440 V and a speed of 1152 rpm. When uncoupled from the load, the readings are found to be: 440 V, 21.3 A, 2.3 kW, and 1199 rpm. The resistance measured between terminals for the stator winding is 0.25 Ω for a Y-connection. The stator core losses and the rotational losses are known to be equal. Determine:

(a) The power factor of the motor when loaded.

(b) The motor efficiency when loaded.

(c) The horsepower rating of the load.

18-16. A three-phase Y-connected 440-V 200-hp induction motor has the following blocked-rotor data: $P_b = 10$ kW, $I_b = 250$ A, $V_b = 65$ V, and $r_1 = 0.02$ Ω. Find the value of the rotor resistance referred to the stator.

18-17. A 500-Hp three-phase 2200-V 25-Hz 12-pole Y-connected wound-rotor induction motor has the following parameters: $r_1 = 0.225$ Ω, $r_2' = 0.235$ Ω, $x_1 + x_2' = 1.43$ Ω, $x_\phi = 31.8$ Ω, $r_c = 780$ Ω. A no-load and a blocked-rotor test are performed on this machine. Neglect rotational losses.

(a) With rated voltage applied in the no-load test, compute the readings of the line ammeters as well as the total wattmeter reading.

(b) In the blocked-rotor test the applied voltage is adjusted so that 228 A of line current is made to flow in each phase. Calculate the reading of the line voltmeter and the total wattmeter reading.

18-18. For the machine of Prob. 18-17 compute the following:

(a) The slip at which maximum torque occurs.

(b) The input line current and power factor at the condition of maximum torque.

(c) The value of the maximum torque.

18-19. Refer to the motor of Prob. 18-17 and find the value of resistance that must be externally connected per phase to the rotor winding in order that the maximum torque be developed at starting. What is the value of this torque?

18-20. From the no-load test on a 10-hp four-pole 230-V 60-Hz three-phase Y-connected induction motor with rated voltage applied, the no-load current was found to be 9.2 A and the corresponding input power 670 W. Also, with 57 V applied in the blocked-rotor test, it was found that the motor took 30 A and 950 W from the line. The stator winding resistance was measured to be 0.15 Ω per phase. When this motor is coupled to its mechanical load, the input to the motor is found to be 9150 W at 28 A and a power factor of 0.82. The stator core losses are equal to the rotational losses.

(a) Compute the rotor current referred to the stator.

(b) Find the developed torque.

(c) What is the value of the slip?

(d) At what efficiency is the motor operating?

GROUP II

18-21. An unbalanced three-phase mmf flows through three coils that are space displaced by 120° as shown in Fig. 18-2. The nature of the unbalance is such that phase *b* lags behind phase *a* by 90° instead of 120°. However, each phase has the same amplitude. Investigate the character of the resultant flux field corresponding to various instants of time.

18-22. Determine the no-load speed of a six-pole wound-rotor three-phase induction motor, the stator of which is connected to a 60-Hz line and the rotor of which is connected to a 25-Hz line, when:

   (a) The stator field and the rotor field revolve in the same direction.

   (b) The stator field and the rotor field revolve in opposite directions.

18-23. A six-pole 60-Hz three-phase Y-connected wound-rotor (three phases also) induction motor has a standstill induced rotor voltage of 130 V per phase. At short circuit with the rotor blocked this voltage produces a current of 80 A at a power factor of 0.3 lagging. At full load the motor runs at a slip of 9%. Find the full-load developed torque.

18-24. A three-phase 2000-V Y-connected wound-rotor induction motor has the following no-load and blocked-rotor test data:

No-load:	2000 V	15.3 A	10.1 kW
Blocked-rotor:	440 V	170.0 A	36.4 kW

The resistance of the stator winding is 0.22 Ω per phase. The rotational losses are equal to 2 kW. Calculate all the necessary data for the approximate equivalent circuit at a slip of 2% and draw the circuit showing all parameter values.

chapter nineteen

Three-Phase Synchronous Machines

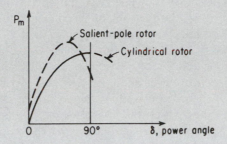

The synchronous generator is universally used by the electric power industry for supplying three-phase as well as single-phase power to its customers. The single-phase power that is brought to homes, shops, and offices originates from one phase of the three-phase system. Moreover, the assignment of commercial load circuits to each phase is made in an effort to keep the phases balanced.

The basic construction features of these machines are described in Sec. 17-3 and illustrated in Fig. 17-8. It is worth noting here that synchronous generators are classified into two types. The first is the *low-speed* (engine- or water-driven) type, which is characterized physically by salient poles, a large diameter, and small axial length. The second is the *turbogenerator,* which uses the steam turbine as the prime mover. The usual salient-pole rotor construction is abandoned in these high-speed generators in favor of the cylindrical (or smooth) rotor because the protruding-pole construction gives rise to dangerously high mechanical stresses. For 60-Hz three-phase power the two-pole generator must operate at 3600 rpm. Moreover, since turbogenerators are invariably designed for two poles, it follows that these machines are further characterized by a small diameter and long axial length.

19-1 GENERATION OF A THREE-PHASE VOLTAGE SYSTEM

How is a three-phase voltage generated? This can be readily understood from a study of Fig. 19-1(a). Depicted here is a two-pole rotor, the field winding of which is assumed energized from a d-c source to create the pole flux. It is further assumed that the pole pieces are shaped to produce a sinusoidal flux field. Appearing in the stator is a balanced three-phase winding with the axis of each phase displaced by 120°. The complete winding of each phase is represented in Fig. 19-1 by a single coil. Now consider that the rotor is driven by the prime mover in a counterclockwise direction at synchronous speed. Applying the $\bar{v} \times \bar{B}$ rule for the field direction shown reveals that the instantaneous voltage induced in coil sides a, b', c' is directed out of the paper while in coil sides a', b, c it is directed into the paper. Furthermore, since coil sides a and a' are located beneath the maximum value of the flux-density wave, the induced emf in phase a is at its maximum value. The corresponding induced emf in phase b, as identified by coil side b, is seen to be of opposite sign and of smaller magnitude. In fact, since coil sides b and b' are both displaced from the maximum-flux-density position by 60°, it follows that this instantaneous voltage is at one-half (cos 60°) its maximum value. A similar reasoning yields the same result for phase c.

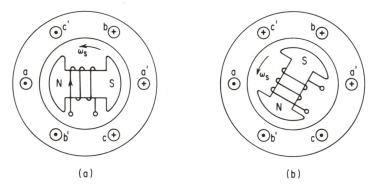

(a) (b)

Fig. 19-1 Induced-voltage distribution in the stator winding of a synchronous generator: (a) phase a at a positive maximum—time t_1 in Fig. 19-2; (b) phase b at a negative maximum—time t_2.

Next consider that the rotor has advanced by 60° in the counterclockwise direction. This puts the amplitude of the north-pole flux directly beneath b', indicating that the induced emf is now a negative maximum in phase b. Also, coil c-c' now finds itself under the influence of flux of reversed polarity as indicated in Fig. 19-1(b). Hence the direction of the emf in c is now out of the paper and into the paper for c'. The instantaneous value of the three phase voltages for the first and second time instants are depicted in Fig. 19-2 and are identified as t_1 and t_2, respectively.

By repeating the foregoing procedure many times and plotting the results

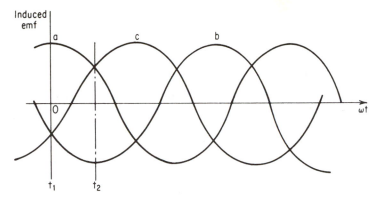

Induced emf

a   c   b

O ωt

t_1 t_2

Fig. 19-2 Time history of three-phase induced emf's for machine of Fig. 19-1.

for each instant of time we obtain the complete curves of Fig. 19-2. Note that the induced-emf variation for each phase is identical to the other two except for time displacements of 120° and 240°, respectively. As a matter of fact this variation is a direct consequence of having the beginning of each phase space-displaced by 120 electrical degrees. Physical connection of the ends of each phase (points a', b', c') gives a Y-connected stator winding.

19-2 SYNCHRONOUS GENERATOR PHASOR DIAGRAM AND EQUIVALENT CIRCUIT

The principles underlying the development of the phasor diagram are similar to those of the transformer. In fact, one may say that the principles are the same as for any magnetic device subjected to a varying flux. Thus, when the synchronous generator is operating at no-load (i.e., no current flowing through the stator winding, which is the armature winding in this case) there exists a pole flux Φ_f produced by the field winding, which in turn induces an rms voltage per phase which we denote by $\overline{E}_f$. Of course this emf lags the flux that induces it by 90° as depicted in Fig. 19-3, $\overline{E}_f$ is frequently called the *excitation voltage* for obvious reasons. It is always the voltage appearing at the armature terminals at no-load with the field winding excited.

Now consider that a balanced three-phase unity-power-factor load is placed across the armature terminals so that in the steady state the load draws an rms

Φ_f

$\overline{E}_f$

Fig. 19-3 No-load phasor diagram of a synchronous generator.

current $\bar{I}_a$ at a terminal voltage per phase of $\bar{V}_t$. This situation is depicted on a per phase basis in Fig. 19-4. The resulting phasor diagram is shown in Fig. 19-5. The presence of a three-phase current in the stator winding accounts for the differences in Figs. 19-3 and 19-5. The development of the phasor diagram starts by placing $\bar{I}_a$ in phase with $\bar{V}_t$ because the load is resistive. However, a three-phase current flowing through the three-phase stator winding gives rise to a rotating field as described in Sec. 18-1. This armature flux field Φ_A combines with the flux field Φ_f, which is produced by the field winding, to yield the resultant flux per pole Φ. If, for simplicity, the stator-winding resistance and leakage reactance are neglected, then this resultant flux Φ must lead V_t by 90°. Note that Φ_A is shown coincident with $\bar{I}_a$, which is consistent with the fact that they are in phase—Φ_A being produced by $\bar{I}_a$.

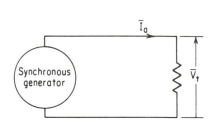

Fig. 19-4 Synchronous generator connected to a unity-pf load.

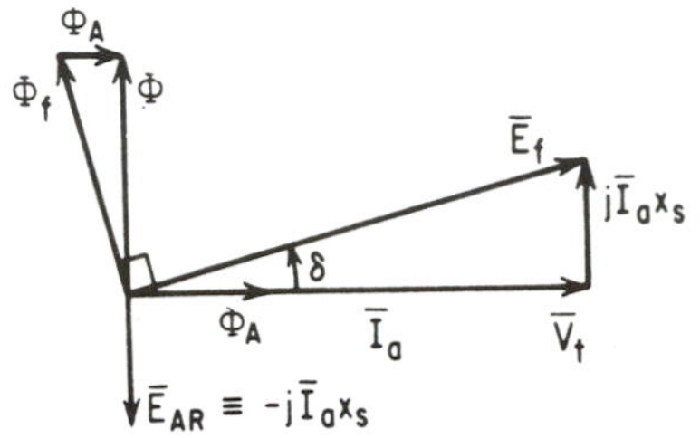

Fig. 19-5 Synchronous generator phasor diagram for unity-pf load.

In analyzing the effect of Φ_A we can resort to superposition and consider that Φ_A is a synchronously rotating field which cuts the stator armature winding, thereby inducing an "armature-reaction" voltage per phase denoted by $\bar{E}_{AR}$. This voltage lags the flux that creates it by 90°. Now the excitation voltage, by Kirchhoff's voltage law, must contain not only a component equal to $\bar{V}_t$ but also one equal and opposite to $\bar{E}_{AR}$. To treat the effect of the armature flux as a voltage drop, we replace the armature reaction voltage by an equivalent reactance drop as was done with leakage fluxes in the case of transformers. Thus

$$\bar{E}_{AR} \equiv -jI_a x_s \qquad (19\text{-}1)$$

The minus sign is used in the definition so that the equal and opposite quantity can be handled with a plus sign. That is, the plus sign is used for the drop and the minus sign for the rise. The parameter x_s is called the *synchronous reactance* per phase of the stator winding. It is a fictitious quantity which replaces the effect of the armature winding mmf on the field flux. By Kirchhoff's voltage law the excitation (or source) voltage must be equal to the sum of the voltage drops. Thus

$$E_f\underline{/\delta} = V_t\underline{/0°} + jI_a\underline{/0°}x_s \qquad (19\text{-}2)$$

Figure 19-5 shows this addition graphically. Note, too, that once $\bar{E}_f$ is found in this manner, Φ_f can also be shown in the phasor diagram by putting it 90° leading

$\overline{E}_f$ and of such a magnitude that when it is added to Φ_A there is yielded the resultant flux Φ.

Appearing in Fig. 19-6 is the phasor diagram of a synchronous generator as it applies to a lagging-power-factor load. Note that the armature flux field Φ_A causes two effects on the original flux field Φ_f. It directly demagnetizes as well as cross-magnetizes (or distorts) the pole-flux field. This is readily apparent if we consider Φ_A in terms of its two components—one along the line of direction of Φ_f and the other in quadrature. A physical picture of this situation is depicted in Fig. 19-7. The complete three-phase armature winding is again, for convenience, represented by a single coil marked e-e'. In fact this coil is located at the center of the emf distribution of the entire winding. (This corresponds to using coil a-a' in Fig. 19-1(a) to represent the emf distribution of the complete winding.) A similar ampere-conductor distribution can be identified for the armature winding and represented by a single coil i-i'. The phasor diagram of Fig. 19-6 shows that this current coil must be located in the lag direction (with respect to direction of rotation) by ψ electrical degrees. It is interesting to note that this angle ψ is the same one that appears in the basic torque expression of Eq. (17-39). Applying the right-hand rule to this coil gives an armature flux direction which with respect to the flux field yields the effects already cited.

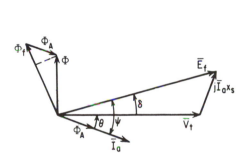

Fig. 19-6 Synchronous generator phasor diagram for lagging-pf load.

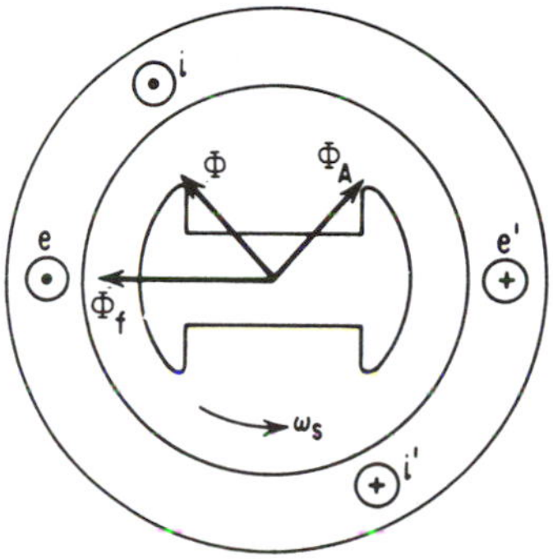

Fig. 19-7 Space picture showing the effect of armature flux at lagging pf.

Equivalent Circuit. The equivalent circuit follows directly from a circuit interpretation of Eq. (19-2). $\overline{E}_f$ is considered the source voltage and x_s is treated as an internal source impedance since it is associated with the armature winding. Figure 19-8 shows the resulting equivalent circuit.

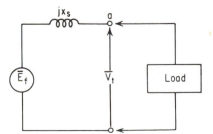

Fig. 19-8 Equivalent circuit of the synchronous generator appears to the left of terminals ab.

19-3 THE SYNCHRONOUS MOTOR

The significant and distinguishing feature of synchronous motors in contrast to induction motors is that they are doubly excited. Electrical energy is supplied both to the field and the armature windings. When this is done torque can be developed at only one† speed—the *synchronous speed*. At any other speed the average torque is zero. The synchronous speed refers to that rotor speed at which the rotor flux field and the armature ampere-conductor distribution (or the armature flux field) are stationary with respect to each other. To illustrate this point consider the case of the conventional synchronous motor equipped with two poles. It has a d-c voltage applied to the rotor winding, and a three-phase a-c voltage applied to the stator winding which is usually 60 Hz. The three-phase voltage produces in the stator an ampere-conductor distribution which revolves at a rate of 60 Hz. The d-c (or zero-frequency) currents in the rotor set up a two-pole flux field which is stationary so long as the rotor is not turning. Accordingly, we have a situation in which there exists a pair of revolving armature poles and a pair of stationary rotor poles. A little thought should make it clear that for a period of half a cyle (or $\frac{1}{120}$ s) a positive torque is developed by the cooperation of the field flux and the revolving ampere-conductor distibution. However, in the next half-cycle this torque is reversed so that the average torque is zero. This is the reason why a synchronous motor per se has no starting torque.

In order to develop a continuous torque, then, it is necessary not only to have a flux field and an appropriately displaced ampere-conductor distribution, *but the two quantities must be stationary with respect to each other.* In singly fed machines such as the induction motor this comes about automatically. The relative speed between the stator field and the rotor winding induces rotor currents at those slip frequencies which yield a revolving rotor mmf. Superposing this upon the rotor speed gives a revolving rotor mmf, which in turn is stationary relative to the stator field. In the synchronous motor this condition cannot take place automatically because of the separate excitation used for the field and armature windings. In light of these comments it is reasonable to expect that if an induction motor is doubly energized, it too will behave as a synchronous motor.‡

On the basis of the foregoing comments it follows that to produce a continuous nonzero torque it is first necessary to bring the d-c excited rotor to synchronous speed by means of an auxiliary device. Sometimes the auxiliary device takes the form of a small d-c motor mounted on the rotor shaft. Most often, however, use is made of a squirrel-cage winding similar to that used in induction motors and embedded in the pole faces. By means of this winding (also called the *amortisseur winding*) the synchronous motor is brought up to almost synchronous

---

† There is a second speed, too, but it is only of academic interest because of the excessive currents that flow.

‡ One such device is the Schrage motor, which is commercially available from the General Electric Company.

speed. Then, if the field winding is energized at the right moment, a positive torque will be developed for a sufficiently long period to allow the armature poles to pull the rotor poles into synchronism.

A general expression may be written to identify that frequency of rotation required of the rotor so that the stator and rotor fields are stationary with respect to each other. Let

$$f_1 = \text{frequency of the currents through the stator winding}$$

$$f_2 = \text{frequency of the currents through the rotor winding}$$

$$f_r = \text{frequency of rotation of the rotor structure}$$

Then the defining equation is

$$f_1 = f_2 + f_r \qquad (19\text{-}3)$$

In the conventional synchronous motor the frequency of the rotor current f_2 is zero, since it is d-c. Hence for an f_1 of 60 Hz the rotor must revolve at 60 Hz in order for nonzero torque to be developed. On the other hand note that if the rotor winding were energized with 20-Hz current in place of d-c, then for an f_1 of 60 Hz the synchronous speed would be 40 Hz.

19-4 SYNCHRONOUS MOTOR PHASOR DIAGRAM AND EQUIVALENT CIRCUIT

The synchronous motor phasor diagram differs from that of the synchronous generator in two respects. The first involves the voltage equation for the stator circuit. In the generator case, $\overline{E}_f$, the excitation voltage, played the role of a source voltage and the terminal voltage $\overline{V}_t$ was dependent upon the value of $\overline{E}_f$ and the synchronous impedance drop. In the motor case the roles are interchanged. Now $\overline{V}_t$ is the source voltage applied to the synchronous motor armature winding and $\overline{E}_f$ is a reaction or counter emf which is internally generated. It is assumed that the terminal voltage originates from an infinite bus system† and so remains invariant. Applying Kirchhoff's voltage law to the synchronous motor then leads us to

$$\boxed{V_t \angle 0° = \overline{E}_f + j\overline{I}_a x_s} \qquad (19\text{-}4)$$

Clearly, this equation states that the applied stator voltage is equal to the sum of the drops. The excitation voltage is treated as a reaction-voltage drop in much the same way as E_1 was treated in the case of the transformer.

The circuit interpretation of Eq. (19-4) leads to the equivalent circuit as it applies to the synchronous motor. This appears in Fig. 19-9.

The second point of view in which the phasor diagram of the motor differs from that of the generator involves the angle δ, which is called the power angle for reasons described soon. Physically, for the generator, the phasor $\overline{E}_f$ is ahead

---

† A power system of tremendous capacity compared to the rating of the synchronous motor.

of $\overline{V}_t$ in time because of the driving action of the prime mover. Keep in mind that $\overline{E}_f$ may be considered associated with the rotor field axis as was done in connection with Fig. 19-7. A similar line of reasoning pertains for the motor. At no-load, δ is zero and so the axis of a field pole and the resultant flux associated with the terminal voltage are in time phase. As shaft load is applied, however, the rotor falls slightly behind its no-load position and in this way causes δ to increase but in a sense opposite to that for the generator.

On the basis of these modifications the phasor diagram of the synchronous motor becomes that shown in Fig. 19-10. Note that $\overline{E}_f$ now lags $\overline{V}_t$ and that it is the phasor sum of $\overline{E}_f$ and $j\overline{I}_a x_s$ which equals $\overline{V}_t$ as called for by Eq. (19-4). Another interesting point about this diagram is revealed upon comparison with Fig. 19-6. In both cases the magnitude of the excitation voltage $\overline{E}_f$ exceeds that of the terminal voltage $\overline{V}_t$. This is referred to as a condition of *overexcitation* (i.e., $|\overline{E}_f| > |\overline{V}_t|$). Note, however, that overexcitation in the synchronous generator is accompanied by a current of *lagging* power factor whereas in the synchronous motor overexcitation is accompanied by a current of *leading* power factor. When the magnitude of $\overline{E}_f$ is equal to $\overline{V}_t$ the condition is referred to as *100% excitation*. A situation where $|\overline{E}_f| < |\overline{V}_t|$ is described as *underexcitation*.

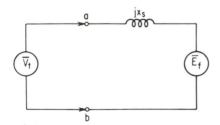

Fig. 19-9 The equivalent circuit of the synchronous motor appears to the right of terminals *ab*.

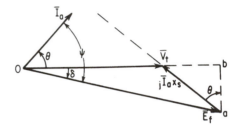

Fig. 19-10 Phasor diagram of the synchronous motor.

19-5 COMPUTATION OF SYNCHRONOUS MOTOR PERFORMANCE

The analysis of the performance of a synchronous motor is readily established in terms of the phasor diagram, the power-flow diagram, and an appropriate expression for the mechanical power developed which is derivable from the phasor diagram. In connection with the phasor diagram one additional item is needed which has not yet been discussed. It is the saturation curve for the magnetic circuit of the synchronous motor. This curve, of course, is a plot of induced armature-winding voltage as a function of field-winding current. Thus corresponding to any computed value of excitation voltage $\overline{E}_f$, the field current needed to produce it can be specified. Conversely, the excitation voltage is known once the field current is named.

The worth of a power-flow diagram has already been demonstrated in connection with the induction motor. Hence it is presented in Fig. 19-11 without very much additional comment. There is one step fewer in this flow graph because the a-c source does not have to supply rotor-winding copper loss since the rotor is traveling at synchronous speed. This is not to say that rotor-winding copper losses do not exist. They do, of course, but the power is supplied from the separate d-c source. Figure 19-11 represents the flow of power from the a-c source to the shaft of the synchronous motor.

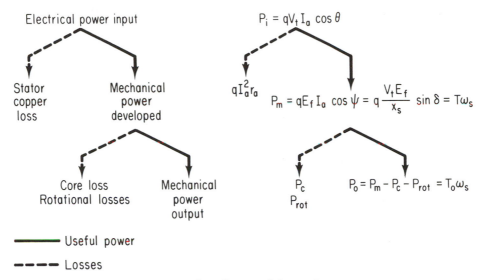

Fig. 19-11 Power-flow diagram of the synchronous motor.

A useful expression for the mechanical power developed can be derived from the phasor diagram of Fig. 19-10, which is drawn for negligible armature-winding resistance. Because r_a is assumed zero, the power-flow diagram indicates that the mechanical power developed is equal to the power input. Thus

$$P_m = qE_fI_a \cos \psi = qV_tI_a \cos \theta \qquad (19\text{-}5)$$

This statement is also apparent from Fig. 19-10 when we recognize that the projection of E_f on the I_a phasor is identical to the projection of V_t on I_a. Furthermore, from the geometry of the phasor diagram of Fig. 19-10, we note that the quantity ab may be expressed in one of two ways. Thus

$$ab = E_f \sin \delta = I_a x_s \cos \theta \qquad (19\text{-}6)$$

or

$$I_a \cos \theta = \frac{E_f}{x_s} \sin \delta \qquad (19\text{-}7)$$

Insertion of Eq. (19-7) into Eq. (19-5) yields the result being sought. Hence the mechanical power developed in a synchronous motor that has negligible armature

resistance may be expressed as

$$P_m = q \frac{V_t E_f}{x_s} \sin \delta \qquad (19\text{-}8)$$

Because of the dependence of the mechanical power developed upon the sine of the angle δ, this angle is called the *power angle*. Equation (19-8) states that if the power angle is zero the synchronous motor cannot develop torque. It further points out that the mechanical power developed is a sinusoidal function of the angle δ as depicted in Fig. 19-12. Note, too, that the maximum developed power for a given excitation occurs at δ equal to 90°.

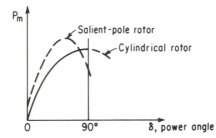

Fig. 19-12 Power-angle curves for synchronous motors.

The application of Eq. (19-8) for computation purposes has a further restriction in addition to that of negligible armature-winding resistance. It assumes that the machine has a uniform air gap. That is, the rotor is assumed to be of the nonsalient-pole type. The use of the phasor diagram of Fig. 19-10 presupposes this, because the diagram applies only to such a machine. For a salient-pole machine the corresponding phasor diagram is more complicated and lies beyond the scope of this book. However, even though synchronous motors are invariably of salient-pole construction, we shall continue to use the phasor diagram of Fig. 19-10 as well as Eq. (19-8) because the general results are the same. Moreover, it is easier to gain insight into the basic operating principles and performance behavior when the secondary effects are mitigated.

The analogous expression for the mechanical power developed by a salient-pole synchronous motor includes a second term besides that appearing in Eq. (19-8). In fact this second term is completely independent of the excitation of the machine; it depends solely upon the terminal voltage and machine parameters, which reflect the nonuniform character of the air gap. In most motors this extra term has a value sometimes as much as 25% of the P_m given by Eq. (19-8). For obvious reasons this term is referred to as *reluctance power*. It causes a change in the power-versus-angle plot as indicated by the broken-line curve in Fig. 19-12.

What is the mechanism by which the synchronous motor becomes aware of the presence of a shaft load, and how is the electric energy source made aware of this so that it proceeds to provide energy balance? To answer this question we start first with a study of the conditions prevailing at no-load. As the power-flow diagram indicates, the only mechanical power needed is that which is required to supply the rotational and core losses. This calls for a very small value of δ, as depicted in Fig. 19-13. The machine is assumed to be overexcited since $\overline{E}_f > \overline{V}_t$. Note that δ_0 is just large enough so that the in-phase component of $\overline{I}_a$ is sufficient to supply the losses. Consider next that a large mechanical load is suddenly applied to the motor shaft. The first reaction is to cause a momentary drop in speed. In turn this appreciably increases the power angle and thereby causes a phasor voltage difference to exist between $\overline{V}_t$ and the excitation voltage $\overline{E}_f$. The result is the flow of an increased armature current at a very much improved power factor compared to no-load. As a matter of fact, the speed changes momentarily by a sufficient amount to allow the power angle to assume that value which enables the armature current and input power factor to assume those values which permit the input power to balance the power required at the load plus the losses. Figure 19-10 is typical of what the final phasor diagram looks like. Note that the power component of the current is considerably increased over the no-load case. The increase in power angle from no-load to the load case can be measured by means of any stroboscopic instrument. The stroboscope furnishes a convincing demonstration of the physical character of the power angle.

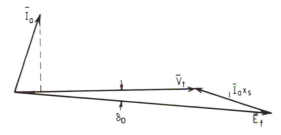

Fig. 19-13 Phasor diagram of synchronous motor at no-load and overexcited.

EXAMPLE 19-1 A 2300-V, three-phase, 60-Hz, Y-connected cylindrical-rotor synchronous motor has a synchronous reactance of 11 Ω per phase. When it delivers 200 hp, the efficiency is found to be 90% exclusive of field loss, and the power angle is 15 electrical degrees as measured by a stroboscope. Neglect ohmic resistance and determine:

(a) The induced excitation voltage per phase, E_f,
(b) The line current, $\overline{I}_a$,
(c) The power factor.

Solution: (a) The power input and the mechanical power developed are the same. Hence

$$3\frac{V_t E_f}{x_s}\sin\delta = \frac{200(746)}{0.9}$$

Inserting $V_t = 2300/\sqrt{3}$, $\delta = 15°$, and $x_s = 11$ yields

$$E_f = 1768 \text{ V/phase}$$

(b) The armature current follows directly from Eq. (19-4). Thus

$$\overline{I}_a = \frac{1328\underline{/0°} - 1768\underline{/-15°}}{11\underline{/90°}} = 54.1\underline{/39.7°}$$

Thus since the angle of $\overline{I}_a$ is positive, the power factor is leading.

(c) The line power factor is merely

$$pf = \cos 39.7° = 0.769 \text{ leading}$$

19-6 POWER-FACTOR CONTROL

For a fixed mechanical power developed (or load) it is possible to adjust the reactive component of the current drawn from the line by varying the d-c field current. This feature is achievable in the synchronous motor precisely because it is a doubly excited machine. Thus, although operation at a constant applied voltage demands a fixed resultant flux, both the d-c source and the a-c source may cooperate in establishing this resultant flux. If the field-winding current is made excessively large, then clearly the resultant air-gap voltage in the motor tends to be larger than that demanded by the applied voltage. Accordingly, a reaction occurs which causes the armature current to assume such a power-factor angle that the armature mmf exerts that amount of demagnetizing effect which is needed to restore the required resultant flux. Similarly, if the field is underexcited, then the resultant gap flux tends to be too small. This also creates a phasor voltage difference between the line voltage and the motor excitation voltage which acts to cause the armature current to flow at that power-factor angle which enables it to magnetize the air gap to the extent needed to provide the necessary resultant flux.

Whether or not the reactive component of current must be leading or lagging readily follows from an investigation of the phasor diagrams depicted in Fig. 19-14 for various values of excitation voltage. The diagram applies for an assumed constant mechanical power developed. Hence as E_f is changed, the sin δ must change correspondingly to keep E_f sin δ invariant. In Fig. 19-14 this means that as the excitation is varied, the locus of the tip of the $\overline{E}_f$ phasor is the broken line. Furthermore, the in-phase component of the current in each case must be the same. In the first case, where the excitation voltage is $\overline{E}_{f1}$, the field current is producing too much flux. This creates a reaction between the motor and the source which calls for a leading current of such a magnitude that it provides that amount and direction of synchronous reactance drop which when added to $\overline{E}_{f1}$ yields the fixed terminal voltage $\overline{V}_t$. Physically, what is happening is that a

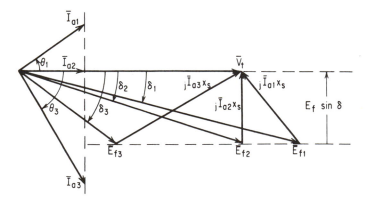

Fig. 19-14 Showing the effect of varying excitation on power factor.

leading reactive current is made to flow which acts to demagnetize the flux field to the extent needed. When the excitation is reduced to $\overline{E}_{f2}$, there is no excess flux produced by the field winding. Consequently, the a-c line current contains no reactive component. It merely has the value of an in-phase component needed to supply power to the load. At the excitation corresponding to $\overline{E}_{f3}$ the machine is greatly underexcited. To compensate for this a reaction occurs which allows the line to deliver a large lagging reactive current, which helps to establish the value of air-gap flux demanded by the terminal voltage. Note, too, that as the excitation is decreased, the power angle must increase for a fixed mechanical power developed.

If a plot is made of the armature current as the excitation is varied for fixed mechanical power developed, the current is observed to be large for underexcitation and overexcitation and passing through a minimum at some intermediate point. The plot in fact resembles a V-shape as demonstrated in Fig. 19-15.

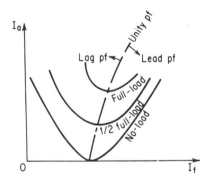

Fig. 19-15 Synchronous motor V-curves.

The ability of the synchronous motor to draw leading current when overexcited can be used to improve the power factor at the input lines to an industrial establishment that makes heavy use of induction motors and other equipment drawing power at a lagging power factor. Many electric power companies charge increased power rates when power is bought at poor lagging power factor. Over the years such increased power rates can result in an appreciable expenditure of money. In such instances the installation of a synchronous motor operated overexcited can more than pay for itself by improving the overall input power factor to the point where the penalty clause no longer applies.

19-7 SYNCHRONOUS MOTOR APPLICATIONS

Synchronous motors are rarely used below 50 hp in the medium-speed range because of their much higher initial cost compared to induction motors. In addition they require a d-c excitation source, and the starting and control devices are usually more expensive—especially where automatic operation is required. However, synchronous motors do offer some very definite advantages. These include constant-speed operation, power-factor control, and high operating ef-

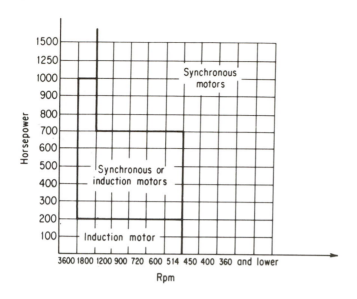

Fig. 19-16 Indicating the general areas of application of synchronous and induction motors.

TABLE 19-1 SYNCHRONOUS MOTOR CHARACTERISTICS AND APPLICATIONS†

Type designation	Synchronous, high speed, above 500 rpm	Synchronous, low speed, below 500 rpm
Starting torque (% of normal)	Up to 120	Low 40
Pull-in torque (% of normal)	100 to 125	30
Pull-out torque	Up to 200	Up to 180
Starting current	500–700	200–350
Slip	Zero	Zero
Power factor	High, but varies with load and with excitation	High, but varies with excitation
Efficiency (%)	Highest of all motors, 92–96	Highest of all motors, 92–96
Typical applications	Fans, blowers, d-c generators, line shafts, centrifugal pumps and compressors, reciprocating pumps and compressors. Useful for power-factor correction. Constant speed. Frequency changers.	Lower-speed direct-connected loads such as reciprocating compressors when started unloaded, d-c generators, rolling mills, band mills, ball mills, pumps. Useful for power-factor control. Constant speed. Flywheel used for pulsating loads.

† By permission from M. Liwschitz-Garik and C. C. Whipple, *Electric Machinery,* Vol. II (Princeton, N.J.: D. Van Nostrand Co., Inc., 1946).

ficiency. Furthermore, there is a horsepower and speed range where the disadvantage of higher initial cost vanishes, even to the point of putting the synchronous motor to advantage. This is demonstrated in Fig. 19-16. Where low speeds and high horsepower are involved the induction motor is no longer cheaper because it must use large amounts of iron in order not to exceed a gap flux density of 0.7 T. In the synchronous machine, on the other hand, a value twice this figure is permissible because of the separate excitation.

Appearing in Table 19-1 are some of the more important characteristics of the synchronous motor along with some typical applications. In addition to the area of application listed there, they are also quite prominently found in the following applications, which are characterized by operation at low speeds and high horsepower: large, low-head pumps; flour-mill line shafts; rubber mills and mixers; crushers; chippers; pulp grinders and jordans and refiners used in the papermaking industry.

Summary review questions

1. What is a turbogenerator?
2. Describe the manner in which a balanced three-phase voltage is produced by a synchronous generator. Give due regard to construction details.
3. Draw the phasor diagram of a synchronous generator per phase delivering electrical power to a resistive load. Restrict the diagram to the inclusion of the excitation voltage and its associated flux, the terminal voltage and its associated flux, and the armature current and the flux it produces. Assume negligible armature leakage impedance.
4. Redraw the phasor diagram of Question 3 by replacing the effect of the rotating armature mmf (or the armature flux it produces) by a reactance drop. Explain why this reactance drop takes the magnitude and direction it does.
5. Describe how the synchronous reactance of a synchronous generator is defined. On what important assumption is this definition based?
6. Describe the effects produced by the armature mmf on the air-gap flux produced by the field mmf when electrical power is delivered to lagging power factor load.
7. Repeat Question 6 when the electrical load is characterized by a leading power factor.
8. Draw the equivalent circuit of a synchronous generator. Neglect armature leakage impedance. Write Kirchhoff's voltage law as it applies to this circuit. Specify magnitudes and phase angles for each of the phasor quantities.
9. Does a synchronous motor of itself have a starting torque? Explain. Describe how the synchronous motor can be made to attain synchronous speed.
10. Draw the phasor diagram of the synchronous motor and indicate the differences it has with that of the synchronous generator. Write Kirchhoff's voltage law for this motor circuit.
11. What is the effect on the field flux of an armature current in the synchronous motor that leads the terminal voltage?
12. What adjustment must be made to the synchronous motor to cause it to draw an

armature current at a leading power factor angle? Give a physical explanation that describes the circumstances which force the synchronous motor to draw the leading current.

13. What adjustment must be made to the synchronous motor to cause it to draw an armature current at a lagging power factor angle while delivering constant output at its shaft?

14. In the synchronous motor is the a-c electrical source applied to the stator or rotor? Does the a-c source supply energy to the field winding? Explain.

15. Name the quantities that determine the mechanical power developed in a synchronous motor. What is the maximum value of this power for the smooth-rotor synchronous motor?

16. Define the following terms: overexcitation, underexcitation, 100% excitation.

17. When the synchronous motor is operating at full load at 100% excitation, explain whether the power factor is lagging or leading.

18. What is reluctance power? Do all synchronous motors have this power? How does it affect the power angle at which maximum power is developed?

19. Describe how the synchronous motor may be made to behave as a large capacitor on the power lines feeding power to a plant. How can the plant take advantage of this feature to reduce its billings for electrical energy supplied to the plant?

20. Why are synchronous motors rarely used in small (<50) horsepower ratings?

21. Why are synchronous motors generally preferable over induction motors for high-horsepower low-speed applications?

Problems

GROUP I

19-1. A three-phase 1732-V (line to line) Y-connected cylindrical-pole synchronous motor has $r_a = 0$ and $x_s = 10 \, \Omega$ per phase. The friction and windage plus the core losses amount to 9 kW. The motor delivers an output of 390 hp. The greatest excitation voltage that may be obtained is 2500 V per phase.
   (a) Calculate the magnitude and power factor of the armature current for maximum excitation at the specified load.
   (b) Compute the smallest excitation for which the motor will remain in synchronism for the given power output.

19-2. A 2300-V three-phase 60-Hz Y-connected cylindrical-rotor synchronous motor has a synchronous reactance of 11 Ω per phase. When it delivers 100 hp, the efficiency is 85% and the power angle is 7 electrical degrees. Neglect r_a and determine:
   (a) The excitation emf per phase.
   (b) The line current.
   (c) The power factor.

19-3. A synchronous motor is operated at half-load. An increase in its field current causes a decrease in armature current. Does the armature current lead or lag the terminal voltage? Explain.

19-4. A three-phase Y-connected synchronous motor is operating at 80% leading power factor. The synchronous reactance is 2.9 Ω per phase and the armature winding resistance is negligible. The armature current is 20 A/phase. The applied line voltage is 440 V.

(a) Find the excitation voltage and the power angle.

(b) It is desired to increase the armature current to 40 A/phase and maintain the power factor at 80% leading. Show clearly in a phasor diagram how the field excitation must be changed. Can this be accomplished if the shaft load remains fixed? Explain.

19-5. A Y-connected three-phase 60-Hz 13,500-V synchronous motor has an armature resistance of 1.52 Ω/phase and a synchronous reactance of 37.4 Ω/phase. When the motor delivers 2000 hp, the efficiency is 96% and the field current is so adjusted that the motor takes a leading current of 85 A.

(a) At what power factor is the motor operating?

(b) Calculate the excitation emf.

(c) Find the mechanical power developed.

(d) If the load is removed, describe (do not calculate) how the magnitude and power factor of the resulting armature current compare with the original.

19-6. As a synchronous motor is loaded from zero to full-load, can the power factor ever become leading if the field excitation is maintained at 75%? Explain.

19-7. A synchronous motor delivers rated power to a load. Is there a lower limit on the excitation current beyond which it may not be reduced? Explain.

19-8. A 440-V three-phase Y-connected synchronous motor has a synchronous reactance of 6.06 Ω per phase. The armature resistance is negligible, and the induced excitation emf per phase is 200 V. Moreover, the power angle between $\overline{V}_i$ and $\overline{E}_f$ is 36.4 electrical degrees.

(a) Calculate the line current and pf.

(b) What values of excitation emf and power angle are necessary to make the power factor unity for the same input?

GROUP II

19-9. A 1250-hp three-phase cylindrical-pole synchronous motor receives constant power of 800 kW at 11,000 V. Armature winding resistance is negligible and the synchronous reactance per phase is 50 Ω. The armature is Y-connected. The motor has a rated full-load current of 52 A. If the armature current shall not exceed 135% of this value, determine the range over which the excitation emf can be varied through adjustment of the field current.

19-10. A synchronous motor at no-load is connected to an infinite bus system. The field circuit is accidentally opened. Explain what happens to the armature current.

19-11. An eight-pole synchronous motor draws 45 kW from the 208-V 60-Hz three-phase power system at a pf of 0.8 lagging. The motor is Y-connected and has a synchronous reactance of 0.6 Ω per phase. Armature resistance is negligible.

Without any further manipulations on the motor, what is the highest possible value of its steady-state torque?

19-12. Answer whether the following statements are true or false. When the statement is false, describe why.

(a) Increasing the air gap of a machine reduces the synchronous reactance and raises the steady-state power limit of the machine.

(b) For a synchronous motor, the developed electromagnetic torque is in a direction opposite to the direction of rotation.

(c) In a synchronous machine, the space-phase relation between the field mmf wave and armature-reaction-mmf wave is determined by the power factor of the load.

(d) The sum of the armature mmfs in the three phases of the stator winding is zero at any given instant.

(e) The sum of the main field flux Φ_f and the armature flux Φ_A always add to give the resultant flux Φ.

(f) A synchronous motor operating at leading power factor is underexcited.

(g) The high currents flowing during a short circuit cause a synchronous machine to operate under saturated conditions.

(h) The load on a synchronous motor is increased. This causes the main field mmf axis to drop further behind the air-gap mmf axis, so as to increase the torque angle.

(i) An overexcited synchronous motor is a generator of lagging kvar. This means that it supplies lagging kvar to the system to which it is connected.

(j) For large machines, the armature resistance and leakage reactance are about 10% of the machine rating and the synchronous reactance is about 100%.

chapter twenty

D-C Machines

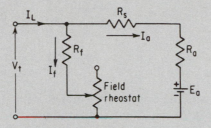

$$E_a = K_E \Phi n$$

$$T = K_T \Phi I_a$$

$$V_t = E_a + I_a(R_a + R_s)$$

$$I_L = I_f + I_a$$

The basic voltage equation, the developed-torque equation, the construction details, and the need for a commutator have all been described in Chapter 14. In this chapter we analyze the performance of these machines and discuss typical applications.

20-1 D-C GENERATOR ANALYSIS

The d-c machine functions as a generator when mechanical energy is supplied to the rotor and an electrical load is connected across the armature terminals. In order to supply electrical energy to the load, however, a magnetic field must first be established in the air gap. The field is necessary because it serves as the coupling device permitting the transfer of energy from the mechanical to the electrical system. There are two ways in which the field winding may be energized to produce the magnetic field. One method is to excite the field separately from an auxiliary source as depicted in the schematic diagram of Fig. 20-1. But clearly this scheme is disadvantageous because of the need of another d-c source. After all, the purpose of the d-c generator is to make available such a source. Therefore, invariably d-c generators are excited by the second method which involves a process of self-excitation. The wiring diagram appears in Fig. 20-2, and the arrangement is called the *self-excited shunt generator*. The word shunt is used

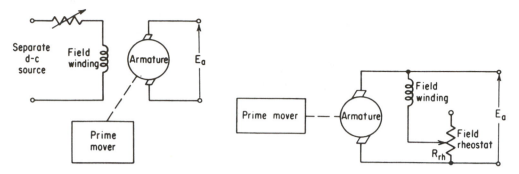

Fig. 20-1 Schematic diagram of a separately excited d-c generator.

Fig. 20-2 Schematic diagram of a self-excited shunt generator.

because the field winding appears in parallel with the armature winding. That is, the two windings form a shunt connection.

To understand how the self-excitation process takes place we must start with the *magnetization curve* of the machine. Sometimes this is called the *saturation curve*. Strictly speaking, the magnetization curve represents a plot of air-gap flux versus field-winding magnetomotive force. However, in the d-c generator where the winding constant K_E is known and the speed n is fixed, the magnetization curve has come to represent a plot of the open-circuit induced armature voltage as a function of the field-winding current. With K_E and n fixed, Eq. (17-47) shows that Φ and E_a differ only by a constant factor. Figure 20-3 depicts a typical magnetization curve, valid for a constant speed of rotation of the armature. It is especially important to note in this plot that even with zero field current an emf is induced in the armature of value Oa. This voltage is due entirely to residual magnetism, which is present because of the previous excitation history of the magnetic-circuit iron. The linear curve appearing on the same set of axes is the *field-resistance line*. It is a plot of the current caused by the voltage applied to the series combination of the field winding and the active portion of the field rheostat. Clearly, then, the slope of the linear curve is equal to the sum of the field-winding resistance R_f and the active rheostat resistance R_{rh}. The voltage Oa due to residual magnetism appears across the field circuit and causes a field current Ob to flow. But in accordance with the magnetization curve this field current aids the residual flux and thereby produces a larger induced emf of value bc. In turn, this increased emf causes an even larger field current, which creates more flux for a larger emf, and so forth. This process of voltage build-up continues until the induced emf produces just enough field current to sustain it. This corresponds to point f in Fig. 20-3. Note that in order for the build-up process to take place three conditions must be satisfied: (1) There must be a residual flux. (2) The field winding mmf must act to aid this residual flux. (3) The total field-circuit resistance must be less than the critical value. The *critical field resistance* is that value which makes the resistance line coincide with the linear portion of the saturation curve.

The performance of the d-c generator can be analyzed in terms of an equivalent circuit, a set of appropriate performance equations, the power-flow

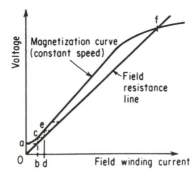

diagram, and the magnetization curve. To generalize the equivalent circuit consider first the schematic diagram of the compound generator in Fig. 20-4. A *compound generator* is a shunt generator equipped with a series winding. The *series winding* is a coil of comparatively few turns wound on the same magnetic axis as the field winding and connected in series with the armature winding. Since the series field winding must be capable of carrying the full armature current, its cross-sectional area is much greater than that used in the shunt field winding. The purpose of the series field is to provide additional air-gap flux as increased armature current flows, in order to neutralize the armature-winding resistance drop as well as the voltage drops occurring in the feeder wires leading to the load. In such cases the generator is usually referred to as a *cumulatively* compounded generator, because the series-field flux *aids* the shunt-field flux.

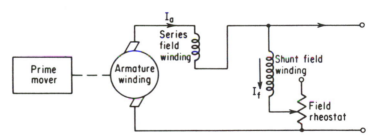

Fig. 20-4 Schematic diagram of a compound d-c generator.

By imposing the appropriate constraint on the connection diagram shown in Fig. 20-4, we can identify any one of the three modes of operation of the d-c generator. Thus, besides the armature winding we have the following:

Compound generator: includes shunt- and series-field windings
Shunt generator: shunt-field winding
Series generator: series-field winding

Since the series generator is rarely used except for special applications, all further treatment of generators is confined to the shunt and compound modes.

Appearing in Fig. 20-5 is the equivalent circuit of the compound generator. The armature winding is replaced by a source voltage having the induced emf

E_a and a resistance R_a, which represents the armature circuit resistance.† The series field is replaced by its resistance R_s; likewise for the shunt field.

The governing equations for determining the performance are the following:

$$E_a = \left(\frac{pZ}{60a}\right)\Phi n = K_E \Phi n \tag{20-1}$$

$$T = \left(\frac{pZ}{2\pi a}\right)\Phi I_a = K_T \Phi I_a \tag{20-2}$$

$$V_t = E_a - I_a(R_a + R_s) \tag{20-3}$$

$$I_a = I_L + I_f \tag{20-4}$$

These include the two basic relationships for induced voltage and electromagnetic torque as developed in Sec. 17-4. Equations (20-3) and (20-4) are merely statements of Kirchhoff's voltage and current laws as they apply to the equivalent circuit

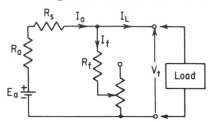

Fig. 20-5 Equivalent circuit of the compound d-c generator.

of Fig. 20-5. For the compound generator the expression for the air-gap flux must include the effect of the series field as well as the shunt field. Thus

$$\Phi = \Phi_{sh} + \Phi_s \tag{20-5}$$

where Φ_{sh} denotes the flux produced by the shunt-field winding and Φ_s denotes the flux caused by the series-field winding. For the shunt generator, of course, Φ_s is zero in Eq. (20-5), and so too is R_s in Eq. (20-3).

The power-flow diagram for the d-c generator is depicted in Fig. 20-6. Note

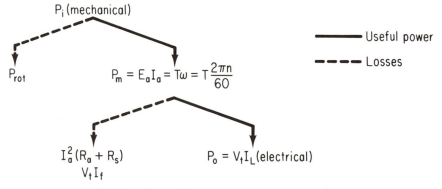

Fig. 20-6 Power-flow diagram of the d-c generator.

† The armature-circuit resistance includes the armature-winding resistance plus the effect of the voltage drop in the carbon brushes.

the similarity it bears to that of the synchronous generator. The field-winding loss is included in the power flow directly because it is assumed that the generator is self-excited. Of course, if the field winding is separately excited, the field losses are not supplied from the prime mover and so must be handled separately. Moreover, note that the field losses are represented in terms of the product of the field terminal voltage and the field current. This ensures that the field-winding rheostat losses are included as well.

EXAMPLE 20-1   The magnetization curve of a 10-kW, 250-V, d-c self-excited shunt generator driven at 1000 rpm is shown in Fig. 20-7. Each vertical division represents 20 V and each horizontal unit represents 0.2 A. The armature-circuit resistance is 0.15 Ω and the field current is 1.64 A when the terminal voltage is 250 V. Also, the rotational losses are known to be equal to 540 W. Find at rated load:

    (a) The armature induced emf.
    (b) The developed torque.
    (c) The efficiency.

Assume constant-speed operation.

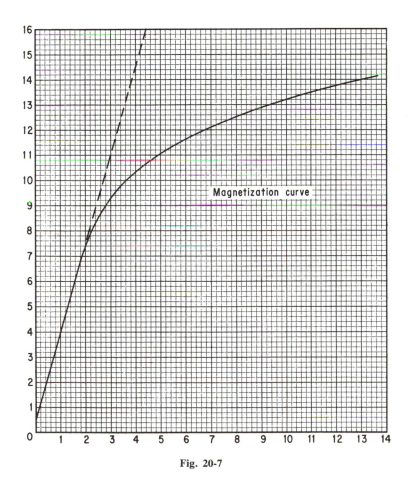

Fig. 20-7

Solution: (a) With the generator delivering rated load current it follows that

$$I_L = \frac{10,000}{250} W = 40 \text{ A}$$

Hence from Eq. (20-4) the armature current is

$$I_a = I_L + I_f = 40 + 1.64 = 41.64 \text{ A}$$

The induced armature voltage then follows from Eq. (20-3):

$$E_a = V_t + I_a R_a = 250 + 41.64(0.15) = 256.25 \text{ V} \qquad (20\text{-}6)$$

(b) To determine the developed torque it is first necessary to compute the electromagnetic power $E_a I_a$. Thus

$$E_a I_a = 256.25(41.64) = 10,670 \text{ W}$$

The power-flow diagram reveals that this power may also be found from

$$E_a I_a = P_o + I_a^2 R_a + V_t I_f = 10 \text{ kw} + (41.64)^2(0.15) + 250(1.64) \qquad (20\text{-}7)$$

$$= 10 \text{ kw} + 261 + 410 = 10,671 \text{ W}$$

which checks closely with the preceding calculation. Hence the developed torque is

$$T = \frac{E_a I_a}{2\pi n/60} = \frac{10,670}{2\pi(1000)/60} = 101.9\text{N-m} \qquad (20\text{-}8)$$

(c) The efficiency is found from Eq. (18-20), which for the d-c generator takes the form

$$\eta = 1 - \frac{\sum \text{losses}}{E_a I_a + P_{rot}} \qquad (20\text{-}9)$$

In this case

$$\sum \text{losses} = P_{rot} + I_a^2 R_a + V_t I_f$$

$$= 540 + 260 + 410 = 1210 \text{ W}$$

Hence

$$\eta = 1 - \frac{1210}{10,670 + 540} = 1 - \frac{1210}{11,210} = 1 - 0.108 = 0.892$$

The efficiency is therefore 89.2%.

20-2 D-C MOTOR ANALYSIS

A d-c motor is a d-c generator with the power flow reversed. In the d-c motor electrical energy is converted to mechanical form. Also, as is the case for the generator, there are three types of d-c motors: the *shunt* motor, the *cumulatively compounded* motor, and the *series motor*. The compound motor is prefixed with the word cumulative in order to stress that the connections to the series-field winding are such as to ensure that the series-field flux *aids* the shunt-field flux. The series motor, unlike the series generator, finds wide application, especially for traction-type loads. Hence due attention is given to this machine in the treatment that follows.

The performance of the d-c motor operating in any one of its three modes can conveniently be described in terms of an equivalent circuit, a set of performance equations, a power-flow diagram, and the magnetization curve. The equivalent circuit is depicted in Fig. 20-8. It is worthwhile to note that now the armature induced voltage is treated as a reaction or counter emf. By imposing constraints similar to those applied to the d-c generator, we obtain the correct equivalent circuit for the desired mode of operation. For example, for a series motor the appropriate equivalent circuit results upon removing R_f from the circuitry of Fig. 20-8.

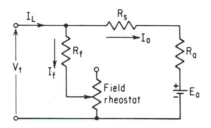

Fig. 20-8 Equivalent circuit of the d-c generator.

The set of equations needed to compute the performance is listed below.

$$E_a = K_E \Phi n \tag{20-1}$$

$$T = K_T \Phi I_a \tag{20-2}$$

$$V_t = E_a + I_a(R_a + R_s) \tag{20-10}$$

$$I_L = I_f + I_a \tag{20-11}$$

The first two equations are identical to those used in generator analysis. Note, however, that the next two equations are modified to account for the fact that for the motor V_t is the applied or source voltage and as such must be equal to the sum of the voltage drops. Similarly the line current is equal to the *sum* rather than the *difference* of the armature current and field currents.

The power-flow diagram, depicting the reversed flow from that which occurs in the generator, is illustrated in Fig. 20-9. The electrical power input $V_t I_L$ originating from the line supplies the field power needed to establish the flux field as well as the armature-circuit copper loss needed to maintain the flow of

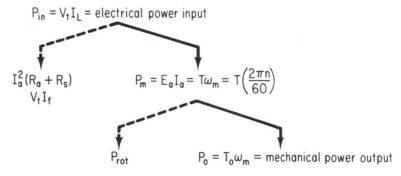

Fig. 20-9 Power-flow diagram of the d-c motor.

I_a. This current flowing through the armature conductors imbedded in the flux field causes torque to be developed. The law of conservation of energy then demands that the electromagnetic power, $E_a I_a$, be equal to $T\omega_m$, where ω_m is the steady-state operating speed. Removal of the rotational losses from the developed mechanical power yields the mechanical output power. The use of the foregoing tools in computing motor behavior is illustrated by the following examples.

EXAMPLE 20-2 A 20-hp 230-V 1150-rpm shunt motor has four poles, four parallel armature paths, and 882 armature conductors. The armature-circuit resistance is 0.188 Ω. At rated speed and rated output the armature current is 73 A and the field current is 1.6 A. Calculate:

    (a) The electromagnetic torque.
    (b) The flux per pole.
    (c) The rotational losses.
    (d) The efficiency.
    (e) The shaft load.

Solution: (a) The torque can be computed from Eq. (20-8), but we first need E_a. Hence

$$E_a = V_t - I_a R_a = 230 - 73(0.188) = 230 - 13.7 = 216.3 \text{ V}$$

Also,

$$\omega_m = \frac{2\pi n}{60} = \frac{2\pi(1150)}{60} = 120 \text{ rad/s}$$

Therefore,

$$T = \frac{E_a I_a}{\omega_m} = \frac{216.3(73)}{120} = 132 \text{ N-m}$$

(b) $E_a = K_E \Phi n = \dfrac{p}{60}\left(\dfrac{Z}{a}\right)\Phi n = \dfrac{4}{60}\left(\dfrac{882}{4}\right)\Phi(1150)$

$$\therefore \Phi = \frac{216.3(60)}{882(1150)} = 0.0128 \text{ Wb}$$

(c) From the power-flow diagram

$$P_{rot} = P_m - P_o = E_a I_a - 20(746) = 15{,}790 - 14{,}920 = 870 \text{ W}$$

(d) First find the sum of the losses. Thus

$$\sum \text{losses} = P_{rot} + I_a^2 R_a + V_t I_f = 870 + (73)^2(0.188) + 230(1.6)$$
$$= 870 + 1002 + 368 = 2240 \text{ W}$$

Hence

$$\eta = 1 - \frac{\sum \text{losses}}{E_a I_a + I_a^2 R_a + V_t I_f} = 1 - \frac{2240}{17{,}160} = 0.869$$

The efficiency is 86.9%.

(e) $T_o = \dfrac{1}{\omega_m} P_o = \dfrac{1}{120}(20 \times 746) = 124 \text{ N-m}$

The difference between this torque and the electromagnetic torque is the amount needed to overcome the rotational losses.

EXAMPLE 20-3 The shaft load on the motor of Example 20-2 remains fixed, but the field flux is reduced to 80% of its value by means of the field rheostat. Determine the new operating speed.

Solution: Information about the speed is available from Eq. (20-1). However, in turn, knowledge of flux and armature induced emf E_a is needed. By the statement of the problem the new flux Φ' is related to the original flux by

$$\Phi' = 0.8\Phi$$

To obtain information about E_a, we must determine the change in I_a, if any. By the constant-torque condition we have

$$K_T\Phi I_a = K_T\Phi' I'_a$$

Hence

$$I'_a = \frac{\Phi}{\Phi'}I_a = \frac{1}{0.8}(73) = 91.3 \text{ A}$$

Consequently,

$$E'_a = V_t - I'_a R_a = 230 - 91.3(0.188) = 212.8 \text{ V}$$

Returning to Eq. (20-1) we can now formulate the ratio

$$\frac{E'_a}{E_a} = \frac{K_E\Phi' n'}{K_E\Phi n}$$

from which the expression for the new operating speed becomes

$$n' = \frac{E'_a}{E_a}\left(\frac{\Phi}{\Phi'}\right)n = \frac{212.8}{216.3}\left(\frac{1}{0.8}\right)1150 = 1414 \text{ rpm}$$

20-3 MOTOR SPEED-TORQUE CHARACTERISTICS. SPEED CONTROL

How does the d-c motor react to the application of a shaft load? What is the mechanism by which the d-c motor adapts itself to supply to the load the power it demands? The answers to these questions can be obtained by reasoning in terms of the performance equations that appear on p. 769. Initially our remarks are confined to the shunt motor, but a similar line of reasoning applies for the others. For our purposes the two pertinent equations are those for torque and current. Thus

and

$$T = K_T\Phi I_a \tag{20-2}$$

$$I_a = \frac{V_t - K_E\Phi n}{R_a} \tag{20-12}$$

Note that the last expression results from replacing E_a by Eq. (20-1) in Eq. (20-10). With no shaft load applied, the only torque needed is that which overcomes the rotational losses. Since the shunt motor operates at essentially constant flux, Eq. (20-2) indicates that only a small armature current is required compared to its rated value to furnish these losses. Equation (20-12) reveals the manner in which the armature current is made to assume just the right value. In this expression V_t, R_a, K_E, and Φ are fixed in value. Therefore the speed is the critical variable. If, for the moment, it is assumed that the speed has too low a value, then the numerator of Eq. (20-12) takes on an excessive value and in turn makes I_a larger than required. At this point the motor reacts to correct the situation. The excessive armature current produces a developed torque which exceeds the opposing torques of friction and windage. In fact this excess serves as an accelerating torque, which then proceeds to increase the speed to that level which corresponds to the equilibrium value of armature current. In other words, the acceleration torque becomes zero only when the speed is at that value which by Eq. (20-12) yields just the right I_a needed to overcome the rotational losses.

Consider next that a load demanding rated torque is suddenly applied to the motor shaft. Clearly, because the developed torque at this instant is only sufficient to overcome friction and windage and not the load torque, the first reaction is for the motor to lose speed. In this way, as Eq. (20-12) reveals, the armature current can be increased so that in turn the electromagnetic torque can increase. As a matter of fact the applied load torque causes the motor to assume that value of speed which yields a current sufficient to produce a developed torque to overcome the applied shaft torque and the frictional torque. Power balance is thereby achieved, because an equilibrium condition is reached where the electromagnetic power, $E_a I_a$, is equal to the mechanical power developed, $T\omega_m$.

A comparison of the d-c motor with the three-phase induction motor indicates that both are *speed-sensitive* devices in response to applied shaft loads. An essential difference, however, is that for the three-phase induction motor developed torque is adversely influenced by the power-factor angle of the armature current. Of course no analogous situation prevails in the case of the d-c motor.

On the basis of the foregoing discussion it should be apparent that the speed-torque curve of d-c motors is an important characteristic. Appearing in Fig. 20-10 are the general shapes of the speed-torque characteristics as they apply for the shunt, cumulatively compounded, and series motors. For the sake of comparison the curves are drawn through a common point of rated torque and speed. An understanding of why the curves take the shapes and relative positions depicted in Fig. 20-10 readily follows from an examination of Eq. (20-1), which involves the speed. For the shunt motor the speed equation can be written as

$$n = \frac{E_a}{K_E \Phi_{sh}} = \frac{V_t - I_a R_a}{K_E \Phi_{sh}} \qquad (20\text{-}13)$$

The only variables involved are the speed n and the armature current I_a. At rated output torque the armature current is at its rated value and so, too, is the speed. As the load torque is removed, the armature current becomes correspondingly smaller, making the numerator term of Eq. (20-13) larger. This results

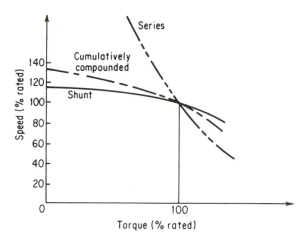

Fig. 20-10 Typical speed-torque curves of d-c motors.

in higher speeds. The extent to which the speed increases depends upon how large the armature circuit resistance drop is in comparison to the terminal voltage. It is usually around 5 to 10%. Accordingly, we can expect the percent change in speed of the shunt motor to be about the same magnitude. This change in speed is identified by a figure of merit called the *speed regulation*. It is defined as follows:

$$\text{percent speed regulation} = \frac{\text{(no-load speed)} - \text{(full-load speed)}}{\text{full-load speed}} 100 \qquad (20\text{-}14)$$

The speed equation as it applies to the cumulatively compounded motor takes the form

$$n = \frac{V_t - I_a(R_a + R_s)}{K_E(\Phi_{sh} + \Phi_s)} \qquad (20\text{-}15)$$

A comparison with the analogous expression for the shunt motor bears out two differences. One, the numerator term also includes the voltage drop in the series-field winding besides that in the armature winding. Two, the denominator term is increased to account for the effect of the series-field flux Φ_s. Starting at rated torque and speed, Eq. (20-14) makes it clear that as load torque is decreased to zero there is an increase in the numerator term which is necessarily greater than it is for the shunt motor. At the same time, moreover, the denominator term decreases because Φ_s reduces to zero as the torque goes to zero. Both effects act to bring about an increase in speed. Therefore the speed regulation of the cumulatively compounded motor is greater than for the shunt motor. Figure 20-10 presents this information graphically.

The situation regarding the speed-torque characteristic of the series motor is significantly different because of the absence of a shunt-field winding. Keep in mind that the establishment of a flux field in the series motor comes about solely as a result of the flow of armature current through the series-field winding. In this connection, then, the speed equation for the series motor becomes

$$n = \frac{E_a}{K_E \Phi_s} = \frac{V_t - I_a(R_a + R_s)}{K_E' I_a} \qquad (20\text{-}16)$$

where K'_E denotes a new proportionality factor which permits Φ_s to be replaced by the armature current I_a. When rated torque is being developed the current is at its rated value. The flux field is therefore abundant. However, as load torque is removed less armature current flows. Now since I_a appears in the denominator of the speed equation, it is easy to see that the speed will increase greatly. In fact, if the load were to be disconnected from the motor shaft, dangerously high speeds would result because of the small armature current that flows. The centrifugal forces at these high speeds can easily damage the armature winding. For this reason a series motor should never have its load uncoupled.

Because the armature current is directly related to the air-gap flux in the series motor, Eq. (20-1) for the developed torque may be modified to read as

$$T = K_T \phi I_a = K'_T I_a^2 \tag{20-17}$$

Thus the developed torque for the series motor is a function of the square of the armature current. This stands in contrast to the linear relationship of torque to armature current in the shunt motor. Of course in the compound motor an intermediate relationship is achieved. It is interesting to note, too, that as the series motor reacts to develop greater torques, the speed drops correspondingly. It is this capability which suits the series motor so well to traction-type loads.

Speed Control. One of the attractive features the d-c motor offers over all other types is the relative ease with which speed control can be achieved. The various schemes available for speed control can be deduced from Eq. (20-13), which is repeated here with one modification:

$$n = \frac{V_t - I_a(R_a + R_e)}{K_E \Phi} \tag{20-18}$$

The modification involves the inclusion of an external armature-circuit resistance R_e. Inspection of Eq. (20-18) reveals that the speed can be controlled by adjusting any one of the three factors appearing on the right side of the equation: V_t, R_e, or Φ. The simplest to adjust is Φ. A field rheostat such as that shown in Fig. 20-8 is used. If the field-rheostat resistance is increased, the air-gap flux is diminished, yielding higher operating speeds. General-purpose shunt motors are designed to provide a 200% increase in rated speed by this method of speed control. However, because of the weakened flux field the permissible torque that can be delivered at the higher speed is correspondingly reduced, in order to prevent excessive armature current.

A second method of speed adjustment involves the use of an external resistor R_e connected in the armature circuit as illustrated in Fig. 20-11. The size and cost of this resistor are considerably greater than those of the field rheostat because R_e must be capable of handling the full armature current. Equation (20-18) indicates that the larger R_e is made, the greater will be the speed change. Frequently the external resistor is selected to furnish as much as a 50% drop in speed from the rated value. The chief disadvantage of this method of control is the poor efficiency of operation. For example, a 50% drop in speed is achieved by having approximately half of the terminal voltage V_t appear across R_e. Accordingly, almost 50% of the line input power is dissipated in the form of heat

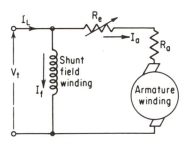

Fig. 20-11 Speed adjustment of a shunt motor by an external armature-circuit resistance.

in the resistor R_e. Nonetheless, armature-circuit resistance control is often used—especially for series motors.

A third and final method of speed control involves adjustment of the applied terminal voltage. This scheme is the most desirable from the viewpoint of flexibility and high operating efficiency. But it is also the most expensive because it requires its own d-c supply. It means purchasing a motor-generator set with a capacity at least equal to that of the motor to be controlled. Such expense is not generally justified except in situations where the superior performance achievable with this scheme is indispensable, as is the case in steel mill applications. Armature terminal voltage control is referred to as the *Ward-Leonard* system.

20-4 SPEED CONTROL BY ELECTRONIC MEANS

The development of reliable thyristors (SCRs) with high current and voltage ratings has made the speed control of d-c motors by armature voltage adjustment more feasible. Keep in mind that adjustment of the armature voltage offers the advantage of speed control below base speed as well as operation at high efficiency coupled with good speed regulation. Moreover, the control of speed can be achieved whether the motor is operating under load conditions or no load.

Single-Phase Half-Wave Thyristor Drive. Figure 20-12 shows the arrangement that is used when a single-phase source is available. It is frequently used for motors which are equipped with permanent magnet fields and rated for 1 hp or less. Because a single thyristor is used in this circuit, control of the firing point, θ_f, by the gate signal can occur anywhere between 0 and π radians. Illustrated in Fig. 20-13 is a typical situation that can prevail for a firing angle of $\theta_f = 90°$ during steady-state conditions. Time in this diagram is assumed to be measured after steady state is reached. Moreover, the load torque to be supplied by the motor is assumed to be at such a level as to require an average armature current of I_a projected over a full cycle. To produce this average armature current, however, it is necessary for the actual variation of the armature current to have such an instantaneous variation that integration over the conduction period yields an average value of I_a. The conduction period is the time during which the thyristor is conducting and generally this is represented by

$$\theta_c = \theta_f - \theta_e \tag{20-19}$$

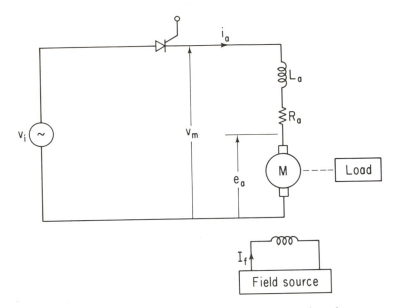

Fig. 20-12 Armature voltage control of a d-c motor through use of a single-phase, half-wave controlled rectifier. The motor may have either a permanent magnetic field or a separate source of excitation.

where θ_f is the angle in the positive excursion of the supply voltage where the gate signal fires the thyristor, and θ_e is the extinction angle of the thyristor.

The extinction angle is determined as that point in the cycle where the actual armature current i_a returns to zero, which in turn depends upon the relationship of the instantaneous value of the supply voltage v_i and the variation of the armature counter emf e_a. This e_a must always assume such a position

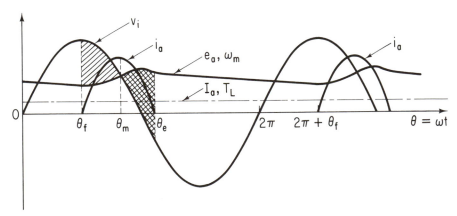

Fig. 20-13 Waveshapes of the various electrical quantities and motor speed and load torque associated with system of Fig. 20-12. The thyristor firing angle is illustrated for $\theta_f = 90°$. Voltage, current, speed, and torque are drawn with different scales.

relative to v_i that it produces a time variation in i_a which yields a value of the average armature current that allows a torque to be developed to meet the needs of the load. It is instructive to note here that the instantaneous armature current i_a undergoes an increase in value from θ_f to θ_m, where θ_m corresponds to the time instant where $v_i = e_a$. In this interval the voltage across the armature winding inductance in Fig. 20-12 is positive. Consequently, an increase in current from zero takes place which can be described by a form of Faraday's law as

$$\Delta i_a = \frac{1}{L_a} \int_{\theta_f}^{\phi_m} (v_i - e_a) \, dt \qquad (20\text{-}20)$$

where L_a denotes the armature winding inductance. Beyond θ_m the quantity $(v_i - e_a)$ becomes negative. By Eq. (20-20) the effect of these negative volt-seconds, appearing across the armature inductance, is to cause the armature current to diminish. In fact, the extinction angle is that point in the cycle where the sum of the negative volt-seconds (i.e., the double-crosshatched area in Fig. 20-13) is equal to the sum of the positive volt-seconds (i.e., the single-crosshatched area).

When the load calls for an increased developed torque by the motor, the motor responds by reducing its speed (and therefore e_a) by an amount that allows a sufficient increase in positive volt-seconds to yield a greater instantaneous current flow, i_a. Correspondingly, a larger average value of I_a is made to exist over a full cycle, thereby meeting the increased torque demand of the load. The variation with time of e_a and the motor speed, ω_m, are represented in Fig. 20-13 by the same curve. It is understood, of course, that the curve is drawn to a different scale. The same applies to the curves marked I_a and T_L. In the interval from θ_f to θ_e observe that the motor is receiving a pulse of energy from the source which manifests itself as a slight increase in speed. The energy stored in the rotating mass, consisting of the rotor of the motor and its attached mechanical load (i.e., the total inertia J), is thereby increased. However, because there is no armature current flow in the interval from θ_e to $(2\pi + \theta_f)$, the motor finds itself coasting along. During this period, energy is delivered to the load by extracting it from the kinetic energy stored in the rotating mass. The result is a dip in speed. When the motor is called on to deliver full load torque at low speeds, the speed dip will be greater than when it operates at speeds close to its base value.

Although small d-c motors do operate successfully in the circuitry of Fig. 20-12, it is not without notable shortcomings. For example, the need to meet load torque requirements in terms of an armature current averaged over a full cycle of the supply frequency means that large pulses of armature current must flow during the conduction period. This leads to much higher heating losses. Unless special arrangements are made for cooling, it may be necessary to operate the motor at a reduced horsepower rating. Another disadvantage is related to the fact that an average d-c current is made to flow through the a-c source. If this source should be a single-phase transformer, it is possible for the transformer to operate in a state of partial saturation. These problems are essentially avoided by operating from a three-phase source.

Chopper Drives.   There is an area of application of d-c motors where d-c energy is available, often in the form of batteries and where operation below base speed is extremely important. This is the case for example with forklifts, golf carts, electrically powered trucks, and even rapid-transit cars. Of course, as pointed out in the preceding section, one workable scheme is to use a series resistor. But we already know that this solution carries with it such serious shortcomings as poor efficiency, poor speed regulation, and high cost. A popular method of providing armature voltage control today in such applications is to use the thyristor in a pulse-width modulation mode or a pulse-frequency modulation mode. In the former the d-c source voltage is switched on and off for a variable part of a cycle in the interest of adjusting the level of average voltage applied to the motor armature winding. In the latter a pulse of voltage of fixed width which is derived from the d-c source is repeated more or fewer times over a given time period. A typical schematic arrangement of the circuitry is shown in Fig. 20-14. Because the principles are the same, attention is directed here to just one of the schemes, namely, pulse-width modulation of the d-c source voltage.

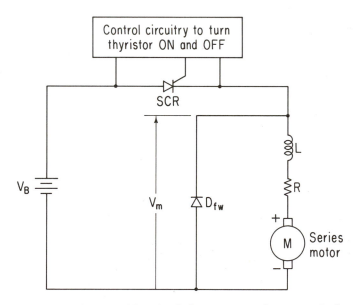

Fig. 20-14   Chopper drive circuit for armature voltage control of a d-c series motor.

The control circuitry that is connected to the thyristor has a twofold purpose. First, it serves to turn the thyristor on over a variable part of a period T_P as illustrated in Fig. 20-15. In the time interval from 0 to T_0 the gate signal puts the thyristor into a conductive state, thereby passing the battery voltage to the armature winding. Second, at time T_0, a reverse voltage is instantaneously (measured in the tens of microseconds) applied to the thyristor, thereby cutting it off. At the end of the period represented by T_P, the process is repeated. In essence, then, the control circuitry acts to "chop" the d-c supply voltage before it is

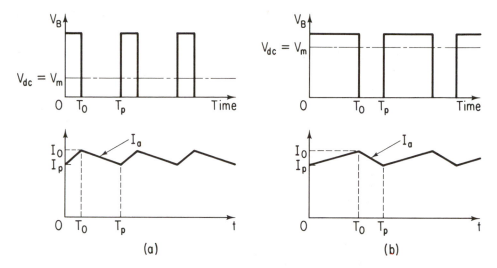

Fig. 20-15 Voltage and current waveshapes appearing at the armature terminals of the motor in Fig. 20-14: (a) thyristor fired for low-speed operation; (b) thyristor fired for high-speed operation.

applied to the motor armature winding. These drive circuits are called *chopper drives*, for obvious reasons. The average value of the d-c voltage that is transferred to the motor is quite simply given by

$$V_{dc} = V_m = \frac{T_0}{T_P} V_B \tag{20-21}$$

where the meaning of the quantities are self-evident from their identities in Fig. 20-15.

The frequency at which the thyristor is pulsed is critical to the design of the control circuitry if large peaks in the armature current are to be avoided. In a properly designed chopper drive circuit, the variation of the armature looks very much like that illustrated in Fig. 20-15 during steady-state operation. For equilibrium to exist the increase in current during time T_0 is equal to the decrease in current during the time $(T_P - T_0)$. In terms of the quantity of volt-seconds applied to the armature inductance, we can write as the increase in current in period T_0:

$$\Delta I_a = \frac{1}{L_a} \int_0^{T_0} [V_B - (V_m - I_a R_a)] \, dt = \frac{[V_B - (V_m - I_a R_a)]T_0}{L_a} \tag{20-22}$$

since the integrand is essentially constant. The motor voltage V_m is given by Eq. (20-21). Of course, the effect of this increased current is to augment the magnetic energy stored in the armature field. Then, following cutoff of the thyristor at time T_0, the energy is released to the motor through the circuit established by the free-wheeling diode D_{fw}. The rate of decay of the armature current during the time $(T_P - T_0)$ is determined by the time constant of the armature circuit, which is often very large compared to the cycle time of the pulses. For this reason the decay is illustrated as a straight line. Because of the important role of the armature inductance in storing sufficient magnetic energy, chopper drives are frequently used with series motors. The series motor has larger armature inductance than the shunt motor, although perfectly acceptable performance can be had when the latter machine is used with an external series inductor.

EXAMPLE 20-4   A d-c motor, which is rated at 550 V and 60 hp, draws 30 A when it delivers rated torque at a speed of 800 rpm. The field is set at a fixed value that gives the motor a torque and speed parameter of 5.94 N-m/A (or volts/rad per second). The armature circuit resistance is 1 Ω. Assume that this motor is to be operated with a chopper drive from a 550-V d-c source.

(a) Find the motor speed when the thyristor is made to operate with a ratio of ON time T_0 to cycle time T_P of 0.4. Assume that the motor delivers rated torque.

(b) Determine the required pulse frequency of the thyristors that serves to limit the total change of armature current during the ON period to 5 A. Assume an armature winding inductance of 0.1 H.

(c) Repeat parts (a) and (b) for $T_0/T_P = 0.7$.

Solution: (a) From Eq. (20-21) the average value of the armature terminal voltage is

$$V_m = \frac{T_0}{T_P} V_B = 0.4(550) = 220 \text{ V}$$

the corresponding value of the induced armature voltage then becomes

$$E_a = V_m - I_a R_a = 220 - (30)(1) = 190 \text{ V}$$

Therefore,

$$\omega_m = \frac{190}{5.94} = 32 \text{ rad/s or 305.4 rpm}$$

(b) The time T_0 is found directly from Eq. (20-22). Thus

$$T_0 = \frac{L_a(\Delta I_a)}{V_B - E_a} = \frac{0.1(5)}{550 - 190} = 1.4 \times 10^{-3} \text{ s}$$

Accordingly,

$$T_P = \frac{T_0}{0.4} = \frac{1.4}{0.4} \times 10^{-3} = 3.5 \times 10^{-3} \text{ s}$$

and this leads to the following required pulses per second:

$$f = \frac{1}{T_P} = \frac{10^3}{3.5} = 286 \text{ Hz}$$

(c) Now Eq. (20-21) yields

$$V_m = 0.7(550) = 385 \text{ V}$$

$$E_a = 385 - 30 = 355 \text{ V}$$

$$\omega_m = \frac{355}{5.94} = 64.81 \text{ rad/s or 618.9 rpm}$$

Also, from Eq. (20-22),

$$T_0 = \frac{(0.1)5}{195} = 2.56 \times 10^{-3}$$

$$T_P = \frac{2.56 \times 10^{-3}}{0.7} = 3.66 \times 10^{-3}$$

$$f = \frac{1}{T_P} = 273 \text{ Hz}$$

20-5 APPLICATIONS OF D-C MOTORS

The d-c motor is often called upon to do the really tough jobs in industry because of its high degree of flexibility and ease of control. These features cannot easily be matched by other electromechanical energy-conversion devices. The d-c motor offers a wide range of control of speed and torque as well as excellent acceleration and deceleration. For example, by the insertion of an appropriate armature-circuit resistance, rated torque can be obtained at starting with no more than rated current flowing. Also, by special design of the shunt-field winding, speed adjustments over a range of 4:1 are readily obtainable. If this is then combined with armature-voltage control, the range of speed adjustment spreads to 6:1. In some electronic control devices that are used to provide the d-c energy to the field and armature circuits, a speed range of 40:1 is possible. The size of the motor being controlled, however, is limited.

Table 20-1 lists some of the salient characteristics and typical applications of the three types of d-c motors. It is interesting to note that the maximum torque in the case of the d-c motor is limited by commutation and not, as with all other motor types, by heating. Commutation refers to the passage of current from the brushes to the commutator and thence to the armature winding itself. The passage from the brushes to the commutator is an arc discharge. Moreover, as a coil leaves a brush the current is interrupted, causing sparking. If the armature current is allowed to become excessive, the sparking can become so severe as to cause flashover between brushes. This renders the motor useless.

Another point of interest in the table is the considerably higher starting torque of the compound motor by comparison with the shunt motor. This feature is attributable to the contribution of the series-field winding. The same comment is valid as regards the maximum running torque. In each case, of course, the limit for the armature current is the same.

20-6 STARTERS AND CONTROLLERS
FOR D-C MOTORS

The limitations imposed by commutation as well as voltage-dip restrictions on the source as set forth by electric utility companies make it necessary to use a starter or controller on all d-c machines whose ratings exceed two horsepower. A glance at Eq. (20-12) discloses that at starting ($n = 0$) the armature current is limited solely by the armature-circuit resistance. Hence if full terminal voltage

TABLE 20-1 CHARACTERISTICS AND APPLICATIONS OF D-C MOTORS†

Type	Starting torque (%)	Max. running torque, momentary (%)
Shunt, constant speed	Medium—usually limited to less than 250 by a starting resistor but may be increased	Usually limited to about 200 by commutation
Shunt, adjustable speed	Same as above	Same as above
Compound	High—up to 450, depending upon degree of compounding	Higher than shunt—up to 350
Series	Very high—up to 500	Up to 400

† By permission from M. Liwschitz-Garik and C. C. Whipple, *Electric Machinery,* Vol. I (Princeton, N.J.: D. Van Nostrand Co., Inc., 1946).

is applied, excessive armature currents will flow. This is especially so where large machines are involved because the armature resistance gets smaller as the rating increases. In addition to limiting the armature current, controllers fulfill other useful functions as described in Sec. 18-8.

Depicted in Fig. 20-16 is a simple line starter which is used for small d-c motors. The operation is straightforward. Pushing the start button energizes the main coil which then closes the main contactors M and the interlock M_a. Note that when the main contactors close, voltage is applied to the armature winding through the starting resistor R and simultaneously to the field winding. This arrangement prevents "shock" starting because some time is needed before the field flux is fully established. Note, too, that this starter has no provisions for removing the starting resistor once the motor has attained its operating speed. In small motors this is of little consequence and it makes for an inexpensive

Speed-regulation or characteristic (%)	Speed control (%)	Typical application and general remarks
5–10	Increase up to 200 by field control; decrease by armature-voltage control	Essentially for constant-speed applications requiring medium starting torque. May be used for adjustable speed not greater than 2:1 range. For centrifugal pumps, fans, blowers, conveyors, woodworking machines, machine tools, printing presses.
10–15	6:1 range by field control, lowered below base speed by armature voltage control	Same as above, for applications requiring adjustable speed control, either constant torque or constant output.
Varying, depending upon degree of compounding—up to 25–30	Not usually used but may be up to 125 by field control	For drives requiring high starting torque and only fairly constant speed; pulsating loads with flywheel action. For plunger pumps, shears, conveyors, crushers, bending rolls, punch presses, hoists.
Widely variable, high at no-load	By series rheostat	For drives requiring very high starting torque and where adjustable, varying speed is satisfactory. This motor is sometimes called the traction motor. Loads must be positively connected, not belted. For hoists, cranes, bridges, car dumpers. To prevent overspeed, lightest load should not be much less than 15 to 20% of full-load torque.

starter. However, the starter resistor in this case does serve another purpose. When the motor is stopped, the field winding is disconnected from the line. The energy stored in the magnetic field then discharges through the starting resistor, preventing possible damage to the field winding.

Three types of magnetic controllers are in use today for controlling the

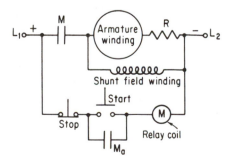

Fig. 20-16 Line starter for low-hp d-c motor.

starting current, the starting torque, and the acceleration characteristics of d-c motors. One type is the *current-limit* controller, which works on the principle of keeping the current during the starting period between specified minimum and maximum limits. It is not too commonly used because its success depends upon a tricky electrical interlock arrangement which must be kept in excellent operating condition at all times. No further consideration is given to this type. The second type is the *counter-emf* type which is illustrated in Fig. 20-17. Two kinds of relays are used in this controller. One is a light, fast-acting unit called an accelerating relay (AR). The other is the strong, heavy-duty type previously discussed. The accelerating relays appearing across the armature circuit in Fig. 20-17 are voltage-sensitive devices. They are designed to close when the voltage across the coil exceeds a preset value. For the controller under discussion the accelerating relay 1AR is usually adjusted to "pick up" at 50% of line voltage and 2AR is adjusted to close at 80 per cent of line voltage.

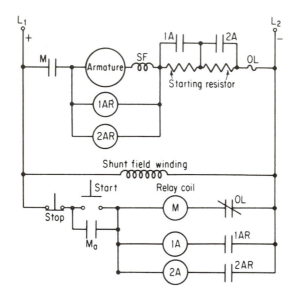

Fig. 20-17 Counter-emf magnetic controller for a d-c motor.

Pressing the start button energizes coil *M* which then closes interlock M_a and the main contactors *M*. Since the armature is initially stationary, the accelerating relays are deenergized so that both steps of the starting resistor are in the circuit. As the armature gains speed and develops an induced emf exceeding 50% of line voltage, coil 1AR snaps closed, closing contactors 1AR. In turn, coil 1A is energized, closing contactors 1A, which short out the first section of the starting resistor. This then applies increased voltage to the armature, which furnishes further acceleration. When the armature induced emf exceeds 80% of line voltage, accelerating relay 2AR closes. This excites coil 2A, which shorts out the second section of the starting resistor. The motor then accelerates to its full-voltage operating speed.

The counter-emf controller has the advantage of providing a contactor closing sequence which adjusts itself automatically to varying load conditions.

Furthermore, this is accomplished in a manner that maintains uniform accelerating current and torque peaks. There can be no question about the desirability of such starting performance; however, there is one disadvantage. The contactor closing sequence is based on the assumption that the motor will start on the first step. If it fails to do so, all subsequent operations cannot take place. Furthermore, the starting resistor is in danger of burning up. To avoid such occurrences general-purpose d-c motors are most often equipped with *definite time-limit controllers*. Figure 20-18 shows the schematic diagram of such a controller. After a relay coil is energized, the corresponding contactors do not close until the elapse of a preset time delay. The time delay is achieved either by means of magnetic flux decay, a pneumatic device, or a mechanical escapement.

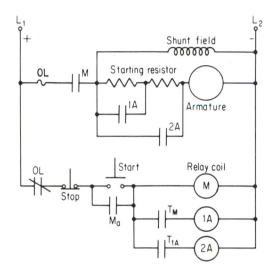

Fig. 20-18 Definite time-limit controller for d-c motor.

Pressing the start button energizes coil M and closes interlock M_a and the main contactors M, thereby applying voltage to the armature winding through the starting resistor. A definite time after the armature of relay coil M closes, contactors T_M in the control circuit close, regardless of whether the rotor is turning. This energizes coil 1A which closes contactors 1A, thus shorting out the first section of the starting resistor. At a preset time delay after coil 1A is energized, contactors T_{1A} close. Coil 2A becomes energized, removing the entire starting resistor. The motor then assumes its normal operating speed.

In the definite time-limit controller the preset time intervals between the closing of contactors are adjusted to obtain smooth acceleration and uniform currents peaks for average load conditions. If a heavy starting condition occurs and the motor fails to start on the first step, the first accelerating (time-delay) contactors close anyway. This allows an increased starting torque to be developed. Thus the motor is made to "work harder" if it does not start on the first step. Accordingly, whenever a controller must be selected for a general-purpose motor, it is wiser to prescribe the definite time-limit type. The reason is that as a rule for general-purpose applications the starting conditions are not well known.

1. Describe the difference between the separately excited shunt generator and the self-excited one.

2. Explain the process of voltage buildup in a self-excited shunt generator.

3. What is the critical field resistance of the shunt generator, and why is it important to know?

4. Describe the role played by the field winding resistance in determining the value of the no-load voltage of a d-c shunt generator.

5. What is a compound d-c generator? What is the purpose of such a compounding mode of operation.

6. Draw the equivalent circuit of the compound d-c generator and explain the meaning of each circuit element and electrical quantity shown.

7. Draw the power flow diagram of the compound d-c generator.

8. Cite the basic torque and voltage formula that can be used both for the study of the motor and generator modes of the d-c machine.

9. Draw the equivalent circuit and the power flow diagram of the compound d-c motor. Explain the meaning of all terms.

10. What is a series d-c motor, and how does it differ from a cumulatively compounded d-c motor?

11. Why is it dangerous to uncouple the mechanical load of a series d-c motor?

12. Describe the mechanism by which the d-c motor accommodates the demand for increased power to be delivered at the shaft to a mechanical load.

13. Both the d-c motor and the induction motor are described as speed-sensitive devices because of the way they respond to increased load demands. Explain this statement.

14. A cumulatively compounded motor is operating at rated horsepower. Explain what happens to the speed as the load torque is reduced to zero.

15. Define *speed regulation*.

16. Which of the following three motors has the poorest speed regulation: shunt motor, series motor, or cumulative compounded motor? Explain.

17. Which of these two motors has the poorer speed regulation: the shunt motor or the cumulatively compounded motor? Explain.

18. Why is the control of the speed of a d-c motor easier to achieve than it is for an induction motor?

19. List the conventional nonelectronic methods of speed control that are available for the d-c motor.

20. Describe how speed control of a d-c motor is obtained using field control and cite the advantages and shortcomings of this scheme.

21. Repeat Question 20 for the use of an external armature resistor.

22. Repeat Question 20 for the case where a variable d-c voltage source is used.

23. Describe the manner in which the speed of a series d-c motor can be achieved electronically through use of a silicon-controlled rectifier (thyristor).

24. Distinguish between the firing angle and extinction angle as it relates to a single-phase half-wave thyristor drive for a series d-c motor.

25. Explain the operation of electronic chopper drives for the speed control of a series d-c motor. What is the role of the thyristor in these chopper drives?

26. What design factor limits the maximum torque of a d-c motor? Of an a-c motor?

27. List some of the features of the d-c motor that makes it an outstanding choice for the really tough jobs in industry.

28. Why is it necessary to use line starters and controllers in the operation of d-c motors?

29. Describe the basic principle on which the counter emf magnetic controller of a d-c motor is based.

30. Describe the basic principle that underlies the operation of a definite time-limit controller for a d-c motor.

Problems

GROUP I

20-1. Explain your answer to each part:
   (a) Can a separately excited d-c generator operate below the knee of its magnetization curve?
   (b) Can a d-c shunt generator operate below the knee of its magnetization curve?

20-2. The magnetization curve for a d-c shunt generator driven at a constant speed of 1000 rpm is shown in Fig. 20-7. Each vertical division represents 20 V; each horizontal division represents 0.2 A.
   (a) Compute the critical field resistance.
   (b) What voltage is induced by the residual flux of this machine?
   (c) What must be the resistance of the field circuit in order that the no-load terminal voltage be 240 V at a speed of 1000 rpm?
   (d) Determine the field current produced by the residual flux voltage when the field circuit resistance has the value found in part (c).
   (e) At what speed must the generator be driven in order that it would fail to build up when operating with the field circuit resistance of part (c)?

20-3. A d-c shunt generator has a magnetization curve given by Fig. 20-7, where each ordinate unit is made equal to 20 V and each abscissa unit is set equal to 0.2 A and the speed of rotation is 1000 rpm. The field circuit resistance is 156 Ω.

   Determine the voltage induced between brushes when the generator is operated at a reduced speed of 800 rpm.

20-4. The no-load characteristic of a 10-kW 250-V d-c self-excited shunt generator driven at 1000 rpm is shown in Fig. 20-7. Each vertical division denotes 20 V and each horizontal division represents 0.2 A. The armature circuit resistance is 0.3 Ω and the field circuit (shunt field winding plus field rheostat) resistance is set at 110 Ω. Determine:
   (a) The critical resistance of the shunt field circuit.
   (b) The voltage regulation.
   (c) The shunt field current under rated load conditions.

(d) The approximate no-load speed at which this machine would run when connected to a 220-V line as a motor.

20-5. A 10-hp, 230-V shunt motor has an armature circuit resistance of 0.5 Ω and a field resistance of 115 Ω. At no-load and rated voltage the speed is 1200 rpm and the armature current is 2 A. If load is applied, the speed drops to 1100 rpm. Determine:

(a) The armature current and the line current.

(b) The developed torque.

(c) The horsepower output assuming that the rotational losses are 500 W.

20-6. A 20-hp 230-V 1150-rpm four-pole d-c shunt motor has a total of 620 conductors arranged in two parallel paths and yielding an armature circuit resistance of 0.2 Ω. When it delivers rated power at rated speed, the motor draws a line current of 74.8 A and a field current of 3 A. Compute:

(a) The flux per pole.

(b) The developed torque.

(c) The rotational losses.

(d) The total losses expressed as a percentage of the rated power.

20-7. A 250-V 50-hp 1000-rpm d-c shunt motor drives a load that requires a constant torque regardless of the speed of operation. The armature circuit resistance is 0.04 Ω. When this motor delivers rated power, the armature current is 160 A.

(a) If the flux is reduced to 70% of its original value, find the new value of armature current.

(b) What is the new speed?

20-8. When a 250-V 50-hp 1000-rpm d-c shunt motor is used to supply rated output power to a constant-torque load, it draws an armature current of 160 A. The armature circuit has a resistance of 0.04 Ω and the rotational losses are equal to 2 kW. An external resistance of 0.5 Ω is inserted in series with the armature winding. For this condition compute:

(a) The speed.

(b) The developed power.

(c) The efficiency assuming that the field loss is 1.6 kW.

20-9. A d-c shunt motor has the magnetization curve shown in Fig. 20-7, where one vertical unit represents 20 V and one horizontal unit represents 2 field amperes. The armature circuit resistance is 0.05 Ω. At a specified load and a speed of 1000 rpm the field current is found to be 12 A when a terminal voltage of 245 V is applied to the motor. The rotational losses are 2.5 kW.

(a) Find the armature current.

(b) Compute the developed torque.

(c) What is the efficiency?

20-10. Depicted in Fig. P20-10 is the reversing controller circuitry for a d-c shunt motor. Identify the unmarked armature contacts and explain the reversing operation. Explain, too, why the auxiliary interlock contacts (F_a and R_a) are placed in series with the REV and FWD switches.

20-11. A current-limit type controller for a d-c shunt motor is shown in Fig. P20-11. The accelerating relays (AR) respond to armature current. These relays have a light-weight armature so that they easily pick up and open their contacts whenever the armature current exceeds the rated value. By means of a step-by-step procedure describe the operation of this controller. Be careful to identify the proper closing and opening sequence wherever two or more contactors appear in series.

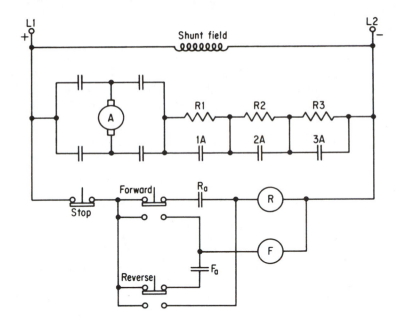

Fig. P20-10

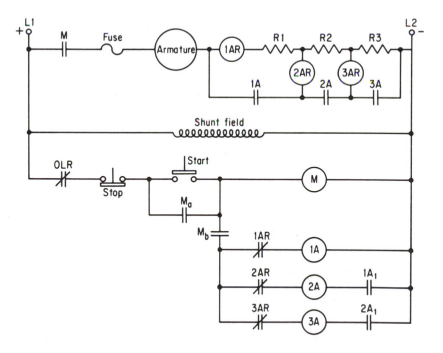

Fig. P20-11

GROUP II

20-12. It is desired to reverse the terminal voltage polarity of a generator that has been operating properly as a cumulative compound generator. The machine is stopped. Residual magnetism is reversed by temporarily disconnecting the shunt field and separately exciting it with reversed current. The connections are then restored exactly as they were before.

(a) Does the terminal voltage build up? Explain.

(b) If answer to part (a) is yes, will the compounding be cumulative or differential? Explain.

20-13. A 230-V 50-hp d-c shunt motor delivers power to a load drawing an armature current of 200 A and running at a speed of 1100 rpm. The magnetization curve is given by Fig. 20-7, where each vertical unit represents 20 V and each horizontal unit represents 2 A.

(a) Find the value of the armature induced emf at this load condition.

(b) Compute the motor field current.

(c) Compute the value of the load torque. The rotational losses are 600 W.

(d) At what efficiency is the motor operating?

(e) At what percentage of rate power is it operating?

20-14. Refer to Prob. 20-13 and assume the load is reduced so that an armature current of 75 A flows.

(a) Find the new value of speed.

(b) What is the new horsepower being delivered to the load?

20-15. A 230-V 10-hp d-c series motor draws a line current of 36 A when delivering rated power at its rated speed of 1200 rpm. The armature circuit resistance is 0.2 Ω and the series field winding resistance is 0.1 Ω. The magnetization curve may be considered linear.

(a) Find the speed of this motor when it draws a line current of 20 A.

(b) What is the developed torque at the new condition?

(c) How does this torque compare with the original value? Why?

20-16. A 25-hp 250-V d-c motor has an armature circuit resistance of 0.1 Ω and a developed torque parameter of 6.25 N-m/A. This motor is driven by the chopper circuitry of Fig. 20-14 from a 250-V d-c source. The thyristor control circuitry is arranged to provide a ratio of on-time to cycle time of 0.5.

(a) Find the operating speed when rated torque is delivered by the motor.

(b) Determine the no-load speed and the speed regulation for the specified switching ratio.

(c) Repeat parts (a) and (b) for $T_0/T_p = 1.0$.

20-17. The motor of Prob. 20-16 is driven by a chopper drive of a 250-V source where the ratio of on-time to cycle time is 0.5. Determine the required pulse frequency of the thyristors in order that the change in armature current during the *on* period be limited to 5 A when operating at rated load. The armature inductance is 0.05 H.

20-18. Repeat Prob. 20-17 for the case where the on-time to cycle time is 0.8.

chapter twenty-one

Single-Phase Induction Motors

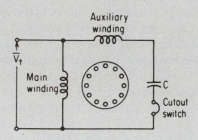

By far the vast majority of single-phase induction motors are built in the fractional-horsepower range. Single-phase motors are found in countless applications doing all sorts of jobs in homes, shops, offices, and on the farm. An inventory of the appliances in the average home in which single-phase motors are used would probably number a dozen and more. An indication of the volume of such motors can be had from the fact that the sum total of all fractional-horsepower motors in use today far exceeds the total of integral horsepower motors of all types.

The treatment of the single-phase motor as developed in the following pages is concerned chiefly with their method of operation, the classification and characteristics of the various types, and their typical applications. The analysis of the performance of such motors is beyond the scope of this book.

21-1 HOW THE ROTATING FIELD IS OBTAINED

In its pure and simple form the single-phase motor usually consists of a distributed stator winding (not unlike one phase of a three-phase motor) and a squirrel-cage rotor. The a-c supply voltage is applied to the stator winding, which in turn creates a field distribution. Since there is a single coil carrying an alternating current, a little thought reveals that the air-gap flux is characterized by being fixed in space and pulsating in magnitude. If hysteresis is neglected, the flux is a maximum when the current is instantaneously a maximum and it is zero when the current is zero. Such an arrangement gives the single-phase motor no starting torque. To understand this in terms of the concepts discussed in Sec. 17-1 refer to Fig. 21-1. As a matter of convenience the distributed single-phase winding is represented by a coil wrapped around protruding pole pieces; some motors do in fact use this configuration. Assume instantaneously that the flux-density wave is increasing in the upward direction as shown. Then by transformer action a voltage is induced in the rotor having that distribution which enables the corresponding rotor mmf to oppose the changing flux. To accomplish this, current flows out of the right-side conductors and into the left-side conductors as illustrated in Fig. 21-1. Note that this resulting ampere-conductor distribution corresponds to a space phase angle of $\psi = 90°$. By Eq. (17-39) the net torque is therefore zero. Of course what this means is that beneath each pole piece there are as many conductors producing clockwise torque as there are producing counterclockwise torque. This condition, however, prevails only at standstill. If by some means the rotor is started in either direction, the motor will develop a nonzero net torque in that direction and thereby cause the motor to achieve normal speed. The problem therefore is to modify the configuration of Fig. 21-1 in such a way that it imparts to the rotor a nonzero starting torque.

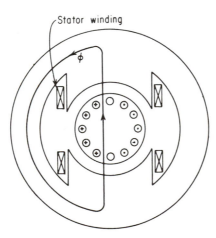

Fig. 21-1 Simplified diagram of the single-phase motor.

The answer to this problem lies in so modifying the motor that it closely approaches the conditions prevailing in the *two-phase* induction motor. In accordance with the general results developed in Sec. 18-1 we know that to obtain a revolving field of constant amplitude and constant linear velocity in a two-phase induction motor two conditions must be satisfied. One, there must exist two coils (or windings) whose axes are space-displaced by 90 electrical degrees. Two, the currents flowing through these coils must be time-displaced by 90 electrical degrees and they must have magnitudes such that the mmf's are equal. If the currents are less than 90° apart in time but greater than 0°, a rotating field will still be developed but the locus of the resultant flux vector will be an ellipse rather than a circle. Hence in such a case the linear velocity of the field varies from one point in time to another. Also, if the currents are 90° apart but the mmf's of the two coils are unequal, an elliptical locus for the rotating field again results. Finally, if the currents are neither 90° time-displaced nor of a magnitude to furnish equal mmf's, a rotating magnetic field will continue to be developed but now the locus will be more elliptical than in the previous cases. However, the important aspect of all this is that a revolving field can be so obtained even if its amplitude is not constant during its time history, and satisfactory performance can be achieved with such a revolving field. Of course such performance items as power factor and efficiency will be poorer than for the ideal case, but this is not too serious because the motors are of relatively small power.

Appearing in Fig. 21-2 is the schematic diagram which shows the modifications needed to give the single-phase motor a starting torque. A second winding called the *auxiliary* winding is placed in the stator with its axis in quadrature with that of the main winding. Usually the main winding is made to occupy two-thirds of the stator slots and the auxiliary winding is placed in the remaining one-third. In this way the space-displacement condition is met exactly. The time displacement of the currents through the two windings is obtained at least partially by designing the auxiliary winding for high resistance and low leakage reactance. This is in contrast to the main winding, which has low resistance and higher leakage reactance. Figure 21-3 depicts the time displacement existing between the auxiliary winding current $\bar{I}_A$ and the main winding current $\bar{I}_M$ at standstill. Frequently in motors of this design the $\bar{I}_A$ and $\bar{I}_M$ phasors are displaced by about 45° in time. Thus with the arrangement of Fig. 21-2 a revolving field results and so the motor achieves normal speed. Because of the high-resistance character of the auxiliary winding, this motor is called the *resistance-start split-phase* induction motor. Also, the auxiliary winding used in these motors has a short time power rating and therefore must be removed from the line once the operating speed is reached. To do this a cut-out switch is placed in the auxiliary winding circuit which, by centrifugal action, removes the auxiliary winding from the line when the motor speed exceeds 75% of synchronous speed.

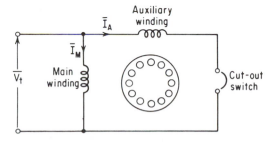

Fig. 21-2 Schematic diagram of the resistance split-phase motor.

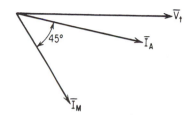

Fig. 21-3 Phasor diagram showing the time-phase displacement between the auxiliary and main winding currents at standstill.

21-2 THE DIFFERENT TYPES OF SINGLE-PHASE MOTORS

Many different types of single-phase motors have been developed primarily for two reasons. One, the torque requirements of the appliances and applications in which they are used vary widely. Two, it is desirable to use the lowest-priced motor that will drive a given load satisfactorily. For example, a high-torque version of the split-phase induction motor just discussed is designed almost exclusively for washing-machine applications. This motor is available in a single speed rating and in just two horsepower ratings. This, coupled with the large volume of sales, enables it to be the least expensive motor in its category. See the second line of Table 21-1 for more details.

The Capacitor-Start Induction-Run Motor. The chief difference between the various types of single-phase motors lies in the method used to start them. In the case of this motor a capacitor is placed in the auxiliary winding circuit so selected that it brings about a 90° time displacement between $\bar{I}_A$ and $\bar{I}_M$. See Fig. 21-4. The result is a much larger starting torque than is achievable with

TABLE 21-1 CHARACTERISTICS AND APPLICATIONS OF SINGLE-PHASE A-C MOTORS†

Type designation	Starting torque (% of normal)	Approx. comparative price	Breakdown torque (% of normal)	Starting current at 115v
General-purpose split-phase motor	90–200 Medium	85%	185–250 Medium	23 1/4 hp
High-torque split-phase motor	200–275 High	65%	Up to 350	32 High 1/4 hp
Permanent-split capacitor motor	60–75 Low	155%	Up to 225	Medium
Permanent-split capacitor motor	Up to 200 Normal	155%	260	
Capacitor-start general-purpose motor	Up to 435 Very high	100%	Up to 400	
Capacitor-start capacitor-run motor	380 High	190%	Up to 260	
Shaded-pole motor	50	—	150	

† By permission from M. Liwschitz-Garik and C. C. Whipple, *Electric Machinery,* Vol. II (Princeton, N.J.: D. Van Nostrand Co., Inc., 1946).

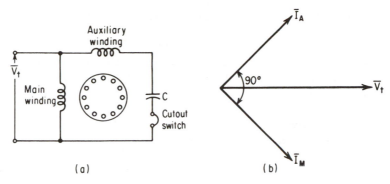

Fig. 21-4 Capacitor-start induction-run motor: (a) schematic diagram; (b) phasor diagram at starting.

resistance split-phase starting. This motor is widely used for general-purpose applications.

The Capacitor-Start Capacitor-Run Motor. As Fig. 21-5 shows, two capacitors are used in the auxiliary circuit of this motor. By keeping one capacitor in during

Power factor (%)	Efficiency (%)	Hp range	Application and general remarks
56–65	62–67	1/20 to 3/4	Fans, blowers, office appliances, food-preparation machines. Low- or medium-starting torque, low-inertia loads. Continuous-operation loads. May be reversed.
50–62	46–61	1/6 to 1/3	Washing machines, sump pumps, home workshops, oil burners. Medium- to high-starting torque loads. May be reversed.
80–95	55–65	1/20 to 3/4	Direct-connected fans, blowers, centrifugal pumps. Low-starting-torque loads. Not for belt drives. May be reversed.
80–95	55–65	1/6 to 3/4	Belt-driven or direct-drive fans, blowers, centrifugal pumps, oil burners. Moderate-starting-torque loads. May be reversed.
80–95	55–65	1/8 to 3/4	Dual voltages. Compressors, stokers, conveyors, pumps. Belt-driven loads with high static friction. May be reversed.
80–95	55–65	1/8 to 3/4	Compressors, stokers, conveyors, pumps. High-torque loads. High power factor. Speed may be regulated.
30–40	30–40	1/300 to 1/20	Fans, toys, hair dryers, unit heaters. Desk fans. Low-starting-torque loads.

normal operation, improved performance is obtained because the motor then behaves more like the balanced two-phase motor. The improved performance is manifested in terms of less noise and higher efficiency and power factor. A second capacitor is needed at starting because the reactive component of the input impedance of the auxiliary circuit is considerably different at standstill than at full speed.

The Permanent-Split Capacitor Motor. In this motor a single capacitor is used both for starting and for running. To take advantage of improved running performance, the value of the capacitor used is that needed at full speed. Consequently, starting torque must be sacrificed. The schematic wiring diagram appears in Fig. 21-6.

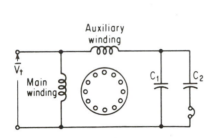

Fig. 21-5 Capacitor-start capacitor-run motor.

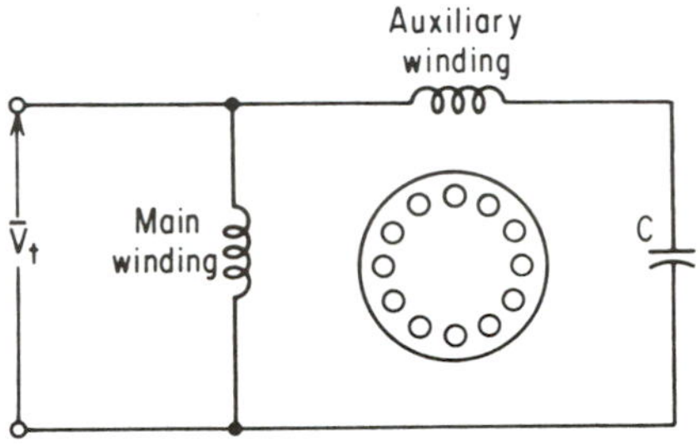

Fig. 21-6 Permanent-split capacitor motor.

The Shaded-Pole Motor. The shaded-pole motor is very extensively used in applications that require 1/20 hp or less. The construction is extremely rugged, as Fig. 21-7 reveals. There is little that can go wrong with this motor aside from overheating. Note that it contains no cut-out switch, which could be a source of trouble, nor does it have an auxiliary winding, which could burn up—especially if the cut-out switch became faulty. The shaded-pole motor consists essentially of copper wound on iron and not much more.

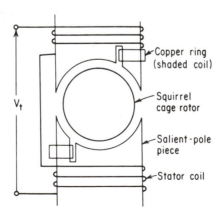

Fig. 21-7 Construction features of the shaded-pole motor.

The manner of obtaining a moving-flux field, however, is different in this motor than it is in those considered previously. Here use is made of a copper ring imbedded in each salient-pole piece. As the gap flux is changing in response to the alternating coil current, transformer action causes currents to be induced in the copper rings, always so directed as to oppose the changing flux. The net effect of this action is that for one portion of the cycle the ring currents cause the flux to concentrate in that part of the pole piece which is free of the ring. At a subsequent portion of the cycle the ring currents act to crowd the gap flux through that part of the pole piece around which the ring is wrapped. In this fashion the air-gap flux undergoes a sweeping motion across the pole face. It appears to be moving from the unshaded to the shaded (or ring) portion of the pole. The sweeping action of the flux occurs periodically, thereby producing a starting as well as a running torque. The starting torque is normally 50% of rated value. The breakdown torque is also relatively low.

21-3 CHARACTERISTICS AND TYPICAL APPLICATIONS

The principal performance features as well as typical applications of single-phase a-c motors appear in Table 21-1. Note that for applications involving less than $\frac{1}{20}$ hp the shaded-pole motor is invariably used. On the other hand, for applications involving $\frac{1}{20}$ to $\frac{3}{4}$ hp the choice of the motor depends upon such factors as the starting and breakdown torque and even quietness of operation. Where a minimum of noise is desirable and low starting torque is adequate, such as for driving fans and blowers, the motor to choose is the permanent-split type. If quietness is to be combined with high starting torque, as might be the case where a compressor drive is needed, then the suitable choice would be the capacitor-start, capacitor-run motor. Of course, in circumstances where the compressor is located in a noisy environment, clearly the choice should be the capacitor-start induction-run motor because it is less expensive.

Summary review questions

1. Explain why a single-phase motor has no starting torque.
2. State the ideal conditions for torque production in the two-phase induction motor and discuss how these conditions can be approached in the single-phase induction motor.
3. What is a resistance split-phase induction motor? What is the function of the auxiliary winding? Why is a cutout switch placed in series with the auxiliary winding?
4. What is the locus of the revolving field vector of the resistance split-phase induction motor? Explain why the locus is not circular.

5. The capacitor-start induction run motor has a much higher starting torque than the resistance split-phase motor. Explain.

6. What is a permanent-split capacitor motor? Its starting torque is lower than that of the resistance split-phase motor. Explain.

7. Describe the construction features of the shaded-pole induction motor and explain how a starting torque is produced.

8. The shaded-pole motor is said to be a rugged, low-cost motor. Why is this?

9. When quietness of operation is essential in a fractional horsepower operation, which single-phase induction motor type is preferred? Why?

10. Under what circumstances is the capacitor-start general-purpose motor the motor of choice?

Problems

GROUP I

21-1. Two field coils are space displaced by 45 electrical degrees. Moreover, sinusoidal currents that are time displaced by 45 electrical degrees flow through these windings. Will a revolving flux result? Explain.

21-2. Two coils are placed in a low reluctance magnetic circuit with their axes in quadrature—the first along the vertical and the second along the horizontal. Sinusoidal currents that are time displaced by 90 electrical degrees and of the same frequency ω are made to flow through the coils. Moreover, the mmf of the first coil is twice that of the second.
   (a) Assuming that at time ωt_0 the mmf of the first coil is at its positive peak value, determine the relative magnitude and direction of the resultant mmf.
   (b) Repeat part (a) for a time ωt_1 that is 45° later than ωt_0.
   (c) Repeat part (a) for a time ωt_2 that is 90° later than ωt_0.
   (d) What conclusion can you draw from parts (a), (b), and (c)?

21-3. The two coils of Prob. 21-2 are now energized with sinusoidal currents that are 45 electrical degrees out of phase but yield equal mmfs.
   (a) Assuming that at time ωt_0 the mmf of the first coil is at its positive peak value, find the relative magnitude and direction of the resultant mmf.
   (b) Repeat part (a) for a time ωt_1 that is 45° later than ωt_0.
   (c) Repeat part (a) for a time ωt_2 that is 90° later than ωt_0.
   (d) What conclusion can you draw from parts (a), (b), and (c)?

GROUP II

21-4. In Fig. P21-4, I, II, and III denote the three coils of a balanced three-phase winding which are space displaced by 120° from one another. If a single-phase voltage is applied to the three-phase squirrel-cage motor in the manner shown, will the motor operate? Justify your answer.

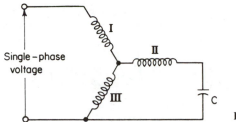

Fig. P21-4

21-5. At standstill, the currents in the main and auxiliary windings of a single-phase induction motor are given by $\bar{I}_M = 10\underline{/0°}$ and $\bar{I}_A = 5\underline{/60°}$. There are five turns per pole on the auxiliary winding for every four turns on the main winding. Moreover, these windings are in space quadrature.

What should be the magnitude and phase of the current in the auxiliary winding to produce an mmf wave of constant amplitude and velocity?

chapter twenty-two

Stepper Motors

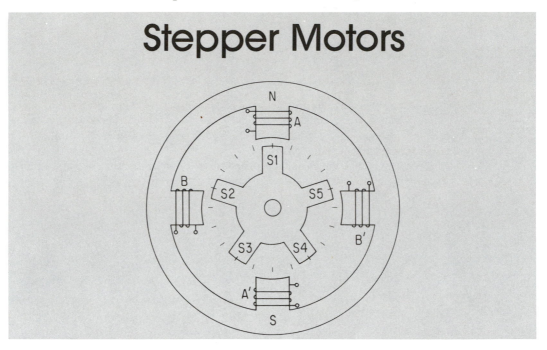

The explosive growth of the computer industry in recent years has also meant an enormous growth for stepper motors because these motors provide the driving force in many computer peripheral devices. They can be found, for example, driving the paper-feed mechanism in line printers and printing terminals. Stepper motors are also used exclusively in floppy disk drives, where they provide precise positioning of the magnetic head on the disks. The x- and y-coordinate pens in expensive plotters are driven by stepper motors. Here they offer good dynamic performance while eliminating the need for a heavy maintenance schedule associated with alternative drive schemes that employ gears, slidewires, and brushes. The stepper motor is especially suited for such applications because essentially it is a device that serves to convert input information in *digital form* to an output that is mechanical. It thereby provides a natural interface with the digital computer.

The stepper motor, however, can be found performing countless tasks outside the computer industry as well. In many commercial, military, and medical applications, the stepper motors perform such functions as mixing, cutting, stirring, metering, blending, and purging. They provide many supporting roles in the manufacture of packaged foodstuffs, commercial end products, and even the production of science-fiction movies. The advantages offered by the stepper motor in these applications can easily be exploited with the use of two pieces of auxiliary equipment: the microprocessor and controlled switches in the form of transistors.

Although the stepper motor is driven by a properly sequenced set of digital signals, the motor itself exhibits the characteristics of a synchronous motor. In one of its most popular forms, it acts as a doubly excited machine, which in turn means that it produces a steady-state torque at one speed.

22-1 CONSTRUCTION FEATURES

The construction of the stepper motor is quite simple. It consists of a slotted stator equipped with two or more individual coils and a rotor structure that carries no winding. The classification of the stepper motor is determined by how the rotor is designed. If it is provided with a permanent magnet attached to its shaft, it is called a *permanent-magnet* (PM) *stepper motor*. If the permanent magnet is not included, it is simply classified as a *reluctance-type stepper motor*. Attention here is directed first to the PM stepper motor. Of course, the presence of the permanent magnet furnishes the motor with the equivalent of a constant d-c excitation. Thus when one or more of the stator coils are energized, the machine behaves as a synchronous motor.

The basic construction details of the PM stepper motor are illustrated in Fig. 22-1 for a four-pole stator and a five-pole rotor structure. Observe that the action of the permanent magnet on the particular orientation of the rotor structure is to magnetize each of the poles (or slots) at one end of the rotor with a N-pole polarity and each of the poles at the other end with a S-pole polarity. Moreover, the N-pole set of rotor poles is arranged to be displaced from the S-pole set of rotor poles by one-half of a pole pitch and is readily evident by a comparison of Figs. 22-1(b) and (c). The symmetry that exists between the stator poles and each set of rotor poles makes it apparent that each rotor pole set behaves in an identical fashion. For example, if coil A-A' is assumed to be energized to yield a N-pole polarity at A and an S-pole polarity at A', the relationship between N of the stator and S1, S2, and S5 of the rotor in Fig. 22-1(b) corresponds exactly to that of S of the stator and N1, N2, and N5 of the rotor in Fig. 22-1(c). A similar statement can be made for the remaining sets of poles.

22-2 METHOD OF OPERATION

The manner in which the PM stepper motor can be used to perform precise positioning is explained by examining the sequence of diagrams that appear in Fig. 22-2. The starting point is represented in Fig. 22-2(a) with coil A-A' energized to carry positive d-c current so that the upper stator pole is given a N-pole polarity and the corresponding lower stator pole is made to behave with a S-pole polarity. Coil B-B' is assumed to be deenergized. The rotor maintains this orientation as long as the stator coil currents remain unchanged. At this position, the net torque developed by the stepper motor is zero. This conclusion is readily

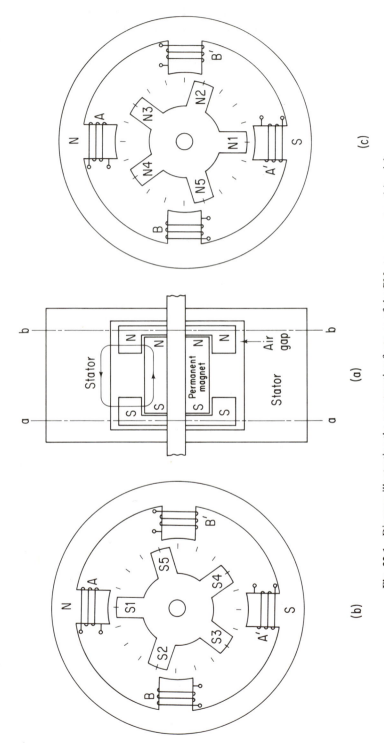

Fig. 22-1 Diagram illustrating the construction features of the PM stepper motor: (a) axial view showing the permanent magnet attached to rotor shaft; (b) cross-sectional view at a-a shows rotor poles of S-pole polarity; (c) cross-sectional view at b-b showing poles of N-pole polarity.

apparent in this case from symmetry considerations. For example, at the S-pole end of the rotor it is seen that S1 is in alignment with the N-pole of the stator. The torque produced by the action of the field distributions of the stator pole and the rotor pole S1 is dependent on the sine of the angular displacement of these two flux fields. With exact alignment the angle is clearly zero and so is the torque between these two fields. Proceeding on the basis of superposition, we note that a torque of attraction exists between the stator N-pole field and the field associated with S5, which acts to cause counterclockwise (ccw) rotation. The magnitude of this torque is proportional to the sin 72°, which is the rotor pole-pitch angle. However, the action between the N-pole stator field and S2 is also one of attraction but it is exactly equal and opposite to that produced between the stator N-pole and S5. A similar result is obtained upon examining the situation that exists under the S-pole stator field. It is useful to note here that a difference in detail does exist in that the torques involved are of a repulsive nature rather than attractive as occurs beneath the N-pole stator field for this orientation. Because the repulsive force has its maximum effect when a rotor flux field is in exact alignment with a stator flux field, the cosine of the displacement angle between the axes of two flux fields of opposite angle must be used. Accordingly, there is a torque existing between the stator S-pole and the rotor S3 pole which is proportional to the strength of these fields as well as the cos 36°, which acts to produce a clockwise rotation. But the action between S4 and the stator S-pole is exactly equal and opposite, thus yielding a resultant of zero torque produced by conditions beneath the stator S-pole. The net torque is zero, so the rotor remains in a position of equilibrium.

Let us now examine the effect of alternately energizing the two quadrature stator coils in sequence through positive and negative values of d-c excitation. The first step in the sequence is represented in Fig. 22-2(b) by placing coil A-A' in a deenergized state and coil B-B' in a fully energized state. The rotor is shown in the final steady-state position that results as long as coil B-B' is energized with positive current (i.e., so that the left pole becomes a north pole and the right pole assumes a S-pole polarity). The net rotor displacement for this switch in coil excitation in this case is 18°. Once the rotor assumes the position depicted in Fig. 22-2(b), the orientation of the rotor poles relative to the newly located stator poles is identical to that appearing in Fig. 22-2(a), so the rotor is again in equilibrium. It is instructive, however, to examine the torque conditions that develop when coil A-A' is deenergized and coil B-B' is energized while the rotor is still in the position corresponding to Fig. 22-2(a). We proceed on the assumption that the currents in these coils change instantaneously, i.e., the time constants of the windings are very small compared to the mechanical time constants involved. This is not an unrealistic assumption. Because the magnitude of the flux field produced by a given coil excitation in the fixed geometry of a given machine is determinable and since the permanent magnet flux field of each rotor pole is also calculable, we can represent the maximum torque produced by these fields as T_m. When the rotor orientation is such that the torque developed is other than maximum, the appropriate sine or cosine function is introduced depending upon whether the torque is one of attraction or one of repulsion. Thus with coil

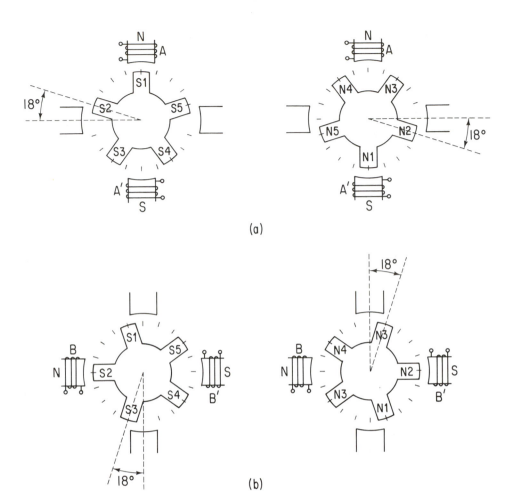

Fig. 22-2 Illustrating the indexing of the rotor in a ccw direction for a full cycle of coil *A-A′*/coil *B-B′* excitation in sequence through positive and negative values. (a) Coil *A-A′* energized for a N-S orientation; coil *B-B′* deenergized. Comment: starting point. (b) Coil *A-A′* deenergized; coil *B-B′* energized for a N-S orientation. Result: Rotor advances ccw by 18° from position in (a). (c) Coil *A-A′* energized with reversed current for a S-N orientation; coil *B-B′* deenergized. Result: Rotor advances ccw an additional 18° from position in (b). (d) Coil *A-A′* deenergized. Coil *B-B′* energized with reverse current for a S-N orientation. Result: Rotor advances ccw an additional 18° from position (c). (e) Coil *A-A′* energized with positive current for a N-S orientation, coil *B-B′* deenergized. Result: Rotor advances an additional 18° for a total of 72° or one rotor pole pitch corresponding to one cycle of ± excitation of stator coils. Observe that the orientation between rotor and stator in (e) exactly corresponds to the starting point in (a), except for notation.

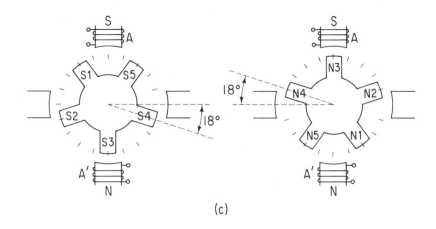

(c)

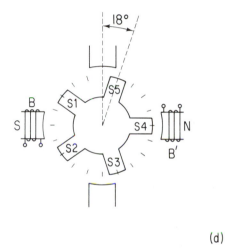

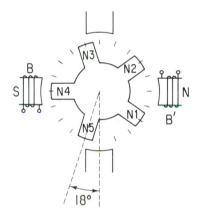

(d)

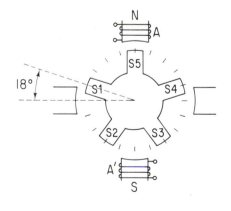

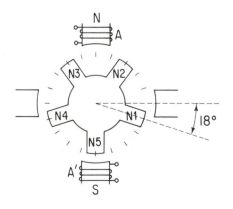

(e)

B-B' energized and the rotor still in the position depicted in Fig. 22-2(a), the expression for the torque employing superposition is

$$T = T_m \left(\sin \left\langle \frac{S2}{N} + \sin \left\langle \frac{S1}{N} - \sin \left\langle \frac{S3}{N} + \cos \left\langle \frac{S5}{S} - \cos \left\langle \frac{S4}{S} \right. \right) \quad (22\text{-}1)$$

where the plus sign is used if the torque acts to produce ccw movement and the negative sign is used if the movement would be cw. Inserting the proper angles corresponding to the stator pole orientation of Fig. 22-2(b) and the rotor position of Fig. 22-2(a), Eq. (22-1) becomes

$$T = T_m(\sin 18° + \sin 90° - \sin 54° + \cos 18° - \cos 54°)$$

$$= T_m(0.309 + 1.0 - 809 + 0.951 = 0.588) \quad (22\text{-}2)$$

$$= 0.863 \ T_m$$

Because this is a positive torque, the rotor in Fig. 22-2(a) moves to the new equilibrium position shown in Fig. 22-2(b).

Consider next that coil B-B' is switched off and coil A-A' is again energized but this time with current flowing in the opposite direction. The upper stator pole assumes a S-pole polarity and the lower stator pole takes on a north-pole polarity. Accordingly, the new stator flux configuration [in Fig. 22-2(c)] exerts a torque on the rotor orientation of Fig. 22-2(b) which is again described by Eq. (22-2). The ensuing torque displaces the rotor by an additional 18° counterclockwise in order to reach the new equilibrium position shown in Fig. 22-2(c). In the subsequent step in the sequence, the current in coil A-A' is switched off while the current in coil B-B' is switched on but in the reversed direction. The stator poles then assume the polarity depicted in Fig. 22-2(d), which in turn produces a torque on the rotor to yield a further 18° displacement in the ccw direction. In other words, the rotor moves to the new equilibrium position that finds S4 in alignment with the N-pole associated with coil B-B'.

It is instructive to observe here that once a sequence is initiated and maintained through the proper positive and negative alternate excitation of the stator coils, the action of the torque produced by the subsequent coil excitation is to displace the rotor in the established direction. This is done in a manner to ensure that the rotor pole having the least angular alignment with a stator pole of opposite polarity is brought into exact alignment with that pole. Accordingly, when coil A-A' is next energized, the result is to move S5 into alignment with the upper stator N-pole. The new equilibrium position is shown in Fig. 22-2(e). Observe that this last orientation is identical to the first one except that the entire rotor structure has been indexed through a total displacement equal to one rotor pole pitch, or 72° in this case. Note too that for a *two-coil* (or two-phase) stator with *two* states (positive and negative), it requires *four* switching operations to complete a switching cycle. Thus this motor is made to *step* four times during each cycle of excitation, which in turn produces a net displacement of one rotor pole pitch. Or, stated differently, each switch of the stator excitation produces a rotor displacement of one-fourth a rotor pole pitch. The *stepping* action exhibited by this motor in response to alternate sequenced excitation of stator coils is the reason it is called a *stepper motor*.

The direction of rotation of the stepper motor can be conveniently reversed by merely reversing the switching sequence. Thus, if in Fig. 22-2(b) the excitation of coil B-B' were reversed, the action of the developed torque would be to displace S5 in Fig. 22-2(a) into a position of exact alignment with the N-pole flux which would now be associated with coil B'. The result is a clockwise displacement of $18°$, in contrast to the counterclockwise displacement that occurs when coil B is made to produce a N-pole polarity.

22-3 DRIVE AMPLIFIERS AND TRANSLATOR LOGIC

The foregoing description of the method of operation of the stepper motor makes it plain that the stepping action of the motor is dependent on a specific switching sequence that serves to energize and deenergize the stator coils. A practical scheme that achieves this objective is the one of a drive amplifier consisting of transistors that are switched on and off sequentially in accordance with the signals that originate from an appropriate translator logic control circuit. A block diagram of the arrangement is shown in Fig. 22-3. The drive amplifier provides the d-c excitation to coils A-A' and B-B' through switched circuitry that is controlled by the translator logic device. Appearing in Fig. 22-4 is a detailed description of the composition of one type of drive amplifier, namely, the *bipolar* type. A total of four switching transistors is needed for each phase coil whenever a single power supply is used and current reversal in the phase coils is desired. The term *bipolar* is used for such a circuit because the circuitry permits either end of the phase coil to be connected to the positive side of the power supply. There is a drive circuit available that employs two power supplies, in which case it becomes necessary to use only two transistors per phase coil. In addition to the eight transistors, the drive amplifier of Fig. 22-4 also includes eight diodes. Of course, the diodes are required to provide paths for the inductive coil currents to flow whenever the transistors are switched off by logic control circuitry. For example, if transistors Q_1 and Q_2 are on during the first quarter of a switching cycle thus

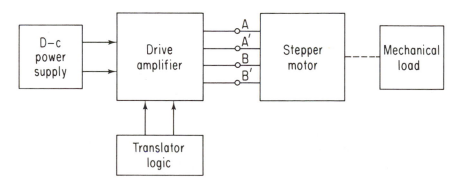

Fig. 22-3 Block diagram of the stepper motor and its switched supply.

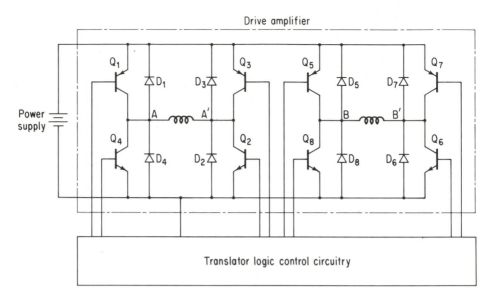

Fig. 22-4 Diagram illustrating the details of a bipolar drive amplifier.

causing current to flow in the phase coil from A to A', the transition into the next quarter of the cycle calls for Q_1 and Q_2 to be switched off. Because of the magnetic storage associated with the current in coil A-A', the current cannot be reduced to zero instantaneously. But as it attempts to do so, a large emf is induced in the coil with terminal A' positive and A negative. This induced emf puts a positive voltage across diodes D_3 and D_4, thus furnishing a path for the switched-off current in coil A-A'. The current decays rapidly because of the small time constant of the discharging circuit.

How is the logic schedule of the translator control circuit determined? This is best explained by developing the logic required to perform the operations of the stepper motor described in connection with Fig. 22-2. In this case a complete switching cycle required that the stator coils be energized in accordance with the schedule in the table below for a counterclockwise rotation.

Coil A-A'	Coil B-B'
ON+	OFF
OFF	ON+
ON−	OFF
OFF	ON−

Here ON+ is distinguished from ON− by a current reversal in the coil. The logic schedule that achieves this switching sequence for the drive amplifier of Fig. 22-4 expressed in binary language, where 1 denotes ON and 0 denotes OFF is as follows:

Stepper Motors   Chap. 22

Coil A-A'				Coil B-B'			
Q_1	Q_2	Q_3	Q_4	Q_5	Q_6	Q_7	Q_8
1	1	0	0	0	0	0	0
0	0	0	0	1	1	0	0
0	0	1	1	0	0	0	0
0	0	0	0	0	0	1	1

Each repetition of this switching cycle provides an advancement of the stepper motor equal to one rotor pole pitch. Hence in this case five cycles of this switching sequence are needed in order to complete one full revolution of the rotor.

A very popular circuit for the drive amplifier of stepper motors is the one illustrated in Fig. 22-5. This circuit clearly has the advantage of fewer electronic components and simpler bias circuitry for the switching transistors. Each phase coil of the stepper motor that is used with these drives is equipped with a center tap and the switching logic for the associated transistors arranged so that current flows in only one section at a time. Reversal of the magnetic pole polarity is achieved by directing the current either to one section or to the other. The disadvantage of such a scheme is that the copper is not made to work continuously in producing useful output. This arrangement is called the *bifilar unipolar drive*. It is referred to as unipolar because the ends of both coil sections stay connected

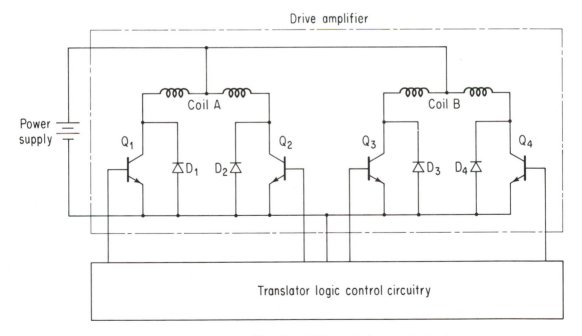

Fig. 22-5 Drive amplifier of the bifilar, unipolar type. Each phase coil of the stepper motor comes equipped with a center tap.

to the negative side of the power supply during operation unlike the bipolar case, where the connection alternates between the high and low side of the source.

The logic schedule to be furnished by the translator control circuitry for each full switching cycle of the operation of a stepper motor such as the one illustrated in Fig. 22-2 is the following:

Coil A		Coil B	
Q_1	Q_2	Q_3	Q_4
1	0	0	0
0	0	1	0
0	1	0	0
0	0	0	1

It is assumed here that coil A is equipped with a center tap by using the point of connection of coil A and coil A' in Fig. 22-1. A similar arrangement is assumed to be used for coil B.

The diodes D_1 through D_4 appearing in the circuitry of the drive amplifier shown in Fig. 22-5 are for the purpose of maintaining a current path following the cutoff of a particular transistor switch. For example, when current is flowing through the left-half section of coil A in Fig. 22-5 and Q_1 is subsequently cut off, a circuit is established by transformer action in the second section of coil A which allows the current in the first section to drop to zero as the current in the second section instantaneously rises to the level of the current in the first section. The latter current then flows through a path that involves the power supply and diode D_2. Diode D_2 is made conductive since a positive voltage appears across it as a result of the transformer voltage induced in the right-half section of coil A. The transfer of current from the left section to the right section of coil A is allowed to occur instantaneously because there is *no change* in magnetic energy in this process. The mmf of both coils act on the same magnetic circuit.

22-4 HALF-STEPPING AND THE REQUIRED SWITCHING SEQUENCE

The description of the operation of the stepper motor so far has been limited to situations where either one stator coil or the other was allowed to be energized at any one time. What would be the consequence if a switching schedule is devised which permits both stator coils to be excited simultaneously at certain periods in a cycle? To examine this matter, let us return to the configuration depicted in Fig. 22-2(a). Recall that a full 18° ccw displacement resulted from switching coil A-A' off and coil B-B' on. However, suppose that coil B-B' is switched on while maintaining the excitation on coil A-A'. In other words, coil A-A' is not deenergized when excitation is applied to coil B-B'. Intuition, based on the symmetry prevalent in this situation, leads us to suspect that the motor takes a *half-step*, i.e., it advances in the ccw direction by one-eighth of a rotor pole-pitch. An examination of the resultant torque on the rotor structure at this displacement angle reveals the motor to be in equilibrium. This is readily verified by writing the expression for the net torque on the rotor corresponding to this condition. The situation is illustrated in Fig. 22-6 for a half-step angle of 9°. Note that all four stator poles are energized with the polarities indicated. Applying superposition, the expression for the net torque can be written as

$$T_R = T_m \left(-\sin \left\langle \frac{N_A}{S1} - \sin \left\langle \frac{N_A}{S2} + \sin \left\langle \frac{N_A}{S5} - \cos \left\langle \frac{S_{A'}}{S3} + \cos \left\langle \frac{S_{A'}}{S4} \right. \right.\right.\right.\right.$$
$$+ \sin \left\langle \frac{N_B}{S2} + \sin \left\langle \frac{N_B}{S1} - \sin \left\langle \frac{N_B}{S3} + \cos \left\langle \frac{S_{B'}}{S5} - \cos \left\langle \frac{S_{B'}}{4} \right) \right.\right.\right.\right.$$

where the first line represents the torque produced by the stator poles created by coil A-A' interacting with the S-pole section of the rotor and the terms on the second line represent the torque contributions associated with the stator

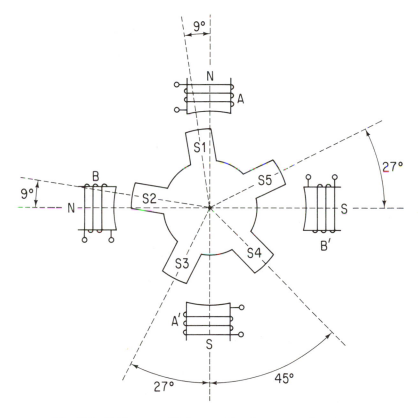

Fig. 22-6 Half-step equilibrium position that occurs when the rotor is initially at the position illustrated in Fig. 22-2(a) and both coils A-A' and B-B' are energized to produce the field orientation shown here. The result is a 9° displacement of the rotor as indicated here.

poles of coil B-B' and the S-pole section of rotor. Upon inserting the angular displacements existing between the axes of the indicated poles, we get

$$T_R = T_m \left(-\sin 9° - \sin 81° + \sin 63° - \cos 27° + \cos 45°\right.$$

$$+ \sin 9° + \sin 81° - \sin 63° + \cos 27° - \cos 45°\left.\right)$$

$$= 0$$

Hence this half-step displacement of the rotor represents an equilibrium position.

If coil A-A' is now deenergized while the excitation of coil B-B' is maintained, the motor steps an additional 9°, thus yielding the originally expected 18° that results when one coil is switched off and the other is switched on. An extension of this analysis through a complete switching cycle executed by half-step increments leads to the following switching sequence for counterclockwise rotation:

Coil A-A'	Coil B-B'	
ON+	OFF	
ON+	ON+	
OFF	ON+	
ON−	ON+	
ON−	OFF	One complete switching cycle
ON−	ON−	
OFF	ON−	
ON+	ON−	
ON+	OFF	
⋮	⋮	

A full switching cycle now includes a total of eight switching states which yield eight half steps to give a total displacement of one rotor pole pitch. The switching schedule required to half-step the motor of Fig. 22-1 arranged to operate in the bifilar mode and driven by the unipolar drive amplifier of Fig. 22-5 is detailed in the following table:

Coil A		Coil B	
Q_1	Q_2	Q_3	Q_4
1	0	0	0
1	0	1	0
0	0	1	0
0	1	1	0
0	1	0	0
0	1	0	1
0	0	0	1
1	0	0	1

The advantage of half-stepping, of course, is that it improves the precision of the stepper motor in applications where more precise positioning is demanded.

22-5 THE RELUCTANCE-TYPE STEPPER MOTOR

The reluctance-type stepper motor is not equipped with a permanent magnet on the rotor. It develops an indexing torque in response to the sequenced d-c excitation of stator coils by virtue of the large difference in magnetic reluctance that exists between a direct-axis path and a quadrature-axis path. The resulting stationary flux fields produced by the d-c current in selected stator coils always exert a force which acts to minimize the reluctance of the flux path in accordance with the characterization of Eq. (15-42). A typical configuration of the reluctance-type stepper motor is illustrated in Fig. 22-7. The stator in this case is designed for eight poles and is equipped with four phase coils each of two sections. The rotor is slotted to produce the effect of six poles corresponding to which is a pole pitch of 60°. The rotor is at a position of equilibrium which occurs when coil A-A' (i.e., phase A) is energized and the remaining phases are left unexcited. Observe that the constant flux field produced by phase A encounters a path of minimum reluctance by passing through rotor poles 1 and 4, which are aligned with the stator pole structures associated with phase A.

For the combination of stator and rotor poles that is designed into this motor, it is instructive to note that the axes of both rotor poles 2 and 5 are 15° displaced from the axes of the two stator poles associated with phase B. Hence,

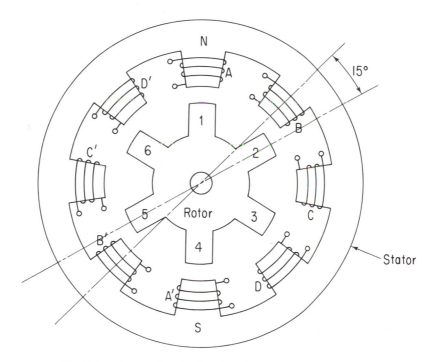

Fig. 22-7 Construction details of a reluctance-type stepper motor equipped with four stator phases (i.e. eight poles) and six rotor poles.

if phase *A* is deenergized and phase *B* is energized to produce a N-pole polarity at coil *B* and a S-pole polarity at coil *B'*, a reluctance torque is developed that acts to align the axes of rotor poles 2 and 5 with the axes of the stator poles at *B-B'*. The rotor advance is 15° ccw, which in this machine too is one-quarter of a rotor pole pitch. Successive excitation of phases *C*, *D*, and *A* yields a total advance in the ccw direction equal to a rotor pole pitch. A sixfold repetition of this switching sequence for the excitation of the stator phases results in one revolution of the motor shaft.

The equilibrium position of the rotor in Fig. 22-7 also prepares the motor to move in a clockwise direction. However, it now requires that coil *D'* be energized for a N-pole polarity and coil *D* for a S-pole polarity following the removal of excitation on phase *A*. A reluctance torque now develops that pulls rotor teeth 6 and 3 into alignment with the poles produced by coil *D' − D*. Clearly, this results in a clockwise rotation.

The design of the drive amplifiers and the translator logic for the reluctance-type stepper motor differ in some details from those reported for the PM stepper motor. This is necessary to account for differences in the design aspects of the two types of motors. The principles, however, remain the same and are easily transferable from one motor to the other.

22-6 RATINGS AND OTHER CHARACTERISTICS

On the basis of the material that is related in the preceding pages concerning the stepper motor, it should come as no surprise that these motors are used in circumstances where the need for large horsepower is not a factor. A motor's output power is dependent upon the extent to which the copper is worked electrically and the iron is worked magnetically. The stepper motor ranks low on both counts. To achieve the characteristics which make the stepper motor unique, it is necessary that during each complete switching cycle each of the phase coils spend time in a deenergized state and that some parts of the magnetic flux paths be essentially free of field flux. Consequently, stepper motors are found to range in output power from about 1 W to a maximum size of 3 hp. In terms of physical dimensions, the largest machines are built with diameters up to 7 in., while the smallest ones are constructed in a pancake form with diameters as small as 1 in.

Many step sizes are also available with stepper motors. They can be purchased with step sizes as small as 0.72° or as large as 90°. The most common step sizes are 1.8°, 7.5°, and 15°. To procure a step size of 1.8° in a PM type, it is necessary to design the stator for 40 poles and the rotor for 50 poles (or slots).

The shape of the torque-speed curve of the stepper motor takes on the general form depicted in Fig. 22-8. The speeds indicated in this particular motor are achieved as a result of programming the switching sequence for appropriate repetition. For example, in order to realize 1 revolution per second, it is necessary for the translator logic to switch a 1.8° step motor at the rate of 200 steps per second. Any attempt to operate this motor at a rate of five revolutions per second

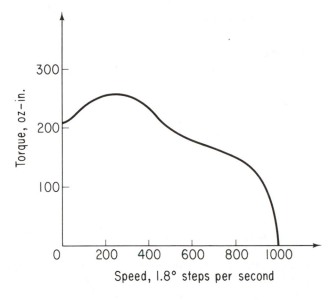

Fig. 22-8 Torque-speed curve of a PM stepper motor operated at full step.

fails because the torque drops to zero. This decrease in torque is attributable to the effects of the speed voltages induced in the phase coils at the higher speeds as well as to the inductance of these coils. Compensation circuits are available which can double the useful speed range, but this is at the cost of increased circuit complexity for the drive amplifiers.

Summary review questions

1. List some of the areas of applications where stepper motors can be found performing important tasks.
2. Name the two types of stepper motors.
3. Describe the essential construction features of the permanent-magnet stepper motor.
4. Describe the basic operating principle of the stepper motor. In particular stress the relationship between the excitation sequence of the stator coils and the response of an appropriately slotted permanent magnet rotor.
5. How many switching operations are needed for a displacement of one rotor slot pitch if the stepper motor is equipped with two phases and two states (positive and negative)?
6. How is the direction of rotation of the stepper motor reversed?
7. The indexing action of the stepper motor is dependent on an appropriate switching sequence to drive the stator coils (phases). Identify the electronic equipment that is used to provide these switching sequences.
8. What is a bipolar drive amplifier as it relates to the stepper motor?
9. How many transistors per phase coil are used in a bipolar drive amplifier equipped with a single power supply? How many diodes?
10. What is a bifilar unipolar drive amplifier? Describe how it differs from the bipolar type.

11. Distinguish between half-stepping and full-stepping in the stepper motor. What modification of the switching sequence is required to achieve half-stepping?

12. What advantage does half-stepping offer over full-stepping?

13. In what essential aspect does the reluctance-type stepper motor differ from the permanent-magnet type?

14. Describe the principle that underlies the operation of the reluctance-type stepper motor.

15. For a given physical size the stepper motor produces much less useful output than conventional motors. Explain.

Problems

22-1. Consider the situation corresponding to the rotor position depicted in Fig. 22-2(b). Assume that coil A-A' is energized so that the magnetic material on which coil A' is wrapped is of a N-pole polarity and that of coil A is of a S-pole polarity.
   (a) Write the expression for the torque produced on the rotor by the S-pole section of the rotor. Assume sinusoidal flux distributions throughout.
   (b) How many degrees of rotation occur before a position of equilibrium is reached?

22-2. Repeat Prob. 22-1 for the conditions that exist at the N-pole section of the rotor.

22-3. Let T_m denote the maximum torque produced by the cooperation of a single PM rotor pole with the stator pole in the configuration of Fig. 22-2. Calculate the net torque produced on the rotor shaft for the condition represented by Fig. 22-2(c).

22-4. The schematic diagram of a four-phase two-pole PM stepper motor is shown in Fig. P22-4. The phase coils are excited in sequence by means of a translator logic circuit.

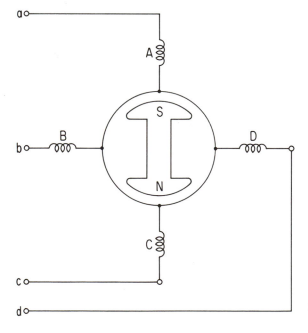

Fig. P22-4

(a) Write the logic schedule for full-stepping of this motor.

(b) What is the displacement angle of the full step?

(c) Write the logic schedule for operating this motor in the half-step mode.

(d) Make a sketch of the stator poles and the relative rotor position when coils B and C are simultaneously energized.

(e) What is the total rotor displacement in part (d) measured from the position depicted in Fig. P22-4?

22-5. The excitation signals used to drive the coils in the motor of Fig. P22-4 originate from digital signals that drive the supply amplifier. A continuous sequencing of these signals will cause the stepper motor to run at a constant speed (thus behaving as a synchronous motor). Determine the frequency rate of these digital signals if the speed of rotation is to be 360 rpm. Assume a half-stepping pattern.

22-6. Determine the frequency rate at which the digital signals that drive the amplifier of the motor of Fig. 22-2 must be set in the half-step mode if the motor is to run at a speed of 360 rpm.

22-7. The drive amplifier of Fig. 22-4 is used to drive the stepper motor of Fig. 22-2 in the full-step mode. Write the logic schedule for switching the transistors in a manner such that it yields a *clockwise* direction of rotation.

22-8. A PM stepper motor is designed to provide a half-step size of 0.9°.

(a) Determine the rotor pole pitch in degrees.

(b) Find the number of stator poles.

22-9. Draw the diagram that illustrates the orientation of the stator poles with the N-pole section of the rotor for the half-step case illustrated in Fig. 22-6.

chapter twenty-three

Principles of Automatic Control

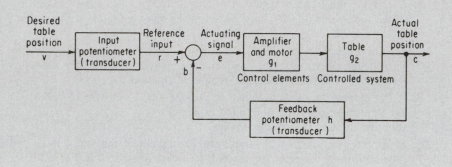

The essential feature of many automatic control systems is feedback. Feedback is that property of the system which permits the output quantity to be compared with the input command so that upon the existence of a difference an actuating signal arises which acts to bring the two into correspondence. This principle of feedback is really not new to us; it surrounds every phase of everyday living. It underlies the coordinated motions executed by the human body in walking, reaching for objects, and driving an automobile. It plays an equally important role in the countless applications of control system engineering in the fields of manufacturing, process industries, control of watercraft and aircraft, special-purpose computers for many types of military equipment, and in many other fields including the home and the office.

In this chapter the mechanism through which the feedback action exhibits itself is studied in control systems ranging from elementary ones to those possessing a high degree of complexity. In each, the same functional behaviors will be observed, and the general nature of the method of analysis will thus be emphasized. The distinction between an elementary system and a complex one lies primarily in the difficulty of the task to be performed. The more difficult the task, the more complex the system. In fact, with many present-day systems this complexity has reached such proportions that system design has virtually become a science. The functional behavior of each system will be treated in terms of a block-diagram notation and its associated terminology.

23-1 DISTINCTION BETWEEN OPEN-LOOP AND CLOSED-LOOP (FEEDBACK) CONTROL

It is important in the beginning that the distinction between closed-loop and open-loop operation be clearly understood. Both terms will often be used throughout this part of the book. The following simple definitions point out the difference.†

An *open-loop system* is one in which the control action is independent of the output (or desired result).

A *closed-loop system* is one in which the control action is dependent upon the output.

The key term in these definitions is *control action*. Basically, it refers to the actuating signal of the system, which in turn represents the quantity responsible for activating the system to produce a desired output. In the case of the open-loop system the input command is the sole factor responsible for providing the control action, whereas for a closed-loop system the control action is provided by the difference between the input command and the corresponding output.

To illustrate the distinction, consider the control of automobile traffic by means of traffic lights placed at an intersection where traffic flows along north-south and east-west directions. If the traffic-light mechanism is such that the green and red lights are on for predetermined, fixed intervals of time, then operation is open-loop. This conclusion immediately follows from our foregoing definition upon realizing that the desired output, which here is control of the volume of traffic, in no way influences the time interval during which the light shines green or red. The input command originates from a calibrated timing mechanism, and this alone establishes how long the light stays green or red. The control action is thereby provided directly by the input command. Accordingly, the timing mechanism has no way of knowing whether or not the volume of traffic is especially heavy along the north-south direction and therefore in need of longer green-light intervals. A closed-loop system of control would provide precisely this information. Thus, if a scheme is introduced which measures the volume of traffic along both directions, compares the two, and then allows the difference to control the green and red time periods, feedback control results because now the actuating signal is a function of the desired output.

To complete the comparison of closed-loop versus open-loop operation, it is instructive next to list briefly some of the performance characteristics of each. Generally, the open-loop system of control has two outstanding features. First, its ability to perform accurately is determined by its calibration. As the calibration deteriorates, so too does its performance. Second, the open-loop system is usually easier to build since it is not generally troubled with problems of instability. One of the noteworthy features of closed-loop operation is its ability faithfully to reproduce the input owing to feedback. This in a large measure is responsible for the high accuracy obtainable from such systems. Since the actuating signal is a function of the deviation of the output from the input, the control action

---

† More precise definitions are given in the next section.

persists in generating sufficient additional output to bring the two into corre-spondence. Unfortunately, this very factor (feedback) is also responsible for one of the biggest sources of difficulty in closed-loop systems, namely, the tendency to oscillate. A second important feature of closed-loop operation is that, in direct contrast to the open-loop system, it usually performs accurately even in the presence of nonlinearities.

23-2 BLOCK DIAGRAM OF FEEDBACK CONTROL SYSTEMS. TERMINOLOGY

Every feedback control system consists of components that perform specific functions. A convenient and useful method of representing this functional char-acteristic of the system is the block diagram. Basically this is a means of representing the operations performed in the system and the manner in which signal information flows throughout the system. The block diagram is concerned not with the physical characteristics of any specific system but only with the functional re-lationship among various parts in the system. In general, the output quantity of any linear component of the system is related to the input by a gain factor and combinations of derivatives or integrals with respect to time. Accordingly, it is possible for two entirely different and unrelated physical systems to be represented by the same block diagram, provided that the respective components are described by the same differential equations.

The general form of the block diagram of a feedback control system is shown in Fig. 23-1.† Since the differential equation or the gain factor of each block is not specifically identified, lowercase letters are used to represent the input and output variables. When this information is known, capital letters are used. This notation is followed throughout this part of the book. The symbols in the block diagram were selected to avoid those which imply the mechanics

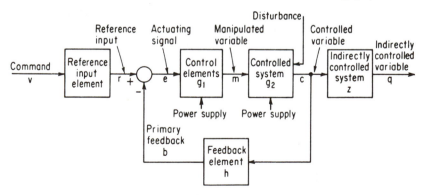

Fig. 23-1 Block diagram of a feedback control system, showing terminology.

† A.I.E.E. Committee Report, "Proposed Symbols and Terms for Feedback Control Systems," *Elec. Eng.*, **70** (1951), pt. 2, pp. 905–909.

of the system such as θ for angle, p for pressure, etc. This helps to preserve the generic nature of the block diagram.

Before we consider the significance and advantages of the block-diagram notation, it is very important that the meanings of the terms used in Fig. 23-1 be clearly understood and remembered. These terms and others are described below:

A *feedback control system* is a control system that tends to maintain a prescribed relationship of one system variable to another by comparing functions of these variables and using the difference as a means of control.

The *controlled variable* is that quantity or condition of the controlled system which is directly measured and controlled.

The *indirectly controlled variable* is that quantity or condition which is controlled by virtue of its relation to the controlled variable and which is not directly measured for control.

The *command* is the input which is established or varied by some means external to and independent of the feedback control system under consideration.

The *reference input* is a signal established as a standard of comparison for a feedback control system by virtue of its relation to the command.

The *primary feedback* is a signal that is a function of the controlled variable and that is compared with the reference input to obtain the actuating signal. (This designation is intended to avoid ambiguity in multiloop systems.)

The *actuating signal* is the reference input minus the primary feedback.

The *manipulated variable* is that quantity or condition which the controller (g_1) applies to the controlled system.

A *disturbance* is a signal (other than the reference input) that tends to affect the value of the controlled variable.

A *parametric variation* is a change in system properties that may affect the performance or operation of the feedback control system.

The *controlled system* is the body, process, or machine, a particular quantity or condition of which is to be controlled.

The *indirectly controlled system* is the body, process, or machine that determines the relationship between the indirectly controlled variable and the controlled variable.

The *feedback controller* is a mechanism that measures the value of the controlled variable, accepts the value of the command, and, as a result of comparison, manipulates the controlled system in order to maintain an established relationship between the controlled variable and the command.

The *control elements* comprise the portion of the feedback control system that is required to produce the manipulated variable from the actuating signal.

The *reference-input elements* comprise the portion of the feedback control system that establishes the relationship between the reference input and the command.

The *feedback elements* comprise the portion of the feedback control system that establishes the relationship between the primary feedback and the controlled variable.

The *summing point* is a descriptive symbol used to denote the algebraic summation of two or more signals.

The reference-input element usually consists of a device that converts control signals from one form into another. Such devices are known as *transducers*. Most transducers have an output signal in the form of electrical energy. Thus a potentiometer can be used to convert a mechanical position to an electrical voltage. A tachometer generator converts a velocity into a d-c or an a-c voltage. A pressure transducer changes a pressure drop or rise into a corresponding drop or rise in electric potential. The transducer provides the appropriate translation of the command into a form usable by the system.

The feedback element also frequently consists of a transducer of the same kind as the reference-input element. When the two signals of the input and feedback transducers are compared, the result is the actuating signal. This portion of the closed-loop system is very important, and it is usually more descriptively identified as the *error detector* (see Fig. 23-2). Error detectors may take on a variety of forms depending upon the nature of the feedback control system. Some of the more frequently used error detectors include the following: two bridge-connected potentiometers, a control-transmitter control-transformer synchro combination, two linear differential transformers, two tachometer generators, differential gear, E-type transformers, bellows, gyroscopes.

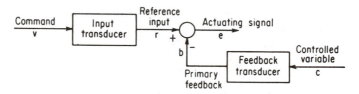

Fig. 23-2 Error-detector portion of a feedback control system.

The primary function of the control element g_1 of the block diagram is to provide amplification of the actuating signal since it is generally available at very low power levels.† Many types of amplifying devices are used either singly or in combination. Some of the more common types are the following: electronic, solid-state (transistor), electromechanical, hydraulic, pneumatic, magnetic. Note that the amplifier must have its own source of power. This is indicated in Fig. 23-1 by the use of an external arrow leading to the g_1 block. It should be emphasized that the *closed-loop block diagram* refers to the flow of the *control power* and not to the main source of energy for the system. In essence, then, the block diagram represents the manner in which the control signal manipulates the main source of power, in order to adjust the controlled variable in accordance with the command.

The block identified as the controlled system g_2 in Fig. 23-1 refers to that portion of the system which generates the controlled variable in accordance with

---

† The control element also frequently serves the function of changing the basic time character of the signal to control the transient system response.

the dictates of the manipulated variable. The controlled system obviously can take on many forms. It may represent an aircraft frame, the body of a ship, a gun mount, a chemical process, or a temperature or pressure system. The milling machine and the lathe are examples where the controlled system represents a machine as distinguished from a process or body.

The operation of any feedback control system can be described in terms of the block diagram of Fig. 23-1. The application of a specific command causes a corresponding signal at the reference input through the action of the input transducer. Since the controlled variable cannot change instantaneously because of the inertia of the system, the output of the feedback element no longer is equal and opposite to the reference input. Accordingly, an actuating signal exists, which in turn is received by the control element and amplified, thereby generating a new value of controlled variable. This means that the primary feedback signal changes in such a direction as to reduce the magnitude of the actuating signal. It should be clear that the controlled system will continue to generate a new level of output as long as the actuating signal is different from zero. Therefore, only when the controlled variable is brought to a level equal to the value commanded will the actuating signal be zero.† The system then will be at the new desired steady-state value.

To have feedback control systems operate in a stable fashion, they must be provided with *negative* feedback, which simply means that the feedback signal is opposite in sign to the reference input. This helps to ensure that, as the controlled variable approaches the commanded value, the actuating signal approaches zero. A moment's reflection reveals that, if the feedback is positive and the controlled variable increases, the actuating signal will also increase. This causes an increased value of controlled variable, which results in a further increase in actuating signal, and so on. The result is an unbounded‡ increase in controlled variable and loss of control by the command source.

Summarizing, we see that the block diagram of a feedback control system is, first of all, a representation of its functional characteristics which permits a description of the manner in which the control-signal energy flows through the system. Second, it is a means of emphasizing that the closed-loop system is composed of three principal parts: the error detector, the control element (amplifier and output devices), and the controlled system.

23-3 POSITION FEEDBACK CONTROL SYSTEM. SERVOMECHANISM

A common industrial application of a feedback system is to control position. These systems occur so frequently in practice and are so important that they are given a special name—*servomechanism*. A servomechanism is a power-amplifying feedback control system in which the controlled variable is mechanical

† There are situations where the steady-state level of the actuating signal is finite.
‡ The nonlinearities inherent in the system would ultimately limit the magnitude of the output.

position. The term also applies to those systems which control time derivatives of position, as, for example, velocity and acceleration. The schematic diagram of a servomechanism is shown in Fig. 23-3.

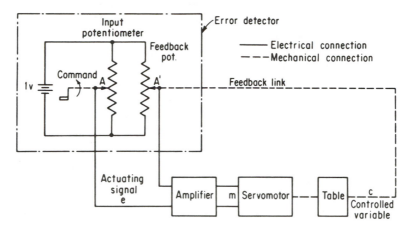

Fig. 23-3 Example of a servomechanism.

Consider that the system of Fig. 23-3 is to be used to control the angular position of a table in accordance with a command signal originating from a remotely located station. The first step in obtaining the block diagram and understanding the operation of the system is to identify the error-detector portion of the system. As pointed out in Fig. 23-2, this requires merely finding the input and feedback transducers and the summing point of their output signals. For the system of Fig. 23-3 the error detector consists of the section blocked off with the dash-dot line. The second principal part of the system—the control element—consists of the amplifier unit and the output motor. The latter is more commonly referred to as a servomotor, since it requires special design considerations. The table, the position of which is being controlled, represents the controlled system. In a specific application it may be a gun mount or the support table for a lathe or milling machine. The block diagram for the system is shown in Fig. 23-4.

To understand the operation of the system, assume that initially the slider arms of the input and feedback potentiometers are both set at $+50$ mV. The voltage of the input potentiometer is the reference input. For this condition the

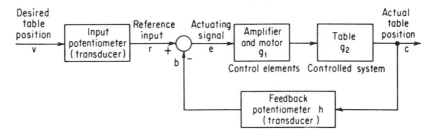

Fig. 23-4 Block diagram of the servomechanism of Fig. 23-3.

actuating signal is zero, and so the motor has zero output torque. Next consider that the command calls for a new position of the table, namely, that corresponding to a potentiometer voltage of $+60$ mV. When arm A is placed at the $+60$-mV position, arm A' remains instantaneously at the $+50$-mV position because of the table inertia. This situation creates a $+10$-mV actuating signal, which is really a measure of the lack of correspondence between the actual table position and the commanded position. A $+10$-mV input to the amplifier applies an input to the servomotor, which in turn generates an output torque, which repositions the table. With negative feedback present the table moves in a direction which causes the potential of A' to increase beyond $+50$ mV. As this takes place, the actuating signal gets smaller and finally reaches zero, at which time A' has the same potential as A (this is referred to as the *null* position). The actual table position is therefore equal to the commanded value. Note that, if the table is not exactly at the commanded position, the actuating signal will be different from zero, and hence a motor output torque persists, forcing the table to take the commanded position.† On the basis of the foregoing explanation, we can begin to appreciate that a system of this type possesses an integrating property; i.e., the motor output angle is proportional to the integral of the actuating signal and thus stops changing only when the actuating signal itself is zero. This follows from the fact that for any given actuating signal there results a corresponding motor velocity. However, the feedback potentiometer is sensitive not to velocity but rather to motor position, which is the integral of velocity and therefore also of the actuating signal.

One distinguishing feature that the amplifier of this type of control system must have is *sign sensitivity*. It must function properly whether the command places arm A at a higher or a lower voltage than the original value. In terms of position this means that the command should be capable of moving the table in a clockwise or counterclockwise direction. Thus, if arm A had been placed at $+40$ mV rather than $+60$ mV by the command, the actuating signal would have been -10 mV. The amplifier must be capable of interpreting the intelligence implied in the minus sign by applying a reversed signal to the servomotor, thereby reversing the direction of rotation.

The manner in which a feedback control system reacts to an externally applied disturbance may be illustrated by considering further the example of the servomechanism. Assume the disturbance to be in the form of an external torque applied to the table. The effect of this torque will be to offset the feedback potentiometer from its null position. Upon doing this, the actuating signal no longer is zero but takes on such a value that when multiplied by the torque constant (units of torque per volt of actuating signal) the developed torque of the servomotor is equal and opposite to the applied disturbance. It should be clear that, as long as the external torque is maintained, the actuating signal cannot be zero. How close it will be to zero depends upon the torque constant

---

† This is true provided that the components are all assumed to be perfect. Actually, imperfections such as motor dead zone cause the commanded and actual position to be out of correspondence by a small amount.

of the servomotor and the magnitude of the external force. Of course, the larger the torque constant, the smaller the actuating signal required. Since the actuating signal can no longer be zero, the potential of arm A' is no longer equal to the commanded value. Hence, for a system of the type described in Fig. 23-3, a steady-state position error exists in the presence of a constantly maintained external disturbance.† This characteristic of the system of readjusting itself to counteract the disturbance is referred to as the self-correcting, or automatic control, feature of closed-loop operation.

On the basis of the foregoing description of the servomechanism, a few general characteristics of this type of control system can be stated. A feedback control system is classified as a servomechanism if it satisfies all three of the following conditions: (1) It is error-actuated (i.e., closed-loop operation). (2) It contains power amplification (i.e., operates from low signal levels). (3) It has a mechanical output (position, velocity, or acceleration). Essentially, the servomechanism is a special case of a feedback control system. In regard to performance features we can say at this point that among its salient characteristics are automatic control, remote operation (of the command station), high accuracy, and fast response. The aspect of remote operation deserves further comment since it is one of the chief motivations for the development of these systems. It should be apparent from Fig. 23-3 that the input potentiometer (often called the command station) can be far removed from the output device. Thus, if the controlled system is the rudder of a ship, its heading can be easily and accurately manipulated from the ship's bridge.

23-4 TYPICAL FEEDBACK CONTROL APPLICATIONS

To assure a more thorough understanding of the principles of feedback control, we shall next apply these principles to several widely different situations. For each system the description will revolve about the block-diagram representation, thus again emphasizing the functional nature of the approach. Also, a thorough comprehension of the procedure should develop within us a facility that will make further thinking in terms of block diagrams a routine matter. It is important to establish such a facility before dealing with the general problem of the dynamic behavior of feedback control systems.

Pressure Control System for Supersonic Wind Tunnel. Of considerable importance in the aircraft and outer space industry nowadays is obtaining information about aircraft stability derivatives and aerodynamic parameters prevailing at Mach‡ numbers ranging from 0.5 to 5 and higher. Several of the leading aircraft companies have built wind tunnels to accomplish this. The wind tunnel is often

---

† This position error may be eliminated by introducing additional control elements.

‡ *Mach number* refers to the ratio of the speed of the aircraft at a given altitude to the speed of sound at the same altitude.

of the blowdown type, which means that compressors pump air to a specified pressure in a huge storage tank and, when a test is to be performed, the stored air is bled through a valve into a settling chamber, where air flow at high Mach numbers is realized. In such tests, keeping the Mach number constant is of the utmost importance. A problem exists since, as the air is bled from the storage tanks, its pressure drops and so too will the Mach number in the test section unless the valve is opened more. The need for a control system therefore presents itself. The nature of the control system depends upon the means available for monitoring the quantity it is desired to control, which in this case is the Mach number. Since the Mach number can be identified in terms of pressure, it turns out that to keep the Mach number fixed requires keeping the settling-chamber pressure constant. Hence, the use of a pressure-activated control system suggests itself. It is interesting to note that the Mach number is now the indirectly controlled variable while the pressure is the directly controlled variable.

A schematic diagram of the wind tunnel† without the control system is shown in Fig. 23-5. The block diagram of the complete system including the components required for control is shown in Fig. 23-6. As indicated in Fig. 23-5, the air to operate the tunnel is stored in six tanks at a pressure of 600 psia at 100°F. The tanks discharge into a 30-in.-diameter manifold, which in turn leads to a 24-in.-diameter rotating-plug control rotovalve. The discharge side of this valve is connected to the settling chamber by means of a duct. In Fig. 23-6 the pressure transducer is a device that converts pressure to a corresponding voltage, which is then compared with the reference-input voltage, and the difference constitutes the actuating signal. The controller refers to an amplifying stage plus electrical networks that integrate and differentiate the input signal. (The reason for integration and differentiation will be discussed in the next chapter.)

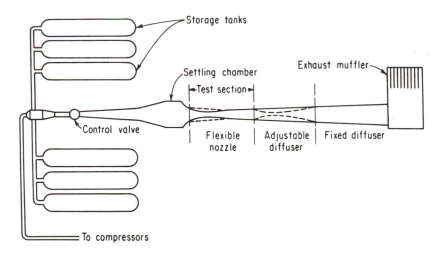

Fig. 23-5 High-speed wind tunnel.

† The wind tunnel discussed here is installed at Convair, a division of General Dynamics Corporation at San Diego, Calif.

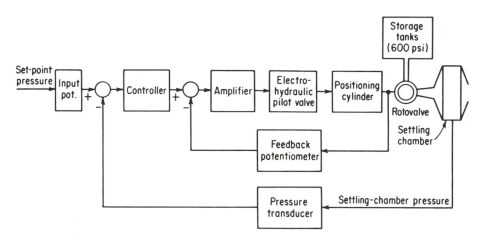

Fig. 23-6 Block diagram of a supersonic wind-tunnel pressure control system.

The electrohydraulic valve is an electrically operated two-stage hydraulic amplifier. It consists of a coil-magnet motor (which receives the electrical output of the preceding amplifier), a low-pressure pilot valve, and a high-pressure pilot valve. The output motion of the latter drives a positioning cylinder, which in turn positions the rotovalve. For improved performance a position feedback is put round the electrohydraulic valve. The controller, the electrohydraulic valve and its feedback path, and the rotovalve represent that part of the feedback control system identified as the *control elements* g_1. The *error detector* consists of the input potentiometer plus the feedback transducer, while the *controlled system* is the process of pressure build-up in the settling chamber.

Assume that a wind-tunnel test is to be performed at Mach 5, to which the settling-chamber pressure of 250 psia corresponds. Operation is started by putting the input potentiometer at a set-point pressure of 250 psia. An actuating signal immediately appears at the controller, which in turn causes the electrohydraulic pilot valve and its positioning cylinder to open the rotovalve and thereby build up pressure in the chamber. As the rotovalve is opened, the input to the pilot-valve amplifier decreases because of the position feedback voltage. Moreover, as pressure builds up, the actuating signal to the controller decreases. When the settling-chamber pressure has reached the commanded 250 psia, the actuating signal will be zero and no further movement of the rotovalve plug takes place. The time needed to accomplish this is relatively small (about 5 s), and the attendant decrease in tank pressure is also very small. Consequently, an air flow of Mach 5 is established in the settling chamber. However, as time passes and more and more air is bled from the storage tanks, the storage pressure will decrease and, unless a further opening of the rotovalve is introduced, the settling-chamber pressure will drop also. Of course, additional rotovalve opening will occur because of the pressure feedback. Specifically, as the settling-chamber pressure decreases, the corresponding feedback voltage drops and, since the reference input voltage is fixed, an actuating signal appears at the controller, which causes the pilot valve to reposition the rotovalve. This, in turn, maintains the desired chamber pressure. It should be clear that this corrective action will prevail as long as there is any tendency for the chamber pressure to drop and provided that the rotovalve is not at its limit position, i.e., not fully open.

Automatic Machine-Tool Control. Application of the principles of feedback control techniques to machine tools, together with the ability to feed the machine tool programmed instructions, has led to completely automatic operation with increased accuracy as well. Fundamentally, three requirements need to be satisfied to obtain this kind of control. First, the machine tool must receive instructions regarding the final size and shape of the workpiece. Second, the workpiece must be positioned in accordance with these instructions. Third, a measurement of the desired result must be made in order to check that the instructions have been carried out. This, of course, is accomplished through feedback.

To understand better the procedure involved, consider the system depicted in Fig. 23-7. It represents the programmed carriage drive for a milling machine. It could just as well represent the vertical and lateral drive of the cutting tool. In practice, for the cutting of a three-dimensional object, three such systems are provided, and so whatever is said about one holds for all three. The information regarding the size and shape of the workpiece usually originates from engineering drawings, equations, or models and is subsequently converted to more usable forms such as punched cards, magnetic tapes, magnetic disks and cams. In Fig. 23-7 this is the box identified as the punched taped input. Frequently, this part of the system is referred to as the input memory. The information contained in this memory is then applied to a computer, whose function is to make available a command signal indicating a desired position and velocity of the carriage. The computer in the case illustrated is a data-interpreting system and decoding servomechanism. Together with the input memory, this constitutes the total programmed instructions, or command input, to the carriage feedback control system.

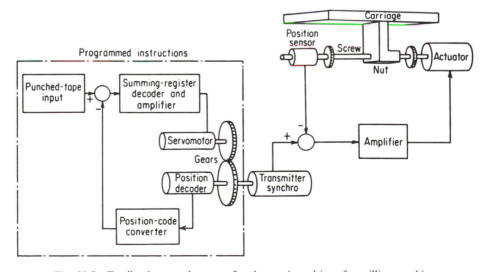

Fig. 23-7 Feedback control system for the carriage drive of a milling machine.

The transmitter synchro is the transducer, which converts the programmed command to a voltage which is the reference input to the control system. This voltage appears at the amplifier, which generates an output to drive the actuator. The actuator may be of the hydraulic, pneumatic, or electric type or perhaps a clutch. The lead screw is then driven and the carriage positioned in accordance with the instruction. Assurance that the instruction has been followed is provided by the feedback position sensor. Should the instruction fail to be fully carried out, the position sensor will allow an actuating signal to appear at the input terminals of the amplifier, thereby causing additional motion of the carriage. When the commanded position is reached, the actuating signal is nulled to zero until the next instruction comes along.

The block diagram for the system of Fig. 23-7 is shown in Fig. 23-8. The programmed command refers to the dashed-line enclosure and is identified as the programmed instructions in Fig. 23-7. The control elements consist of the amplifier and actuator. The controlled system is the carriage. The error detector is made up of the input and feedback transducers as usual. It is interesting to note, too, that the data-interpreting system is itself a servomechanism.

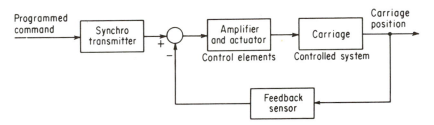

Fig. 23-8   Block diagram of Fig. 23-7.

Automobile Power-Steering Servomechanism.   One of the most common servomechanisms is the power-steering unit found in the automobile. A simplified schematic diagram of the system appears in Fig. 23-9. The corresponding block diagram is given in Fig. 23-10. The purpose of the system is to position the wheels in accordance with commands applied to the steering wheel by the driver. The inclusion of the hydraulic amplifier means that relatively small torques at the steering wheel will be reflected as much larger torques at the car wheels, thereby providing ease of steering.

The operation is simple and can be explained by applying the same approach that has been applied to the foregoing systems. Initially, with the steering wheel at its zero position (i.e., the crossbar horizontal), the wheels are directed parallel to the longitudinal axis of the car. In this position, the control-valve spool is centered so that no pressure differential appears across the faces of the power ram. When the steering wheel is turned to the left by an amount θ_i, the control-valve spool is made to move toward the right side. This opens the left side of the power cylinder to the high-pressure side of the hydraulic system and the right side to the return, or low-pressure, side. Accordingly, an unbalanced force appears on the power ram, causing motion toward the right. Through proper

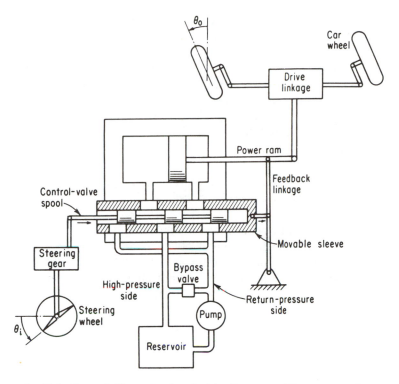

Fig. 23-9 Example illustrating the principle of an automobile power-steering servomechanism.

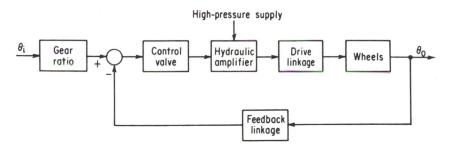

Fig. 23-10 Block diagram of Fig. 23-9.

drive linkage, a torque is applied to the wheels, causing the desired displacement θ_0. As the desired wheel position is reached, the control valve should be returned to the centered position in order that the torque from the hydraulic unit will be returned to zero. This is assured through the action of the feedback linkage mechanism. The linkage is so arranged that, as the power ram moves toward the right, the movable sleeve is displaced toward the right also, thereby sealing the high-pressure side. Such action signifies that the system has negative feedback.

The control valve and the power cylinder are part of the same housing, and these, together with the mechanical advantage of the linkage ratio, constitute the control element of the system. The controlled system in this case refers to the wheels. The error detector consists of the net effect of the feedback output position of the movable sleeve and the displacement introduced by the steering wheel at the control-valve spool. A centered valve position means no lack of correspondence between the command and the output.

Figure 23-10 is a representation of the position feedback loop only. Actually there are several more loops involved, such as the velocity loop and the load loop, which account for such things as car dynamics and tire characteristics. These are omitted for simplicity.

23-5 FEEDBACK AND ITS EFFECTS

Besides the feature of automatic control described in the preceding sections, the feedback principle has several other advantages, effects, and uses. However, before we turn to these matters, it is helpful first to obtain mathematically the relationship between input and output quantities both for the open- and closed-loop systems. Depicted in Fig. 23-11(a) is the block-diagram notation of an open-loop system which has a *direct transmission path function* denoted by G. The symbol G represents the *transfer function* of the system. It is that quantity which when multiplied by the input E yields the output C. Thus

$$C = GE \qquad (23\text{-}1)$$

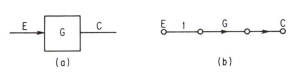

Fig. 23-11 Open-loop system: (a) the block-diagram representation in terms of the input and output variables E and C and the direct transmission function G; (b) the corresponding signal-flow graph.

The transfer function of the system, which is defined as the ratio of output C to input E, is clearly given by

$$\boxed{\frac{C}{E} = G} \qquad (23\text{-}2)$$

G is characterized entirely by the components that make up the system. In general, in addition to a gain factor, it also contains operations of differentiation and integration. In such instances the transfer function G can be expressed algebraically in terms of the differential operator p.

Appearing in Fig. 23-11(b) is an alternative way of representing the block diagram. Here use is made of nodes (the small circles) and straight lines marked by the appropriate corresponding transfer function which relates the output and input quantities of any two successive nodes. Appearing at the nodes are the variables of the system. The arrows emphasize the unilateral nature of the signal flow. Transmission in block G can occur from E to C and not vice versa. This representation is called a *signal-flow graph.*

The general block diagram of a closed-loop system is shown in Fig. 23-12(a) and the corresponding signal-flow graph in Fig. 23-12(b). The feedback function is denoted by H. In general H, too, can be expressed as an algebraic function of the differential operator p. From the point of view of the direct function G, the output C still depends upon the input signal E. However, the significant difference in the closed-loop case is that the actuating signal E is dependent upon a portion of the output signal, HC, as well as the reference input, R. That is,

$$E = R - HC \tag{23-3}$$

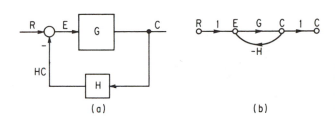

Fig. 23-12   Closed-loop system: (a) block-diagram representation; (b) signal-flow graph. The feedback function is denoted by H.

The feedback signal is opposite in sign from the reference input because, for reasons previously cited, *negative* feedback is desired. Therefore the expression for the output quantity with feedback becomes

$$C = G(R - HC) \tag{23-4}$$

or

$$C = \frac{G}{1 + HG}R \tag{23-5}$$

Moreover, it follows from Eq. (23-5) that the transfer function of the closed-loop system is

$$T \equiv \frac{C}{R} = \frac{G}{1 + HG} \tag{23-6}$$

where T denotes the closed-loop transfer function. The quantity HG is called the *loop transfer function* since it is obtained by taking the product of the transfer functions appearing in a traversal of the closed loop. A little thought indicates that a more general way of expressing Eq. (23-6) is to write

$$\text{closed-loop transfer function} = \frac{\text{direct transfer function}}{1 + \text{loop transfer function}} \tag{23-7}$$

The manner in which these transfer functions are found, once the system components are known, is treated in Chapter 24.

Effect of Feedback on Sensitivity to Parameter Changes.   A particularly noteworthy feature of feedback lies in its ability to reduce significantly the sensitivity of a system's performance to parameter changes. To understand how this is achieved, we define the sensitivity S of the closed loop system T to parameter change p in the direct transmission function G as

$$\text{sensitivity} = \frac{\text{per unit change in closed-loop transmission}}{\text{per unit change in open-loop transmission}} \qquad (23\text{-}8)$$

or, expressed mathematically,

$$S_p^T = \frac{\partial T/T}{\partial G_p/G_p} \qquad (23\text{-}9)$$

where G_p is used to denote that the derivative is to be taken with respect to the parameter p of the function G.

The partial derivative of the closed-loop transfer function with respect to the p parameter of the direct transfer function follows directly from Eq. (23-6)

$$\partial T = \frac{-HG_p\partial G_p}{(1 + HG_p)^2} + \frac{\partial G_p}{1 + HG_p} \qquad (23\text{-}10)$$

or

$$\partial T = \frac{\partial G_p}{(1 + HG_p)^2} \qquad (23\text{-}11)$$

Inserting Eqs. (23-11) and (23-6) into Eq. (23-9) yields

$$S_p^T = \frac{\partial T/T}{\partial G_p/G_p} = \frac{[\partial G_p/(1 + HG_p)^2]\,[(1 + HG_p)/G_p]}{\partial G_p/G_p} = \frac{1}{1 + HG_p}$$

or

$$\frac{\partial T}{T} = \left(\frac{1}{1 + HG_p}\right)\frac{\partial G_p}{G_p} \qquad (23\text{-}12)$$

A study of the last expression reveals that any per unit (or percentage) change in the direct transmission function brought about by a change in the p parameter carries over as a change in the closed-loop system that is only $1/(1 + HG_p)$ as large. Clearly, by making HG_p a large number, it is possible to reduce changes in G to insignificant proportions in T.

Let us illustrate these ideas by applying the procedure to the grounded-source amplifier shown in Fig. 23-13. The amplifier is enclosed by a dashed-line box called G to denote that this is the direct transmission function. The input

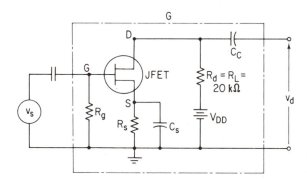

Fig. 23-13 Grounded-source JFET amplifier.

signal in this instance is v_s and the varying component of the corresponding output is v_d. From Eq. (9-174) the output is related to the input by

$$v_d = A_V v_s = \frac{-\mu}{1 + \dfrac{r_d}{R_L}} v_s \tag{23-13}$$

A comparison with Eq. (23-1) indicates that

$$G = \frac{-\mu}{1 + \dfrac{r_d}{R_L}} \tag{23-14}$$

Appearing in Fig. 23-14 is the same JFET amplifier of Fig. 23-13 but modified to include a portion of the output in the gate-to-source voltage. The feedback connection is achieved by applying the output quantity across the resistor R_f and then taking a portion, H, of this amount and placing it in series with the signal voltage. The net input to the G portion of the feedback amplifier of Fig. 23-14 thus becomes $(v_s + v_f)$. The varying component of the output voltage, v_d, is related to the net input by G. Expressed mathematically we have

$$(v_s + v_f)G = v_d \tag{23-15}$$

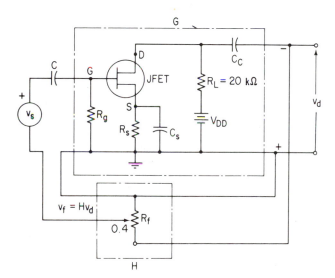

Fig. 23-14 JFET amplifier provided with voltage feedback.

It is important to note here that v_f is not preceded with a minus sign (which is often used to denote the negative feedback feature) because G already includes the minus sign owing to the phase reversal feature of the electronic amplifier. This is evident from Eq. (23-14) which upon insertion into Eq. (23-15) yields

$$(v_s + v_f) \frac{-\mu}{1 + r_d/R_L} = v_d \tag{23-16}$$

In the feedback circuit we have

$$v_f = H v_d \tag{23-17}$$

Introducing the last expression into Eq. (23-16) and collecting terms leads to

$$v_s \frac{-\mu}{1 + r_d/R_L} = \left(1 + \frac{\mu H}{1 + r_d/R_L}\right)v_d \qquad (23\text{-}18)$$

Finally, upon formulating the ratio of output to input, there results the transfer function of the complete feedback amplifier.

$$T = \frac{v_d}{v_s} = \frac{\dfrac{-\mu}{1 + r_d/R_L}}{1 + \dfrac{\mu H}{1 + r_d/R_L}} \qquad (23\text{-}19)$$

The feedback quantity H is simply a numeric in this case. Equation (23-19) discloses that the feedback amplifier also possesses the phase reversal feature, i.e., a *positive* change in input signal yields an amplified *negative* change in output signal. Furthermore, it is instructive to observe that the presence of the plus sign in the denominator corroborates the use of *negative* feedback.

For ease of illustration, assume that the JFET employed in Fig. 23-14 is the 2N4222 and that for the specified load resistor of 20 kΩ the operating point has the following associated parameters: $\mu = 40$ and $r_d = 20$ kΩ. Then when H is taken to be 0.4, the expression for the gain of the feedback amplifier becomes

$$T = \frac{v_d}{v_s} = \frac{-0.5\mu}{1 + 0.2\mu} \qquad (23\text{-}20)$$

This result can also be obtained directly from the block-diagram representation of Fig. 23-14 (which is depicted in Fig. 23-15) and the use of Eq. (23-6).

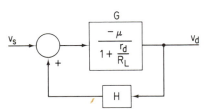

Fig. 23-15 Block-diagram representation of the feedback amplifier of Fig. 23-14. Note the use of the + sign at the output of H. No physical reversal at H is necessary because the phase reversal inherent in the transistor amplifier furnishes the required negative feedback.

The amplification factor μ is left explicit in order that we might now proceed to investigate the effect of changes in μ on the output quantity. The amplification factor can readily change with age as well as vary from transistor to transistor. Variations of the order of 20 per cent or more are not uncommon. When the JFET amplifier is used without feedback, the direct transmission function expressed in terms of μ is simply

$$G = -0.5\mu \qquad (23\text{-}21)$$

Accordingly, if there occurs a 20% drop in μ, the output quantity drops correspondingly by 20%. Thus, the per unit change in the direct transmission of the open-loop arrangement becomes

$$\frac{\partial G}{G} = \frac{-0.5\partial\mu}{-0.5\mu} = \frac{\partial\mu}{\mu} = -0.2 \text{ for a 20\% drop} \qquad (23\text{-}22)$$

On the other hand, with feedback the situation is significantly different. The transmission function is now described by Eq. (23-20). Hence the per unit decrease in the closed loop case now becomes, by Eq. (23-12),

$$\frac{\partial T}{T} = \frac{1}{1 + 0.5\mu H}\frac{\partial G}{G} = \frac{1}{1 + 0.2\mu}\frac{\partial \mu}{\mu} \tag{23-23}$$

$$= \frac{1}{1 + 0.2(40)}(-0.2) = -0.022$$

Thus, as a feedback amplifier a 20% change in the μ parameter of the direct transfer function reflects only as a 2.2% decrease in the closed-loop transmission.

The insensitivity to parameter changes associated with negative feedback is not obtained without a price. There always occurs a reduction in gain for a fixed G. This conclusion is immediately apparent upon comparing Eqs. (23-1) and (23-6). For the same G the gain as a closed-loop system is $1/(1 + HG)$ of the value as an open-loop system. For the example at hand the open-loop gain has a value of $G = -0.5\mu = -20$. But as a closed-loop system the over-all gain by Eq. (23-20) is

$$T = \frac{-0.5\mu}{1 + 0.2\mu} = \frac{-20}{1 + 8} = -2.2$$

Clearly, if it is desirable to maintain the same nominal gain of 20 for the closed-loop system between the output and input terminals, it becomes necessary to increase the direct transfer function accordingly. This matter is illustrated by the example that follows.

EXAMPLE 23-1   A voltage amplifier without feedback has a nominal gain of 400. Deviations in the amplifier parameters, however, cause the gain to vary in the range from 380 to 420, thus yielding a total per unit change in open-loop transmission of 0.1. It is desirable to introduce feedback to reduce the per unit change to 0.02 while maintaining the original gain of 400. Determine the new direct transfer gain G and the feedback factor H needed to achieve this result.

Solution: Let G' be the gain without feedback. Then

$$\frac{\partial G'}{G'} = \frac{420 - 380}{400} = 0.1$$

From Eq. (23-12)

$$\frac{\partial T}{T} = \frac{\partial G}{G}\left(\frac{1}{1 + HG}\right) = 0.02 \tag{23-24}$$

where G denotes the new direct transmission gain with feedback. Also, from Eq. (23-6)

$$T = \frac{G}{1 + HG} = 400 \tag{23-25}$$

From Eq. (23-24)

$$1 + HG = \frac{0.1}{0.02} = 5 \tag{23-26}$$

Inserting this into Eq. (23-25) yields

$$G = 400(1 + HG) = 400(5) = 2000 \tag{23-27}$$

Finally from Eq. (23-26)

$$H = \frac{5 - 1}{2000} = 0.002 \qquad (23\text{-}28)$$

Effect of Feedback on Dynamic Response and Bandwidth. The effect of feedback on the transient response of a circuit or system can be conveniently described by showing how it modifies the behavior of a system containing one energy-storing element. In this connection assume the direct transfer function† of the system to be given by

$$\frac{C}{E} = G = \frac{K}{1 + p\tau} = \frac{K/\tau}{p + 1/\tau} \qquad (23\text{-}29)$$

Figure 23-16(a) depicts the open-loop system. The presence of the p term indicates that there exists a time derivative relating the input and output variables. Since the denominator of Eq. (23-29), when equated to zero, is the characteristic equation of the system, it follows that the transient solution of the system described by G must be of the form

$$c_t = A\varepsilon^{-t/r} \qquad (23\text{-}30)$$

The transient in this system thus decays in accordance with a time constant of τ seconds.

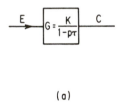

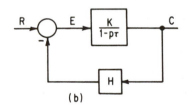

Fig. 23-16 Studying the effect of feedback on dynamic response and bandwidth of the open-loop system shown in (a) and the closed-loop system of (b).

(a) (b)

Consider next the case where a feedback path having the transfer function H is wrapped around the function G in the manner illustrated in Fig. 23-16(b). Now the transmission between input and output is represented by the closed-loop transfer function. Hence

$$T = \frac{C}{R} = \frac{\dfrac{K/\tau}{p + 1/\tau}}{1 + \dfrac{HK/\tau}{p + 1/r}} = \frac{K/\tau}{p + \dfrac{1 + HK}{\tau}} \qquad (23\text{-}31)$$

It follows therefore that the corresponding transient solution of the closed-loop system has the form

$$c_{tf} = A_f\varepsilon^{-[(1 + HK)/\tau]t} \qquad (23\text{-}32)$$

where c_{tf} denotes the transient response of the output variable with feedback.

† This transfer function is readily obtained from the differential equation describing the relationship between the input (E) and output (C) variables of the system. It merely requires writing the differential equation in operator form and then formulating the ratio of output to input variables (see Sec. 24-5).

A comparison of the last equation with the corresponding equation of the open-loop case, Eq. (23-30), immediately reveals that the time constant with feedback is smaller by the factor $1/(1 + HK)$. Hence, the transient decays faster.

The effect of feedback on the bandwidth of the system of Fig. 23-16(a) can also be determined from the foregoing equations. To speak of bandwidth is to speak of a frequency description of the system transfer function. This is readily accomplished by looking upon p as the sinusoidal frequency variable $j\omega$. Proceeding on this basis, then, for the direct transfer function of Eq. (23-29), it follows that the bandwidth spreads over a range from zero to a frequency of $1/\tau$ radians per second. Note that when $p = j\omega = j(1/\tau)$ is inserted in Eq. (23-29) the value of the gain is $0.707K$. On the other hand, for the system employing feedback Eq. (23-31) reveals that the bandwidth spreads from zero to $(1 + HK)/\tau$ radians per second. Thus the bandwidth has been augmented by increasing the upper frequency limit by $(1 + KH)$.

Effect of Feedback on Stability. By the judicious choice of the feedback transfer function H, it is possible to convert an inherently unstable system to a stable one. Appearing in Fig. 23-17(a) is the block-diagram representation of a system that is unstable. The transfer function is

$$G = \frac{C}{E} = \frac{K}{1 - p\tau} = \frac{-K/\tau}{p - 1/\tau} \qquad (23\text{-}33)$$

To understand why this system is unstable, assume that the input quantity E is a unit-step function. Then the expression for the output becomes

$$C(p) = \frac{-K/\tau}{p - 1/\tau}E(p) = \frac{-K/\tau}{p(p - 1/\tau)} \qquad (23\text{-}34)$$

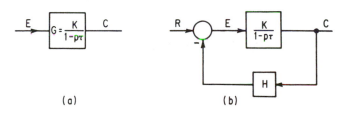

(a)

(b)

Fig. 23-17 (a) Direct transfer function of an unstable first-order system; (b) the same system with a feedback path wrapped around it.

By employing the procedure described in Chapter 4 we find the corresponding time solution to be

$$c(t) = K(1 - \varepsilon^{t/\tau}) \qquad (23\text{-}35)$$

The second term on the right is the transient component of the solution and, because the exponent has a positive power, it increases with time. Since the response increases without limit as time passes, a constant steady-state output consistent with the steady-state input is never reached. For this reason the system is said to be unstable. A stable system must have negative powers of the exponential terms associated with the transient components.

By placing a feedback path H around the direct transfer function in the manner depicted in Fig. 23-17(b) and applying the feedback formula of Eq. (23-

6) we obtain

$$T = \frac{C}{R} = \frac{\dfrac{K}{1 - p\tau}}{1 + \dfrac{KH}{1 - p\tau}} = \frac{K}{1 - p\tau + HK} \qquad (23\text{-}36)$$

The plus sign for HK is used in the denominator because negative feedback is assumed. A comparison of Eqs. (23-33) and (23-35) reveals that the increasing time response is attributable to the presence of the minus sign in the denominator of the direct transfer function [Eq. (23-33)]. We may describe this more elegantly by saying that the direct transfer function has a pole located in the right half-plane at $p = 1/\tau$. Examination of Eq. (23-36), however, indicates that by choosing

$$H = ap \qquad (23\text{-}37)$$

The closed-loop transfer function is modified to

$$T = \frac{K}{1 - p\tau + paK} \qquad (23\text{-}38)$$

Then, if we impose the condition that

$$aK > \tau \qquad (23\text{-}39)$$

it follows that the pole of the closed-loop transfer function is located in the left-half p plane. Accordingly, the time response of the closed-loop system, when subjected to a unit-step becomes

$$c_f(t) = K(1 - \varepsilon^{-t/(aK-\tau)}) \qquad (23\text{-}40)$$

where $c_f(t)$ denotes the response with feedback. With the constraint of Eq. (23-39) imposed, the exponential term of Eq. (23-40) decays to zero so that the response reaches steady state. The insertion of feedback has thus caused the unstable direct transmission system represented by G to become stable. This technique is frequently used to stabilize space rockets and vehicles, which are inherently unstable because of their large length-to-diameter ratios.

Summary review questions

1. Describe the notion of feedback.
2. Distinguish between open-loop control and closed-loop control. Use illustrations.
3. Name some disadvantages of open-loop control.
4. Define the following terms: actuating signal, reference input element, transducer, controlled variable, feedback element, error detector, negative feedback, servo-mechanism, sign sensitivity.
5. Name two practical automatic control systems and describe the manner of operation of each in response to an applied command.
6. Draw the block diagram and the signal-flow graph of a single-loop feedback system and explain the meaning of each symbol that is used.
7. Define in words and by equation the sensitivity of a feedback system.

8. Describe and illustrate the meaning of each of the following: direct transfer function, loop transfer function, closed-loop transfer function.

9. When the sensitivity to parameter changes of a device is improved by the addition of feedback, explain what happens to the gain between the input and output.

10. When a device with a single-energy storing element is equipped with feedback, explain how the dynamic behavior is affected, if at all.

11. Can feedback be used to impart stability to an inherently unstable device? Explain and illustrate.

Problems

GROUP I

23-1. The control system for remotely positioning the rudder of a ship is shown in Fig. P23-1. The resistor coil of the potentiometer is fastened to the frame of the ship. The desired heading is determined by the gyroscope setting (which is independent of the actual heading). Explain how the system operates in following a command for a northbound direction.

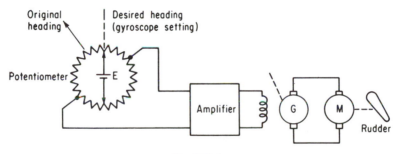

Fig. P23-1

23-2. Devise an elementary temperature-control system for a furnace fed by a valve-controlled fuel line. Use a Wheatstone-bridge error detector and a temperature-sensitive resistor such as the thermistor. Assume that the disturbance takes the form of heat losses to the ambient temperature. Describe how control is obtained. Also indicate how the desired reference temperature is established.

23-3. Negotiating a turn in the process of driving an automobile is an example of feedback control. In this instant the eyes, the brain, the arms, and the automobile come into play. Draw a block diagram depicting the interrelationhip of these elements and describe the operation.

23-4. Describe in block-diagram form a control system that allows trucks to be weighed as they pass over a platform. The readings are to be recorded at a station remotely located from the platform.

23-5. Because of aging, the parameters of the FET in a grounded-source amplifier undergo a change that causes a net per-unit decrease in gain of 25%. Several identically constructed amplifier circuits (called stages) having a nominal gain of 80 are available.

It is desirable, however, that despite aging an amplifier be designed that yields a gain between the input-output terminals of 80 with the change at no time exceeding 0.1%.

(a) Determine the minimum number of stages required to meet the specifications.

(b) Compute the corresponding feedback factor H.

23-6. It is desirable to have an amplifier with an overall gain of 1000 ± 20. The gain of any one stage is known to drift from 10 to 20. Find the required number of stages and the feedback function needed to meet these specifications. [*Hint*: Treat the closed-loop system as one that has an initial value of 980 and a final value of 1020. Similarly, consider each stage as having an initial value of 10 and a final value of 20.]

23-7. A system has a direct transmission function expressed as

$$G(p) = \frac{10}{p^2 + p - 2}$$

(a) Explain clearly why the system as it stands is unstable.

(b) A feedback path having the transfer function H is put around $G(p)$. Is it possible to make the feedback system stable? Explain. Be specific.

23-8. Refer to the circuitry of Fig. 23-14.

(a) Draw the block diagram for this feedback amplifier.

(b) Derive the general expression for the voltage gain using the feedback formula of Eq. (23-6). Express the result in terms of the feedback factor H where $0 \leqslant H \leqslant 1$.

GROUP II

23-9. A heavy mass rests on a horizontal slab of steel. The slab is hinge-supported on one side and rests on an adjustable support on the other side. The hinged support is subjected to random vertical displacements. The mass is to be kept level by changing the height of the adjustable support. Devise a feedback control system to achieve this task.

23-10. Devise a control system to regulate the thickness of sheet metal as it passes through a continuous rolling mill (see Fig. P23-10). The control system should not only keep the thickness within suitable tolerances but permit the adjustment of the thickness as well. Show a clearly labeled diagram of the system.

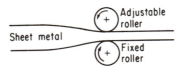

Fig. P23-10

23-11. The speed of a gasoline engine is to be controlled in accordance with a command that is in the form of a voltage. Devise a control system that is able to provide this kind of control. Describe the operation of your system.

23-12. A medical clinic calls upon you to design a servo anesthetizer for the purpose of regulating automatically the depth of anesthesia in a patient in accordance with the level of energy output of brain-wave activity as measured by an electroencephalograph (EEG). The anesthesia is to be administered by means of a hypodermic syringe which is activated by a stepping relay fed from an appropriate amplifying source. Show a diagram of the control system, and state clearly what constitutes the error detector, the feedback, and the corrector. Explain the way the system operates.

23-13. Repeat Prob. 23-7 for the case where

$$G(p) = \frac{10}{p(p^2 + p - 2)}$$

chapter twenty-four

Dynamic Behavior of Control Systems

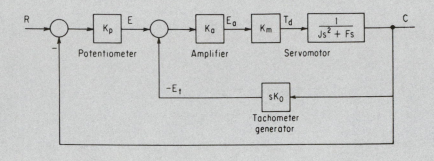

In the course of the development of the subject matter of this part of the book the dynamic behavior of feedback control systems is analyzed by several methods, each of which has a place and purpose in the over-all picture. A necessary starting point in each is to obtain the differential equation describing the system. The first approach, which constitutes the chief subject matter of this chapter, involves a direct solution of the differential equation by using the procedure outlined in Chaps. 4 and 6.

Usually the complete solution of the differential equation provides maximum information about the system's dynamic performance. Consequently, whenever it is convenient, an attempt is made to establish this solution first. Unfortunately, however, this is not easily accomplished for high-order systems, and, what is more, the design procedure that permits reasonably direct modification of the transient solution is virtually impossible to achieve. In such cases we are forced to seek out other, easier, more direct methods, such as the frequency-response method of analysis.

The existence of transients is characteristic of systems that possess energy-storing elements and that are subjected to disturbances. Usually the disturbance is applied at the input or output end or both. Often oscillations are associated with the transients, and, depending upon the magnitude of the system parameters, such oscillations may even be sustained.

844

24-1 DYNAMIC RESPONSE OF THE SECOND-ORDER SERVOMECHANISM

The system to be analyzed is shown in Fig. 24-1. It contains two energy-storing elements in the form of inertia and shaft stiffness, making it a second-order system. Here stiffness refers to the torsional elasticity of the servoamplifier-servomotor combination which causes the system to act as though an elastic shaft connected the input to the inertial load. The application of a command in the form of a displacement of the slider arm of the input potentiometer causes an actuating signal e to exist. The amplified version of e is then applied to the servomotor, which in turn develops a torque and causes movement of the load.

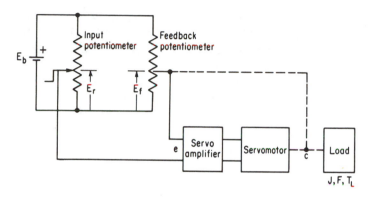

Fig. 24-1 Second-order servomechanism. A direct connection is assumed between the servomotor and the feedback potentiometer as well as the load.

The motor developed torque, T_d, may thus be represented by

$$T_d = eK_aK_m \qquad (24\text{-}1)$$

where K_a = amplifier gain factor, V/V
K_m = motor developed torque constant, lb-ft/V

Moreover, the actuating signal e may be expressed in terms of the potentiometer displacements as

$$e = K_p(r - c) \qquad (24\text{-}2)$$

where K_p is the potentiometer transducer constant expressed in V/rad, r is the input command in radians, and c is the output displacement in radians. For simplicity it has been assumed that there is no gear reduction between the output shaft and the load or between the output shaft and the feedback potentiometer. Introducing Eq. (25-2) into Eq. (25-1) allows the motor developed torque to be expressed as

$$T_d = (r - c)K_pK_aK_m = (r - c)K \qquad (24\text{-}3)$$

where

$$K \equiv K_p K_a K_m \, \frac{\text{lb–ft}}{\text{rad}} \tag{24-4}$$

If, as indicated in Fig. 24-1, it is assumed that there appears at the motor shaft an effective inertia J, an equivalent viscous friction F, and a resultant load torque T_L, it then follows that the developed motor torque must be equal to the sum of the opposing torques represented by these quantities. Thus

$$T_d = \sum \text{opposing torques} \tag{24-5}$$

or

$$(r - c)K = J\frac{d^2c}{dt^2} + F\frac{dc}{dt} + T_L \tag{24-6}$$

The J parameter accounts for the inertia of the load itself as well as the inertia of the rotating member of the servomotor. The viscous friction parameter F is very often determined to a large extent by the drooping slope of the torque-speed characteristic of the servomotor.

Rearranging Eq. (24-6) leads to the preferred form of expressing the defining differential equation for obtaining the dynamic behavior of the system. Thus

$$\boxed{J\frac{d^2c}{dt^2} + F\frac{dc}{dt} + Kc + T_L = Kr} \tag{24-7}$$

Step-Position Input. It is assumed that the command r applied to the system described by Eq. (24-7) is a step of magnitude r_0 radians. Furthermore, it is assumed that the system is resting at null prior to the application of the command. Hence the displacement and velocity immediately after the input is applied must be zero, because no movement of the output member can occur in infinitesimal time in the presence of a finite forcing function. The two required initial conditions for the solution of Eq. (24-7) may thus be written as

$$c(0^+) = 0 \quad \text{and} \quad \frac{dc}{dt}(0^+) = \dot{c}(0^+) = 0 \tag{24-8}$$

By Laplace transforming Eq. (25-7), it becomes

$$J[s^2C(s) - sc(0^-) + c(0^-)] + F[sC(s) - c(0^-)] \\ + KC(s) + T_L(s) = KR(s) \tag{24-9}$$

where $C(s) = \mathscr{L}c(t)$ and $c(t)$ denotes the desired time solution of the controlled variable. Moreover, $T_L(s)$ and $R(s)$ represent the Laplace transforms of the disturbances applied at the output and input ends of the servomechanism. When a step change in load of magnitude T_L and a step command of magnitude r_0 are assumed applied simultaneously for the initial conditions of Eq. (24-8), the preceding equation becomes

$$C(s)[s^2J + sF + K] = \frac{Kr_0}{s} - \frac{T_L}{s} \tag{24-10}$$

where $R(s) = r_0/s$ and $T_L(s) = T_L/s$. Thus the solution for the controlled variable in the s domain is

$$C(s) = \frac{(K/J)r_0}{s\left(s^2 + s\frac{F}{J} + \frac{K}{J}\right)} - \frac{T_L/J}{s\left(s^2 + s\frac{F}{J} + \frac{K}{J}\right)} \tag{24-11}$$

It is interesting to note in passing that, with the Laplace transform method of solving the system differential equation, the solution for two disturbances is found with little more effort than that required for one.

Even at this early point in the solution some useful information may be obtained from an inspection of Eq. (24-11). For example, if T_L is zero, then at steady state $c_{ss} = r_0$ so that exact correspondence exists between the command and the controlled variable. On the other hand, in the presence of a fixed load torque the steady-state value of the controlled variable becomes

$$c_{ss} = r_0 - \frac{T_L}{K} \tag{24-12}$$

Therefore an error in position exists which is directly proportional to the load torque and inversely proportional to the gain K. It will be recalled that these are the same conclusions arrived at in Sec. 23-3 through a process of physical reasoning. Another notable observation is that the denominators of both terms on the right side of Eq. (24-11) involve the same quadratic expression. In fact this quadratic is precisely the characteristic equation of the system which follows from Eq. (24-7) when the external disturbances are set equal to zero. Consequently the source-free modes are of the same character whether the disturbance with which they are associated originates at the input or at the output end.

To determine the time-domain solution corresponding to the s-plane solution of Eq. (24-11), we must find the roots of the characteristic equation. Because of the frequent occurrence of such quadratic terms in control systems work, particular attention is directed to this case with a view toward generalizing the results as was done for the second-order RLC circuit in Sec. 5-3. Starting with the characteristic equation as it appears in Eq. (24-11), we have

$$s^2 + \frac{F}{J}s + \frac{K}{J} = 0 \tag{24-13}$$

The roots are

$$s_{1,2} = -\frac{F}{2J} \pm \sqrt{\left(\frac{F}{2J}\right)^2 - \frac{K}{J}} \tag{24-14}$$

Depending on the radical term, the response can be any one of the following: (1) overdamped if $(F/2J)^2 > K/J$; (2) underdamped if $(F/2J)^2 < K/J$; (3) critically damped if $(F/2J)^2 = K/J$. Often a desirable feature of a servomechanism is that it be fast-acting. Unless it is otherwise specified, therefore, we shall confine our attention to the underdamped case because it responds by moving quickly to its new commanded state in spite of the fact that it may then oscillate about the new level before settling down.

Critical damping occurs whenever the damping term (in this case F) is related to the gain and inertia parameters by

$$\boxed{F_c = 2\sqrt{KJ}} \tag{24-15}$$

For the underdamped situation the roots may be written as

$$s_{1,2} = -\frac{F}{2J} \pm j\sqrt{\frac{K}{J} - \left(\frac{F}{2J}\right)^2} \tag{24-16}$$

To put these roots in a form that will give greater significance in terms of the resulting time response, we introduce the following definition for the damping ratio:

$$\boxed{\zeta \equiv \frac{\text{total damping}}{\text{critical damping}} = \frac{F}{2\sqrt{KJ}}} \tag{24-17}$$

and the following definition for the system natural frequency:

$$\boxed{\omega_n \equiv \sqrt{\frac{K}{J}}} \tag{24-18}$$

Accordingly,

$$\frac{F}{2J} = \frac{\zeta F_c}{2J} = \frac{2\zeta\sqrt{KJ}}{2J} = \zeta\omega_n \tag{24-19}$$

and

$$\sqrt{\frac{K}{J} - \left(\frac{F}{2J}\right)^2} = \sqrt{\omega_n^2 - \zeta^2\omega_n^2} = \omega_n\sqrt{1 - \zeta^2} \tag{24-20}$$

Therefore Eq. (24-16) may be written as

$$\boxed{s_{1,2} = -\zeta\omega_n \pm j\omega_n\sqrt{1 - \zeta^2} = -\zeta\omega_n \pm j\omega_d} \tag{24-21}$$

where

$$\boxed{\omega_d \equiv \omega_n\sqrt{1 - \zeta^2} = \text{damped frequency of oscillation}} \tag{24-22}$$

A distinguishing feature of Eq. (24-21) is that both the real and j parts have the unit of inverse seconds, which is frequency. This in fact gives rise to the term *complex frequency*. The practical interpretation is that the real part of the complex frequency represents the damping factor associated with the decaying transients, while the imaginary (or j part) refers to the actual frequency of oscillation at which the underdamped response decays.

In terms of the quantities ζ and ω_n Eq. (24-11) can be rewritten

$$C(s) = \frac{\omega_n^2 r_0}{s(s^2 + 2\zeta\omega_n s + \omega_n^2)} - \frac{T_L/J}{s(s^2 + 2\zeta\omega_n s + \omega_n^2)} \tag{24-23}$$

To determine the time solution corresponding to Eq. (24-23) we need a partial fraction expansion. First, however, we shall set $T_L = 0$ in order to simplify the algebraic manipulations. Nothing is thereby lost because the forms of the two terms on the right side of the last equation are similar. Thus the equation to be solved is reduced to

$$C(s) = \frac{\omega_n^2 r_0}{s(s^2 + 2\zeta\omega_n s + \omega_n^2)} \tag{24-24}$$

As a consequence of the simplicity of this expression, a formal partial fraction expansion is unnecessary. A little thought discloses that the right side of Eq. (24-24) may be rearranged as

$$C(s) = \frac{r_0}{s} - \frac{(s + 2\zeta\omega_n)r_0}{s^2 + 2\zeta\omega_n s + \omega_n^2} \tag{24-25}$$

Moreover, in the interest of putting Eq. (24-25) in a form each term of which is immediately identifiable in Table 6-1, it can be written as

$$C(s) = \frac{r_0}{s} - r_0 \left[\frac{(s + \zeta\omega_n)}{(s + \zeta\omega_n)^2 + \omega_d^2} + \zeta\frac{\omega_n}{\omega_d}\frac{\omega_d}{(s + \zeta\omega_n)^2 + \omega_d^2} \right] \tag{24-26}$$

The first term in brackets is readily seen to be a damped cosine function and the second term is the damped sine function. Accordingly, the time solution corresponding to Eq. (24-26) is

$$c(t) = r_0 - r_0\varepsilon^{-\zeta\omega_n t}\left(\cos \omega_d t + \frac{\zeta}{\sqrt{1 - \zeta^2}} \sin \omega_d t \right) \tag{24-27}$$

This last equation can be further simplified as outlined in Sec. 5-4 to

$$\boxed{c(t) = r_0 \left[1 - \frac{\varepsilon^{-\zeta\omega_n t}}{\sqrt{1 - \zeta^2}} \sin (\omega_d t + \theta) \right]} \tag{24-28}$$

where

$$\theta = \tan^{-1}\frac{\sqrt{1 - \zeta^2}}{\zeta} \tag{24-29}$$

and is valid for a step input, $0 \le \zeta \le 1$, and $T_L = 0$.

Undamped Solution. One extreme condition of Eq. (24-28) occurs when $\zeta = 0$. This makes $\theta = 90°$ so that

$$c(t) = r_0[1 - \sin (\omega_d t + 90°)] = r_0(1 - \cos \omega_n t) \tag{24-30}$$

The solution for the controlled variable is therefore a sustained oscillation. This result is entirely expected because the system has two energy-storing elements and zero damping (see Fig. 24-2). By Eq. (24-22) the frequency of oscillation is

$$\omega_d = \omega_n = \sqrt{\frac{K}{J}} \tag{24-31}$$

which is the natural frequency of the system. It depends upon the two energy-storing elements K (stiffness) and J (inertia).

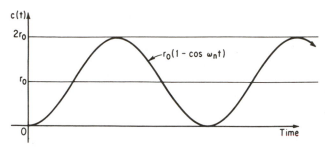

Universal Curves for Second-Order Systems. It is useful to depict the response of second-order systems graphically for specific values of damping ratios independently of the magnitude of the input step command or the system parameters. The response of Eq. (24-28) can readily be made independent of the magnitude of the step input by plotting $c(t)/r_0$ rather than $c(t)$. Thus the solution is normalized with respect to the value of the step input. This means that for any given system with a fixed damping ratio a single curve may be used to represent the response irrespective of the magnitude of the input signal r_0.†

A germane question at this point, however, is: Can this same curve be used for a different second-order system having the same ζ but another value of ω_n? A little thought makes it clear that this cannot be done because of the role played by $\omega_n t$ in the exponential and in the argument of the sine term in Eq. (24-28). In order to use a single response curve so that it applies for all linear second-order systems, it is necessary to plot time in a nondimensional form which masks the effect of the natural frequency parameter. Thus, if the magnitude-normalized response is plotted not as a function of time t directly but rather as a function of $\omega_n t$, then the resulting plot is truly universal, since it applies to all linear second-order systems. These plots are shown in Fig. 24-3 for various values of ζ. They describe the response to step input commands.

Figures of Merit Identifying the Transient Response. As was pointed out in the analysis of the *RLC* circuit in Sec. 5-3, the dynamic behavior of any linear second-order system is readily described in terms of two figures of merit: ζ and ω_n. To understand what each quantity conveys about the transient response, let us determine the time at which the maximum overshoot occurs. Differentiating Eq. (24-28) with respect to time yields upon simplification

$$\frac{d}{dt}c(t) = \frac{\omega_n}{\sqrt{1 - \zeta^2}}\varepsilon^{-\zeta\omega_n t}\sin\omega_n t \qquad (24\text{-}32)$$

Clearly, then, the first and, therefore, the peak overshoot occurs when

$$\omega_d t = \pi \qquad (24\text{-}33)$$

or

$$\omega_n t = \frac{\pi}{\sqrt{1 - \zeta^2}} \qquad (24\text{-}34)$$

† It is assumed of course that saturation effects do not exist. In any practical system this is a real restriction.

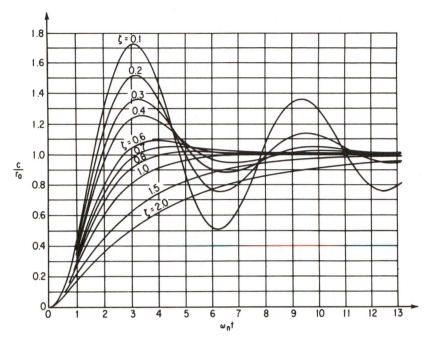

Fig. 24-3 Universal transient response curves for second-order systems subjected to a step input.

If this result is put into Eq. (24-28), the maximum instantaneous value of the controlled variable is found to be

$$\frac{c_{max}}{r_0} = 1 + \varepsilon^{-\zeta\pi/\sqrt{1-\zeta^2}} \qquad (24\text{-}35)$$

Hence the maximum percent overshoot becomes

$$\boxed{\frac{c_{max} - c_{ss}}{c_{ss}} = \frac{c_{max} - r_0}{r_0} = 100\varepsilon^{-\zeta\pi/\sqrt{1-\zeta^2}} \%} \qquad (24\text{-}36)$$

where c_{ss} is the steady-state value of the output. Figure 24-4 shows the plot of Eq. (24-36). It is important to note that the maximum overshoot depends solely upon the value of ζ. Therefore, the damping ratio is looked upon as a figure of merit which provides information about the maximum overshoot in the system when it is excited by a step forcing function. This is a common and very useful way of obtaining information about the dynamic behavior of systems.

The information conveyed by the value of the system's natural frequency ω_n can be deduced from an examination of the output response to a step input plotted nondimensionally for a given ζ. Such a plot is shown in Fig. 24-5. Note the inclusion of a $\pm 5\%$ tolerance band. A system can often be considered as having reached steady state once the response stays confined within the limits of this band. The time it takes to do this is called the *settling time*. As a practical matter the response time of all systems is necessarily gauged in terms of a suitable

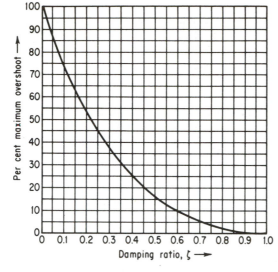

Fig. 24-4 Percent maximum overshoot versus damping ratio for linear second-order system.

tolerance band, however large or small the particular specifications may demand. For the situation illustrated in Fig. 24-5 it follows that the settling time is inversely proportional to the natural frequency. Thus, if the response curve shown in Fig. 24-5 applies to two different systems having the same ζ, then clearly the one with the larger natural frequency will have the smaller settling time in responding to input commands or load disturbances. Consequently, the natural frequency of a system may be interpreted as a figure of merit which provides a measure of the settling time.

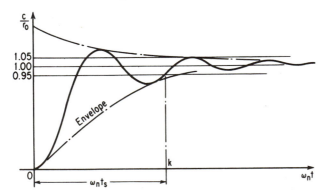

Fig. 24-5 Settling time of a step response identified in terms of a ± 5% tolerance band.

Another way to arrive at this conclusion is to refer to Eq. (24-28) and to note that the magnitude of the transient term is controlled by the value of the exponential $\varepsilon^{-\zeta\omega_n t}$. It is possible to speak of a time constant for the second-order system:

$$T = \frac{1}{\zeta\omega_n}$$ (24-37)

Equation (24-37) reveals too that for a fixed ζ, the time constant is inversely proportional to ω_n. Expressed in terms of the time constant, it can then be said that the response to a step input will reach 95 percent of its steady-state value within three time constants. That is, the settling time is no greater than $t_s = 3/\zeta\omega_n$. For a 1 percent tolerance band the settling time will be no greater than $t_s = 5/\zeta\omega_n$.

Transfer Function Analysis. For the sake of completeness and in the interest of applying the results described in Sec. 23-4, we next analyze the closed-loop response of the system of Fig. 24-1 by developing the appropriate block diagram and then applying Eq. (23-6). The general procedure in finding the transfer function of a component is to identify first the equation that relates the output variable to the input variable. Then in those instances where time derivatives and time integrals are involved, the next step is to Laplace-transform the defining integrodifferential equation assuming all initial conditions are zero. This assumption is made because in determining the transfer function we are looking to identify the characteristic equation, which depends solely upon the parameters and which establishes the nature of the dynamic response. Recall that the initial conditions affect only the magnitudes of the transient terms but not their basic decaying characteristics. Finally, the transfer function is obtained by formulating the ratio of the output to input variable.

Let us apply this procedure to each of the components that make up the system of Fig. 24-1. The input command takes the form of a displacement of r radians applied to the slider arm of the input potentiometer. In turn this makes available at the electrical wire attached to the slider arm of the potentiometer a new level of voltage. For a given r displacement the amount of voltage change depends, of course, upon how much excitation voltage E_b is used. It should be apparent, then, that the potentiometer is a transducer that converts a mechanical input variable r expressed in radians to a corresponding output voltage expressed in volts. Accordingly we can say that the potentiometer has a transfer function

$$\frac{E_b}{R} = K_p \qquad \text{V/rad} \qquad (24\text{-}38)$$

where E_r is the output voltage at the slider arm corresponding to displacement R. Capital letters are used to stress that we are dealing with the Laplace transform versions of the variables. The potentiometer in the feedback path has the same transfer constant except that the output and input quantities are called E_f and C, respectively. Figure 24-6 shows both potentiometer transfer constants at the proper places in the block diagram.

Continuing with the development of the block diagram, note in Fig. 24-1 that the electrical connections to the servoamplifier are such that its input quantity E is obtained as the difference between E_r and E_f. This is represented in the block diagram of Fig. 24-6 by the circle, which is the symbol for a summing point. The transfer function of the amplifier in this case is merely a gain factor which raises the level of the input voltage E to E_a. Thus

$$\frac{E_a}{E} = K_a \qquad \text{V/V} \qquad (24\text{-}39)$$

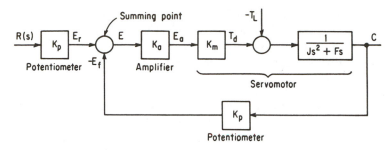

Fig. 24-6 Block diagram of the servomechanism of Fig. 24-1, showing the transfer function of each component.

If the application of a voltage E_a to the input terminals of the servomotor is assumed to develop an output which is the developed torque T_d, the gain factor associated with the motor is clearly

$$\frac{T_d}{E_a} = K_m \qquad \frac{\text{lb–ft}}{\text{V}} \qquad (24\text{-}40)$$

Finally, the manner in which this developed torque manifests itself as a displacement angle c is described by the differential equation relating these two variables, namely,

$$T_d = J\frac{d^2c}{dt^2} + F\frac{dc}{dt} + T_L \qquad (24\text{-}41)$$

The presence of a load disturbance may be conveniently treated by placing it on the left side in Eq. (24-41).

$$(T_d - T_L) = J\frac{d^2c}{dt^2} + F\frac{dc}{dt} \qquad (24\text{-}42)$$

Again a small circle representing the summing operation called for on the left side of the last equation is used in the block-diagram notation. Laplace transforming Eq. (24-42) for zero initial conditions leads to

$$T_d(s) - T_L(s) = Js^2C(s) + FsC(s) \qquad (24\text{-}43)$$

Formulating the ratio of output to input yields the transfer function, which describes how torque is carried over to output displacement. Thus

$$\frac{C(s)}{T_d(s) - T_L(s)} = \frac{1}{Js^2 + Fs} \qquad (24\text{-}44)$$

By interconnecting the output of one component to the input of the following component in the system, we arrive at the complete block diagram of Fig. 24-6. This block diagram can be simplified by observing that, since the actuating signal is given by $E = K_p(R - C)$, the potentiometer transfer factor may be removed from the input and feedback paths and placed instead in the direct transmission path as shown in Fig. 24-7. Also, for simplicity, T_L is assumed to be zero.

The direct transmission function for the system of Fig. 24-7 can clearly be

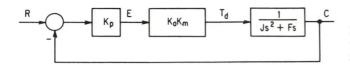

Fig. 24-7 Simplified form of Fig. 24-6 with the load torque assumed to be zero.

identified as

$$G(s) = \frac{K_p K_a K_m}{Js^2 + Fs} = \frac{K}{Js^2 + Fs} \tag{24-45}$$

By Eq. (23-2) the corresponding closed-loop transfer function is then

$$T(s) = \frac{C(s)}{R(s)} = \frac{G(s)}{1 + HG(s)} = \frac{\dfrac{K}{Js^2 + Fs}}{1 + \dfrac{K}{Js^2 + Fs}} = \frac{K}{Js^2 + Fs + K}$$

or

$$\frac{C(s)}{R(s)} = \frac{\dfrac{K}{J}}{s^2 + \dfrac{F}{J}s + \dfrac{K}{J}} \tag{24-46}$$

It is important to note that the output variable, as well as the input command, is represented in terms of their Laplace transforms. To stress the point capital letters are employed along with the "of s" notation; i.e., $C(s)$ and $R(s)$. This procedure is entirely consistent with the fact that the transfer function often involves operations of time, which in the Laplace transform language means that the transfer function is an algebraic function of s. A glance at the right side of Eq. (24-46) corroborates this for the case at hand. Another notable observation in Eq. (24-46) is that the denominator expression of the closed-loop transfer function, when equated to zero, yields the characteristic equation of the closed-loop system. Thus it is that the transfer function is the means through which information about the dynamic behavior of systems can be readily and conveniently accounted for in the process of analysis. It should also be apparent that the use of the Laplace transformation not only lends great facility to the algebraic manipulations involved in arriving at solutions but also makes available a convenient procedure of converting from the s domain to the time domain and vice versa.

Once the transfer function of a system is determined and put in a form similar to that of Eq. (24-46), the transformed solution can be written down forthwith. For the system described by Eq. (24-46) this becomes

$$C(s) = R(s) \frac{\dfrac{K}{J}}{s^2 + \dfrac{F}{J}s + \dfrac{K}{J}} = R(s) \frac{\omega_n^2}{s^2 + 2\zeta\omega_n s + \omega_n^2} \tag{24-47}$$

The particular form of the forced solution depends upon the type of forcing function used. If a step input is applied of magnitude r_0, then $R(s) = r_0/s$ so that the complete transformed solution becomes

$$C(s) = \left(\frac{r_0}{s}\right)\frac{\omega_n^2}{s^2 + 2\zeta\omega_n s + \omega_n^2} \tag{24-48}$$

This is the same result shown in Eq. (24-24), which was found directly from the differential equation describing the closed-loop operation. Inspection of Eq. (24-48) reveals that the transient terms resulting from a partial fraction expansion are associated with the poles of the denominator of the closed-loop transfer function. On the other hand the steady-state solution is generated by the pole associated with $R(s)$.

EXAMPLE 24-1 A feedback control system has the configuration shown in Fig. 24-8. The relationship between the output variable c and the input to the direct transmission path e is

$$\frac{d^2c}{dt^2} + 8\frac{dc}{dt} + 12c = 68e \tag{24-49}$$

(a) When the system is operated open-loop (i.e., the feedback path removed), find the complete solution for the output variable for $e = u(t)$. Assume the system is initially at rest.

(b) Describe the nature of the dynamic response of the controlled variable when the feedback loop is connected and the forcing function is again a unit step, i.e., $R(s) = 1/s$. Find the solution by working with the differential equation for the closed-loop system.

(c) Identify the closed-loop behavior by working solely in terms of transfer functions.

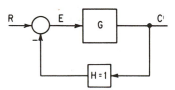

Fig. 24-8 Block diagram for Example 24-1.

Solution: (a) The complete solution of the open-loop response is the solution to Eq. (24-49). If we Laplace transform this equation, we obtain for zero initial conditions

$$s^2C(s) + 8sC(s) + 12C(s) = \frac{68}{s} \tag{24-50}$$

Hence the transformed solution is

$$C(s) = \frac{68}{s(s^2 + 8s + 12)} = \frac{68}{s(s + 2)(s + 6)} = \frac{K_0}{s} + \frac{K_1}{s + 2} + \frac{K_2}{s + 6} \tag{24-51}$$

Evaluating the coefficients of the partial fraction expansion in the usual fashion yields

$$C(s) = \frac{17}{3}\left(\frac{1}{s}\right) - \frac{17}{2}\left(\frac{1}{s + 2}\right) + \frac{17}{6}\left(\frac{1}{s + 6}\right) \tag{24-52}$$

The corresponding time solution is then

$$c(t) = \frac{17}{3} - \frac{17}{2}\varepsilon^{-2t} + \frac{17}{6}\varepsilon^{-6t} \tag{24-53}$$

Equation (24-53) indicates that the transient response of this open-loop system takes the form of two exponentially decaying terms—one having a time constant of $\frac{1}{2}$ s and the other a time constant of $\frac{1}{6}$ s. This general information could be deduced directly from Eq. (24-51) once the root factors of the denominator are known.

(b) The essential difference between the differential equation describing open-loop operation and that describing closed-loop operation is that in the latter the variable e is replaced by $(r - c)$. Thus

$$\frac{d^2c}{dt^2} + 8\frac{dc}{dt} + 12c = 68(r - c) \tag{24-54}$$

Now the differential equation describing the relationship between the output and input variables of the closed-loop system takes the form

$$\frac{d^2c}{dt^2} + 8\frac{dc}{dt} + 80c = 68r \tag{24-55}$$

To describe the nature of the dynamic response of the closed-loop system as represented by Eq. (24-55) it is no longer necessary to obtain a formal solution for $c(t)$. Instead, use is made of the information conveyed by the two figures of merit ζ and ω_n. Both these quantities follow almost immediately from the characteristic equation of the closed-loop system, which for this case is

$$s^2 + 8s + 80 = 0 \tag{24-56}$$

Of course Eq. (24-56) comes from Eq. (24-55) when the input is set equal to zero and the left side is Laplace transformed with zero initial conditions. The general form of this equation, as it applies to all linear second-order systems, is

$$s^2 + 2\zeta\omega_n s + \omega_n^2 = 0 \tag{24-57}$$

Comparing coefficients leads to

$$\omega_n = \sqrt{80} = 8.95 \text{ rad/s} \tag{24-58}$$

and

$$\zeta = \frac{8}{2\omega_n} = \frac{4}{8.95} = 0.448 \tag{24-59}$$

From Fig. 24-4 we find the maximum overshoot to be 17%. Also, since $\zeta\omega_n = 4$, it follows that the controlled variable reaches within 1% of its steady-state value after the elapse of

$$t_s = \frac{5}{\zeta\omega_n} = \frac{5}{4} = 1.25 \text{ s} \tag{24-60}$$

(c) The transfer function of the direct transmission path follows from Eq. (24-49). Thus

$$\frac{C(s)}{E(s)} = G(s) = \frac{68}{s^2 + 8s + 12} \tag{24-61}$$

The closed-loop transfer function then becomes

$$T(s) = \frac{C(s)}{R(s)} = \frac{G(s)}{1 + HG(s)} = \frac{\dfrac{68}{s^2 + 8s + 12}}{1 + \dfrac{68}{s^2 + 8s + 12}} \tag{24-62}$$

or

$$\frac{C(s)}{R(s)} = \frac{68}{s^2 + 8s + 80} \tag{24-63}$$

Because the denominator expression of Eq. (24-63) is identical to that of Eq. (24-56), the damping ratio and the natural frequency have the same values as found in part (b). Hence the dynamic behavior can again be described as oscillatory, having a maximum overshoot of 17%, and requiring 1.25 s to reach with 1% of steady state.

24-2 ERROR-RATE CONTROL

To meet the requirements for the steady-state performance as well as the dynamic performance of the feedback control system of Fig. 24-1, it is necessary to provide independent control of both performances. In the system of Fig. 24-1 the viscous friction coefficient F is associated with the speed-torque characteristic of the output actuator and is generally not adjustable. Furthermore, the inertia term is kept as small as is physically possible in the interest of achieving a large natural frequency. Therefore the sole adjustable parameter as the system now stands is the loop gain K. This is not sufficient to yield satisfactory steady-state and dynamic performance. The reason is apparent from an examination of Eqs. (24-12) and (24-17). If a command r_0 is applied to the control system in the presence of an external torque, Eq. (24-12) indicates that in the steady state the output displacement will lag behind the command by the amount T_L/K. Clearly, to diminish this position lag error to tolerable limits, it becomes necessary to use high values of the loop gain K. In such circumstances Eq. (24-17) shows that the dynamic behavior may significantly deteriorate because of the inverse relationship between ζ and K. Therefore, if K is to be adjusted to meet accuracy requirements in the steady state, another parameter—the damping coefficient— must be made adjustable to furnish acceptable dynamic behavior at the same time.

The need to control the accuracy performance in steady state is often present even in cases where there are no external disturbances to speak of, but where very high accuracy is to be achieved. Under such circumstances the imperfections of the system components loom large—such as coulomb friction, bearing friction, and dead-zone effects in potentiometers, servomotors, and the like. The use of high loop gains can greatly minimize these effects.

Although there are several methods that can be used to furnish the control being sought, only two are truly practical and have thereby gained widespread acceptance. The first of these makes use of error-rate damping to control the dynamic response, while the second employs output-rate damping and is discussed in the next section.

A system possesses error-rate damping when the generation of the output in some way depends upon the rate of change of the actuating signal. For the system of Fig. 24-1 a convenient way of introducing error-rate damping is to design the amplifier so that it provides an output signal containing a term proportional to the derivative of the input as well as one proportional to the input itself. Thus, if the input voltage to the amplifier is $e = K_p(r - c)$, the output voltage of the amplifier e_a will then be

$$e_a = \left(K_a + K_e\frac{d}{dt}\right)e = K_ae + K_e\frac{de}{dt} \qquad (24\text{-}64)$$

where K_e denotes the error-rate gain factor of the amplifier. The developed torque of the servomotor is found by multiplying Eq. (24-64) by K_m, the motor torque constant. Thus the differential equation for the system becomes

$$K_aK_me + K_eK_m\frac{de}{dt} = J\frac{d^2c}{dt^2} + F\frac{dc}{dt} + T_L \qquad (24\text{-}65)$$

If we next insert $e = K_p(r - c)$ and rearrange terms, Eq. (24-65) may be expressed as

$$J\frac{d^2c}{dt^2} + (F + Q_e)\frac{dc}{dt} + Kc = Kr + Q_e\frac{dr}{dt} - T_L \qquad (24\text{-}66)$$

where

$$K = \text{loop proportional gain factor} = K_pK_aK_m \qquad (24\text{-}67)$$

$$Q_e = \text{loop error-rate gain factor} = K_pK_eK_m \qquad (24\text{-}68)$$

Equation (24-66) is the governing differential equation for the system when error-rate damping is included. A comparison with Eq. (24-7), which applies to the linear second-order system without error-rate damping, reveals that the significant difference lies in the makeup of the coefficient of dc/dt. For the system with error-rate control this coefficient is augmented by the amount Q_e and is separately adjustable by K_e. Accordingly the characteristic equation is modified from that of Eq. (24-13) to the following:

$$s^2 + \frac{F + Q_e}{J}s + \frac{K}{J} = 0 \qquad (24\text{-}69)$$

yielding the roots

$$s_{1,2} = -\frac{F + Q_e}{2J} \pm j\sqrt{\frac{K}{J} - \left(\frac{F + Q_e}{2J}\right)^2} \qquad (24\text{-}70)$$

for the case of underdamping. From the last expression the critical viscous friction is found to be

$$(F + Q_e)_c = 2\sqrt{KJ} \qquad (24\text{-}71)$$

which shows that, whether the system is just viscously damped or viscously damped plus error-rate damped, the critical value of the damping coefficient continues to be dependent upon the loop proportional gain K and the system inertia J. Moreover, since the damping ratio of a second-order system is always expressed as the ratio of the total damping coefficient to the critical damping coefficient, the expression for the damping ratio in the viscous plus error-rate case is

$$\zeta = \frac{F + Q_e}{2\sqrt{KJ}} \qquad (24\text{-}72)$$

A comparison with Eq. (24-17) shows that now the damping ratio can be independently adjusted through Q_e. Of course, the expression for the natural frequency remains as before: $\omega_n = \sqrt{K/J}$. This is obvious from Eq. (24-69).

The steady-state solution for a step input r_0 in the presence of a load torque T_L is seen from Eq. (24-66) to be the same whether or not error-rate damping is present. Hence the use of high loop gains can effect a closer correspondence between the actual and commanded values of the controlled variable. However, the advantage offered by the error-rate system is that this can be achieved without deteriorating the transient response because of the influence of the Q_e term in establishing the damping ratio [see Eq. (24-72)]. Herein lies the strength of error-rate damping—*it allows higher gains to be used without adversely affecting the damping ratio and in this manner makes it possible to satisfy the specifications for the damping ratio as well as for the steady-state performance.* It offers the additional advantage of increasing the system's natural frequency, which in turn means smaller settling times.

The viscous plus error-rate damped system can also be analyzed through use of the block diagram and the appropriate transfer functions of the system components. For the system that includes error-rate, the only modification required concerns the expression for the transfer function of the servoamplifier. Instead of just the gain K_a, Eq. (24-64) indicates that now the transfer function is

$$\frac{E_a(s)}{E(s)} = K_a + sK_e \tag{24-73}$$

The complete block diagram is depicted in Fig. 24-9.

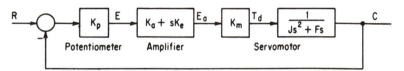

Fig. 24-9 Block diagram of a viscous plus error-rate damped system.

The direct transmission function in this case takes the form

$$G(s) = \frac{K_p K_m (K_a + sK_e)}{Js^2 + Fs} \tag{24-74}$$

This yields the closed-loop transfer function

$$\frac{C(s)}{R(s)} = T(s) = \frac{G(s)}{1 + HG(s)} = \frac{K + sQ_e}{Js^2 + (F + Q_e)s + K} \tag{24-75†}$$

where K and Q_e are defined by Eqs. (24-67) and (24-68). Note again the recurrence of the proper form of the characteristic equation in the denominator term of the closed-loop transfer function.

† Equation (24-75) also shows an additional numerator term when compared with Eq. (24-46), which applies for a system with only viscous damping. Although at first glance it seems that the additional terms appearing in the solution for the controlled variable may result in a poor dynamic response, this is in fact not the case.

24-3 OUTPUT-RATE CONTROL

A system is said to have output-rate damping when the generation of the output quantity in some way is made to depend upon the rate at which the controlled variable is changing. Its introduction often involves the creation of an auxiliary loop, making the system multiloop. For the servomechanism group of feedback control systems a common way of obtaining output-rate damping is by means of a tachometer generator, driven from the servomotor shaft. In order to illustrate the general results and to demonstrate the method of analysis, this study is developed with a particular system in mind, namely, that of a second-order servomechanism employing tachometric feedback damping. Although most of the derived results apply specifically to this system, it should be understood that the procedures and conclusions apply to feedback control systems in general, irrespective of the particular system composition. Moreover, the system performance will again be analyzed by the differential equation method and the transfer function approach. Attention is first directed to the former method.

Let the output-rate signal be denoted by $K_0(dc/dt)$, where K_0 is the output-rate gain factor and is expressed in volts per radian per second. This signal with reversed polarity is combined with the signal e from the error detector to make up the input signal to the servoamplifier. Thus

$$\text{servoamplifier input} = e - K_0\frac{dc}{dt} = K_p(r - c) - K_0\frac{dc}{dt} \qquad (24\text{-}76)$$

It is important to subtract these signals, otherwise the output-rate term will deteriorate rather than improve the dynamic response as will be shown presently. If the servoamplifier input signal is then multiplied by the amplifier gain K_a and the motor torque constant K_m, the expression for the developed torque is obtained. Upon equating this to the opposing torques, we obtain

$$\left(e - K_0\frac{dc}{dt}\right)K_aK_m = J\frac{d^2c}{dt^2} + F\frac{dc}{dt} + T_L \qquad (24\text{-}77)$$

or

$$(r - c)K_pK_aK_m = J\frac{d^2c}{dt^2} + (F + K_0K_aK_m)\frac{dc}{dt} + T_L \qquad (24\text{-}78)$$

Rearranging terms leads to the final form of the governing differential equation of the viscous plus output-rate system, namely,

$$\boxed{J\frac{d^2c}{dt^2} + (F + Q_o)\frac{dc}{dt} + Kc = Kr - T_L} \qquad (24\text{-}79)$$

where

$$Q_o = \text{loop output-rate gain factor} = K_oK_aK_m \qquad (24\text{-}80)$$

$$K = \text{loop proportional gain factor} = K_pK_aK_m \qquad (24\text{-}81)$$

The characteristic equation is therefore

$$s^2 + \frac{F + Q_o}{J}s + \frac{K}{J} = 0 \qquad (24\text{-}82)$$

which yields as the expression for the damping ratio

$$\zeta = \frac{F + Q_o}{2\sqrt{KJ}} \qquad (24\text{-}83)$$

As was the case with error-rate damping, output-rate damping adds a term to the numerator of the expression for the damping ratio, thus providing control over ζ when K is used to meet accuracy requirements.

The manner in which output-rate damping asserts itself in controlling the transient response is most easily demonstrated by assuming a step input applied to the system. If for reasons of accuracy the gain is assumed to be very high, then immediately upon applying the input, a large developed torque is produced due to the proportional gain. Explained in terms of Eq. (24-77), at $t = 0^+$ the term dc/dt is zero, and so torque is produced only by the proportional term. This torque acts to bring the controlled variable quickly to the commanded steady-state level. However, as it does so with the elapse of time, the system begins to generate a large rate of change of output. In turn this means that the output-rate signal, $K_o(dc/dt)$, in Eq. (24-77) appears in opposition to the proportional signal, thus removing the tendency for an excessively oscillatory response. This discussion should also make clear why the sign of the output-rate signal must be connected in opposition to the proportional signal. Certainly, greater overshoots would result if the output-rate signal were made to aid the proportional signal. Improper polarity means a reversal in the sign of Q_o in Eq. (24-83), from which it then becomes obvious that the transient behavior deteriorates rather than improves.

The complete block diagram of the servomechanism employing output-rate damping is depicted in Fig. 24-10. The output voltage of the tachometer generator that furnishes the output-rate signal is given by

$$e_t = K_o \frac{dc}{dt} \qquad \text{V} \qquad (24\text{-}84)$$

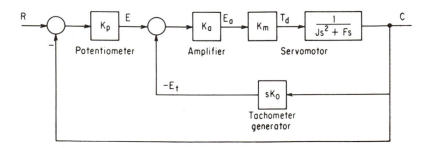

Fig. 24-10 Block diagram of a viscous plus output-rate damped system.

Hence the corresponding transfer function is

$$\frac{E_t(s)}{C(s)} = sK_o \qquad (24\text{-}85)$$

In Fig. 24-10 the output of the tachometer generator is negatively summed with the signal originating from the error detector as called for by Eq. (24-76). The feedback relationship of Eq. (20-6) may be applied to this minor loop to obtain the functional relationship between C and E. Thus

$$\frac{C(s)}{E(s)} = \frac{\dfrac{K_aK_m}{Js^2 + sF}}{1 + \dfrac{sK_oK_aK_m}{Js^2 + Fs}} = \frac{K_aK_m}{Js^2 + (F + Q_o)s} \qquad (24\text{-}86)$$

Accordingly the block diagram of Fig. 24-10 reduces to that shown in Fig. 24-11.

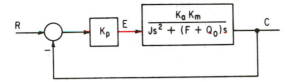

Fig. 24-11 Reduced form of the block diagram of Fig. 24-10.

The closed-loop transfer function for the complete system results when the feedback relationship is applied to Fig. 24-11. Thus

$$\frac{C(s)}{R(s)} = T(s) = \frac{\dfrac{K}{Js^2 + (F + Q_o)s}}{1 + \dfrac{K}{Js^2 + (F + Q_o)s}} = \frac{K}{Js^2 + (F + Q_o)s + K} \qquad (24\text{-}87)$$

Once again note the equivalence of the denominator expression with the characteristic equation. See Eq. (24-82).

EXAMPLE 24-2 A feedback control system, having the configuration depicted in Fig. 24-1, has the following parameter values:

$$K_p = 0.5 \text{ V/rad} \qquad F = 1.5 \times 10^{-4} \text{ lb-ft/rad/s}$$
$$K_a = 100 \text{ V/V} \qquad J = 10^{-5} \text{ slug-ft}^2$$
$$K_m = 2 \times 10^{-4} \text{ lb-ft/V}$$

(a) Describe the dynamic response of this system when a step input command is applied. Assume the system is initially at null.

(b) A load disturbance of 10^{-3} lb-ft is present on the system when a step command is applied. Find the position lag error in radians.

(c) To what value must the amplifier gain be changed in order that the position lag error of part (b) be no greater than 0.025 rad?

(d) Compute the damping ratio for the gain of part (c). What is the maximum percent overshoot?

(e) Find the value of the output-rate gain factor which for the gain of part (c) makes the maximum overshoot 25%.

Solution: (a) By Eq. (24-4) the loop proportional gain is

$$K = K_p K_a K_m = \tfrac{1}{2}(100)(2 \times 10^{-4}) = 10^{-2} \text{ lb–ft/rad}$$

$$\therefore \zeta = \frac{F}{2\sqrt{KJ}} = \frac{1.5 \times 10^{-4}}{2\sqrt{10^{-2} \cdot 10^{-5}}} = 0.236$$

Hence from Fig. 24-4 the maximum overshoot is found to be 47%. Also

$$\omega_n = \sqrt{\frac{K}{J}} = \sqrt{\frac{10^{-2}}{10^{-5}}} = 31.8 \text{ rad/s}$$

Thus the commanded value of the controlled variable reaches within 1% of its final value in

$$t_s = \frac{5}{\zeta \omega_n} = \frac{5}{0.236(31.8)} = 0.67 \text{ s}$$

The damped oscillations occur at a frequency of

$$\omega_d = \omega_n \sqrt{1 - \zeta^2} = 30.8 \text{ rad/s}$$

(b) From Eq. (24-12) the position lag error is seen to be

$$\frac{T_L}{K} = \frac{10^{-3}}{10^{-2}} = 0.1 \text{ rad}$$

(c) The loop gain must be increased by a factor of 4. Hence the new value of amplifier gain is

$$K_a' = 4K_a = 4(100) = 400 \text{ V/V}$$

(d) Equation (24-17) shows that when the gain is quadrupled the damping ratio is halved. Hence

$$\zeta' = \frac{\zeta}{2} = 0.118$$

This yields a maximum overshoot of 70%.

(e) For a 25% overshoot Fig. 24-4 reveals that $\zeta = 0.4$. Then by Eq. (24-83) we have

$$0.4 = \frac{F + Q_o}{2\sqrt{KJ}} = \frac{1.5 \times 10^{-4} + Q_o}{12.72 \times 10^{-4}}$$

$$\therefore Q_o = 0.4(12.72)10^{-4} - 1.5 \times 10^{-4} = 3.6 \times 10^{-4}$$

Hence

$$K_o = \frac{Q_o}{K_a' K_m} = \frac{3.6 \times 10^{-4}}{4 \times 10^2 \times 10^{-2}} = 0.9 \times 10^{-4} \text{ V/rad/s}$$

24-4 INTEGRAL-ERROR (OR RESET) CONTROL

A control system is said to contain integral-error control when the generation of the output in some way depends upon the integral of the actuating signal. Integral control is readily achieved by designing the servoamplifier so that it makes available an output voltage that contains an integral term as well as a proportional term. Expressed mathematically we have

$$e_a = eK_a + K_i \int_0^t e \, dt \tag{24-88}$$

where e = input to the servoamplifier

e_a = output of the servoamplifier, which is also the motor input voltage

K_i = proportionality factor of the integral-error component

Since the motor developed torque is $T_d = K_m e_a$, the equation of motion for the system can be written as

$$eK_a K_m + K_i K_m \int_0^t e \, dt = J\frac{d^2c}{dt^2} + F\frac{dc}{dt} + T_L \qquad (24\text{-}89)$$

Differentiating both sides to remove the integral leads to

$$K_a K_m \frac{de}{dt} + K_i K_m e = J\frac{d^3c}{dt^3} + F\frac{d^2c}{dt^2} + \frac{dT_L}{dt} \qquad (24\text{-}90)$$

Inserting $e = K_p(r - c)$ yields

$$K_a K_m K_p \left(\frac{dr}{dt} - \frac{dc}{dt} \right) + K_i K_p K_m(r - c) = J\frac{d^3c}{dt^3} + F\frac{d^2c}{dt^2} + \frac{dT_L}{dt} \qquad (24\text{-}91)$$

Rearranging gives

$$J\frac{d^3c}{dt^3} + F\frac{d^2c}{dt^2} + K\frac{dc}{dt} + Q_i c + \frac{dT_L}{dt} = Q_i r + K\frac{dr}{dt} \qquad (24\text{-}92)$$

where

$$Q_i = K_i K_p K_m \qquad (24\text{-}93)$$

and K is as defined before.

A glance at Eq. (24-89) reveals the interesting fact that the integral term does not combine with the viscous-friction term as was the case when error-rate and output-rate signals were introduced. Instead the integral term stands alone in the differential equation, thus stressing that its influence on system performance differs basically from the previous compensation schemes. Equation (24-92) shows that integral-error compensation has changed the order of the system from second to third. The inclusion of the integral term means therefore that a third independent energy-storing element is present. This stands in sharp contrast to the influence that error rate or output rate exerts on the same basic proportional system. For rate control the effect is confined to altering the coefficient of the first derivative term in the governing differential equation, and for this reason does not appear as an additional independent energy-storing element.

It follows from Eq. (24-92) that the characteristic equation now takes the form of a cubic, namely

$$s^3 + \frac{F}{J}s^2 + \frac{K}{J}s + \frac{Q_i}{J} = 0 \qquad (24\text{-}94)$$

Finding the solution of this equation is more involved than it is for the second-order system. Moreover, if standard time-domain procedures are used, it becomes considerably more difficult to design *directly* for a specific dynamic behavior as is done in Example 24-2. In third- and higher-order systems it is laborious and difficult to see what the effect on the dynamic behavior is when the value of a particular parameter is modified. However, if the analysis is done in the frequency domain, the solution of the problem is easier and more direct.

In view of the fact that integral compensation increases the order of the system and thus makes satisfactory dynamic behavior more difficult to achieve, what advantage is to be gained by its inclusion? The answer readily follows from an examination of Eq. (24-92). Note that in the steady state, even in the presence of a fixed external load, exact correspondence exists between the actual and commanded values of the output. That is, the position lag error, which exists with error-rate and output-rate control, disappears with integral-error control. In fact it is characteristic of integral control that it greatly improves the steady-state performance (system accuracy). At the same time, however, it makes the dynamic behavior more difficult to cope with successfully.

The analysis of integral-error control by means of the transfer function approach leads, of course, to the same results. Since the servoamplifier is the only component in the block diagram that is modified in the integral-error controlled system, the new transfer function is readily found from Eq. (24-88). Thus

$$\frac{E_a(s)}{E(s)} = K_a + \frac{K_i}{s} \tag{24-95}$$

The complete block diagram then takes the form depicted in Fig. 24-12, from which it immediately follows that the direct transmission function is

$$G(s) = \frac{K_p K_m (K_i + s K_a)}{J s^3 + F s^2} \tag{24-96}$$

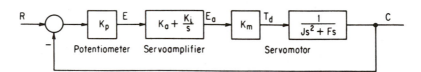

Fig. 24-12 Block diagram of a proportional plus integral-error controlled system.

The corresponding closed-loop transfer function then becomes

$$\frac{C(s)}{R(s)} = \frac{Q_i + sK}{J s^3 + F s^2 + Ks + Q_i} \tag{24-97}$$

This expression is equivalent to Eq. (24-92) for the case where T_L equals zero. The characteristic equation is the same.

24-5 TRANSFER FUNCTIONS OF SYSTEM COMPONENTS

The typical system components whose transfer functions are developed here represent a necessary selection. Because of the nature of this book they are primarily chosen from the electrical field; a few are from the mechanical area. Examples from the pneumatic, hydraulic, and process field are omitted, not only because of space considerations but also because insofar as the procedure is concerned nothing new is gained. No matter what type of component we consider, the general procedure for deriving the transfer function is always the same. It involves the following three steps: (1) determine the governing equation for the component expressed in term of the output and input variables; (2) Laplace-transform the governing equation, assuming all initial conditions are zero; (3) rearrange the equation to formulate the ratio of the output to input variable.

Lag Network (*or Integrating Circuit*).   It often is necessary to reshape the frequency-response characteristic of a system's transfer function in order to meet the performance specifications. Integrating and differentiating circuits are very commonly used to achieve these results. Effectively they provide the system with integral- and error-rate control in a simple and economical fashion.

The circuit shown in Fig. 24-13 is called an integrating circuit because under certain conditions the output signal e_o is effectively the integral of the input signal e_i. To understand this, recall that the output voltage across the capacitor terminals is

$$e_o = \frac{1}{C} \int i \, dt \qquad (24\text{-}98)$$

If the frequency of e_i is assumed to be such that the capacitive reactance is negligible compared to the resistance, it follows that the capacitor current is given approximately by

$$i \approx \frac{e_i}{R} \qquad (24\text{-}99)$$

Inserting Eq. (24-99) into Eq. (24-98) yields

$$e_0 \approx \frac{1}{RC} \int e_i \, dt \qquad (24\text{-}100)$$

which shows that the output signal is the integral of the input signal. However, this circuit ceases to be an integrating circuit at very low frequencies because of the correspondingly high capacitive reactance.

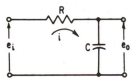

Fig. 24-13   An integrating circuit or lag network.

An exact description of the properties of the circuit of Fig. 24-13 is readily had from the transfer function. The governing equations involving the input and output variables are

$$e_i = iR + \frac{1}{C} \int i \, dt = R\frac{dq}{dt} + \frac{q}{C} \qquad (24\text{-}101)$$

$$e_o = \frac{1}{C} \int i \, dt = \frac{q}{C} \qquad (24\text{-}102)$$

The corresponding Laplace transforms are

$$E_i(s) = RsQ(s) + \frac{Q(s)}{C} \tag{24-103}$$

$$E_o(s) = \frac{Q(s)}{C} \tag{24-104}$$

where $Q(s) = \mathscr{L}q(t)$. Formulating the ratio of output to input then yields

$$\frac{E_o(s)}{E_i(s)} = \frac{\dfrac{1}{sC}}{R + 1/sC} = \frac{1}{1 + sRC} \tag{24-105}$$

Note that $Q(s)$ cancels out in the formulation. In fact the middle expression of Eq. (24-105) shows that, where networks are involved, the transfer function can be found directly by using the voltage-divider rule and the operational form of the impedance of the circuit elements.

For sinusoidal input functions the frequency s is replaced by $j\omega$ so that the final expression for the transfer function becomes

$$\boxed{\frac{E_o(j\omega)}{E_i(j\omega)} = \frac{1}{1 + j\omega RC} = \frac{1}{1 + j\omega\tau}} \tag{24-106}$$

where $\tau = RC$. It is important to understand that the frequency ω appearing in the transfer function always refers to the frequency of variation of the input signal. When this frequency has a very low value (i.e., $\omega \to 0$), the capacitor behaves as an open circuit and so a direct transmission of the signal occurs. Thus the circuit does not integrate. On the other hand, as the frequency increases (i.e., $\omega\tau \gg 1$), the transfer function of Eq. (24-106) becomes approximately $1/j\omega\tau$. The presence of the factor $1/j\omega$ in the transfer function means that the sinusoidal input function appears across the capacitor terminals as an integrated quantity. Recall that integration is algebraically denoted by $1/s$ or $1/j\omega$. Finally, it should also be noted that, when this circuit performs an as integrator, the input signal is appreciably attenuated because $\omega\tau \gg 1$. In such circuits the integration property goes hand in hand with attenuation of the signal level.

Attention is next focused on the graphical representation of the transfer function of Eq. (24-106) as a function of the signal frequency ω. Two methods are generally preferred—the polar plot and the Bode plot. The polar plot is merely a plot of the amplitude and phase of Eq. (24-106) at each frequency in the complex plane. The result is shown in Fig. 24-14(a). Note that the locus of the output-to-input phasor is a semicircle in the fourth quadrant. The location in this quadrant stresses the fact that the phase angle is always a *lagging* one. In fact, integration in the time domain is characteristically described by phase lag in the frequency domain. When the lag angle in sinusoidal steady-state analysis of the output of a linear component is 90°, the corresponding behavior in the time domain is one of pure integration. When the lag angle lies between 0 and 90°, the result in the time domain involves a proportional term as well as an integral term.

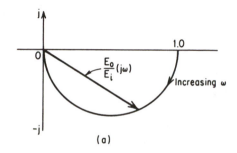

(a)

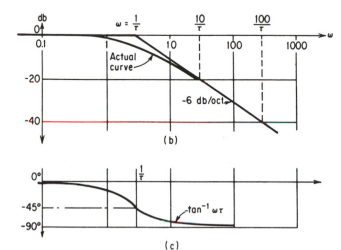

(b)

(c)

Fig. 24-14 Graphical representation of the lag network transfer function: (a) polar plot; (b) attenuation in db versus frequency; (c) phase versus frequency.

In the Bode-diagram representation of the transfer function the amplitude and phase characteristics are individually plotted as a function of frequency on a logarithmic scale. Moreover, it is customary to plot the amplitude expressed in units of *decibels*, which is defined as twenty times the logarithm of the voltage ratio in this case and is frequently called the log-modulus characteristic. The log-modulus characteristic as a function of frequency can be expressed mathematically by

$$A = 20 \log \frac{1}{|1 + j\omega\tau|} = 20 \log 1 - 20 \log |1 + j\omega\tau|$$
$$= -20 \log |1 + j\omega\tau| \tag{24-107}$$

The absolute signs are included to emphasize that we are dealing only with the magnitude of the complex number for each value of ω. The phase characteristic is given by

$$\phi = -\tan^{-1} \omega\tau \tag{24-108}$$

Usually the attenuation characteristic can be quickly plotted by drawing two straight lines originating from the point $\omega\tau = 1$ or $\omega = 1/\tau$. To understand this, consider the behavior of Eq. (24-107) at low and high frequencies. In this instance a low frequency is one that is small compared to $1/\tau$. Examination of

Eq. (24-107) discloses that when $\omega\tau \ll 1$, the log-modulus attenuation characteristic is just a straight horizontal line at the 0-db level. That is,

$$A = -20 \log 1 = 0 \text{ dB} \qquad \text{for } \omega\tau \ll 1 \qquad (24\text{-}109)$$

At high frequencies, Eq. (24-106) becomes

$$A \approx -20 \log |j\omega\tau| \qquad \text{for } \omega\tau \gg 1 \qquad (24\text{-}110)$$

Accordingly, at

$$\omega_1\tau = 10 \qquad A_1 = -20 \log |j10| = -20 \text{ dB} \qquad (24\text{-}111)$$

$$\omega_2\tau = 20 \qquad A_2 = -20 \log |j20| = -26 \text{ dB} \qquad (24\text{-}112)$$

$$\omega_3\tau = 100 \qquad A_3 = -20 \log |j100| = -40 \text{ dB} \qquad (24\text{-}113)$$

A plot of these high-frequency values of the amplitude identifies a straight line having a slope of -6 dB/octave or -20 dB/decade. Since $\omega_2 = 2\omega_1$, ω_2 is said to be an *octave* larger than ω_1. Similarly, ω_3 is said to be a *decade* higher than ω_1 because it is ten times greater. The extension of this sloping line intersects the frequency axis at the value $1/\tau$. This frequency is a key quantity in plotting the log-modulus characteristic because it marks the frequency value at which the asymptotic representation of the transfer function changes slope. Hence, to distinguish it from other frequencies, it is called the *breakpoint* or *corner* frequency of the transfer function. It is readily found by setting the j part of the transfer-function equal to unity and solving for ω. The log-modulus characteristic is depicted in Fig. 24-14(b). Note that the actual curve and the asymptotic approximations are in very close agreement everywhere in the frequency spectrum except for the small region about the breakpoint frequency. A study of Eq. (24-107) shows that the maximum deviation between these plots occurs at the breakpoint frequency and has a value of

$$20 \log |1 + j1| = 20 \log |\sqrt{2}| = 3 \text{ dB}$$

Since this deviation is a determinable quantity, the procedure in plotting the log-modulus characteristic of a transfer function is to use the straight line asymptotic approximations and then to introduce corrections only when the situation warrants it.

The phase characteristic of Eq. (24-108) plots simply as the arc tangent curve. This is shown in Fig. 24-14(c). Note that at the breakpoint frequency the phase shift is $-45°$.

Lag Network with Fixed Maximum Attenuation. At high frequencies the capacitor in the circuit of Fig. 24-13 appears almost as a short circuit so that the output signal level is practically zero. In the interest of limiting the high-frequency attenuation this circuit is modified to the practical form shown in Fig. 24-15. Taking advantage of the validity of using the voltage-divider rule in finding network transfer functions, we get for this circuit

$$\frac{E_o(s)}{E_i(s)} = \frac{R_2 + \dfrac{1}{sC}}{R_1 + R_2 + 1/sC} = \frac{1 + sR_2C}{1 + s(R_1 + R_2)C} \qquad (24\text{-}114)$$

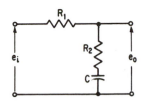

Fig. 24-15 Practical form of integrating circuit to limit attenuation at high frequencies.

Calling

$$\tau \equiv R_2 C \tag{24-115}$$

$$\alpha \equiv \frac{R_1 + R_2}{R_2} \tag{24-116}$$

Equation (24-114) may be rewritten as

$$\frac{E_o(s)}{E_i(s)} = \frac{1 + s\tau}{1 + s\alpha\tau} \tag{24-117}$$

For sinusoidal inputs this becomes

$$\boxed{\frac{E_o(j\omega)}{E_i(j\omega)} = \frac{1 + j\omega\tau}{1 + j\omega\alpha\tau}} \tag{24-118}$$

The corresponding polar plot is shown in Fig. 24-16(a). Note that at $\omega = 0$ the transfer function ratio is unity. At $\omega \to \infty$ the value is $1/\alpha$. Also, since the radius of this semicircle is

$$\left(1 - \frac{1}{\alpha}\right)\frac{1}{2} = \frac{\alpha - 1}{2\alpha}$$

and the distance from the origin to this center is

$$\left(\frac{1}{\alpha} + \frac{\alpha - 1}{2\alpha}\right) = \frac{\alpha + 1}{2\alpha}$$

it follows that the maximum phase lag angle possible with this circuit is

$$\phi_m = \sin^{-1}\frac{\alpha - 1}{\alpha + 1} \tag{24-119}$$

The quantity ϕ_m is solely dependent upon α.

The Bode diagrams of attenuation and phase are depicted in Figs. 24-16(b) and (c). The asymptotic attenuation characteristic is found by superposing the effects of the numerator and denominator terms in Eq. (24-118). The denominator is handled as before. Its breakpoint frequency is $\omega_{b1} = 1/\alpha\tau$. Hence we have a zero-db line originating from zero frequency and going up to ω_{b1}. At this point we have a negatively sloping line of -20 dB/decade which continues over the remainder of the frequency range. The numerator term is treated in a similar fashion. A zero-db line proceeds from zero frequency to the breakpoint frequency $\omega_{b2} = 1/\tau$. This frequency occurs higher in the scale than ω_{b1}. Beyond ω_{b2} the numerator term provides an increase in decibels at the rate of $+20$ dB/decade. Since we are using logarithms to express the amplitude variations, the resultant attenuation characteristics is found by simply adding the contributions of the

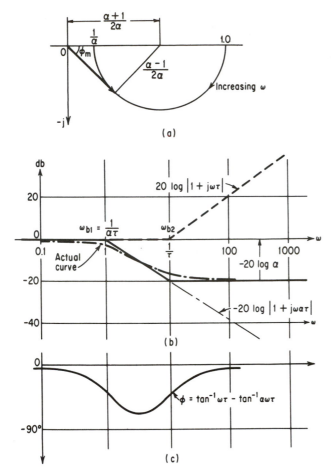

(a)

(b)

$\phi = \tan^{-1} \omega\tau - \tan^{-1} \alpha\omega\tau$

(c)

Fig. 24-16 Graphical representation of the transfer function of the network of Fig. 24-15: (a) polar plot; (b) and (c) Bode plots of attenuation and phase versus frequency.

numerator and denominator terms. This leads to the resultant asymptotic curve shown in Fig. 24-16(b). It is worth noting that below ω_{b2} the resultant curve is the same as that of the denominator term. Above ω_{b2} the $+20$ dB/dec slope of the numerator term cancels the effect of the -20 dB/dec slope of the denominator term, thus yielding a horizontal line placed at a value corresponding to the fixed high-frequency attenuation of $-20 \log \alpha$.

The phase characteristic is illustrated in Fig. 24-16(c). It is a plot of the equation

$$\phi = \tan^{-1} \omega\tau - \tan^{-1} \alpha\omega\tau \qquad (24\text{-}120)$$

The first term is the angle of the numerator and the second is that of the denominator. Since α is defined always greater than unity, the resultant phase angle at any finite frequency is negative, indicating that the output lags the input. Another notable observation regarding the Bode diagram is that the phase angle follows the changes in attenuation. Thus when the slope of the attenuation curve is changing rapidly so too does the phase curve. In fact, mathematical relationships are available that allow one characteristic to be derived from the other.

Lead Network (or Differentiating Circuit). Rate control can be obtained by using the differentiating circuit of Fig. 24-17. Like the integrating circuit, the differentiating circuit provides the derivative of the input only in a limited portion of the frequency spectrum. For example, if e_i varies at relatively high frequencies where the capacitive reactance is negligibly small, the circuit of Fig. 24-17 shows that the input signal appears directly at the output terminals without undergoing a time operation. On the other hand, at low frequencies, where the capacitive reactance predominates, the current can be written approximately as

$$i \approx C\frac{de_i}{dt} \tag{24-121}$$

Hence the output voltage as obtained across the resistor becomes

$$e_o = iR \approx RC\frac{de_i}{dt} \tag{24-122}$$

Clearly, then, this circuit behaves as a differentiating circuit in the restricted frequency range.

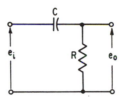

Fig. 24-17 Differentiating circuit or lead network.

A more exact description of the properties of the circuit can be obtained from the transfer function, which in this case is

$$\frac{E_o(s)}{E_i(s)} = \frac{R}{R + 1/sC} = \frac{sRC}{1 + sRC} = \frac{s\tau}{1 + s\tau} \tag{24-123}$$

For sinusoidal inputs Eq. (24-123) becomes

$$\frac{E_o(j\omega)}{E_i(j\omega)} = \frac{j\omega\tau}{1 + j\omega\tau} \tag{24-124}$$

This transfer function makes it clear that when $\omega\tau \ll 1$, the ratio of output to input signals is approximately $j\omega\tau$. Since the term $j\omega$ is the operational representation of differentiation where sinusoids are involved, the circuit behaves as a differentiator for the restriction specified. Again note that the output signal level is attenuated when the circuit provides differentiation. Finally, in terms of a frequency-domain description, a differentiating circuit can be described as a lead network because at any finite frequency the output phasor *leads* the input phasor by the resultant phase angle of the transfer function. Equation (24-124) shows this phase lead angle to be

$$\phi = +90° - \tan^{-1}\omega\tau \tag{24-125}$$

A polar plot of Eq. (24-124) appears in Fig. 24-18(a). The locus in this case is a semicircle located in the first quadrant, which is consistent with the fact that the phase angle is a positive or lead angle. The log-modulus attenuation

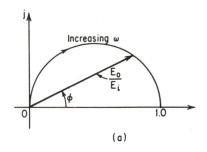

(a)

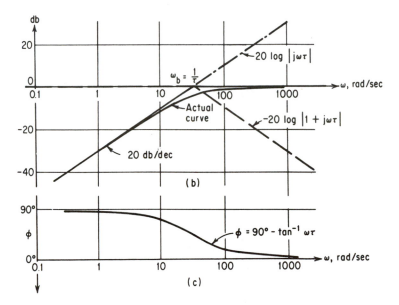

(b)

(c)

Fig. 24-18 Graphical representation of the transfer function of the circuit of Fig. 24-17: (a) polar plot; (b) attenuation versus frequency plot (note the high-pass character of the circuit above ω_b); (c) phase versus frequency plot.

characteristic in decibels is

$$A = 20 \log \left| \frac{E_o}{E_i} \right| = 20 \log |j\omega\tau| - 20 \log |1 + j\omega\tau| \qquad (24\text{-}126)$$

A little thought discloses that the first term plots as a straight line passing through $\omega_b = 1/\tau$ with a slope of 20 dB/decade. The second term is plotted as two straight lines—a horizontal line at zero db spreading from zero frequency to $\omega_b = 1/\tau$, and another straight line with a slope of -20 dB/decade spreading over the frequency range from ω_b to infinity. The resultant asymptotic amplitude curve is the solid curve in Fig. 24-18(b). The phase characteristic, given by Eq. (24-125), is depicted in Fig. 24-18(c).

Lead Network with Fixed Low-Frequency Attenuation. A serious limitation of the circuit of Fig. 24-17 is that it permits no signal transfer when the system reaches steady state (i.e., $\omega = 0$). Accordingly, if a system is so composed that it demands a finite signal level at steady state to generate the output quantity, the use of this circuit forbids it. To avoid the difficulty the practical form of the lead network shown in Fig. 24-19 is commonly used. Its transfer function can be derived as follows:

$$\frac{E_o(s)}{E_i(s)} = \frac{R_2}{R_2 + \dfrac{R_1}{1 + sR_1C}} = \frac{R_2}{R_1 + R_2}\left[\frac{1 + sR_1C}{1 + s\left(\dfrac{R_1R_2}{R_1 + R_2}\right)C}\right] \quad (24\text{-}127)$$

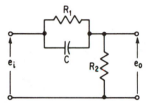

Fig. 24-19   Practical form of a rate network.

Calling

$$\alpha \equiv \frac{R_1 + R_2}{R_2} \quad (24\text{-}128)$$

$$\tau \equiv \frac{R_1R_2}{R_1 + R_2}C = \frac{R_1C}{\alpha} \quad (24\text{-}129)$$

and inserting into Eq. (22-127) yields

$$\frac{E_o(s)}{E_i(s)} = \frac{1}{\alpha}\left(\frac{1 + s\alpha\tau}{1 + s\tau}\right) \quad (24\text{-}130)$$

The expression for the transfer function for sinusoidal inputs then becomes

$$\boxed{\frac{E_0(j\omega)}{E_i(j\omega)} = \frac{1}{\alpha}\left(\frac{1 + j\omega\alpha\tau}{1 + j\omega\tau}\right)} \quad (24\text{-}131)$$

The polar plot of Eq. (24-131) is depicted in Fig. 24-20(a). It is identical to Fig. 24-16(a) with the exception that the semicircle is now located in the first quadrant because lead angles are involved. The expressions for the log-modulus and phase characteristics are given by

$$A = 20 \log \frac{1}{\alpha} + 20 \log |1 + j\omega\alpha\tau| - 20 \log |1 + j\omega\tau| \quad (24\text{-}132)$$

and

$$\phi = \tan^{-1} \omega\alpha\tau - \tan^{-1} \omega\tau \quad (24\text{-}133)$$

By plotting each term of Eq. (24-132) and summing, we obtain the resultant curve of Fig. 24-20(b). Note that with this circuit the attenuation at low frequencies (i.e., $\omega \ll 1/\alpha\tau$) is fixed at $1/\alpha$ where α is defined by Eq. (24-128). A plot of Eq. (24-133) yields the phase characteristic shown in Fig. 24-20(c).

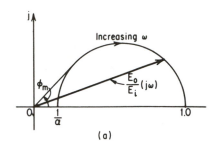

(a)

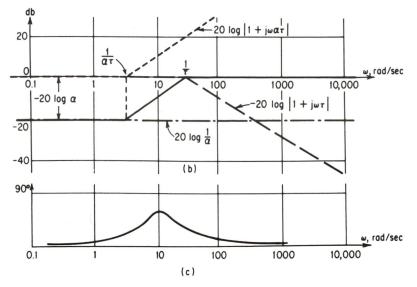

Fig. 24-20 Graphical representation of the transfer function of the circuit of Fig. 24-19: (a) polar plot; (b) attenuation versus frequency; (c) phase versus frequency.

The Linear Accelerometer. The linear accelerometer has gained wide-spread use in high-speed aircraft and space vehicle applications. In these applications it can be employed either as a measuring device, which provides valuable data on the extent to which the spacecraft or jet aircraft is subjected to vibrations and shock, or else as an important, inherent part of the spacecraft or aircraft guidance system.

Our attention is directed to a linear accelerometer having a single degree of freedom as shown in Fig. 24-21. The frame of the accelerometer is assumed attached to the missile frame. Let

y = motion of M relative to inertial space†

M = accelerometer mass (24-134)

† Inertial space is the reference frame in which a force-free body is unaccelerated; it is the celestial space determined by the "fixed stars."

$$x_i = \text{motion of frame relative to inertial space}$$
$$x_0 = y - x_i = \text{motion of } M \text{ relative to frame}$$

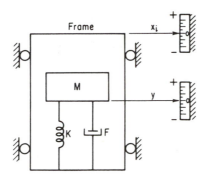

Fig. 24-21 Construction features of a linear accelerometer constrained for a single degree of freedom.

The quantity x_0 is considered the output of the accelerometer. The input quantity may be taken to be either the displacement x_i or the input acceleration $s^2 x_i$. The equation of motion for the system is found by equating to zero the sum of the forces associated with an assumed displacement of the mass and the frame. Thus

$$M\frac{d^2y}{dt^2} + F\left(\frac{dy}{dt} - \frac{dx_i}{dt}\right) + K(y - x_i) = 0 \qquad (24\text{-}135)$$

Three transfer functions can be identified for the accelerometer. One of these follows directly from Eq. (24-135). Thus, rearranging and taking the Laplace transform yields

$$s^2 Y + \frac{F}{M}sY + \frac{K}{M}Y = \frac{F}{M}sX_i + \frac{K}{M}X_i \qquad (24\text{-}136)$$

Introducing

$$\frac{F}{M} = 2\zeta\omega_n \qquad \text{and} \qquad \omega_n = \sqrt{\frac{K}{M}}$$

and formulating the ratio of output to input results in the expression

$$\frac{Y(s)}{X_i(s)} = \frac{1 + (2\zeta/\omega_n)s}{(s/\omega_n)^2 + (2\zeta/\omega_n)s + 1} \qquad (24\text{-}137)$$

The transfer function of Eq. (24-137) applies whenever the output is measured relative to inertial space.

A more useful form of the output of the accelerometer, however, occurs when the displacement of the mass M is measured relative to the frame and not to inertial space. To provide this transfer function, Eq. (24-134) is inserted in Eq. (24-135) to yield

$$\frac{d^2x_o}{dt^2} + 2\zeta\omega_n\frac{dx_o}{dt} + \omega_n^2 x_o = -\frac{d^2x_i}{dt^2} \qquad (24\text{-}138)$$

Upon taking the Laplace transform and formulating the ratio of output to input, we obtain the desired result. Thus

$$\frac{X_o(s)}{X_i(s)} = -\frac{s^2}{s^2 + 2\zeta\omega_n s + \omega_n^2} = -\frac{1}{\omega_n^2}\frac{s^2}{(s/\omega_n)^2 + (2\zeta/\omega_n)s + 1} \qquad (24\text{-}139)$$

The attenuation and phase characteristics of this transfer function are shown in Fig. 24-22.

Examination of Eq. (24-139) or of its corresponding Bode plots reveals some useful information about the application limitations of linear accelerometers. For example, the question may arise regarding the advisability of using the accelerometer as a measuring device for detecting low-frequency oscillations in the spacecraft air frame. Equation (24-139) shows that the transfer function under such circumstances is approximately

$$\frac{X_o}{X_i} \approx -\frac{1}{\omega_n^2}s^2 \tag{24-140}$$

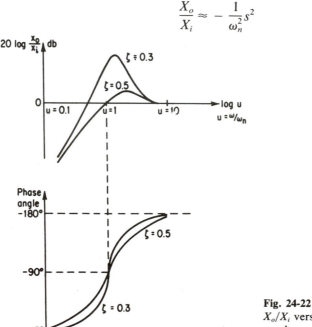

Fig. 24-22 Attenuation and phase of X_o/X_i versus frequency for a linear accelerometer.

However, the accelerometer's natural frequency is designed to be very high by virtue of a small mass and a very stiff spring. Accordingly, there is very little observable output displacement if it is used for such a purpose. In fact, the output signals may very well fall within the noise level of the system, thereby making such measurements unreliable. By the same token, however, it follows that to detect and measure an oscillation having a frequency comparable with that of the natural frequency of the accelerometer, the use of the accelerometer is satisfactory.

Of most importance is the transfer function that relates the output displacement to input acceleration. The right side of Eq. (24-138) represents the input acceleration, which we shall now call a_x. Consequently the desired transfer function becomes

$$\frac{X_o}{a_x} = -\frac{1}{s^2 + 2\zeta\omega_n + \omega_n^2} = -\frac{1}{\omega_n^2}\frac{1}{(s/\omega_n)^2 + (2\zeta/\omega_n)s + 1} \tag{24-141}$$

which is readily recognized as the standard quadratic form. Since the input quantity is acceleration, it follows that the output displacement X_o is proportional to input acceleration. Recalling the nature of the second-order response curves, we can easily understand why the damping ratio in these units is kept between 0.4 and 0.7. In this way large overshoot and therefore false indications of peak accelerations are avoided. Furthermore, the large natural frequency is designed into the system to ensure the ability of the accelerometer to measure acceleration pulses with a high degree of accuracy.

Since the magnitude of X_o provides the means of measuring spacecraft acceleration, the accelerometer serves an important function in the spacecraft autopilot. Through the inclusion of an accelerometer in a feedback path of the spacecraft air frame, it follows that, whenever the spacecraft output acceleration deviates from the command value, an actuating signal arises which acts to remove this lack of correspondence. Thus, high performance is ensured.

The Gyroscope. The gyroscope has achieved wide use in the aircraft industry as an indicating instrument and as a component in fire control, automatic flight control, and navigational systems. The principle of the gyroscope is well known and is based upon the law of conservation of angular momentum.

When this instrument is used as a practical component, the motion of the spinning wheel of the gyroscope is usually restrained by mounting it in a set of bearings called *gimbals*. It is customary to classify gyroscopes in terms of the number of degrees of freedom of angular motion of the spinning wheel. Figure 24-23(a) shows a system with one degree of freedom and Fig. 24-23(b) a system with two degrees of freedom. There are many types of gyroscopes in practical use, including the rate gyro, the integrating rate gyro, and the directional gyro of the one-degree-of-freedom class, and the displacement gyro, the vertical gyro, and the gyrocompass of the two-degrees-of-freedom class. Because of the general complexity of the subject and the limited treatment possible here, only the transfer functions of some simple one-degree-of-freedom components are derived.

Figure 24-24 illustrates a general one-degree-of-freedom system with an elastic restraint (spring K) and a viscous damper (dashpot F) about the free, or

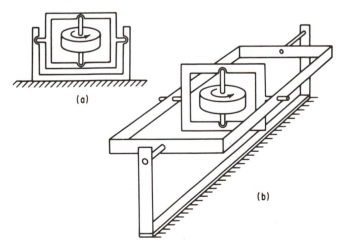

(a)

(b)

Fig. 24-23 Gyroscope degrees of freedom: (a) one-degree-of-freedom gyroscope; (b) two-degrees-of-freedom gyroscope.

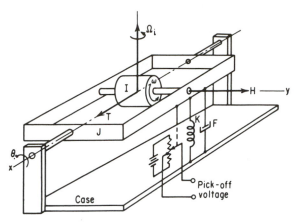

Fig. 24-24 Generalized one-degree-of-freedom gyroscope.

output, axis x. If the gyro case is rotated about the z axis with an angular velocity Ω_i, then a torque $T = H\Omega_i$ is developed as shown, which tends to rotate the gyro through an angle θ about the gimbal axis x. The unit tends to rotate so as to align its angular-momentum vector H with the input angular-velocity vector Ω_i. This torque is opposed by the inertia of the system about the output x axis, the spring restraint K, and the viscous damping F. If the moment of inertia about the x axis is denoted by J and the angular displacement by θ, then the equation for the dynamic balance of the system is given by

$$H\Omega_i = J\frac{d^2\theta}{dt^2} + F\frac{d\theta}{dt} + K\theta \tag{24-142}$$

The transfer function between the output angle θ and the input angular velocity is therefore given by

$$\frac{\theta}{\Omega_i} = \frac{H}{Js^2 + Fs + K} \tag{24-143}$$

Usually the output angle is transformed to a voltage by means of a "pick-off" transducer, which is pictured as a potentiometer in Fig. 24-23 although differential transformers and other types of pick-offs are also used.

This type of gyro is known as a *rate gyro* and is commonly used for measuring aircraft motion and for computing rates of turn for use in autopilots and gunfire control systems. The application of this device as a rate-of-turn indicator may be illustrated by considering the steady-state output deflection for a constant angular-velocity Ω_{io} input. The steady-state deflection about the output axis is therefore given by (letting s go to zero)

$$\theta_{ss} = \frac{H}{K}\Omega_{io} \tag{24-144}$$

Thus the output angular displacement is proportional to the constant input angular velocity. Some damping is always necessary in this gyroscope to prevent an oscillatory response.

It is important to note that a constant deflection θ_{ss} introduces an error into the system because it causes the H vector to move away from the y axis, and the input vector Ω_i is therefore no longer in line with the z axis. The resulting error in angular-rate measurement is said to be due to *geometric cross coupling*. In order to keep its effect to a minimum, the angular rotation about the output axis must be restrained to small values by increasing the spring constant K. This requires that the pick-off transducer be extremely sensitive, which in turn gives rise to problems of drift and noise. The effects of frictional torques in the output bearings and pick-off transducer are also a major source of error in the system. Thus designing a precision gyroscope becomes a difficult engineering problem.

If the elastic restraint of the rate gyro is removed, then the instrument becomes an *integrating rate gyro*. This unit is also known as an integrating gyro or HIG (hermetically sealed integrating gyro). In this case the equation for dynamic balance about the output axis is given by

$$H\Omega_i = J\frac{d^2\theta}{dt^2} + F\frac{d\theta}{dt} \qquad (24\text{-}145)$$

The transfer function corresponding to Eq. (24-145) depends on how the input and output variables are defined, which, of course, depends upon the particular application. If the output pick-off transducer is sensitive to angular deflection about the output axis, then θ becomes the output variable. Usually the input to the instrument is taken as the angular displacement about the input z axis (denoted by the letter θ_i) rather than the input angular velocity Ω_i (note that $\Omega_i = d\theta_i/dt$). For these variables the transfer function becomes

$$\frac{\theta}{\theta_i} = \frac{H}{Js + F} \approx \frac{H}{F} \qquad \text{for negligible } J \qquad (24\text{-}146)$$

If the gyroscope input consists of a constant angle θ_{io}, then the steady-state deflection about the output axis is given by

$$\theta_{ss} = \frac{H}{F}\theta_{io} \qquad (24\text{-}147)$$

Thus the steady-state output angular displacement is proportional to the input angular displacement. The problem of geometric cross coupling also exists in the integrating rate gyro. Typical values for H/F are of the order of 1 to 10 so that the problem cannot be neglected. Because of this, integrating rate gyros are usually restricted to applications such as stabilized platforms, where the output angle is returned close to the zero position by feedback.

Output Actuators. Very often control systems contain components that generate the controlled variable or the manipulated variable in a manner that exhibits a pure integration property. In such cases the transfer function of the component assumes the general form given by Eq. (24-148):

$$\frac{Y(s)}{X(s)} = \frac{K}{s(1 + s\tau)} \qquad (24\text{-}148)$$

where $Y(s)$ is the Laplace transform of the output variable and $X(s)$ is the Laplace transform of the input variable. Our knowledge of the Laplace transform of a time integral tells us that the presence of s as a root factor in the denominator means that the output variable is a function of the integral of the input variable. The pure integration property is a very useful one in control systems because it eliminates the need for errors in the steady state. This matter was briefly treated in Sec. 23-4.

The servomotor in the servomechanism of Fig. 23-3 has a transfer function the form of which is described by Eq. (24-148), provided the output quantity is considered to be the motor displacement angle c and the input is the voltage to the servomotor control winding m. Thus

$$\frac{C(s)}{M(s)} = \frac{K}{s(1 + s\tau_m)} \tag{24-149}$$

where K is an appropriate motor constant expressed in inverse volt-seconds and τ_m is the mechanical time constant of the motor expressed as the ratio of its inertia to viscous friction coefficient. The same is true for the output actuator of Fig. 23-7.

Another output device commonly found in control systems is the hydraulic actuator. It is used, for example, in the systems of Figs. 23-9 and 23-6. For the system of Fig. 23-6 both the electrohydraulic pilot valve and the positioning cylinder have transfer functions that are expressed by Eq. (24-148). In these cases, however, $Y(s)$ denotes the output power-ram piston position and $X(s)$ denotes the input-valve-spool displacement.

Summary review questions

1. Draw the schematic diagram of the servomechanism with a connected inertial load and write the governing differential equation using the controlled variable as the dependent variable.
2. Name the various modes of dynamic behavior associated with a system that has two independent energy-storing elements.
3. Define the critical damping coefficient as it applies to the servomechanism.
4. How is the natural frequency defined for a system with two independent energy-storing elements? What is this expression for the servomechanism?
5. Explain in physical terms the distinction between the natural frequency and the damped frequency of oscillation.
6. What *physical* meaning can be given to the complex frequency as it arises in the characteristic equation of an underdamped second-order system?
7. What is meant by the characteristic equation of a second-order system, and how is it obtained?
8. State the important information that is obtained from the characteristic equation of a system. Illustrate.
9. What is a critically damped response?
10. What is a universal curve, and what advantage does it offer?
11. Describe carefully how it is possible to characterize the dynamic behavior of a second-order system without resorting to a formal solution of the defining differential equation.
12. Why is the unit-step function used to characterize the dynamic behavior of the second-

order system? Can other source functions be used such as the impulse function? Explain.

13. Describe how it is possible to use a single time constant to represent the exponential decays associated with an underdamped second-order system. What useful purpose does this time constant serve?

14. Describe the essential features of a transfer function analysis of the servomechanism. What advantages can you cite for this approach over that of the differential equation?

15. What is error-rate damping as it applies to feedback control system?

16. Can error-rate damping be used to improve the accuracy of a system? Explain.

17. Explain what is meant by proportional plus derivative control.

18. Assume that a second-order system is operating at a fixed loop gain. Explain the effect that the introduction of error-rate damping has on the dynamical and steady-state behavior of the system.

19. Write the transfer function of a proportional plus derivative controller.

20. What is output-rate damping? What is the mechanism by which such damping is introduced into a servomechanism?

21. Describe how output-rate damping affects the dynamical behavior of a second-order system.

22. When output-rate damping is inserted into a system, state the precaution that must be taken to ensure an improvement in dynamical behavior.

23. Describe the circumstances that call for the use of error-rate or output-rate damping.

24. Give a *physical* explanation to describe how output-rate damping exerts influence in controlling the transient response of an underdamped system.

25. Draw the block diagram of the servomechanism equipped with output-rate damping. Show all component transfer functions.

26. Explain integral-error control. How is it put into a system?

27. What is the chief benefit to the system when integral-error control is introduced?

28. In what significant way is the feedback system changed with integral-error control compared to error-rate control?

29. Name two important disadvantages that are associated with the use of integral-error control.

30. Draw an integrating circuit using a single resistor and capacitor and indicate the output terminals. Which of these circuit elements must be the dominant one in order that the circuit perform as an integrator?

31. Why is the integrating circuit called a lag network when a frequency-domain description is used?

32. Sketch the Bode diagrams for magnitude and phase for the lag network without fixed high-frequency attenuation.

33. Repeat Question 32 for the case where the lag network is modified to include a fixed high-frequency attenuation.

34. Why is it desirable to include a resistor to limit high-frequency attenuation in the lag network? Is there a penalty associated with this modification? Explain.

35. Draw a differentiating circuit using a simple resistor and capacitor. Which parameter must exert the dominant influence in order that the circuit behave as a differentiator?

36. Why is the differentiating circuit called a lead network when a frequency-domain description is used?

37. Sketch the Bode diagrams for magnitude and phase for the lead network.

38. Repeat Question 37 for the case where the capacitor is shunted with a resistor.

39. In the practical form of the differentiating circuit why is it necessary to shunt the capacitor with a resistor?

40. Show the construction features of the linear accelerometer. Write the transfer function and indicate how it is made to convey information about the acceleration of the vehicle to which it is attached.

41. Derive the transfer function of the gyroscope.

Problems

GROUP I

24-1. Measurements conducted on a servomechanism show the output response to be

$$\frac{c(t)}{r_0} = 1 - 1.66\varepsilon^{-8t} \sin{(6t + 37°)}$$

when the input is a step displacement of magnitude r_0.
  (a) Determine the natural frequency of the system in radians per second.
  (b) What is the damped frequency of oscillation?
  (c) Find the value of the damping ratio.
  (d) The inertia of the output member is known to be 0.01 lb-ft-s², and the viscous coefficient is 0.16 lb-ft/rad/s. What is the loop gain?
  (e) If the damping ratio is not to be less than 0.4, how much can the loop gain be increased?

24-2. A feedback control system is represented by the following differential equation:

$$\frac{d^2c}{dt^2} + 6.4\frac{dc}{dt} = 160e$$

where $e = r - 0.4c$ and c denotes the output variable.
  (a) Find the value of the damping ratio.
  (b) What information does this convey about the transient behavior of the system? Be specific.

24-3. A feedback control system has the configuration shown in Fig. P24-3.
  (a) Obtain the expression for the closed-loop transfer function.
  (b) Find the complete expression for the output response as a function of time when the input is a unit step.
  (c) Plot each component of the output response as a function of time. Also sketch the resultant response and find graphically the time required to reach 95% of the final value.

24-4. A control system is composed of components whose transfer functions are those specified in the block diagram of Fig. P24-4.

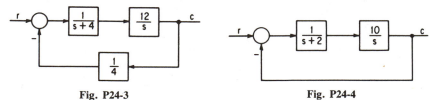

Fig. P24-3   **Fig. P24-4**

(a) What is the expression for the closed-loop transfer function?
(b) At what frequency does the output variable oscillate in responding to a step command before reaching steady state?
(c) What is the maximum percent overshoot in part (b)?
(d) How many seconds are required for the output to reach within 99% of steady state in part (b)?

24-5. The step response of a unity feedback control system is given by

$$\frac{c(t)}{r_0} = 1 - 1.66\varepsilon^{-8t} \sin{(6t + 37°)}$$

where r_0 is the magnitude of the input quantity.
(a) Find the closed-loop transfer function.
(b) What is the corresponding open-loop transfer function?
(c) Determine the complete output response for a unit-step input when the system is operated open-loop.

24-6. The governing differential equation of a feedback speed control system is given by

$$\left(\frac{J}{F}\right)\frac{d\omega_e}{dt} + (1 + K)\omega_e = \left(\frac{J}{F}\right)\frac{d\omega_r}{dt} + \omega_r$$

where ω_e is the error in speed between the reference speed ω_r and the output speed ω_o, i.e., $\omega_e = \omega_r - \omega_o$. Moreover, J is the system inertia, F is the system viscous-friction coefficient, and K is the system loop gain.
(a) What is the expression for the steady-state speed error?
(b) Determine the complete expression for the speed error when the reference speed is applied to the system in the form of a step command.
(c) Can any setting of the gain K cause sustained oscillations in the response of the system? Explain.

24-7. The dynamic behavior of a device is described by the equation

$$\frac{dc}{dt} + 10c = 40e$$

(a) Find the expression for the complete response of this system when $e = u(t)$.
(b) What is the expression for the direct transmission function?
(c) A unity-feedback path is placed around the device. Find the response of the feedback system to a unit step command.
(d) Compare the solutions to parts (a) and (c) and draw appropriate conclusions.

24-8. Solve Prob. 24-7 for the case where a unit ramp function, i.e., $e(t) = tu(t)$, is used as the driving force in parts (a) and (c).

24-9. A servomechanism has the following parameters:

K = open-loop gain = 18×10^{-4} lb-ft/rad

J = system inertia = 10^{-5} slug-ft^2

F = system viscous-friction coefficient = 161×10^{-6} lb-ft/rad/s

(a) Find the value of the damping ratio for the constants specified.
(b) To meet the accuracy requirements during steady-state operation, it is found that the loop gain must be increased to 185.5×10^{-4} lb-ft/rad. The damping ratio is to remain the same as in part (a). Determine the error-rate damping coefficient needed to achieve this result.

24-10. Depicted in Fig. P24-10 is the system shown in Fig. P24-4 modified to include error-rate damping. Determine the value of the error-rate factor K_e so that the damping ratio of the modified characteristic equation is 0.6.

24-11. For the system depicted in Fig. P24-11 find the value of the output-rate factor that yields a response to a step command having a maximum overshoot of 10%.

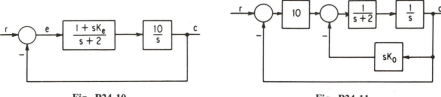

Fig. P24-10   **Fig. P24-11**

24-12. Repeat Prob. 24-11 for the system whose block diagram appears in Fig. P24-12.

24-13. Find the transfer function for the circuit shown in Fig. P24-13. What is the purpose of the unity-gain amplifier?

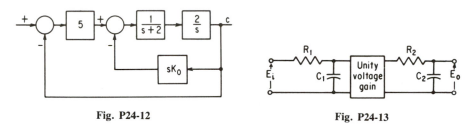

Fig. P24-12   **Fig. P24-13**

24-14. Derive the transfer function for the network shown in Fig. P24-14. Manipulate the result in the form of either a phase lag or a phase lead (whichever applies), and define an appropriate α and τ of the element values of the circuit.

24-15. Derive the transfer function for the circuit of Fig. P24-15. Express the result in terms of an appropriate α and τ.

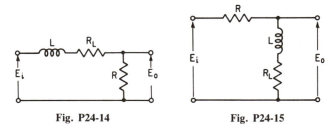

Fig. P24-14   **Fig. P24-15**

24-16. Find the transfer function for the circuit of Fig. P24-16, and express the result in terms of an appropriate α and τ.

24-17. For the network depicted in Fig. P24-17, show that

$$\frac{E_o(s)}{E_i(s)} = \frac{1}{\tau_1 \tau_2 s^2 + (\tau_1 + \tau_2 + \tau_{12})s + 1}$$

where $\tau_1 = R_1 C_1$, $\tau_2 = R_2 C_2$, and $\tau_{12} = R_1 C_2$.

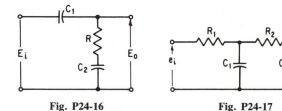

| Fig. P24-16 | Fig. P24-17 |

24-18. Determine the transfer function for the circuit shown in Fig. P24-18.

24-19. Derive the transfer function $X_o(s)/X_i(s)$ for the mechanical system shown in Fig. P24-19. Put the transfer function in the form of a phase lead or phase lag by using the α and τ notation.

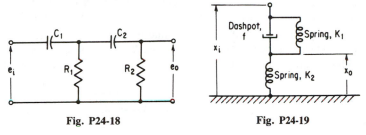

| Fig. P24-18 | Fig. P24-19 |

24-20. Find the transfer function of the mechanical system depicted in Fig. P24-20, and express the result in terms of an appropriate α and τ.

24-21. Determine the transfer function $X_o(s)/F(s)$ for the mechanical system of Fig. P24-21. To which standard form does this transfer function belong?

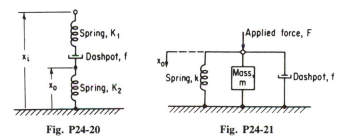

| Fig. P24-20 | Fig. P24-21 |

24-22. Sketch the attenuation and phase characteristics of the system components whose transfer functions are given by

(a) $T(s) = \dfrac{10(1 + j\omega\tau)}{(j\omega\tau)^2}$, where $\tau = 1$ s.

(b) $T(s) = \dfrac{1}{3 + j2\omega}$.

(c) $T(s) = \dfrac{1}{(1 + j\omega)(1 + j10\omega)}$.

24-23. A system component is described by the differential equation

$$c(t) = r(t) + \tau \frac{dr(t)}{dt}$$

where $c(t)$ is the output quantity and $r(t)$ is the input variable. This device is said to provide proportional plus derivative control. Sketch the log-modulus and phase-angle characteristics as well as the polar plot.

24-24. A proportional-plus-integral controller is described by the equation

$$c(t) = r(t) + \frac{1}{\tau} \int r(t)\, dt$$

where $c(t)$ and $r(t)$ are the output and input variables, respectively. Sketch the attenuation and phase-angle characteristics as well as the polar plot.

24-25. The asymptotic log-modulus for a system component is shown in Fig. P24-25.
(a) Find the corresponding transfer function.
(b) Sketch the polar plot.

24-26. Sketch the log-modulus and phase-angle characteristics described by the transfer function $T(j\omega) = (j\omega)^2 + j2\zeta\omega + 1$ for $\zeta = 0.1$, 0.3, and 1.0.

24-27. In the system of Fig. P24-27 determine the expression for the transformed output $C(s)$ in terms of the inputs $R_1(s)$, $R_2(s)$, and $R_3(s)$ and the specified system components. What is the form of the characteristic equation for closed-loop operation?

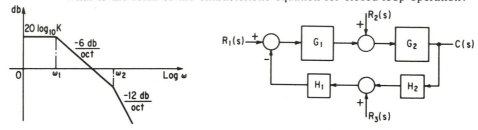

Fig. P24-25   Fig. P24-27

GROUP II

24-28. A feedback system employing proportional plus error-rate control is depicted in Fig. P24-28. The output actuator is a servomotor that develops a torque in accordance with

$$T_d = -D\frac{dc}{dt} + Be_a$$

where $D = 10^{-5}$ lb-ft/rad/s
$\quad\ B = 0.375 \times 10^{-5}$ lb-ft/V
$\quad\ e_a =$ amplifier output = servomotor input voltage

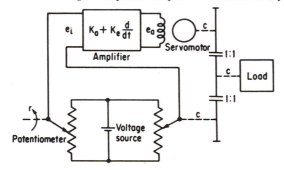

Fig. P24-28

The output voltage of the amplifier is given by $e_a = K_a e_i + K_e(de_i/dt)$. The other constants of the system are

$$K_p = \text{potentiometer transducer constant} = 1.615 \text{ V/rad}$$

$$J = \text{system inertia} = 1.56 \times 10^{-8} \text{ slug-ft}^2$$

$$F_L = \text{load viscous-friction coefficient} = 0.625 \times 10^{-5} \text{ lb-ft/rad/s}$$

(a) Determine the complete differential equation of this system relating the output variable c to input r.

(b) Obtain the transfer function of the output actuator relating the output variable $C(s)$ to the input variable $E_a(s)$.

(c) The amplifier gain is set at a value $K_a = 40$ to meet accuracy requirements. Find the value of K_e that allows the dynamic behavior to be critically damped.

24-29. A second-order feedback control system has the configuration shown in Fig. 24-10 and the following parameters:

$$K_p = \text{transducer constant} = 0.5 \text{ V/rad}$$

$$K_m = \text{motor torque constant} = 2 \times 10^{-4} \text{ lb-ft/V}$$

$$F = \text{viscous-friction coefficient} = 1.5 \times 10^{-4} \text{ lb-ft/rad/s}$$

$$J = \text{system inertia} = 10^{-5} \text{ slug-ft}^2$$

A load torque of 10^{-3} lb-ft appears constantly on the output shaft. It is required that in the steady state the effect of this load torque is not to cause a position-lag error between the command and the response exceeding 0.01 rad. Moreover, the dynamic behavior of the system is to be characterized by a maximum overshoot that does not exceed 25% when subjected to a step input. Specify the amplifier gain and the output-rate factor that will enable the system to meet these specifications.

24-30. A control system has the log-modulus attenuation characteristic depicted in Fig. P24-30.

(a) Determine the transfer function.

(b) Sketch the corresponding polar plot.

24-31. (a) Derive the transfer function of the circuit shown in Fig. P24-31.

(b) Sketch the log-modulus attenuation characteristic for the case where $C_1 = 0.1$ μF, $C_2 = 4.86$ μF, $R_1 = 12.8$ MΩ, and $R_2 = 0.103$ MΩ.

(c) Comment on the nature of the phase-angle characteristics.

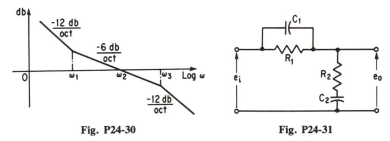

Fig. P24-30   Fig. P24-31

24-32. Determine the closed-loop transfer function for the system depicted in Fig. P24-32.

24-33. Find the closed-loop transfer function for the system shown in Fig. P24-33.

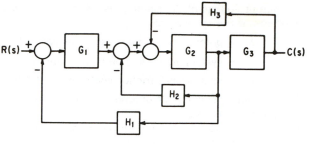

Fig. P24-32

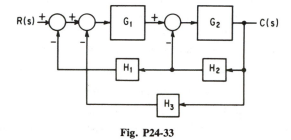

Fig. P24-33

appendices

Appendix A: units and conversion factors

TABLE A-1 UNITS

Quantity	Symbol	Name of unit	Dimensions
Fundamental			
Length	l, L	meter	L
Mass	m, M	kilogram	M
Time	t	second	T
Current	i, I	ampere	I
Mechanical			
Force	F	newton	MLT^{-2}
Torque	T	newton-meter	ML^2T^{-2}
Angular displacement	θ	radian	—
Velocity	v	meter/second	LT^{-1}
Angular velocity	ω	radian/second	T^{-1}
Acceleration	a	meter/second^2	LT^{-2}
Angular acceleration	α	radian/second^2	T^{-2}
Spring constant (translation)	K	newton/meter	MT^{-2}
Spring constant (rotational)	K	newton-meter	ML^2T^{-2}
Damping coefficient (translational)	D, F	newton-second/meter	MT^{-1}
Damping coefficient (rotational)	D, F	newton-meter-second	ML^2T^{-1}
Moment of inertia	J	kilogram-meter^2	ML^2
Energy	W	joule (watt-second)	ML^2T^{-2}
Power	P	watt	ML^2T^{-3}
Electrical			
Charge	q, Q	coulomb	TI
Electric potential	v, V, E	volt	$ML^2T^{-3}I^{-1}$
Electric field intensity	$\mathscr{E}$	volt/meter (or newton/coulomb)	$MLT^{-3}I^{-1}$
Electric flux density	D	coulomb/meter^2	$L^{-2}TI$
Electric flux	ψ, Q	coulomb	TI
Resistance	R	ohm	$ML^2T^{-3}I^{-2}$
Resistivity	ρ	ohm-meter	$ML^3T^{-3}I^{-2}$
Capacitance	C	farad	$M^{-1}L^{-2}T^4I^2$
Permittivity	ε	farad/meter	$M^{-1}L^{-3}T^4I^2$
Susceptance	S	siemens (ampere/volt)	$M^{-1}L^{-2}T^3I^2$
Magnetic			
Magnetomotive force	$\mathscr{F}$	ampere(-turn)	I
Magnetic field intensity	H	ampere(-turn)/meter	$L^{-1}I$
Magnetic flux	ϕ	weber	$ML^2T^{-2}I^{-1}$
Magnetic flux density	B	tesla	$MT^{-2}I^{-1}$
Magnetic flux linkages	λ	weber-turn	$ML^2T^{-2}I^{-1}$
Inductance	L	henry	$ML^2T^{-2}I^{-2}$
Permeability	μ	henry/meter	$MLT^{-2}I^{-2}$
Reluctance	$\mathscr{R}$	ampere/weber	$M^{-1}L^{-2}T^2I^2$

TABLE A-2 CONVERSION FACTORS

Quantity	Multiply number of:	By:	To obtain:
Length	meters	100	centimeters
	meters	39.37	inches
	meters	3.281	feet
	inches	0.0254	meters
	inches	2.54	centimeters
	feet	0.3048	meters
Force	newtons	0.2248	pounds
	newtons	10^5	dynes
	pounds	4.45	newtons
	pounds	4.45×10^5	dynes
	dynes	10^{-5}	newtons
	dynes	2.248×10^{-5}	pounds
Torque	newton-meters	0.7376	pound-feet
	newton-meters	10^7	dyne-centimeters
	pound-feet	1.356	newton-meters
	dyne-centimeters	10^{-7}	newton-meters
Energy	joules (watt-seconds)	0.7376	foot-pounds
	joules	2.778×10^{-7}	kilowatt-hours
	joules	10^7	ergs
	joules	9.480×10^{-4}	British thermal units
	foot-pounds	1.356	joules
	electron-volts	1.6×10^{-19}	joules
Power	watts	0.7376	foot-pounds/second
	watts	1.341×10^{-3}	horsepower
	horsepower	745.7	watts
	horsepower	0.7457	kilowatts
	foot-pounds/second	1.356	watts

Appendix B: periodic table of the elements

	I	II	III	IV	V	VI	VII	VIII
1	H 1 1.0081							He 2 4.002
2	Li 3 6.940	Be 4 9.02	B 5 10.82	C 6 12.01	N 7 14.008	O 8 16.000	F 9 19.00	Ne 10 20.183
3	Na 11 22.997	Mg 12 24.32	Al 13 26.97	Si 14 28.06	P 15 31.02	S 16 32.06	Cl 17 35.457	Ar 18 39.994
4	K 19 39.096	Ca 20 40.08	Sc 21 45.10	Ti 22 47.90	V 23 50.95	Cr 24 52.01	Mn 25 54.93	Fe 26 55.84 Co 27 58.94 Ni 28 58.69
4	Cu 29 63.57	Zn 30 65.38	Ga 31 69.72	Ge 32 72.6	As 33 74.91	Se 34 78.96	Br 35 79.916	Kr 36 83.7
5	Rb 37 85.48	Sr 38 87.63	Y 39 88.92	Zr 40 91.22	Cb 41 92.91	Mo 42 96.0	Te 43 ...	Ru 44 101.7 Rh 45 102.91 Pd 46 106.7
5	Ag 47 107.880	Cd 48 112.41	In 49 114.76	Sn 50 118.70	Sb 51 121.76	Te 52 127.61	I 53 126.92	Xe 54 131.3
6	Ce 55 132.91	Ba 56 137.36	La 57 138.92	Hf 72 178.6	Ta 73 180.88	W 74 184.0	Re 75 186.31	Os 76 191.5 Ir 77 193.1 Pt 78 195.23
6	Au 79 197.2	Hg 80 200.61	Tl 81 204.39	Pb 82 207.21	Bi 83 209.00	Po 84 ...	At 85 ...	Rn 86 222
7	Fr 87 ...	Ra 88 226.05	Ac 89 ...	Th 90 232.12	Pa 91 231	U 92 238.07		

Note: The number to the right of the symbol for the element gives the atomic number. The number below the symbol for the element gives the atomic weight. This table does not include the rare earths and the synthetically produced elements above 92.

Appendix C: conduction properties of common metals

TABLE C-1 RESISTIVITY AND RESISTANCE TEMPERATURE COEFFICIENTS

Material	Resistivity, ρ		Resistance temp. coefficient at 20°C, α
	microhm-cm at 20°C	ohm-cir. mils per foot at 20°C	
Aluminum	2.828	—	0.0039
Brass	—	40	0.0017
Copper (std. annealed)	1.724	10.37	0.00393
Nichrome	100	—	0.0004
Silver	1.63	—	0.0038
Tungsten	—	33.2	0.0045

TABLE C-2 ROUND COPPER-WIRE DATA

AWG number	Area (cir. mils)	Resistance (Ω/1000 ft)	Weight (lb/1000 ft)	Allowable current[a] (A)
0000	212,000	0.0490	640	358
000	168,000	0.0618	508	310
00	133,000	0.0779	402	267
0	106,000	0.0983	319	230
1	83,700	0.1240	253	196
2	66,400	0.156	201	170
3	52,600	0.197	159	146
4	41,700	0.248	126	125
5	33,100	0.313	100	110
6	26,300	0.395	79.5	94
8	16,500	0.628	50	69
10	10,400	0.999	31.4	50
12	6,530	1.59	19.8	37
14	4,110	2.52	12.4	29

[a] For type RH insulation—National Electrical Code®.

Appendix D: selected transistor characteristics

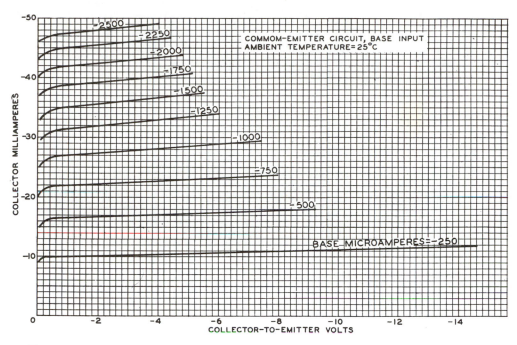

Fig. D-1. Average collector characteristics of RCA transistor 2N104 in common-emitter mode.

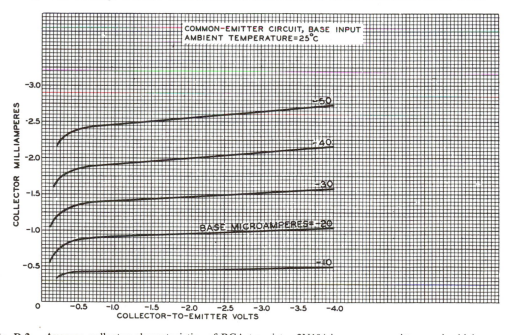

Fig. D-2. Average collector characteristics of RCA transistor 2N104 in common-emitter mode, high current.

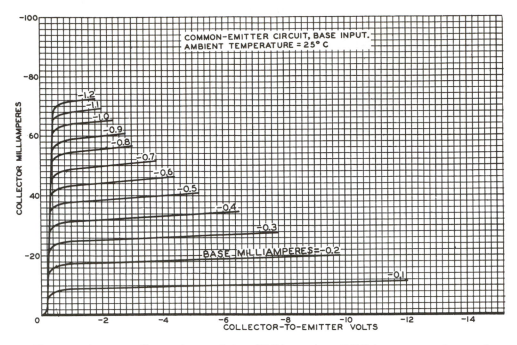

Fig. D-3. Average collector characteristics of RCA transistor 2N109 in common-emitter mode.

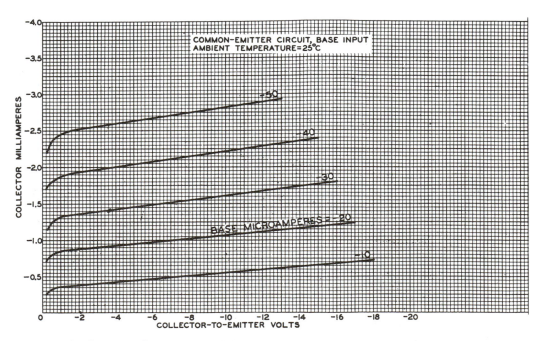

Fig. D-4. Average collector characteristics, common-emitter connection, RCA transistor 2N139.

Appendices

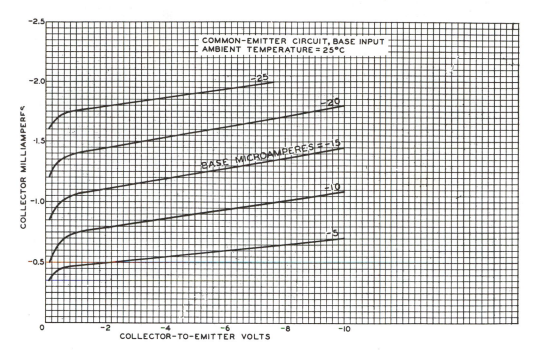

Fig. D-5. Average collector characteristics, common-emitter mode, RCA transistor 2N175.

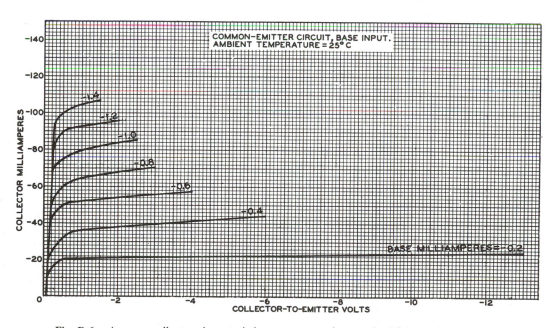

Fig. D-6. Average collector characteristics, common-emitter mode, RCA transistor 2N270.

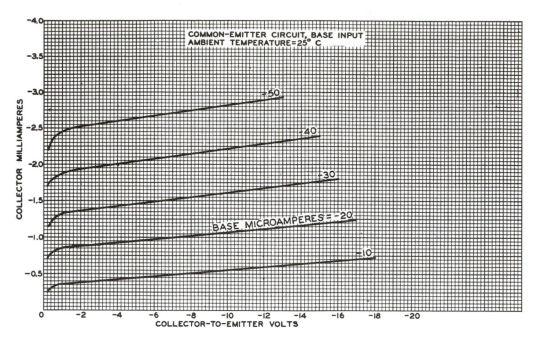

Fig. D-7. Average collector characteristics of RCA transistor 2N410 in common-emitter mode.

Appendix E: answers to selected problems

Chapter 1

1-2. (a) $\varepsilon = 0.018$ V/m; (b) $W = 0.0576 \times 10^{-19}$ J; (c) $F = 115.2 \times 10^{-22}$ N.
1-6. $V_A = 12$ V, $V_B = 4$ V, $V_c = 6$ V, $V_D = 14$ V. **1-9.** $R_3 = 5\ \Omega$. **1-11.** $R = 17.5\ \Omega$.
1-13. $e = 20$ V.

Chapter 2

2-1. 2 A, 12 V. **2-3.** 1 A, 8 Ω. **2-5.** (a) 6 V; (b) 0 V. **2-6.** 62.5 Ω. **2-7.** (a) 21¢;
(b) 129,600 C; (c) 8.08×10^{23}; (d) 1380 W. **2-9.** (a) 556 Ω; (b) 556 Ω. **2-11.** 55 °C.
2-13. 79.4 ft. **2-15.** (a) 0.5 H; (b) 0.0025 J. **2-17.** (a) 1 mA; (b) 0. **2-20.** $R_f = 27$ kΩ,
$R_1 = 3$ kΩ. **2-23.** (a) $v_o = -3x + 12y - 2z$; (b) $v_o = -2.77x + 10.91y - 1.89z$.
2-25. (a) 8 W; (b) 4 V; (c) $P_{3A} = -12$ W. **2-27.** (a) $V_a = 8$ V, $V_b = 5$ V; (b) $V_a =$
10 V, $V_b = 10$ V; (c) $V_a = 10$ V, $V_b = 0$. **2-31.** 10 H. **2-32.** 0.89 H. **2-33.** (a) 1.03 μF;
(b) 0.00462 J; (c) 1 mA; (d) 0. **2-34.** Capacitance of 1 μF. **2-36.** (b) $i(1) = 20$A,
$i(2) = 32.72$ A, $i(4) = 6.7$ A.

Chapter 3

3-4. (a) 70 μF; (b) 14×10^{-3} C; (c) $Q_i = 2 \times 10^{-3}$, $Q_2 = 4 \times 10^{-3}$, $Q_3 = 8 \times$
10^{-3} C. **3-5.** (a) 40/7 μF; (b) 2×10^{-3}; (c) $V_1 = 200$ V, $V_2 = 100$ V, $V_3 = 50$ V.
3-8. (a) 4/75 H; (b) 50 V. **3-12.** 165 V. **3-13.** 4.78 Ω. **3-15.** 18.75 Ω. **3-16.** (a) 150
V; (b) 40 V. **3-17.** 15.44 Ω. **3-20.** 7 Ω. **3-21.** 1 A. **3-23.** $I_{3\text{-}\Omega} = I_{8\text{-}\Omega} = I_{6\text{-}\Omega} = 0$,
$I_{12\text{-}\Omega} = 0.835$ A. **3-24.** (a) 12: 9 resistors and 3 voltage sources; (b) 10; (c) 3; (d) 5;
(e) 3. **3-27.** 1.67 A from bottom to top. **3-29.** 0.448 A. **3-32.** $V_{oc} = 130$ V,
$R_i = 22\ \Omega$. **3-37.** 2.86 A. **3-40.** (a) $Q_1 = 240\ \mu$C, $Q_2 = 180\ \mu$C, $Q_3 = 60\ \mu$C;
(c) $W_1 = 4.8 \times 10^{-3}$ J, $W_2 = 5.4 \times 10^{-3}$ J, $W_3 = 1.8 \times 10^{-3}$ J. **3-42.** 4.5 and 14.5 H.
3-44. 300 W. **3-45.** $V_{oc} = 12$ V, $R_i = 48\ \Omega$.

Chapter 4

4-1. $Z(p) = 2p^2 + 6p + 3/2p^2 + 3$. **4-3.** (a) $3e = 3i_L + 6\dfrac{di_L}{dt} + 2\dfrac{d^2i_L}{dt^2}$; (b) $3e =$

$2p^2i_L + 6pi_L + 3i_L$. **4-6.** (a) $4\dfrac{d^2v_b}{dt^2} + 14\dfrac{dv_b}{dt} + 14v_b = 4\dfrac{de_i^2}{dt^2} + 10\dfrac{de_i}{dt} + 4e_i + 2e_2$;

(b) The right side of part (a). **4-7.** (a) $Z_{bd} = \dfrac{6p}{3 + 2p}$, $Z_{bc} = \dfrac{20}{2p + 10}$; (b) $Z_{ad} =$

$\dfrac{3p^2 + 55p + 15}{3p^2 + 25p + 15}$, $Z_{cd} = \dfrac{12p^2 + 220p + 60}{16p^2 + 80p + 30}$. **4-9.** $6\dfrac{d^2i_2}{dt^2} + 110\dfrac{di_2}{dt} + 30i_2 = 8\dfrac{d^2e_2}{dt^2} +$

$43\dfrac{de_2}{dt} + 15e_2$. **4-11.** (a) $\dfrac{d^2v}{dt^2} + 50\dfrac{dv}{dt} + 100v = \dfrac{d^2e}{dt^2}$; (b) $\dfrac{di_R}{dt} + 2i_R = \dfrac{di_c}{dt}$.

4-13. $3\dfrac{d^3v}{dt^3} + 6\dfrac{d^2v}{dt^2} + 16\dfrac{dv}{dt} + 8v = 8e$. **4-15.** $2\dfrac{di_2}{dt} + 4i_2 = I$. **4-17.** (a) $\dfrac{d^2i}{dt^2} +$

$6\dfrac{di}{dt} + 8i = 4E$; (b) $i_f = E/2$; (c) capacitor is an open circuit and inductor is a short

circuit, hence $i_f = E/2$. **4-23.** (a) $\dfrac{d^2v}{dt^2} + 50\dfrac{dv}{dt} + 20v = 20e$; (b) $-20/29\varepsilon^{-t}$; (c) $-\frac{1}{19}\varepsilon^{-10t}$;

(d) $\dfrac{\varepsilon^{10t}}{31}$.   **4-24.** (a) $2\dfrac{d^2i}{dt^2} + 6\dfrac{di}{dt} + 3i = 0$; (b) $s^2 + 3s + 1 = 0$; (c) $i = K_1\varepsilon^{-2.6t} +$

$K_2\varepsilon^{-0.4t}$.   **4-25.** (a) $\dfrac{p + 10}{p}$; (b) $\dfrac{di}{dt} + 10i = 12e$; (c) 0; (d) $12\varepsilon^{-10t}$.   **4-31.** $\cos 10t$.

4-35. $\dfrac{-5}{4} + 2.5t, \tfrac{5}{4} + 7.5t$.   **4-38.** (a) 0; (b) 12; (c) 12; (d) -48; (e) See 4-17(a); (f) 6;

(g) $s^2 + 6s + 8 = 0$; (h) $i = 6(1 + 2\varepsilon^{-4t} - \varepsilon^{-2t})$.   **4-40.** (a) 0.02 A; (b) $K_1\varepsilon^{-33t}$,

$K_2\varepsilon^{-3t}$.

Chapter 5

5-1. $i(t) = 2 + 1 - \varepsilon^{-25t}$ A.   **5-2.** (a) $i(t) = 2(1 - \varepsilon^{-25/3t})$ A; (b) 2 A; (c) $\tfrac{9}{25}$ s.
5-3. (a) $v(t) = \tfrac{1}{2}(1 - \varepsilon^{-4t})$ m/s; (b) $\tfrac{1}{2}$ m/s; (c) 1.25 s.   **5-5.** $i(t) = 5(1 - \varepsilon^{-200t})$ A.
5-6. (a) $2\varepsilon^{-10t}$; (b) 6 J; (c) 6 J.   **5-10.** $10 - 6\varepsilon^{-10t}$ A.   **5-11.** (a) $i(t) = 15 - 3\varepsilon^{-5.4t}$ A;
(b) 0.185 s.   **5-13.** $i(t) = \tfrac{3}{2}(1 - \varepsilon^{-32t})$ A.   **5-14.** $4 \times 10^{-3}\varepsilon^{-t}$ CCW.   **5-15.** (a) $s^2 +$
$1.25s + 0.25 = 0$; (b) 1 s, 4 s.   **5-19.** (a) 20 V; (b) 20V; (c) $v(t) = 7.5 +$
$12.5\varepsilon^{-4/15t}$V; (d) 3.75 s.   **5-20.** (a) 0; (b) $2 \times 10^{-3}\varepsilon^{-t}$ A; (c) 5 s; (d) $i(t) = q_0\omega \sin \omega t$.
5-23. (a) ε^{-30t}; (b) 1000 V; (c) tenfold increase.   **5-27.** (a) 0.4; (b) 25%; (c) 7.32 rad/s;
(d) 0.936 s.   **5-28.** (a) $2 \cos (20t - 53°)$; (b) 1.2 A; (c) $-1.2\varepsilon^{-15t}$.   **5-31.** $i(t) = 1.6 -$
$0.4\varepsilon^{-20/3t}$ A.   **5-33.** 1.15 s, 2.77 s.   **5-38.** (a) A/RT; (b) $i(t) = \dfrac{A}{RT}\varepsilon^{-R/Lt}$.

5-44. (a) 24.8 Ω, 12.4H; (b) positive; (c) peak value.

Chapter 6

6-1. (c) $(s^2 + 3s + 2)C(s) = \dfrac{1}{s^2} - s - 1$.   **6-2.** (a) $\mathscr{L}f(t) = \dfrac{1}{s^2(s + \alpha)}$, (c) $\mathscr{L}\varepsilon^{-\alpha t} \sin$

$(\omega t + \theta) = \dfrac{\omega \cos \theta}{(s + \alpha)^2 + \omega^2} + \dfrac{(s + \alpha) \sin \theta}{(s + \alpha)^2 + \omega^2}$.   **6-3.** $\mathscr{L}f(t) = \dfrac{A}{s}(\varepsilon^{-sT_1} - 2\varepsilon^{-sT_2} + \varepsilon^{-sT_3})$.

6-4. $\mathscr{L}f(t) = \dfrac{(1 - \varepsilon^{-s})^2}{s^2}$.   **6-5.** $\mathscr{L}f(t) = \dfrac{1}{s^2T_1} - \dfrac{\varepsilon^{-sT_1}}{s} - \dfrac{\varepsilon^{-sT_1}}{s^2T_1}$.   **6-7.** $x(t) = \tfrac{1}{8} + 1.79\varepsilon^{-4t}$

$- 0.92\varepsilon^{-10t}$.   **6-8.** $f(0^+) = 2$.   **6-9.** $f(t) = 2\varepsilon^{-t} - t\varepsilon^{-2t} - 2\varepsilon^{-2t}$.   **6-11.** $f(t) = 5\varepsilon^{-2t}$
$\sin 2t$.   **6-24.** $f(t) = 3\varepsilon^{-3t} \cos 4t - \tfrac{1}{4}\varepsilon^{-3t} \sin 4t$.

Chapter 7

7-1. (a) 0, I_m; (b) $0.5I_m$, $0.707I_m$.   **7-2.** (a) 0, $0.577V_m$; (b) $0.25V_m$, $V_m/\sqrt{6}$.
7-3. (a) $0.5I_p$, $0.577I_p$; (b) 0, $0.577I_p$; (c) $0.25I_p$, $I_p/\sqrt{6}$.   **7-5.** 862 W.
7-6. (a) $8.63\angle -5.1°$; (b) $12.2 \sin (377t - 5.1°)$; (c) 863 VA; (d) 0.966.
7-10. (a) $15 \angle 53.2°$; (b) $300\sqrt{2} \sin (377t + 53.2°)$; (c) 0.0317 H.   **7-14.** (a) 39.4 sin
$(377t + 15.1°)$.   **7-20.** $\overline{V}_{ab} = 84\angle 28°$, $\overline{V}_{bc} = 47.3\angle -56.9°$.   **7-21.** 33.3 Ω.   **7-22.** $X_1 =$
4, $X_2 = 3$, and $\overline{I} = 39.6\angle -45°$.   **7-23.** $11.8\angle -80.2°$.   **7-26.** (a) $11.72 + j47.1$;
(b) $-j47.1$; (c) 53.2 μF; (d) $6.8\angle 0°$.   **7-27.** (a) $5\angle -36.8$; (b) $L = 1.5$ H;
(c) $12.5\angle -90°$ V.   **7-34.** (a) 1000 rad/s; (b) 100; (c) 10 rad/s; (d) 995 and 1005 rad/s;
(e) 100 V.   **7-36.** (a) $0.233\angle 90°$ mA; (b) $0.567\angle -90°$ mA.   **7-39.** (a) $17.32\angle -60°$;
(b) 5710 W.   **7-40.** (a) $15.21\angle -83.14°$, $15.21\angle -203.14°$, $15.21\angle 36.7°$; (b) 1160 W;
(c) 0.   **7-43.** (a) $i(\omega t) = \dfrac{4}{\pi}I_m (\sin \omega t + \tfrac{1}{3} \sin 3\omega t + \tfrac{1}{5} \sin 5\omega t + \cdots +)$; (b) $i(\omega t) =$

$\dfrac{I_m}{2} + \dfrac{2}{\pi}I_m (\sin \omega t + \tfrac{1}{3} \sin 3\omega t + \tfrac{1}{5} \sin 5 \omega t + \cdots +)$.   **7-44.** $v(\theta) = \dfrac{8}{\pi^2}\bigg(\sin \theta -$

$$\frac{1}{3^2} \sin 3\theta + \frac{1}{5^2} \sin 5\theta - \frac{1}{7^2} \sin 7\theta + \cdots + \Big).$$ **7-46.** $i(\theta) = \dfrac{Ip}{2} + \dfrac{1}{\pi}I_p \left(\sin \theta + \tfrac{1}{2} \sin 2\theta + \right.$ $\tfrac{1}{3} \sin 3\theta + \cdots + \right).$ **7-49.** 80 V. **7-50.** (a) $6\underline{/36.8°}$; (b) $6\sqrt{2} \sin (\omega t + 36.8°)$; (c) $\overline{V}_{ab} = 24\underline{/-53°}$, $\overline{V}_{bc} = 107.4\underline{/10.3°}$. **7-53.** $47.8\underline{/-83.14°}$.

Chapter 8

8-1. (a) False; (b) false; (c) false; (d) false. **8-2.** (1) 4.4×10^{14} holes/cm³; (b) 6.6×10^{12} electrons/cm³; (c) see Figs. 8-13 and 8-16 of text. **8-4.** (a) 418 Ω; (b) 108 Ω. **8-5.** 150 mA. **8-6.** 472 mA and 0.235 mA. **8-7.** 7.6 Ω. **8-9.** (a) -8.12 V; (b) -0.388 V; (c) 149. **8-12.** 2 mA/V and 240. **8-13.** (a) 4 mA; (b) 16.6 kΩ; (c) 1.55 mA/V; (d) 25.7.

Chapter 9

9-1. (a) 6.3 V; (b) 80; (c) 24.5; (d) 1960. **9-2.** (a) Yes; (b) 0.2 V; (c) 0.74 V. **9-4.** (a) 5 mV; (b) 400; (c) 50; (d) 400. **9-5.** (a) 2550 Ω. **9-7.** (a) 0.46 mA; (b) 30 kΩ. **9-9.** 25,180 Ω. **9-10.** $R_E = 500\ \Omega$, $R_B = 44$ kΩ. **9-12.** (b) 1.07 μA; (c) 68.4 μA; (d) 34.2. **9-13.** (a) 34.1; (b) 17.3; (c) 590; (d) 1.92 kΩ; (e) 2.46 kΩ. **9-15.** (b) 241.5 mV; (c) 46.7, 80.5, and 2185. **9-16.** (a) 356 μA; (b) 291 μA; (c) 4.1 V and 20.5 mA; (d) 2420; (e) 205; (f) 496,000. **9-18.** (a) 36.6; (b) 1.26 Hz; (c) 89,300 Hz; (d) -109.8 μA.

Chapter 10

10-1. (b) 5 V; (c) 7.07 V. **10-2.** (b) 1.1 A. **10-3.** (b) 0.022 A; (c) full-wave rectification. **10-5.** $3E_m$. **10-7.** $2E_m$. **10-8.** (a) $\dfrac{V_o}{V_b} = h_{fe}\dfrac{R_L}{h_{ie} + (1 + h_{fe})R_E}$; (c) place a capacitor in parallel with R_E, which yields $\dfrac{V_o}{V_b} = h_{fe}\dfrac{R_L}{h_{ie}}$. **10-9.** (a) 660 k$\Omega$; (b) 56 Ω; (d) 5 V. **10-13.** Carrier frequency, 10^8 rad/s; first lower-side frequency $= \omega_C - \omega_{m1} = 10^8 - 10^3$; first upper-side frequency $= \omega_C + \omega_{m1} = 10^8 + 10^3$; second lower-side frequency $= \omega_C - \omega_{m2} = 10^8 - 1.8 \times 10^3$; second upper-side frequency $= \omega_C + \omega_{m2} = 10^8 + 1.8 \times 10^3$; (b) For ω_{m1}, $m_1 = 0.4$; for ω_{m2}, $m_2 = 0.2$. **10-16.** $4E_m$.

Chapter 11

11-3. $Z = (D \cdot S_D + P \cdot S_P) \cdot K$. **11-4.** **11-6.** $L = A \cdot B + \overline{A} \cdot \overline{B}$. **11-8.** $L = A \cdot (E + (B \cdot (C + D)))$.

11-10.

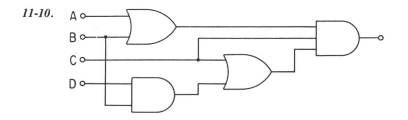

11-13. (a) 77_{10}; (b) 227_{10}; (c) 682_{10}. **11-14.** (a) 1101; (b) 100011; (c) 101101; (d) 1101101011. **11-17.** (a) $N_{10} = 255$; (b) $N_{10} = 4011$; (c) $N_{10} = 4077$; (d) $N_{10} = 65{,}261$. **11-18.** $N_2 = .10100110$; error $= 0.0015625$. **11-21.** $N_2 = 10.101111$. **11-23.** $N_7 = 23.4125$. **11-25.** $N_3 = 122.12101$. **11-28.** $Z = \overline{A + B}$.

Chapter 12

12-1. $A \cdot A = A \cdot A + 0 = A \cdot A + A \cdot \overline{A} = A(A + \overline{A}) = A \cdot 1 = A$. **12-2.** $A + AB = A \cdot 1 + AB = A(1 + B) = A \cdot 1 = A$.

12-1.

A	B	C	$B + C$	$A(B + C)$	$A \cdot B$	$A \cdot C$	$AB + AC$
0	0	0	0	0	0	0	0
0	0	1	1	0	0	0	0
0	1	0	1	0	0	0	0
0	1	1	1	0	0	0	0
1	0	0	0	0	0	0	0
1	0	1	1	1	0	1	1
1	1	0	1	1	1	0	1
1	1	1	1	1	1	1	1

Observe that columns 5 and 8 are identical.

12-5.

A	B	C	Z
0	0	0	1
0	0	1	1
0	1	0	0
0	1	1	0
1	0	0	0
1	0	1	1
1	1	0	1
1	1	1	1

12-6. (a) $Z = B$; (b) $Z = AB + BD$.

12-9. (a) $Z = \overline{A}\,\overline{B}\,\overline{C} + AB$.

12-10.

A	B	C	Z
0	0	0	0
0	0	1	1
0	1	0	1
0	1	1	0
1	0	0	0
1	0	1	1
1	1	0	1
1	1	1	0

12-11. (a) $Z = \overline{A}BC + A\overline{B} + A\overline{C}$;

(b) $Z = \overline{A}BC + A\overline{B}C + A\overline{B}\,\overline{C} + AB\overline{C}$;

(c)

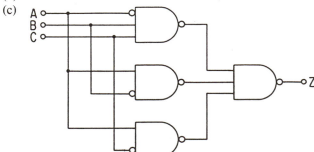

12-13.

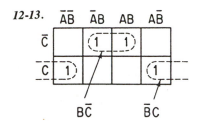

12-15. (a) $Z_1 = \overline{A}\,\overline{B}C + \overline{A}BC + A\overline{B}C + ABC$, $Z_2 = \overline{A}\,\overline{B}\,\overline{C} + \overline{A}B\overline{C} + A\overline{B}\,\overline{C} + AB\overline{C}$; (b) $Z_1 = (A + B + C)(A + \overline{B} + C)(\overline{A} + B + C)(\overline{A} + \overline{B} + C)$, $Z_2 = (A + \overline{B} + \overline{C})(\overline{A} + B + C)(\overline{A} + B + \overline{C})(\overline{A} + \overline{B} + \overline{C})$; (c) $Z_1 = C$, $Z_2 = \overline{A}B + B\overline{C}$; (d) $\overline{Z}_1 = \overline{C}$ or $Z_1 = C$, $Z_2 = (\overline{B} + \overline{C})(\overline{A} + B)$. **12-17.** (a) $Z = (A + B + C)(A + B + \overline{C})(A + \overline{B} + \overline{C})(\overline{A} + B + \overline{C})$; (b) $Z = (A + C)(B + \overline{C})$; (c)

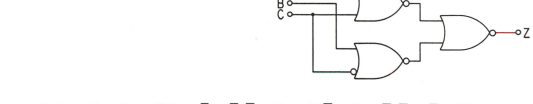

12-18. (a) $Z = (A + B + C)(A + \overline{B} + \overline{C})(\overline{A} + B + C)(\overline{A} + B + \overline{C})(\overline{A} + \overline{B} + \overline{C})$; (b) $Z = (\overline{A} + B)(B + C)(\overline{B} + \overline{C})$. **12-19.** $Z = AC$. **12-22.** $Z = A + BC + D$. **12-23.** (a) $Z = AC\overline{D} + \overline{B}D$; (b)

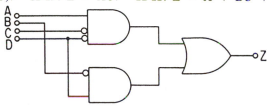

Chapter 13

13-2. $Z = D_1 + D_3 + D_5 + D_7 + D_9 + D_B + D_D + D_F$ for least significant bit

    $Y = D_2 + D_3 + D_6 + D_7 + D_A + D_B + D_E + D_F$

    $X = D_4 + D_5 + D_6 + D_7 + D_C + D_D + D_E + D_F$

    $W = D_8 + D_9 + D_A + D_B + D_C + D_D + D_E + D_F$ for most significant bit

13-3. $Z = D_1 + D_3$, $Y = D_2 + D_3$.   **13-6.** (a) 10001; (b) 10101000.   **13-7.** $M = 0$, adder; $M = 1$, subtractor.

13-8.

X	Y	B	D
0	0	0	0
0	1	1	1
1	0	0	1
1	1	0	0

$B = \overline{X}Y$

$D = \overline{X}Y + X\overline{Y}$ (this logic is exactly the same as the sum for the half-adder).

13-11. 10001, 1000.   **13-12.** (a) 1000111; (b) 10001; (c) $-(10001)$; (d) $-(1000111)$.

13-14.

A	B	D_0	D_1	D_2	D_3	Minterms
0	0	1	0	0	0	$m_0 = \overline{A}\,\overline{B}$
0	1	0	1	0	0	$m_1 = \overline{A}B$
1	0	0	0	1	0	$m_2 = A\overline{B}$
1	1	0	0	0	1	$m_3 = AB$

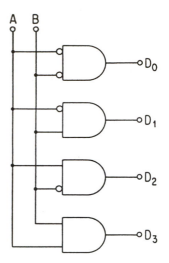

13-16.

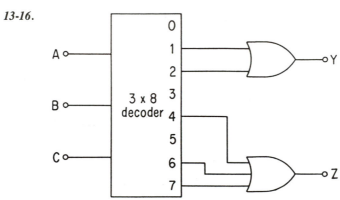

13-19. $2^n = 2^2 = 4$ control lines.

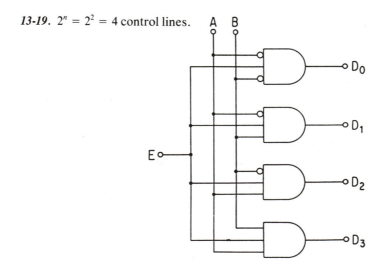

13-21. $I_7I_6I_5I_4I_3I_2I_1I_0 = 00110011.$ **13-23.**

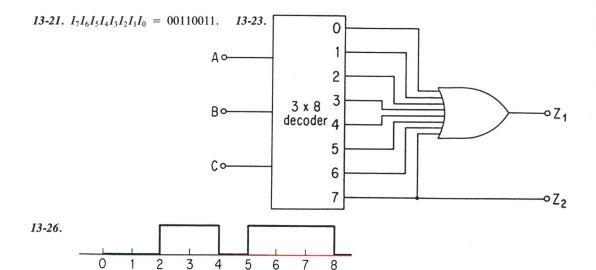

13-26.

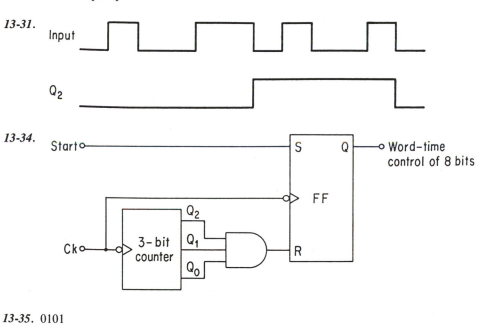

13-29. The D flip-flop.

13-31.

13-34.

13-35. 0101
1011
0110
1101
1010
0101

13-39. 14 bits.   **13-41.** (a) 16; (b) 8 FF.   **13-42.** When $S_3S_2S_1S_0 = 0$, $A = B$; when $C = 1$, $A \geq B$; when $C = 0$, $A < B$.

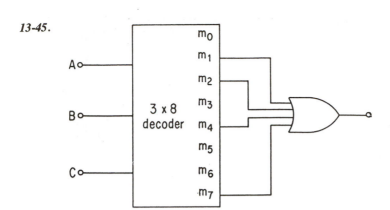

13-45.

Chapter 14

14-1.

MR address in binary	Contents of RAM Op code	Address code	Instruction
0000	11	1010	Transfer A to MR10 (ten)
0001	10	1010	SUB contents of MR10 form A

14-2.

Step	Memory address	Contents of RAM Op code	Address code	Instruction
Instruction				
0	0000	01	1010	ADD MR
1	0001	01	1011	ADD MR
2	0010	10	1100	SUB MR
3	0011	01	1101	ADD MR
4	0100	11	1111	TR MR
5	0101	00	xxxx	HALT
⋮	⋮	⋮	⋮	⋮
Data				
10	1010	00	0010	2
11	1011	00	0111	7
12	1100	00	0100	4
13	1101	00	0001	1
Output				
15	1111	00	0110	6
14-4. 0	0000	11	1010	TR MR
⋮	⋮	⋮		
Data section				
10	1010	00	0000	

14-6. 1. LXI H
      <0001 1001>
      <0000 0000>
  2. MOV A, M
  3. ANI
      <1111 0000>
  4. INX H
  5. MOV M, A

14-8. 1. LDI H
      <0010 0011>
  2. MOV A, M
  3. CMA
  4. INR A
  5. MOV M, A

14-7. 1. MOV H, B
  2. DAD D
  3. MOV B, D

14-11. 1. SUB A
  2. LXI H
      <0000 1100>
  3. MOV A, M
  4. INR H
  5. ADD M
  6. INR H
  7. ADD M
  8. INR H
  9. ADD M
  10. INR H
  11. MOV M, A

14-13. Introduce an OR gate at the output of A4 and modify as follows:

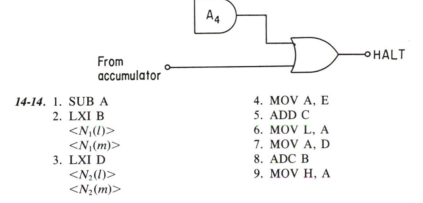

14-14. 1. SUB A
  2. LXI B
      $<N_1(l)>$
      $<N_1(m)>$
  3. LXI D
      $<N_2(l)>$
      $<N_2(m)>$

  4. MOV A, E
  5. ADD C
  6. MOV L, A
  7. MOV A, D
  8. ADC B
  9. MOV H, A

Chapter 15

15-1. (a) $4/\pi$; (b) 32×10^{-7} N/m; (c) 0.32 N/m.　**15-5.** 20 A.　**15-7.** (a) 98 At; (b) 1.225 A.　**15-8.** 4.96 A.　**15-9.** 183.6 At.　**15-11.** 0.0045 Wb.　**15-14.** (a) 1500 At; (b) 2300 At/m; (c) 2.9 W/m²; (d) 1.25×10^6; (e) 1.26×10^{-3}; (f) 1010.　**15-16.** 20.1 W. **15-18.** 0.003125 Wb.　**15-19.** 361 At applied in direction to cause flux in right leg to flow CCW.　**15-21.** At 60 Hz: $P_h = 1400$ W, $P_e = 400$ W; at 90 Hz: $P_h = 2100$ W and $P_e = 900$ W.　**15-22.** $F = \dfrac{B^2}{\mu} bg$.　**15-23.** (a) 5 J; (b) 2 W-s; (c) 20 H; (d) 3.35 W-s; (e) no, it comes from electrical source.

Chapter 16

16-1. (a) 0.0108 Wb; (b) 1.44 in.².　**16-4.** A 50% reduction.　**16-5.** (a) 114 W; (b) 360 W (c) 97.7%; (d) 0.055%.　**16-6.** (a) 2340.8 V; (b) 98%.　**16-8.** 4.82%.　**16-10.** 103 A on high side, 206 A on low side.　**16-11.** 87.8 V.　**16-14.** $150\angle-30°$. **16-15.** $906\angle-44°$.　**16-17.** A lead pf angle of 28.6° with respect to V_1.

Chapter 17

17-1. (a) 0.0024 Wb; (b) 0.144 Wb-turns; (c) $\lambda = 0.144 \cos 5\pi t$; (d) 2.26 V; (e) 1.96 V.
17-2. (a) $e = \pi \sin 10\pi t$; (b) πV. **17-3.** (a) $\Phi = Blr(\pi - 2\theta_0)$; (b) $e = 2NBlv = $ 4.8 V; (c) 7.54 V; (d) 4.8 V. **17-4.** 531 At/pole. **17-5.** (a) 0.0925 Wb; (b) 34.7°.
17-7. (a) 8640 At; (b) 53.1 N-m; (c) 104 V; (d) 0.00122 Wb.

Chapter 18

18-3. 1.5 $\phi_m\angle 30°$, $1.5\phi_m\angle -30°$. **18-5.** (a) 1800 rpm; (b) 3 Hz; (c) 90 rpm; (d) 1800 rpm; (e) 0; (f) yes, in direction of revolving field. **18-7.** 900 rpm, 1.8 Hz. **18-9.** 87.45%.
18-10. (a) 44.7 A at 0.87 pf lag; (b) 17.85 hp; (c) 125 N-m. **18-12.** (a) 6770 N-m; (b) 127 A at 0.942 pf lag. **18-14.** (a) 88.3 A at a pf of 0.958 lag; (b) 2260 N-m; (c) 94%.
18-16. 0.0333 Ω/phase. **18-18.** (a) 0.162; (b) 90.3 A at 0.505 pf lag; (c) 560 N-m.
18-20. (a) $22.4\angle -18°$ A; (b) 45 N-m; (c) 3.56%; (d) 86.1%. **18-22.** (a) 700 rpm; (b) 1700 rpm. **18-23.** 69.5 N-m.

Chapter 19

19-1. (a) $163.5\angle 52.8°$; (b) 1000 V/phase. **19-2.** (a) 1790 V/phase; (b) $55.4\angle 41°$; (c) 0.755 lead. **19-4.** (a) $293\angle -9.1°$; (b) no. **19-8.** (a) 24.85 A at 0.79 pf lag; (b) $295\angle -30.7°$ V. **19-9.** 4120 V to 9408 V.

Chapter 20

20-2. (a) 366.7 Ω; (b) 12 V; (c) 176.5 Ω; (d) 0.068 A; (e) 617 rpm. **20-3.** 201 V.
20-4. (a) 375 Ω; (b) 5.08%; (c) 2.27 A; (d) 880 rpm. **20-6.** (a) 0.00905 Wb; (b) 129 N-m; (c) 580 W; (d) 15.2%. **20-7.** (a) 229 A; (b) 1414 rpm. **20-9.** (a) 240 A; (b) 535 N-m; (c) 86.5%. **20-13.** (a) 226 V; (b) 207.8 A; (c) 388 N-m; (d) 93.33%. **20-16.** (a) 176.93 rpm; (b) 7.9%; (c) 3.8%. **20-17.** 269 Hz.

Chapter 21

21-1. Yes. **21-4.** Yes. **21-5.** $8\angle 90°$.

Chapter 22

22-1. (a) $0.863T_m$ CCW; (b) 18°. **22-3.** 0.
22-4. (a)

A	B	C	D
1	0	0	0
0	1	0	0
0	0	1	0
0	0	0	1

(b) 90°; (c)

A	B	C	D
1	0	0	0
1	1	0	0
0	1	0	0
0	1	1	0
0	0	1	0
0	0	1	1
0	0	0	1
1	0	0	1

(d)

(e) 135°.

22-6. 240 bits/s.　**22-9.** (a) 7.2°; (b) 40.

Chapter 23

23-5. (a) Three stages; (b) 0.487×10^{-3}.　**23-6.** Five stages.

Chapter 24

24-1. (a) 10; (b) 6 rad/s; (c) 0.8; (d) unity; (e) four times.　**24-2.** (a) 0.4; (b) 25%.

24-4. (a) $\dfrac{10}{s^2 + 2s + 10}$; (b) 3 rad/s; (c) 35%; (d) 5 s.　**24-6.** (a) $\dfrac{\omega_r}{1 + K}$; (b) $\omega_e(t) =$

$\dfrac{1}{1 + K} + \dfrac{K}{1 + K} \varepsilon^{[(1+K)/(J/F)]t}$; (c) No.　**24-8.** (a) $c(t) = 4t - 0.4(1 - \varepsilon^{-20t})$; (b) $\dfrac{40}{s + 10}$;

(c) $c(t) = \frac{4}{5}t - \frac{4}{250}(1 - \varepsilon^{-50t})$.　**24-9.** (a) 0.6; (b) 355×10^{-6}.　**24-10.** $K_e = 0.179$.

24-12. $K_o = 0.9$.　**24-13.** $1/(1 + s\tau_1)(1 + s\tau_2)$.　**24-15.** $\dfrac{E_o}{E_i} = \dfrac{1}{\alpha}\left(\dfrac{1 + s\alpha\tau}{1 + s\tau}\right)$ where $\alpha =$

$\dfrac{R + R_L}{R}$ and $\tau = \dfrac{L}{R + R_L}$.　**24-19.** $\dfrac{X_o}{X_L} = \dfrac{1}{\alpha}\left(\dfrac{1 + s\alpha\tau}{1 + s\tau}\right)$ where $\alpha = \dfrac{K_1 + K_2}{K_1}$ and $\tau =$

$\dfrac{f}{K_1 + K_2}$.　**24-21.** $\dfrac{X_o}{F} = \dfrac{1/M}{s^2 + (f/M)s + k/M}$ where $\omega_n = \sqrt{\dfrac{k}{M}}$ and $\zeta = \dfrac{f}{2\sqrt{kM}}$.

24-25. $T(j\omega) = \dfrac{k}{\left(1 + j\dfrac{\omega}{\omega_1}\right)\left(1 + j\dfrac{\omega}{\omega_2}\right)}$.　**24-29.** $K = 1000$ V/V, $K_o = 3.25 \times 10^{-3}$.

24-30. $T(j\omega) = \dfrac{1 + j\dfrac{\omega}{\omega_1}}{\dfrac{(j\omega)^2}{\omega_1\omega_2}\left(1 + j\dfrac{\omega}{\omega_3}\right)}$.　**24-32.** $\dfrac{C}{R} = \dfrac{G_1 G_2 G_3}{1 + H_2 G_2 + H_1 G_1 G_2 + H_3 G_2 G_3}$.

24-33. $\dfrac{C}{R} = \dfrac{G_1 G_2}{1 + H_3 G_1 G_2 + H_2 G_2 + H_1 H_2 G_1 G_2}$.

index

Absorption rule, Boolean algebra, 500
Accelerometer, 876
Acceptor impurity, 337
Accumulator, 588
A-c load line, 384
Actuators, output, 881
Adders,
   carry ripple, 531
   full-adder, 526
   half-adder, 526
   parallel type, 4-bit, 530
Addition, op-amps, 57
Admittance, complex, 286
Algebraic multiplier, 133
Alternating current, 256
ALU, 588, 597
Amortisseur winding, 750
Ampere, definition, 3
Ampère's circuital law, 619
Ampère's law, 19, 615
Amplification factor,
   of common-base amplifier, 353
   of common-emitter amplifier, 377
   of JFET, 361
Amplifiers (*see also* Transistor amplifiers and
       Operational amplifiers)
   BJT amplifier, 407
   differential, 50
   drive, 809
   emitter follower, 448
   feedback, 835
   JFET, 385
   operational, 49
Amplitude demodulation (*see* Demodulation)
Amplitude modulation (*see* Modulation)
Amplitude, sinusoid, 254
AND, 466
Apparent power, 264
Angular velocity, 255
Armature reaction, 748
Armature winding, 616
Assembler, 593
Assembly language programming,
   elementary computer, 593
   Intel 8080, examples, 606–610
Asynchronous circuits, 545
Asynchronous machines, 714
Average power, 263
Average value, 257
AWG (American wire gage), 36

Bandwidth,
   BJT amplifier, 426
   resonant circuit, 298
   JFET amplifier, 428
   with feedback, 839
Biasing methods,
   collector-to-base, 397
   fixed, 396
   self, 398
Binary digits (bits)
   least significant bit, 483
   most significant bit, 483
Binary addition, 491
Binary logic, 465
Binary number system, 483
Biot-Savart law, 19
Bipolar junction transistor amplifier (BJT) (see
      Transistor amplifier)
Bits (*see* Binary digits)
Block diagram, 820
Bode diagram,
   accelerometer, 878
   lag network, 869, 872
   lead network, 873
Boltzmann relation, 342
Boolean algebra, 468
   absorption rule, 500
   postulates and theorems, 498–504
Branch, of a circuit, 87
Breakdown torque, 731
Breakover voltage, SCR, 367
Breakpoint (Corner) frequency, 870
Buffer, 54
Byte, 483

Capacitance,
   circuit viewpoint, 44
   definition, 44
   diffusion, 350
   energy viewpoint, 46
   geometrical viewpoint, 48
   initial condition property, 46
   interterminal, 428
   parallel-connected, 77
   series-connected, 76
   units, 44
Capacitive reactance, 279
Capacitor,
   definition, 44
   integrated circuit, 432

Capacitor motor, 794
Carrier frequency, 454
Carry ripple, 531
Characteristic equation, 141
    RC circuit, 166
    RL circuit, 157
    RLC circuit, 181, 242
Charge, 4, 5, 10
Charge neutrality, law of, 344
Chopper drive, d-c motor, 775
Circuit,
    asynchronous, 545
    combinational, logic, 522
    clamping, 446
    definition, 5
    differentiating, 873
    diffused, 432
    dual, 173
    ground, 29
    integrating, 867
    sequential, 544
    series-parallel, 80
    terminal, definition, 25
    three-phase, 305, 311
Circuit analysis,
    with complex impedances, 291
    delta-Y transformation, 109
    duality, 171
    evaluation of initial conditions, 158
    forced solutions, 130
    formulation of differential equations, 129
    maximum power transfer, 104
    mesh method, 87
    network reduction, 81
    nodal method, 94
    Norton's theorem, 104
    superposition, 84
    Thévenin's theorem, 100
    transient solutions, 137
    via matrices, 90
Clamping circuit, 446
Circuit elements, diffused, 432
Circular mil, 35
Closed-loop system, 819
CMOS, 367, 479
Coercivity, 627
Color-coded resistors, 37
Combinational logic circuits, 522
Common-emitter mode, transistor, 375
Commutator, 701
Compiler, 610
Complement, 467
Complete response,
    RC circuit to step input, 237
    RL circuit to sinusoids, 195, 245
    RL circuit to steps, 233
    RLC circuit to sine input, 199
    RLC circuit to step input, 241
Complex admittance, 286
Complex frequency, 183, 848
Complex impedance, 281, 288, 289

Complex numbers,
    addition, 265, 272
    multiplication and division, 270
    powers and roots, 271
Complex translation, Laplace transform, 224
Compound generator, 765
Compound motor (see D-c motors)
Computer architecture, 587 (see also Computer, elementary)
Computer, elementary,
    basic architecture, 587
    assembly language programming, 590
    controller, 592
    fetch and execute modes, 592
    input/output, 596
    mnemonic codes, 592
Conductance, 36
    mutual, 97
    RL circuit, 286
    self, 97
Conduction band, semiconductor, 338
Construction features,
    d-c machine, 701
    induction motor, three-phase, 698
    single-phase induction motor, 794–797
    synchronous machine, 700
Control instructions, Intel 8080, 603–606
Controller, elementary computer, 592, 594
Controllers,
    d-c motors, 781–785
    induction motors, 733
Conversion factors, Table, 891
Conversions, number systems, 484–489
Core loss, 642
Coulomb,
    definition, 5
    law, 6
Counters, 555
    decade, 565
    program, 588
    ring, 560
    ripple, 556
    switchtail, 563
    synchronous, 558
    up-down, 560
Cramer's rule, 92
Critical damping, 190
Critical field resistance, 764
Curie point, 625
Current,
    definition, 5
    as a fundamental unit, 3
    by conduction, 338
    by diffusion, 344
    by injection, 347
    positive direction, defined, 29
    reverse saturation, 346
    source, 26
    recombining, 347
Current amplification factor, transistor, 377
Current-divider rule, 108

Current element, 615
Current gain, 382, 385, 406, 412, 415, 417
Current impulse, 179
Current source, 27, 59
Current transfer ratio, 353, 390

Damped frequency, 183
Damping ratio, 182
    as a figure of performance, 188, 851
    effect on maximum overshoot, 189, 851
Data distributer, 539
Data latch, 545
Data selectors, 540
D-c generator,
    compound, 765
    critical field resistance, 764
    equivalent circuit, 766
    performance equations, 766
    power flow diagram, 766
    saturation curve, 764
    self-excited, 763
D-c load line, transistor, 379
D-c machine, construction features, 701
D-c motor,
    applications, 701
    controllers, 781
    equivalent circuit, 769
    performance equations, 769
    power flow diagram, 769
    speed control, 774
    speed regulation, 773
    speed-torque curves,
        series motor, 773
        shunt motor, 772
Decade counter, 565
Decibels, 869
Decoders, 536
    BCD-to-decimal, 489
Decoding, 490
D flip-flop, 551
Delta-connection, three-phase, 308
Delta-Y transformation, 109
Demodulation, amplitude, 458
    diagonal clipping, 459
De Morgan's theorem, 500
DE MOSFET, 365
    v-i characteristic, 366
Demultiplexers, 539
Depletion region, 342
Detection (see Demodulation)
Diagonal clipping, 459
Diamagnetism, 623
Difference amplifier, 50
Differential operator, 122
Differentiating circuit, 873
Diffused circuit elements, 432
Diffusion capacitance, 350
Diffusion current, 344
Dimension, definition, 3
Diode clamper, 446

Diode detector, 458–459
    transistor type, 460
Diode limiter, 447
Donor impurity, 336
Driving point impedance, 124
Drive amplifiers, stepper motors, 808, 809
Duality, 171
Dual circuit, by graphical procedure, 173
Dynamic response (see Natural response and
        Complete response)

ECL, 473–478
Eddy-current loss, 642
Effective value, definition, 259
Efficiency,
    d-c generator, 768
    d-c motor, 770
    single-phase induction motors, 795
    synchronous motors, 758
    three-phase induction motors, 728
    transformers, 678
Electric charge (see Charge)
Electric field intensity, definition, 6
Electric flux density, 11
Electricity,
    definition, 4
    frictional, 4
Electromagnet, 643
Electromagnetic torque, 692 (see also Torque)
Electromotive force (see also Voltage),
    of self-induction, 15
Electron volt, 340
Electronic speed control, d-c motor,
    chopper drive, 778–781
    thyristor drive, 775–777
Electronics, definition, 333
Electronic switching, 472
Emitter-coupled logic, 473
Emitter-follower, 448
    input resistance, 450
    output resistance, 450
    voltage gain, 451
Encoders, 524
Energy bands, 338
Energy defined, 31
Enhancement-mode FET, 366
EPROM, 542
Equivalent circuits,
    BJT, 388
    d-c generator, 766
    d-c motor, 769
    hybrid, 423
    induction motor, three-phase, 722–725
    JFET, 394
    JFET, high frequency, 395
    synchronous generator, 748
    transformer, 664, 447, 671
Error detector, 822
Error-rate control, 858

Excitation voltage, 748
Exclusive-OR, 469

Farad, 44
Faraday's law, 14
  in thyristor drives, 777
  in transformers, 656
Feedback,
  amplifier, 835
  definition, 818
  effects, 832–840
    on bandwidth, 838
    on parameter changes, 833
    on stability, 839
  negative, 823
  sign sensitivity, 825
Feedback amplifier, 835
Feedback control applications,
  automatic machine tool, 829
  automatic power steering, 830
  high-speed wind tunnel, 826
Ferromagnetism, 624
FET (see Junction field effect transistor)
Field resistance line, 764
Field winding, 616
Figures of merit, second order, 850
Final-value theorem, Laplace transform, 225
Flags, status register, 597, 603
Flip-flops, 545
  basic circuit, 546
  clocked SR type, 548
  D type, 551
  glitch, 550
  JK type, 552
  latch, 545
  master-slave, 558
  T type, 554
  toggling, 553
  trailing-edge triggering, 549
Flux, leakage in transformers, 660, 661
Flux density (see Magnetic flux density)
Flux leakage, 14
  maximum, 697
Forced solutions, 130
  to constant sources, 131
  to exponential sources, 135
  to polynomial sources, 136
  of RC circuit, 165
  of RL circuit, 157
  to sinusoidal sources, 135
Forcing function, in circuits, defined, 128
Fourier series, 314–321
  fundamental, 314
  harmonics, 314
  mean square error, 321
  periodic functions, 314
  symmetry properties, 318
Forbidden energy bands, 338
Forward bias, semiconductor diode, 341
  transistor, 352
Frequency,
  breakpoint (or corner), 870

carrier, 454
complex, 183, 848
damped, 183
definition, 255
half-power, 298
modulation, 454
natural, 182
resonant, 297
sideband, 456
Frequency divider, 555
Frequency modulation, 454
Frequency response,
  BJT amplifier, 420–427
  JFET amplifier, 428
  transformer, 670
Full-adder, 526
Fundamental units, 2

Galvanometer, 14
Gate, logic, (see Logic gates)
Gauss, 634
Gauss' law, 10
Gauss' elimination method, 91
Generator (see also D-c generator and Syn-
      chronous generator)
  compound, 765
  direct-current, 763
  self-excited, 763
  shunt, 763, 765
  synchronous, 745
Generator, definition, 688, 689
Gilbert, 634
Giorgi, G., 2
Glitch, 550
Ground reference, 29, 94
Gyroscope, 879

Half-adder, 526
Half-power frequency, 298
Half-stepping logic control, stepper motors,
    810
Half-wave rectifier, 444
Half-wave symmetry, 318
Harmonics, 314
Henry, definition, 39
Hexadecimal number, 483
High-frequency response, BJT amplifier, 423
High-pass filter, 873
Hole, definition, 337
Hybrid parameters, BJT amplifier, 389–390
  characteristics, 392
Hybrid π equivalent circuit, 423
Hysteresis, 627
  losses, 641
Hysteretic angle, 658

Immittance (see Impedance and Admittance)
Impedance,
  complex, RL circuit, 281
  RLC circuit, 289
  driving-point, 124

Impedance (*cont.*)
  input, of transformers, 670
  operational, 123
Impulse, definition, 177
Impulse response, RC circuit, 178, 240
Impurities,
  n-type, 336
  p-type, 337
Incremental resistance, diode, 349
Independent current source, 26
Independent voltage source, 26
Induced voltage (*see* Voltage)
Inductance,
  circuit viewpoint, 39
  definition, 38
  energy viewpoint, 41
  geometrical viewpoint, 42
  initial condition property, 40
  mutual, transformers, 680
  parallel-connected, 79
  series-connected, 78
  unit, 39
Induction motor (*see also* Single-phase
          motors),
  applications and ratings, 733–735
  as a transformer, 719
  construction features, 698
  controllers, 733
  equivalent circuits, 722–725
  maximum torque, 731–732
  no-load tests, 731
  performance calculations, 726–729
  power flow diagram, 727
  revolving magnetic field, 715
  slip, 721
  slip-frequency, 724
  starting torque, 729
  torque-speed curve, 736
Inductive reactance, 277
Inductor, defined, 39
  linear, 39
  nonlinear, 40
Initial conditions, evaluation, 158
Initial-value theorem, Laplace transform, 225
Injection current, 347
Instruction register, 588
Insulated-gate field effect transistor, 363
Insulator, 339
Integral-error control, 864
Integrated circuits, 429–435
  diffused circuit elements, 432
  fabrication, 429
  transistor amplifier, 434
Integrated transistor amplifier, 434
Integrating circuit, 867
Intel 8080 microprocessor, 597
Interface, input-output, 596
Intrinsic silicon semiconductor, 339
Inverse Laplace transformation, 228
  complex poles, 230
  first-order poles, 229
  multiple poles, 231

Inversion, CMOS, 480
Iron loss (*see* Core loss)
Isolation amplifier (*see* Voltage follower)

JFET amplifier,
  biasing, 403
  circuit, 385
  equivalent circuits, 393–395
  frequency response, 428
  performance evaluation, 418–419
  power gain, 387
  voltage gain, 387, 428
JK flip-flop, 552
Johnson counter (*see* Switchtail counter)
Joule's law, 34
Jump instruction, Intel 8080, 603, 606
Junction, definition, 87
Junction diode, 341
Junction field effect transistor, JFET, 356
  depletion mode, 358
  enhancement, 365
  parameters, 360
  pinch-off voltage, 359
  transconductance, 362
  transfer characteristic, 361
  voltage amplification factor, 361
  volt-ampere characteristic, 359

Karnaugh maps, 505, 508
Kilogram, definition, 3
Kirchhoff's laws,
  current law, 16
  voltage law, 17

Lag network, 867
Laplace transform,
  common functions, 217
  complex translation, 224
  definition, 214
  final-value theorem, 225
  impulse function, 221
  initial value theorem, 225
  inverse, 228
  operational impedance, 235
  partial fraction expansion, 228
  real differentiation, 215
  real integration, 215
  sinusoidal response, RL circuit, 245
  step response, RC circuit, 237
  step response, RL circuit, 233
  step response RLC circuit, 241
  table, 218–219
  time-displacement theorem, 224
Latch gate, 545
Lead network, 873
Lenz's law, 15, 656
Lines of force, 616
Logic,
  half-stepping, 810
  negative, 471

Logic (*cont.*)
  positive, 471
  translator, for stepper motors, 807–810
Logic gates, 469
  CMOS, 479
  ECL, 473–478
  MOS, 478
  TTL, 473–475
    Schottky, 475
Logical functions,
  algebraic simplification, 504–508
  product of sums, maxterms, 513
  simplification by Karnaugh maps, 508–513
  sum of products, minterms, 513
Logical operations,
  AND, 466
  NOT, 467
  OR, 466
  principle of duality, 499
  XOR, 469
Loop,
  of a circuit, 88
  of a feedback system, 819
Low frequency response, BJT amplifier, 421

Magnetic circuit, 629
Magnetic domains, 624
Magnetic field intensity, 616, 619
Magnetic field, revolving, 715
Magnetic flux, 618
Magnetic flux density, 616
Magnetic moment, 623
Magnetic saturation, 624
Magnetic units, 633
Magnetization curve, 627–628
Magnetomotive force, 620
Majority carriers, 340
Mapping, Karnaugh, 505, 508
Master-slave flip-flop, 558, 583
Matrix formulation, circuit equations, 90, 98
Maximum percent overshoot, vs. damping
    ratio, 189
Maximum power transfer, 104
Maxterms, 513
Maxwells, 632
Mean square error, 321
Memory,
  address, 573, 588
  RAM, 573
  ROM, 542
Memory address register, 573, 588
Mesh current, 89
Mesh, definition, 88
Mesh method, 87
  with complex impedances, 293
  number of independent equations, 89
  using matrices, 90, 98
Meter, 3
Microcomputer system, 596
Microprocessor, 588
Microprocessor, Intel 8080, 597

control instructions, 603–606
operation instructions, 602
registers, general purpose, 599
transfer instructions, 599–601
Microresistor, 432
Midband frequency gain, BJT amplifier, 420
Miller effect, 425
Minority carriers, 340
Minterm, 508
Mnemonic codes, 593
Modulation,
  amplitude, 454
  carrier, 454
  frequency, 454
  modulation index, 454
  sideband frequencies, 456
Modulation frequency, 454
Modulation index, 454
Modulator,
  base-injection type, 457
  square law type, 457
MOS, 478
MOSFET, 363
Motor, definition, 688, 689
Motor (*see also* D-c, Induction, Single-phase,
    and Synchronous Motors)
  capacitor, 794
  compound, 768
  direct-current, 768
  induction, three-phase, 714
  series, 768
  shaded-pole, 796
  shunt, 768
  split-phase, 793
  squirrel-cage, 699
  stepper, 800
  synchronous, 750
  wound-rotor, 719
MSI chips, 524
Multiplexers, 540
Mutual inductance, 680
Mutual resistance, 90

NAND, 469, 473
  with CMOS gate, 481
Natural frequency, 182
Natural response, 137
  characteristic equation, 141
  critically damped, 190
  generalized solution, 141
  natural frequency, 182
  overdamped, 143, 193
  RC circuit, 166
  RL circuit, 138, 157
  RLC circuit, 143, 145, 181
  time constant, 160
  underdamped, 145
Negative feedback, 823
Negative logic, 471
Network analysis (*see* Circuit analysis)

Network, definition, 81, 87
    operational function, 130
Network reduction, 81
Newton's law, 3
Nibble, 483
Nodal method, 94
    with complex impedances, 294
    number of independent node-pair equations,
        96
    using matrices, 98
Node, definition, 87
Nomenclature, transistors, 378
NOR, 469, 473
Normal magnetization curve, 627–628
Norton's theorem, 104
    with complex impedances, 296
NOT, 467
N-type impurity, 336
Number systems,
    binary, 483
    conversions, 484–489
    decimal, 482
    general, 482
    hexadecimal, 483
    octal, 483
Octal number, 483
Oersted, 634
Ohm's law,
    linear form, 11
    nonlinear, 13
    microscopic form, 632
Op-amp (operational amplifier)
    as a current-controlled current source, 58
    as a difference amplifier, 50
    features, 51
    inverting mode, 55
    as an isolation device, 54
    noninverting mode, 52
    physical IC layout, 49
    transfer characteristic, 51
    as a voltage-controlled voltage source, 53, 56
    as a voltage follower, 54
Open-circuit test, transformer, 674
Open-loop system, 819
Operation instructions, Intel 8080, 602
Operational impedance, 123, 235
Operational network function, 130
OR, 466
Output conductance, BJT amplifier, 390
Output-rate control, 861
Overdamped response, 193
Overexcitation, synchronous motor, 752
Overshoot, maximum, 189

Parallel resonance, 302
Parameter changes, 833
Paramagnetism, 623
Particular solution (see Forced solution)
Peak inverse voltage, 446
Peak rectifier, 445
Pentavalent impurity (see Donor impurity)

Periodic function, 314
Periodic table, 892
Permeability, 617
    relative, 617
Permittivity,
    definition, 7
    relative, 7
Phase angle, 255
Phase inversion, transistor, 380
Phase lag, 255, 277
Phase lead, 255, 280
Phase, relative, 255
Phase sequence, three-phase, 306
Phasor, definition, 267
Phasor representation of sinusoids, 264
Phasor diagram, transformer, 660–661
Pinch-off voltage, JFET, 359
Polyphase induction motor (see Induction
        motor)
Positive logic, 471
Potential difference,
    definition, 8
    as work per unit charge, 8, 9
Potential energy barrier, 342
Power,
    average, sinusoids, 263
    defined, 31
    instantaneous, sinusoids, 262
    gain, transistor, 354
    reactive, in RL circuit, 284
    three-phase system, 309
Power factor, 264, 562
Power flow diagrams,
    d-c generator, 766
    d-c motor, 769
    synchronous motor, 753
    three-phase induction motor, 727
Power gain, CE transistor amplifier, 382, 385
Power transfer, maximum, 104
Product of sums, logical functions, 513
Program counter, 588
Programming, assembly language, 593, 606
PROM, 542
P-type impurity, 337
Pull-out torque (see Breakdown torque)
Pulse response, RC circuit, 174
Push-pull amplifier, 451

Quality factor, resonance, 299
Quiescent operating point, 380

Radix, 482
RAM, 573
    memory cell, 574
Rate gyro, 880, 881
Ratio of transformation, 659
RC circuit (see Complete response)
Reactance,
    capacitive, 279
    inductive, 277
    synchronous, 748

Reactance (*cont.*)
  in transformers,
    leakage, 661, 662
    equivalent, 668
    magnetizing, 666
Reactive power, 284
Reactor, ideal, iron core, 654
Read-and-write memory, 573
Read-only-memory, 542
Recombining electron current, 347
Rectifier,
  half-wave, 444
  peak, 445
  peak inverse voltage, 446
  SCR, 367
Reference directions,
  current, 29
  power, 29
  voltage, 29
Registers, 567
  general purpose, 599
  instruction, 588
  memory address, 573
  parallel, 567
  PROM, 542
  ROM, 542
  shift, 567, 568, 572
  status, 597
  storage, 531
Regulation,
  speed, 773
  voltage, 679
Relay, 643
Reluctance, 620
Reluctance force, 645
Reluctance power, 754
Residual flux density, 627
Resistance,
  armature circuit, 766
  circuit viewpoint, 32
  definition, 12, 32
  energy viewpoint, 34
  geometrical viewpoint, 35
  incremental, 349, 350
  incremental drain, 362
  mutual, 90
  parallel-connected, 75
  physical aspects, 36
  referred, 667, 725
  self, 89
  semiconductor diode,
    apparent, 349
    incremental, 349
  series-connected, 74
  temperature coefficient, 33
  variation with temperature, 33
Resistivity, 12, 35
Resistor, 32
  wire-wound, 36
  carbon type, 36
  color-code, carbon types, 37
  integrated type, 432

metal film type, 38
power-rating, 34
standard values, 38
Resonance, 296
  bandwidth, RLC circuit, 298
  energy description, 302
  quality factor, 299
  parallel, RLC circuit, 302
  resonant voltage rise, 301
Resonant frequency, 297
Retentivity, 627
Reverse amplification factor, 389
Reverse bias, transistor, 352
Reverse saturation current, 346
Revolving magnetic field, 715, 791
Right-hand rule, 657
Ring counter, 560
Ripple counter, 556
RL circuit (*see* Complete response)
RLC circuit (*see* Second-order system)
Rms value, periodic function (*see* Effective
    value)
ROM, 542–544
Rotor, 690
Round-off error, 487

Saturation, 624
Schottky TTL gate, 475
SCR, 367
  breakover voltage, 367
  firing, 368
  v-i characteristic, 368
Second, definition, 3
Second-order system,
  critically damped response, 190
  damping ratio, 851
  figures of merit, 850
  maximum percent overshoot, 188
  natural frequency, 849, 182
  RLC circuit, 180
  settling time, 185, 851
  time constant, 184, 852
  overdamped response, 147, 193
  underdamped response, 144, 181, 241, 848
  undamped response, 849
  universal curves, 850
Selectivity, 299
Self-bias, BJT amplifier, 398
Self inductance, 681
Self-excited shunt generator, 763
Self resistance, 89
Semiconductor, 335
  energy band description, 338
  intrinsic type, 339
  n-type impurity, 336, 340
  p-type impurity, 337, 341
  thermal agitation in, 339
Semiconductor junction diode,
  carrier densities in, 343
  depletion region, 342
  potential energy barrier, 342
  resistance, 348

Semiconductor junction diode (*cont.*)
 reverse saturation current, 346
 v-i characteristic, 345
 Zener breakdown voltage, 346
Semiconductor triodes (*see* Transistor)
Sensitivity, 834
Sequential logic circuits, 544
Series motor (see D-c motors)
Servomechanism, 823
 dynamic response, 845
 transfer function analysis, 853
Settling time, 155
 of RLC circuit, underdamped, 184, 190
Shaded-pole motor, 796
Shift register (see Registers)
Short-circuit test, transformer, 675
Shunt motor (*see* D-c motors)
Sideband frequency, 456
Sign sensitivity, 825
Signal flow graph, 833
Silicon-controlled rectifier (*see* SCR)
Single-phase motors,
 applications and ratings, 794–795, 797
 method of operation, 791–793
 resistance-start, split-phase, 793
 starting torque, 792
 types, 794
Sinusoidal function,
 addition of sinusoids, 265
 amplitude, 254
 average value, 257
 cycle, 255
 effective value, 259
 frequency, 255
 phasor representation, 264
 power, 262, 263
 relative phase, 255
 terminology, 254
Sinusoidal response,
 capacitor, 278
 inductor, 275
 RC circuit, 288
 resistor, 273
 RL circuit, 280
 RLC circuit, 289
SI units, 2, 632
Slip, 721
Software, 599
Space displacement angle, induction motor,
  729
Specific dielectric constant (*see* Permittivity)
Speed control, d-c motor, 774 (*see also* Elec-
  tronic speed control)
Speed regulation, d-c motors, 773
Split-phase motor, 793
Square-law modulator, 457
Squirrel-cage induction motor, 699
SR flip-flop, 548
SSI chips, 522–523
Stability factor, BJT amplifier, 397
Stability, effect of feedback, 839
Starters (*see* Controllers)

Starting torque,
 induction motor, three-phase, 729
 single-phase, 792
Stator, 690
Status register, 597
Steady-state solutions (*see* Forced solutions)
Stepper motors,
 construction features, 801–807
 drive amplifiers,
  bipolar, 808
  bifilar, 809
 half-stepping, 810
 method of operation, 801
 permanent-magnet type, 801
 ratings, 814
 reluctance type, 813
 translator logic, 807–810
Subtraction, by 2s complement, 533
 by op-amps, 57
Subtractors, 532
Sum of products, logical function, 508
Superposition method, 84
Susceptance, 286
Switch, electronic, 473
Switchtail counter, 563
Symmetry properties, Fourier series, 318
Synchronous counter, 558
Synchronous Generator,
 armature reaction voltage, 748
 construction features, 700
 equivalent circuit, 749
 excitation voltage, 748
 phasor diagram, 747
 synchronous reactance, 748
 voltage generation, 746
Synchronous motor,
 amortisseur winding, 750
 applications and ratings, 758
 equivalent circuit, 752
 overexcitation, 752
 performance calculations, 752
 phasor diagram, 751
 power angle, 754
 power flow diagram, 753
 reluctance power, 754
 underexcitation, 752
 V-curves, 757
Synchronous reactance, 748
Synchronous speed, 720, 750
System (*see* Second-order system)

Tachometer generator, 861
Terminology, feedback control, 820–823
Tesla, 616
T flip-flop, 554
Thermal agitation, semiconductors, 339
Thermal runaway, 397
Thévenin's theorem, 100
 with complex impedances, 295
Three-phase circuits, 305
 delta connection, 308

Three-phase circuits (*cont.*)
  line voltages, 306
  neutral, 306
  phase rotation, 306
  phase voltages, 306
  power, 309
  relationship, phase and line quantities, 308, 309
  wye connection, 306
Thyristor, 367
Thyristor drive, d-c motor, 775
Time constant,
  RC circuit, 160
  RL circuit, 166, 239
  RLC circuit, underdamped, 184
  servomechanism, 852
Time-displacement theorem, Laplace transform, 224
Toggling, 553
Torque,
  a-c machines, 693, 705, 706
  d-c machines, 695, 709
  developed, 692
  maximum, induction motor, 731
  starting, induction motor, 729
Trailing-edge triggering, flip-flops, 549
Transconductance, JFET, 362
Transducer, 822
Transfer functions,
  accelerometers, 876
  actuators, 881
  circuit, 125, 236
  gyroscope, 879
  lag network, 867
  lead network, 873
  open-loop, 832
  system, 833
  transformer, 669
Transfer instructions, Intel 8080, 599–601
Transformer,
  air-core, 654
  balance current, 663
  construction, 662, 663
  efficiency, 678
  equivalent circuits, 664, 667, 671
  equivalent reactance, 668
  equivalent resistance, 668
  frequency response, 670
  iron core, 654
  leakage flux,
    primary, 660
    secondary, 662
  leakage reactance,
    primary, 661
    secondary, 662
  mutual inductance, 680
  nameplate rating, 669
  no-load tests, 673
  phasor diagram
    full load, 661
    no load, 660
  ratio of transformation, 662, 663

transfer function, 669
voltage regulation, 679
windings, 654
Transient response (*see* Natural response)
Transistor,
  common-base mode, 351
  current transfer ratio, 353
  definition, 354
  fabrication, 429
  forward bias, 352
  insulated gate, 363
  junction field effect type, 356
  parameters, 356
  reversed bias, 352
  v-i characteristic, 355
Transistor amplifier, (BJT)
  a-c load line, 384
  bandwidth, 426
  biasing methods, 395–403
  common-base mode, 352
  common-emitter mode, 375
  current gain, single stage, 382, 385, 406
  current gain, multistage, 412, 415, 417
  cutoff, 472
  d-c load line, 379
  emitter follower, 448
  equivalent circuit, 388
  frequency response, 420–427
  h-parameters, 389
  integrated, 434
  nomenclature, 378
  performance evaluation, 404–418
  power gain, 382
  push-pull type, 451
  phase inversion, 380
  quiescent point, 380
  self bias, 398
  stability factor, 397
  voltage gain, 381, 385, 407
Translator logic, stepper motors, 807–810
Tree, of a circuit, 88
Triac, 369
Trivalent impurity (*see* Acceptor impurity)
Truth table, 466, 468
TTL, 473–475
  Schottky, 475
Turbogenerator, 745

Undamped response, 849
Underdamped response, 181, 848
Underexcitation, synchronous motor, 752
Units,
  definition, 2
  farad, 44
  fundamental, 2
  henry, 39
  magnetic, 633
  SI, 2, 632
  table, 890
  tesla, 616
  weber, 619

Universal curves, second-order systems, 850
Up-down counter, 560

Valence electrons, 336
V-curves, 757
Voltage,
    a-c machines, induced, 697, 703
    breakover, SCR, 367
    d-c machines, induced, 706, 707
    definition, 8
    excitation, 748
    pinch-off, JFET, 359
    source, 26
Voltage amplification factor, JFET, 361
Voltage controlled voltage source, 53
Voltage-divider rule, 81
Voltage doubler, 447
Voltage follower, 54
Voltage gain,
    BJT amplifier, 381, 385, 407
    JFET, 387, 428
Voltage regulation, 679
Voltage-source conversion, 105
Voting machine, 506–508

Weber, 619
Winding,
    amortisseur, 750
    Gramme-ring, 702
    squirrel-cage, 699
    three-phase, 746
Word, 542
Wordtime, 565
Wound-rotor motor, 698
Wye connection, three-phase, 306

XOR, 469

Y-connection of three-phase load, 306
Y-delta transformation, 109

Zener breakdown voltage, 346
Zilog Z80 microprocessor, 597